Advances in Fracture and Damage Mechanics XI

Edited by
Li Qingfen
Li Yulong
M.H. Aliabadi

Advances in Fracture and Damage Mechanics XI

Selected, peer reviewed papers from the
11th International Conference on
Fracture and Damage Mechanics
(FDM 2012),
September 18-12, 2012, Xian, China

Edited by

Li Qingfen, Li Yulong, M.H. Aliabadi

Copyright © 2013 Trans Tech Publications Ltd, Switzerland

All rights reserved. No part of the contents of this publication may be reproduced or transmitted in any form or by any means without the written permission of the publisher.

Trans Tech Publications Ltd
Kreuzstrasse 10
CH-8635 Durnten-Zurich
Switzerland
http://www.ttp.net

Volumes 525-526 of
Key Engineering Materials
ISSN 1013-9826

Full text available online at *http://www.scientific.net*

Distributed worldwide by

Trans Tech Publications Ltd
Kreuzstrasse 10
CH-8635 Durnten-Zurich
Switzerland

Fax: +41 (44) 922 10 33
e-mail: sales@ttp.net

and in the Americas by

Trans Tech Publications Inc.
PO Box 699, May Street
Enfield, NH 03748
USA

Phone: +1 (603) 632-7377
Fax: +1 (603) 632-5611
e-mail: sales-usa@ttp.net

printed in Germany

Preface

This volume is a collection of edited papers presented at the 11th International Conference on Fracture and Damage Mechanics (FDM2012), held in Xian, China. Previous meetings were held in London, UK(1999), Milan, Italy (2001), Paderborn, Germany (2003), Mallarca, Spain (2005), Harbin, China (2006), Madeira, Portugal (2007), Seoul, Korea (2008) and St. Julian's Bay, Malta (2009) and Nagasaki, Japan (2010), Dubrovnik, Croatia (2011).

The conference which was attended by researchers from 18 countries served as a forum to promote and exchange latest theoretical, computation and experimental research works on fracture and damage mechanics as well as structural integrity and durability.

The FDM2012 covers a wide range of topics: Fracture Mechanics, Failure analysis, Composites, Multiscale Modelling, Micromechnaics, Structural Health Monitoring, Damage Tolerance, Corrosion, Creep, Non-linear problems, Dynamic Fracture, Residual Stress, Environmental effects, Crack Propagation, Metallic and Concrete Materials, Probabilistic Aspects, Computer Modelling Methods (Finite Elements, Boundary Elements and Meshless), Microstructural and Multiscale Aspects.

The Editors would like to thank the members of the Steering Committee and International Scientific Committee for their help and support that were essential to the success of the conference. Furthermore, we express our gratitude to the Northwestern Polytechnic University and Harbin Engineering University for help in promoting the meeting. Finally, we would like to thank all those who helped to organize this scientific event, especially member of The Centre for Foreign Talents Introduction and Academic Exchange of Material Behaviour of Advanced Structure and Materials.

Editors
MH Aliabadi, LI Qingfen, LI Yulong,

International Conference on Fracture and Damage Mechanics XI

18-21 September 2012, Xian, China

Steering Committee
Prof. M.H.Aliabadi Imperial College, London
Prof. B.Abersek University of Maribor, Slovania
Prof. J.K.Kim KAIST, Korea
Prof. M.Guagliano Politecnico di Milano, Italy
Prof. S-I.Nishida Saga University, Japan
Prof. S.W.Shin Hanyang University, Korea
Prof. YANG Zhichun School of Aeronautics, Northwestern Polytechnic University
Prof. LI Qingfen, Harbin Engineering University, China

Conference Organisers
Prof. LI Qingfen, Harbin Engineering University, China
Prof. LI Yulong, Northwestern Polytechnic University, China
Professor Ferri M H Aliabadi, Department of Aeronautics, Imperial College London,UK

Local Organizing Committee
Prof. YANG Zhichun School of Aeronautics, Northwestern Polytechnic University
Prof. YUE Zhufeng Northwestern Polytechnic University (Dean of School of Mechanics and Civil & Architecture)
Prof. MA Cunbao Northwestern Polytechnic University (Vice Dean of Scholl of Aeronautics)
Prof. XU Fei Northwestern Polytechnic University (Vice Dean of Scholl of Aeronautics)
Prof. ZHAO Hongzhang Northwestern polytechnic University (Vice Dean of Scholl of Aeronautics)
Prof. LIN Nan Northwestern Polytechnic University (Director of School Office of Aeronautics)
Prof. WANG Zhenqing Harbin Engineering University (Dean of College of Aerospace and Civil Engineering)
Prof. ZOU Guangping Harbin Engineering University (Vice Dean of College of Aerospace and Civil Engineering)

International Organizing Committee

LI Huaxing Northwestern polytechnic University (Director of International Office)
ZHANG Qingbin Harbin Engineering University (Director of International Office)
GOU Xingwang Northwestern polytechnic University
LI Ziwei Northwestern polytechnic University
CHI Miao Harbin Engineering University
ZHAO Yan Harbin Engineering University

International Scientific Advisory Committee

Abersek,B University of Maribor, Slovenia

Alfaiate,J Instituto Superior Tecnico, Portugal

Ashcroft,I.A Loughborough University, UK

Baragetti,S Universita degli Studi di Bergamo, Italy

Cerny,I SVUM , Czech Rep

Chen, D-H Science University of Tokyo, Japan

Constantin,N University Politehnica of Bucharest, Romania

De Corte, W Ghent University, Belgium

Dhondt,G MTU Aero Engines, Germany

Doblare,M University of Zaragoza, Spain

Ebara,R Hiroshima Institute of Technology, Japan

Espinosa,C (ISAE), TOULOUSE, France

Flewitt,P.E.J BNFL Magnox Generation, UK

Fujimoto,T Kyushu Sangyo University, Japan

Galvez,J.C Universidad Politécnica de Madrid, Spain

Guagliano,M Politecnico di Milano, Italy

Hadavinia,H Kingston University, UK

Hashida,T Tohoku University, Japan

Hornikova,J Brno University of Technology, Czech, Republic

Horst,P IFL, Braunschweig, Germany

Ivankovic,A University College Dublin, Ireland

Jankowski,R Gdansk University of Technology, Poland

Kawagoishi,N Kagoshima University, Japan

Kikuchi,M Science University of Tokyo, Japan

Kim, Jeong Guk (Korea Railroad Research Institute, Korea)

Kisu,H Nagasaki University, Japan

Kolednik,O Austrian Academy of Sciences, Austria

Korsunsky,A.M Oxford University, UK

Kotoul,M Brno University of Technology, Czech Republic

Kuna,M Technische Universität Bergakademie Freiberg, Germany

Lee, Jinyi (Chosun University, Korea)

Li Li Harbin Engineering University, China

Li,Q.F Harbin Engineering University, China

Makabe,C University of Ryukus, Japan

Marsavina,L Politechnica University of Timisoara, Romania

Nobile,L University of Bologna, Italy

Pappalettere,C Politecnico di Bari, Italy

Pastor,J ESIGEC, University of Savoie, France

Piat,R Univeristy of Karlsruhe, Germany

Providakis,C Technical University of Crete, Greece

Rezaeepazhand,J Ferdowsi University of Mashhad, Iran

Schlangen,E Delft University of Technology, The Netherlands

Sluys,L.J Delft University, The Netherlands

Soprano,A Second University of Naples, Italy

Takuda,H Kyoto University, Japan

Teranishi,T Kyushu Sangyo University, Japan

Theilig,H Univ. Appl. Sci. Zittau/Gorliz, Germany

Tohgo,K Tokyo Institute of Technology, Japan

Torii,T Okayama University, Japan

Toribio,J University of Salamanca, Spain

Tsukrov, Igor University of New Hampshire, USA

Watanabe, K Yamagata University, Japan

Wang,Z.Q Harbin Engineering University, China

Yan,S.Y. Harbin Engineering University, China

Yigeng Xu. University of Hertfordshire, United Kingdom

Yoon,I.S. Induk Institute of Technology, South Korea

Yufeng Zheng Harbin Engineering University, China

ZHANG,J-Z Harbin University of Science and Technology, China

Zhu,S.F. Harbin Engineering University, China

Table of Contents

Preface   v
Committees   vi

A Grain Boundary Formulation for the Analysis of Three-Dimensional Polycrystalline Microstructures
I. Benedetti and M.H. Aliabadi 1

Comparisons of High-Temperature Structural Analysis Results on the Medium-Scale PHE Prototype under the Steady-State and Trip Conditions of a Small-Scale Gas Loop
K.N. Song, S.D. Hong and H.Y. Park 5

Effect of Niobium Film on Corrosion Resistance of AZ91D Magnesium Alloy
E.B. Liu, X.F. Cui, G. Jin, Q.F. Li and T.M. Shao 9

The Measurement of Mechanical Properties of Thermal Barrier Coatings by Micro-Cantilever Tests
D. Liu and P.E.J. Flewitt 13

Crack Growth by Dimensional Reduction Methods
P.H. Wen and M.H. Aliabadi 17

Nondestructive Evaluation of Polyurethane Materials Using Transient Thermography
M. Amarandei, K. Berdich, I. Szigyarto, L. Kun and L. Marşavina 21

Numerical Modelling of Particulate Composite with a Hyperelastic Matrix
B. Máša, L. Náhlík and P. Hutař 25

Fracture Toughness of PIR Foams Produced from Renewable Resources
J. Andersons, E. Spārniņš, U. Cābulis and U. Stirna 29

Modelling of Discontinuous Environment
J. Boštík and K. Weiglová 33

Influence of Finite Element Modeling Choices in the Assessment of Stress Concentrations at Fatigue Prone Locations in Orthotropic Bridge Decks
W. De Corte and A. Jansseune 37

Numerical Analysis of Femoral Neck Angle Influence on Stress Distribution of Cemented Austin Moore Hip Prosthesis
L. Bogdan, C.S. Nes, N. Faur, M. Amarandei and A. Enkelhardt 41

Advanced Natural Stitched Composite Materials in Skin-Stiffener of Wind Turbine Blade Structures
F. Papadopoulos, D. Aiyappa, R. Shapriya, E. Sotirchos, H. Ghasemnejad and R. Benhadj-Djilali 45

Interfacial Shear Strength of Carbon Fiber Reinforced Polypropylene
K. Takemura and H. Katogi 49

Effect of Surface Treatment on Creep Property of Jute Fiber Reinforced Green Composite under Environmental Temperature
K. Takemura, S. Miyamoto and H. Katogi 53

Micro-Scale Cantilever Testing of Linear Elastic and Elastic-Plastic Materials
J.E. Darnbrough, S. Mahalingam and P.E.J. Flewitt 57

The Influence of the Epoxy Interlayer on the Assessment of Failure Conditions of Push-Out Test Specimens
J. Klusák, P. Helincks, S. Seitl, W. De Corte, V. Boel and G. De Schutter 61

Fracture and Damage Characterization of Natural Fiber Composites
H. Takagi, Y. Hagiwara and A.N. Nakagaito 65

Modeling of Damage and Failure of Dual Phase Steel in Nakajima Test
J.H. Lian, P.F. Liu and S. Münstermann 69

Micromechanical Modeling for the Evaluation of Elastic Moduli of Woven Composites
D. Abe, O. Bacarreza and M.H. Aliabadi 73

Characterization of Metal Magnetic Memory of Thermal Damage Correlated with Initial Cumulative Fatigue Damage for Laser Cladding Remanufacturing Technology
C. Zeng, W. Tian and L. Hua 77

SEM *In Situ* Study on Pre-Corrosion and Fatigue Cracking Behavior of LY12CZ Aluminum Alloy
X.D. Li, Z.T. Mu and Z.G. Liu 81

Analysis of Stress Intensity Factor for Cracked Flattened Brazilian Disk: Part I – Analysis Method and Pure Mode I Crack
S.M. Dong and Q.Y. Wang 85

Analysis of Stress Intensity Factor for Cracked Flattened Brazilian Disk: Part II – Mixed-Mode Crack
S.M. Dong and Q.Y. Wang 89

Thermo-Mechanical Stress near Apex of a Bi-Material Wedge by a Novel Finite Element Analysis
X.C. Ping, L. Leng and S.H. Wu 93

Simulation of Spall Fracture Based on Material Point Method
W.D. Chen, F. Zhang and W.M. Yang 97

Buckling Reliability Analysis for the Cylindrical Shell with Initial Defects
H. An, W.G. An, Y.Y. Gao and X.H. Song 101

To Research Residual Stress Using Finite Element Analysis
Z.F. Wang and Y.Y. Hu 105

Numerical Simulation of Underwater Explosion Based on Material Point Method
W.D. Chen, W.M. Yang and F. Zhang 109

Numerical Simulation of Fatigue Behavior of Honeycomb Sandwich Panels
J. Lu, G.P. Zou and P.F. Yang 113

The Debond Fracture of Sandwich Plate with Corrugated Core Using Cohesive Zone Element
G.P. Zou, P.F. Yang, J. Lu and Y.G. Li 117

Study on a Numerical Calculation Method of Fatigue Behavior in Glare Panel
Y.E. Ma and S. Yun 121

Investigation of Residual Stress Distribution in Clad 2024 T3 after Laser Peening
Z.H. Wang and Y.E. Ma 125

Investigation on the Corrosion Effect of Friction Stir Welded AA2024 T3 Aluminum Alloy Joints
Y.E. Ma and Z.Q. Zhao 129

Distortion Induced Fatigue Analysis of Steel Bridges
C.S. Wang, S.L. Yan and J. Cheng 133

Material Properties and Fatigue Safety Evaluation of Old Metal Bridges
C.S. Wang, H.J. Sheng, J.Y. Hu, S.L. Yan and L. Duan 137

Damage Monitoring and Evaluation Using AE Sensors for Existing Concrete Bridges
C.S. Wang, M.S. Zhai and L. Tian 141

Repair and Service Safety Evaluation of a Fire Damaged Concrete Bridge
C.S. Wang, L.P. Liu and M.S. Wei 145

Scaling Effects on the Tensile Strength of Fibrous Composites
F. Wang, L. Li and Z.Q. Chen 149

Analysis of Fatigue and Cumulative Damage Characteristics for Concrete in Freezing-Thawing Environment
Q.H. Xiao 153

Residual Fatigue Life Analysis of Submarine Pressure Structure Based on Probabilistic Fracture Mechanics
Z.Y. Zhang, T.S. Song and Y. He 157

Geometrical Parameters Influencing a Hybrid Mechanical Coupling
G. Lamanna, F. Caputo and A. Soprano 161

Fatigue Test of Small Sized AZ31 Magnesium Alloy Using Micropillar Specimen
S. Mizuno, T. Kakiuchi and Y. Uematsu 165

Fatigue Behavior of A356 Cast Aluminum Alloy Microstructurally Modified by Friction Stir Processing under Low Strain Rate Condition
T. Kakiuchi, Y. Uematsu and Y. Tozaki 169

Load Influence on the Behavior of Micro-Crack in the Particulate Composite
Z. Majer 173

Pressure Pipe Damage: Numerical Estimation of Point Load Effect
M. Zouhar, P. Hutař, M. Ševčík and L. Náhlík 177

Quasi-Static Simulation of Crack Growth in Elastic Materials Considering Internal Boundaries and Interfaces
P. Judt and A. Ricoeur 181

Deterioration of FRC Plate due to Explosion and Change of Temperature
P.P. Prochazka 185

Monitoring Fatigue Cracks and Stress Intensity Factors Based on the Distributed Dislocation Technique
R. Boukellif and A. Ricoeur 189

Computational Characterization of Micro-To Macroscopic Deformation Behavior of Double Network Hydrogel
I. Riku and K. Mimura 193

Multilevel/Multiobjective Design of Composite Structures
O. Bacarreza, M.H. Aliabadi and A. Apicella 197

The Role of Grain Size on Deformation of 316H Austenitic Stainless Steel
S. Mahalingam, P.E.J. Flewitt and A. Shterenlikht 201

Experimental Investigation of the Influence of the Bond Conditions on the Shear Bond Strength between Steel and Self-Compacting Concrete Using Push-Out Tests
P. Helincks, W. De Corte, J. Klusák, V. Boel and G. De Schutter 205

Experimental Study of the Influence of the Initial Notch Length in Cubical Concrete Wedge-Splitting Test Specimens
S. Korte, V. Boel, W. De Corte, G. De Schutter and S. Seitl 209

Micro-Analyses of Small Cracks in 6061-T6 Aluminium Alloy Subjected to High-Cycle Fatigue
Y. Takahashi, T. Shikama and H. Noguchi 213

Fatigue Properties of Solution-Treated Type 304 Stainless Steel after Nitriding
Y. Nakamura, M. Nakajima, T. Shimizu, K. Suzuki, Y. Bai and Y. Uematsu 217

Fatigue Crack Growth of Alloy 7050-T7451 Plate in L-S Orientation
R. Bao, X.C. Zhao, T. Zhang and J.Y. Zhang 221

Structural Design and Finite Element Analysis of Composite Wind Turbine Blade
S.F. Zhu and I. Rustamov 225

Determination of Intensity Factors for Interfacial Cracks in TIP Materials
H.J. Zhong and J. Lei 229

Failure of Composite T-Joints in Bending with Through-the-Thickness Reinforcement: Stitching Vs Z-Pinning
H. Cui and Y.L. Li 233

Investigation on T_{11}-Stress for Semi-Elliptical Surface Cracks in Finite Thickness Plates under Remote Tension
W. Xie 237

Investigation of Thermal Damage and the Consequent Fatigue Property of Laser Surface Heat Treated Component
W. Tian, C. Zeng and L. Hua 241

A 3-Dimensional Model Related to Stress-Strain-Time of Rock Salts
H.B. Zhang, Z.Y. Wang and J.W. Ma 245

Experimental Research on Interlaminal Shear Strength of GFRP Bridge Decks under Simulated Concrete Environment
W.C. Xue and K. Fu 249

Vibration Fatigue Behavior of 2024-T62 Aluminum Alloy Cantilever Beam under Different Vibration State
H.T. Hu, Y.L. Li and J.L. Wang 253

Simulation of Pylon Emergency Break-Away of Large Commercial Aircraft
X. Zhang and Y.L. Li 257

Dynamic Mechanical Behavior of Two Fiber-Reinforced Composites
Y.Z. Guo, X. Chen, X.Y. Wang, S.G. Tan, Z. Zeng and Y.L. Li 261

Numerical Simulation of Compression-After-Impact Process of Composite Laminates
B. Li, Y.Z. Li, X. Li and Z.H. Yao 265

Deposition Process and Interface Properties of Electro-Thermal Explosion Sprayed WC/Co Coating
X.F. Cui, G. Jin and Q.F. Li 269

Grain-Boundary Segregation of Phosphorus and Inter-Granular Fracture Behavior under Low Tensile Stresses
Y.D. Fu, Q.F. Li and W.X. Sun 273

Effect of the Neodymium Content on Mechanical Properties of the Electro-Brush Plated Nano-Al_2O_3/Ni Composite Coating
G. Jin, X.F. Cui, E.B. Liu and Q.F. Li 277

FE Simulation Methods to Predict Welding Residual Stresses
L. Li, K. Asifa, H. Li and S. Khurram 281

Study on the Pin-Load Distribution of Multiple-Bolted Composite to Metal Joints
X.D. Liu, Y.Z. Li, Z.H. Yao and H. Shu 285

The Sensitivity of the Foam-Core Parameters under Impact Loading
F. Xu and M.G. Duan 289

Dynamic Anti-Plane Behaviors on Two Dissimilar Piezoelectric Media with an Interfacial Non-Circular Cavity
T.S. Song, D. Li, M.J. Zhang and Y.F. Zhou 293

Analysis of Crack Tip Field in Materials with Creep Behavior
Q.H. Meng, W.Y. Liang and Z.Q. Wang 297

The Conservation Laws and Path-Independent Integrals for Piezo-Magnetic Media
H.Y. Song, J.S. Zhou and Z.M. Liu 301

Scattering of Anti-Plane SH-Wave by Multiple Cylindrical Inclusions in Elastic Semi-Space
H.L. Li and Y. Yang 305

Multiscale Simulation of the Size Effect of Nano-Indentation
Z.Q. Wang, Q.H. Meng and Z.J. Yang 309

J-Integral and its Dual Form Based on Finite Deformation Theory
Z.M. Liu, J.Z. Mao and H.Y. Song 313

Study on the Diffusion Coefficient in NGS under Low Tensile Stress
Y.J. Qiao, H.B. Fu and C.K. Li 317

Pressureless Sintering and Properties of Boron Carbide-Titanium Diboride Composites by *In Situ* Reaction
A.D. Liu, Y.J. Qiao and Y.Y. Liu 321

Galvanic Corrosion of Titanium/Cu-Ni Alloy/High Strength Steel Multiphase Material System in Seawater
C.L. Wang, Q.F. Li and J.H. Wu 325

Wear Failure Behavior of Steel Surface with Palygorskite Powders as Lubricant Additives
L.M. Wang, B.S. Xu, Y. Xu, B. Zhang, H.L. Yu and Q.F. Li 329

Fatigue Assessment of Trimaran Structure Based on Simplified Procedure
H.L. Ren, S. Khurram, C.B. Zhen and K. Asifa 333

A Comparative Study between Steel Grillage and SPS Stiffening Plate Based on FEA Eigen Value Analysis
Q.C. Xue, G.P. Zou, Y. Wu, J. Li and L. Shang 337

Yield Criterion and Crack Tip Plastic Zone of Nickel-Based Single Crystal
L.H. Yang, G.P. Zou and J. Qu 341

Interaction of Elliptical Inclusion and Crack in Half-Space under SH-Waves
Z.L. Yang, H.N. Xu, B.P. Hei and Y. Yang 345

Influence of Welding Conditions on Residual Stresses of Multi-Pass Tube Sheet Welds
H. Li, Y. Zheng and L. Li 349

Study on Mode I Quasi-Static Growing Crack in a Rigid-Viscoelastic Material Interfacial Crack
Y. Yang, Z.L. Yang and L.Q. Tang 353

Analysis of Interface Deformation of Steel-Concrete-Steel Sandwich Beam
P.X. Xia, G.P. Zou and Z.L. Chang 357

Reliability Analysis for Stiffened Plate on the Maximum Entropy Method
J. He and X.Y. Chen 361

An Enhanced Time-Reversal Method for Impact Damage Monitoring on Plate-Like Structures
C.L. Chen, Y.L. Li and F.G. Yuan 365

Experimental Study on Residual Strength of Panels with Multiple Site Damage
X.Z. Yan, S.N. Wang and W. Wang 369

Damage Analysis of 2D Woven Composite Laminates Containing an Open-Hole under Tensile Loadings
X.Q. Zhang, W.G. Guo and D.S. Kong 373

A New Approach to Determine Dynamic Strength Model Parameters under Taylor Impact Test
F. Xu, W.G. Guo, Q.J. Wang and Z.Y. Zeng 377

A Novel Testing Method for Measuring Through-Thickness Properties of Thick Composite Laminates
Y.T. Li, X.T. Zheng and G. Luo 381

Numerical Simulation of Compressive Residual Strength for Damaged Composite Laminates
T.J. Qu, X.T. Zheng and D. Zhang 385

A Research on Characterization for Damage Tolerance of Composite Laminates
D. Zhang, X.T. Zheng and L.N. Cheng 389

The Simulation of Low-Velocity Impact on Composite Laminates with the Damage Model Based on Strain
C.H. Ma and F. Xu 393

Investigation into Damage of Stainless Steel Mesh/ALPlate Multi-Shock Shield under Hypervelocity AL-Spheres Impact
G.S. Guan, D.D. Pu and Y. Ha 397

Numerical Simulation of Hypervelocity Impact on Mesh Bumper Causing Fragmentation and Ejection
G.S. Guan and R.T. Niu 401

Analysis of Crack Arrest by Electromagnetic Heating in Metal with Oblique-Elliptical Embedding Crack
Y.M. Fu, H.M. Zhou, J.L. Wang and L.J. Zheng 405

Effect of Characteristic Parameters of Exponential Cohesive Zone Model on Mode I Fracture of Laminated Composites
G.W. Zhu, Y.X. Jia, P. Qu, J.Q. Nie and Y.L. Guo 409

Extended Finite Element Method for Fracture Mechanics and Mesh Refinement Controlled by Density Function
Y. Cen 413

Effects of Surface Roughness on the Fatigue Life of Alloy Steel
W.L. Xiao, H.B. Chen and Y. Yin 417

Correlation between Alternating Temperature Accelerated Aging and Real World Storage of Composite Propellant
B. Ding, P. Shi and X. Qiu 421

Experimental Research on Fracture Toughness of High Grade Line Pipe
Q.R. Xiong, J.X. Zhang, Y.R. Feng, H.T. Wang, Q.R. Ma and Z.Z. Xu 425

Stress-Strain Analysis of Composite Propellant under Cyclic Temperature Loads
C.L. Zhang, B. Ding, C.S. Ai and F. Huang 429

Damage Detection for Structural Health Monitoring Using Ultrasonic Guided Waves
H.Y. Li and H. Xu 433

The Research on Wood Fiber/Stainless Steel Net Electromagnetic Shielding Composite Board
C.W. Su, Q.P. Yuan, W.X. Gan, J.D. Huang and Y.Y. Huang 437

Low Cycle Fatigue Behaviors of TI-6AL-4V Alloy Controlled by Strain and Stress
R.F. Wang, Y.T. Li and H.P. An 441

Analysis of Stress Singularity near the Tip of Artificial Crack
Y.T. Li and H.Q. Li 445

An Evaluation on the Restrained Shrinkage of Ultra-High Performance Concrete
J.J. Park, D.Y. Yoo, S.W. Kim and Y.S. Yoon 449

Structural Strength Evaluation of a Stainless Railroad Car
S.C. Yoon, J.G. Kim, D.B. Choi, D.H. Koo and G.S. Park 453

Study of the Characteristic of Composite Sandwich Panel with Cut-Out under Compression Load
J.W. Lee, C.W. Kong, Y.S. Jang and Y.S. Lee 457

Size Effect of PHE Prototype on High-Temperature Structural Integrity
K.N. Song 461

High-Temperature Structural Analysis on the Small-Scale PHE Prototype
K.N. Song 465

Characterization of Ductile Failure Behavior of the Ferritic Steel Using Damage Mechanics Modeling Approach
P. Kucharczyk and S. Münstermann 469

Thermal and Static Analysis of an Insulation Block from Recycled Polymer HDPE for Solution of Thermal Bridges in Wall-Footing Detail
J. Pěnčík, L. Matějka and L. Matejka 473

Diagnosis of Damage in a Steel Tank Model by Shaking Table Harmonic Tests
D. Burkacki and R. Jankowski 477

Numerical Analysis of a Steel Frame Building with Soft-Storey Failure under Ground Motion Excitation
W. Migda and R. Jankowski 481

Relation between Structural Size and the Discretization Density of Brittle Homogeneous Lattice Models
M. Vořechovský and J. Eliáš 485

Mathing Asymptotic Method in Propagation of Cracks with Dugdale Model
D.T.B. Tuyet, J.J. Marigo and L. Halpern 489

Formability Evaluation of Non-Crimp Carbon Fabric by Non-Contact 3D Deformation Measurement System
K. Tanaka, K. Kanazawa, S. Enoki and T. Katayama 493

Modelling Micro-Damage in Granular Solids
M. Buonsanti, G. Leonardi and F. Scoppelliti 497

Load and Environmental Effects on the Corrosion Behavior of a Ti6Al4V Alloy
S. Baragetti and A. Medolago 501

Fatigue Crack Nucleation and Growth Mechanisms for Ti6Al4V in Different Environments
S. Baragetti, C. Foglia and R. Gerosa 505

Modeling Delamination of Interfacial Corner Cracks in Multilayered Structures
B. Veluri and H.M. Jensen 509

Crack Modelling Using the Material Point Method and a Strong Discontinuity Approach
I. Guiamatsia and G. Nguyen 513

Numerical Simulation of Crack Tip Behavior under Fatigue Loading
Y.G. Xu, W. Tiu and Y.Z. Xu 517

Delamination Threshold Load of Composite Laminates under Low-Velocity Impact
Y.G. Xu, Z. Shen, W. Tiu, Y.Z. Xu, Y. Chen and G. Haritos 521

Inelastic Seismic Behavior without and with Over-Resistance Effects of 10-Story Building RC Damaged due to the 1985 Earthquakes in Mexico City
J.A. Avila 525

Accurate Description of Near-Crack-Tip Fields for the Estimation of Inelastic Zone Extent in Quasi-Brittle Materials
V. Veselý, J. Sobek, L. Šestáková and S. Seitl 529

Application of the Mesh Superposition Technique to the Study of Delaminations in Composites Thin Plates
A. Sellitto, R. Borrelli, F. Caputo, A. Riccio and F. Scaramuzzino 533

Deterministic and Probabilistic Earthquake Scenarios for the Seismic Risk Analysis of URM Buildings
J.A. Avila-Haro, J.R. González-Drigo, Y.F. Vargas, L.G. Pujades and A.H. Barbat 537

Micromechanics Damage Analysis in Fiber-Reinforced Composite Material Using Finite Element Method
C.Y. Kimyong, S. Aimmanee, V. Uthaisangsuk and W. Wechsatol 541

On the Direction of a Crack Initiated from an Orthotropic Bi-Material Notch Composed of Materials with Non-Uniform Fracture Mechanics Properties
T. Profant, J. Klusák, O. Ševeček and M. Kotoul 545

Fracture Behaviour of Thin Sheet Stainless Steel
N. Gubeljak, D. Jagarinec, J. Predan and J. Landes 549

Transient and Steady State Regimes of Fatigue Crack Growth in High Strength Steel
J. Toribio, B. González and J.C. Matos 553

Estimation of Shear Behavior of Ultra High Performance Concrete I Girder without Shear Stirrups
J.W. Lee, C. Joh, E.S. Choi, I.J. Kwak and B.S. Kim 557

A Local Compression Tests of UHPC Anchor Blocks for Post-Tensioning Tendons
E.S. Choi, J.W. Lee, C. Joh, J.W. Kwark, J.S. Kim and Y.S. Choi 561

Impact Identification in Composite Stiffened Panels
M. Ghajari, Z. Sharif Khodaei and M.H. Aliabadi 565

Electro-Mechanical Impedance Technique for Structural Health Monitoring of Composite Panels
M. Schwankl, Z. Sharif Khodaei, M.H. Aliabadi and C. Weimer 569

Dynamic Response Analysis of Lining Structure with Primary Defects for Tunnel under Moving Train Loading
Y.P. Cui, W.Z. Sun, J. Dong and F. Dong 573

Monitoring Analysis of Tunnel Construction in a New Subway Line Down-Through an Existed Line
F. Dong, J. Dong, Y.P. Cui, W.Z. Sun and D.Y. Li 577

SMART Platform for Structural Health Monitoring of Sensorised Stiffened Composite Panels
Z. Sharif Khodaei, M. Ghajari, M.H. Aliabadi and A. Apicella 581

Micromechanical Characterisation of Microstructure in Weld Heat Affected Zone of Structural Steel
Y. Shimada, Y. Kayamori, S. Nishida, M. Matsuda and K. Takashima 585

Two Scale-Based Continuum Damage Model for Brittle Materials under Thermomechanical Loading
A. Ricoeur and D. Henneberg 589

The Applicability Study on the FRP-Concrete Composite Bridge Deck for Cable-Stayed Bridges
S.T. Kim, S.Y. Park, K.H. Cho, J.R. Cho and B.S. Kim 593

Improvement of Fatigue Crack Growth Behavior in the Case of the Cracked Specimen with Relatively Narrow Width
M.S. Ferdous, C. Makabe and T. Miyazaki 597

Dynamic Crack Problems Using Meshless Method
P.H. Wen and M.H. Aliabadi 601

Investigation of Damping Characteristics of Metals by the Use of Inverted Torsion Pendulum Method
T. Asahina, T. Shioya, T. Toguri and M. Sekine 605

Mechanical Characteristics of Bamboo
T. Shioya and T. Asahina 609

Relationship between Crack Extension Behavior and Fatigue Life of C/C Composites
S.A. Setyabudi, C. Makabe and M. Fujikawa 613

Influence of Adhesive Layer on Actuation of Lamb Wave Signals
Z. Sharif Khodaei, Q. Liu and M.H. Aliabadi 617

Evaluation of Setting Time in Ultra High Performance Concrete
S.W. Kim, J.J. Park, D.Y. Yoo and Y.S. Yoon 621

Keyword Index 625

Author Index 631

A grain boundary formulation for the analysis of three-dimensional polycrystalline microstructures

I. Benedetti[1,2,a], M.H. Aliabadi[1,b]

[1]Department of Aeronautics, Imperial College London, South Kensington Campus, SW7 2AZ, London, UK.

[2]Dipartimento di Ingegneria Civile, Ambientale, Aerospaziale e dei materiali, Università degli Studi di Palermo, Viale delle Scienze, Edificio 8, 90128, Palermo, Italy

[a]i.benedetti@imperial.ac.uk, [a]ivano.benedetti@unipa.it, [b]m.h.aliabadi@imperial.ac.uk

Keywords: Polycrystalline materials, Microstructural Modelling, Material Homogenization, Boundary Element Method.

Abstract. A 3D grain boundary formulation is presented for the analysis of polycrystalline microstructures. The formulation is expressed in terms of intergranular displacements and tractions, that play an important role in polycrystalline micromechanics, micro-damage and micro-cracking. The artificial morphology is generated by Hardcore Voronoi tessellation, which embodies the main statistical features of polycrystalline microstructures. Each crystal is modeled as an anisotropic elastic region and the integrity of the aggregate is restored by enforcing interface continuity and equilibrium between contiguous grains. The developed technique has been applied to the numerical homogenization of cubic polycrystals and the obtained results agree well with available data.

Introduction

Macroscopic material properties are determined by the features of material microstructure. The investigation on this relation is an important and technologically relevant aim of modern materials science. The estimation of the effective material properties, or *material homogenization*, can be carried out at different levels [1]. A modern approach to material homogenization is the use of numerical models for the simulation of the material behavior at the microstructural scale [2].

Polycrystalline materials constitute an important class of heterogeneous materials [3]. Many materials of technological interest (metals, ceramics) present a polycrystalline microstructure. The internal structure of polycrystals is determined by the size and shape of the grains, by their crystallographic orientation and by different types of defects. A crucial role in the determination of the polycrystalline aggregate properties, especially when damage is taken into account, is played by the intergranular interfaces and their defects [4].

The microstructure of polycrystalline materials can be investigated by using different experimental techniques [5,6]. These provide fundamental information but require sophisticated equipment, material manufacturing and preparation and complicated postprocessing, resulting then generally expensive and time consuming. A viable alternative, or complement, to the experimental characterization is offered by the computational micromechanics [2]. The dramatic increase in computational power and the formulation of reliable mathematical models allow to simulate the response of complex microstructures at little cost, thus complementing and accelerating the experimental campaigns when, for example, the design of a new material is pursued.

In the present study, a 3D grain boundary integral formulation for the analysis of polycrystalline microstructures is presented. The technique is alternative to the more used FEM and its typical features are: *a*) the simplification in the artificial microstructure generation and modelling, especially in relation to the meshing of the artificial microstructure, since only the discretization of the grains surface is required; *b*) the microstructural problem is formulated directly in terms of intergranular displacements and tractions, which play an important role in polycrystalline micromechanics, especially when damage and microcracking are involved [7].

Generation of artificial polycrystalline microstructures

A reliable model of the microstructure must retain the main topological, morphological and crystallographic features of the aggregate. For polycrystalline materials, Voronoi tesselations are widely used for the generation of the microstructural models [4,8,9]. The Voronoi cells are convex polyhedra bounded by flat polygonal convex faces. Some authors have pointed out that Voroni tessellations underestimate the distribution of grain size and overestimate the number of faces per grain [5]. Some of these issues can be corrected through the application of correction schemes [9]. In any case, Voroni tessellations have the advantage of being analytically defined, relatively simple to generate and possess some features that make them suitable for numerical treatment, (straight edges and flat faces). Here the *Hardcore* Voronoi tessellation is adopted for generating the microstructure: the additional hardcore constraint produces more regular grains and tessellations.

To complete the microstructure model, it is necessary to assign a specific orientation to each crystal of the aggregate. In this work, each grain is assigned a random orientation from a uniform distribution in the group of rotations in the 3D space.

Modelling of polycrystalline microstructures

Material modelling. Each grain of the aggregate is modeled as a 3D linearly elastic orthotropic domain with arbitrary spatial orientation. The hypothesis of orthotropic material is not restrictive, as the majority of single metallic and ceramic crystals present general orthotropic behavior.

Grain boundary element formulation. Each crystal is modeled by using the Boundary Element Method (BEM) for 3D anisotropic elasticity [10]. The polycrystalline aggregate is seen as a multi-region problem, so that different elastic properties and spatial orientation can be assigned to each grain [11]. Given a volume bounded by an external surface and containing N_g grains, two kinds of grains can be distinguished: the *boundary grains*, intersecting the external boundary, and the *internal grains*, completely surrounded by other grains. Boundary conditions are prescribed on the surface of the boundary grains lying on the external boundary, while interface continuity and equilibrium conditions are forced on interfaces between adjacent grains, to restore the integrity of the aggregate. In general, the boundary integral equation for a generic grain G_k is written

$$c_{ij}^k(x)u_j^k(x) + \int_{B_C \cup B_{NC}} T_{ij}^k(x,y)u_j^k(y)dB^k(y) = \int_{B_C \cup B_{NC}} U_{ij}^k(x,y)t_j^k(y)dB^k(y), \tag{1}$$

where u_i^k and t_i^k denote boundary displacements and tractions of the grain G_k, and U_{ij}^k and T_{ij}^k represent the components of the 3D anisotropic displacement and traction fundamental solutions. The integrals appearing in Eq. 1 are extended over the entire surface of the grain, given by the union of *contact* surfaces B_C, in common with other grains, where interface conditions apply, and *non-contact* surfaces B_{NC}, where boundary conditions apply. Eq. 1 is complemented by the

$$\begin{cases} u_i^k = \bar{u}_i^k \\ t_i^k = \bar{t}_i^k \end{cases} \text{Boundary conditions} \quad \begin{cases} \tilde{u}_i^k + \tilde{u}_i^j = \delta\tilde{u}_i^{kj} = 0 \\ \tilde{t}_i^k = \tilde{t}_i^j \end{cases} \text{Interface conditions}, \tag{2}$$

where the overbar denotes prescribed quantities, while the tilde represents quantities expressed in an interface local reference system, more suitable for the interface conditions. The interface conditions involve surface displacements and tractions from two different grains, G_k and G_j.

After discretization and integration of Eq. 1 for each grain, the final system of equations for the aggregate can be written

$$\begin{bmatrix} \mathbf{A}_1 & 0 & 0 \\ 0 & \ddots & 0 \\ 0 & 0 & \mathbf{A}_{N_g} \\ & [\mathbf{I}] & \end{bmatrix} \begin{bmatrix} x_1 \\ \vdots \\ x_{N_g} \end{bmatrix} = \begin{bmatrix} y_1 \\ \vdots \\ y_{N_g} \\ 0 \end{bmatrix}, \qquad (3)$$

where the vectors x_k contain the unknown components of displacements and tractions, the matrix blocks $\mathbf{A}_k$ are the grain boundary element matrices, the matrix $\mathbf{I}$ contain the coefficients of the interface conditions and the terms y_k stem from the boundary conditions. System 3 is highly sparse and the use of specialized sparse solvers is then desirable to speed up the numerical solution.

Grain boundary element discretization. The proposed formulation has the remarkable advantage that only meshing of the grain surfaces is required. The artificial microstructure is, in the context of meshing, a collection of flat convex polygons, that represent the grain boundaries and interfaces. Plane triangular linear elements are used to discretize such faces. Linear discontinuous triangular elements are used for representing the unknown boundary fields. The mesh generator Triangle (http://www.cs.cmu.edu/~quake/triangle.html, [12]) is used for the creation of a two-dimensional high-quality mesh of each plane cell face. Since the Voronoi tessellations used for microstructure modelling have stochastic nature, care must be taken to ensure mesh consistency and homogeneity to the greatest extent. This is achieved by introducing a discretization parameter governing the mesh density, so to create meshes as homogeneous as possible. Fig. 1 shows the mesh of a tessellation with 100 grains and the representative mesh of a single grain.

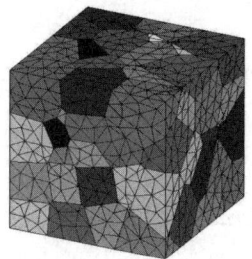
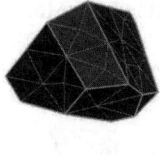

Figure 1: Mesh of a tessellation containing 100 grains and representative mesh of a single grain.

Numerical estimation of elastic properties of FCC nickel

The effective elastic properties of nickel polycrystals are determined. The effective or overall isotropic modula E and G are determined from the knowledge of the properties of the single crystals in a linear framework. The nickel crystals elastic constants are: $C_{11} = 251\ GPa$, $C_{12} = 150\ GPa$, $C_{44} = 124\ GPa$. The effective properties are obtained by taking the ensemble average of the apparent properties over a certain number of microstructure realizations with the same number of grains, see e.g. [13]. Aggregates with $N_g = 10,\ 20,\ 50,\ 100,\ 150$ have been tested. For a given number of grains, 100 different realizations have been generated and simulated. Each realization differs from the others in terms of both geometry and grains orientation. Linear displacement boundary conditions have been enforced on each simulated realization. Fig. 2 shows the mean value and scatter of the apparent modula. First and third order bounds, as taken from [8], are also reported. It is worth noting how the scatter is reduced, and how the computed mean values get closer to the third order bounds, when an higher number of grains is considered.

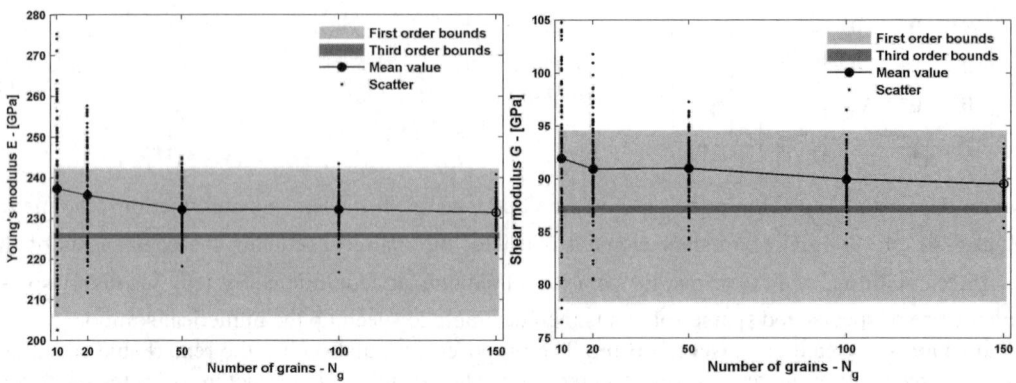

Figure 2: Apparent elastic properties of aggregates of nickel based on ensemble averages taken over 100 realizations.

Summary

A 3D grain boundary formulation has been developed for the analysis of polycrystalline microstructures. The technique allows a remarkable simplification in the generation of the artificial microstructure model and it is directly formulated in terms of intergranular displacements and tractions, that play an important role in polycrystalline micromechanics. The developed method has been applied to the determination of the effective properties of cubic polycrystals. Results show good agreement with literature data in the framework of numerical homogenization.

Acknowledgements

This research was supported by a Marie Curie Intra European Fellowship within the 7th European Community Framework Programme.

References

[1] S. Nemat-Nasser, M. Hori: *Micromechanics: overall properties of heterogeneous materials* (North-Holland, Elsevier, The Netherlands, 1999).
[2] T.I. Zohdi and P. Wriggers: *An introduction to computational micromechanics*, (Springer, Berlin, 2005).
[3] B.L. Adams and T. Olson: Prog Mater Sci, Vol. 43, (1998), p. 1.
[4] A.G. Crocker, P.E.J. Flewitt, and G.E. Smith: Int Mater Rev, Vol. 50(2), (2005), p. 99.
[5] K.M. Döbrich, C. Rau, and C.E. Krill III: Metall Mater Trans A, Vol. 35A, (2004), p. 1953.
[6] M.A. Groeber et al.: Mater Charac, Vol. 57, (2006), p. 259.
[7] G.K. Sfantos and M.H. Aliabadi: Int J Numer Meth Eng, Vol. 69, (2007), p. 1590.
[8] F. Fritzen, T. Böhlke and E. Schnack: Comput Mech, Vol. 43, (2009), p. 701.
[9] R. Quey, P.R. Dawson, and F. Barbe: Comput Meth App Mech Eng, Vol. 200, (2011), p. 1729.
[10] I.Benedetti, A. Milazzo A, MH Aliabadi, International Journal for Numerical Methods in Engineering, Vol:80, (2009), P.1356-1378
[11] M.H. Aliabadi: *The boundary element method: applications in solids and structures*, (John Wiley & Sons Ltd, England, 2002).
[12] J.R. Shewchuk: in Applied Computational Geometry: Towards Geometric Engineering, ed. M.C. Lin and D. Manocha, vol. 1148 of Lecture Notes in Computer Science, Springer-Verlag, Berlin, (1996).
[13] T. Kanit et. al: Int J Solids Struct, Vol. 40, (2003), p. 3647.

Comparisons of high-temperature structural analysis results on the medium-scale PHE prototype under the steady-state and trip conditions of a small-scale gas loop

Kee-nam Song[1,a], S-D Hong[1,b], H-Y Park[2,c]

[1]P.O.Box 105, Yuseong, Daejeon, 305-600, Korea

[2]#1101 Hanjin Officetel, 535-5 Bongmyoung-dong, Yuseong-gu, Daejeon, Korea

[a]knsong@kaeri.re.kr, [b]sdhong1@kaeri.re.kr, [c]hypark@adsolution.co.kr

Key Words: Medium-Scale Process Heat Exchanger (PHE), High-temperature Structure Analysis, Small-scale Gas Loop, Very High Temperature Reactor (VHTR), Steady-State and Trip Condition

Abstract. PHE (Process Heat Exchanger) is a key component for transferring the high-temperature heat generated from a VHTR (Very High Temperature Reactor) to a chemical reaction for massive production of hydrogen. Recently, Korea Atomic Energy Research Institute (KAERI) has manufactured a medium-scale PHE prototype made of Hastelloy-X of high-temperature alloy and a performance test on the PHE prototype is scheduled in a small-scale nitrogen gas loop established at KAERI. In this study, in order to evaluate the high-temperature structural integrity of the PHE prototype under the steady-state and trip conditions of the gas loop before the performance test on the PHE prototype, elastic and elastic-plastic structural analyses on the PHE prototype were carried out and the analyses results were compared each other.

Introduction

Hydrogen is considered a promising future energy solution because it is clean, abundant, and storable, and has high-energy density. One of the major challenges in establishing a hydrogen economy is how to produce massive quantities of hydrogen in a clean, safe, and economical way. Among the various hydrogen production methods, nuclear hydrogen production is garnering attention worldwide since it can produce hydrogen, a promising energy carrier, without an environmental burden. Research demonstrating the massive production of hydrogen using a VHTR designed for operation at up to 950°C has been actively carried out worldwide [1,2,3].

The nuclear hydrogen program in Korea is strongly considering producing hydrogen by employing a Sulfur-Iodine (SI) water-splitting hydrogen production process [1]. Recently, KAERI (Korea Atomic Energy Research Institute) has manufactured a medium-scale PHE prototype made of Hastelloy-X and a performance test on the PHE prototype is scheduled in a small-scale nitrogen gas loop at KAERI. In this study, in order to evaluate the high-temperature structural integrity of the PHE prototype under the steady-state and trip conditions of the gas loop before the performance test on the PHE prototype, structural analyses on the PHE prototype were carried out and the analyses results were compared with each other.

Finite Element (FE) Modeling on a Medium-Scale PHE Prototype

Overall structure. The medium-scale PHE prototype which is composed by primary and secondary flow plates as shown in Fig. 1 is designed as a hybrid concept [4]. That is to say, the hot nitrogen gas channel has a compact semicircular shape, similar to a printed circuit heat exchanger, and is designed to withstand the high pressure difference between loops, while the sulfuric acid gas channel has a plate fin shape with sufficient space to install and replace the catalysts for sulfur trioxide decomposition. All parts of the PHE prototype are made from Hastelloy-X of high-temperature alloy. Grooves of 1.0 mm diameter are machined into the flow plate for the primary coolant. Waved channels are bent into the flow plate for the secondary coolant (SO_3 gas). Forty flow plates for the primary and secondary coolants are stacked in turn, and are bonded along the edge of the flow plate using a solid-state diffusion bonding method. After stacking and bonding the flow plates, the outside of the PHE is covered with the Hastelloy-X plate of 3.0 mm thickness.

FE Modeling. Figure 2 shows the overall dimensions and each part of the PHE prototype via the 3-D CAD modeling. Based on Fig. 2, FE modeling using I-DEAS/TMG Ver. 6.1 [5] was carried out and analyses such as a thermal analysis and structural analysis are carried out using ABAQUS Ver. 6.8 [6]. For the sake of simplicity and understanding the overall behavior of the PHE prototype, the FE model is composed of 680,772 2-D linear quadrilateral shell elements and 870,696 3-D linear solid elements including 66,456 tetrahedron elements. Figure 3 shows the boundary conditions of the primary/secondary flow plates for a thermal analysis under a steady-state and trip condition of the gas loop. The material properties of Hastelloy-X [7] are used in the FE model.

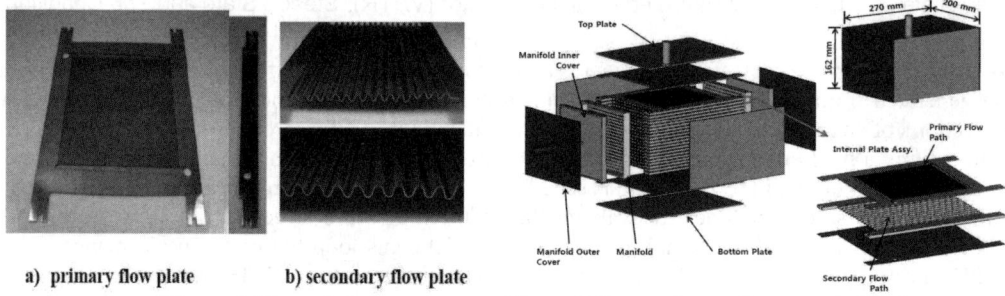

a) primary flow plate   b) secondary flow plate
Figure 1 Flow plates   Figure 2 Parts of a medium-scale PHE prototype

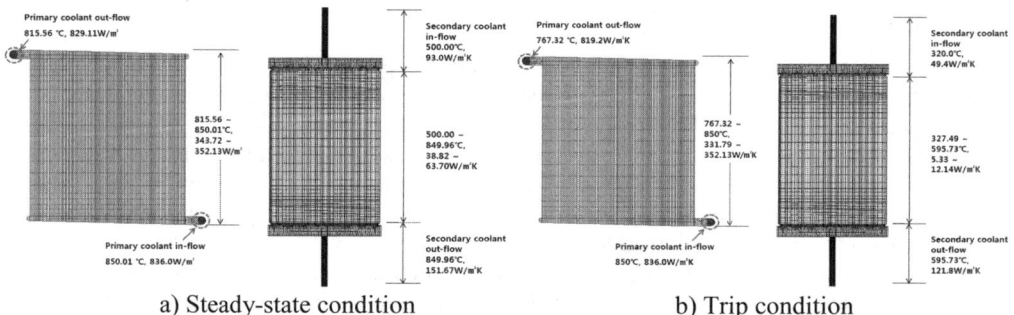

a) Steady-state condition   b) Trip condition
Figure 3 Thermal boundary condition

Analysis

Thermal analysis. Figure 4 shows the thermal analysis results of the outside surface of the PHE prototype under the steady-state and trip conditions of the gas loop. From Fig. 4, the maximum temperature of the outside surface is about 836.26°C and 787.69°C for a steady-state and trip condition of the gas loop, respectively.

Coolant pressure. Coolant pressures of 3.0MPa and 0.1MPa are acting on the primary and secondary flow plate, respectively.

Boundary condition for structural analysis. As a displacement constraint condition, equivalent spring stiffness is imposed on each end of in-flow/out-flow pipelines connected to the PHE prototype [8] to simulate the pipelines in the gas loop. Figure 5 shows the displacement constraint condition.

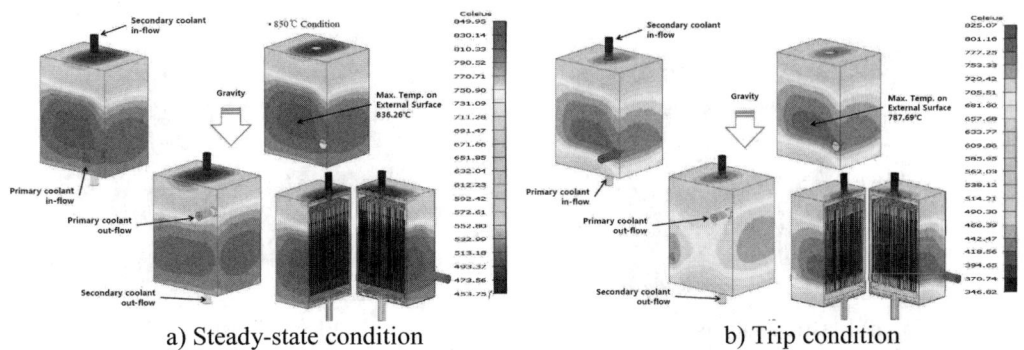

a) Steady-state condition   b) Trip condition

Figure 4 Temperature contour

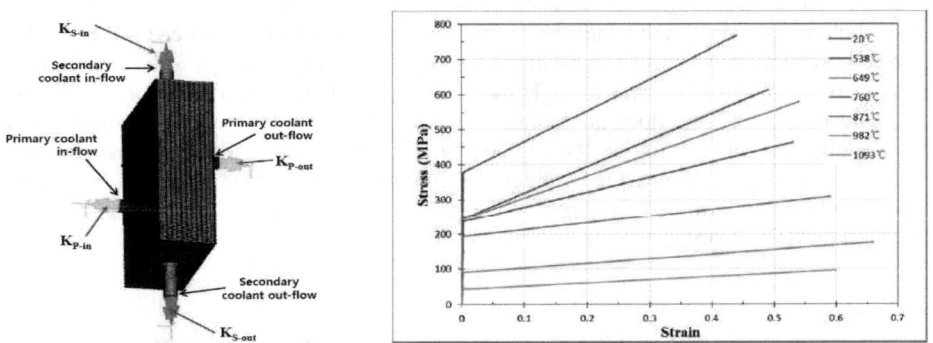

Figure 5 Boundary condition   Figure 6 Bilinear stress-strain curve for elastic-plastic analysis

Structural analysis. Based on the thermal analysis result, high-temperature elastic and elastic-plastic structural analyses were performed. Stress-strain curve shown in Fig. 6 is used for the elastic-plastic structural analysis. Figures 7 and 8 represent the stress contours at the pressure boundary of the PHE prototype under the steady-state and trip condition, respectively. From the analysis results shown in Fig. 7 under the steady-state condition, the maximum local stress of 331.23 MPa occurs near the edge between the top plate and side plate in the elastic analysis, while 263.65 MPa does in the elastic-plastic analysis. The maximum stresses exceed the yield stress of Hastelloy-X under both analyses. Meanwhile, from the analysis results in Fig. 8 under the trip condition, the maximum local stress of 441.73 MPa occurs in the elastic analysis, while 263.65 MPa does in the elastic-plastic analysis. The maximum stresses also exceed the yield stress of Hastelloy-X under both analyses.

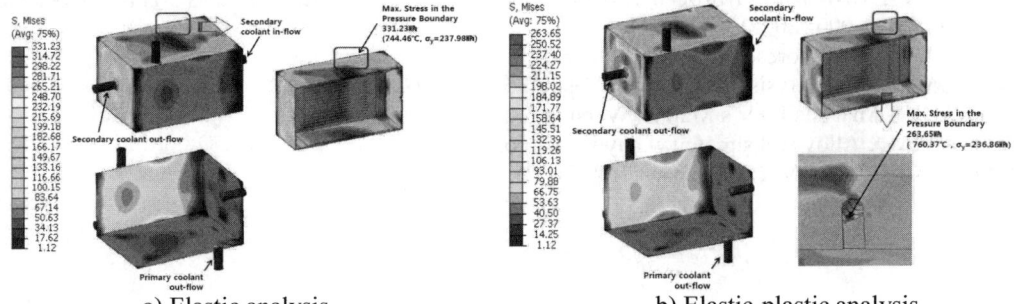

a) Elastic analysis   b) Elastic-plastic analysis

Figure 7 Stress contour under the steady-state condition of the small-scale gas loop

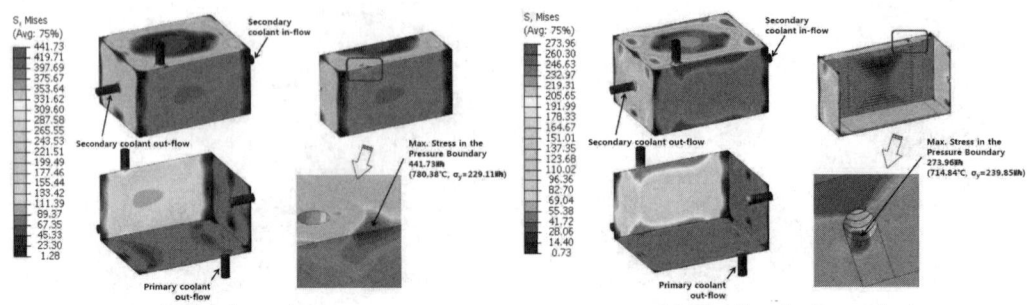

a) Elastic analysis   b) Elastic-plastic analysis
Figure 8 Stress contour under the trip condition of the small-scale gas loop

Since edges of the PHE prototype are chamfered realistically, the maximum stress will be decreased to some extent when considering the chamfered edges. Due to the chamfering effect, the high-temperature structural integrity of the medium-scale PHE prototype seems to be maintained under the steady-state condition of the gas loop. However, the high-temperature structural integrity of the medium-scale PHE prototype does not seem to be maintained under the trip condition of the gas loop, even though the chamfering on the edge of the PHE prototype lead to decrease the maximum stress level. Therefore, some measure to modulate the thermal expansion of the medium-scale PHE prototype should be determined for both re-designing the pipelines of the gas loop to be more flexible and the PHE prototype.

Summary

To understand the macroscopic behavior of a medium-scale PHE prototype, FE modeling, thermal analysis, and high-temperature elastic and elastic-plastic structural analyses were carried out under the steady-state and trip condition of the gas loop. As a result of these analyses, we draw the following conclusions.
1. Due to the chamfering effect, the high-temperature structural integrity of the medium-scale PHE prototype seems to be maintained under the steady-state condition of the gas loop.
2. Meanwhile, the high-temperature structural integrity of the medium-scale PHE prototype does not seem to be maintained under the trip condition of the gas loop. Therefore, some measure should be determined for both re-designing the pipeline of the gas loop and the PHE prototype.

References

[1] Chang, J. H. *et al.*, Nuclear Engineering and Technology, Vol. 39, No. 2 (2007), p. 111.
[2] US DOE, *Financial Assistance Funding Opportunity Announcement*, NGNP Program (2009).
[3] AREVA, *NGNP with Hydrogen Production Pre-conceptual Design Studies Report*, Doc. No. 1209052076-000 (2007).
[4] Kim, Y. W., R.O.Korea patent # 10-0877574 (2008).
[5] I-DEAS/TMG Analysis User Manual Version 6.1 (2009).
[7] ABAQUS Analysis User's Manual, Version 6.8 (2009).
[7] Hastelloy-X Alloy website, www.haynesintl.com.
[8] Song, K. N., ICFDM 2012 Papae #152 (2012).

Effect of Niobium Film on Corrosion Resistance of AZ91D Magnesium Alloy

Liu Er-bao [1,a], Cui Xiu-fang [1], Jin Guo [1], Li Qing-fen [1,b], Shao Tian-min [2]

[1] College of Material Science and Chemical Engineering, Harbin Engineering University, Harbin 150001, China

[2] State Key Laboratory of Tribology, Tsinghua University, Beijing 100084, China

[a]liutong_05@163.com, [b]qingfli@yahoo.com.cn

Keywords: niobium film; corrosion resistance; AZ91D magnesium alloy

Abstract. The niobium film is prepared by magnetron sputtering on the surface of the AZ91D magnesium alloy. The morphology, phase structure, roughness, nano-hardness and elastic modulus of the niobium films were studied by filed emission scanning electron microscope, X-ray diffraction, atomic force microscope and nano-indentation respectively. The influences of film deposition parameters, such as substrate temperature, negative bias and power on the properties of films were investigated. The corrosion resistance of niobium films on magnesium alloy was investigated by electrochemical system. Results show that the microstructure, phase structure, roughness, nano-hardness and elastic modulus of the niobium films are determined by power, negative bias and substrate temperature. And the corrosion resistance of magnesium alloy improved obviously when coated with the niobium films.

Introduction

Corrosion failure exit in many industries and the cost of such damage may reach billions of dollars annually. Magnesium alloys have many excellent properties. However, their poor corrosion resistance has greatly restricted their application in practical industries [1-4]. Corrosion and protection of magnesium has therefore been being one of the most important fields in corrosion research [5-10]. Niobium has excellent corrosion resistance and good performance at high temperature and is widely used in the chemical and aerospace industry [11, 12].

In this paper, the niobium film is prepared by magnetron sputtering on the surface of the AZ91D magnesium alloy. The morphology, phase structure, roughness, nano-hardness and elastic modulus of the niobium films were studied, and effect of niobium film on corrosion resistance of AZ91D magnesium alloy was investigated.

Experimental

The niobium films are prepared on the surface of AZ91D magnesium alloy by direct current (DC) magnetron sputtering (JGP450) . The AZ91D substrates (15mm ×10mm ×5mm) were polished by using waterproof abrasive paper from 360 grits to 2500 grits, and then fine polished using diamond paste of 3.5 μm. They were whereafter degreased with absolute ethanol in ultrasonic bath for 15 min and subsequently dried by cold air in room temperature. The sputtering target is niobium (99.99%) with the size of Φ 60 ×3mm. The detailed films deposition parameters are shown in Table 1.

The morphology and phase structure of the films were observed by field emission scanning electron microscopy (FESEM) , atomic force microscope (AFM) and X-ray diffraction (XRD) . The nano-mechanical properties, such as hardness, elastic modulus, were examined by nano-mechanical testing system.

The anticorrosion performance was evaluated by recording the corrosion potential and corrosion current density of the samples immerged in 3.5% sodium chloride (NaCl) solution, which was conducted on a CHI660B electrochemical workstation. The electrochemical cell used a classic three-electrode system which consists of a reference electrode, a counter electrode and a working electrode. The potentiodynamic polarization curves were performed at a scanning rate of 1 mV/s. OCP and EIS measurements were carried out at corrosion potential in a frequency range between 0.01 Hz and 100000 Hz using a 10 mV amplitude perturbation.

Table 1 Deposition parameters of niobium films by DC magnetron sputtering

Sample number	power (W)	temperature (°C)	negative bias (V)
1#	100	300	100
2#	100	25	100
3#	150	25	100
4#	50	25	100
5#	100	25	50
6#	100	25	150

Results and discussion

Morphology of Niobium Films. The morphologies of niobium films on the AZ91D magnesium alloys for different deposition parameters are illustrated in Fig. 1. It can be seen that niobium films are all uniform and dense. The lower deposition temperature leads to the compact and fine microstructure of niobium films as shown in Fig. 1(a) and (b). The lower sputtering power helps forward the fine microstructure formation of niobium films as shown in Fig. 1(b), (c) and (d). The higher negative bias contributes to the dense microstructure of niobium films as shown in Fig. 1(b), (e) and (f).

Fig. 1 SEM photographs of niobium films: (a) 1#, (b) 2#, (c) 3#, (d) 4#, (e) 5#, (f) 6#

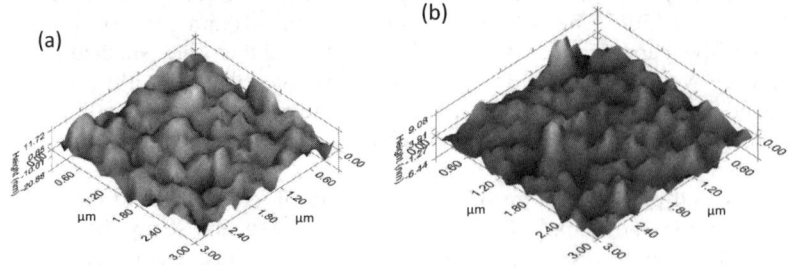

Fig. 2 Three-dimensional AFM images of niobium films deposited on different sputtering powers: (a) 150 W, (b) 50 W

Fig. 2 shows the three-dimensional AFM images of niobium films deposited at different sputtering powers. It can be seen that the particles of niobium films prepared at 150 W are larger than that at 50 W.

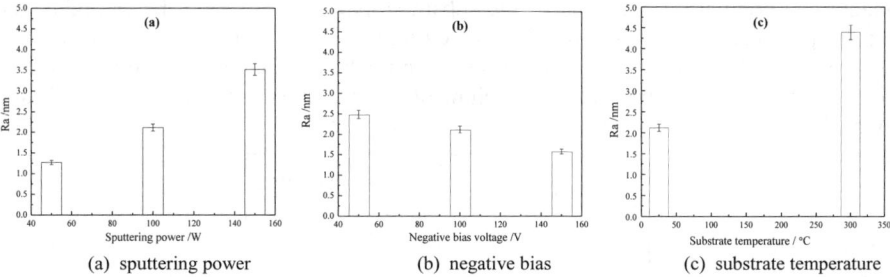

(a) sputtering power (b) negative bias (c) substrate temperature

Fig. 3 The influence of deposition parameters on the surface roughness of niobium films

Fig. 3 The influence of deposition parameters on the surface roughness of niobium films

The influence of deposition parameters on the surface roughness of niobium films is shown in Fig. 3. Where, it can be seen that the lower sputtering power, the higher negative bias and the lower substrate temperature are all in favor of decreasing the surface roughness of niobium films.

Phase Structure of Niobium Films. The influence of sputtering power on the phase structure of niobium films is shown in Fig. 4. The niobium films deposited at different sputtering power exhibit (110), (200) and (211) planes. However, the crystallization of the films increases with increasing sputtering power, since the niobium peaks especially for (211) gradually become obvious as the sputtering power increases.

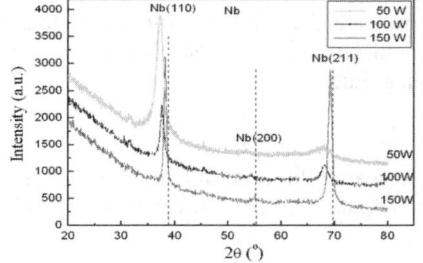

Fig. 4 Influence of sputtering power on the films phase structure of niobium films

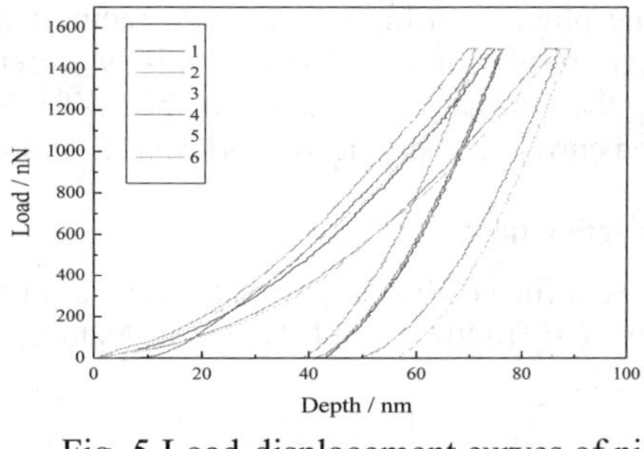

Fig. 5 Load-displacement curves of niobium deposited at different parameters

Nano-mechanical Properties of Niobium Films. The load-displacement curves of niobium films deposited at different parameters are shown in Fig. 5 and the nano-mechanical properties (such as elastic modulus and hardness) are shown in Table 2. It can be seen that the hardness increases with increasing substrate temperature and decreases obviously with increasing sputtering power. While both the elastic modulus and hardness decreases at first and then increases with increasing negative bias.

Table 2 Nano mechanical properties of niobium films deposited at different parameters

Sample number	Er /GPa	H /GPa
1#	156.0	9.7
2#	132.0	7.6
3#	133.9	7.1
4#	153.2	9.9
5#	158.9	9.7
6#	172.3	11.0

Corrosion Resistance of Niobium Films. Potentiodynamic polarization curves of the AZ91D magnesium alloys coated with niobium films immerged in 3.5 wt% NaCl aqueous are shown in Fig. 6. It can be seen that niobium films can decrease the corrosion current density (Icorr) about four orders of magnitude comparing with that of the substrate, partially blocking the cathodic reaction and shifting the polarization curves toward lower current density values. This indicates that the niobium films on AZ91D restrain the anodic reaction and enhance the electrochemistry stability of AZ91D. These results demonstrate that the niobium films can improve the corrosion resistance of AZ91D obviously.

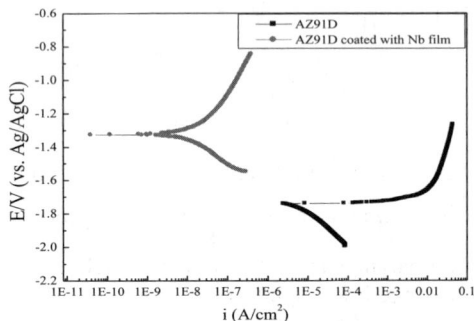

Fig. 6 Potentiodynamic polarization curves of AZ91D coated with niobium films

Summary
The microstructure and mechanical properties of niobium films are significantly affected by the deposition parameters. The lower sputtering power, the lower substrate temperature and the higher negative bias are all in favor of decreasing the surface roughness of niobium films. Besides, lower sputtering power and higher substrate temperature is helpful to increase the hardness of niobium films. The results of the electrochemistry experiments indicate that niobium films can evidently improve the corrosion resistance of the AZ91D magnesium alloy in 3.5% NaCl solution and the corrosion current density decreases about four orders of magnitude.

Acknowledgement

This work is financially supported by the National Basic Research Program of China (973 Program) (No. 2011CB013404) and National Natural Science Foundation of China (Nos. 50905038, 50875053).

References

[1] H. Wang, R. Akid and M. Gobara. Corros. Sci. 52 (2010), p. 2565.
[2] V. Barranco, N. Carmona, J.C. Galván, M. Grobelny, L. Kwiatkowski and M.A. Villegas. Prog. Org. Coat. 68 (2010), p. 347.
[3] Y. Gao, C. Liu, S. Fu, J. Jin, X. Shu and Y. Gao. Surf. Coat. Technol. 204 (2010), p. 3629.
[4] G.S. Wu, X.Q. Zeng and G.Y. Yuan. Mater. Let. 62 (2008), p. 4325.
[5] X.F. Cui, Q.F. Li, Y. Li, F.H. Wang, G. Jin and M.H. Ding. Appl. Surf. Sci. 255 (2008), p. 2098.
[6] C.D. Gu, J.S. Lian, G.Y. Li, L.Y. Niu and Z.H. Jiang. J. Alloys Compd. 391 (2005), p. 104.
[7] J.N. Balaraju, S.M. Jahan and C. Anandan. Surf. Coat. Technol. 200 (2006), p. 4885.
[8] Z.H. Li, Z.Y. Chen, S.S. Liu and F. Zheng. Trans. Non. Met. Soc. China, 18 (2008), p. 819.
[9] C.D. Gu, J.S. Lian, G.Y. Li, L.Y. Niu and Z.H. Jiang. Surf. Coat. Technol. 197 (2005), p. 61.
[10] W.X. Zhang, J.G. He, Z.H. Jiang and Q. Jiang. Surf. Coat. Technol. 201 (2007), p. 4594.
[11] A. Robin and J.L. Rosa. Int. J. Refract. Met. H. 18 (2000), p. 13.
[12] R. Günzel, S. Mändl, E. Richter, A. Liu, B.Y. Tang, and P.K. Chu. Surf. Coat. Technol. 1 16-119 (1999), p. 1107.

The Measurement of Mechanical Properties of Thermal Barrier Coatings by Micro-cantilever Tests

Dong Liu[1,2,a] and Peter E J Flewitt[1,3,b]

[1]Interface Analysis Centre, University of Bristol, Bristol, BS2 8BS, UK

[2]Dept. of Mechanical Engineering, University of Bristol, Bristol, BS8 1TR, UK

[3]School of Physics, University of Bristol, Bristol, BS8 1TL, UK

[a]dong.liu@bristol.ac.uk, [b]peter.flewitt@bristol.ac.uk

Keywords: Micro-cantilever test, air plasma sprayed thermal barrier coating, force measurement system, brittle fracture, rupture

Abstract Micro-scale cantilever beam specimens have been created in air plasma sprayed thermal barrier coatings (APS-TBC) by focus-ion beam milling and tested in-situ using a force measurement technique. The elastic modulus, fracture toughness and the flexural strength of the specimens are calculated from the loading-deflection curve. In addition, the failure modes of the tested TBC are analysed.

Introduction

Plasma-spraying is a commonly used deposition process to apply ceramic coatings to a substrate. The solid ceramic particles are injected into a high temperature gas where they are heated, melted, then accelerated and sprayed at relatively high velocity onto the substrate where they impact and form micro-scale multilayers of overlapping thin lenticular splats. The properties of the coating depend on (i) the starting material, (ii) microstructure produced, (iii) residual stresses induced while spraying, and (iv) the porosity of the coating (open or closed) [1]. The mechanical properties and their variability within a given coating are important when seeking to demonstrate the overall integrity of the coating. Air plasma-sprayed (APS) yttria stabilized zirconia (YSZ) ceramic coatings are used extensively for thermal barrier coatings (TBC) on first stage blades and vanes of advanced land-based gas turbines to protect the underlying superalloy substrate from oxidation and extreme operating temperatures [2].

Cracking or eventual delamination of the ceramic coating can expose the underlying superalloy to extreme temperatures leading to the failure in the function of turbines. The residual stresses and mechanical properties within the APS-TBC interact with each other and have been extensively investigated due to their important role of leading to failure [2]. The commonly encountered approach for mechanical property measurements is indentation methods carried out on bulk samples [3, 4]. The interpretation and deconvolution of the indentation data, however, is not straightforward due to the influence from the residual stress in the samples, the effects from the underlying substrates and most importantly the complex stresses created under the indent [5, 6].

This paper describes the data obtained using standard micro-scale specimens created at selected locations within the APS-TBC. The correlation between the loading-deflection curves and the internal microstructure, crack propagation and failure of the specimens are explored and discussed. Due to the small size of the specimens created, it is possible to separately examine the individual defects created by the manufacturing routes of the TBC and therefore improve the understanding of the flaw-sized sensitivity of the fracture strength of the material [7] to enhance the stability during service.

Experimental

The experimental arrangement for the mechanical property testing is a combination of a FEI Helios NanoLab 600i Dualbeam workstation and a compact force measurement system supplied by Kleindiek which allows in-situ loading and force readout. The dualbeam workstation provides ion beam milling and in-situ SEM imaging. Before commencing the measurements, a calibration of the force sensor is carried out by loading a standard spring embedded in the system to provide the resistance conversion references and zero the load reading. The specimen used is APS-7 wt.% Y_2O_3-stabilised ZrO_2 (YSZ) of thickness ≤ 250 µm. Micro-cantilever specimens with the size of ~$2\times2\times10$ µm are created at preferred locations of this coating. Details of the force measurement system and the cantilever specimen preparation are described elsewhere [8, 9].

Results and discussion

Failure is a process by which a material changes from one state of behaviour to another including stages of crack initiation and propagation. The specimens were found to divide into two: (i) those containing no external flaws and (ii) those containing visible external flaws. In the first case, the failure was evaluated in terms of elastic modulus and flexural strength whereas for the latter the fracture toughness could be derived.

Fig. 1a shows an example of a micro-cantilever with no surface flaws where the dimension is approx. $2.3\times2.3\times10$ µm. The loading was applied first from one side of the beam for three sequences (load 1, 2 and 3), and the fourth sequence was undertaken by applying force from the other side. The force-deflection curve for each sequence is plotted in Fig. 1b and a fractured surface in Fig.1c. The loading/unloading rates are listed in Table 1. Load 1 and 2 have a similar loading/unloading rate, whereas load 3 has a higher rate, and load 4 has the lowest rate of 9.3 µN/s. For the sequences 1, 2 and 3, the elastic modulus was calculated from the linear gradient of the curve to be 75±4 GPa, 90±5 GPa, and 87±7 GPa. For sequence 4, the elastic modulus decreased to 60±6 GPa.

There are three characteristic features associated with the loading/unloading sequences: (i) no residual plasticity was observed. The material showed linear elasticity during loading but with steps in the unloading, Fig. 1b; (ii) the elastic modulus increased by ~15 GPa after the first sequence and then remained constant during the second and third. This indicates that there may be internal modification of the microstructure after the initial loading so that the bending resistance of the beam is enhanced; (iii) for the sequences 1, 2 and 3, during unloading the deflection of the beam lags behind the force which indicates the dissipated energy in the micro-cantilever is history dependent, whereas for loading sequence 4, the force lags behind the deflection indicating gain of energy. Close to the end of unloading (~0.1 µm of deflection), the opposite occurred, hence little energy introduced by the loading process was stored in the specimen. On the other hand, this irregular hysteresis of the micro-cantilever may be caused by the contact between the probe and the specimen, but this is unlikely because the hysteresis loops switched systematically with the direction of loading, Fig. 1b. Therefore, the micro-cantilever has a complementary mechanical response when loaded from opposite directions. The flexural strength calculated is 28±6 MPa, which is consistent with the macroscopic values in the range of 30 to 40 MPa [10].

Table 1 Loading and unloading rate of the loading sequences

Loading rate (µN/s)	Load 1	Load 2	Load 3	Load 4
Loading (±2)	11.5	11.6	12.6	9.3
Unloading (±2)	11.1	11.5	12.6	9.3

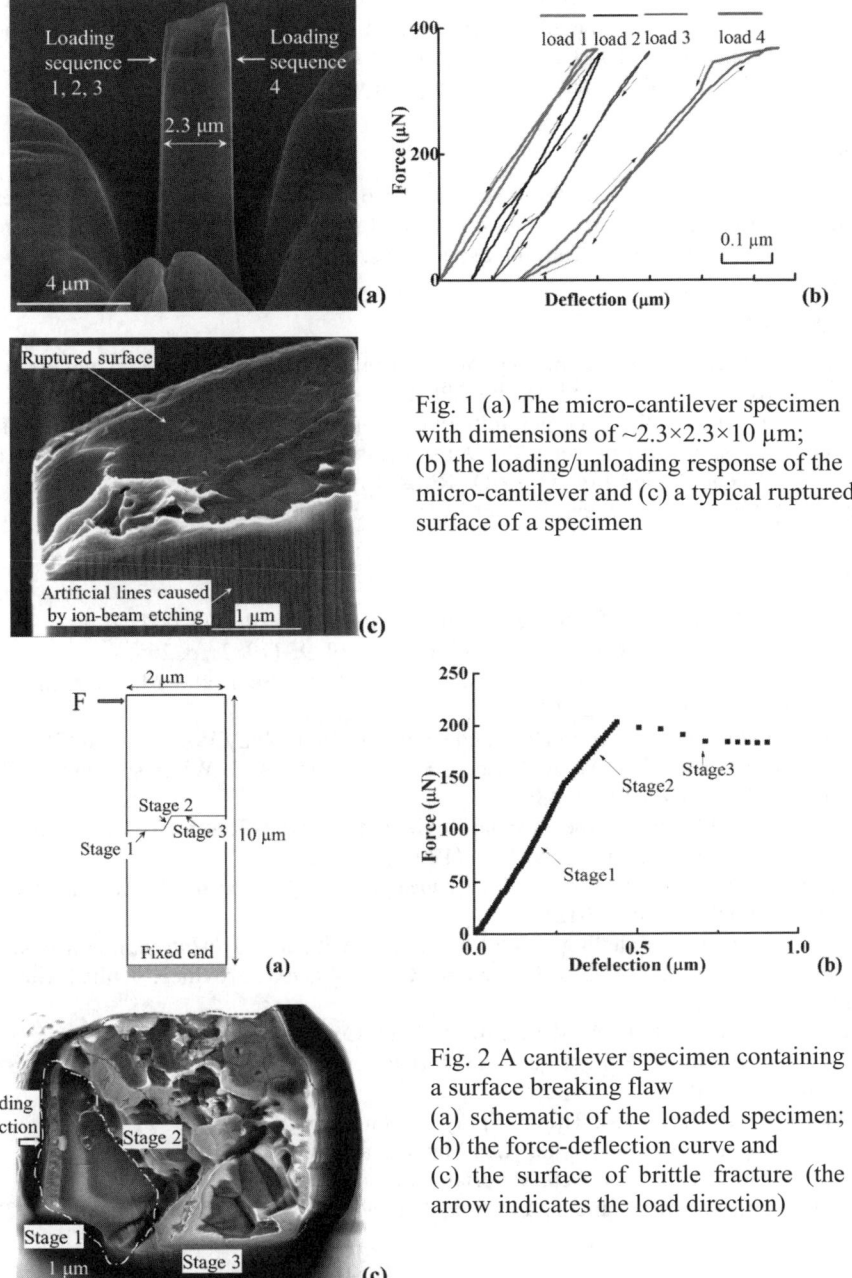

Fig. 1 (a) The micro-cantilever specimen with dimensions of ~2.3×2.3×10 μm; (b) the loading/unloading response of the micro-cantilever and (c) a typical ruptured surface of a specimen

Fig. 2 A cantilever specimen containing a surface breaking flaw
(a) schematic of the loaded specimen;
(b) the force-deflection curve and
(c) the surface of brittle fracture (the arrow indicates the load direction)

In the case of cantilever specimens containing surface breaking flaws, it can be considered as a bend specimen containing a pre-crack, Fig. 2a. The force-deflection curve, Fig. 2b, contains three district stages 1 to 3. During stage 1 there is elastic deflection and at the end of this stage (~150 μN) the gradient changes. Stage 2 reaches a maximum at ~200 μN. This is followed by a period of stable deflection as cracks propagate stably during stage 3. Examination of the final fracture surface, Fig. 2c, reveals the initial pre-existing flaw. There is evidence for some crack extension during stage 2 and then final mainly intergranular brittle fracture in stage 3. Compare the failed surfaces in Fig. 1c and Fig. 2c, the differences between the two failure modes can be distinguished. The elastic modulus calculated from these specimens varied from 30 to 40 GPa, which falls into the range of macroscopic values of 10 to 100 GPa [11-13]. The fracture toughness calculated is ~5 MPam$^{-1/2}$, which is in the higher end of the macroscopic range of values, 0.7 to 5 MPam$^{-1/2}$ [14, 15].

Conclusions

The micro-cantilever tests provide the mechanical properties for the relatively thin layer of APS-TBC coating. The values for elastic modulus, flexural strength and fracture toughness are in broad agreement with those determined from bulk specimens.

Acknowledgement

We would like to acknowledge the support of The Energy Programme, which is a Research Councils UK cross council initiative led by EPSRC and contributed to by ESRC, NERC, BBSRC and STFC, and specifically the Supergen initiative (Grants GR/S86334/01 and EP/F029748) and the following companies; Alstom Power Ltd., Doosan Babcock, E.ON, National Physical Laboratory, Praxair Surface Technologies Ltd, QinetiQ, Rolls-Royce plc, RWE npower, Siemens Industrial Turbomachinery Ltd. and Tata Steel, for their valuable contributions to the project.

References

[1] X. Q. Cao, R. Vassen and D. Stoever: J. Eur. Ceram. Soc. Vol. 24 (2004), p. 1.
[2] R. A. Miller and C. E. Lowell: Thin. Solid. Films. Vol. 95 (1982), p. 265.
[3] D. J. Kim, S. K. Cho, J. H. Choi, J. M. Koo, C. S. Seok and M. Y. Kim: J. Nanosci. Nanotechnol. Vol. 9 (2009), p. 7271.
[4] M. Eskner and R. Sandström: Surface and Coatings Technology Vol. 177-178 (2004), p. 165.
[5] D. E. J. Armstrong, A. S. M. A. Haseeb, S. G. Roberts, A. J. Wilkinson and K. Bade: Thin Solid Films Vol. 520 (2012), p. 4369.
[6] K. J. Hemker and W. N. Sharpe: Annual Review of Materials Research Vol. 37 (2007), p. 93.
[7] P. F. Becher: J. Am. Ceram. Soc Vol. 74 (1991), p. 255.
[8] D. Liu and P. E. J. Flewitt, in: *19th European Conference on Fracture*, Kazan, Russia. Submitted for publication. (2012).
[9] J. E. Darnbrough, S Mahalingam and P. E. J. Flewitt, in: *11th International Conference on Fracture and Damage Mechanics*, Xi'an City, Shaanxi Province, China. Submitted for publication. (2012).
[10] S. R. Choi, D. Zhu and R. A. Miller: Int. J. Appl. Ceram. Tec. Vol. 1 (2004), p. 330.
[11] K. W. Schlichting, N. P. Padture, E. H. Jordan and M. Gell: Mater. Sci. Eng. A-Struct. Vol. 342 (2003), p. 120.
[12] J. Wallace and J. Ilavsky: J. Therm. Spray. Techn. Vol. 7 (1998), p. 521.
[13] J. A. Thompson and T. W. Clyne: Acta Materialia Vol. 49 (2001), p. 1565.
[14] A. Rabiei and A. G. Evans: Acta Materialia Vol. 48 (2000), p. 3963.
[15] Y. Yamazaki, S. I. Kuga and T. Yoshida: Acta Metall. Sin. (Engl. Lett.) Vol. 24 (2011), p. 109.

Crack Growth by Dimensional Reduction Methods

P.H. Wen[1,a] and M.H. Aliabadi[2,b]

[1] School of Engineering and Material Sciences, Queen Mary, University of London, UK

[2] Department of Aeronautics, Imperial College, London, UK

[a] p.h.wen@qmul.ac.uk, [b] m.h.aliabadi@imperial.ac.uk

Keywords: Dual boundary element method, element-free Galerkin method, stress intensity factors, fatigue crack growth.

Abstract This paper presents a new fatigue crack growth prediction by using the dimensional reduction methods including the dual boundary element method (DBEM) and element-free Galerkin method (EFGM) for two dimensional elastostatic problems. One crack extension segment, i.e. a segment of arc, is introduced to model crack growth path. Based on the maximum principle stress criterion, this new prediction procedure ensures that the crack growth is smooth everywhere except the initial growth and the stress intensity factor of mode II is zero for each crack extension. It is found that the analyses of crack paths using coarse/large size of crack extension are in excellent agreement with analyses of the crack paths by the tangential method with very small increments of crack extension.

Introduction

This paper aims to develop a more accurate and simple technique to predict the crack growth path with a fixed length arc crack extension. The curve is smooth everywhere and the radius of arc is determined by the condition that the mode II stress intensity factor is zero at the new crack tip. This new crack growth prediction is implemented by the use of the dual boundary element method and element-free Galerkin method. The proposed crack growth prediction and predictions described in the literature are demonstrated by solving two different examples. In the two examples it is found that analyses of the crack paths with the proposed technique using big increments of crack tip extension are in excellent agreement with analyses of the crack paths using very small increments of crack tip extension.

Dimensional reduction methods

1. *Dual boundary element method*

Based on the displacement boundary integral equation, direct boundary element formulation is written as [1-3]

$$c_{ij}u_j(\mathbf{x}') = \int_\Gamma U_{ij}(\mathbf{x}',\mathbf{x})t_j d\Gamma(\mathbf{x}) - \int_\Gamma T_{ij}(\mathbf{x}',\mathbf{x})u_j d\Gamma(\mathbf{x}) \quad (1)$$

where U_{ij} and T_{ij} represent the displacement and traction fundamental solutions, $\mathbf{x}'$ and $\mathbf{x}$ are the source point and boundary integral field point respectively, $\int_\Gamma$ stands for the Cauchy principal-value integral and coefficient c_{ij} depends on the boundary geometry, which is equal to $\delta_{ij}/2$ for a smooth boundary. The dual boundary element method incorporates two independent boundary integral equations which are the displacement equation (1) applied for source points on one of crack surfaces and the traction boundary integral equation

$$\frac{1}{2}t_j(\mathbf{x}') = n_i(\mathbf{x}')\int_\Gamma D_{kij}(\mathbf{x}',\mathbf{x})t_k d\Gamma(\mathbf{x}) - n_i(\mathbf{x}')\int_\Gamma S_{kij}(\mathbf{x}',\mathbf{x})u_j d\Gamma(\mathbf{x}) \quad (2)$$

applied for source points on the other crack surface. In which, $\int_\Gamma$ stands for the Hadamard principal-value integral, D_{kij} and S_{kij} contain the derivatives of the displacement and traction fundamental solutions U_{ij} and T_{ij}, n_i denotes a unit outward normal vector. The numerical implementation of dual boundary element method can be found in [2-3]. By introducing continuous and discontinuous element shape functions, the boundary integral equations (1) and (2) are transformed into a system of linear algebraic equations relative to the nodal field variables.

2. *Element-free Galerkin method*

With the shape functions, the displacements $\mathbf{u}(\mathbf{y})$ at domain point $\mathbf{y}$ can be approximated in terms of the nodal values in a local domain, called as support domain [4-5], as

$$u_i(\mathbf{y}) = \sum_{k=1}^{n} \phi_k(\mathbf{y},\mathbf{x}_k)\hat{u}_i^k = \overline{\Phi}(\mathbf{y},\mathbf{x})\hat{\mathbf{u}}_i \qquad (4)$$

where $\overline{\Phi}(\mathbf{y},\mathbf{x}) = \{\phi_1(\mathbf{y},\mathbf{x}_1), \phi_2(\mathbf{y},\mathbf{x}_2),...,\phi_n(\mathbf{y},\mathbf{x}_n)\}$ (5)

and the nodal value $\hat{\mathbf{u}}_i = \{\hat{u}_i^1, \hat{u}_i^2,..., \hat{u}_i^n\}^T$, $i = 1,2$ (6)

at collocation point $\mathbf{x}_k = \{x_1^{(k)}, x_2^{(k)}\}$, where $k = 1,2,...,n(\mathbf{y})$, ϕ_k the shape function and $n(\mathbf{y})$ the number of nodes in the local supported domain. To capture the singular stresses at crack tip, an enriched radial basis function has been selected as following [5]

$$R_k(\mathbf{y},\mathbf{x}) = R(y,x) = \sqrt{c^2 + |\mathbf{y}-\mathbf{x}_k|^2} + \sqrt{r}\, e^{-\alpha r} \qquad (7)$$

where $r = |\mathbf{y}-\mathbf{y}_c|$; c and α are free parameters; $\mathbf{y}_c\,(y_1^{(c)}, y_2^{(c)})$ denotes the location of the crack tip. For the two dimensional elasticity, we can rearrange the above relation in a matrix form as

$$[\mathbf{K}]_{2N \times 2N}\, \hat{\mathbf{u}}_{2N} = \mathbf{f}_{2N} \qquad (8)$$

Crack growth with arc increment

In general case, the crack path should be a curved smooth path. There are few methods to determine the crack growth path which is simulated by successive linear increments. For these methods, several criteria have been used to describe the local direction of mixed-mode crack growth. The maximum principal stress criterion is most popular adopted in the crack propagation analysis. That is, at each crack tip, the local direction of crack growth is determined by condition that the shear stress is zero, i.e.

$$K_I \sin\theta_c + K_{II}(3\cos\theta_c - 1) = 0 \qquad (9)$$

$$\theta_c = 2\arctan\left[\left(\psi \pm \sqrt{\psi^2 + 8}\right)/4\right] \qquad (10)$$

Obviously the direction of crack growth depends on the ration ψ only near the crack tip.

Assume that the location of crack tip is $a_n(x_1^n, x_2^n)$ and the crack local coordinate system is (x_1', x_2'). At crack tip, we hold $K_{II} = 0$ and therefore the crack extension must be exist along direct of coordinate x_1'. An arc crack extension with length of Δa and radius ρ_n is introduced in this approach. Considering the smooth crack growth path, the centre of the arc extension should be allocated on the axis x_2' and its coordinate $c_n\,(0,\pm\rho_n)$ is unknown for certain length of crack extension as shown in Figure 1(a). For each crack extension, the length of the arc Δa is specified for each problem. In addition, the arc is divided into several straight elements for DBEM or several collocation points are distributed on the arc for EFGM as shown in Figure 1(b). To determine the radius of arc, the condition of mode II stress intensity factor $K_{II}(a_{n+1}) = 0$ at new crack tip $a_{n+1}(x_1^{n+1}, x_2^{n+1})$ is considered. The angle of rotation between these two local coordinate systems (x_1^n, x_2^n) and (x_1^{n+1}, x_2^{n+1}) is $\varphi_n = \pm\Delta a/\rho_n$ (see Figure 1). Two conditions are satisfied by this approach: (a) the crack growth path is smooth; (b) the mode II stress intensity factor at crack tip for each extension is zero. In fact, the stress intensity factor $K_{II}(s)$ is relatively small compared with $K_I(s)$ for any crack length l apart from each crack tips a_n, $n = 1,2,...,N$. An iteration procedure is introduced to determine radius ρ_n, then the new crack tip a_{n+1} with new local coordinate system (x_1^{n+1}, x_2^{n+1}) are determined consequently. It needs to be point out that, for the first crack growth element (at initial crack tip), the direction of local coordinate axis x_1^1 is determined by θ_c in Eq. (10).

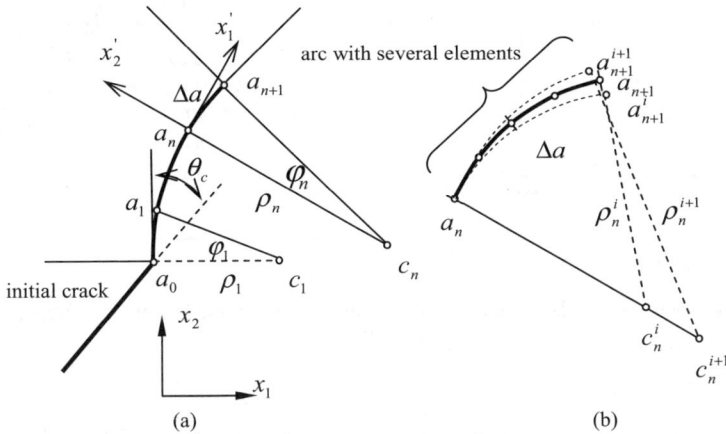

Figure 1. Crack growth path and local crack tip coordinate system: (a) crack growth path; (b) iteration procedure to determine position of arc crack extension.

A square plate under a mode II fatigue load

A square plate (20cm×20cm) was investigated for a mode II fatigue load by Kim and Lee [7]. Constant cyclic mode II traction in the range from 0 to 165MPa is applied on the upper and lower left edge of the plate in the opposite direction. Possion ratio $v = 0.3$. By using DBEM, there are 90 boundary elements collocated on the outer boundary and 20 discontinuous elements on each crack surface. In the case of $a = 2cm$, all discontinuous elements on the crack surface are of the same length of 0.1cm. In addition, EFGM is used for this problem. For element-free Galerkin method, total 41×41 notes are uniformly distributed in the domain and 17 extra nodes around the crack tip to achieve more accurate results. The fatigue crack growth is simulated with crack element of length 0.25cm. The radius of the support domain r_y centered at field point $\mathbf{y}$ is determined such that the minimum number of nodes in the support domain $n(\mathbf{y}) = 8$. The integration is performed on 80×80 cells with 4×4 Gauss points in each cell. The growth paths are shown in Figures 2 for different initial crack lengths. Two crack extension lengths are considered too, i.e. $\Delta a = 6cm$ and $\Delta a = 3cm$ respectively. It has been observed that very similar results can be obtained with different crack extensions by DBEM. To compare with other crack growth simulation, the numerical results given by Kim and Lee [7] are plotted in the same figure. Apparently the agreement between these different approaches has shown to be satisfactory.

Conclusion

This paper presented a new crack growth correct implemented with the dual boundary element method and element-free Galerkin method. The determination of new crack tip with an arc extension ensures that the mode II stress intensity factor is zero and the crack growth path is smooth everywhere. Both dimensional reduction methods do not need regeneration of mesh for crack propagation analysis. The proposed method can be extended to other types of cracks growth as reported in [8-18].

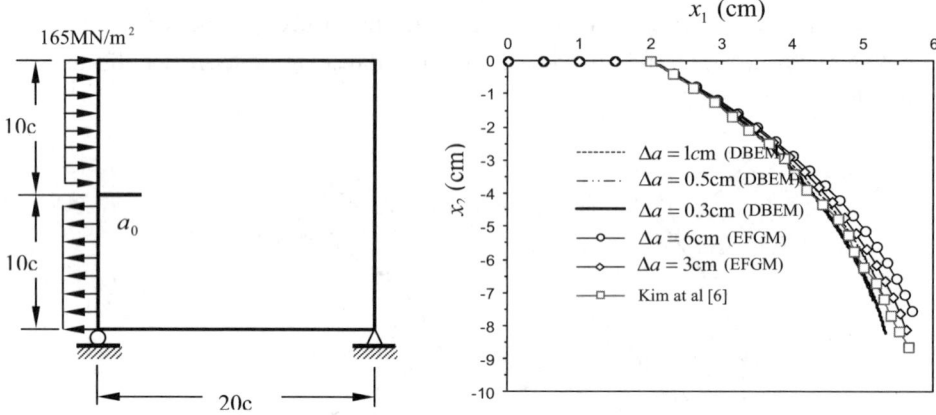

Figure 2. Geometry of an edge cracked square plate under a mode II load and paths of crack growth.

References

[1] Portela, A., Aliabadi, M.H., Rooke, D.P., Computer and Structures, 46 (2), 237-247, 1993.
[2] Portela, A., Aliabadi, M.H., Rooke, D.P INTERNATIONAL JOURNAL FOR NUMERICAL METHODS IN ENGINEERING Vol33 Issue: 6 P 1269-1287, 1992
[3] Aliabadi, M.H., The boundary element method: Applications in solids and structures, John Wiley & Sons, Ltd., West Sussex, England, 2002.
[4] Nayroles, B., Touzot, G and Villon, P., 1992. Computational Mechanics 10, 307-318.
[5] Belytschko, T., Lu, Y.Y & Gu, L., 1994. Int. J. Numerical Methods in Engineering 37, 229-256.
[6] Wen, P.H. and Aliabadi, M.H., Structural Durability and Health Monitoring, **3** (2), 107-119, 2007.
[7] Kim, K and Lee, H., Int. J. Numerical Methods in Engineering 72, 697-721, 2007.
[8] Wen PH; Aliabadi MH; Rooke DP, Computer Methods in Applied Mechanics and Engineering, Vol 167 Issue: 1-2 P139-151, 1998
[9] AL Saleh, MH Aliabadi Engineering Fracture Mechanics 51 (4), 533-545, 1995.
[10] A Portela, MH Aliabadi, DP International Journal of Fracture 55 (1), 17-28, 1992
[11] P Sollero, MH Aliabadi Composite Structures 31 (3), 229-233, 1995
[12] PH Wen, MH Aliabadi International Journal of Solids and Structures
[13] Fedelinski P; Aliabadi MH; Rooke DP Computers & Structures Vol59 Issue: 6 P 1996
[14] Wen P. H.; Aliabadi M. H.; Lin Y. W. CMES-COMPUTER MODELING IN ENGINEERING & SCIENCES Vol 30   Issue: 3   P 133-147   2008
[15] Albuquerque EL; Sollero P; Aliabadi MH, International Journal for Numerical Methods in Engineering Vol: 59 Issue: 9   P 1187-1205   2004
[16] Dirgantara T; Aliabadi MH International Journal of Fracture Vol 105   Issue: 1   P: 27-47, 2000
[17] Sfantos G. K.; Aliabadi M. H. Source: Computer Methods in Applied Mechanics and Engineering, Vol **196**   Issue: **7** P **1310-1329   2007**
[18] Wen P. H.; Aliabadi M. H., Communications in Numerical Methods in Engineering Vol: 24   Issue: 8   P: 635-651   2008

Nondestructive Evaluation of Polyurethane Materials Using Transient Thermography

Mihaela Amarandei[1,a], Karla Berdich [1,b], Izabella Szigyarto[2,c], Lorand Kun[1,d] and Liviu Marşavina[1,e]

[1]"Politehnica" University of Timişoara, Mechanical Engineering Faculty, Mechanics and Strength of Materials Department, Bd. Mihai Viteazul 1, RO-300222 Timişoara, Romania

[2] "Politehnica" University of Timişoara, Faculty of Civil Engineering, Hidrotechnics Department, Str. Traian Lalescu 2, RO-300223 Timişoara, Romania

[a]amarandei_mihaela@yahoo,com, [b]karla.berdich@gmail,com, [c]izabela_sz@yahoo.com, [d]kunlori@yahoo.com, [e]lmarsavina@yahoo.com

Keywords: thermography, nondestructive evaluation, defect, necuron.

Abstract. The aim of this work is to investigate the potential of transient thermography in the nondestructive evaluation of structural defects of NECURON 1001 using the FLIR thermographic system. Necuron is a polyurethane material used for applications like: fixture and gauges, master and copy models, models with high mechanical stress, etc. Transient thermography is a thermographic method which implies the investigation of materials that are of a different (often higher) temperature than the ambient. The heat flow into the sample is altered in the presence of a subsurface defect or feature, creating a temperature contrast at the surface that is recorded by the infrared system. Results show that this method of evaluation can indicate, in necuron, defects of small sizes that can be overlooked in the manufacturing process. Also, it was shown that the transient thermography method presented can be an important tool in evaluating structural defects of materials.

Introduction

Every object with a temperature above zero absolute (0 K) emits infrared radiation (IR). Thus, an excellent way of performing quality assurance and evaluation of materials in a nondestructive manner is to use a infrared vision device such as an infrared camera capable of detecting radiation in the two of the high transmittance infrared bands (MWIR and LWIR) [1].

In the last decade, the industrial interest is more and more oriented towards using and developing new nondestructive testing techniques in order to obtain cost efectiveness and high accuracy [2].

Thermography is one of these nondestructive testing techniques and is widely used for detection of defects impact damages, fatigue degradation of different materials [3] [4], as a fast inspection method and for the characterization of many engineering materials [2] [5].

For the evaluation of materials using thermography there are two aproaches one can use: passive [6] and active [7].

Passive thermography is used when the features of interest are of different temperature than the background. When an energy source is introduced in order to obtain a contrast between the features of interest and the ambient we speak of active thermography. The transfer of energy during active thermography can occur in different ways, one of which is transient thermography that uses short or long pulses of energy from different sources (i.e. optical flash lamps, heat lamps, hot or cold air guns) [8].

Our study used active long-pulse transient thermography in order to determine and assess defects of various depths and sizes in a block of NECURON 1001.

NECURON 1001 is a polyurethane material usually used in application like: fixtures and gauges, master and copy models, models with high mechanical stress and tooling jigs, tools for serial production [9].

Lately necuron was used in different industrial applications and research such as airplane wing tesing in wind tunnels or in bioengineering research for heart valve damage [10] [11].

It is also very important to mention that characterization research of necuron is still in its first years. Many data sheets of manufacturers lack reliable information for material properties and they usualy differ from one producer to another. It can be found that in the description of some polymeric materials, necuron included, the impact strength refers to notched specimens and for some to un-notched specimens. This and the fact that polymeric materials have different chemical composition and are produced in different conditions shows how important research is that can provide new characterization information and determines mechanical properties in different experimental conditions [12] [13].

Materials and methods

Images were obtained using the FLIR A40M Infrared Camera System, a laser thermomether and a weather station to determine ambient conditions. The A40M IR camera was specially created to be mounted in any position and to depend only on the ambiental factors. Measurements are flexible and it provides images with a resolution of 320x240 pixels (FLIR A40M manual). In order for temperature contrast to occur, an infrared lamp and a refrigerator were used.

A NECURON 1001 parallelipipedic sample was drilled and 9 holes were obtained: 3 holes of 1 mm, 3 of 3 mm and 3 of 5 mm, each having a different depth of 5, 10 and 14 mm (going through the entire thickness of the necuron block) as shown in Fig. 1.

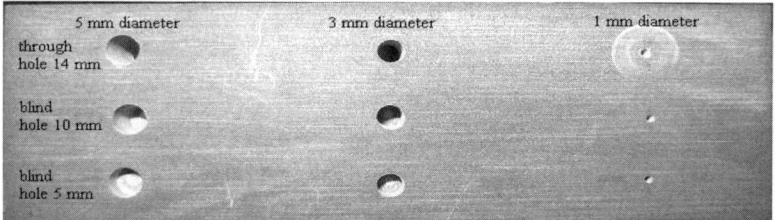

Fig. 1 Sample of Necuron 1001 with the 9 drilled holes

All test parameters are shown in Table 1, as well as the emissivity of the NECURON 1001 sample. When using an IR camera and processing its data, it is important to take into account that the emissivity differs depending on the temperature of the material.

The sample of necuron underwent 5 cycles of heating and 5 cycles of cooling. Both heating and cooling began at ambient temperature (Table 1).

Table 1 Test parameters

Parameter	Cold	Hot
Object emissivity[-]	0,4 for -20 to +20°C	0,8 for +20 to +60°C
Distance from sample to IR camera [m]	0.3 m	0.5 m
Ambient temperature [°C]	24.5	24.5
Relative humidity [%]	25	25
Optic transmission of IR camera[-]	1	1
External temperature of camera [°C]	25	25

The heating test was done until the sample reached 65°C. The dilatation of the material was observed beyond this temperature. The cooling test was done keeping the sample for 10 minutes in the refrigerator (2.5°C).

After heating or cooling, the sample's return to the ambient temperature was visualized with the IR camera. At different steps, images were taken in order to observe the defects in the material.

Results

Processed data showed that all holes exhibit an accumulation of energy inside, especially for the heating tests. Thus Fig. 2, for the heating cycles, shows in sections A, B and C the accumulation of higher temperatures for the 5, 3 and 1 mm holes, regardless of depth.

Similarly, for the cooling cycles, we can observe (Fig. 3) that section B and C show peaks of temperature, regardless of depth, for the 5 and 10 mm depths. Section A describes here also the 14 mm depth holes, but in this case we can not conclude that these defects were identified in the graphic. All 10 tests, 5 cycles of cooling and 5 of heating, showed similar results.

In Fig. 4 one can observe that on the left, the 1 mm holes are more difficult to observe. This happened for all cooling cycles regardless of the temperature. On the right, we can see the block of necuron after it was heated and left to return to ambient temperature. Results showed that for every heating cycle, the 1 mm holes are visible above 40°C.

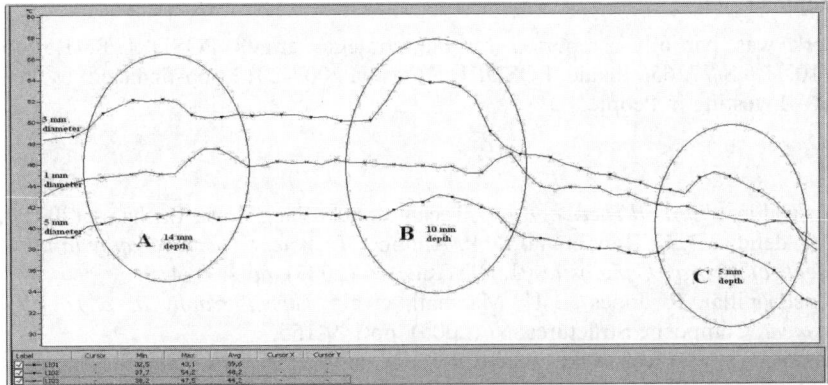

Fig. 3 Recorded data of the returning to ambient temperature process after a heating cycle

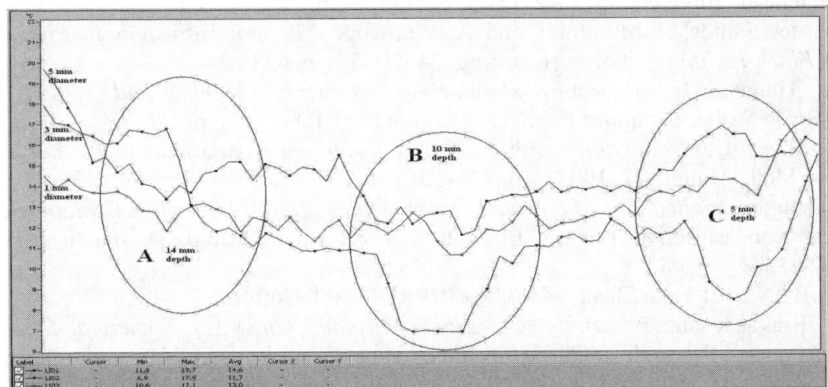

Fig. 3 Recorded data of the returning to ambient temperature process after a cooling cycle

9.2 °C 45.5 °C
Fig. 4 The necuron block: left - after cooling; right - after heating.

Conclusions

It is known that through a cross section of homogeneous material heat propagates uniformly. If there is a defect in the structure, the temperature distribution is disturbed due to heat concentrators [1]. This was the case of our tests also. For both, heating and cooling cycles, the defects were visualized when inspecting images with the naked eye and in the processed data because of the energy retained inside. Thus the method chosen showed good and reliable results.

It was shown that heating tests were more reliable in providing information about defects of the material. All of the holes were visible with the IR camera for the heating tests and only about 90 percent were visible for the cooling tests. This was similar for the visual inspection of the thermographic images.

One very important future research for our team is to determine defects from these images using pattern recognition software.

Acknowledgment

This work was partially supported by the strategic grants POSDRU/88/1.5/S/50783 and POSDRU 107/1.5/S/77265, inside POSDRU Romania 2007-2013, co-financed by the European Social Fund – Investing in People.

References

[1] M. Alexandrina: *Infrared Thermography*, Technical Publishing House Bucharest (2005), pp. 23-28.
[2] N. P. Avdelidis, B. C. Hawtin and D. P. Almond: *Transient Thermography in the Assessment of Defects of Aircraft Composites*, J. NDT&E., 36 (2003), pp. 433–439.
[3] M. Krishnapillai, R. Jones, I. H. Marshall, et al.: *Thermography as a Tool for Damage Assessment*, Composite Structures, 67 (2005), pp 149-155.
[4] G. Muzia, Z. M. Rdzawski, M. Rojek, et al.: *Thermographic Diagnosis of Fatigue Degradation of Epoxy-Glass Composites*, Journal of Achievements in Materials and Manufacturing Engineering, 24(2) (2007), pp.123-126.
[5] W. N. dos Santos, P. Mummery and A. Wallwork: *Thermal Diffusivity of Polymers by the Laser Flash Technique*, Polymer Testing, 24 (2005), pp. 628-634.
[6] R. A. Thomas: *Thermography Monitoring Handbook, Machine and Systems Condition Monitoring Series*, Coxmoor Publishing Company (UK) (1999). pp. 70-72.
[7] S. M. Shepard: *Introduction to Active Thermography for Nondestructive Evaluation*, J. Anti-Corros. Meth. Mater., 44 (1997), pp. 236–239.
[8] X. Maldague: *Applications of Infrared Thermography in Nondestructive Evaluation*, Trends in Optical Nondestructive Testing, Elsevier Science, P.K. Rastogi, D. Inaudi (Eds.) (2000), pp.112-123.
[9] NECURON 1001 Data Sheet, NECUMER-PRODUCT GmbH.
[10] K. H. Brakhage and P. Lamby: *Generating Airplane Wings for Numerical Simulation and Manufacturing*, Proceedings of the 9th International Conference on Numerical Grid Generation in Computational Field Simulations, San Jose, California, USA (2005), pp.85-89.
[11] M. Vermeulen, R. Kaminsky, Van Der Smissen et al.: *In Vitro Flow Modelling for Mitral Valve Leakage Quantification*, 8th International symposium on particle image velocimetry, Melbourne, Victoria, Australia (2009), pp.221-224.
[12] L. Marsavina, A. Cernescu, E. Linul, et al.: *Experimental Determination and Comparison of Some Mechanical Properties of Commercial Polymers*, Materiale plastice, 47(1) (2010), pp.85-89.
[13] I. M. Ward and J. Sweeney: *An Introduction to the Mechanical Properties of Solid Polymers*, Second edition, John Wiley and Sons, Cheichester (2004), pp. 25-28.

Numerical Modelling of Particulate Composite with a Hyperelastic Matrix

B. Máša[1,2,a], L. Náhlík[1,2,b] and P. Hutař[1,2,c]

[1]CEITEC IPM, Institute of Physics of Materials, Academy of Sciences of the Czech Republic, Žižkova 22, 616 62 Brno, Czech Republic

[2]Faculty of Mechanical Engineering, Brno University of Technology, Technická 2
616 69 Brno, Czech Republic

[a]masa@ipm.cz, [b]nahlik@ipm.cz, [c]hutar@ipm.cz

Keywords: hyperelasticity, particulate composite, FEM, unit cell model

Abstract. The main aim of the paper is an estimation of the macroscopic mechanical properties of particulate composites using numerical methods. Matrix of the considered composite was cross-linked polymethyl methacrylate - PMMA in a rubbery state, which exhibits hyperelastic behaviour. The three parameter Mooney Rivlin material model, which is based on the strain energy density function, was chosen for description of the matrix behaviour. Alumina based particles (Al_2O_3) were used as a filler. Numerical modelling based on the finite element method (FEM) was performed to determine stress-strain curve of the considered particulate composite. Representative volume element (RVE) model was chosen for FE analyses as a modelling approach of a composite microstructure. Various geometry arrangements of particles and various directions of loading have been considered and composite anisotropy has been investigated. A good agreement between numerical calculations with damage model and experimental data has been found and the described method may have a great potential for numerical modelling of composite behaviour and design of new particulate composite materials.

Introduction

The main advantage of composite materials is generally a usage of various materials in a different geometric configuration; the resulting composite is optimally designed for the expected function. Laminates can be obtained by stacking various plates for the required rigidity, weight, improved thermal and fracture properties. Likewise, long-fibre composites can exhibit better mechanical properties e.g. by winding fibres in the principal stress directions. In both cases the anisotropy is apparent. However, the interesting question is the anisotropy of particle composites. From the macroscopic point of view the composite seems to be homogenous and isotropic. However, the greatest influence on the degree of anisotropy has just distribution of particles in the matrix.

Nowadays, the main way for determination of mechanical properties of materials is the experimental testing. Nevertheless, for some applications the numerical methods are also suitable and more progressive. Some approaches have already been published. However, formerly performed two-dimensional calculations do not provide satisfactory results and have a large number of shortcomings see e.g. [1-5]. The necessity to carry out 3D computations has become obvious. Based on previous assumptions 3D FE models based on periodically repeated unit cells have been successfully developed, see e.g. [4,6-11] to determine the macroscopic behaviour of a composite material.

Material Model

The material of composite matrix was a cross-linked PMMA (polymethyl methacrylate), mechanical properties were determined experimentally by a standard tensile test using a typical dog bone specimen. The temperature of 180 °C has been applied to reach the rubbery state of the matrix.

Mooney-Rivlin material model is often used for the elastic response of rubber-like material, e.g. [12]. For this application the hyperelastic three parameter Mooney-Rivlin material model was chosen, where the strain energy density function W is described as:

$$W = C_{10}(\overline{I}_1 - 3) + C_{01}(\overline{I}_2 - 3) + C_{11}(\overline{I}_1 - 3)(\overline{I}_2 - 3) + \frac{1}{d}(J - 1)^2, \qquad (1)$$

where $\overline{I}_1$ is the first deviatoric strain invariant ($\overline{I}_1 = J^{-2/3} I_1$, where $I_1 = \lambda_1^2 + \lambda_2^2 + \lambda_3^2$), $\overline{I}_2$ is the second deviatoric strain invariant ($\overline{I}_2 = J^{-4/3} I_2$, where $I_2 = \lambda_1^2 \lambda_2^2 + \lambda_2^2 \lambda_3^2 + \lambda_3^2 \lambda_1^2$), λ_1, λ_2, λ_3 are principal stretches, J is the determinant of the deformation gradient, C_{10}, C_{01} and C_{11} are material constants and d is the parameter of incompressibility of the material. The relationship between strains and stretches is as follows: $\varepsilon_i = \lambda_i - 1$, where i = 1,2,3. Rubber-like materials are nearly incompressible (Poisson's ratio $v = 0.5$) and their bulk modulus is infinite. This fact complicates numerical calculations. Due to that in the numerical calculations performed the value of Poisson's ratio of the matrix $v_m = 0.4995$ was considered. The material constants of the cross-linked polymer matrix were fitted from experimental data and identified as: $C_{10} = 0.18560$ MPa, $C_{01} = 0.12356$ MPa, $C_{11} = 0.00351$ MPa and parameter d = 0.00500 MPa^{-1}. The alumina based particles were used as a filler of the particulate composite with volume fraction of 17.3% (30 weight %). Particles had a spherical shape and their typical size was approximately 10μm. Material model of a filler was considered as a linear, homogenous and isotropic with elastic constants: Young's modulus $E_p = 380$ GPa and Poisson's ratio $v_p = 0.23$. The limit strain (strain at break) of the composite was experimentally identified as ε = 0.49.

Numerical Model of the Composite

The distribution of particles in composite was inspired by atomic arrangement in crystal lattices. Thus, three different basic types of regular particle arrangements were used in the numerical models: simple cubic (SC), body-centred cubic (BCC) and face-centred cubic (FCC). The rotation of these arrangements by 45 degrees has been investigated to describe the degree of anisotropy (models SC-45, BCC-45 and FCC-45, see Figure 1). Extended numerical analyses were also performed for differences in particle sizes: body-centred cubic arrangement with two sizes of particles 2 μm and 10 μm (BCC-VAR and rotated BCC-VAR-45).

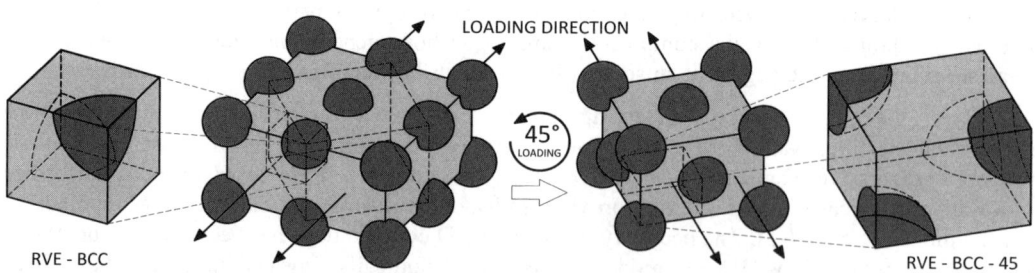

Figure1. Particles positions for BCC arrangement model for both considered loading directions and differences in RVE for BCC and BCC-45 models.

No clustering of particles was observed in the real composite - the assumption of regular distribution of particles in the matrix is close to reality. The regular distribution of particles in the matrix was modelled using unit cells. Every cell has a unique geometry characteristic for the distribution of particles and considered direction of loading. For example, rotated BCC model has the same configuration of particles in the cell like FCC, but different ratios of cell sides and size of particles. For finite element calculations the commercial software ANSYS [13] was used. The number of elements exceeded 100 000 3D 10-node tetrahedral elements SOLID187 and models have been solved by computer cluster with 12 x 2.4 GHz cores, GPU NVIDIA Tesla and 48 GB RAM.

Boundary conditions were applied to describe the behaviour of the composite in uniaxial tension test in terms of its microstructure. Three perpendicular faces of the cubic geometry were fixed in the direction of their normal, another two faces were constrained to remain parallel and planar after loading and normal displacement was applied to the top face (Figure 2). The macroscopic applied (engineering) strain was obtained as $\varepsilon_{appl} = u_{appl} / h$, where h is the height of the elementary cell. The maximal value of macroscopic strain exceeded $\varepsilon_{appl} = 0.45$, i.e. almost the value of the failure strain of the real composite. The engineering stress was obtained as a sum of particular reactions in the nodes on the top face of the model divided by the area of this face.

The experimental stress-strain data were also expressed in engineering coordinates. Ideal adhesion between particle and matrix was presumed. For all above mentioned models have been performed numerical solutions. The obtained results are shown in Figure. 3.

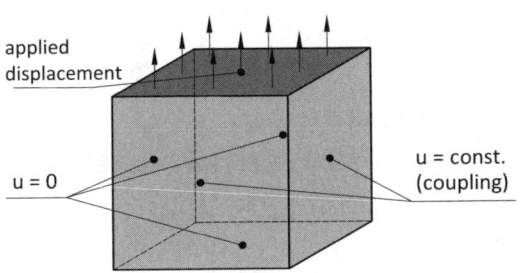

Figure 2. Applied boundary conditions on faces of the unit cell (RVE).

It is obvious that the distribution of particles in the matrix has a large influence on the stiffness of the composite, although the volume fraction of particles was retained. The best agreement with experimental data was obtained by SC and BCC-VAR models up to macroscopic strain $\varepsilon_{appl} = 0.18$, at higher strains the numerical results are rather overestimated.

This is due to begin of damage processes in the matrix, see [14]. Moreover, it may also be noted that much smaller particles in the RVE have a smaller effect than would be expected. Rotated modifications SC-45 and BCC-VAR-45 have stiffness decreased to about 20%.

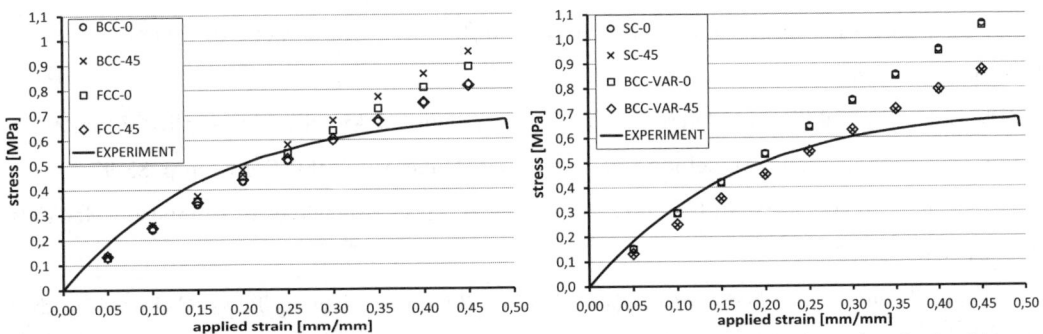

Figure 3. Comparison of experimental data (smooth curve) and numerical results obtained by models: body-centred cubic (BCC), rotated body-centred cubic by 45 degrees (BCC-45), face-centred cubic (FCC) and rotated face-centred cubic by 45 degrees (FCC-45), simple cubic (SC), rotated simple cubic by 45 degrees (SC-45), body-centred cubic arrangement with two sizes of particles (BCC-VAR) and rotated BCC-VAR (BCC-VAR-45).

BCC and FCC models exhibit from the beginning lower stiffness than a real composite, the FCC model in the basic configuration has a stiffness of 10% higher than the BCC model. However, BCC-45 model (BCC arrangement after rotation turns to the FCC arrangement but the ratios of RVE edges and size of particles are different) increase in the rigidity of the BCC arrangement by about 20%, but the FCC model after rotation by 45 degrees (FCC-45) has a stiffness of 10% lower.

Summary

The problem of anisotropy of particulate composites with hyperelastic matrix was studied by a numerical modelling. It was found that distribution of particles significantly influences the macroscopic stiffness of the composite. This effect was quantified for a various arrangements of particles in the matrix and different direction of loading. The differences in stiffness are presented and discussed in wider context. The results and described method may have a great potential for numerical modelling of composite behaviour and design of new particulate composite materials.

Acknowledgment: this work was supported through the Specific academic research grant of the Ministry of Education, Youth and Sports of the Czech Republic No. FSI-S-11-11/1190 provided to Brno University of Technology, Faculty of Mechanical Engineering and grant No. 106/09/H035 of Czech Science Foundation. The research was realised in CEITEC - Central European Institute of Technology with research infrastructure supported by the project CZ.1.05/1.1.00/02.0068 financed from European Regional Development Fund.

References

[1] T. Iung and M. Grange: Mater. Sci. Engng. A201 (1995), L8-L11

[2] H.P. Gänser, F.D. Fischer and E.A. Werner: Comp. Mater. Sci. Vol. 11 (1998), p. 221-226

[3] N. Chawla, R.S. Sidhu and V.V. Ganesh: Acta Mater. 54 (2006), p. 1541-1548

[4] P.R. Marur: Acta Mater. 52 (2004), p. 1263-1270

[5] W. Han, A. Eckschlanger and H.J. Böhm: Compos. Sci. Technol. 61 (2001), p. 1581-1590

[6] X. Wang, K. Xiao, L. Ye, Y. W. Mai, C.H. Wang and L.R.F. Rose: Acta Mater. 48 (2000), p. 579-586

[7] X. Zeng, H. Fan and J. Zhang: Comp. Mater. Sci. 40 (2007), p. 395-399

[8] J. Cho, M.S. Joshi and C.T. Sun: Compos. Sci. Technol. 66 (2006), p. 1941-1952

[9] P. Hutař, Z, Majer, L. Náhlík, L. Šestáková and Z. Knésl: Mech. Compos. Mater., 45, No. 3 (2009), p. 281-286

[10] P. Hutař, L. Náhlík, Z. Majer, Z. Knésl: Computational Methods in Applied Sciences, 1, Volume 24, Computational Modelling and Advanced Simulations (2011) Springer, p. 83-97

[11] L. Náhlík, P. Hutař, M. Dušková, K. Dušek and B. Máša: Mechanics of Composite Materials, 47 (2012), p. 627-634

[12] L.R.G. Treloar: *The Physics of Rubber Elasticity* (Oxford University Press, Oxford, UK, third edition 2005).

[13] Ansys Release 12.1 Documentation (ANSYS, Inc. USA 2009)

[14] B. Máša, L. Náhlík and P. Hutař: Proceedings of MCM, Riga, 2012

Fracture toughness of PIR foams produced from renewable resources

J. Andersons[1,a], E. Spārniņš[1,b], U. Cābulis[2,c] and U. Stirna[2,d]

[1] Institute of Polymer Mechanics, University of Latvia, Aizkraukles iela 23, LV-1006, Rīga, Latvia

[2] Latvian State Institute of Wood Chemistry, Dzērbenes iela 27, LV-1006, Rīga, Latvia

[a]janis.andersons@pmi.lv, [b]sparns@pmi.lv, [c]cabulis@edi.lv, [d]stirna@edi.lv

Keywords: polyisocianurate foams, fracture toughness, rapeseed oil.

Abstract. Rigid low-density closed-cell polyisocyanurate (PIR) foams are used primarily as a thermal insulation material. Traditionally, they are manufactured from constituents produced by petrochemical industry. Introducing renewable materials in PIR formulation brings definite economical and environmental benefits. Fracture toughness of PIR foams obtained from renewable resources (with the polyol system comprising up to 80% of rapeseed oil esters) and petrochemical PIR foams has been characterized experimentally, by compact tension tests, for mode I crack propagation along the rise direction of the foams.

Introduction

Polyurethanes (PU) represent a class of polymers that have found a widespread use as insulating material in building construction and in the global appliances (refrigerators, freezers, etc.) industry, but they also are used in furniture, packaging, automotive and shoe industry, agriculture and medicine. Petrochemical resources, used intensively in the worldwide chemical industry, are limited and are rapidly decreasing. The polymer industry is making big efforts to find an alternative for petrochemical recourses due to decreasing reserves and increasing price. Nowadays rigid polyurethane (PUR) and PIR foams are the main thermal insulation materials used in construction industry, especially PIR, due to their thermal stability and low flammability. The use of PIR foams is expected to grow due to the need for high energy efficiency as well as drive toward reduced emission of greenhouse gasses [1]. The production of polymers, including PUR and PIR foams, from renewable resources is being actively investigated by research teams from different countries. Polyols used in foam production can be obtained from different natural oils – rapeseed, castor, soya, sunflower, etc. Apart from the primary functional characteristics, e.g. low thermal conductivity, a sufficient level of resistance to cracking should be reached in order to ensure the mechanical integrity of foam materials in service. In the absence of test standards developed specifically for the evaluation of toughness of polymer foams, a standard for plastics [2] is routinely applied, see e.g. [3, 4].

The principal aim of this work is to utilize rapeseed oil (RO) for synthesis of natural oil polyols as starting raw materials for PU chemistry, and to evaluate fracture toughness of the resulting PIR foams.

Experimental

Materials. Polyols of rapeseed oil were prepared by amidisation with diethanolamine. Temperature: 140°C, catalyst: zinc acetate. The diethanolamine / natural oil molar ratio was 2.9 / 1.0 [5].

PIR samples were obtained by mechanically mixing appropriate amounts of IsoPMDI 92140 and the polyol system (polyols, surfactant, catalysts, blowing agent) for 10-15 s. The unreacted mixture was poured into a plastic mould (20 x 30 x 10 cm) for free foaming. The polymerization reaction took place at room temperature for all the obtained samples and was completed in about 3-5 min; the samples were conditioned before tests for 24 h. The content of different components in the composites was calculated on the basis of initial weights.

Foams with apparent density in the rage of 35 to 65 kg/m^3 were obtained.

Specimens. The experimental procedure applied for evaluation of the fracture toughness of PIR foams generally followed that described in the standard [2]. Compact tension (CT) specimen geometry was chosen for the test program, with a few tests performed on single-edge-notch bending (SENB) specimens for the purpose of comparison. The nominal dimensions of the specimens are shown in Fig. 1. The nominal thickness of the specimens amounted to 20 mm. The direction of crack propagation was chosen along the foam rise direction. The pre-crack consisted of a 35 mm (CT) or 15 mm (SENB) long prenotch plus a 5 mm long razor notch. All the dimensions of each specimen were measured with 0.01 mm accuracy. After weighing, the density of each specimen was calculated.

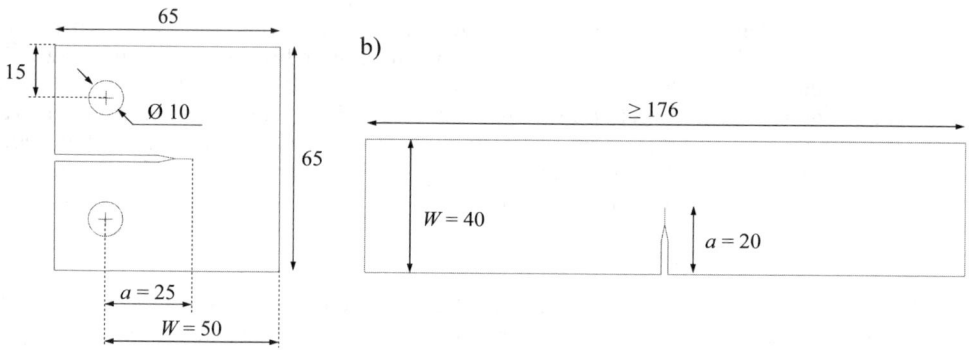

Fig. 1. Schematic of a compact tension (a) and single-edge-notch bending (b) specimens. The dimensions are expressed in mm.

Tests. Mechanical tests were performed using Zwick/Roell 2.5k electromechanical testing machine equipped with a load cell of 100 N capacity. The grips were similar to those described in the standard [2]. Pins made of polished bronze were used in order to reduce the friction of rotation of a CT specimen. Loading rate was 2 mm/min for CT specimens. Thus the maximum load was reached in 1-2 minutes. A test was continued until the load decreased below a half of the maximum value. The load-displacement curves were electronically recorded during the test.

In order to determine the system compliance and indentation displacement, one or few CT specimens of each batch without pre-cracks were tested. A reduced cross-head speed of 0.5 mm/min was applied for the unnotched specimens in order to have a comparable loading rate (and test time) with the CT specimen tests. The obtained indentation displacement data were used to correct the CT load-displacement curves, by subtracting them from the CT specimen displacement data.

Data reduction

According to [2], fracture toughness K_{Ic} is expressed via the critical load P_Q and specimen dimensions as follows:

$$K_{Ic} = \frac{P_Q}{\sqrt{BW}} f(x) \tag{1}$$

where B and W designate the thickness and width of a specimen, a is the pre-crack length (see Fig. 1), and $x = a/W$. The non-dimensional geometrical correction factor $f(x)$ in Eq. (1) is given in [2] by analytical relations for CT specimens:

$$f(x) = (2+x)(0.886 + 4.64x - 13.32x^2 + 14.72x^3 - 5.6x^4)/(1-x)^{\frac{3}{2}},$$

and for SENB specimens:

$$f(x) = 6\sqrt{x}\left(1.99 - x(1-x)(2.15 - 3.93x + 2.7x^2)\right) / (1+2x)(1-x)^{\frac{3}{2}}.$$

The critical load P_Q is determined using the following procedure:
a) first, the compliance of a specimen is evaluated based on a linear fit of the load-displacement data within a load interval of $0.25 P_{max}$ to $0.5 P_{max}$, where P_{max} designates the maximum load;
b) then the intersection point of the experimental load-displacement curve and the line that corresponds to a 5 % higher compliance is found. The load value at this point is denoted by P_Q;
c) if thus obtained P_Q has been reached at a smaller displacement than that at P_{max}, then it is retained; otherwise P_Q is set equal to P_{max};
d) finally, validity of the test is checked. In terms of loads, $P_{max}/P_Q < 1.1$ should hold for the test to be valid, while the size criteria ensure appropriate stress state in the specimen [2].

Results and discussion

The load-displacement curves of CT specimens of petrochemical foams and RO PIR foams exhibited a consistent qualitative difference. Fig. 2 shows typical loading diagrams for both foam types (with comparable densities of ca. 63 kg/m^3). The fracture of RO PIR foam specimens was brittle, with load-deflection curve being linear up to the point of maximum load. By contrast, the fracture of petrochemical foam was markedly more ductile. It is also seen in Fig. 2 that not only the maximum load, but also the stiffness is higher for the petrochemical foam specimens.

All the tests of RO PIR specimens were found to be valid due to their linear response up to the maximum load. The dependence of fracture toughness, evaluated by Eq. (1) from CT tests, on the foam density is shown in Fig. 3. Note that the SENB tests, performed for 53 kg/m^3 density foams, yielded toughness of 14.3 ± 1.5 kPa·m$^{1/2}$ confirming the CT results.

The apparent density of specimens of the same batch is seen to exhibit moderate scatter, with the coefficient of variation less than 3.5%. Such density fluctuations could result from the presence of a few air bubbles (larger than the typical cell size) in some specimens and the higher foam density in the vicinity of edges of the foam blocks. However, the specimens were cut out so that both the bubbles and the block edge zones, if present, were far from the energy dissipation zone.

The validity of the CT tests of petrochemical PIR foams was doubtful since the criterion ratio P_{max}/P_Q was larger than 1.1 in most cases. Therefore, only the provisional toughness [2], K_Q, based on the 5% offset, was evaluated by Eq. (1) and plotted in Fig. 3. Nevertheless, the obtained toughness estimates are in reasonable agreement with the results for similar foams of comparable density [3], see Fig. 3. Note that the fracture toughness of both types of foams is proportional to the foam density within the range studied, as demonstrated by the best-fit dashed lines in Fig. 3. Such a linear relation has been inferred in [6] allowing for the non-singular stresses in the vicinity of the crack front.

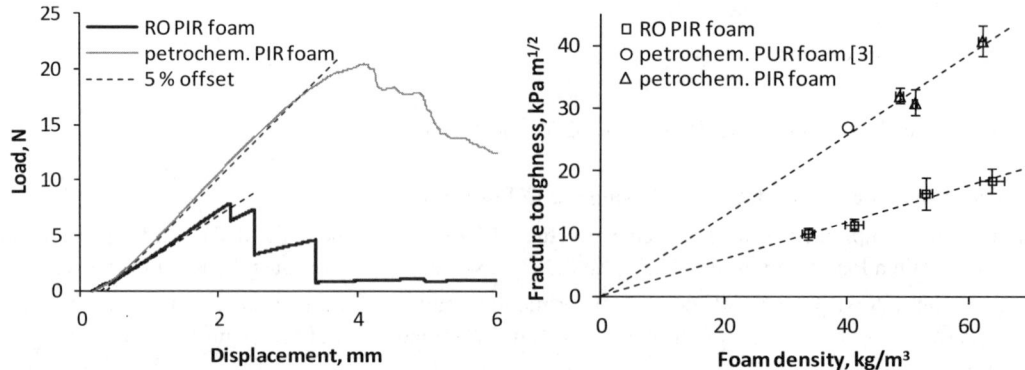

Fig. 2. Typical load-displacement curves of CT specimens produced from RO and petrochemical PIR foams.

Fig. 3. The dependence of fracture toughness on the apparent density of RO and petrochemical PIR foams.

The toughness of RO PIR foams is thus markedly, by a factor of two, lower than that of their petrochemical counterparts. However, the principal functional characteristics of thermal insulation, hydrophobicity, and fire resistance (not reported here) of RO PIR foams are comparable or superior to those of petrochemical foams, therefore they can be used in applications not requiring high resistance to cracking. Should the latter be needed, foam toughness can be improved by adding appropriate fillers [4] or by introducing long-chain petrochemical polyols in the RO PIR foam composition, but the latter would decrease the renewable material content in the foams.

Summary

Rigid, low-density PIR foams have been developed that incorporate natural oil polyols, synthesized from rapeseed oil. Fracture toughness of such RO PIR foams has been experimentally characterized by compact tension tests, for mode I crack propagation along the rise direction of the foams. The toughness was found to vary proportionally to the foam density in the range of 35 to 65 kg/m^3 studied, reaching up to 17 kPa·m$^{1/2}$. Similar PIR foams from petrochemical constituents exhibited more ductile fracture and higher toughness.

Acknowledgement

This work has been funded by ERDF via project 2010/0290/2DP/2.1.1.1.0/10/APIA/VIAA/053.

References

[1] Protecting the Ozone Layer, Foams, *United Nations Publication*, **4** (Sales No. 92-III-D.10.) (1992)

[2] Standard Test Methods for Plane-Strain Fracture Toughness and Strain Energy Release Rate of Plastic Materials, ASTM Standard D5045-99, *Annual Book of ASTM Standards* (1999)

[3] L. Marsavina and E. Linul, in: *Proc of 18th Eur. Conf. on Fracture* (2010)

[4] M.C. Saha, Md.E. Kabir and S. Jeelani: Polym. Compos. Vol. 30 (2009), p. 1058

[5] U. Stirna, U. Cabulis, I. Beverte: J. Cell. Plast. Vol. 44 (2008), p.139

[6] J.B. Choi and R.S. Lakes: Int. J. Fract. Vol. 80 (1996), p. 73

Modelling of Discontinuous Environment

Jiří Boštík[1,a] and Kamila Weiglová[1,b]

[1]Brno University of Technology, Faculty of Civil Engineering, Institute of Geotechnics,
Veveří 331/95, CZ-602 00 Brno, Czech Republic

[a]bostik.j@fce.vutbr.cz, [b]weiglova.k@fce.vutbr.cz

Keywords: underground structures, experimental modeling, dislocation zones

Abstract. Deformation process and strength of rock environment is significantly influenced by a presence of discontinuity planes. The paper deals with experimental modeling of underground structures in such a rock environment. Parametric analysis presents results from twin circular tunnels simulated in a scale model. The cases with and without zone of dislocations in between the tunnels were observed. Another variable factor studied was the distance between tunnels. The relation between model surface displacement and excavated length of the tunnel, which was monitored during the simulations, was used for mutual comparison of individual cases.

Introduction

Prediction of the behavior of geotechnical structures is a complex engineering problem. In order to get reasonable results it is necessary to take into account fundamental aspects of the rock mass behavior using appropriate methods for analyzing its behavior which is significantly influenced by presence of discontinuity planes. This prediction can be performed on *scale experimental (physical) models* [1].

For the *scale models* of discontinuous environment (numerical calculations) based on continuum mechanics, there are the following approaches (see e.g. [2]): a) rock environment is approximated by a continuum and the effect of discontinuities is incorporated in the constitutive (material) model of the equivalent continuum; b) each discontinuity is modeled individually (contact elements, the finite layer thickness); c) combination of the two previous cases (sometimes called hybrid models).

Numerical analyses are currently the most common way of prediction of the geotechnical structures behavior. Examples of their applications can be found in numerous publications, e.g. 2D numerical analysis of the circular tunnels in the rock mass impaired by the tectonic fault zone is described in [3]. In this model the effect of selected factors on the stability of tunnels is studied. It involves the occurrence of dislocations in the rock mass, their slopes, thickness and distance from the tunnels. The analysis was carried out by the FEM and dislocations were modeled as finite layer thickness.

Numerical analysis of the retaining structure in the rock environment disturbed by joints can be found in [4]. An influence of some factors on displacements and bending moments of diaphragm wall was studied. Particularly an inclination, thickness and location of the joint from the retaining wall were analyzed. Joints were modeled by clusters, instead of using interface elements, due to significant fault thickness of up to 3 m.

The equivalent model of rock environment was prepared in [5], wherein the parametric analysis of the influence of discontinuity orientations on the response of rock environment during driving an underground structure in the direction of discontinuity planes is presented. Rock environment was created by one kind of stratified rock. It was assumed that single layers are of the same width, which was relatively small as compared to the transverse dimensions of the underground structure. Further, it was supposed that the discontinuities were not filled. Three alternatives were considered: the underground structure in rock mass with one-sided dip of discontinuities, the underground structure situated in synclinal bend and the underground structure situated in anticline bend. In all three cases, the variability of discontinuities dip angle in the range of $0 - 90°$ was considered.

This paper is focused on an analysis of underground structure in discontinuous environment using *experimental models*. Attention is concentrated on the rock environment response to the tunneling and simulation of the excavation of a pair of circular tunnels.

Simulation of system consisting of two tunnels

In this section experimental models (without the application of dimensional analysis) of two circular parallel tunnels are presented. Tunnels were modeled in the rock mass with and without zones of dislocations in the space between the tunnels. Another variable factor having been monitored was the distance between tunnels, which was successively set as 1.75, 1.20 and 0.65 times of the diameter of tunnel.

Models were created in the rectangular parallelepiped modeling box with the bottom size of 280 x 160 mm and the height of 200 mm. In the rear wall, there were two circular holes cut out. Two paper tubes with size of 43/0.5 mm and 39/1 mm were used for modeling of excavation of each of both tunnels. Colored dry sand was used as the modeling material. The modeling box was filled by sand in layers and each layer was compacted. In the zone of dislocations the pillar between the tunnels was modeled. Its width was 20 mm and the deviation from the horizontal plane was approximately 65 deg. Its lower level corresponded to the tunnel bottom level and its upper level was located approximately 1D (the diameter of tunnels, see Fig. 1) below the surface of the models. Fault zone in between the tunnels was modeled by polystyrene balls.

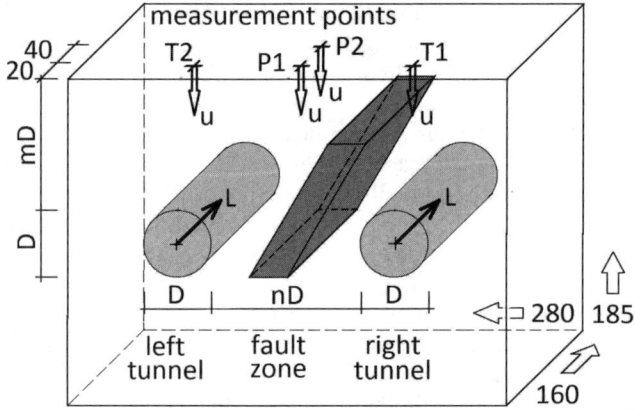

Fig. 1 Experimental model: dimensions in mm, m = 1.9, n = 1.75, 1.20, 0.65, D = 39 mm

Response of the models during the excavating the tunnels was evaluated based on the vertical surface displacement in the specific measurement points. These points were located in the axes of both tunnels and in the axis of the pillar (Fig. 1). The results obtained were graphically worked out in the form of relation of these normalized vertical displacements to normalized progress of the excavation. The relations for selected tunnel arrangements described above are presented in Fig. 2 - 4, where the excavation effect of the particular tunnels is separated. Excavation of the tunnels was simulated in the order of the right tunnel and then the left tunnel.

By comparison of the results obtained, the influence of fault zone between tunnels was evaluated. Vertical model surface displacement in the rock pillar axis is contributed to by excavation of the both left and right tunnels – this applies to the both fault and faultless zone cases. For the fault zone case, the ratio of model surface vertical displacement in the rock pillar axis and the model surface vertical displacement in the tunnels' axes is greater than for the faultless zone case.

Increase of vertical displacement increment in the axis of the tunnel caused by excavating the other tunnel is influenced by presence of the fault zone which apparently attenuates the interaction. The final vertical surface deflection above the axis of tunnel was induced only by excavating the tunnel itself was practically reached for excavated length of 3D. This applies to the both fault zone and faultless zone cases, and also to the both left and right tunnels.

The increment of vertical deflections at the surface points above the axis of pillar induced between the tunnels, which was brought about by the excavation of a single tunnel was observed in both cases (with and without fault zone) and higher in excavating the second (left) tunnel. In the case of considering the fault zone the increase in deflection is higher than that observed in the case of faultless zone – this applies to the deflection induced by excavating both left and right tunnel.

The final surface deflection of the right tunnel axis for the case of fault zone is greater than that attained in the case of the faultless zone (mutual distance between tunnels equals 0.65D). For a greater rock pillar width, the deflection is greater in the case of faultless zone. In the latter case above the axis of the left tunnel this deflection is greater than that for the case of faultless zone (mutual distance between tunnels equals 0.65D and 1.75D) or the displacement is same in both cases (pillar width equals 1.20D).

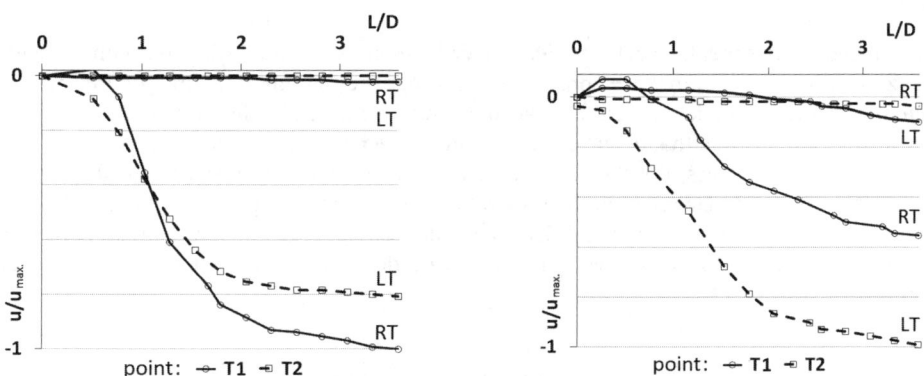

Fig. 2 Normalized deflection at the surface points for excavation of only right (RT) and only left (LT) tunnel: without fault zone (left), with fault zone (right).
Tunnels with mutual distance 1.75D

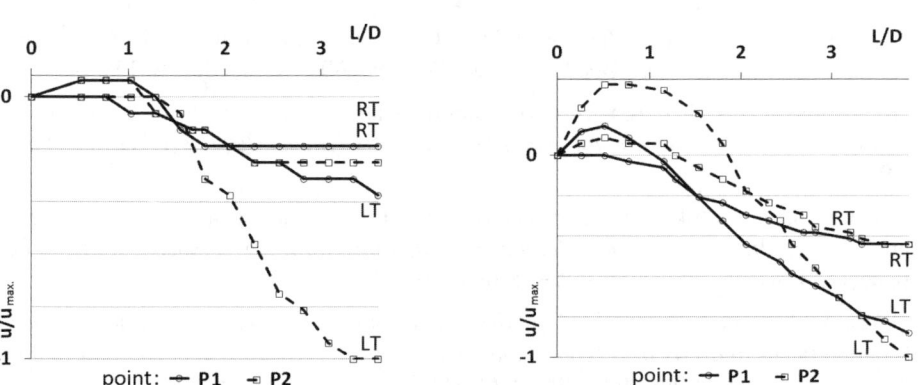

Fig. 3 Normalized deflection at the surface points for excavation of only right (RT) and only left (LT) tunnel: without fault zone (left), with fault zone (right).
Tunnels with mutual distance 1.75D

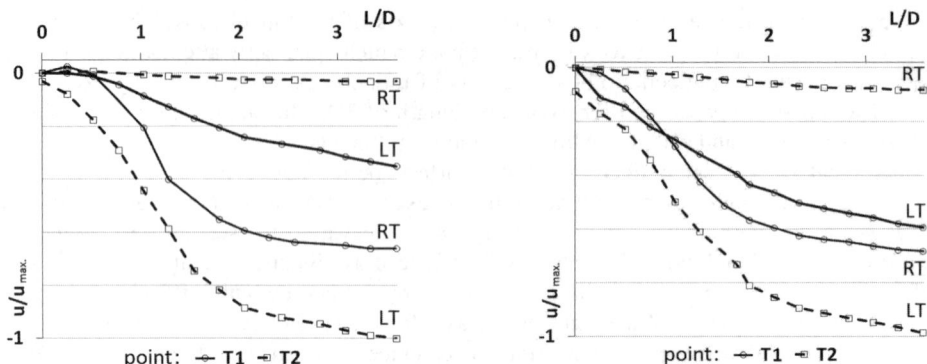

Fig. 4 Normalized deflection at the surface points for excavation of only right (RT) and only left (LT) tunnel: without fault zone (left), with fault zone (right).
Tunnels with mutual distance 0.65D

Conclusions

In this paper, experimental scale models of twin tunnels in the rock environment with and without zones of dislocations are presented. During the simulations of progressive tunnel excavation, the vertical model surface displacement was measured in the measurement points. The obtained results demonstrate that occurrence of a fault zone between the tunnels significantly affects the response of the rock mass. For the case where the zone of dislocations was considered, the final vertical model surface displacement in the rock pillar axis is significantly bigger than in the faultless zone case. The total difference of vertical displacement between the fault and faultless zone case is a function of the mutual distance between tunnels – with decreasing pillar width the total difference increases.

This contribution was financially supported by the project of the Ministry of Education, Youth and Sports (MŠMT ČR) No. MSM0021630519. Authors acknowledge this support.

References

[1] K. Weiglová, P.P. Procházka, Increase of stability of underground works. In *1 st International Conference on Underground Spaces - Design*. Wessex, WIT Press, 2008. p. 139-147.

[2] T. Kawamoto, Ö. Aydan, A review of numerical analysis of tunnels in discontinuous rock masses, *International Journal for Numerical and Analytical Methods in Geomechanics*, 23, 1999. p. 1377-1391.

[3] K. Weiglová, J. Boštík, Mutual effect of tectonic dislocations and tunnel linings during tunnelling. In *Analysis of discontinuous deformation: New developments and applications*. Singapore, Research Publishing. 2009. p. 693-701.

[4] L. Miča, V. Račanský, J. Grepl, Numerical analysis of deep excavation affected by tectonic discontinuity. In *Analysis of discontinuous deformation: New developments and applications*. Singapore, Research Publishing. 2009. p. 661-667.

[5] J. Boštík, Underground structure in discontinuous rock mass. In *Proceeedings of the Fifth International Symposium "Compunational Civil Engineering 2007"*. Iasi, Editura Societatii Academice Matei - Teiu Botez, 2007. p. 66-74.

Influence of finite element modeling choices in the assessment of stress concentrations at fatigue prone locations in orthotropic bridge decks

Wouter De Corte[1,2,a] and Arne Jansseune[1,2,b]

[1]Department of Construction, Faculty of Applied Engineering Sciences, University College Ghent, Valentin Vaerwyckweg 1, 9000 Ghent, Belgium

[2]Department of Structural Engineering, Faculty of Engineering Sciences, Ghent University, Technologiepark-Zwijnaarde 904, 9000 Ghent

[a]wouter.decorte@hogent.be, [b]arne.jansseune@hogent.be [c]email

Keywords: orthotropic, bridge, deck, fatigue, finite element.

Abstract. In orthotropic bridge decks, the rib to floorbeam connection is a major source of fatigue problems. Commonly, the trapezoidal ribs cross the floorbeam continuously necessitating clearance holes in its web, and frequently additional web cutouts are foreseen to relieve the ribs lower edges. This solution is favorable for rib cracking but will generate stress concentrations in the web itself. The shape of the additional cutout has a major influence on the sizes of the concentrations rendering differences of a factor 2 or 3 for corresponding overall geometries and loading schemes. Various authors have studied cutout shapes through full scale testing or by computation by finite element modeling. This paper presents such a study, but focuses on the influence of the finite element modeling itself. It is shown that the mesh density, the element type, the choice between shell and volume elements. This is an important finding and should not be overlooked when comparing finite element based results to coded values or measured results. In order to do so, the results of the numerical work are compared to strain results from full scale tests.

Introduction

Steel orthotropic bridge decks contain at least 5 fatigue critical locations [1], among which the rib to floorbeam connection is probably the hardest to tackle mathematically, given the high complexity of the connection and the occurring stresses. The behavior of such a when combined with closed continuous ribs is substantially different from that of an average beam, especially if additional cutouts are used around the bottom flange of the longitudinal ribs. When comparing measured or FE calculated stresses to results from classical beam theory it becomes clear that the stress distribution in the web in no way relates to the one in a simple beam. It relates much more to the distribution found in a special truss without diagonal members commonly referred to as a Vierendeel truss. Based on this observation, a calculation model was proposed by Haibach [2], and further developed by De Corte [3][4].This configuration allows comparing various types of cutout geometries independent from the floorbeam characteristics. However, when assessing finite element results it is clear that the modeling parameters equally influence the stress results and should be chosen with care, especially when comparing results from strain gauge measurements to finite element results. This paper presents results of a comparison of measured results on a test specimen with finite element results from shell and solid models with various choices in mesh size.

Experimental program

The test panel is shown in Fig. 1. It comprises of a 12mm deckplate measuring 1800 by 2000 mm. There are 3 trapezoidal 200mm high 5 mm thick ribs with a 100 mm lower rib spaced 600 mm apart. The ribs are supported at midspan by a 500 mm high and 10 mm thick floorbeam. As can be seen from Fig. 1, 2 different web cutouts are used. On the right side a cutout according to Haibach [2] is present whereas on the left hand side an improved cutout shape is present [3]. By doing so, two cutout geometries can be tested on one test specimen. In this setup forces are applied by hydraulic jack, while strains are measured inside the cutout edges along the curved edge using small base strain gauges (See Fig. 1). The gauges are placed at the positions with largest stress, as found by the finite element calculation, and recorded during loading. For this the hydraulic actuator is moved transversely across the specimen width and stresses are recorded at 50 mm intervals of actuator position.

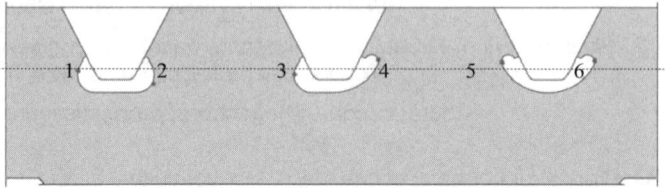

Figure 1. Test setup, floorbeam cross section and strain gauge locations

Numerical calculations

For comparison purposes, calculations were done by a full 3D linear FE analysis. Tetrahedron (C3D4) and brick shaped (C3D8) solid elements were used for the solid model. For the shell model, triangular (S3R) and quadrilateral (S4) shell elements were used. The mesh of the floorbeam was generated by a free meshing technique whereby the element size along the edge of the cutout in the floorbeam was varied to consider its influence on the stress distribution.

Figure 2. Cross section of Abaqus model (Left : solid / Right : shell)

Results and discussion

In order to assess the influence of calculation method a 50 kN load is moved transversely over the test specimen, both mathematically and in the laboratory. The results are shown in Fig. 3 for the stress at gauge n°1 (Fig. 1), and at the corresponding nodes of the FE model. For this, it is important to notice that the strain gauges (6mm grid length) are located directly adjacent to the cutout edge, and consequently do not record the strain at the edge itself. Rather, a strain averaged over the strain gauge width is recorded. In Fig. 3, both the effective edge stress, as well as the stress average over the strain gauge width is given. The figure indicates that for this gauge, the recorded values shown excellent agreement with the gauge-averaged values of the shell model. Values found from the solid model are generally much lower. This observation is further proven in Fig. 4 which displays the recorded stresses at all 6 gauges for the most negative load

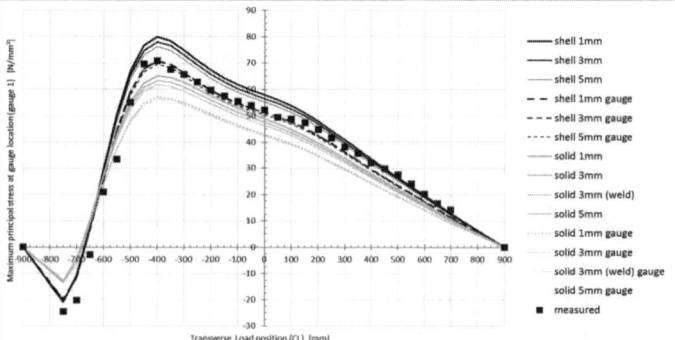

Figure 3. Maximum principal stress results for gauge n°1

location, together with the calculated values from the shell (left) and solid model (right). For 5 of the 6 gauges, the shell model provides much better agreement with the measured values. Overall, the average values of the relation of calculated to measured values are given in Fig. 5. Clearly, the shell model results coincide much better to the measured results. Within the shell results, the average difference between 1mm and 5mm meshing is 3% (max 5%), this value being practically equal for the solid results : 3% (max 10%). However, the average difference between the corresponding shell and solid values is much larger, averaging 29% (max 44%). The difference between the shell and solid results become larger when nodes closer to the surface are considered as can be seen from Fig. 6. This figure depicts the maximum principal stresses in the horizontal cut from Fig. 1. Clearly, the difference between the shell model results and the solid model results are especially large in the vicinity of the cutout edges.

A similar horizontal cut made near the lower edge of the floorbeam web reveals a less that 5% average difference between the shell and solid results. Other overall calculation results such as vertical displacements show excellent agreement between both models. This validates both models but indicates that care should be taken when comparing results of FE analysis on orthotropic bridge decks by various authors.

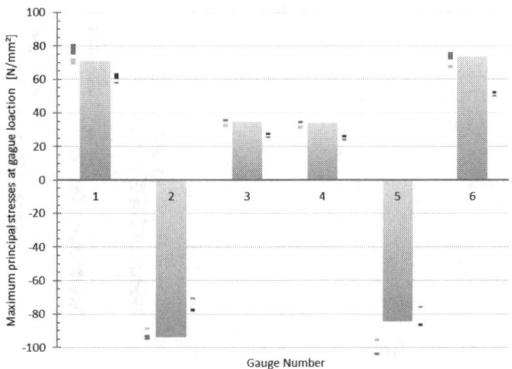

Figure 4. Comparison of measured to shell (L) and solid (R)

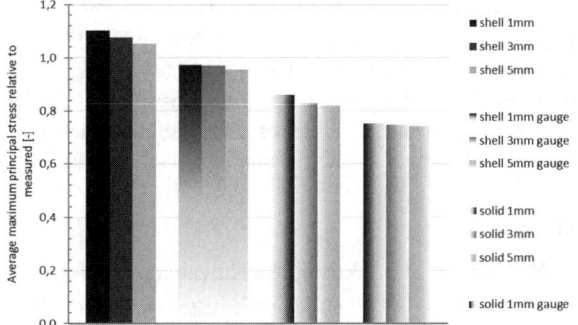

Figure 5. Average FEA values relative to measured

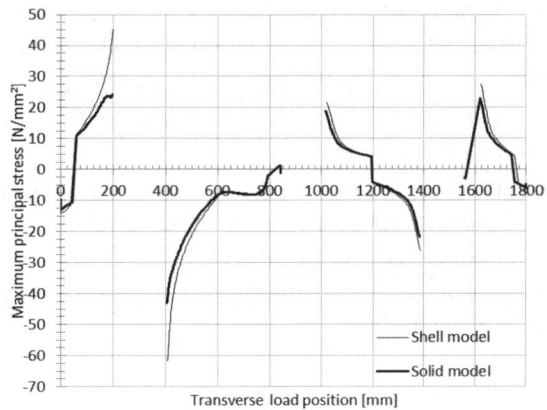

Figure 6. Comparison of stresses in transverse cut

From the above calculations, the reader could get the impression that a shell model inevitably results in larger and more accurate stress results. This cannot be confirmed however, since a parametric study with the geometry of the test specimen (Fig. 1), but with the alternative cutout geometries as described in [3] and [4]. The results of the overall maximum compressive principal stresses for all load cases (Fig. 3) and all possible locations around the cutout edges are given in Fig. 7 for the 10 alternatives from [4], adapted to the test specimen geometry. Clearly the substantial differences in FE results are found for 5 out of the 10 shapes only. In 4 cases, the difference is acceptable, and in 1 case the solid result is even larger compared to the shell result. For the maximum tensile values, the result is completely different, providing larger values for the solid model for all cutout alternatives. In any case, care is required when analyzing orthotropic details by FEA.

Summary

A study focusing on the influence of the finite element modeling is given in this contribution. It is shown that the mesh density and the choice between shell and volume elements have a substantial influence on the stress results at fatigue prone locations along the additional cutout edge of rib to floorbeam connections in orthotropic decks. Since such values are often used to predict fatigue life care should be taken and a study of the FE analysis itself might be advisable.

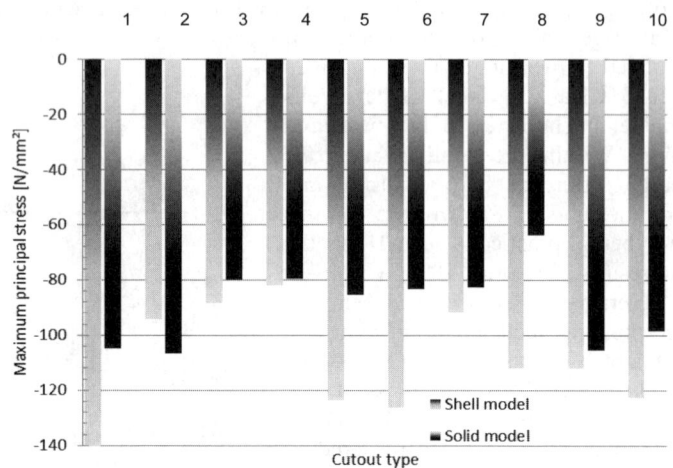

Figure 7. Overall comparison of maximum compressive stresses for various cutout geometries from [4]

References

[1] D. Uchida, S. Inokuchi, A. Kawabata, M. Ishio and T. Tamakoshi: *Field Investigations and Measurements of Orthotropic Steel to Draft Efficient Method of Stock Management*, Proc. of the International Orthotropic Bridge Conference, Sacramento, CA, USA, August 25-29, (2008), CD-ROM.

[2] E. Haibach and I. Pläsil, *Untersuchungen zur Betriebsfestigkeit von Stahlleichtfahrbahnen met Trapezhohlsteifen im Eisenbahnbrückenbau*. Der Stahlbau Vol.9 (1983), p. 269–74.

[3] W. De Corte and P. Van Bogaert, *Improvements to the analysis of floorbeams with additional web cutouts for orthotropic plated decks with closed continuous ribs*. Steel and Composite Structures Vol.7(1) (2007), p. 1-18.

[4] W. De Corte, *Parametric Study of floorbeam cutouts for orthotropic bridge decks to determine shape factors*. Bridge Structures, Vol.5 (2) 2009, p. 75-85.

[5] C.S. Wang, Q. Zhang, T. Zhang, Y.C. Feng, *Floor-Beam Web Cutout Shape Analysis to Improve the Fatigue Resistance in Orthotropic Steel Bridge Decks*, Key Engineering Materials Vol. 452-453, p.161-164.

Numerical analysis of femoral neck angle influence on stress distribution of cemented Austin Moore hip prosthesis

L. Bogdan[1a], C.-S. Nes[1b], N. Faur[1c], M. Amarandei[1d], A. Enkelhardt[1e]

[1]"Politehnica" University of Timisoara, Romania

[a]blucian85@gmail.com, [b]cristianedonis@yahoo.com, [c]ruaf2001@yahoo.com,
[d]amarandei_mihaela@yahoo.com, angelica_enkelhardt@yahoo.de

Keywords: hip, prosthesis, numerical, stress, cement, Austin Moore

Abstract. This paper presents a finite element analysis regarding the stress distribution in a cemented Austin Moore type hip prosthesis. The 3-D model was obtained using a Roland PICZA 3-D laser scanner. The applied loads simulate the normal gait cycle. The prosthesis is made from stainless steel with a femoral head of 45mm diameter. The numerical analysis was performed using the ABAQUS code. The results showed that the stress level in the cement is sensitive to the femoral neck angle. Starting with a standard, 125° angle, and increasing the angle with up to 5°, the resulting stress can be reduced with more than 10%. The proposed angle increase produces a more uniform stress distribution in the cemented section, increasing the durability of the arthoplasty.

Introduction

Total hip arthoplasty (THA) is a widely used clinical procedure to restore the normal function of hip joint disrupted by disease or fracture [1]. About 200000 interventions/year are performed in The United States and 80000 in the United Kingdom; they are estimated to increase about 170% by 2030 [2]. From an engineering point-of-view hip implants are not a complete success and still need further improvements [3-4].

From the most popular cemented hip prosthesis - Charnley, Austin Moore, Thompson - to the newer ones like Filler and Lutecia, two main critical issues for implant success are agreed: the implant fixation/loosening related to the implant/bone interaction and the wear of the articulating surfaces [3]. Fatigue failure of hip prosthesis seems to be reduced significantly in the past two decades [5].

Bone resorption and implant loosening is initiated by a biological adverse response due to mechanical failure of the implant or bone cement, wear debris into interface region, stress shielding in the bone [6], bone quality, patients body weight and activity level must be mentioned as well [7]. These processes can be minimized using downsized prosthesis components, advanced materials with increased biocompatibility and design improvements.

Reducing the stresses in the prosthesis/cement and cement/bone wall interfaces by optimizing the femoral neck angle is a cheap and efficient improvement of the THA.

In this paper we used a finite element analysis to examine the influence of the femoral-neck angle of an Austin Moore hip prosthesis on the cement used in THA. This angle is formed between the

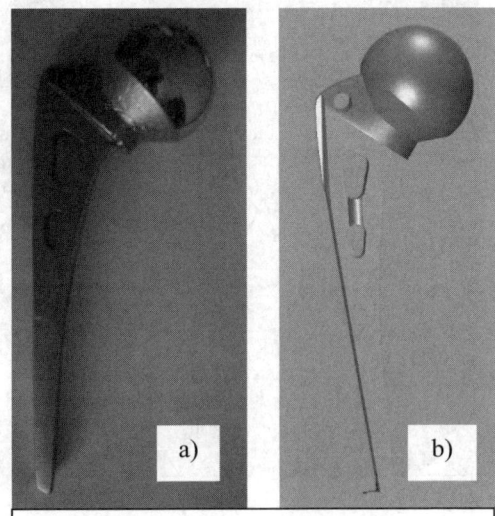

Fig. 1. Physical and 3-D models of Austin-Moore hip prosthesis

axes of the neck and stem of the hip prosthesis with a normal value of 125° [8]. This angle is necessary for normal function of the hip joint, if this value is exceeded it may disrupt normal mechanic and function of the joint [9].

Materials and methods

Finite element analysis is an alternative approach to pre-clinical tests and it is commonly used because it offers a first aspect of the expected mechanical behavior of implants [10-12].

For the FEM analysis, carried out with the Abaqus commercial software, a 3D prosthesis model was build based on direct measurement of an Austin Moore hip prosthesis, with a femoral head of 45 mm (Excel-Broad-Stem S.S. model) [13] made from AISI 316-I grade stainless steel (Fig. 1.a). The Roland PICZA 3D scanner, SolidWorks and CATIA P3 V5R17 software were used for CAD modeling of the prosthesis and bone cement (Fig. 1.b). Mechanical properties of AISI 316-I stainless steel and bone cement used in the simulation are indicated in Table 1.

Table 1. Mechanical properties of materials used

Materials properties	AISI 316-I grade stainless steel	Cement
Elastic modulus (GPa)	205	2.8
Poisson's ratio	0.3	0.33

The prosthesis-cement assembly (Fig. 2) was created using ABAQUS default contact controls. Considering that the cement fills the cavities within the bone, its outer surface had all the degrees of freedom supressed.

The assembly was subjected to a concentrated force applied in the center of the femoral head sphere (FAP).

Linear tetrahedral elements (C3D4) were used for the model, in order to provide sufficient accuracy and the possibility to mesh complicated geometries.

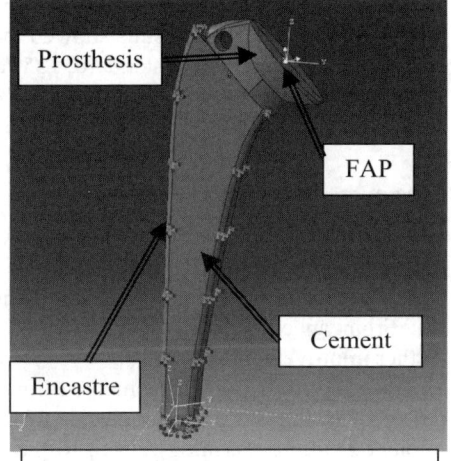

Fig. 2. The prosthesis/cement assembly and boundary conditions

The prosthesis contains 11333 elements, while the cement part contains 271722 elements (Fig. 3).

A total of 9 models were analyzed.

The angle of the femoral neck of the first model was 125°. This angle was then increased with 0.5° steps, up to 130°.

Applied loads (Fig. 4) simulate the hip joint reactions produced during the normal gait cycle. The reaction forces were determined using the AnyBody software. A body weight of 840 N (about 85.6 kg) was considered.

Each component was independently defined with its own amplitude. Considering the low level of dynamic loads produced during the normal gait cycle, the analysis was performed under static conditions.

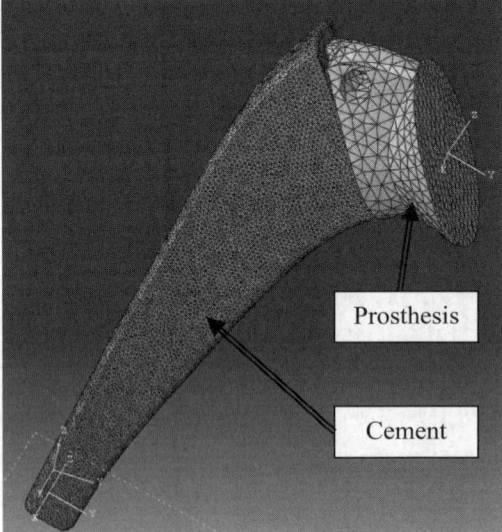

Fig. 3. The meshed hip prosthesis/bone cement model

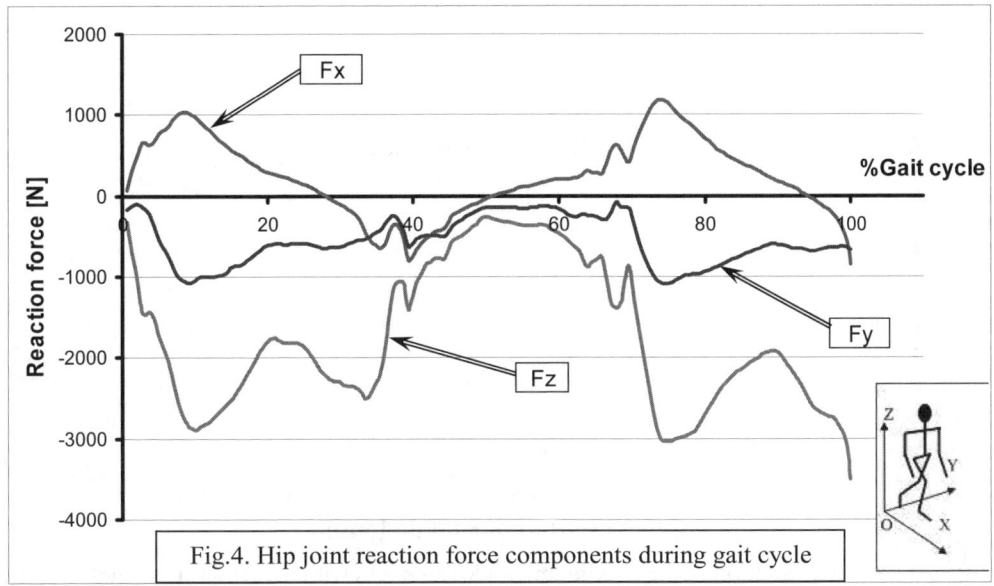

Fig.4. Hip joint reaction force components during gait cycle

Results and discussions

The finite element analysis showed stress concentration around the two cement tubes that connect the prosthesis' sides (Fig. 5). These tubes are designed to lock the cement mantle onto the stainless steel prosthesis and take much of the applied load.

The stress level was high, suggesting inappropriate design for the locking holes. Nevertheless, in reality, these holes may not be filled completely; in this case, the load is distributed primarily on the side surfaces and the resulting stresses are significantly reduced. Ultimate tensile strength of commercial acrylic bone cement available in literature is reported from 35 to 45 MPa and 31.7 to 51.4 MPa, with the fatigue strength hovering around 8 – 10 MPa.

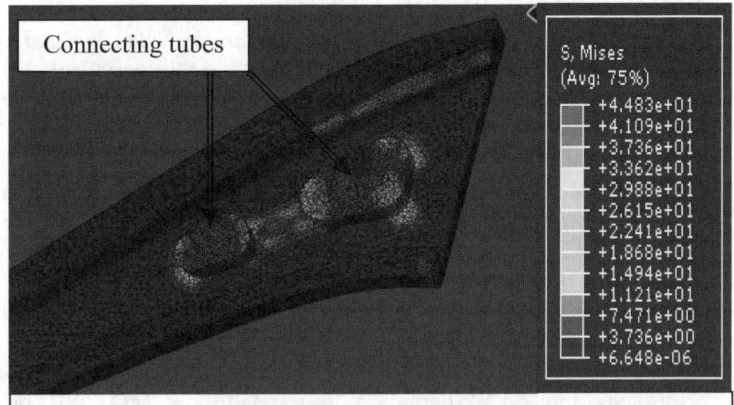

Fig. 5. The stress distribution around the cement connecting tubes

The results also showed that an increase of the femur neck angle produces a decrease of the equivalent von Mises stress (Fig. 6). The dependence is linear, and a 130° angle reduces the maximum stress with more than 10%. This shows that cheap, optimized prosthesis design can be very efficient in providing reliable improvements in THA procedures.

Acknowledgment

This work was partially supported by the strategic grants POSDRU/88/1.5/S/50783, Project ID50783 (2009) and POSDRU 107/1.5/S/77265 co-financed by the European Social Fund – Investing in People, within the Sectoral Operational Programme Human Resources Development 2007-2013.

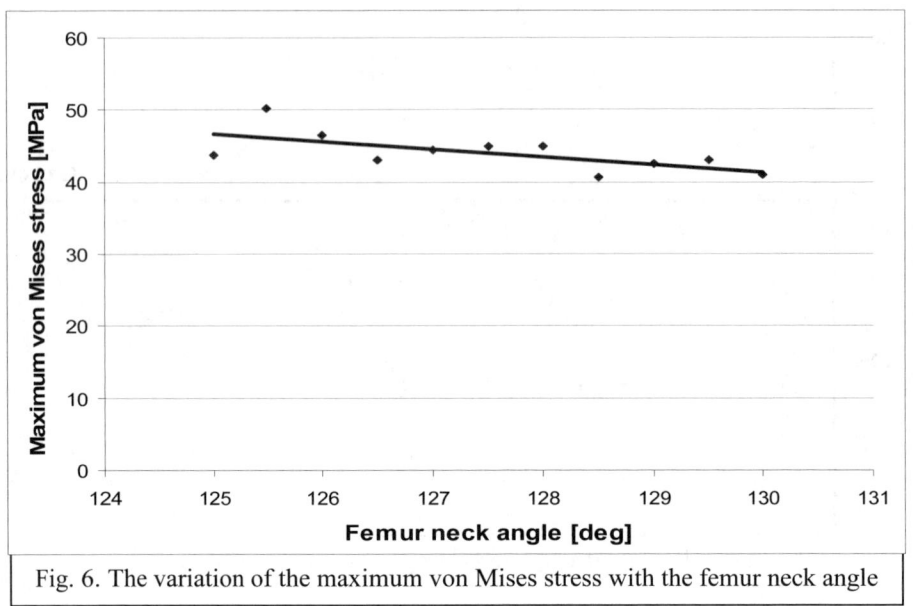

Fig. 6. The variation of the maximum von Mises stress with the femur neck angle

References

[1] Griza S., et al. – *Fatigue failure analysis of a specific total hip prosthesis stem design*, International Journal of Fatigue, Vol. 30, 2008, pp. 1325-1332.
[2] Kurtz S., et al. – *Projections of primary and revision hip and knee arthoplasty in the United States from 2005 to 2030*, Journal of Bone Joint Surgery, Vol. 89, 2007, pp.780-785.
[3] Mattei L., et al. – *Lubrication and wear modeling of artificial hip joints: A review*, Tribol Int. (2010), doi:10.2016/jtriboint.2010.06.010
[4] Hip and Knee Survey 2009
[5] Senalp A.Z., et al. – *Static, dynamic and fatigue behavior of newly designed stem shapes for hip prosthesis using finite element analysis*, Materials and Design, Vol. 28, 2007, pp. 1577-1583.
[6] Joshi M.G., et al. – *Analysis of a femoral hip prosthesis designed to reduce stress shielding*, Journal of Biomechanics, Vol. 33, 2000, pp.1655-1662.
[7] Gravius S., et al. – *In vitro interface and cement mantle analysis of different femur stem designs*, Journal of Biomechanics, Vol. 41, 2008, pp. 2021-2028.
[8] Nordin M., et al. – *Biomechanics of the hip, Basic Biomechanics of the Musculoskeletal System*, third ed. Lippicott Williams & Wilkins, Baltimore, MD, 2001; pp. 202-221.
[9] Peng T.P., et al. – *Review of osteoporosis*, J.Jiangxi Univ., Tradit. Chinese med., Vol. 8.
[10] Prendergast P.J. – *Finite element models in tissue mechanics and orthopedic implant design*, Clinical Biomechanics, Vol. 12, 1997, pp.343-366.
[11] Fialho J.C., et al. – *Computational hip joint simulator for wears and heat generation*, Journal of Biomechanics, Vol. 40, 2007, pp. 2358 – 2366.
[12] Qian J. G., et al. – *Examination of femoral-neck structure using finite element model and bone mass density using dual-energy X-ray absorption*, Clinical Biomechanics, Vol. 24, 2009, pp. 47-52.
[13] http://www.ortho.in

Advanced natural stitched composite materials in skin-stiffener of wind turbine blade structures

F. Papadopoulos[1,a], D. Aiyappa[1,b], R. Shapriya[1,c], E. Sotirchos[1,d], H. Ghasemnejad[1,e], R. Benhadj-Djilali[1,f]

[1]Faculty of Science, Engineering & Computing, Kingston University London, SW15 3DW, UK

[a]F. Papadopoulos @kingston.ac.uk , [b]D. Aiyappa@kingston.ac.uk, [c]R. Shapriya@kingston.ac.uk, [d]E. Sotirchos@kingston.ac.uk, [e]h.ghasemnejad@kingston.ac.uk [f]R. Benhadj-Djilali@kingston.ac.uk

Keywords: Wind; Failure, Joint; Stitched; Composite; CZM;

Abstract: In this paper the failure behaviour of natural stitched composite materials in the skin-stiffener of wind turbine blade structures are investigated. For this study, the laminated composite beams were stitched using Flax yarns before curing process. Two stiffener structures of T-beam and Box-beam are studied in this paper. These specimens were tested under quasi-static loading condition to compare the failure resistance of adhesive and stitched bonding methods. Furthermore, the cohesive zone modelling (CZM) which is known as a variation in the cohesive stresses with the interfacial opening displacement along the localised fracture process zone is used to predict bonding failure in the skin-stiffener of wind turbine blade structures.

Introduction

Adhesive joints are one of the most important joining techniques used for bonding composite structures. Joint design technology has become a main factor in structural integrity to design of composite sub-structures in various engineering disciplines such as aerospace structures, marine engineering, automotive structural parts, micro-electro-mechanical systems, wind turbine blades and also civil structures for strengthening.
Fibre-reinforced polymer (FRP) composite materials are widely used because of their high strength-to-weight and stiffness-to-weight ratios compared with many traditional materials [1,2]. Wind turbine blades are typically manufactured from FRP composites and joint failure can be an important issue in these structures. In extreme conditions, like ice impacting, failure of the adhesive joints is found in different parts of a blade, introducing local damage, which can cause catastrophic failure under various loading conditions including fatigue loading.

In the research reported here, experimental and numerical studies of failure in skin-stiffener composite beams were carried out. Laminated composite skin-stiffeners were stitched through their thickness using natural flax yarns to improve joint failure in wind turbine blade structures.

Manufacturing: Design of the Moulds

The most common through-thickness reinforcement techniques are 3D weaving, stitching and braiding. More advanced techniques include embroidery, tufting and z-anchoring, hereafter known as z-pinning. These techniques are effective at increasing the delamination resistance and impact damage tolerance. The manufacturing of natural stitched composite specimens is similar to the normal composite specimens except that subsequent to stitch the laminated composite with natural Flax yarn before curing process (see Figure 1). Four different moulds of T-beam with adhesive bonding, T-beam with natural fibre stitching, box beam with adhesive bonding and box beam with natural fibre stitching were studied in this research. All specimens were manufactured from carbon fibre-reinforced plastic (CFRP) material of density 1.8 g/cm^3 with epoxy resin.

Fig. 1. Stitching of a) T-joints and b) box-beam in skin-stiffener using Flax yarns.

Experimental studies

The performed test was a three point Quasi-static (2 mm/min) pull out test; the equipment used was a Universal Tensile Testing machine. The three point test consisted of the surface plate simply supported at both ends at a distance of 127mm, to keep this length constant a jig was used to locate the specimens shown in Figures 2 and 3. A vice grip was attached to the machine which was common for all tests; different jigs were used for the T-beam and Box-beam, which would firstly prevent the specimens from slipping and secondly to centralise the load. In order to keep to standards, three tests were carried for each specimen.

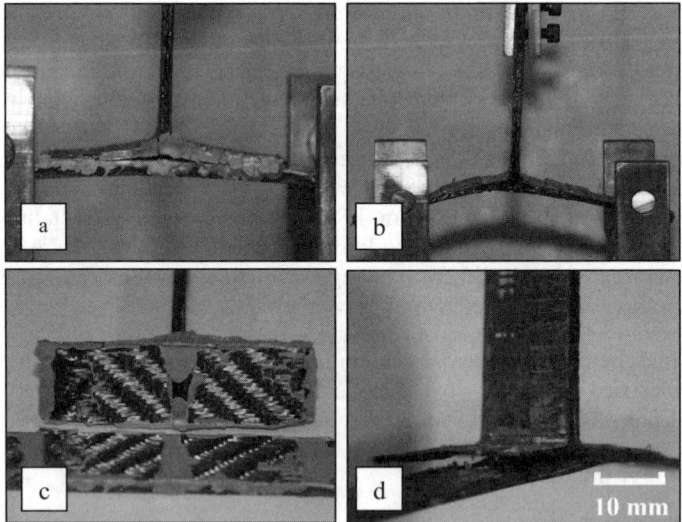

Fig. 2. Quasi-static pull out test a) adhesively bonded T-joint, b) natural fibre stitched T-joint, c) de-bonded adhesively bonded joint after test and d) de-bonded area in stitched T-joint specimen.

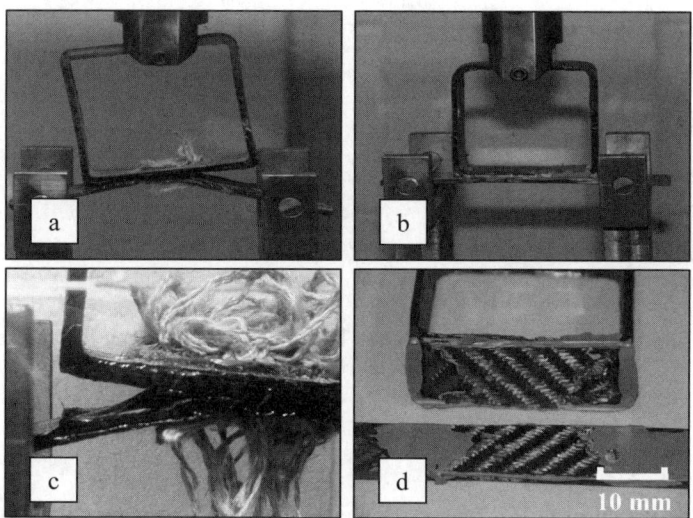

Fig. 3. Quasi-static pull out test a) adhesively bonded box-beam joint, b) natural fibre stitched box beam, c) de-bonded area in stitched box-beam joints and d) de-bonded interfaces after the pull out test.

Discussion and Conclusion

The tests were executed until rapture and the results were plotted on a force (N) versus displacement (mm) graphs (see Figure 4). In this case the pull out tests were followed by two failure mechanisms of intra-laminar and inter-laminar (delamination) failures. The intra-laminar failure which is related to failure through the lamina consists of some fibre breakage and matrix cracking. In addition delamination propagation is an inter-laminar failure mode which occurred in all specimens. In this regard, natural Flax yarns which were stitched through the thickness of multi delaminated specimens can significantly arrest the crack propagation in the specimens. In this case there is more resistance within the specimens which consequence to increase the failure resistance in the skin-stiffeners (see Figures 2 and 3).

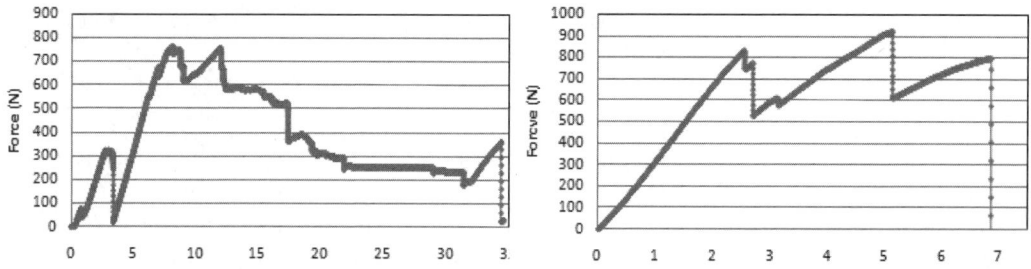

Fig. 4. Force-displacement diagrams a) adhesively bonded in T-joint specimen and b) natural fibre stitched joint in box-beam specimen.

The composite skin-stiffeners were also modelled with the same design and lay-ups using finite element software ANSYS. The CZM element was located along the interface between two adherends (box-beam and beam-plate). Automatic solution procedure was adapted which led to relatively small displacement increments. In the CZM/FEA the 20-node Non-layered/Layered SOLID 186 was used to model the composite beam (see Figure 5). The CZM/FEA results were in reasonable agreement with experimental results.

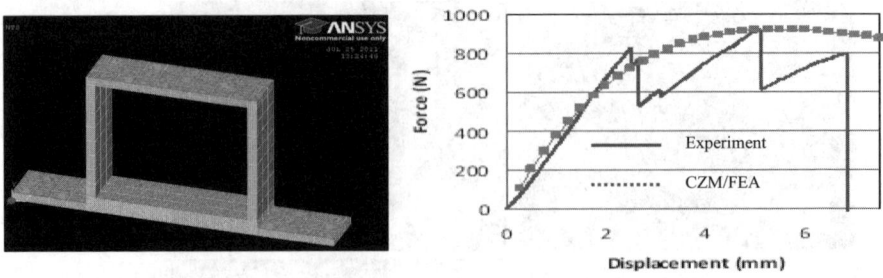

Fig. 5. CZM/FEA of Box-beam in ANSYS and b) comparison between FE and relevant experimental results.

References

[1] H. Ghasemnejad, V.R. Soroush, P.J. Mason, B Weager, To improve impact damage response of single and multi-delaminated FRP composites using natural Flax yarn, in 'Materials and Design, 2012;36:865–873.

[2] H. Ghasemnejad, L. Occhineri, D.T. Swift-Hook, Post-buckling failure in multi-delaminated composite wind turbine blade materials, in 'Materials and Design, 2011;32(10): 5106–5112.

Interfacial Shear Strength of Carbon Fiber Reinforced Polypropylene

Kenichi Takemura[1, a] and Hideaki Katogi[1, b]

[1]Kanagawa University, Department of Mechanical Engineering, Faculty of Engineering

3-27-1 Rokkakubashi, Kanagawa-ku, Yokohama 221-8686, Japan

[a]takemura@kanagawa-u.ac.jp, [b]katougi@kanagawa-u.ac.jp

Keywords: CFRP, PP, Interfacial strength, Bending Property, Contact angle, Micro debonding test

Abstract. In this study, interfacial shear strength of carbon fiber reinforced polypropylene were investigated. Two kinds of reinforcements are used. One is non-treated carbon fiber, another is acetone-treated carbon fiber. And two kinds of matrices are used. One is non-treated polypropylene, another is maleic anhydride -polypropylene. Three point flexural tests and micro debonding tests are conducted. As a result, following conclusions are obtained. Acetone treatment and maleic anhydride are effective to the adhesives on the surface between fiber and matrix. But simultaneous treatments are not effective. The shear strength is not dependent on fiber embedded length. The contact angle and fracture load are dependent on fiber embedded length. The interfacial strength is dependent on the contact angle. As the contact angle increases, the interfacial strength increases.

Introduction

CFRP(Carbon Fiber Reinforced Plastics) are widely used for airplane etc. Most matrix of CFRP is thermosetting type like epoxy. This thermosetting type resin has a problem from the viewpoint of waste. So, thermoplastic resin is focused as matrix of CFRP[1,2]. When themoplastic resin is used, interfacial strength is important.
Gamstedt et al. [3] showed that synthesis of unsaturated polyesters for improved interfacial strength in carbon fiber composite. As a result, interfacial shear strength was increased with increasing degree of unsaturation of resin, which is controlled by the relative amount of maleic anhydride. Tanaka et. al. [4] reported interfacial property of carbon fiber/polyamide model composite after water absorption. The interfacial shear strength was decreased by water absorption. The water absorbed through the interface degraded the interfacial shear strength. Then, absorbed water diffuses to the resin and resin was degraded. But there are few reports about the interfacial strength and interfacial fracture.
In this study, the micro-debonding test apparatus was made in order to examine the interfacial strength. Two kinds of fiber interface and two kinds of matrix are used. Three point bending test and microdebonfing test are conducted.

Specimen

Materials. Torayca cloth CO6343 (Toray Co.) which is plain woven type is used as reinforcement. A sheet type of polypropylene(Shin-kobe electric Co.) is used as matrix. Maleic asid modified polypropylene (Sanyo kasei kogyo Co.) are used for maleic modified matix. The density is 5 %.
Chemical Treatment of Carbon Fiber. Chemical treatment using acetone was conducted to the surface of carbon fiber to improve the adhesive property. The length and width of carbon fiber cloth are 150mm and 120 mm, respectively. The immersion time is 15 minutes in acetone solution. After immersion, the carbon cloth are washed with water, and dried.
Maleic Acid Modified PP. Generally, the adhesion between untreated PP and carbon fiber is not enough. So, maleic acid modified PP was used as matrix to improve the adhesive properties.

Geometry and Dimensions of Specimen. For static bending test, the laminates are made by hot press method. The specimens are cut out on the base of JIS(Japanese Industrial Standard)7074 using band saw.
For micro debonding test, specimen is the filament of carbon fiber with a resin ball in a center position. The diameter of resin ball is less than 0.3mm. The reason is that if the diameter is more than 0.3mm, the resin ball drops by gravity and it can not keep the ball shape.

Experimental Method

Static Three Point Bending Test. The gage length is 80±0.2mm and crosshead speed is 5±1mm/min. Autograph AG-IS (Shimadzu Co.) was used as testing machine.

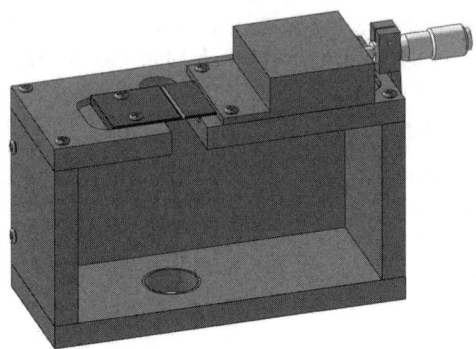

Fig.1 Micro debonding test apparatus.

Micro Debonding Test. Micro debonding test apparatus(Figure 1) was made in the laboratory. The apparatus is attached with tensile test machine(Tensilon RTC-1250A, Orientec Co.). Crosshead speed is 0.1mm/min. Interfacial shear strength is calculated by Equation(1) using maximum tensile load. Here, τ is interfacial shear strength, F_{max} is maximum tensile load, d is fiber diameter and l is embedded fiber length.
Eq(1)

$$\tau = \frac{F_{max}}{\pi d \ell} \quad \cdots \quad (1)$$

Results and Discussion

Static Three Point Bending Test. Figure 2 shows the result of static bending test. When the results are compared each other, the strength of acetone treated specimen is 22 % more than untreated specimen. The stiffness of acetone treated specimen is also 26 % more than untreated specimen. So, it is thought that the adhesive properties at the interface between untreated carbon fiber and PP resin. When they are compared from the viewpoint of maleic acid modification, both of strength and stiffness are twice more than untreated ones.
When both treatments are conducted, the bending properties are improved compared with aceton treatment. But it was not better than maleic acid treatment. So, it is thought that Maleic acid treatment is suitable to untreatment carbon fiber than acetone treated one.
Micro DebondingTest. Figures 3, 4 and 5 show the results of micro debonding tests. The scatter is relatively big. The reason is thought that the cross section is not a real circle, the adhesive area is not constant, the dimension of the specimen is very small order.

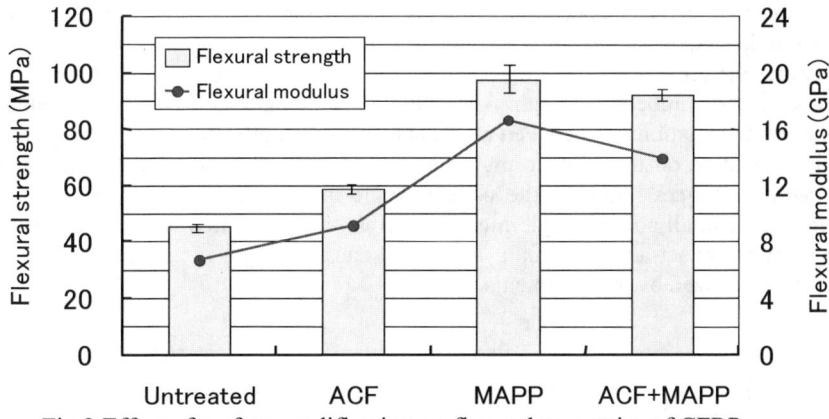

Fig.2 Effect of surface modification on flexural properties of CFRP.

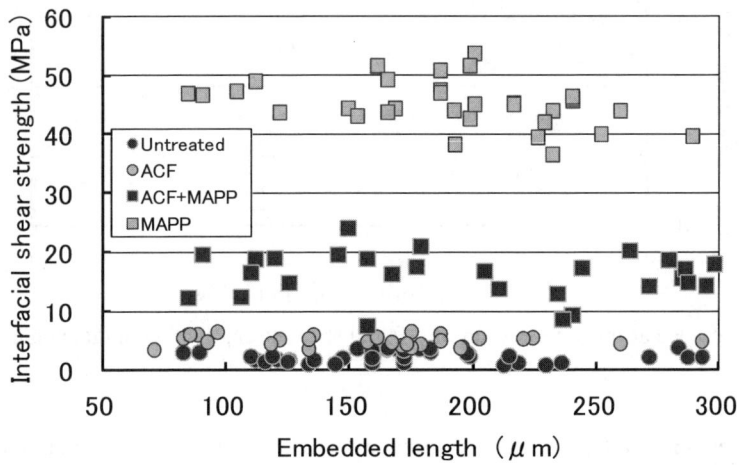

Fig.3 Relationship between interfacial shear strength and embedded length.

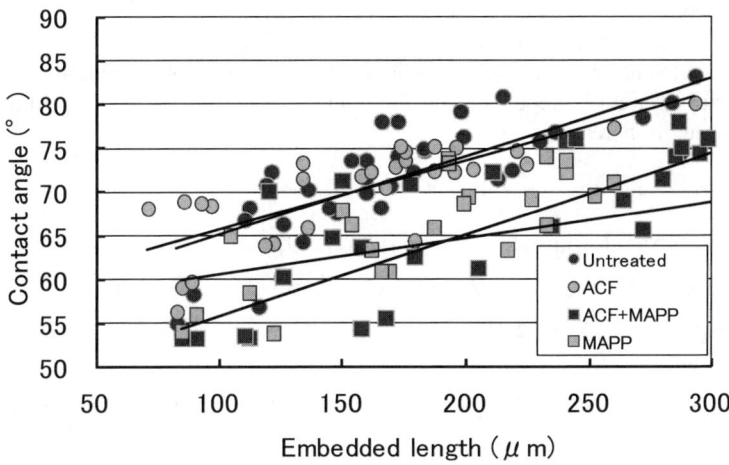

Fig.4 Relationship between contact angle and embedded length.

Figure 3 shows that interfacial shear strength has s flat shape to embedded length. So, the interfacial strength is not dependent on the embedded length.

Figure 4 shows the all slopes of the lines are positive. So, the contact angle at the resin and carbon fiber is dependent on embedded length. And as embedded length increases, contact angle increases.

Figure 5 shows the relationship between interfacial shear strength and contact angle. Interfacial shear strength is dependent on the contact angle. When the contact angle is small, the interfacial shear strength becomes bigger. Because the contact angle is dependent on embedded length which is mentioned above, small contact angle means the small adhesive area which is equal the embedded length. So when contact angle is small, interfacial shear strength is big. That means the adhesive properties are improved by the treatments.

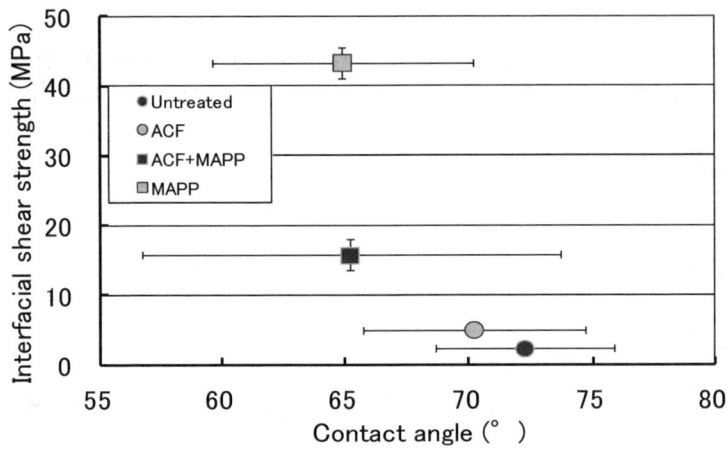

Fig.5 Relationship between interfacial shear strength and contact angle.

Conclusions

In this study, interfacial strength of carbon fiber reinforced polypropylene was examined. As a result, following conclusions were obtained.
(1) Acetone treatment and maleic anhydride are effective to the adhesives on the surface between fiber and matrix. But simultaneous treatments are not effective.
(2) The shear strength is not dependent on fiber embedded length.
(3) The contact angle and fracture load are dependent on fiber embedded length.
(4) The interfacial strength is dependent on the contact angle. As the contact angle increases, the interfacial strength increases.

References

[1] S.W. Beckwith, SAMPE Journal, Vol.4, 1 (1996) p.497.

[2] K.Uzawa, J. Soc. Mat. Sci., Japan, Vol.55, 1 (2006), p.131 (in Japanese).

[3] E.K. Gamstedt, M. Skrifvars, T.K. Jacobsen and R. Pyrz, Composites: Part A, Vol.33 (2002) p.1239.

[4] K. Tanaka, Y.Masabe and T.Katayama, J.Soc. Mat. Sci., Japan, Vol.58, 7 (2009), p.635 (in Japanese).

Effect of Surface Treatment on Creep Property of Jute Fiber Reinforced Green Composite under Environmental Temperature

Kenichi Takemura[1,a], Satoshi Miyamoto[1,b] and Hideaki Katogi[1,c]

[1] Kanagawa University, Department of Mechnical Engineering, Faculty of Engineering

3-27-1 Rokkakubashi,Kanagawa-ku, Yokohama, Kanagawa 221-8686, Japan

[a]takemura@kanagawa-u.ac.jp, [b]r201170052jw@kanagawa-u.ac.jp, [c]katougi@kanagawa-u.ac.jp

Keywords: Green composite, Natural fiber, PLA, Creep property, Surface treatment

Abstract. In this study, effect of surface treatment on creep property of green composite under environmental temperature was investigated. Jute fiber was used as reinforcement. PLA (polylactic acid) was used as matrix. Surface treatments were conducted using 5 % solution of silane coupling agent and PVA (polyvinyl alcohol). The flexural test was conducted under 25°C environment. The flexural creep test was conducted for 50 hours at 25, 40 and 50°C environment. As a result, the flexural property of composite increased by the surface treatments. And the surface treatments affected the adhesion of fiber/resin interface. Creep strains of surface-treated jute fiber/resin composites were lower than that of virgin composite. The creep strain was decreased by the treatments. The effects are confirmed under various temperature conditions.

Introduction

Green composites, consisting of an association of a biodegradable polymer matrix and a natural fiber as reinforcement have become popular due to both increasing social and economic pressure to conserve petroleum resource. Green composites offer environmental benefits such as biodegradability, less greenhouse gas emissions, and renewability of the base material.

Most of the natural fibers and reinforcement used for composites are hydrophilic nature, whereas synthetic polymers are hydrophobic. Poor adhesion between the natural fibers and polymer matrix often prevents the possibility of natural fibers to act as fillers, resulting in poor dispersion, inadequate reinforcement, and low mechanical properties [1-3]. Therefore, natural fibers require the addition of coupling agents or the chemical treatment for industrial application in composites. Many research efforts have been made to green composites with high strength, modulus, durability estimated by tensile, flexural and fatigue tests, with extensive attention being focus on the improvement of fiber-matrix interface compatibility, such as the use of coupling agent and surface treatments [4-6].

Creep is one of the principal properties which serve to be good reference when developing and using composite materials. It was well known that most materials respond differently depending on the time required to complete the mechanical test [7]. The creep behavior of composite materials is particularly important and a critical issue for many modern engineering application.

In spite of many reports on green composites, the analysis of creep behavior is still rare. Vazquez et al. [8] have performed flexural-creep test on bagasse fiber-polypropylene composites. The surface treated fiber improved the creep behavior of the composites due to the higher adhesion between the fiber and the matrix.

In this study, effect of surface treatment of creep property of jute fiber reinforced green composite under environmental temperature was investigated.

Experimental Details

Materials. The woven fabric jute fiber was supplied by Kawashima Selkon (Kyoto, Japan). PLA sheet (Terramac SS300) was supplied by Unitika Ltd. (Tokyo, Japan). The interfacial adhesion between fibers and matrix was modified using a silane coupling agent Z-6040 (Glycidoxypropyl-trimetho- xysilane) for PLA and PVA supplied by Wako pure chemical industries, Ltd.

Surface Treatment and Composites Fabrication. To obtain an improved fiber matrix adhesion, the fibers were modified silane coupling agent and PVA treatment. Silane coupling agent treatment was carried out in distilled water with 5 % silane content for 1hr at room temperature, and dried in the oven for 24hrs. PVA treatment was performed at same process with silane coupling agent.

Prior to composite fabrication, the woven jute fiber was completely dried at $50 °C$ in the oven. The fiber weight fraction of jute fiber green composites was 35 wt% and all green composites were made by a compression molding technique with vacuum. The woven jute fibers and PLA sheets was placed in an aluminum matched-die mold. The molding temperature, pressure and holding time were $190 °C$, 10 MPa and 10 min, respectively. And then the mold was cooled down to room temperature by water.

Static Flexural and Creep Properties. The flexural tests were performed according to JIS K7017 standard using tensile testing machine, and the crosshead speed was 2 mm/min. The flexural creep tests were performed with handmade equipment using dry oven according to JIS K 7017. Creep tests were carried out with a constant load level of 40 N, temperatures were room temperature(R.T.), 40 and 50 $°C$.

SEM Observation. The surface treated jute fiber and fracture surface of flexural tested samples were investigated by scanning electron microscopy (S-4000 FE-SEM, Hitachi High-Tech Technologies co.).

Results and Discussion

Flexural Property and SEM Observation. Figure 1 shows the relationship between the flexural properties with PVA and silane treated jute/PLA composites. The flexural properties, especially in modulus, of the surface treated jute fiber reinforced PLA composites were compared with these of the untreated jute fiber. The flexural modulus of untreated jute/PLA composites were increased about 21% from 3.8 GPa to 4.6 GPa by PVA treatment. The flexural modulus of virgin composites was improved after silane treatment, but the improvement rate of modulus was lower than PVA treatment. Figure 2 shows SEM image of surface treated fibers. The coated surface of jute fiber and the connection between fiber and fiber by PLA treatment were observed in this study. It can be explain that the connection of fibers lead to enhanced interfacial adhesion between the fibers and the matrix. Fracture surface of surface treated jute/PLA composites were shown in Fig. 3. The PVA treated jute/PLA composites show the better retention of resin on failed fiber. Therefore, their results implied that flexural property of composite was probably improved by using PVA treated fiber as reinforcement.

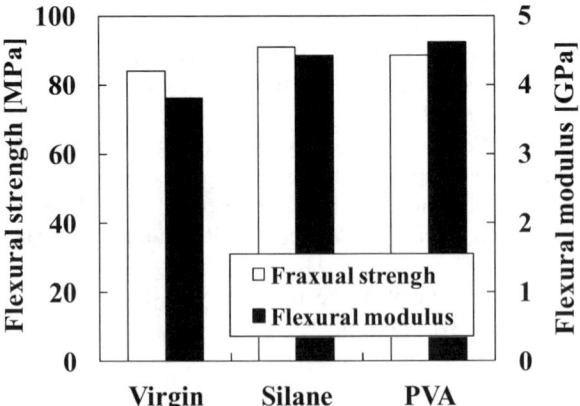

Fig. 1 Flexural properties of jute/PLA composites.

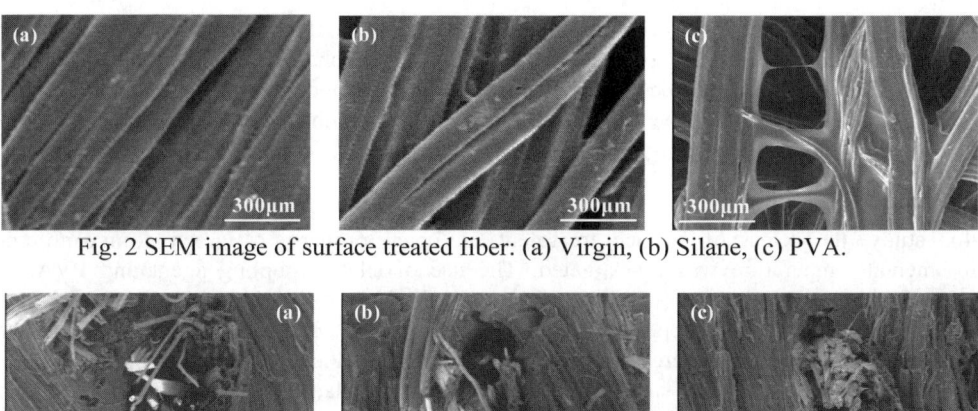

Fig. 2 SEM image of surface treated fiber: (a) Virgin, (b) Silane, (c) PVA.

Fig. 3 Fracture morphologies of composites: (a) Virgin, (b) Silane, (c) PVA.

Creep Behavior. Figure 4 shows the short-term creep behavior of jute/PLA composites using PVA and silane treatment. In case of creep test at R.T., the creep strains of jute/PLA composites using PVA and silane treatment were approximately lower than that without surface treatment, the coupling agent (Z-6040) was more effective than PVA treatment in jute/PLA composites. This could be utilized in chemical interaction between jute fiber and PLA at the interface leading to better bonding. This behavior can be directly related with the interfacial adhesion. As compatibility of hydrophobic polymer and hydrophilic cellulose fiber by surface treatment is improved, the fiber-matrix adhesion is increased. It leads to the enhancement of creep behavior.

As expected, as the temperature increases, the creep strain of composite increases. Nevertheless, creep strain of composites at 40 and 50°C were noticeably reduced by using silane coupling agent and PVA treatment. The creep strain of surface treated jute /PLA composite at 50°C was higher than that of virgin composite at 40°C. In spite of higher creep strain at 50°C, their jute/PLA composites were not failed.

In spite of temperature increasing, the reason that the creep strain of composite using PVA treated fiber is lower is due to enhanced compatibility and thermal stability between the surface treated fiber and the

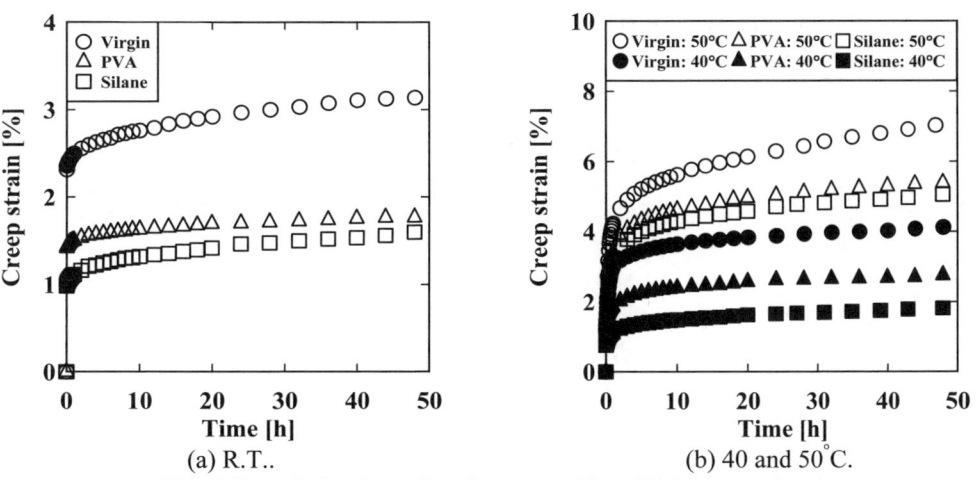

Fig. 4 Creep behaviors of surface treated jute/PLA composites.

matrix. Q. Liu et al. [9] and L.E. Millon et al. [10] revealed that the PVA was increased the number of hydrogen bonding between fiber and matrix, it leads to high mechanical and thermal properties for composites. In this case, it can be explained that the PVA treated fiber is increased by chemical cross-linking between the jute fiber and matrix, it occurred the enhanced compatibility and thermal stability between fiber and matrix.

Conclusions

In this study, the effect of surface treatment on creep property of green composite under environmental temperature was investigated. The use of silane coupling agent and PVA were effective in improving the compatibility of jute fiber with the matrix, and resulted in further enhancements of mechanical properties of jute/PLA composites. The interfacial adhesion between fiber and matrix was probably increased by surface treatment and then the flexural property of the jute fiber composites increased. After surface treatment by PLA and silane coupling agent, creep strains of surface treated jute/PLA composites under environmental temperature was improved. Therefore, the surface treated jute fiber composites show the lower creep strain than virgin composite due to the enhanced thermal stability by chemical cross-linking between the fiber and the matrix.

References

[1] J. Konnerth, A. Jager, J. Eberhardsteiner, U. Muller and W. Gindl: J. Appl. Polym. Sci., Vol. 102 (2006), p.1234.

[2] J. Gassan and A.K. Bledzki: Polym. Compos., Vol. 18 (1997), p.179.

[3] M. Bera, R. Alagitusamy and A. Das: J. Reinforced Plast. Compos., Vol. 29 (2010), p. 3155.

[4] O.A. Khondker, U.S. Ishiaku, A. Nakai and H. Hamada: J.Polym. Environment, Vol. 13 (2005), p.115.

[5] P.S. Razi, R. Portie and A. Raman: J. Compos. Mater., Vol. 33 (1999), p.1064.

[6] A. Espert, W. Camacho and S. Karlson: J.Appl. Polym. Sci., Vol. 89 (2003), pp. 2353.

[7] E.J. Barbero: *Introduction to composite materials design*, CRC Press (2011).

[8] A. Vazquez, V. Dominguez and J.M. Kenny: J. Thermoplast. Compos. mater., Vol. 12 (1999), p. 477.

[9] Q. Liu, T. Stuart, M. Hughes, H.S.S. Sharma and G. Lyons: Compos.: Part A, Vol. 38 (2007), p.1403.

[10] L.E. Millon and W.K. Wan: J. Biomed. Mater. Res. Part B: Appl. Biomater., Vol. 79B (2006), p.245.

Micro-scale cantilever testing of linear elastic and elastic-plastic materials

J E Darnbrough[1,2*], S Mahalingam[1] and P E J Flewitt[1,2]

[1] Interface Analysis Centre, University of Bristol,

Oldbury House, 121 St. Michael's Hill Bristol, BS2 8BS, UK

[2] School of Physics, University of Bristol,

H. H. Wills Physics Laboratory, Tyndall Avenue, Bristol, BS8 1TL, UK.

*j.e.darnbrough@bristol.ac.uk

Keywords: micro-scale testing, cantilever specimens, elastic deformation, elastic-plastic deformation, silicon, nanocrystalline nickel

Abstract. It is increasingly a requirement to be able to determine the mechanical properties of materials: (i) at the micro-scale, (ii) that are in the form of surface coatings and (iii) that have nano-scale microstructures. As a consequence micro-scale testing is an important tool that has been developed to aid the evaluation of the mechanical properties of such materials. In this work cantilever beam specimens (typically 2μm by 2μm by 10μm in size) have been prepared by gallium ion milling and then deformed in-situ within a FEI Helios Dual Beam workstation. The latter is achieved using a force probe with a geometry suitable for loading the micro-scale test specimens. Thus force and displacement can be measured together with observing the deformation and fracture of the individual specimens. This paper considers the evaluation of the mechanical properties in particular elastic modulus, yield strength and fracture strength of materials that result in relatively large deflections to the micro-scale cantilever beams. Two materials are considered the first is linear elastic single crystal silicon and the other elastic-plastic nanocrystalline (nc) nickel. The results are discussed with respect to the reproducibility of this method of mechanical testing and the evaluated properties are compared with those derived by alternative procedures.

Introduction

For many years micro-scale testing has been utilized to investigate both the mechanical properties and deformation and fracture characteristics of a range of materials. Typical examples are tensile tests and punch tests [1]. These tests were developed to allow mechanical properties to be evaluated where there was either a limited amount of material available or where the material was subjected to irradiation and therefore active. However, there is a requirement to determine the mechanical performance of materials at the micro-scale to investigate specific microstructures including nc materials. This has resulted in a range of techniques being adopted including micro-scale tension specimens prepared by focused ion beam milling and nano-indentation [2]. The work of Roberts et al has been directed primarily to preparing horizontal cantilever beam specimens by focused ion milling and then testing ex-situ using a nano-indenter to apply the load to cause deformation and fracture of those specimens [3]. In this paper we explore a modified technique where vertical cantilever beams are produced by focused ion beam milling and loaded within a dual beam workstation. The dual beam system is used to both prepare the specimens and view, using scanning electron microscopy imaging, during deformation. Here we consider the technique applied to both elastic single crystal silicon and elastic-plastic nc nickel.

Test Procedure

The mechanical testing experiments were conducted within a FEI Helios 500i Dual Beam Workstation, fitted with a gallium ion source. The specimens were deformed using a specially adapted force probe and a nano-manipulator supplied by Kliendiek and housed in the work chamber. Thus, the material to be tested is placed in the chamber and the focused ion beam is used to prepare the vertical cantilever beams at an edge of the sample. This edge preparation is selected to guarantee the uninhibited movement of the nano-manipulator used to deform the cantilever. However there is a restriction of the force that can be applied through the tip which means that careful consideration of the dimensions of the cantilever is required to allow deformation of a material. The loading tip fitted to the nano-manipulator is supplied by Kleindiek but the area used for application of force to the cantilever has been modified to match the specimen geometry, figure 1(a). The silicon tip was modified by ion beam milling to create a suitable geometry to apply force to the cantilever specimen, at a defined point of contact. A piezo-electric device is attached to the tip and this records the change in voltage with applied force. Displacements are measured directly from the recorded image. The raw data collected is in the form of force-time graphs and images of the displacement at the position of load application on the cantilever. This leads to the creation of force-displacement curves and the corresponding images of the specimen as deformation progresses. The silicon {100} single crystal wafer used is a well characterized linear elastic material. The elastic-plastic material selected is nc nickel produced by pulsed electrodeposition Integran Ltd. as plates with a thickness of 0.5 mm with a grain size of 20-80 nm. Vertical cantilever beam specimens with a square cross-section were produced in two stages. First initial trenches are cut at 90 degrees to the surface using gallium ions with a high ion beam current to remove material, typically 20 nA, creating the volume that the cantilever can be deformed into, and leaving a column of material. The initial focused ion beam intensity distribution is Gaussian and as a result the beam created has sloping sides. Therefore a second stage cut was performed just off perpendicular with a slightly defocused ion beam. Here the beam current is lower, typically 2.6 nA, which allows a finishing cut to provide square section parallel sided cantilever beams typically 2 μm x 2 μm x 10 μm, figure 1(a). It also removes the severe gallium ion damage introduced into the specimen by the first milling sequence.

This method of preparation does not produce a right angle between the base of the cantilever and the bulk of the material, due to the nature of the ion milling. A finite element (FE) model was used to consider the implication of this. The FE model shows that tapering of more than 12.5° below the base of the cantilever specimen has a negligible effect on deformation, seen by the stress map shown in figure 1(b). As a consequence it is possible to use the small deflection cantilever beam analysis without any corrections to the evaluated properties. The cantilever bend geometry under deformation introduces a stress distribution across the section of the specimen from tension to compression. Given a cantilever of known length, and cross-section (aspect ratio 5:1) and the force required to produce a deflection is proportional to the elastic modulus of the material. Thus given a force, F, and a deflection, δ, with a known length, L, and cross-section, A, which is used to find the second moment of inertia, I, an elastic modulus, E, can be calculated.

$$E = \frac{FL^3}{\delta I} \qquad (1)$$

Departure from linearity in the force-displacement curve, provides a measure of the yield or flow stress.

Results

The silicon specimens were initially used to provide a measure of the reproducibility of the technique. Five specimens were prepared with a common orientation with respect to the {100} plane of the silicon single crystal wafer; thus the faces of the cantilever beams were milled with a common crystallographic surface planes. The results of these tests are shown in figure 2(a) as force-

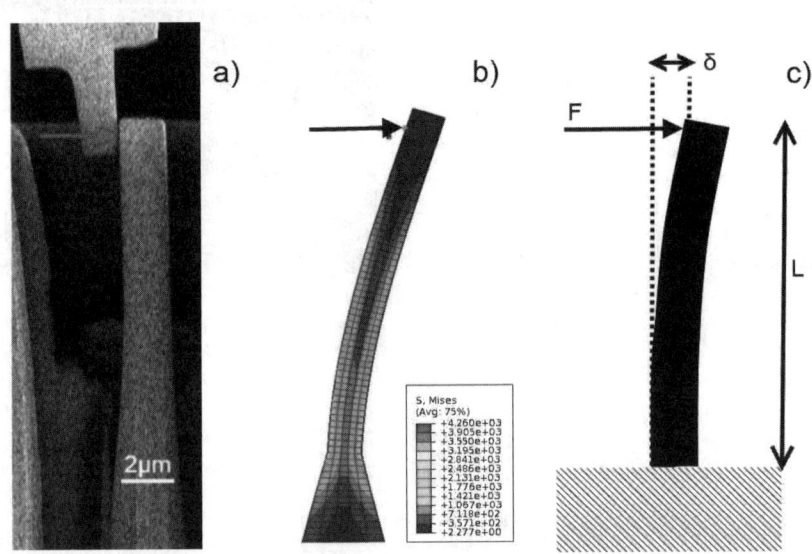

Figure 1: Diagram of a cantilever beam in bending (a) typical specimen (b) Finite Element analysis including tapering and (c) small deflection cantilever beam analysis.

time plots. These show extremely good repeatability in the peak force for the five replicated tests of silicon specimens to fracture. Specimen 4 had a smaller cross-section when observed in comparison with the other specimens which lead to the lower force at failure. When combined with the images force-displacement curves where prepared allowing both elastic modulus and fracture strength to be obtained. These give and elastic modulus of 128±17 GPa which is within the range given by Hopcroft for {100} orientation single crystal silicon loaded in this orientation [4]. Figure 2(b) shows a typical fracture surface for these silicon cantilever beams viewed at an angle. This surface is inclined at ~45° to the {001} plane and is therefore consistent with {110} cleavage fracture where the crack propagates in the $<\bar{1}10>$ as predicted from pseudo-potential calculations [5]. This is the product of a measured room temperature cleavage fracture stress of 72±6.8 MPa, which is below that recorded by Petersen [6].

The typical force-displacement curve for nc nickel shows the elastic-plastic nature through the departure from a linear gradient in the load-displacement curve, figure 2(c). Here departure from linearity occurs at a displacement of ~2 μm and corresponds to a yield stress of 340±32 MPa which lies within the range found by previous workers [7]. The result for elastic modulus of as-received nc nickel is 349±48 GPa; above that expected for the material [7].

Concluding comments

Micro-scale testing is a powerful tool for characterizing the deformation and fracture of specific mechanical microstructural features. In the case of the silicon specimens the force-displacement results, figure 2(a), demonstrate good repeatability both in preparing the specimens with a consistent geometry and for the loading system. The elastic modulus obtained for the silicon single crystal was 128±17 GPa which is consistent with the values evaluated previously [4]. The brittle fracture initiated and propagated on the {110} cleavage plane in the <111> direction. This is in agreement with previously observed cleavage on either {111} or {110} planes in silicon. However the fracture energy of 72±6.8 MPa is significantly lower than previously observed [6]. It is also noteworthy that this crack path was followed for about 0.75 of the cross-section and then the crack

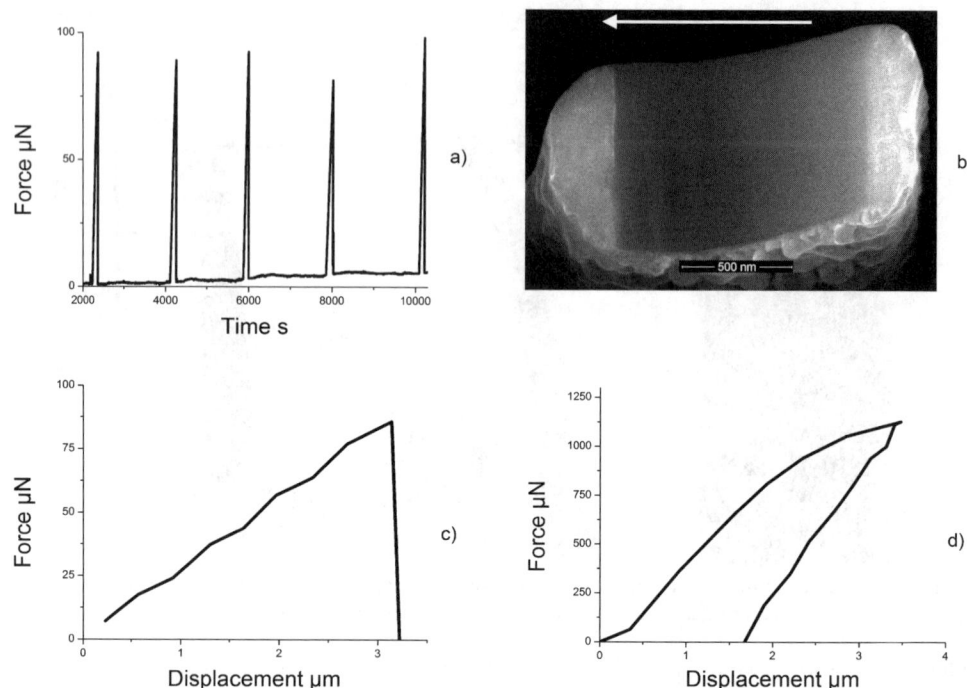

Figure 2: (a) Repeatability force-time curves of silicon (b) typical fracture surface of silicon specimen with the direction of fracture labled (c) typical force-displacement graph for silicon and (d) nc nickel.

deviated to a {001} plane when it reached the compression strained region, likely due to the redistribution of stresses as the crack progressed through the speicmen. In the case of the elastic plastic nc nickel, figure 2(d), the elastic modulus of 349±48 GPa is consistent with, and the yield strength of 340±32 MPa is higher than previous values [7]. It is noteworthy that these nc nickel specimens contain typically 40,000 grains and ~18% of the volume is grain boundary. Hence at this size the micro-scale specimen should provide representative values. In this case the overall plastic deformation is probably a combination of dislocation motion within the grains and grain boundary sliding.

Acknowledgements. The authors acknowledge EPSRC grant EP/H006729/1 for funding.

References

[1]- V. Vorlicek, L.F. Exworthy and P.E.J. Flewitt: Journal of Mat. Sci. vol 30 pp 2936-2943 1995
[2]- K. Fujii, K. Fukuya: Materials Transactions vol 52 no 1 pp 20-24 2011
[3]- D. DiMaio and S.G. Roberts; Journal of Materials Research vol 20 no 2 pp 299-302 2005
[4]- M.A. Hopcroft, W.D. Nix and T.W. Kenny: Journal of Microelectromechanical Systems vol 19 no 2 pp 229-238 2010
[5]-R. Perez and P. Gumbsch: Phys. Rev. Lett. Vol 84 pp 5347-5350 2000
[6]- K.E. Petersen: Proceedings of the IEEE vol 70 no 5 pp 420-457 1982
[7]- U. Erb, K.T. Aust and G. Plaumbo: Electrodeposited Nanocrystalline Metals, Alloys and Composites in 'Nanostructured Materials: processing, properties and applications Second Edition by C. C. Koch 2007

The influence of the epoxy interlayer on the assessment of failure conditions of push-out test specimens

Jan Klusák[1,a], Peter Helincks[2,3,b], Stanislav Seitl[1], Wouter De Corte[2,3], Veerle Boel[2,3] and Geert De Schutter[3]

[1]Institute of Physics of Materials, Academy of Sciences of the Czech Republic, Žižkova 22, 616 62 Brno, Czech Republic

[2]Department of Construction, Faculty of Applied Engineering Sciences, University College Ghent, Schoonmeersstraat 52, BE-9000 Ghent, Belgium

[3] Department of Structural Engineering, Faculty of Engineering Sciences, Ghent University, Technologiepark-Zwijnaarde 904, BE-9000 Ghent, Belgium

[a]klusak@ipm.cz, [b]peter.helincks@hogent.be

Keywords: push-out test, generalized fracture mechanics, failure initiation, steel-concrete joint, epoxy adhesive layer.

Abstract. Connection between steel and concrete parts is frequently required in constructions where the steel-concrete joints are often realized by welded shear studs. In order to avoid stress concentrations, corrosion proneness, and other negative consequences of the welding process, steel-concrete connection without welded mechanical shear connectors is sought nowadays.
Connection can be realized via an epoxy adhesive layer and gritted with granules. In the paper, the assessment of the push-out test configuration was performed from the generalized fracture mechanics point of view. The numerical-analytical modelling of a steel-concrete connection is performed without and with the epoxy interlayer, while 2D and 3D modelling is used. Thus conditions of crack initiation can be predicted from knowledge of the standard mechanical and fracture-mechanics properties of particular materials. The model of a bi-material notch with various geometry, and material properties is used to simulate various singular stress concentrators that can be responsible for failure initiation. Various manners of preparation of the epoxy interlayer are tested experimentally. Results of the fracture-mechanics studies are compared with each other and with experimental results. On the basis of the comparison, the 2D simulation of the steel-concrete connection without the epoxy interlayer is shown to be suitable for the estimation of failure conditions.

Introduction

Steel concrete joints used in constructions require good connection between steel and concrete. The quality of the adhesion is a substantial condition to exploit the advantages of both components (the tensile strength of steel and the compressive strength of concrete). As mechanical shear connectors (studs) welded to steel parts of constructions degrade both the mechanical and corrosion properties of steel, an epoxy adhesion layer gritted with granules is used. In cases of that kind, a push-out test is used for experimental determination of shear bond strength. The experimental program is supplemented by assessment of conditions of failure initiation based on generalized fracture mechanics approaches. The methods utilise a model of a bi-material notch, which is suitable to simulate geometric and/or material discontinuities in constructions. In the paper the push-out specimen is modelled in three levels of model difficulty and the results of the simulations are used for determination of the most satisfactory model.

Generalized fracture mechanics methods. The methods of evaluation of the singular stress concentrators based on generalized fracture mechanics are described in many previous papers [1,2]. The inputs of the assessment procedure are geometry of the specimen (see Fig. 1) as well as known elastic and fracture-mechanics parameters of materials. From the geometry and material properties, the stress singularity exponents p_1 and p_2 are analytically determined. Thus the order of the stress singularity is given. To know the absolute level of the stresses in the stress concentrator vicinity, the generalized stress intensity factors (GSIFs) H_1 and H_2 must be estimated based on the results of finite element methods (FEM) simulations. Possible crack initiation directions are determined from the ratio $\Gamma_{21} = H_2/H_1$ where in the paper the criterion of the direction works on the presumption of the maximum of the mean value of tangential stress. Then the global and local maxima must be evaluated within the stability criterion. The conditions of possible crack initiation are ascertained from the stability criterion based on the mean value of tangential stress as well. In the criterion of stability, the value of GSIF H_1 gained from a numerical solution is compared with its critical value H_{1C} following from analytical relations. The critical loading F_{Crit} is given by the following relation:

$$F_{Crit} = F_{appl} \frac{H_{1C}}{H_1(F_{appl})}$$

where F_{appl} represents applied force for which the $H_1(F_{appl})$ value is calculated. The critical applied force F_{Crit} must be calculated for every direction of possible crack initiation in each singular stress concentrator. From this set of critical forces we take the minimum one, which is considered as the loading under which failure is initiated in a particular stress concentrator's tip.

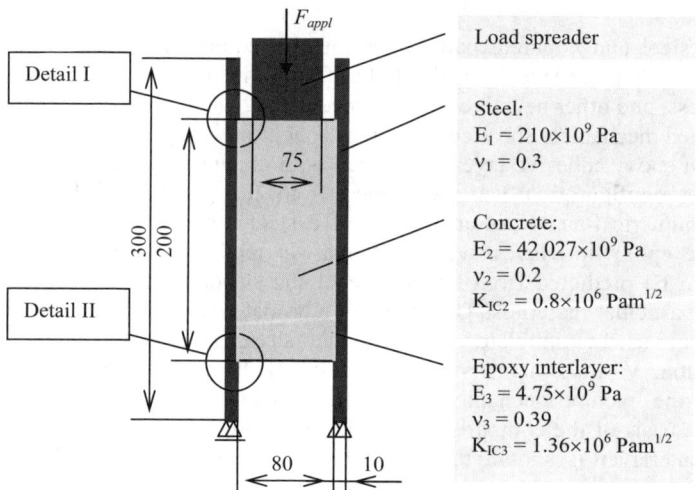

Fig. 1 Geometry and material characteristics of the specimen

Numerical models

The push-out specimen is symmetrical according to its vertical axis. Within the FEM modelling, half of the specimen is modelled in three levels. First, the 2D model of the steel and concrete parts without an epoxy interlayer is performed. Second, the 2D model of the steel-epoxy-concrete connection is used. Finally, the 3D model respecting the grooved surface of the epoxy interlayer was analyzed. The three models exhibit different stress concentrators in the Details I and II (see Fig. 1) and they are shown in Fig. 2. Note that the FEM models in Fig. 2 are rotated 90° anticlockwise. The stress concentrators in the specimen are evaluated individually.

2D steel-concrete FEM model. This plane-strain model is the simplest one and it contains two singular stress concentrations caused by connection of major material components: steel and concrete. Both bi-material notches A and B (see Fig. 2a) were analyzed in [2]. Due to pressure stresses near the notch B, the assumed crack initiation point is supposed to be the notch A. The critical applied force F_{Crit} resulting from the fracture mechanics assessment was determined in the interval (309; 343) kN, depending on the averaging distance in the criterion, see [2] for details.

2D steel-epoxy-concrete FEM model. In order to cover the presence of the epoxy interlayer, a simple 2D model of the specimen with an interlayer is suggested, see Fig. 2b. In this model four bi-material notches occur. The distribution of the tangential stress $\sigma_{\theta\theta}$ around the notches A – D is shown depending on the polar coordinate θ in Fig. 3 ($\theta = 0$ matches to the interface). It is seen that the presence of the interlayer in the FEM model redistributes the stresses. Only in the notches B and C the stresses are positive. The negative stresses $\bar{\sigma}_{\theta\theta}$ around the notches A and D imply pressure which does not lead to failure initiation (from the supposed fracture-mechanics point of view). The notches B and C were evaluated where the results are stated in Table 1. In the case of notch B, the crack initiation angle is taken 0° as there is the maximum of tangential stress in the region of epoxy. In the case of notch C, crack initiation conditions are evaluated both in the direction of 150° (into concrete) and 180° (into epoxy). Note that in the case of notch C only one singular term occurs, thus the value H_2 is not stated. The most probable conditions for crack initiation are in the notch B with F_{Crit} = 1275 kN. Nevertheless the value is considerably far from the experimental results.

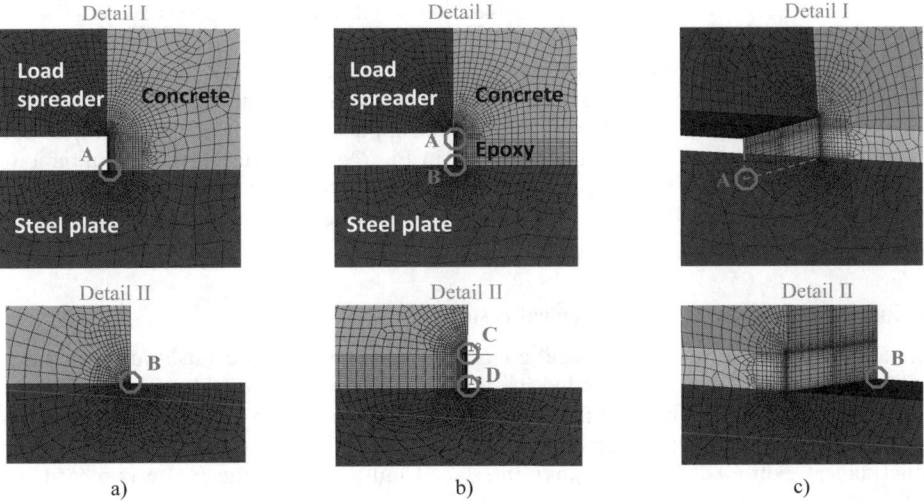

Fig. 2 FEM models: a) 2D steel-concrete, b) 2D steel-epoxy-concrete, c) 3D steel-epoxy-concrete

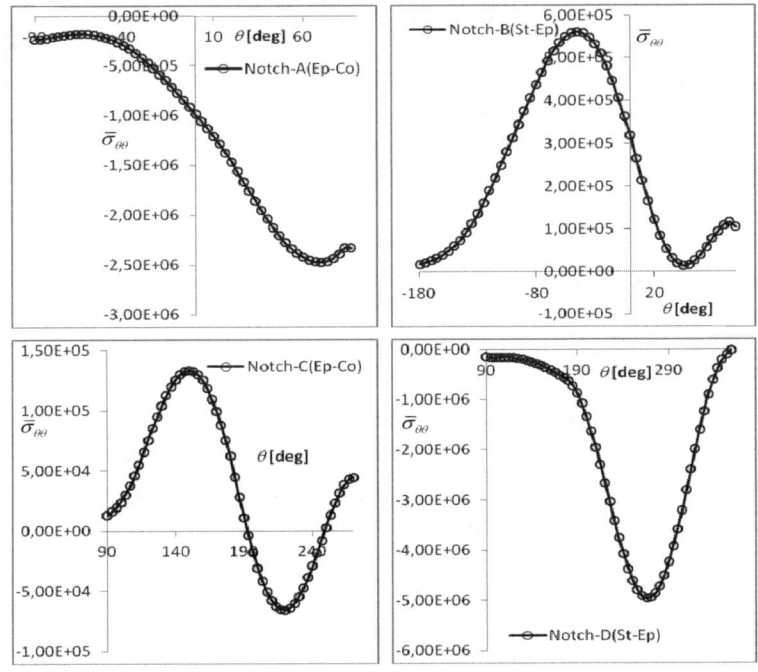

Fig. 3 Mean tangential stress $\bar{\sigma}_{\theta\theta}$ around notches A, B, C and D (2D steel-epoxy-concrete model)

Table 1: Results of a 2D study of a steel-epoxy-concrete joint

		Notch B	Notch C	Notch C
θ_0	[°]	0	150	180
H_1	[Pam^{p_1}]	5,2383E+04	-1,3984E+02	-1,3984E+02
H_2	[Pam^{p_2}]	-1,0321E+04		
H_{1C}	[Pam^{p_1}]	4,6855E+06	-2,1956E+04	-3,6202E+04
F_{Crit}	[N]	1,2746E+06	2,2372E+06	3,6889E+06

3D steel-epoxy-concrete model. The results of the previously realized models lead us to survey conditions of failure initiation in the steel-concrete notch in the 3D model. The epoxy interlayer is modelled locally (see Fig. 2c) in order to represent the application of the epoxy resin with a toothed trowel. From the analysis of the notches A and B in Fig. 2c, the resulting failure initiation forces F_{Crit} are even higher than in the previous steel-epoxy-concrete 2D analysis. This is caused by the larger and more complicated surface of the epoxy layer, where the ideal adhesion conditions are supposed.

Discussion of experimental and numerical results

Within the experimental study, the steel-concrete connections in the push-out specimens were realized in three main ways: a thick adhesive layer applied with a toothed trowel and gritted with granules, a thick smooth layer, and a thin layer of an epoxy resin with higher fluidity applied on the steel plates with a paint roller and gritted with steel grit [3]. The fracture surface of the specimens with thick layers exhibited concrete-epoxy interfacial failure, where the F_{crit} was measured in the interval ⟨177; 227⟩ kN. The case of interfacial failure cannot be assessed without knowledge of interfacial fracture toughness. On the other hand the fracture surface of the specimens with a thin layer exhibited failure in concrete (it corresponds to good adhesion) and the mean value of F_{crit} = 353 kN, see [3]. This case well corresponds to the 2D steel-concrete model, see Fig. 2a. In this case the estimated F_{crit} is in the interval ⟨309; 343⟩ kN depending on the averaging distance in the criterion. Thus the simplest model without modelling the epoxy interlayer is the most suitable. Due to its simplicity, it can be used in engineering practice. The 2D and 3D models with thick interlayers lead to redistribution of the stresses and together with the assumption of ideal adhesion between all components these models overestimate the experimentally measured values.

Acknowledgements

The authors thankfully acknowledge the Research Fund University College Ghent and Czech Science Foundation (grants P108/10/2049 and P105/11/1551) for financial support.

References

[1] Klusák, J., Knésl, Z.: Reliability assessment of a bi-material notch: Strain energy density factor approach, Theor. Appl. Fract. Mech. (2010), doi:10.1016/j.tafmec.2010.03.001

[2] Klusák J., Seitl S., De Corte W., Helincks P., Boel V., De Schutter G., *Failure conditions from push-out tests of a steel-concrete joint: fracture mechanics approach*, Key Engineering Materials Vols. 488-489, 2012, pp. 710-713.

[3] Helincks P., De Corte W., Klusák J., Boel V., De Schutter G., *Experimental investigation of the influence of the bond conditions on the shear bond strength between steel and self-compacting concrete using push-out tests*, Key Engineering Materials - Advances in Fracture and Damage Mechanics XI, 2013.

Fracture and Damage Characterization of Natural Fiber Composites

Hitoshi Takagi[1, a], Yuji Hagiwara[2] and Antonio N. Nakagaito[1, b]

[1]Institute of Technology and Science, The University of Tokushima,
2-1 Minamijosanjima-cho, Tokushima 770-8506, Japan

[2]Graduate School of Engineering, The University of Tokushima,
2-1 Minamijosanjima-cho, Tokushima 770-8506, Japan

[a]takagi@me.tokushima-u.ac.jp, [b]norio@me.tokushima-u.ac.jp

Keywords: Green composites, natural fiber, biodegradable resin, acoustic emission, fiber fracture, interface, adhesion, debonding.

Abstract. This paper reports the microscopic fracture behavior of natural fiber-reinforced green composites. The acoustic emission (AE) method of nondestructive and real-time testing was applied to detect small-scale energy release phenomena during tensile deformation of the green composites. The unidirectional abaca fiber was embedded in a starch-based biodegradable resin matrix. Two kinds of pre-damaged abaca fibers as well as as-received (i.e. undamaged) fiber were used to examine the effect of the pre-damaged abaca fiber on the overall fracture behavior of the unidirectional green composites. In the case of the green composites reinforced with as-received abaca fiber, both of the tensile strength and fracture strain were relatively high. In the case of the green composites reinforced with pre-damaged abaca fiber, however, showed relatively smaller tensile strength and fracture strain. In addition, a wide range of amplitude AE events were measured during the tensile deformation. This tendency was enhanced in the composites reinforced with heavily damaged abaca fiber. The experimental results showed that the AE activity in the early deformation stage was associated with such the microscopic fracture of pre-damaged abaca fibers.

Introduction

In recent years, the research and development of bio-based engineering materials have been done in U.S., Europe, Japan and elsewhere [1-8], as public attention has been focused on various environmental issues such as global warming, waste problem, etc. In order to obtain more sustainable composite materials than conventional glass fiber reinforced plastics (GFRP), many researches have tried to combine natural fiber with bio-based, biodegradable resin matrix, resulting in so-called *green composites*. Almost of the green composites are therefore made from yearly renewable raw materials, and furthermore their disposing treatment becomes easier.

It has been shown that the starch-abaca green composites can be biodegraded faster than neat starch resin [8], and that some kinds of the green composites have acceptable mechanical properties comparable to those of GFRP [8]. It should be noted that the natural fiber composites have not only better mechanical performance [4,6,7] but also interesting functionalities such as biodegradability [8] and thermal insulation property [5,9,10].

However, the detailed information on the micro-fracture behavior of the green composites is still lacking, because there are many complicated factors affecting the fracture behavior of the green composites; such as fiber fracture, debonding and matrix cracking. In our previous study, we reported that relatively high amplitude AE events were frequently observed even at the beginning of the deformation process of unidirectional composites and that this fracture strain range was much smaller than that of abaca fiber [11].

It was therefore suggested that these AE events were derived from the microfracture of pre-damaged abaca fibers, because some extent of damage might be introduced into the natural fibers during its extraction process or composites' fabrication process [11]. In order to clarify this point, we prepared two kinds of pre-damaged abaca fiber-reinforced green composites and carried out their tensile tests using the AE method.

Experimental Method

Raw Materials. The materials used in this study is the same ones as previous work [4,11], namely a starch-based biodegradable resin (CP-300, Miyoshi Oil and Fat Co. Ltd., Japan) was used as a matrix resin [4,11], and long abaca fiber without any surface treatment was used as a reinforcing phase [11].

Fabrication Method of Unidirectional Green Composites. Preliminary unidirectional composite sheets of approximately 110 x 12 mm were prepared by putting the dispersion-type, biodegradable resin (CP-300) on the surface of one layer of unidirectionally aligned abaca fiber. This preliminary composite sheets as well as neat resin sheets were dried at 105°C for 1 hour in an oven. Then, both of the preliminary composite sheets and resin sheets were cut into the exact size of the mold (100 x 10 mm). Some preliminary composite sheets were bent at 90-degree or 180-degree angles at the center point of the sheet (50 mm from both ends), and then bent reversely to the original flat sheet. After that one preliminary composite sheet was sandwiched between four neat resin sheets, then hot-pressed at 10 MPa and 140°C for 10 min. The final tensile specimen size was 100.0 x 10.0 x 1.9 mm. The fiber content of all specimens was fixed to be 3.3wt.%.

Tensile Testing. Details of the tensile test method have been reported elsewhere [11]. Two-channel AE measurement was performed to evaluate the microscopic damaging processes of natural fiber-reinforced composites. Two piezoelectric AE sensors (F217M, Showa Electric Laboratory Co. Ltd., Japan) were attached on the specimen. The distance between the two AE sensors was 30 mm. The AE signals was amplified by using a wide-range pre-amplifier (1220A, Physical Acoustics Co., U.S.A) with a gain of 40 dB, and then recorded in an AE analyzing system (MISTRAS 2001, Physical Acoustics Co., U.S.A). Simultaneously the tensile force information was also recorded in the AE analyzing system [11].

Results and Discussion

Typical stress-strain relationship with AE activities of the unidirectional abaca composite (fiber content is 50wt.%) is shown in Fig. 1[11]. Many AE events with a wide amplitude distribution are measured from the early stage of tensile deformation to final composite fracture. It can be seen that relatively high amplitude AE events are frequently observed even at the beginning of the deformation process. This strain range is much smaller than that associated with fiber fracture. It is therefore suggested that this AE event is derived from the microfracture of pre-damaged abaca fibers. Some damages might be introduced into natural fiber during the extraction process or composite fabrication process. Low amplitude AE events less than 60 dB are also found in the middle stage of the tensile deformation. These low AE events seem to be derived from fiber-matrix debonding, because no AE event was detected during tensile deformation of neat resin specimen [11, 12].

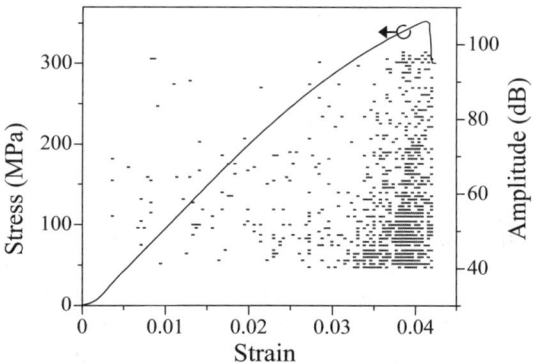

Fig. 1 Stress–strain diagram of unidirectional abaca composite (fiber content=50wt.%) [11].

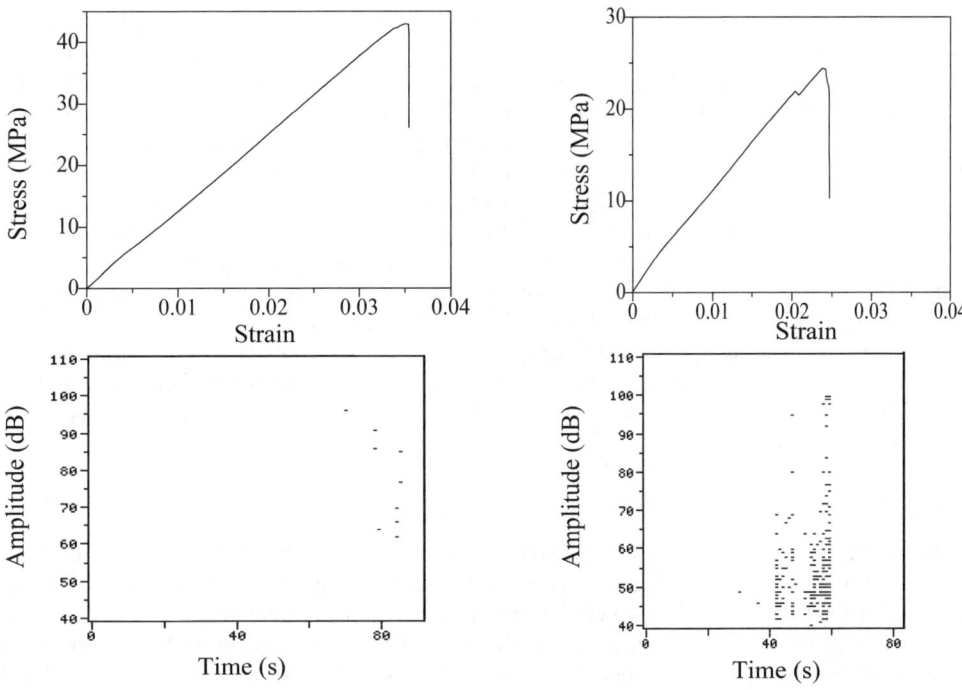

Fig. 2 Stress–strain diagram of undamaged abaca fiber composite (fiber content = 3.3wt.%).

Fig. 3 Stress–strain diagram of 90°-damaged abaca fiber composite (fiber content = 3.3wt.%).

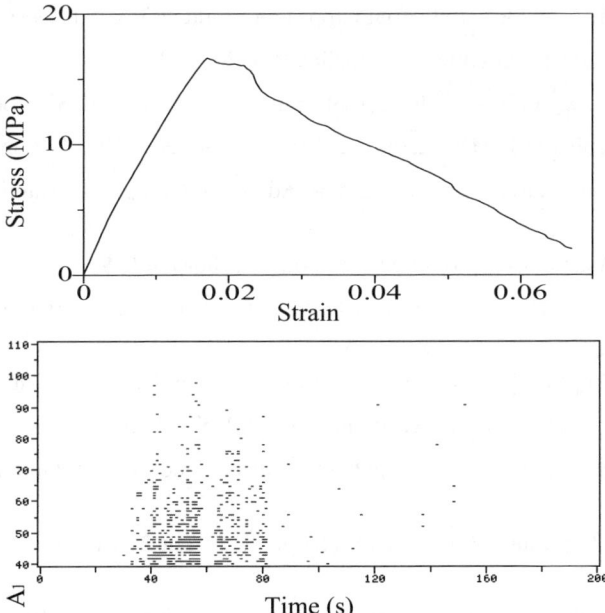

Fig. 4 Stress–strain diagram of 180°-damaged abaca fiber composite (fiber content = 3.3wt.%).

Figure 2 presents a typical stress-strain relationship of a unidirectional composite (fiber content is 3.3wt.%) reinforced with undamaged abaca fiber, and AE activity is also indicated below. The tensile strength is much lower than that in Fig. 1, because of low fiber content. However the fracture strain is approximately 0.035, and this value is almost same as that of abaca fiber [11]. We cannot see any AE events in the early stage of deformation (the strain range lower than 0.02), and relatively higher amplitude AE events appear before final composite fracture (the strain range higher than 0.03).

Figure 3 indicates a typical stress-strain relationship and AE activity of a unidirectional composite (fiber content is 3.3wt.%) reinforced with 90°-damaged abaca fiber. It can be seen that both tensile strength and fracture strain become smaller than those of undamaged green composites (Fig. 2). Both high and low amplitude AE events appear in the latter half of the deformation, suggesting the occurrence of debonding between abaca fiber and starch-based resin.

In the case of heavier damaged abaca fiber composites, namely 180°-damaged sample, its tensile strength and fracture strain become smallest among the all samples tested here. It is therefore suggested that the amount of pre-damage affect the mechanical performance of the natural fiber-reinforced composites.

Summary

In the case of the green composites reinforced with as-received abaca fiber, both of the tensile strength and fracture strain were relatively high. However, in the case of the green composites reinforced with pre-damaged abaca fiber showed relatively lower tensile strength and fracture strain. In addition, this tendency was enhanced in the composites reinforced with heavily damaged abaca fiber; namely 180-degree damaged sample. The experimental results showed that the AE activity observed in the early stage of deformation was associated with microscopic fracture of pre-damaged abaca fibers.

References

[1] M. Wollerdorfer and H. Bader: Industrial Crops and Products Vol. 8 (1998), p. 105.

[2] S. Luo and A.N. Netravali: Polymer Composites Vol. 20 (1999), p. 367.

[3] D.H. Mueller and A. Krobjilowski: Journal of Industrial Textiles Vol. 33 (2003), p. 111.

[4] H. Takagi and Y. Ichihara: JSME International Journal, Series A Vol. 47 (2004), p. 551.

[5] H. Takagi, S. Kako, K. Kusano, and A. Ousaka: Advanced Composite Materials Vol. 16 (2007), p. 377.

[6] H. Takagi and A. Asano: Composites Part A Vol. 38 (2008), p.685.

[7] R. Tokoro, D.M. Vu, K. Okubo, T. Tanaka, T. Fujii, and T. Fujiura: Journal of Materials Science Vol. 43 (2008), p. 775.

[8] H. Takagi: Proceedings of Third International Conference on Eco-Composites (2005), p. 14-1.

[9] K. Liu, H. Takagi, and Z. Yang: Materials and Design Vol. 32 (2011), p. 4586.

[10] K. Liu, H. Takagi, R. Osugi, and Z. Yang: Composites Science and Technology Vol. 72, (2012), p. 633.

[11] H. Takagi and Y. Hagiwara: WIT Transactions on The Built Environment Vol. 112 (2010), p. 221.

[12] I.M. De Rosa, C. Santulli, and F. Sarasini: Composites Part A Vol. 40 (2009), p. 1456.

Modeling of damage and failure of dual phase steel in Nakajima test

Junhe Lian[1,a], Pengfei Liu[1] and Sebastian Münstermann[1]

[1]Department of Ferrous Metallurgy, RWTH-Aachen University,

Intzestr. 1, 52072, Aachen, Germany

[a]junhe.lian@iehk.rwth-aachen.de

Keywords: Formality, Forming limit diagram, DP steel, Localization, Crack initiation, Damage mechanics models

Abstract. For modern high strength steels, instead of metal instability, ductile damage triggered by the formation of microvoids or microcracks resulting from the complex material microstructure, has become the key factor responsible for the final failure in the forming process of such steels. The target of this study is to describe the initiation and evolution of damage in a dual-phase (DP) steel (DP600). By applying a newly proposed approach that is able to indicate the onset of damage in an engineering sense and quantify the subsequent damage evolution, to predict the forming limits for DP600 are predicted by simulating Nakajima test. Accordingly, two forming limit curves (FLC) are numerically computed to characterize two moments: when damage becomes pronounced and when the final failure is triggered by the accumulation of damage. Comparing with the conventional experimentally calibrated FLC at necking, the limit at crack initiation predicted by modeling gives a lower but defect-free forming boundary. The forming limit at final fracture is well captured by allowing the subsequent damage evolution to a critical value.

Introduction

Dual phase (DP) steels consisting of two phases, ferrite and dispersed martensite, offer an attractive combination of strength and stretchability, which is a result of the strong distinctions of these constituents in mechanical properties. On the other hand, in addition to the damage caused by the inclusions or hard particles in the conventional highly ductile steels, the possibility for damage is increased to a large extent by the debonding of the ferrite-martensite phase boundaries and the inner-cracking of martensite islands in DP steels. It was shown by Tasan et al. [1] that an extensive damage scale has been reached before the strain localization in DP steel. Under some specific strain path, the final fracture is triggered even with minor or no strain localization. The importance of damage and damage evolution is therefore highly emphasized to guide the forming process of such steels instead of metal instability.

To address this new feature for sheet metal forming of DP steel, Lian et al. [2] proposed a hybrid damage mechanics approach consisting of three constituents: a plasticity model to characterize the material behavior before onset of damage, a criterion to indicate the initiation of damage, and a damage-induced softening part to describe the behavior of damaged material. One distinction in the approach is that the introduction of onset of damage, which is specified as crack initiation that causes the qualitative change of the material behavior in an engineering sense. In practice, this moment is captured by the direct current potential drop (DCPD) method. Based on the formulation of the model, two criteria are given as guideline for forming process: the forming limit for defect-free condition and the forming limit indicating the final fracture. The first one is given by the crack initiation criterion being reached in the simulation of Nakajima test, while the latter one is defined once the accumulated damage becomes critical to trigger the final fracture. In this study, both limits are calculated from the simulation of Nakajima test and transferred into the forming limit diagram (FLD). Accordingly, they are compared with the experimentally measured FLCs at necking and final fracture respectively.

Model description and material parameters

Under the assumption of isotropic materials and the plastic incompressibility, the Bai–Wierzbicki [3] yield criterion excluding the pressure effect coupled with an isotropic hardening law, defined in Eq. 1, is employed for the description of the plasticity behavior of DP600.

$$\Phi(J_2, J_3) = \bar{\sigma}_{eq} - \bar{\sigma}(\bar{\varepsilon})\left[c_\theta^s + \left(c_\theta^{ax} - c_\theta^s\right)\left(\gamma - \frac{\gamma^{m+1}}{m+1}\right)\right] \leq 0. \quad (1)$$

Where $\bar{\sigma}_{eq}$ is equivalent von Mises stress and $\bar{\sigma}(\bar{\varepsilon})$ is the flow curve; γ and c_θ^{ax} are defined by:

$$\gamma = \frac{\cos(\pi/6)}{1-\cos(\pi/6)}\left[\frac{1}{\cos(\theta-\pi/6)}-1\right]. \quad (2)$$

$$c_\theta^{ax} = \begin{cases} c_\theta^t, & \theta \geq 0 \\ c_\theta^c, & \theta < 0 \end{cases}. \quad (3)$$

where c_θ^s, c_θ^t, c_θ^c and m are material parameters. Note that compared to the conventional J2 plasticity, $\Phi(J_2) = \bar{\sigma}_{eq} - \bar{\sigma}(\bar{\varepsilon}) \leq 0$, the stress-state effect is taken into account by the dependency of the Lode angle, θ, which is related to the third invariant of deviatoric tensor. The detailed description concerning the flow rule and convexity analysis is refereed to Ref. [3] and Ref. [2].

To characterize the crack initiation, a strain-based criterion taking the form of the failure locus in Eq. 4 postulated by Bai and Wierzbicki [3] is employed. Note that the criterion here is used to identify the onset of damage, crack initiation, instead of failure.

$$\bar{\varepsilon}_i = \left[\frac{1}{2}\left(C_1 e^{-C_2\eta} + C_5 e^{-C_6\eta}\right) - C_3 e^{-C_4\eta}\right]\bar{\theta}^2 + \frac{1}{2}\left(C_1 e^{-C_2\eta} - C_3 e^{-C_4\eta}\right)\bar{\theta} + C_3 e^{-C_4\eta}. \quad (4)$$

where $C_1 - C_6$ are material parameters.

After the onset of damage, the isotropic damage evolution is assumed. As an analogy to the CDM based models, an internal variable D is introduced to measure the accumulated damage. Accordingly, the yield function accounting for the effect of damage on plasticity is defined by:

$$\Phi(J_2, J_3, D) = \bar{\sigma}_{eq} - \bar{\sigma}(\bar{\varepsilon})\left[c_\theta^s + \left(c_\theta^{ax} - c_\theta^s\right)\left(\gamma - \frac{\gamma^{m+1}}{m+1}\right)\right](1-D) \leq 0. \quad (5)$$

To describe the evolution of damage, the dissipation-energy-based damage evolution law applied in commercial FE code Abaqus [4] is adopted. The accumulated damage is therefore defined by:

$$D = \int_{\bar{\varepsilon}_i}^{\bar{\varepsilon}} \frac{\sigma_{y0} L}{2G_f} d\bar{\varepsilon}. \quad (6)$$

where G_f is the energy required to open a unit area of a crack, L is the characteristic length associated with an integration point to reduce the mesh dependency in the FE model, and σ_{y0} is the value of the yield stress at the onset of damage. The final fracture is reached when the damage D is accumulated to a critical value D_{cr}. The material parameters for DP600 are listed in Table 1, and the calibration procedure is shown in detail in Ref. [2].

Table 1. Calibrated material parameters in the fourth-described damage model.

Plasticity parameters				Crack initiation parameters						Damage parameters	
c_θ^t	c_θ^c	c_θ^s	m	C_1	C_2	C_3	C_4	C_5	C_6	G_f	D_{cr}
1.0	0.97	0.93	5	0.43	1.14	0.12	0.95	0.43	1.14	300	0.1

Experimental and numerical procedures of Nakajima tests

The Nakajima test is performed according to the standard DIN EN ISO 12004. Nine specimens are prepared with length of 190 mm and thickness of 1.5 mm but different web width from 20 mm to 160 mm, as shown in Fig. 1 (a). During the test, local strains on the surface of the specimens are computed with the aid of the electrochemically etched grids. In this study, a visual analysis method instead of the position-dependent method recommended by the standard is employed to capture the forming limits. In the finite element simulation of Nakajima test, only a quarter of the experimental setup is modeled due to the symmetry as shown in Fig. 1. In the model the drawing die, blank holder and the punch are considered as rigid bodies, and 3D 8-node brick elements with reduced integration point (C3D8R) is employed for the specimen. According to the experimental setup, a constant velocity is applied on the punch while the die and blank-holder are fixed in all freedoms.

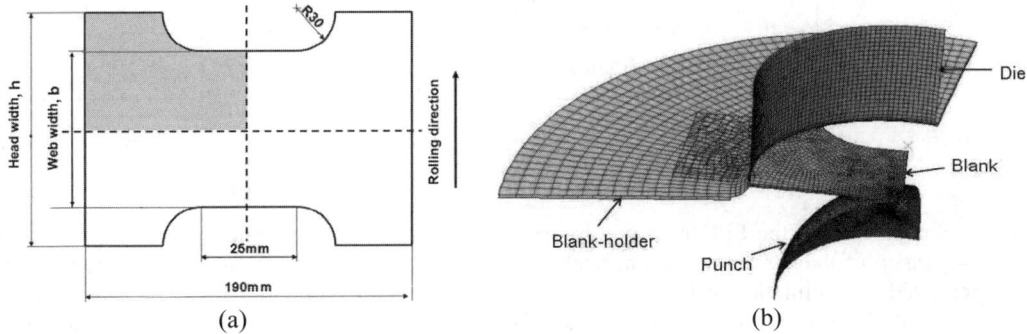

Fig. 1. (a) Drawing and dimensions of the Nakajima test specimens and the colored area is referred to as the quarter of the blank which is modeled in simulations, (b) Mesh of the FE model for Nakajima test after applying reflection symmetry twice.

Results and discussion

The local in-plane strains measured from Nakajima tests at two instants, necking and final fracture are both plotted in Fig. 2 (a) together with the FLC predicted by the simulation at the instant of crack initiation. Comparing with the experimental FLC at necking, the latter one gives a much lower estimation of the forming limits. It proves that the characteristic of forming DP600 steel sheet, i.e. damage has become more critical than the strain localization. As the crack initiation criterion is describing the forming limit corresponding to the occurrence of damage in an engineering sense, one can conclude that forming process under this limit is defect-free. On the contrary, the FLC at necking measured from experiments involves already certain amount of damage depending on the stain path. Taking the experimental one as a criterion in the component forming process would increase the risk of the failure in the subsequent service of the components.

By allowing the damage evolution after the crack initiation criteria is reached, the FLC for final fracture is calculated when the maximum damage accumulation occurs and plotted in Fig. 2 (b). Comparing with the experimental FLC measured at the final fracture, the numerical one gives an overall good prediction. Despite the gap between the numerically predicted FLC at crack intuition and the experimentally measured FLC at necking, the match of the results between two FLCs at the final fracture verifies that the applicability of the hybrid damage mechanics approach to simulate and describe the damage and failure behavior of DP600 steel sheet. On the right branch of the FLC, the numerical prediction shows lower values than the experimental ones. One reason for this is tracked back to the experimental program as well as the analysis technique. Note that the FLCs measured from experiments are not covering the complete strain path from uniaxial to equibiaxial tension. A certain range of forming limits is not given in the vicinity of equibiaxial tension condition. The uncertainty of the analysis technique in conjunction with the missed experimental data produces a certain scatter on the right side of the FLD. In addition, the damage evolution law employed in this study describes damage only proportional to the equivalent plastic strain, not depending on stress states. More accurate and advanced damage evolution laws are under development to improve the predictive capacity of the proposed model.

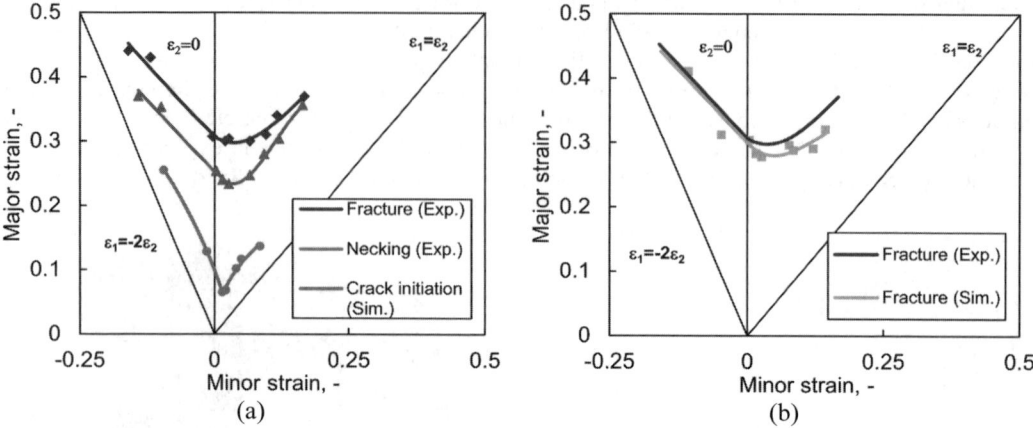

Fig. 2. Comparison of the FLD obtained by experimental measurement and numerical prediction, (a) comparison of the forming limits at necking and fracture measured from experiments with the one predicted by simulation with the criterion of ductile crack initiation, (b) comparison of the forming limit at fracture from experiments and the one predicted by simulation with the criterion of final fracture.

Summary

In this study, a hybrid damage mechanics approach is applied to describe the damage and failure behavior of a damage-dominant DP steel (DP600) in the forming process by simulating Nakajima test. The following conclusions can be drawn:

- By introducing the concept of crack initiation, the moment when damage causes a qualitative irreversible derogation of material behavior is identified in an engineering length scale. The corresponding FLC therefore provides a guideline with which a defect-free forming process is realized.
- Despite the mismatch of the numerally predicted FLC based on the concept of crack initiation with the experimentally measured one at necking, the predicted FLC at final fracture is in an overall good agreement with the experimental one.
- With additional experimental data at equibiaxial tension condition and more advanced damage evolution law expressing the stress state influences on damage, the predictive capacity of the applied damage model can be improved.

References

[1] C.C. Tasan, J.P.M. Hoefnagels, C.H.L.J. ten Horn and M.G.D. Geers: Mech. Mater. 41 (2009), p. 1264.

[2] J. Lian, M. Sharaf, F. Archie and S. Münstermann: Int. J. Damage Mech. (2012), accepted.

[3] Y.L. Bai and T. Wierzbicki: Int. J. Plast. 24 (2008), p. 1071.

[4] ABAQUS User's Manual (Version 6.11) (2011). Hibbit, Karlsson and Sorensen Inc.

Micromechanical Modeling for the Evaluation of Elastic Moduli of Woven Composites

D. Abe[1, a], O. Bacarreza[2, b], M. H. Aliabadi[1, c]

[1]Department of Aeronautics, Imperial College London, SW7 2AZ, London, UK
[2]Imperial Consultants, 58 Prince's Gate, SW7 2PG, London, UK

[a]daisei.abe10@imperial.ac.uk, [b]o.bacarreza-nogales@imperial.ac.uk,
[c]m.h.aliabadi@imperial.ac.uk,

Keywords: Micromechanics, FEM, Effective properties, Composites.

Abstract. Textile composites have increasingly been used as a structural material because of their balanced properties, higher impact resistance, and easier handling and fabrication compared with unidirectional composites. However, the complex architecture of textile composites leads to difficulties in predicting the response in spite of the fact that there is the need to determine mechanical properties in product design. Micromechanical analysis, using the Finite Element Method, was conducted in order to evaluate the effective mechanical properties of plain woven and 3D woven composites. In this study, numerical models of unit cells were used and it is shown that the predicted values of homogenized mechanical properties using the developed procedure were in good agreement with experimental results.

Introduction

Fiber reinforced plastic (FRP) has been employed in various applications in the aircraft and automobile industries to reduce the overall weight of the end product because of a high strength-to-weight ratio. In particular, Unidirectional (UD) composites have been widely used in many applications due to high specific stiffness and strength. On the other hand: woven, braided, knitted and stitched fabric composites have been increasingly used as a structural material in industrial applications because they are efficient at reinforcing all directions within a single layer and their ability to conform to surfaces with complex curvatures. In addition, textile composites, especially woven composites, provide improved impact resistance, better-balanced properties, and easier handling and fabrication compared with UD composites. However, it is known that the complex architecture of woven composites leads to difficulties in predicting the mechanical behavior although there is the need to determine of mechanical properties in product design.

The predictions of mechanical properties for textile composites have been heavily researched with most of studies focusing on plain woven composites which are composed of two fiber tows: fill and warp yarn in each layer. Analytical methods were proposed by Ishikawa and Chou [1], Jiang and Tabiei [2]. FEM has been also used to predict the mechanical properties on the micro-level or constituent level, Woo and Whitcomb [3] proposed a two-dimensional model and Chapman and Whitcomb [4] proposed a three-dimensional model. A meshfree method has also recently been proposed for 2D plain woven composite [5]. This research shows that the FE model of homogenization of the RVE is a powerful tool to predict effective mechanical properties.

For three-dimensional (3D) woven composites in which tows are interlaced in multi-directions, the methods for predicting mechanical properties are more complex than for plain woven composites because 3D woven composites have more complicated architecture. In addition, it is reported that the lack of mechanical characterization data and difficulties in testing methods have been significant issues for 3D woven composites. The level of complexity and lack of mechanical characterization have led to slow progress of research in the field of predicting mechanical behavior for 3D woven composites. However, some prediction approaches for elastic properties using FEM was introduced in recent years; Bogdanovich [6] proposed an FE approach using a three-dimensional mosaic model and the numerical model in FE analysis under low velocity impact conditions was proposed by Ji and Kim [7]. However, the research in the area of 3D woven composites has not reached an advanced stage yet.

In this paper, the procedure for determining the elastic mechanical properties in the case of plain and 3D woven composites is presented in order to provide accurate predictions of mechanical properties. The elastic properties for carbon/epoxy and graphite/epoxy plain woven composites and 93 oz. orthogonal fabric composite are calculated using micromechanical analysis from the results of FE analysis of the unit cell. The mechanical properties obtained from the results of the multi-scale analysis are compared with experimental data and the values obtained from analytical methods. The procedure developed in this study is then assessed and the characteristics of the mechanical properties for plain and 3D woven composites are discussed.

Calculation of Effective Elastic Properties Using FE Analysis

The procedure of determining the effective elastic properties using FE analysis is shown in Fig. 1. Displacement boundary conditions have been commonly used in the homogenization approach (Tang and Whitcomb [8]) and it is reported that the stress–driven approaches, such as traction boundary condition, are not suitable in micro-macro homogenization approaches due to macroscopic rotation effects (Galvanetto and Aliabadi [9], Kouznetsova et al.[10]). Therefore, six types of displacement boundary conditions were selected in order to determine stiffness matrix: each case can determine a row or column of the stiffness matrix. The values of initial volume, all components of stress and strain for each element are the output from FE analysis and the macroscopic stresses and strains are calculated by homogenization of the micro-scale fields.

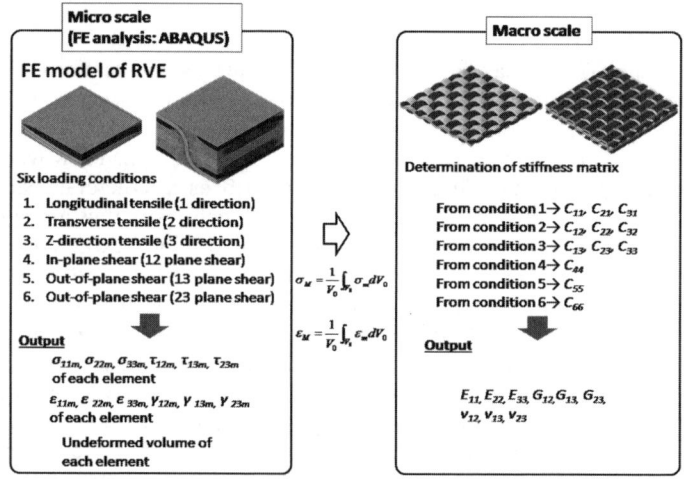

Fig. 1 - Procedure for calculation for elastic properties

Verification by Experimental Data

The direct numerical FE models were made in order to evaluate the elastic properties for both plain and 3D woven composites. The following conditions in FE analysis determined using the sample model was used to obtain the mechanical properties:

The displacement on the supported surface in all directions was fixed and the displacement on the other surfaces was fixed in all directions with the exception of the loading direction.

The section shape of yarn was based on an ellipse.

The fiber orientation of yarn neglects the detailed definition of the transverse direction aligned with the surface of yarn.

The mesh pitch was approximately 0.025 mm.

Plain Woven Composite. Two types of plain woven composites were selected in order to compare the experimental results with micromechanical analysis. One was a carbon/epoxy composite and the other was a graphite/epoxy composite, experimental values of elastic properties for this material were obtained from literature.

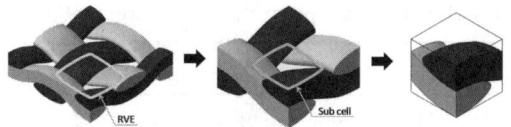

Fig. 2 - Plain Woven Composite

The comparison of elastic properties obtained from FE analyses and experiments on AS4 3k/3501-6 are shown in Table 1. The experimental results reported in the Composite Material Handbook MIL-HDBK-17-2F [11] were used and compared to the values obtained using FE analyses. The longitudinal and transverse elastic modulus (E_{11} and E_{22}) obtained from FE analyses were in good agreement with the experimental value; the values obtained from FE analyses were slightly higher than the experimental value (about 3%).

Table 1. Effective mechanical properties for carbon/epoxy

	$E_{11}=E_{22}$ (GPa)	E_{33} (GPa)	G_{12} (GPa)	$G_{13}=G_{23}$ (GPa)	v_{12}	$v_{13}=v_{23}$
Direct numerical model	69.646	11.289	7.007	4.372	0.042	0.417
Experiment	67.5	N/A	N/A	N/A	N/A	N/A

In order to compare with more experimental results, the graphite/epoxy woven fabric composite introduced in the literature was selected [2]. The comparison of elastic properties obtained from FE analysis, analytical methods and experiments for graphite/epoxy composite are shown in Table 2. The longitudinal and transverse elastic moduli (E_{11} and E_{22}) obtained in the experiments were a bit smaller than the numerical one (about 9%). while the in-plane shear modulus (G_{12}) obtained from FE analysis was slightly smaller than the experimental value.

Table 2. The values of mechanical properties for graphite/epoxy

	$E_{11}=E_{22}$ (GPa)	E_{33} (GPa)	G_{12} (GPa)	$G_{13}=G_{23}$ (GPa)	v_{12}	$v_{13}=v_{23}$
Direct numerical model	45.105	9.975	3.641	2.780	0.055	0.471
Experiment	49.8	N/A	3.83	N/A	0.068	N/A

3D Woven Composite. An FE models for 93 oz. 3D woven S-2 Glass/Dow Derakane 8084 Epoxy-Vinyl Ester resin composite, shown in Fig. 3. The parameters of input material properties and geometry were decided in accordance to Bogdanovich [6]. The angle of the z-yarn had to be smooth in order to make the mesh of brick elements. When the angle from x_3 axis was low, the model was not able to consist of only hexahedra brick elements or the model would have self-penetrating elements which may cause inaccurate results in the FE analysis.

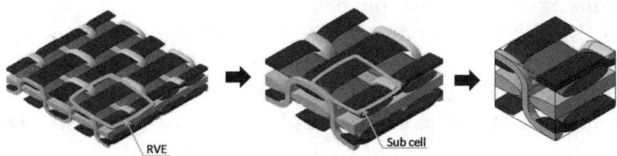

Fig. 3 - 3D Woven Composite

The comparison of elastic mechanical properties obtained from FE analysis and experiments for S-2 glass 93 oz. fabric composite are shown in Table 3. The values obtained from the numerical model correlate well to the experimental results. The differences between experimental values and predicted values of longitudinal and transverse moduli (E_{11} and E_{22}) in the numerical model were approximately 6% and 4%, respectively.

Table 3. The values of mechanical properties for S-2 glass 93 oz. fabric composite

	E_{11} (GPa)	E_{22} (GPa)	E_{33} (GPa)	G_{12} (GPa)	G_{13} (GPa)	G_{23} (GPa)	v_{12}	v_{13}	v_{13}
Direct numerical model	23.22	21.52	9.00	3.01	2.66	2.70	0.128	0.344	0.339
Experiment	24.68	20.75	N/A	N/A	N/A	N/A	0.11	N/A	N/A

Conclusions

Micromechanical modeling analysis using FE analysis was used for predicting the homogenized (effective) elastic mechanical properties in both plain and 3D woven composites. Evaluation and comparison of mechanical properties obtained from the numerical models with experimental data showed good agreement in the accuracy of the predictions. The longitudinal and transverse moduli and Poisson's ratios can be predicted accurately while the out-of-plane shear properties in the numerical model were slightly smaller than the values of the experimental and analytical approach due to the weave and shape of yarn.

The process of making FE mesh, especially in 3D woven composite, had some problems which may lead to inaccurate results in FE analysis, such as distorted elements.

As a result of this study, overall mechanical properties for woven composites were determined and the predicted mechanical properties show more balanced properties than seen in UD composites. Moreover, the results of comparing the mechanical properties between plain and 3D woven composites indicate that 3D woven composites may allow the z-yarn to prevent delamination.

References

[1] T. Ishikawa and T. W. Chou, "Elastic behavior of woven hybrid composites," *Journal of Composite Materials,* vol. 16, pp. 2-19, 1982.
[2] Y. Jiang, A. Tabiei, and G. J. Simitses, "A novel micromechanics-based approach to the derivation of constitutive equations for local/global analysis of a plain-weave fabric composite," *Composites Science and Technology,* vol. 60, pp. 1825-1833, 2000.
[3] K. Woo and J. Whitcomb, "Global/local finite element analysis for textile composites," *Journal of Composite Materials,* vol. 28, pp. 1305-1321, 1994.
[4] C. Chapman and J. Whitcomb, "Effect of assumed tow architecture on predicted moduli and stresses in plain weave composites," *Journal of Composite Materials,* vol. 29, pp. 2134-2159, 1995.
[5] L.Li, P.H.Wen, M.H.Aliabadi "Meshfree modeling and homogenization of 3D orthogonal woven composites" *Composite Science and Technology* Vol **71**, PP **1777-1788, 2011**
[6] A. Bogdanovich, "Multi-scale modeling, stress and failure analyses of 3-D woven composites," *Journal of Materials Science,* vol. 41, pp. 6547-6590, 2006.
[7] K. H. Ji and S. J. Kim, "Dynamic direct numerical simulation of woven composites for low-velocity impact," *Journal of Composite Materials,* vol. 41, pp. 175-200, 2007.
[8] X. Tang and J. D. Whitcomb, "General techniques for exploiting periodicity and symmetries in micromechanics analysis of textile composites," *Journal of Composite Materials,* vol. 37, pp. 1167-1189, 2003.
[9] U. Galvanetto and M. Aliabadi, *Multiscale modeling in solid mechanics: computational approaches* vol. 3: Imperial College Pr, 2010.
[10] V. Kouznetsova, M. Geers, and W. Brekelmans, "Multi-scale constitutive modelling of heterogeneous materials with a gradient-enhanced computational homogenization scheme," *International Journal for Numerical Methods in Engineering,* vol. 54, pp. 1235-1260, 2002.
[11] "MIL-HDBK-17-2F Polymer Matrix Composites - Material Properties," in *Composite Materials Handbook* vol. 2, ed: Department of Defense, 2002.

Characterization of Metal Magnetic Memory of Thermal Damage Correlated with Initial Cumulative Fatigue Damage for Laser Cladding Remanufacturing Technology

C. Zeng[1, a], W. Tian[1, b] and L. Hua[1,2,c]

[1]College of Mechanical and Electrical Engineering, Nanjing University of Aeronautics and Astronautics, Nanjing 210016, China

[2]Nanjing Institute of Railway Technology, Nanjing 210031, China

[a]poetal@126.com , [b]tw_nj@nuaa.edu.cn , [c]hua672@163.com

Keywords: metal magnetic memory , thermal damage , laser cladding , remanufacturing .

Abstract. Laser cladding act as a remanufacturing technology used more and more but the study of the fatigue properties of such products falls relatively behind. To work out the influence fatigue damage and thermal damage have on the fatigue life of laser cladding remanufactured products, this paper deals with the study of the characteristic of metal magnetic memory (MMM) signals of components experienced with fatigue test and a succeeding laser surface burning. Statistic investigation of the experiment result showed that the magnetic signal was closely related to the state of fatigue damage and thermal damage of the specimen. Based on such result, it is expected to quantify the study by statistical analysis of large samples so as to find a correlationship with the last fatigue life.

Introduction

Laser cladding is the high-energy and high-productivity technology in which the cladding material is deposited by some method on the surface of the specimen or component and subsequently melted by high-power laser radiation [1, 2]. It is usually used for directly forming of a component or reconditioning of a condemned product. But, as a remanufacturing technology, it is long before widely put into production as the fatigue life of these products is not that clear, which is affected not only by the consequent working load but also the initial cumulative fatigue damage and thermal damage.

To ensure the quality of the deposited zone and to estimate the remaining life of the remanufactured component, many nondestructive testing methods have been found and in use. Although each method has their overwhelming superiority over their shortcomings [3,4], there is a drawback in common of these method is that, these technology can only diagnose relatively lager defects while early fatigue damage cannot be detected. Been first put forward by the Russian researcher Dubov in 1997 [5], metal magnetic memory (MMM) test is a relatively new approach which can monitoring the flaws even in the early times of the fatigue crack initiation stage. Based on the magneto-mechanical effect, the MMM test method is implemented by analyzing the magnetic signal line Hp(y) of its changing polarity to acquire zero value or the change of the line gradient to achieve its maximum value to diagnose the stress concentration and micro-cracks [6], where the parameter Hp(y) denotes the normal component of spontaneous stray field. Around this technic and its application, researchers have carried out with mass of work elaborated in references [7-12].In this paper, the MMM test method was employed to identify the initial cumulative fatigue damage and thermal damage for the laser cladding remanufacturing technology.

Experiment

Preparation of Specimen. The test material is Q235A steel, according to the Chinese national standard GB/T 700-2006, its chemical composition is given in Table 1 and its mechanical property is given in Table 2 by static tension experiment. Fig. 1 shows the geometry and dimension of specimen which is machined in accordance with the Chinese national standard GB 3075-82. The specimen is of rectangular section, while its thickness is 10 *mm* which is not remarked in the figure. Three lines act as MMM signal test scanning route are arranged parallel to the center line at intervals of 7 *mm*.

Table 1 Chemical composition (wt%)

Steel no.	C	Si	Mn	S	P
Q235A	≤0.22	≤0.35	≤1.40	≤0.050	≤0.045

Table 2 Mechanical properties

Steel no.	Yield strength σ_s [MPa]	Tensile strength σ_b [MPa]	Elongation δ [%]	Rate of size contraction Ψ [%]
Q235A	241	400	52	66.7

Experiment Procedure. In this work, a group of 10 specimens were tested. Data acquisition of parameter Hp(*y*) was conducted for three times, that is at the original state, after fatigue test and after laser treatment, respectively, by a metal magnetic memory apparatus along the scan lines (shown in Fig.1) point by point with interval of 5 *mm*. The tension-compression fatigue test was carried out on a MTS810 servo hydraulic fatigue testing machine in air at room temperature with stress ration R= -1 and frequency *f* = 2 *Hz*. They were tested at the same load level 45 *kN* (150*MPa*) for 1000 times to introduce the fatigue damage. The thermal damage was introduced by a CO_2 laser beam scanning transversely to the loading direction through the middle point of the MMM scan lines. The controlled parameters are laser power from 1 *kW* to 2.6 *kW* at intervals of 0.4 *kW*, the same laser spot diameter 3.5 *mm* and the same scanning speed 180 *mm/min*. To simplify the study, cladding material was not used in this work.

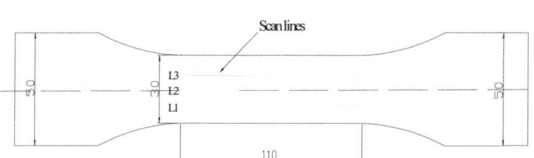

Fig. 1. Geometry and dimensions of specimen (in *mm*) and measuring lines

Experiment Result and Discussion

To study what was brought to the specimens after fatigue test in terms of leakage magnetic signal, the value of Hp(y) measured before fatigue was subtracted from that after. Similarly, the difference between the Hp(y) value after laser surface treatment and before it was studied. If the original MMM signal value, after fatigue value and after laser treatment value was recorded Hp1(*y*), Hp2(*y*) and Hp3(*y*), respectively, then the above statement can be written in equations as below:

$$\Delta Hp_{fatigue}(y) = Hp2(y) - Hp1(y) \qquad (1)$$
$$\Delta Hp_{laser}(y) = Hp3(y) - Hp2(y) \qquad (2)$$

Fig. 2 shows the statistic result of the difference between MMM signal collected when all the specimens were experienced with a certain fatigue test and the initial value. In this picture it's not difficult to find that the dominant features of these lines is that there is a linear relationship between the value $\Delta Hp_{fatigue}(y)$ and the displacement x. Furthermore, most curves are almost close to straight lines, yet a few lines provide a localized value volatility, e.g. signal line A13GHp31, which probably suggest that imperfections exist inner part. Another information Fig.2 tells us is that, although

$\Delta Hp_{fatigue}(y)$ and x has been in a linear relation, the slope of the lines was not approximately the same. Statistic study of the 30 slope values indicates that a majority of the values are less than 10. Whether this difference was caused by the insufficient cumulative fatigue damage or something, it is still a question remained.

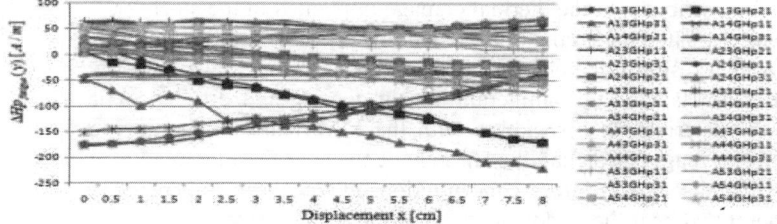

30 lines are included in this figure each with a name AijGHpm1, where Aij stands for specimen ID and m in accordance with the scan line number, as is the situation in Fig.3.

Fig. 2. Statistic regularity of $\Delta Hp_{fatigue}(y)$

What laser burning was done on the after-fatigue pieces reflected on the MMM signal is presented in Fig.3, still, it's the difference of Hp(y) value after laser surface scanning and before. Different from the previous study is that, the specimens carried fatigue damage embedded in them now. Therefore, this discussion is based on the fatigue damage introduced before.

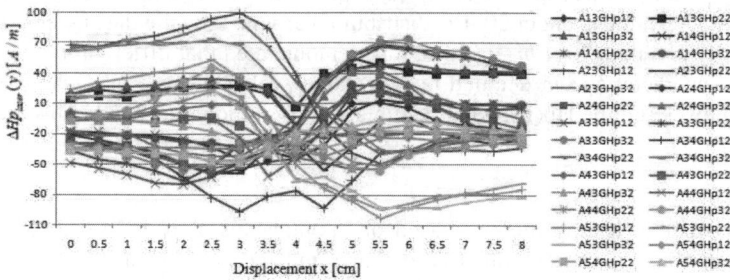

Fig. 3. Statistic regularity of $\Delta Hp_{laser}(y)$

In this figure, it is found that the linear relationship is completely destroyed which exist before. Most of the lines are in a shape similar to "Z" tilted, in spite the values are not agree and some are in the opposite direction. Nevertheless, exceptions always exist, like the three lines on specimen A13. While few lines have the general shape of "Z" though locally fluctuate violently. It is also found that, even lines come from the same specimen, for few cases like A44, their shapes differ greatly in local position, which mainly reveals a fact that this component suffers a transversely heterogeneous microstructure. Since the data rise and fall remarkably especially near the middle point of the routes (where laser beam scanned), it is necessary to investigate the gradient change of Hp(y) of the contiguous point, as is shown in Fig.4.

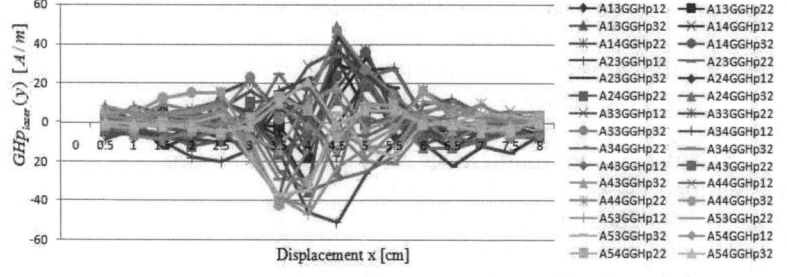

The meaning of the item name AijGGHpm2 is almost the same as AijGHpm2 in Fig.3, while it refers to the gradient of the latter.

Fig.4. Statistic regularity of the gradient of $\Delta Hp_{laser}(y)$

As has been noticed before, it is clearly to see, in Fig.4, practically all the signals show a relatively great saltation near the laser scanned track while far from the middle point they keep a smooth and stable trend. It just so happens that there is a similar case, all these curves reach their maximum peak values at middle point of displacement x = 4 cm or around. Beyond all doubt, this change was induced by the laser treatment which contributes to the thermal damage implanted in the steel. The influence thermal damage perform on the treated material in a laser cladding process mainly act as the change of metallographic structure, which result in the change of hardness of the material, and residual stress been generated in the parent metal due to contraction during cooling and differences in thermal expansion between the two materials. In association with the MMM theory, the position on the inspected component where the magnetic signal line Hp(y) have the change of gradient to achieve its maximum value is often diagnosed with stress concentration or defects in the material, it is no argument to come to the point that near the middle location of the scan lines stress concentration has come into being, as there's no cladding material to cause pores, micro-cracks, etc.

However, it is also recognized that the points where the peak values appear are not agree with each other very well that be in the position of displacement x = 4 cm. This is likely to be caused by that, for the ten specimens it's very hard to ensure the beginning point of these scan lines start from the same place as well as to ensure the probe center, whose diameter is approximately 6 mm, coincide strictly with beginning points of these scan route lines. Fig.4 also reveals that, of most points the magnetic values their absolute value is under 20 A/m distributing around 0, while the absolute value of the peak values are mostly less than 40 A/m. Besides, it is demonstrated that different laser power can lead to different signal characteristic in detailed but the same profile in general. Quantitative relationship between laser power and the Hp(y) signal features still need more work to be discussed in our future work.

Conclusions

As a nondestructive testing method based on spontaneous stray field signals, this approach is feasible for describing fatigue cumulative damage and the subsequent thermal damage for laser cladding remanufactured products. According to the experiment result, it is believed that in the early stage of fatigue the change of the intrinsic scattering magnetic field signals is in a highly linear relationship with the displacement the probe passed, while laser surface scanning can cause fierce sudden change of the signal. However, it is difficult to quantitatively describe the relationship between MMM signals feature and fatigue damage, as well as thermal damage, since some of the mechanism of MMM theory is not clearly yet. Moreover, the spontaneous signal was very weak and after amplification error cannot be avoid taking into account the environmental factors. Further work will be focused on this problem correlating with the fatigue property by statistic analysis methodology.

References

[1] F. Bruckner, D. Lepski and E. Beyer:J. Therm. Spray Technol. Vol.16(2007), p. 355-73
[2] E.M. Birger, G.V. Moskvitin, A.N. Polyakov and V.E. Arkhipov: Weld. Int. Vol.25(2011), p. 234-43
[3] G. Dobmann, L. Debarberis and J.F. Coste: Nucl. Eng. and Des. Vol.206(2001), p. 363-374
[4] J. Spanner and G. Selby: in *The 2002 ASME Pressure Vessels and Piping Conference: NDE Engineering Applications,* Vancouver, 2002.
[5] A.A. Dubov: 7th European Conference on Non-destructive Testing. Copenhagen,1997
[6] A.A. Dubov: Tyazheloe Mashinostroenie. 2005(6): p. 13-15
[7] A.A. Dubov: Therm. Eng. Vol.50(2003), p. 935-8
[8] A.A. Dubov:Therm. Eng. (English translation of Teploenergetika), vol.54(2007), p. 712-5
[9] A.A. Dubov and S.M. Kolokol'nikov: Weld. Int. Vol.21(2007), p. 821-5
[10] A.A. Dubov: Therm. Eng. (English translation of Teploenergetika), vol.56(2009), p. 120-3
[11] A.A. Dubov: Therm. Eng. Vol.57(2010), p. 16-21

SEM *in-situ* study on pre-corrosion and fatigue cracking behavior of LY12CZ aluminum alloy

Li Xudong[1,a], Mu Zhitao[1,b] and Liu Zhiguo[1,c]

[1] Naval Aeronautical Engineering Academy Qingdao Branch, Qingdao, 266000, People's Republic of China

[a]xdli236415064@gmail.com, [b]mzt63@163.com, [c]qdnuaalzg@163.com

Keywords: fatigue crack, aluminum alloy, prior corrosion, microstructure, SEM *in-situ* technology

Abstract. Corrosion fatigue is a form of degradation subjected to combined damage of mechanical stress and corrosive medium, which is an issue in aircraft industry. Experimental investigations on prior corrosion fatigue cracking behavior of LY12CZ were conducted with scanning electron microscope (SEM). Results indicate corrosion damage is important for the fatigue small cracking behavior of LY12CZ aluminum alloy. The effect of corrosion pit on fatigue crack can be characterized by the depth of corrosion pit. Based on small crack, another way to evaluate crack growth rate for AALY12CZ is proposed.

1. Introduction

For the relatively high strength and low density, LY12CZ aluminum alloy is still widely used in aerospace industry which calls for a good combination of light weight and high strength. The natural environmental factors such as moist air, natural water, etc. as well as the substances used in industry, frequently intensify the fatigue crack growth process, thus do great harm to the safety of airplane[1-2].

Models closely related to the microstructure provide a more reliable basis of structure life prediction. Hence, there is increasing interest in fatigue tests combined with high-resolution microscopic techniques [2-4]. The potential of *in-situ* studies has been realized as scanning electron microscope (SEM) combined the servo-hydraulic loading system. SEM is an effective method to capture the microstructure behavior during fatigue. Lots of researches have been done on the corrosion pit-to-crack transition. However, the mechanisms are not sufficient and clear due to the complexity [5, 6].

In this work, we focus mainly on the effect of corrosion features on the fatigue cracking behavior of high-strength LY12CZ aluminum alloy. Fatigue crack initiation and propagation were investigated by SEM. Corrosion pits would often nucleate multiple fatigue cracks, which contributed to the final fatigue failure.

2. Experimental setup

Tests were conducted on AA LY12CZ, whose chemical compositions and mechanical properties are listed in **Table.1**. Specimens of dog-boned shaped with dimension 15mm×5mm were made from rolled sheet with 1mm thickness, which is shown in **Fig. 1**. A notch with a radius of about 0.05 mm and a depth of about 0.1 mm, located at the center of the specimen in order to create a local stress concentration and favor crack capture.

Prior corrosion damage was introduced to the specimen, based on the actual corrosion damage the material suffered during service to make it of practical significance. Corrosion spectrum is made to establish an equivalent relationship between experimental accelerated corrosion damage and the environmental damage based on equivalent electrochemical principle, which were conducted in the ZJF-45G chamber with a constant temperature of $(45\pm2)°C$. The corrosive solution is made by acid 3.5 wt. % NaCl solution with PH=4.0 ± 0.2. One corrosion cycle includes immersion in the solution for 6.5 minutes and 12.5 minutes drying in the air, respectively. 200 such corrosion cycles, of which the cumulative time are about 63.3 hours can make damage to specimen suffered during one calendar year in natural marine environment in China. Specimens were corroded to different calendar corrosion years from 0 year to 19 years. One of the five states subjected to cyclic loadings later is 0 year, 7 years, 11years, and 15 years, 19 years, which are abbreviated as SC0, SC7, SC11, SC15 and SC19, respectively.

To simulate the actual status of aluminum alloy in aircraft structure, the oxide layer on the exhibiting surface of specimens was not polished. All fatigue tests were conducted in high vacuum environment (10^{-4} Pa) and at room temperature(26°C) with SS550 SEM combining a servo-hydraulic testing system (Shimadzu, Japan), which provides pulsating (sin wave) loads at 10 Hz of ± 1 kN maximum capacity and a displacement range of ± 25 mm. The signal of SEM was transferred to a computer, sample 960×1280 pixels image. Fatigue tests were loaded controlled at stress ratio $R=0.1$ under applied stresses of 220, 240 and 250 MPa respectively for every corrosion states. Observations of the whole process were performed at a frequency of 0.2 Hz to get clear images, but between observations the frequency was increased to 6 Hz to accelerate the fatigue damage.

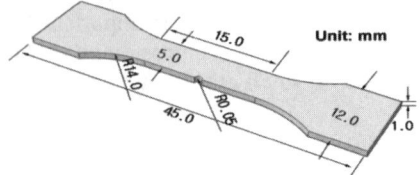

Fig.1 Shape and size sketch of specimen

Table 1 Nominal composition and mechanical properties of LY12CZ aluminum alloy

Composition (Wt. %)	Cu	Mn	Mg	Ti	Al	$\sigma_{0.2}$ (MPa)	σ_b (MPa)
LY12CZ	4.0	0.7	0.60	0.12	Balance	275	415

3. Results and discussion

3.1 Micro characteristics of corrosion pit

Fig. 2 gives a typical case of surface morphology in three-dimension (3-D) of SC7 specimen. All these corrosion pits or concave-convex configurations are potential fatigue crack nucleation sites that promote localized flow in their vicinity, where crack nucleation is caused by morphological changes taking place as a result of localized and cyclical plastic deformation. The pit can diminish the fatigue performance. Depth of corrosion pit is a key parameter to characterize pits. The relationship between average pit depth and the corrosion exposure time are plotted as shown in **Fig.3**.

$$\overline{D} = 0.75 + 18.41 T_{eq.} \quad (1)$$

where $\overline{D}$ denotes the average pit depth and $T_{eq.}$ denotes the equivalent exposure time. Unlike previous report, based on the present corrosion spectrum, the corrosion pits increases linearly with exposure time within 19 years [3-5].

3.2 Fatigue crack initiation behavior

Figs. 4a & 4b displays two cracks originate from pits of SC11 under 240MPa. Pit with diameter of about 20 μm nucleated fatigue crack with length of 140 μm and cracks nucleated from smaller corrosion pits are much shorter as shown in **Figs. 4b**. Big pit indicates more severe local corrosion damage. Different crack length under the same cycle means different crack growth rates. This demonstrates that local corrosion damage is important to micro crack initiation and propagation [5, 6]. Multi-cracks were found as shown in **Figs. 4c**. A portion of cracks would gradually join together to form a main crack, which dominates the final failure. Most micro-cracks would stop growing once the main crack forms.

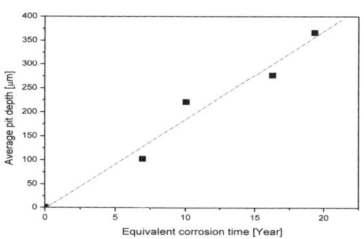

Fig. 2 Typical surface image for SC7   Fig.3 Average pit depth versus equivalent corrosion time

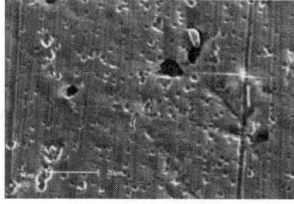 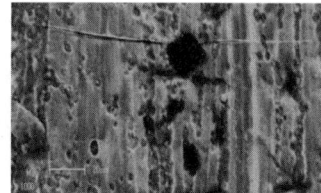 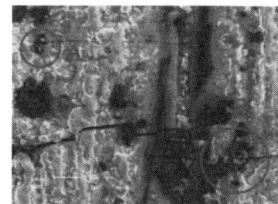

(a) N=4280 cycles   (b) N=4280 cycles   (c) N=12712 cycles

Figs.4 SEM images of pit-nucleated cracks under 220MPa for SC19, (a. b) N=4280 cycles, (c) N=12712 cycles

3.3 Evaluation of small fatigue growth rate with corrosion damage

The effective main crack can be easily decided based on SEM *in-situ* observation. An effective way to estimate the fatigue lives is to evaluate the main crack growth rate, which can be easily obtained by measure the main crack length between two succedent observations.

The relationship between crack length and cycling number in semi logy coordinates is shown in **Fig. 5**. Due to the linear relationship ($\log a \propto N$), the relationship of crack growth rate da/dN with unit of μm/cycle versus a with unit of μm can be obtained when applied stress is constant. Given the fatigue crack length, the fatigue growth rate can be identified.

When the fatigue crack length a is constant (such as $a=1\mu m$), the relation of small crack growth rate da/dN and the maximum applied stress σ_{max} can be obtained as shown in **Fig. 6**. These curves are approximately linear, indicating that $da/dN \propto \sigma_{max}^n$, where n is the slope of each curve. Values of n for different corrosion state are 5.16, 5.17, 5.20, 5.27 and 5.22, respectively, which are around 5.20. The difference is slight, indicating n is mainly dependent the material itself. Considering the combined influences of stress levels and fatigue crack propagation length (σ_{max}^n, a) on da/dN, the relationships between da/dN and the product of $\sigma_{max}^n a$ are shown in **Fig. 7**. Slope values for these linear curves are approximately one, which indicates that da/dN is proportional to $\sigma_{max}^n a$ for each corrosion state (i.e. $\log da/dN \propto \log \sigma_{max}^n a$, then $da/dN \propto \sigma_{max}^n a$). The proportional constant is defined as k. Thus, relationship as follows can be obtained.

$$da/dN = k\sigma_{max}^n a \qquad (2)$$

Unit of k can be identified as (MPa)^n·cycle through dimensional analysis. The dependence of k on the corrosion damage can be characterized linearly as shown in **Fig. 8**. Therefore, the fatigue small crack growth rate of LY12CZ can be described uniquely. Since a lot of phenomenon in fatigue is closely related to the characteristic of small crack growth, the proposed law is valuable in evaluating the fatigue life with corrosion damage for AA LY12CZ.

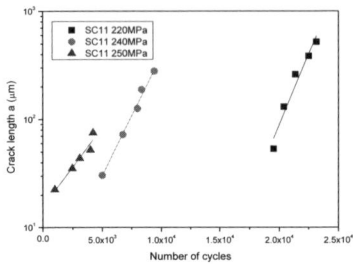
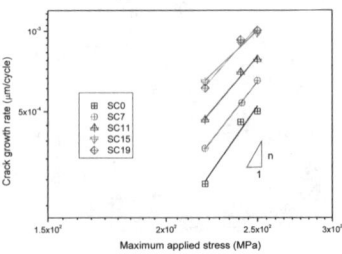

Fig.5 Crack growth rate versus stress cycles for SC11

Fig.6 Crack growth rate versus σ_{max} for specimen with different corrosion damage

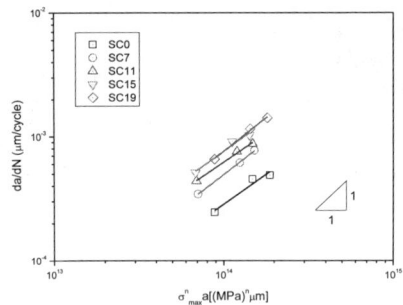
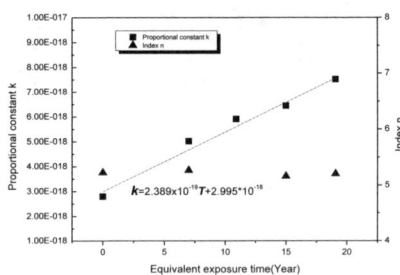

Fig.7 Crack growth rate versus the term of $\sigma_{max}^n a$. All slopes of curves are approximately one.

Fig.8 Proportional constant k and index n versus equivalent exposure time. Index n is approximately constant while k increases with corrosion damage.

4. Conclusions

1. Corrosion pit depth varies with exposure time based on the proposed spectrum, and the pit depth is used to characterize the corrosion damage severity.

2. A novel method of evaluating the fatigue small crack growth rate for LY12CZ aluminum alloy has been proposed.

3. SEM *in-situ* technology is an effective tool for evaluation of fatigue cracking behavior of LY12CZ aluminum alloy with corrosion damage.

References

[1] Biallas G, Maier H J. in-situ fatigue in an environmental scanning electron microscope- potential and current limitations. Int J Fatigue 2007, 29: 1413-1425

[2] Li Xu-Dong, Wang Xi-Shu, Ren Huai-Hui, Chen Yin-Long, Mu Zhi-Tao. Effect of prior corrosion state on the fatigue small cracking behaviour of 6151-T6 aluminum alloy. Corros. Sci. 55(2) (2012) 26-33.

[3] Wang X S, Li Y, Meng X K. An estimation method on failure stress of micro thickness Cu film-substrate structure. Sci China Ser E-Tech Sci 2009, 52(8): 2210-2215

[4] Xi-Shu Wang, Xu-Dong Li, Huai-Hui Ren, Hai-Yan Zhao, Ryosuke Murai, SEM in-situ study on high cyclic fatigue of SnPb-solder joint in the electronic packaging. Microelectronics Reliability. 51(2011) 1377-1384

[5] Wang X S, Yan C K, Li Y, et al. SEM in-situ study on failure of nanocrystal metallic thin films and substrate under three point bending. Int J Facture 2008, 15: 269-279

[6] Maier H J, Gabor P, Karaman I. Cyclic stress-strain response and low-cycle fatigue damage in ultrafine grained copper. Mater Sci Eng A 2005, 410-411: 457-461

Analysis of stress intensity factor for cracked flattened Brazilian disk: Part I—analysis method and pure mode I crack

Shiming Dong*, Qingyuan Wang

College of Architecture and Environment, Sichuan University, Chengdu 610065, China
E-mail: smdong@scu.edu.cn , wangqy@scu.edu.cn
* Corresponding author. Tel.: +86-28-85416486.

Keywords: Cracked flattened Brazilian disk (CFBD); Stress intensity factor (SIF); Central crack; Numerical method

Abstract. In order to solve the problem how to calculate the stress intensity factor for a cracked flattened Brazilian disk under mode I loading, the finite element method was employed to analyze the stress intensity factor for the cracked flattened Brazilian disk under mode I loading, based on the closed-form expression of the stress intensity factor for a cracked Brazilian disk subjected to pressure. The analyzed result shows that within the certain range of the load distribution angle, the formula of the stress intensity factor for the cracked Brazilian disk can be directly used to calculate the stress intensity factor for the cracked flattened Brazilian disk under mode I loading.

1. Introduction

In the fracture toughness experimentation, the cracked Brazilian disk specimen subjected to the diametric compressive load has been widely used to study the mixed mode fracture behavior of brittle materials [1-7]. One advantage of this disk configuration is that the complete mode combinations ranging from pure mode I to pure mode II can be easily achieved by selecting the relative crack length α (the ratio of crack length to disk diameter) and the loading angle θ (Fig.1). The other advantage of using this disk specimen is that the closed-form solutions of the stress intensity factors can be achieved [1, 8].

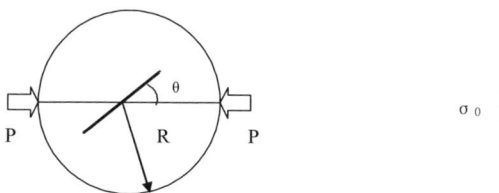

Fig.1 Schematic diagram of the mixed-mode loading   Fig.2 CFBD under mixed-mode loading

However, in the Brazilian testing, the local cracking, breakage or yielding around the loading point are probably occurred because of stress concentration. It is inconsistent with the theoretical analysis result that the crack is firstly occurred in the disk centre, then propagated along with the loading direction. Therefore, it is necessary to improve the Brazilian testing. Guo et al [9] firstly proposed to use the uncracked flattened Brazilian disk to determine the mode I fracture toughness. This configuration was revisited by Wang et al [10] by introducing two parallel flat loading planes, and finally developed into the flattened Brazilian disk with a central straight-through crack (Fig.2), for which there is no stress intensity factor solution. Thus, Wang et al used the finite-element method to compute the stress intensity factor, and gave the following formula to calculate the fracture toughness (K_{Ic}): $K_{IC} = P_{min} \phi_{max} /(\sqrt{Rt})$. Where t is the thickness of the disk, R is the radius, P_{min} is determined by the load-displacement record, ϕ_{max} is determined by the numerical analysis. If the load distribution angle $2\gamma = 20°$ (where $\sin \gamma = h/R$, see Fig.2), $\phi_{max}=0.7997$, If the load distribution angle $2\gamma = 30°$, $\phi_{max}=0.5895$. If the load distribution angle is other value, or the crack length is changed, how to calculate the stress intensity factor for the flattened Brazilian disk is an unsolved problem. In the present paper the explicit expressions of the stress intensity factors for cracked Brazilian disk specimen under pressure loading conditions is firstly introduce. Then, the finite element method is employed to discuss the validity of the expressions of the stress intensity factors for cracked Brazilian disk in calculating the stress intensity factor for the flattened Brazilian disk under mode I loading condition.

2. SIF for cracked Brazilian disk loaded by pressure

Suppose the total force P (Fig.1) is distributed symmetrically and uniformly into the pressure σ_0 over the angle 2γ along the circumference for loading. The schematic diagram of the cracked Brazilian disk subjected to uniformly distributed pressure is shown in Fig. 3. The angle between the crack plane and the loading line of total force P is defined as θ_0. The stress intensity factors for the cracked Brazilian disk loaded by pressures can be expressed as follows [8],

$$K_I = \sigma\sqrt{\pi a}F_I = \sigma\sqrt{\pi a}\left[\frac{\gamma}{\sin \gamma}f_{11} + \sum_{i=1}^{n} B_{1i}f_{1i}\alpha^{2(i-1)}/\sin \gamma\right] \quad (1)$$

$$K_{II} = \sigma\sqrt{\pi a}F_{II} = \sigma\sqrt{\pi a}\sum_{i=1}^{n} B_{2i}f_{2i}\alpha^{2(i-1)}/\sin \gamma \quad (2)$$

Where $\quad \alpha = \dfrac{a}{R}, \quad \sigma = \dfrac{P}{\pi R t} \quad (3)$

$$B_{1i} = \cos(2i\theta_0)\sin(2i\gamma) - \frac{i}{i-1}\cos[2(i-1)\theta_0]\sin[2(i-1)\gamma] \quad (4)$$

$$B_{2i} = \sin(2i\theta_0)\sin(2i\gamma) - \sin[2(i-1)\theta_0]\sin[2(i-1)\gamma] \quad (5)$$

$$f_{ji} = \frac{(2i-3)!!}{(2i-2)!!}\left[1 + \frac{c_{j1}}{2i} + \frac{3c_{j2}}{4i(i+1)}\right] \quad (j=1,2; i=1,2,\ldots n) \quad (6)$$

$$c_{11} = \frac{8 - 4\alpha + 3.8612\alpha^2 - 15.9344\alpha^3 + 24.6076\alpha^4 - 13.234\alpha^5}{\sqrt{1-\alpha}} - 8 \quad (7)$$

$$c_{12} = \frac{-8 + 4\alpha - 0.6488\alpha^2 + 14.1232\alpha^3 - 24.2696\alpha^4 + 12.596\alpha^5}{\sqrt{1-\alpha}} + 8 \quad (8)$$

$$c_{21} = \frac{5 - 2.5\alpha + 1.4882\alpha^2 - 2.376\alpha^3 + 1.1028\alpha^4}{\sqrt{1-\alpha}} - 5 \quad (9)$$

$$c_{22} = \frac{-4 + 2\alpha + 0.4888\alpha^2 + 0.81112\alpha^3 - 0.7177\alpha^4}{\sqrt{1-\alpha}} + 4 \quad (10)$$

Here a is the half length of the crack, α is the relative crack length, γ is the load distribution angle, F_I and F_{II} are the normalized stress intensity factors.

3. Numerical analysis method

3.1 Model parameters. The finite element code ANSYS [11] is employed to calculate the stress intensity factors for the cracked Brazilian disk. The specimen material is PMMA. The relevant material parameters and geometric parameters are listed in table 1.

Table 1 Material parameters and geometric parameters

Specimen	Young's modulus, E	Poisson's ratio, v	Density, ρ	Radius, R	Thickness, t
PMMA	4.87×10^9 Pa	0.388	1170 kg/m^3	0.01m	0.005m

In FEM model, the specimen is presented by 6-node triangular solid element PLANE2. Singular element is employed at the first row of elements around the crack tip.

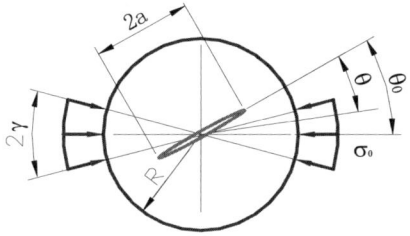

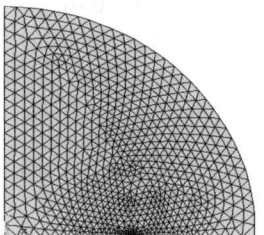

Fig.3 Cracked Brazilian disk subjected to pressure   Fig.4 FEM model for CFBD

3.2 Numerical solution of the stress intensity factor. Based on the numerical results obtained from finite element analysis, the stress intensity factors for crack, which are called the numerical solutions of
the stress intensity factors, can be directly calculated with the following formula[11]:

$$K_I = \pm \frac{2G\sqrt{2\pi}}{1+k} \frac{u_y}{\sqrt{r}} \qquad K_{II} = \pm \frac{2G\sqrt{2\pi}}{1+k} \frac{u_x}{\sqrt{r}} \tag{11}$$

Where '+' corresponds to the upper crack face, and '-' corresponds to the lower crack face. G is the shear modulus of the specimen material. $k = (3-\nu)/(1+\nu)$ (for plane stress state) or $k = 3-4\nu$ (for plane strain state), ν is the Poisson's ratio. r is the distance from the crack tip to a defined FE node.

After the stress intensity factors K_I and K_{II} are obtained, the normalized stress intensity factors F_I and F_{II} can be calculated by the following formula:

$$F_I = F_I(\alpha, \theta) = \frac{K_I}{\sigma\sqrt{\pi a}} \qquad F_{II} = F_{II}(\alpha, \theta) = \frac{K_{II}}{\sigma\sqrt{\pi a}} \tag{12}$$

4. Numerical analysis of SIF for the flattened Brazilian disk under mode I loading condition

Due to the symmetry, only 1/4 part of the whole specimen is considered under mode I loading condition, as is shown in Fig. 4. The symmetrical boundary conditions and the pressure load are applied.

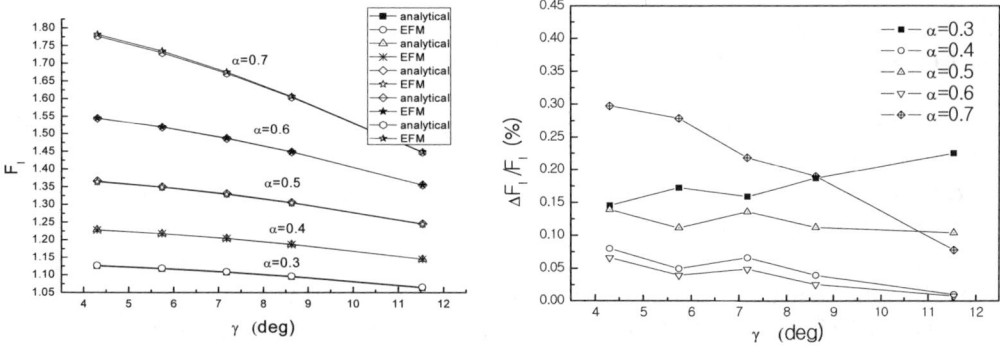

Fig.5. F_I for CFBD under pure mode I loading   Fig.6. Percentage error of F_I for CFBD

Based on the finite element analysis and Eqs.(11) - (12), the normalized stress intensity factor F_I can be obtained for the CFBD with the relative crack length α =0.3-0.7 at different load distribution angles, which are shown in Fig.5. In addition, the analytical solutions of F_I at same conditions can be yielded by using Eqs. (1) - (10), which are also shown in Fig.5.

It is obvious from Fig.5 that the FEM values of F_I coincide very well with the analytical values. The further calculating results show that the maximum percentage error of F_I is less than 0.3% (Fig.6). The relative error (or percentage error) of F_I is defined as follows (the definition of a relative error (or percentage error) of F_{II} is similar to F_I):

$$\Delta F_I / F_I = (F_I^{(analytical)} - F_I^{(EFM)}) / F_I^{(analytical)} \tag{13}$$

Fig.6 shows that we can directly use the explicit expressions of the stress intensity factors for the cracked Brazilian disk loaded by pressure to calculate the stress intensity factor for the CFBD under pure mode I loading.

5. Conclusion

Based on the explicit expressions of the stress intensity factor for the cracked Brazilian disk loaded by pressure, the numerical analysis results of the stress intensity factor for the cracked flattened Brazilian disk under mode I loading condition is compared with the analytical calculating results at different relative crack length and different load distribution angle. The calculating results confirm that within the certain range of the load distribution angle, the explicit expressions of the stress intensity factor for the cracked Brazilian disk loaded by pressure can be directly used to calculate the stress intensity factor for the flattened Brazilian disk under mode I loading condition.

Acknowledgement

This project was supported by China Postdoctoral Science Foundation (Project No.20070420491) and the National Natural Science Foundation of China (Project No. 11172186), and the Program for Changjiang Scholars and Innovative Research Team (Project No.IRT1027).

References
[1] C Atkinson, RE Smelser, J Sanchez. Combined mode fracture via the cracked Brazilian disk test, Int J Fracture (1982), 18:279-291
[2] H Awaji and S Sato, Combined mode fracture toughness measurement by the disk test. J Engng Mater Tech(1978),100:175-182.
[3] DK Shetty, AR Rosenfield and WH Duckworth, Mixed-mode fracture in biaxial stress state: application of the diametral-compression (Brazilian disk) test. Engng Fract Mech(1987), 26: 825-840.
[4] C Liu, Y Huang, ML Lovato and MG Stout, Measurement of the fracture toughness of a fiber-reinforced composite using the Brazilian disk geometry. Int J Fract(1997), 87: 241-263.
[5] RJ Fowell, Suggested method for determining Mode I fracture toughness using cracked chevron notched Brazilian disc (CCNBD) specimens. Int J Rock Mech Min Sci Geomech Abstr(1995),32 (1):57-64.
[6] J Zhou, Y Wang, YM Xia, Mode-I fracture toughness of PMMA at high loading rates. J. Mater. Sci.(2006),41(24):8363-8366.
[7] MR Ayatollahi and MRM.Aliha, Mixed mode fracture in soda lime glass analyzed by using the generalized MTS criterion, Int. J. Solids Struct(2009), 46:311-321.
[8] SM Dong, Y Wang and YM Xia, Stress intensity factors for central cracked circular disk subjected to compression. Engng Fract Mech(2004), 71(7-8): 1135-1148.
[9] H.GUO, N.I.AZIZ, and L.C.SCHMIDT, Rock fracture toughness determination by Brazilian test, Engng. Geology (1993), 33:177-188.
[10] QZ WANG, XM JIA, SQ KOU, ZX ZHANG, and PA LINDQVIST, The flattened Brazilian disc specimen used for elastic modulus, tensile strength and fracture toughness of brittle rocks: analytical and numerical results, Int. J. Rock. Mech. and Min. Sci.(2004), 41: 245-253.
[11] ANSYS9.0, USER'S MANUAL. Swanson Analysis Systems Inc.

Analysis of stress intensity factor for cracked flattened Brazilian disk: Part II—mixed- mode crack

Shiming Dong*, Qingyuan Wang

College of Architecture and Environment, Sichuan University, Chengdu 610065, China

E-mail: smdong@scu.edu.cn , wangqy@scu.edu.cn

*Corresponding author. Tel.: +86-28-85416486.

Keywords: Cracked flattened Brazilian disk (CFBD); Stress intensity factor (SIF); Central crack; Mixed -mode; Finite-element method (FEM)

Abstract: This paper presents a new method to conveniently calculate the stress intensity factors for the cracked flattened Brazilian disks under mixed-mode loading. The finite-element method is employed to confirm an assumption that the formula of the stress intensity factors for the cracked Brazilian disk subjected to pressure can be directly used to calculate the stress intensity factors for the cracked flattened Brazilian disk. The calculated results show that the assumption is valid and reliable. The calculated results also confirm that the Saint-Venant's principle is still valid in fracture mechanics. In addition, the present paper proposes a concept of optimum load distribution angle.

1. Introduction

In the fracture toughness experimentation, the cracked Brazilian disk specimen subjected to the diametric compressive load has been widely used to study the mixed mode fracture behavior of brittle materials. However, during the Brazilian testing, the local cracking, breakage or yielding around the loading point are probably occurred because of stress concentration. It is inconsistent with the theoretical analysis result that the crack is firstly occurred in the disk centre, and then propagated along with the loading direction. Therefore, some improved Brazilian testing methods were proposed. The cracked flattened Brazilian disk (CFBD) with a central straight-through crack (Fig.1) is one of the improved Brazilian disk specimens. Because there is no stress intensity factor solution for the CFBD specimen, how to conveniently calculate the stress intensity factors for CFBD at any relative crack length and any loading angle as well as any load distribution angle is a new problem. Dong et al [1] proposed to use the formula of the stress intensity factors for the cracked Brazilian disk subjected to pressure to calculate the stress intensity factors for CFBD. This assumption has been confirmed that it is valid and reliable under pure mode I loading. But, it needs to further confirm whether the assumption is still valid and reliable under mixed-mode loading. In the present paper, we will use the same analysis methodology as the literature [1] to investigate the calculating method of the stress intensity factors for the CFBD under mixed-mode loading.

2. SIF for the CFBD under mixed-mode loading

The analysis method of SIF for the CFBD under mixed-mode loading is similar to the method under pure mode I loading [1]. Nevertheless, the FEM model for the CFBD under mixed-mode loading is the whole disk, which is different from the model under pure mode I loading. For example, the FEM model for the relative crack length $\alpha=0.5$ and the loading angle $\theta_0=10$ degree is shown in Fig.2. The fixed boundary condition and pressure load are applied on the left and right sides, respectively.

The investigated range of the relative crack length for the CFBD under mixed-mode loading is still 0.3 to 0.7. For the sake of simplification, we take the CFBD with a relative crack length $\alpha=0.5$ as a typical sample to discuss the stress intensity factors at different loading angles and different load distribution angles.

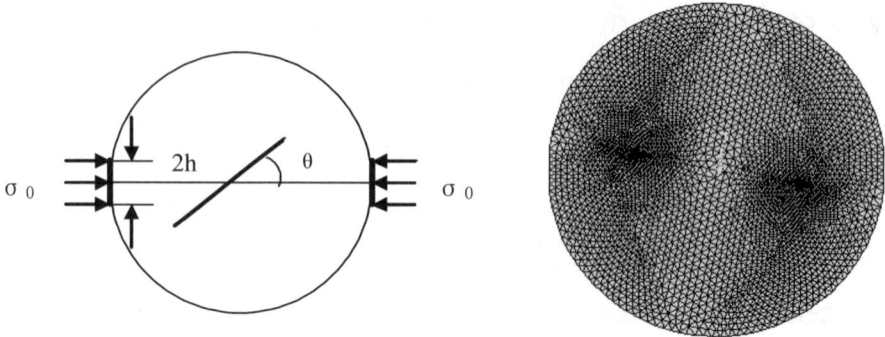

Fig.1 CFBD under mixed-mode loading   Fig.2 FEM model for CFBD (α= 0.5, θ_0=10°)

Based on the finite element analysis and Eqs.(11) - (12) in the literature [1], the normalized stress intensity factor F_I and F_{II} can be obtained for the CFBD with a relative crack length α =0.5 at different load distribution angles and different loading angles, which are shown in Fig.3. In addition, the analytical values of F_I and F_{II} at same conditions can be yielded from Eqs. (1) - (10) in the literature [1], which are also shown in Fig.3.

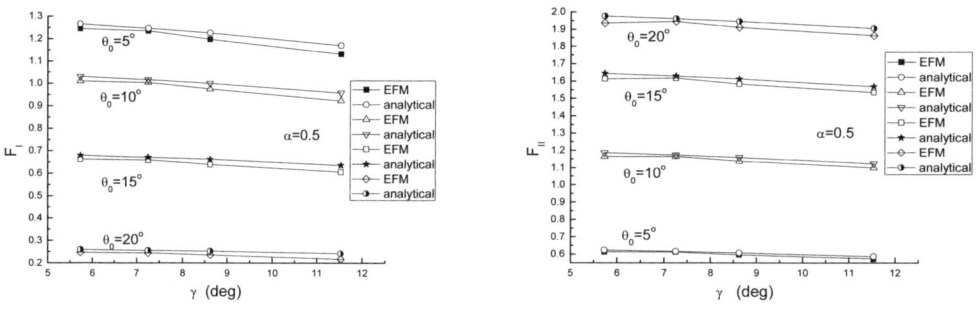

(a) Normalized stress intensity factor F_I (b) Normalized stress intensity factor F_{II}

Fig.3. F_I and F_{II} for CFBD with α= 0.5

It is clear from Fig.3 that the FEM values of F_I and F_{II} for the CFBD agree very well with the analytical values. In order to quantitatively estimate the relative errors of F_I and F_{II}, the percentage errors of F_I and F_{II} are calculated, and illustrated in Fig.4.

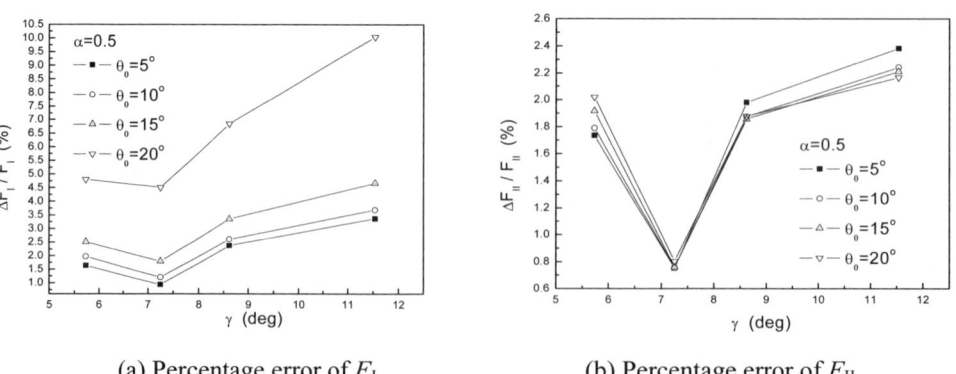

(a) Percentage error of F_I (b) Percentage error of F_{II}

Fig.4. Percentage error of F_I and F_{II} for CFBD with α= 0.5

Similarly, the normalized stress intensity factor F_I and F_{II} can be obtained for the CFBD with the relative crack lengths $\alpha = 0.3$ and $\alpha = 0.7$ at different load distribution angles and different loading angles, and the percentage errors of F_I and F_{II} can also be yielded, which are shown in Figs.5-6, respectively.

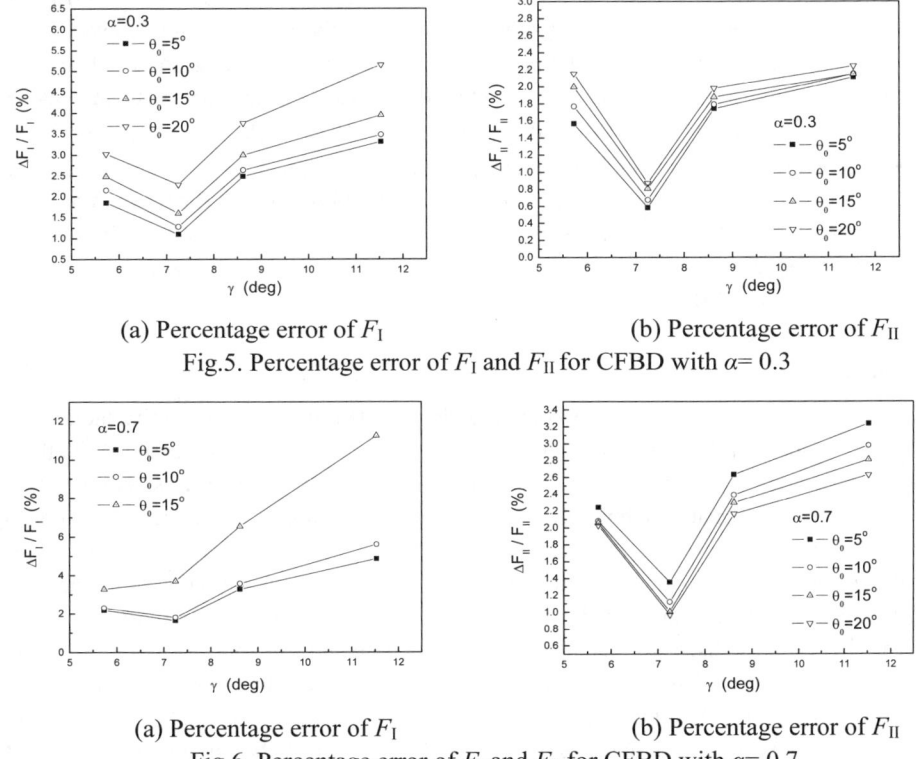

(a) Percentage error of F_I (b) Percentage error of F_{II}
Fig.5. Percentage error of F_I and F_{II} for CFBD with $\alpha = 0.3$

(a) Percentage error of F_I (b) Percentage error of F_{II}
Fig.6. Percentage error of F_I and F_{II} for CFBD with $\alpha = 0.7$

Based on the Figs.3-6, the following conclusions can be drawn:

(1) The FEM values of F_I and F_{II} for the CFBD are always less than the analytical values at same load distribution angles, with other parameters fixed.

(2) Compared the EFM values with the analytical values of F_I and F_{II} for the CFBD, we can find that the absolute errors of F_I and F_{II} are very small; and the relative errors of F_I and F_{II} are generally less than 5% except for some special loading angles which are located at near critical loading angles [2], which means that the explicit expressions of the stress intensity factors for the cracked Brazilian disk loaded by pressure can also be directly used to calculate the stress intensity factors for the CFBD under mixed-mode loading. The reason why the relative error of F_I is bigger when the loading angle is near the critical loading angle is because F_I is very small when the loading angle is equal to 20 degrees, which is near the critical loading angle for $\alpha = 0.5$ [2], hence, a small absolute error of F_I will bring a big relative error based on Eq.(13) in the literature [1].

(3) Generally speaking, the percentage errors of F_I and F_{II} for the CFBD increase with increasing the relative crack length. It means that the effects of the load type on the stress intensity factors increase with decreasing the distance between the crack tip and the load location. This conclusion coincides very well with the Saint-Venant's principle. Therefore, the present paper confirms the validity of the Saint-Venant's principle in fracture mechanics.

(4) The percentage errors of F_I increase with increasing the loading angle when the load distribution angle is fixed; but the relationship between the percentage error of F_{II} and the loading angle is more complicated when the load distribution angle is fixed. If the relative crack length is less than 0.5, the percentage errors of F_{II} increase with increasing the loading angle (Fig.5(b)); if the relative crack length is more than 0.5, the percentage errors of F_{II} decrease with increasing the loading angle (Fig.6(b)); if the relative crack length is equal to 0.5, the percentage errors of F_{II} increase with increasing the loading angle when the load distribution angle is smaller, but the percentage errors of F_{II} decrease with increasing the loading angle when the load distribution angle is bigger (Fig.4(b)).

(5) When the load distribution angle is about 7.25 degree, the percentage errors of F_I and F_{II} are the smallest. We call this load distribution angle as an optimum load distribution angle when the explicit expressions of the stress intensity factors for the cracked Brazilian disk loaded by pressure are used to calculate the stress intensity factors for the CFBD under mixed-mode loading during the CFBD testing.

3. Conclusion

In order to solve the problem how to conveniently calculate the stress intensity factors for the cracked flattened Brazilian disk at any relative crack length and any loading angle as well as any load distribution angle, the finite-element method is employed to confirm a assumption that the formula of the stress intensity factors for the cracked Brazilian disk subjected to pressure are directly used to calculate the stress intensity factors for the CFBD. The calculating results show that within the calculated range of the load distribution angle, the assumption is valid and reliable, so the explicit expressions of the stress intensity factors for the cracked Brazilian disk subjected to pressure can be directly used to calculate the stress intensity factors for the CFBD under mixed-mode loading. The calculating results also confirm that the Saint-Venant's principle is still valid in fracture mechanics. In addition, the present paper proposes the conception of the optimum load distribution angle. We recommend using the optimum load distribution angle to carry out the CFBD test if you use the formula of the stress intensity factors for the cracked Brazilian disk subjected to pressure to calculate the stress intensity factors during the CFBD testing.

Acknowledgement

This project was supported by China Postdoctoral Science Foundation (Project No.20070420491) and the National Natural Science Foundation of China (Project No.11172186), and the Program for Changjiang Scholars and Innovative Research Team (Project No.IRT1027).

References
[1] SM Dong, QY Wang, Analysis of stress intensity factor for cracked flattened Brazilian disk: Part I—Analysis method and pure mode I crack.
[2] SM Dong, Y Wang and YM Xia, Stress intensity factors for central cracked circular disk subjected to compression. Engng Fract Mech (2004), 71(7-8): 1135-1148.

Thermo-mechanical stress near apex of a bi-material wedge by a novel finite element analysis

Xue-Cheng Ping[a], Lin Leng[b] and Si-Hai Wu[c]

School of Mechatronics Engineering, East China Jiaotong university, 330013, Nanchang, P. R. China

[a] xuecheng_ping@yahoo.com.cn, [b] 1055120958@qq.com, [c] 1069080064@qq.com

Keywords: Thermo-mechanical stress, stress intensity factor, bi-material wedge, super singular element

Abstract. A super wedge tip element for application to a bi-material wedge is develop utilizing the thermo-mechanical stress and displacement field solutions in which the singular parts are numerical solutions. Singular stresses near apex of an arbitrary bi-material wedge under mechanical and thermal loading can be obtained from the coupling between the super wedge tip element and conventional finite elements. The validity of this novel finite element method is established through existing asymptotic solutions and conventional detailed finite element analysis.

Introduction

In order to capture the exact singular behavior of the stress field, special elements such as enriched elements [1, 2] and a hybrid crack element [3] have been successfully developed. Enriched elements usually employ transition (overlap) elements, and inter-element compatibility between the special and conventional elements is not satisfied. In general, for a given combination of materials, few analytical eigensolutions are available, thus the hybrid crack element [3] can only be applied to some typical problems. In order to eliminate the aforementioned shortcoming while addressing a junction of dissimilar materials, Chen and Sze [4], Ping and Chen [5] and Barut et al. [6] developed a hybrid global (special) element utilizing the exact analytical solution for the stress and displacement field based on the eigenfunction expansion method. It is noticed that exact analytical solution for the stress and displacement field of any junction of dissimilar materials used by Barut et al [6] cannot be easily solved, elaborate mathematical derivation cannot be avoided for every combination of materials. In contrast, numerical eigensolutions such as orders of stress singularity and angular variations of stress and displacement fields in a bi-material wedge can be easily obtained from the *ad hoc* finite element eigenanalysis method [7]. In this paper, a super wedge tip element with numerical asymptotic solutions is developed to predict the thermo-mechanical stress near apex of bi-material wedge.

Element stiffness matrix of the super wedge tip element

As shown in Fig. 1, a domain composed of bi-material sectors can be partitioned into inner and outer regions. The accurate solution to the entire domain requires coupling of the numerical solution in the inner region with that of the approximate solution through the finite element in the outer region. The coupling can be achieved by developing a super wedge tip element (as shown in Fig. 1) whose interpolation functions satisfy the governing equations exactly near the apex enforcing the inter-element displacement continuity along the common boundary and the nodes between the super and conventional elements. The stiffness matrix for the super wedge tip element of bi-material wedge under mechanical and thermal loading is obtained by considering the total potential expresses in the form

$$\Pi = \sum_{k=1}^{2}\left\{-\frac{1}{2}\int_{\hat{\Gamma}_k}{}^m_\lambda\boldsymbol{\sigma}^{(k)T}\boldsymbol{n}^T{}^m_\lambda\boldsymbol{u}^{(k)}d\Gamma - \int_{\hat{\Gamma}_k}{}^m_\lambda\boldsymbol{\sigma}^{(k)T}\boldsymbol{n}^T{}^t_{\lambda+c}\boldsymbol{u}^{(k)}d\Gamma\right\}$$
$$+ \sum_{k=1}^{2}\left\{\int_{\hat{\Gamma}_k}{}^m_\lambda\boldsymbol{\sigma}^{(k)T}\boldsymbol{n}^T\hat{\boldsymbol{u}}^{(k)}d\Gamma - \int_{\hat{\Gamma}_k}{}^*\boldsymbol{t}^{(k)T}\hat{\boldsymbol{u}}^{(k)}d\Gamma\right\} + \Pi_0 \quad (1)$$

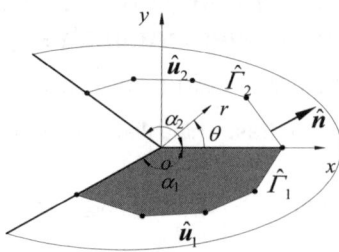

Fig. 1 Description of a bi-material wedge containing a super wedge tip element

The terms ${}_{\lambda}^{m}\sigma_{p}^{(k)}$, ${}_{\lambda}^{m}u_{p}^{(k)}$ and ${}_{\lambda}^{t}\sigma_{p}^{(k)}$, ${}_{\lambda}^{t}u_{p}^{(k)}$ associated with the homogeneous solutions of mechanical and thermal loadings, respectively [4, 5], and in the case of thermal loading, ${}_{c}^{t}\sigma_{p}^{(k)}$ and ${}_{c}^{t}u_{p}^{(k)}$ represent the known complementary solutions for non-singular stress and displacement fields. These vectors can be defined as

$${}_{\lambda}^{m}\sigma_{p}^{(k)} = \Sigma^{(k)}\,{}^{m}\beta, \quad {}_{\lambda}^{m}u_{p}^{(k)} = U^{(k)}\,{}^{m}\beta \tag{2, 3}$$

$${}_{\lambda}^{t}\sigma_{p}^{(k)} = \Sigma^{(k)}\,{}^{t}\beta, \quad {}_{\lambda}^{t}u_{p}^{(k)} = U^{(k)}\,{}^{t}\beta \tag{4, 5}$$

$${}_{c}^{t}\sigma_{p}^{(k)} = F_{c}^{(k)}(r,\theta)q_{c}^{(k)}, \quad {}_{c}^{t}u_{p}^{(k)} = G_{c}^{(k)}(r,\theta)q_{c}^{(k)} + r\alpha_{k}\Delta T e_{r} \tag{6, 7}$$

The details of $\Sigma^{(k)}$ and $U^{(k)}$ refer to Ping and Chen [5], and the definition of $F_{c}^{(k)}$, $G_{c}^{(k)}$, $q_{c}^{(k)}$ and e_{r}^{T} are listed in Barut et al. [6]. ΔT denotes the uniform temperature change. The unknown components of displacement vector along the common boundary segments, $\hat{\Gamma}_{k}$ (shown in Fig. 1), are denoted by $\hat{u}^{(k)}$. ${}^{*}t^{(k)}$ is the known applied traction components along the common boundary segments. n contains the components of the unite normal along $\hat{\Gamma}_{k}$. Π_{0} represents the total potential associated with the known initial strain and stress components arising from thermal loading only. The vector of displacement components, $\hat{u}^{(k)}$, along the common boundary between the global element and the conventional elements can be expressed in terms of the nodal displacement of the conventional elements as

$$\hat{u}^{(k)} = Lv \tag{8}$$

in which the matrix L contains the linear interpolation function compatible with those of the conventional elements. The vector v contains the nodal degrees of freedom associated with the conventional elements located on the boundary segment $\hat{\Gamma}_{k}$ of the super notch tip element.

Substituting for ${}_{\lambda}^{m}\sigma_{p}^{(k)}$, ${}_{\lambda}^{m}u_{p}^{(k)}$ and $\hat{u}^{(k)}$ from Eqs. (2) – (8) into the expression for the total potential leads to

$$\Pi = -\frac{1}{2}{}^{m}\beta^{T}H\,{}^{m}\beta - {}^{m}\beta^{T}\,{}^{t}f + {}^{m}\beta^{T}G\hat{v} - v^{T}\,{}^{*}f + \Pi_{0} \tag{9}$$

in which

$$H = \sum_{k=1}^{2}\frac{1}{2}\int_{\hat{\Gamma}_{k}}\left[\Sigma^{(k)T}Z_{\sigma}^{(k)T}n^{T}Z_{u}^{(k)}U^{(k)} + U^{(k)T}Z_{u}^{(k)T}nZ_{\sigma}^{(k)}\Sigma^{(k)}\right]dS, \quad G = \sum_{k=1}^{2}\int_{\Gamma_{w}}\Sigma^{(k)T}Z_{\sigma}^{(k)T}nL\,dS$$

$${}^{t}f^{(k)} = \sum_{k=1}^{2}\int_{\hat{\Gamma}_{k}}\Sigma^{(k)T}Z_{\sigma}^{(k)T}n^{T}Z_{u}^{(k)}\,{}_{\lambda+c}^{t}u_{p}^{(k)}d\Gamma, \quad {}^{*}f^{(k)} = \sum_{k=1}^{2}\int_{\hat{\Gamma}_{k}}L^{(k)}\,{}^{*}t^{(k)}d\Gamma$$

$Z_{\sigma}^{(k)}$ and $Z_{u}^{(k)}$ are the coordinate syserm transformation matrices. In order to express the total potential in terms of one unknown vector, v, the first variation of the total potential with respect to ${}^{m}\beta$ is taken. While noting that $\delta\Pi_{0} = 0$, enforcing the first variation with respect to ${}^{m}\beta$ to vanish results in

$$^m\beta = H^{-1}\{Rv - {}^tf\} \tag{10}$$

Substituting for $^m\beta$ in the expression for the total potential, and enforcing the first variation of the total potential to vanish results in the nodal equations of equilibrium for the global element as

$$\delta\Pi = v^T\{Kv - {}^tF - {}^*f\} = 0 \tag{11}$$

leading to

$$Kv = {}^tF + {}^*f \tag{12}$$

in which K and ^tF are defined as $K = G^T H^{-1} G$ and $^tF = G^T H^{-1} {}^tf$. The vector $^*f$ represents the internal loading vector at a node common to both the global and conventional elements. If a node is free of conventional elements, the components of the load vector represent the external force components. The vector ^tF represents the reaction force that suppresses the deformation, resulting from thermal loading only, at the common nodes of global and conventional elements.

A benchmark example

As shown in Fig.2, the parameter L describe the geometry of two bonded rectangular plates. The position of the super wedge tip element surrounded by conventional elements in the finite element model is also shown in Fig. 2. It is assumed the first and second materials have the same Poisson's ratio of $\upsilon_1 = \upsilon_2 = 0.2$. Their thermal expension coefficients are specified as $\alpha_1 = 15 \times 10^{-6}\ °C^{-1}$ and $\alpha_2 = 5 \times 10^{-6}\ °C^{-1}$. The first material has a Young's modulus of $E_1 = 280$GPa. The Young's modulus of the second material varies as $E_2 = 14.737, 31.111, 70.0, 120.0, 186.67$GPa. The applied uniform temperature change is $\Delta T = 100°C$. Owing to the symmetry, only is the left half plane modeled.

Under plane strain conditions, the present results are also compared against Munz and Yang [8]'s results and finite element predictions with an extremely refined mesh of only conventional elements using Ansys. In the finite element analysis with conventional elements only, the sub-modeling feature of Ansys is utilized to achieve acceptable mesh refinement near the junction. The model consisted of 1730 elements. In contrast with the conventional model, the present model with a super 5-node wedge tip element used consists of only 82 conventional elements. On the other hand, the solution reported by Munz and Yang [8] was obtained through curve fitting of a fifth-order polynomial in terms of the leading order of the singular behavior.

The comparison of the results of the tangential stress along the interface near point o is presented in Fig. 3. As observed in these figures, the predictions from the present analysis and the finite element analysis with an extremely refined mesh are in remarkable agreement up to a very small distance away from the point o. As is well known, the results of the conventional finite element analysis is strongly dependent on the mesh density and don't converge at a point very close to the point o. The results reported by Munz and Yang [8] deviate from the present analysis and Ansys predictions significantly as the magnitude of the leading-order singular term decreases.

Summary

A super bi-material wedge tip elementhas been developed utilizing the numerical solution for the stress and displacement fields based on the finite element eigenanalysis method. The super wedge tip element for arbitrary geometrical and material configurations is coupled with traditional elements while satisfying the inter-element continuity. The solution method is validated through existing asymptotic solutions and conventional detailed finite element analysis, and it is proved that the present method does not need the refinement of meshes around bi-material wedge tip as the conventional finite element method does

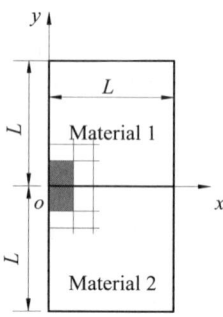

Fig. 2 Two bonded rectangular plates

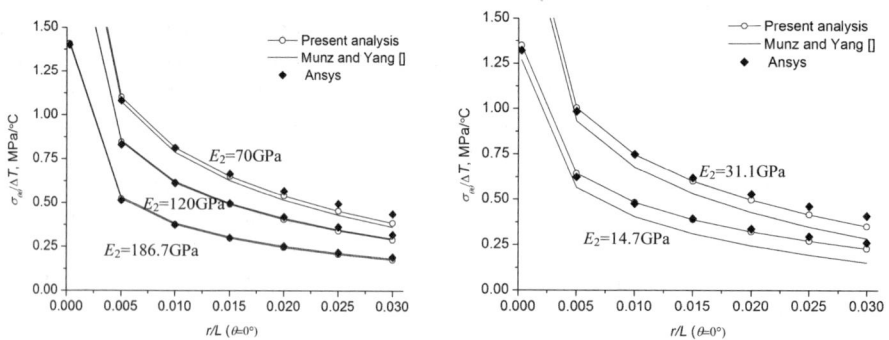

Fig. 3 Variation of the peeling stress away from the junction along the interface for the 90°-90° bonded wedge under thermal loading

Acknowledgement

This study is sponsored by the National Natural Science Foundation of China through grant No.51065008, the Natural Science Foundation of Jiangxi Province through grants No.2010GZW0013 and No. 2007GZW0862, the Jinggang-Star training Plan for Young Scientists of Jiangxi Province through grant No. 20112BCB23013 and the Scientific Research Project of Jiangxi Provincial Department of Education through grant No. GJJ10444.

References
[1] S.S. Pageau, Jr. S.B. Biggers: AIAA J. Vol. 34(1996), p. 1927-1933.
[2] S.S. Pageau, Jr. S.B. Biggers: Int. J. Num. Meth. Eng. (1997), 40, p. 2693-2713.
[3] K.N. Lin, J.W. Mar: Int. Fract. Vol. 12(1976), p.521-531.
[4] Chen MC, Sze KY: Eng Fract Mech Vol. 13(2001), p.1463-1476.
[5] Ping XC, Chen MC: Eng Fract Mech Vol. 75(13)(2008): 3819–3838.
[6] A. Barut, I. Guven, E. Madenci: Int. J. Solids & Struct. Vol. 38 (2001), p. 9077-9109.
[7] K.Y. Sze, H.T. Wang: Finite Elem. Anal. Des. Vol. 35(2) (2000), p. 97-118.
[8] D. Munz, Y.Y. Yang: J. Appl. Mech. (1992), p. 857-881.

Simulation of Spall Fracture Based on Material Point Method

Chen Wei-Dong[1,a], Zhang Fan[1,b] and Yang Wen-Miao[1]

[1] College of Astronautics and Civil Engineering, Harbin Engineering University, Harbin 150001, China

[a]chenweidong@hrbeu.edu.cn, [b]zhangfan3141@yahoo.cn

Keywords: Spall fracture, shock wave, material point method, hypervelocity impact.

Abstract. The spall fracture is a shock wave induced dynamic fracture phenomenon, it's difficult to capture the features of the spall fracture when traditional finite element method based on continuum mechanics is applied. In this paper, a new and flexible meshless method, material point method, is used to study the spall fracture of metal material in the case of hypervelocity impact. Firstly, a computational process is given in which Johnson-cook plasticity model, Mie-Grüneisen equation of state and several failure models including hydrodynamic tensile failure model, effective plastic strain model and Johnson-cook failure model are considered. Then a 3D simulation of spall fracture of an armco iron target under impact loads by a 2024-T351aluminum projectile is carried out. At last, the numerical results show that the material point method can accurately capture important features of spall fracture such as the arrival times, magnitudes and shapes of both the compressive waves and tensile reflections in the spall region, and it's proven that material point method is suitable to simulate the spall fracture in engineering applications.

Introduction

The spall fracture is a shock wave induced dynamic fracture phenomenon, which is due to the tensile stress generated by interaction of two rarefaction waves[1]. Spall fracture was firstly presented by Hopkinson in the beginning of the 20th century. His result indicated the increased brittleness of steel under dynamic conditions, and he also described the brittle appearance of the fractures produced dynamically and very small amount of plastic deformation associated them. Because spall fracture is practically important in high speed impact problems such as automobiles, aircrafts, and military vehicles, a large number of experiments on various kinds of material were carried out. The experimental results show that the macroscopic spall damage is strongly depend on the material kinematics at the microscopic level (0.1-10μm). At this specific scale level, the solid is formed by crystal grains, heterogeneous impurities, material defects and so on. During the shock wave propagating, the microscopic voids will be nucleated first, and grow at dislocation sites. These growing microscopic voids expand its cavity, coalesce each others, finally form macroscopic spall fracture cracks[2].

In this paper, a new and flexible meshless method, material point method, is used to study the spall fracture of metal material in the case of hypervelocity impact. Firstly, a computational process is given in which Johnson-cook plasticity model, Mie-Grüneisen equation of state and several failure models including hydrodynamic tensile failure model, effective plastic strain model and Johnson-cook failure model are considered. Then a 3D simulation of spall fracture of an armco iron target under impact loads by a 2024-T351aluminum projectile is carried out.

Material point method

Material point method, MPM, is a meshfree method for solid mechanics applications,which was presented by Sulsky et al[3]firstly in 1994 ,this mehtod is an extension of FLIP and PIC [4]methods .

In MPM method, material domain is represented by a collection of material points. During a dynamic analysis procedure, the deformation is tracked on the material points which carry all material properties such as position, velocity, acceleration, stress state, etc. At each time step, the equations of motion for the particles are solved on a fixed background grid which may be viewed as a temporary computational scratch pad; this solution is used to update the particles, and the background mesh can be discarded or reused for the next time step in its initial form. The material point method, which takes advantage of both Eulerian and Lagrangian methods and possesses the capability of handling large deformation, can be applied to simulate dynamics problems without special treatment, such as impact/contact, penetration, cracks and fracture.

For a continuous body under purely mechanical loading, momentum equation is given as

$$\rho \mathbf{a} = \nabla \cdot \mathbf{\sigma} + \mathbf{b} \tag{1}$$

Through the variation principle, the equation (1) may be written

$$\int_\Omega \rho \mathbf{a} \cdot \delta \mathbf{v} dx + \int_\Omega \mathbf{\sigma} : \nabla \delta \mathbf{v} dx = \int_\Omega \mathbf{b} \cdot \delta \mathbf{v} dx + \int_{\partial \Omega_\tau} \mathbf{\tau} \cdot \delta \mathbf{v} dS \tag{2}$$

Where $\mathbf{\sigma} = \mathbf{\sigma}(\mathbf{x},t)$ is the Cauchy stress tensor, $\mathbf{b}$ is body force density, $\delta \mathbf{v}$ is an admissible velocity field, Ω is the entire current volume.

Since the whole continuum body is described into a collection of material points, the mass density can be written as

$$\rho(x,t) = \sum_{p=1}^{N_p} M_p \delta(x - x_p^t) \tag{3}$$

The governing equation (2) can eventually be written for each node i as

$$\dot{\mathbf{p}}_i = \mathbf{f}_i^{int} + \mathbf{f}_i^{ext} \tag{4}$$

$$\dot{\mathbf{p}}_i = \sum_{p=1}^{N_p} \overline{S}_{ip} \dot{\mathbf{p}}_p \tag{5}$$

$$\mathbf{f}_i^{int} = -\sum_{p=1}^{N_p} \mathbf{\sigma}_p \cdot \overline{\nabla S_{ip}} V_p \tag{6}$$

$$\mathbf{f}_i^{ext} = \sum_{p=1}^{N_p} V_p \mathbf{b} \overline{S}_{ip} + \int_{\partial \Omega_\tau} \mathbf{\tau} S_i(\mathbf{x}) dS \tag{7}$$

Where $\dot{\mathbf{p}}_i$ is the change rate of momentum on the grid, $\mathbf{f}_i^{int}$ is the vector of nodal internal force and $\mathbf{f}_i^{ext}$ is the vector of nodal external force.

Material constitutive model and damage model

The Johnson-Cook model is empirically based and represents the flow stress with an form:

$$\sigma_y = \left(A + B\varepsilon_p^n\right)\left(1 + C \ln \dot{\varepsilon}_p^*\right)\left(1 - T^{*m}\right) \tag{8}$$

Where A, B, n, C, m are material parameters, σ_y is yield stress; ε_p is the effective plastic strain and ; $\dot{\varepsilon}_p^*$ is the normalized effective plastic strain rate, and $\dot{\varepsilon}_p^* = \dot{\varepsilon}_p / \dot{\varepsilon}_0$ ($\dot{\varepsilon}_0$ is usually constant 1s^{-1}); T^* is the homologous temperature, which is defined as $T^* = (T - T_{room})/(T_{melt} - T_{room})$, T_{room} and T_{melt} represent the room temperature and melting temperature, respectively.

Damage model proposed by Johnson and Cook is used in conjunction with J-C yield model. According to classical damage law, damage (fracture) of an particle is defined by

$$D = \sum \frac{\Delta \varepsilon_p}{\varepsilon^f} \tag{9}$$

Where D can change in a range of 0 to 1, and is set to 0 initially. Fracture is then allowed to occur when $D=1.0$. the general expression for the fracture strain ε^f is given by:

$$\varepsilon^f = \left[D_1 + D_2 \exp(D_3 \sigma^*)\right]\left[1 + D_4 \ln \dot{\varepsilon}_p^*\right]\left[1 + D_5 T^*\right] \tag{10}$$

Where $\sigma^* = p/\bar{\sigma}$ is stress triaxiality, p is hydrostatic pressure, $\bar{\sigma}$ is the Von Mises equivalent stress. In the material point method, failure of a particle may take place when $D=1.0$, and the failure particle can not withstand the shear stress, the deviatoric stress components must be set to zero. For the hydrostatic pressure, p, the particle of failure can not bear tension, but only the pressure. In a process of the pressure update, normal pressure is updated under compression, or set to 0 under tension for a failure particle.

The pressure-volume-energy behavior for the material is computed from Mie-Grüneisen equation of state for solid, and the press of particle is updated by

$$p(\rho,e) = \begin{cases} p_H(\rho)(1-\frac{1}{2}\Gamma u) + \Gamma_0 \rho e & \mu \geq 0 \\ \rho_0 c_0^2 \mu + \Gamma_0 \rho e & \mu < 0 \end{cases} \qquad (11)$$

Where subscript "H" refers to the Hugoniot curve, Γ is the Grüneisen parameter, and $\Gamma\rho = \Gamma_0\rho_0$. In equation 11,

$$p_H = \frac{\rho_0 c_0^2 \mu(1+\mu)}{[1-(s-1)\mu]^2} \qquad (12)$$

$$\mu = \frac{\rho}{\rho_0} - 1 \qquad (13)$$

Simulation of Spall Fracture

Figure 1 shows diagrammatic sketch of typical spall fracture at hypervelocity impact, which consists of a 2024-T351 aluminum projectile impacting an Armco iron target at 0.4cm/μs. Geometric parameters are consistent with the Johnson's article [5], the projectile has a diameter of 0.127 cm and a thickness of 0.064 cm, and the target has a diameter of 10.2 cm and a thickness of 2.54 cm. The aluminum projectile is only as a high-speed external load. For simplicity, tensile failure model, effective plastic strain model are applied in aluminum projectile, and $\varepsilon^p_{max}=2.0$, $p_{min}=-0.04$Mbar. The system is discretized into 1718880 particles in total, other parameters are shown in Tab. 1-2.

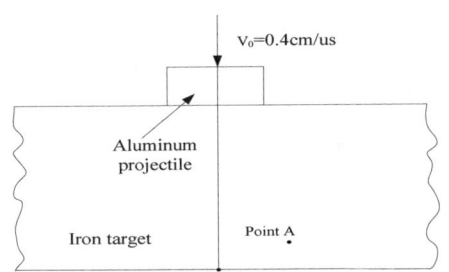
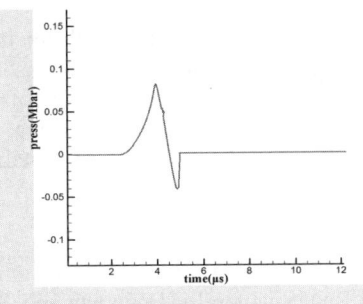

Fig.1 Impact schematic of spall fracture   Fig.2 Pressure versus time at point A

As found in Figs.3-4, a hemispherical shock wave is produced and propagates in the target plate, when equivalent plastic failure begins to form and spread from the center to the edges of plate. The shock wave continues to reach at the lower plane of the target plate, and the compression wave transforms into a tensile wave by the reflection of the free surface, that causes tensile failure destruction on the part of the area of the target plate as in Fig.4. The MPM solution retains all of the material including the failure particles and ejected particle. The Fig. 2 shows pressure versus time at point A, which is in the spall region. The responses are essentially identical. the MPM results with the arrival times, magnitudes and shapes of both the compressive waves and tensile reflections are consistent with the Johnson's results.

Table. 1 Metal material parameters

	Basic material parameters					Mie-Grüneisen EOS		
	ρ_0 [g/cm^{-3}]	E [Mbar]	v	T_{melt} [K]	T_{room} [K]	Γ	c_0 [cm/μs]	s
2024-T351 AL	2.77	0.75	0.33	775	294	2.0	0.533	1.73
Armco iron	7.89	2.10	0.3	1811	294	1.8	0.3570	1.92

Table. 2 Johnson-Cook plastic model and damage model

	Johnson-Cook plastic model					Johnson-Cook damage model				
	A [Mbar]	B [Mbar]	n	c	m	D_1	D_2	D_3	D_4	D_5
2024-T351AL	0.00265	0.00426	0.24	0.015	1.00	-	-	-	-	-
Armco iron	0.00175	0.00380	0.32	0.06	0.55	-2.20	5.43	-0.47	0.016	0.63

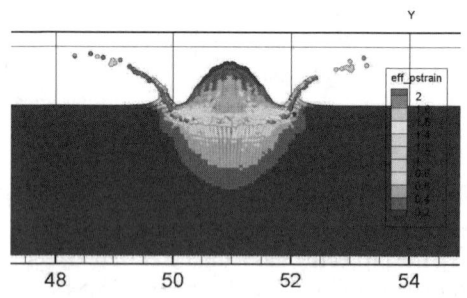 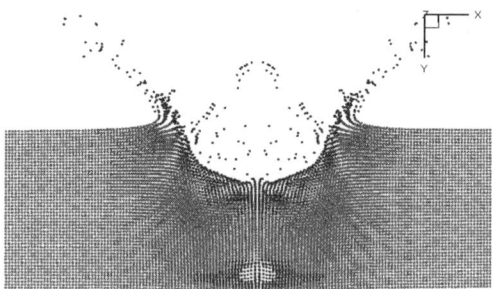

Fig.3 Contour of effective plastic strain at t=3μs   Fig.4 Failure particle and spall fracture at t=12μs

Summary

MPM is used to study the spall fracture in the case of hypervelocity impact. In this paper, Mie-Grüneisen equation of state, Johnson-Cook plastic model and failure model are considered. Then a 3D simulation of spall fracture of an armco iron target under impact loads is carried out. The results of simulation give the right arrival times, magnitudes and shapes in spall fracture process. It's proven that material point method is suitable to simulate the spall fracture in engineering applications.

References

[1] M. A. Meyers: *Dynamic Behavior of Materials* (John Wiley & Sons, Canada 1994).
[2] Bo Ren, Shaofan Li, Jing Qian and Xiaowei Zeng: Computer Methods in Applied Mechanics and Engineering Vol.200 (2011),p 797
[3] D. Sulsky, Z. Chen and H. Schreyer: Computer Methods in Applied Mechanics and Engineering Vol. 118(1994) ,p.179
[4] J. Brackbill and H Ruppel: Journal of Computational Physics Vol. 65(1986),p.314
[5] G. R. Johnson, S. R. Beissel and R. A. Stryk: International Journal for Numerical Methods in Engneering Vol.53(2002),p.875

Buckling Reliability Analysis for the Cylindrical Shell with Initial Defects

Hai An [1,a], Weiguang An [1,a], yuanyin Gao [2,b], Xianghua Song [1,a]

[1] College of Aerospace and Civil Engineering, Harbin Engineering University, Harbin 150001, Heilongjiang, China

[2] Technology Research & Economy Development Institute, CSSC, Beijing 100081, China

[a] anhai@hrbeu.edu.cn, [b] gaoyy1977@163.com

Key words: buckling reliability analysis; local axial symmetry initial defects; buckling critical load coefficient; AFOSM

Abstract. In this paper, the asymptotic expression of the buckling critical load coefficient of the thin cylindrical shell with local axial symmetry initial defects under the axial loads is deduced by used the Karman-Donnel Equation. The buckling safety margin equation of the cylindrical shell with initial defects is constructed. Furthermore, the buckling reliability index is solved by used AFOSM (Advanced First-Order Second Moment) method. In the end, a numerical example is given to analyze the influence of the band width and amplitude of local axial symmetry initial defects on the structural buckling reliability index.

Introduction

The structures with the thin cylindrical shell are widely used in engineering. As concerns these structures, the buckling failure is one of the major failure modes of them. It is known from the experiment that the critical buckling load of cylindrical shell considering initial defects is a fraction of that not considering initial defects. So, the buckling failure of cylindrical shell is highly sensitive to initial geometry defects. For this reason, this paper presents an approach to the buckling reliability analysis for the cylindrical shell with initial defects. The approach proposed in this paper has implications for the structure design, the rational maintenance and renewing strategy in engineering.

The critical load coefficient of the cylindrical shell with local axial symmetry defects

We consider a long circular cylindrical shell of radius, R, length, L and thickness, h, which is made of an homogeneous, isotropic elastic material with Young's modulus E and Poisson's ratio, γ. It is subjected to an axial compressive load P. The coordinate system is shown in Fig. 1 and the displacement components will be denoted by u, v and w. Within the Donnell theory and if the pre-buckling rotations are neglected, in the presence of local initial defects w_0, which is assumed to be on exponential decay, can be expressed

$$w_0 = \xi\, h \exp\left[-\frac{|x|}{\sqrt{C}}\right]. \tag{1}$$

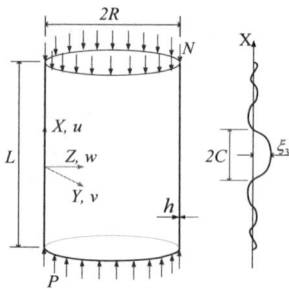

Fig.1. Compressive cylinder with localized axisymmetrical defects

The asymptotic Karman-Donnel Equation with local initial defects [1] can be expressed as follows

$$\frac{\partial^4 w}{\partial x^4} + 2\rho \frac{\partial^2 w}{\partial x^2} + w = -2\rho \frac{\partial^2 w_0}{\partial x^2}. \tag{2}$$

Where $\rho = \frac{P}{P_{cr}}$ is the critical load coefficient. $P_{cr} = \frac{Eh}{R\sqrt{3(1-v^2)}}$ is the classical buckling critical load of the perfect cylindrical shell.

By Fourier Integral Transform to Eqs.(2), we can obtain

$$(i\omega)^4 \overline{w} + 2\rho(i\omega)^2 \overline{w} + \overline{w} = -2\rho(i\omega)^2 \overline{w}_0. \tag{3}$$

Where $\overline{w} = \int_{-\infty}^{\infty} w(x)\exp(-i\omega x)dx$.

We can obtain from Eqs.(3)

$$\overline{w} = \frac{2\rho\omega^2 \overline{w}_0}{\omega^4 - 2\rho\omega^2 + 1}. \tag{4}$$

By the inverse Fourier Transform, it can be obtained

$$w = \frac{1}{2\pi} \int_{-\infty}^{\infty} \frac{2\rho\omega^2 \overline{w}_0}{\omega^4 - 2\rho\omega^2 + 1} \exp(i\omega x) d\omega$$

$$= \frac{\rho}{\pi} \int_{-\infty}^{\infty} \frac{\omega^2 \overline{w}_0}{(\omega^2 - 1)^2 + 2(1-\rho)\omega^2} \exp(i\omega x) d\omega. \tag{5}$$

First order poles of the integrand are

$$\left. \begin{array}{l} \omega_1 = \sqrt{\frac{1+\rho}{2}} - i\sqrt{\frac{1-\rho}{2}}, \quad \omega_2 = -\sqrt{\frac{1+\rho}{2}} - i\sqrt{\frac{1-\rho}{2}}, \\ \omega_3 = -\sqrt{\frac{1+\rho}{2}} + i\sqrt{\frac{1-\rho}{2}}, \quad \omega_4 = \sqrt{\frac{1+\rho}{2}} + i\sqrt{\frac{1-\rho}{2}}. \end{array} \right\} \tag{6}$$

According to Jordan's Lemma and Cauchy Integral Theorem, when $x > 0$, we can obtain

$$w = \frac{\rho}{2}\left[\frac{1}{\sqrt{2(1-\rho)}} - \frac{i}{\sqrt{2(1+\rho)}}\right]\bar{w}_0(\omega_1)\exp(-i\omega_1 x)$$
$$+ \frac{\rho}{2}\left[\frac{1}{\sqrt{2(1-\rho)}} + \frac{i}{\sqrt{2(1+\rho)}}\right]\bar{w}_0(\omega_2)\exp(-i\omega_2 x). \quad (7)$$

When $x < 0$,

$$w = -\frac{\rho}{2}\left[\frac{1}{\sqrt{2(1-\rho)}} - \frac{i}{\sqrt{2(1+\rho)}}\right]\bar{w}_0(\omega_3)\exp(-i\omega_3 x)$$
$$- \frac{\rho}{2}\left[\frac{1}{\sqrt{2(1-\rho)}} + \frac{i}{\sqrt{2(1+\rho)}}\right]\bar{w}_0(\omega_4)\exp(-i\omega_4 x). \quad (8)$$

When it is around a critical state, $\rho \to 1$. Eqs.(8) is simplified as follows

$$w = \frac{\rho}{2\sqrt{2(1-\rho)}}\left[\bar{w}_0(1)\exp(-ix) + \bar{w}_0(-1)\exp(ix)\right]\exp\left(-|x|\cdot\frac{\sqrt{1-\rho}}{2}\right)$$
$$= \frac{\rho}{\sqrt{2(1-\rho)}}|\bar{w}_0(1)|\exp\left(-|x|\sqrt{\frac{1-\rho}{2}}\right)\cos\left(x - \frac{\pi}{2}\right). \quad (9)$$

The critical load coefficient of the imperfect cylindrical shell can be obtained by taking the extreme value of the critical load coefficient to the displace

$$[2(1-\rho_s)]^{3/2} = (3)^{3/2}\sqrt{1-v^2}|\bar{w}_0(1)|\rho_s. \quad (10)$$

Where $|\bar{w}_0(1)| = \left|\int_{-l/2}^{l/2} \xi e^{-\frac{|x|}{\sqrt{C}}} e^{ix} dx\right|$, $l = L\left[\frac{12(1-v^2)}{R^2 h^2}\right]^{1/4}$ is the nondimensional longth of the shell.

The safety margin of the cylindrical shell with initial defects

According to the definition of the critical load coefficient, the buckling safety margin of the cylindrical shell with initial defects can be constructed as follows

$$M = P_{cr} - P = P\left(\frac{1}{\rho_s} - 1\right) = g(P, h, R, L, v, C, \xi). \quad (11)$$

The buckling reliability index of the cylindrical shell can be solved by using AFOSM [2] considering randomness of variables of the equation above.

Numerical example

A thin cylindrical shell with local axial symmetry initial defects under the axial loads is shown in Fig.1. It is assumed that random parameters and defects of the cylindrical shell all obeyed the normal distribution. Their mean value and coefficient of variation are respectively $\mu_L = 134.37mm$,

$CV_L = 0.05$, $\mu_R = 101.6$mm, $CV_R = 0.05$, $\mu_h = 134.37$mm, $CV_h = 0.04$, $\mu_e = 2 \times 10^5$ MPa, $CV_E = 0.03$, $\mu_v = 0.3$, $CV_v = 0.04$, $CV_C = 0.1$, $CV_\xi = 0.1$, $CV_P = 0.05$. The changes of the buckling reliability of the structure with different localized axisymmetrical defects under different axial compressive loads are shown in Fig.2.

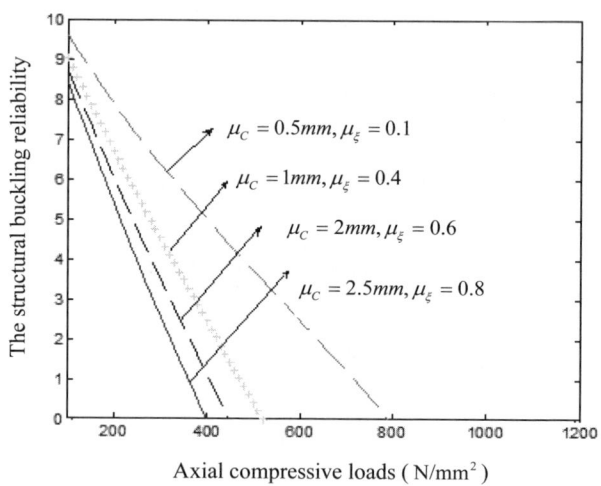

Fig.2. The buckling reliability of the cylindrical shell with different localized axisymmetrical defects under different axial compressive loads

It can be known from Fig.2. that the band width of local defects has a significantly influence on the structural buckling reliability of the cylindrical shell which normally decays with axial compressive loads linearly. The trend of the decay is slower when ξ is smaller. As soon as ξ exceeds a specific value, the decay trend has been basically stable.

Conclusions

This paper proposed an approach to estimating the buckling reliability for the thin cylindrical shell with local axial symmetry initial defects under the axial loads. According to the results of numerical example, the buckling reliability of the thin cylindrical shell can be estimated by analyzing the band width and amplitude of local defects, which provides a reference for formulating maintenance strategy beforehand.

References

[1] Amazigo, J. C. and Budiansky, B., "Asymptotic formulas for the buckling stresses of axially compressed cylinders with localized or random axisymmetric imperfections", J. Appl. Mech., 1972, Vol. 39, P. 179-184

[2] An Wei-guang, Cai Yin-lin, Chen wei-dong: *Reliability analysis and optimum design for stochastic structure systems*, (Harbin Engineering University Press, China 2005). (in chinese)

To Research Residual Stress Using Finite Element Analysis

Zhefeng Wang[1,a], Yaoyang Hu[1,b]

Department of Aerospace Engineering, Shenyang Aerospace University, Shenyang 110136, People's Republic of China

[a]e-mail: zhefeng_w@126.com

[b]e-mail: huyaoyang@126.com

Keywords: residual stress, aluminum alloy thick plate, numerical simulation, theoretical analysis

Abstract. The residual stress distributions of 7075 aluminum alloy rectangular thick plates after quench-hardening had been simulated firstly, then all the results were presented and compared with each other. Some deep theoretical analyses were also carried out. The results show that complicated residual stress distribution regularities in aluminum alloy thick plates can be obtained by the finite element analysis successfully.

1. Introduction

It is impossible to avoid that lots of residual stress is always brought into aluminum alloy thick plates in forming and heat-treatment process. As we all know, residual stress usually plays a negative role which can cause serious distortion and spring–back in kinds of aeronautic components, such as monolithic panels, costae, structure beams and so on [1, 2]. According to some concerned researches show that aluminum alloy thick plates always involve much residual stress in quench-hardening process, which may even approach to the material's yield limit [3, 4]. So the investigation on inner residual stress distributions and levels of quenched aluminum alloy plates is very significant. The research results can contribute to following residual stress relief or control.

Not only carried out finite element analyses of 7075 aluminum alloy thick plates in quench-hardening process, but also gave deep expatiation on residual stress distribution regularities emphatically.

2. Finite Element Analysis Model

2.1 Assumed conditions
To simplify the numerical analyses, a few of basic hypotheses had been made as follow:
   (1)The material of research objects is continuous and isotropic.
   (2)The aluminum alloy thick plates have uniform initial temperature and no stress.
   (3)Quench bath is enough so that its temperature keeps invariable.
   (4)Take no account of the influence from the alloy's microstructure variation.

2.2 Material properties
The alloy's quenching temperature is 465℃, the quench bath temperature is 20℃.The alloy's Poisson ratio is 0.33 and always the same. On the contrary, elastic modulus (E), density (ρ), thermal conductivity (λ), specific heat (Cp), thermal expansion coefficient (α) and yield limit (σs) change a lot with different temperatures (TEMP), as shown in Table 1. Heat transfer coefficient (HF) also changes with surface temperatures of aluminum alloy thick plates, as shown in Table 2.

Table 1 Material properties of 7075 aluminum alloy thick plates [5, 6, 7, 8]

TEMP (℃)	E (10^{10}Pa)	ρ (Kg/m³)	λ (W·m⁻¹k⁻¹)	Cp (J·Kg⁻¹k⁻¹)	α (10^{-6}·k⁻¹)	σ s (10^8Pa)
0	7.333	2800	155	830	22.6	2.867
25	7.2	2788	156	860	23.5	2.773
50	7.12	2781	158.3	870	24.0	2.533
100	6.907	2775	161	900	24.9	2.2
200	6.187	2750	175	970	28.4	1.253
300	5.387	2725	185	1020	29.9	0.773
400	4.853	2700	193	1120	31.4	0.32
500	4.450	2675	197	1320	31.7	0.2

Table 2 Heat transfer coefficient of 7075 aluminum alloy in immerse quench-hardening [6, 7]

TEMP (℃)	30	50	75	100	150	200	250	300	400	500
HF(W·m⁻²k⁻¹)	2500	2660	3500	5000	10000	13857	13857	10000	3000	700

2.3 Establishment of finite element model

The shape of research objects is rectangular block. Its length is 2000mm, width is 220mm, and thickness is different from 12mm to 100mm. According to the symmetries of geometry and boundary condition, 1/4 of the thick plate was established as finite element model to improve the calculation efficiency. To be specific, the model's length size and width size are all half of actual value, thickness size is actual value. It is shown in Fig.1.

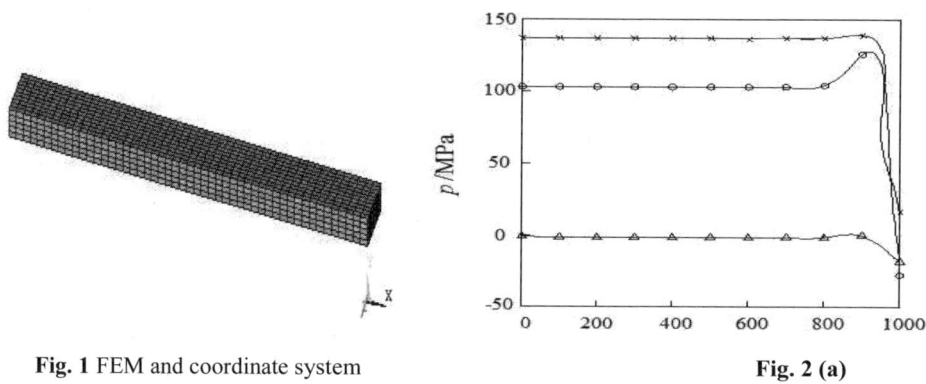

Fig. 1 FEM and coordinate system   Fig. 2 (a)

3. Simulation results

3.1 The residual stress distribution regularity in 50mm thick alloy plate

The residual stress distributions of 50mm thick aluminum alloy plate are shown in Fig. 2.

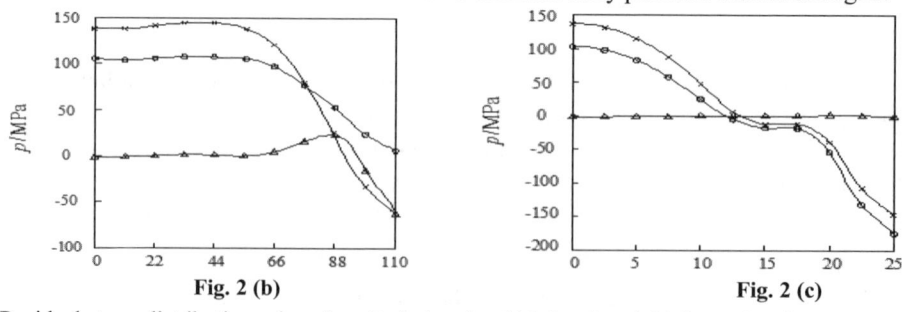

Fig. 2 (b)   Fig. 2 (c)

Fig. 2 Residual stress distributions along longitude (mm), width (mm) and thickness (mm) are respectively shown in fig. 2 (a), fig. 2 (b) and fig. 2 (c). "+" means that the residual stress is tensile state, "-" means that the residual stress is compressive state. "×" means p_x (residual stress in X direction), "△" means p_y (residual stress in Y direction) and "⊙" means p_z (residual stress in Z direction).

From Fig. 2, following information can be obtained.

(1) Along longitude, the residual stress distributions are relatively smooth and they are all tensile stress. Near the cooling surface (800 ~ 1000mm), tensile stress p_y increases by 25MPa. Near the location of 900mm, the values of the tensile stress p_x, p_y and p_z all decrease sharply and change into compressive stress.

Along width, near the location of 55mm, tensile stress both p_x and p_z begins to decrease and changes into compressive stress slowly. Tensile stress p_y is very small, and increases a little from the 55mm location, then changes into compressive stress as well.

Along thickness, a gradual transition happens, which is that tensile stress in central part changes into compressive stress near down cooling surface and top cooling surface.

(2) In all these three figures, p_x and p_y have similar change regularity and p_z is always smaller. So residual stress in 50mm thick aluminum alloy plate can be approximated to 2-dimensional state.

(3) After quench-hardening, residual stress shows tensile state near thick plate surfaces, and shows compressive state in central part. The central zone along width is less than 55mm on both sides of the X-Y coordinate plane. The central zone along thickness is less than 12.5mm on both sides of the middle plane.

3.2 The residual stress distribution regularities of different thickness sizes
With different alloy plate thickness sizes, the residual stress distribution regularities are different.

(1) Table 3 lists the maximum tension stress (Max STE), the maximum compressive stress (Max SCO), the maximum Von Mises equivalent stress (Max SEQV) and minimum Von Mises equivalent stress (Min SEQV) when aluminum alloy plate thickness size (THK) varies from 12mm to 100mm.

As it shows in Table 3, with plate thickness increases, residual stress increases. However, when plate thickness overtakes 80mm, the increase speed of residual stress slows down. When plate thickness increases to 100mm, residual stress has already been very close to the material's yield limit.

Table 3 Residual stress in different aluminum alloy plate thickness sizes

THK(mm)	12	16	25	40	50	70	80	90	100
Max STE(MPa)	149×10^{-5}	839×10^{-5}	34.4	96.2	145	187	244	261	286
Max SCO(MPa)	10.9×10^{-5}	1.4×10^{-5}	1.84	7.45	10.5	14.1	7.11	7.72	7.7
Max SEQV(MPa)	182×10^{-5}	5.7	43.7	129	179	195	224	241	269
Min SEQV(MPa)	2.2×10^{-5}	587×10^{-5}	2.22	5.59	5.68	11.6	13.1	12.1	14.3

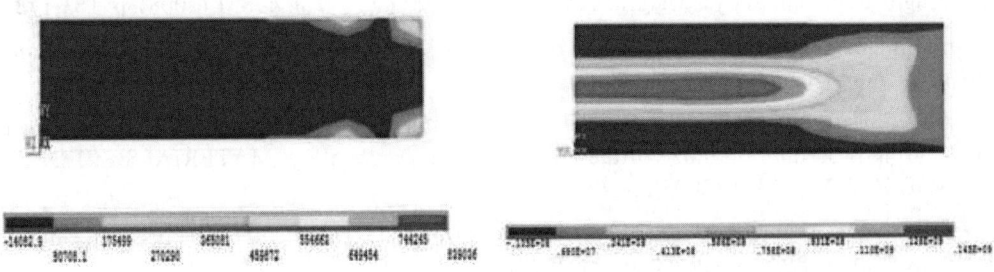

Fig. 3(a)　　　　　　　　　　　　　Fig. 3(b)

Fig. 3 Tensile-compressive stress distribution in different aluminum alloy plate thickness sizes. In fig. 3(a), the plate thickness is 16mm. In fig. 3(b), the plate thickness is 50mm. Both two figures show the middle plane along longitude.

(2) Fig. 3 shows the residual stress distribution difference. When plate thickness is less than 16mm, the residual stress values are so small that we can even ignore them. But, it's worthwhile to note that their distribution regularities are different obviously when plate thickness size is bigger.

From Fig. 3, residual stress in 16mm thick plate shows compressive state in almost the whole volume and there's only a small zone filled tensile stress near the cooling surfaces along width. The main reason: At the beginning of quench-hardening, the rate of cooling shrinkage from alloy plate surfaces is bigger than that from central, so the surface alloy is tied down by central alloy and shows tensile state. Correspondingly, the central alloy is tied down by surface alloy and shows compressive state [6]. For much thicker alloy plates (≥16mm), with the development of quench-hardening, the rate of cooling from central alloy gradually exceeds that from surface alloy, which causes a gradual transition from compressive stress to tensile stress in central part and a gradual transition from tensile stress to compressive stress near alloy surface. Finally, the much thicker alloy plates show tensile state in center and compressive state near surface. However, for little thinner alloy plates (≤16mm), due to more efficient heat exchange, the transition only happens along width.

4. Conclusions

(1) For 50mm thick alloy plate, the central part shows compressive stress and the central zone occupies about half the size in both width and thickness.
(2) With plate thickness increases, residual stress increases. However, when plate thickness overtakes 80mm, residual stress grows slowly. When plate thickness increases to 100mm, residual stress has already been very close to material's yield limit.
(3) After quench-hardening, the residual stress distribution regularities are different when the aluminum alloy plate thickness sizes are different. For little thinner alloy plates (≤16mm), the residual stress shows compressive state in almost the whole volume, and there's only a small zone filled tensile stress near cooling surface along width. For much thicker alloy plates (≥16mm), they show tensile stress in center and compressive stress near surface.

References

[1] Heinz A, Haszler A, Keidel C, and so on: Recent development in aluminum alloys for aerospace applications. Materials Science Engineering A. Vol. 280. 2 (2000), p. 102-107

[2] Dixit M, Mishra R S, sankaran K K: Structure-property correlations in Al 7050 and Al 7055 high-strength aluminum alloys. Materials Science Engineering A. Vol. 478. 1 (2008), p. 163-172.

[3] Tang Zhitao: *Residual Stresses and Deformations of Aerospace Aluminum Alloy in Machining.* (Press of Shandong University, Jinan 2008).

[4] WANG Gui-wei, FANG Hong-yuan, FAN Cheng-lei, NIE Bo: Numerical simulation of the manufacturing procedure optimize of the thick 7B04 aluminum alloy. MATERIALS SCIENCE & TECHNOLOGY. Vol. 13. 1 (2005), p. 70-74.

[5] Wang Zhutang, Tian Rongzhang. *Aluminum Alloy Handbook of Machining.* (Press of Central South University of Technology, Changsha 2000).

[6] YUAN Wang-jiao, WU Yun-xin: Coupled thermal-mechanical simulation on quenching of aluminum alloy thick-plate based on ANSYS. Journal of Central South University (Science and Technology). Vol. 46. 6 (2010), p. 2207-2212.

[7] Cao Jinrong: *Study on Minimizing the Residual Stresses in the 7075 alloy Die-Forged Shell.* (Press of Central South University of Technology, Changsha 2000).

[8] Wang Qiucheng: *Evaluation and Relief of Residual Stresses in Aluminum Alloys for Aircraft Structures.* (Press of Zhengjiang University, Zhejiang 2003).

Numerical simulation of underwater explosion based on material point method

CHEN Wei-Dong[1,a], YANG Wen-Miao[1,b], ZHANG Fan[1,c]

[1]College of Astronautics and Civil Engineering, Harbin Engineering University, Harbin, China

[a]chenweidong@hrbeu.edu.cn, [b]yangwenmiao2008@163.com, [c]zhangfan@hrbeu.edu.cn

Keywords: underwater explosion. material point method. multi-material coupling. numerical simulation.

Abstract: Underwater explosive load calculation and numerical simulation is the key issues of the design of underwater conventional weapons and protection of ships and submarine. Underwater explosion involves strong nonliner problems and multi-material coupling problems. The method of underwater explosion calculation base on the material point method (MPM) is presented. The MPM takes the advantages of the both Euler and Lagrangian methods and overcomes the shortcomings of them. The problems in the underwater explosive simulation such as large deformation, moving material interfaces and deformable boundaries can be solved effectively by the MPM. At last, blast wave produced by TNT exploding under similar infinite water region is computed. The calculated results are in good agreement with the results of empirical formula of Cole and SPH. The simulation results show that the MPM is an effective tool for underwater explosion calculation.

Introduction

As one of the meshfree methods, the MPM (material point method)[1] is an extension to solid mechanics problems of the FLIP(Fluid-implicit particle) method of Brackbill, which itself is an extension of the PIC(particle in cell) method dating back to the pioneering work of Harlow. The MPM takes advantage of both the Eulerian and Lagrangian methods. its feature allows simulations of multi-material interaction problem without the need for a special contact algorithm. The MPM has been widely applied to hypervelocity impact and explosion problems[2,3]. In this paper, the MPM is extended to simulate underwater explosion . The motivation of the work is to probe the feasibility of using MPM to simulate this problem. A MPM computational code for simulation of underwater explosion was developed in Fortran 95 .The process of TNT exploding under similar infinite water region and the propagation of shock wave underwater were simulated. it shows that the MPM is an efficient tool to simulate underwater explosion .

Theory

In MPM, the continuum body is discretized with the use of a finite set of N_p material points.

Each material point at time t is assigned a mass m_p, density ρ_p^t, velocity v_p^t and any other internal state variables necessary for the constitutive model. Thus, these material points provide a Lagrangian description of the continuum body. The governing differential equations can be derived from the conservation equation for mass , momentum and energy

$$\frac{d\rho}{dt} + \rho \nabla \cdot \mathbf{v} = 0 .\tag{1}$$

$$\rho \mathbf{a} = \nabla \cdot \boldsymbol{\sigma} + \rho \mathbf{b} .\tag{2}$$

$$\rho \frac{de}{dt} = \boldsymbol{\sigma} : \dot{\boldsymbol{\varepsilon}} + \mathbf{v} \cdot \mathbf{b} .\tag{3}$$

Where e is the specific internal energy in the current configuration, $\dot{\boldsymbol{\varepsilon}} = \dot{\boldsymbol{\varepsilon}}(\mathbf{x},t)$ is the strain rate, Where $\rho(\mathbf{x},t)$ is the mass density, $\mathbf{v}(\mathbf{x},t)$ is the velocity, $\mathbf{a}(\mathbf{x},t)$ is the acceleration, $\boldsymbol{\sigma}(\mathbf{x},t)$ is Cauchy stress tensor, $\mathbf{b}(\mathbf{x},t)$ is the specific body force, $\mathbf{x}$ is the current position at time of any material point.

Multiplied Eq.2 by the test function and integrate over the current configuration, The weak form of the conservation for momentum is given as

$$\int_\Omega \rho \mathbf{a} \cdot w dV + \int_\Omega \rho \boldsymbol{\sigma}^s : \nabla w dV - \int_\Omega \rho \mathbf{b} \cdot w dV - \int_\Gamma \tau \cdot w dS = 0 \tag{4}$$

Where dV and dS denote the differential volume and surface elements, respectively. $\boldsymbol{\sigma}^s$ is specific stress, $\boldsymbol{\sigma}^s = \boldsymbol{\sigma}/\rho$, Γ the boundary with a prescribed traction, τ is the prescribed traction, w is zero on the boundary with a prescribed displacement.

Since the whole continuum body is described into a collection of material points, the mass density can be written as

$$\rho(x,t) = \sum_{p=1}^{N_p} M_p \delta(x - x_p^t) \tag{5}$$

Where δ is Dirac delta function, x_p^t is the current position of material point p at time t.

In order to calculate the gradient terms in the Eq.4, a background computational grid is constructed. Using this grid, the quantity associated with the material point, ψ_p^t, can be expressed as a piecewise continuous function by using the nodal basis functions assembled from conventional finite element shape functions.

$$\psi_p^t = \sum_{i=1}^{N_n} \psi_i N_i(x_p^t) \tag{6}$$

Where ψ_i^t is the quantity associated with the grid nodes, N_i is the nodal basis function, N_n is the number of nodes in the background grid.

Substituting Eq.5 and Eq.6 in to the Eq.4, and since is arbitrary, Eq.4 becames

$$m_i^t a_i^t = (f_i^t)^{int} + (f_i^t)^{ext} \tag{7}$$

Where $m_i^t, (f_i^t)^{int}, (f_i^t)^{ext}$ denote the lumped nodal mass, the internal force vector and external force vector, respectively. The expressions of them are as follows

$$m_i^t = \sum_{p=1}^{N_p} M_p N_i(x_p^t) \tag{8}$$

$$(f_i^t)^{int} = -\sum M_p \boldsymbol{\sigma}^s(x_p^t, t) \cdot \nabla N_i \big|_{x_p^t} \tag{9}$$

$$(f_i^t)^{ext} = \sum_p m_p b N_i(x_p) + \int_\Gamma N_i(x_p) t d\Gamma \qquad (10)$$

Eq.7 is the discrete form of the conservation equation for momentum on grid nodes. An explicit time integrator is used to solve for the nodal accelerations, with the time step satisfying the stability condition. More details of the algorithm are given in[1,4].

Equation of State

An equation of state(EOS) is a constitutive equation which provides a mathematical relationship between two or more state functions associated with the matter , such as its temperature, pressure, volume, or internal energy. Selecting the correct form of EOS and the appropriate value of the parameters in the EOS is essential to the simulation.

In this paper, the EOS for water takes a polynomial form[5]

$$p = a_1\mu + a_2\mu^2 + a_3\mu^3 + (b_0 + b_1\mu + b_2\mu^2 + b_3\mu^3)\rho_0 e \qquad (11)$$

Where $\mu = \left(\rho/\rho_0\right) - 1$, ρ_0 and ρ are the initial and current density of water, respectively.

e Is the internal energy per unit mass.

The JWL EOS for TNT is given by

$$p = A\left(1 - \frac{\omega}{R_1 V}\right)\exp(-R_1 V) + B\left(1 - \frac{\omega}{R_2 V}\right)\exp(-R_2 V) + \frac{\omega E}{V} \qquad (12)$$

Where V is the ratio of density, e is the internal energy of the high explosive per mass. A, B, R_1, R_2, ω are coefficients obtained by fitting experimental data.

Numerical examples

TNT exploding under similar infinite water region is simulated by MPM. The square symmetric model is used. The size of TNT is 8cm×8cm and the size of water is 400cm×400cm and the simulation model is shown in Fig.1. The number of discrete material points is 160000. The total computation time is 1ms.

In order to comparison the same model is computed by empirical formula of Cole and SPH. Fig.2 shows the curve of peak pressure at different location obtained by the three methods.

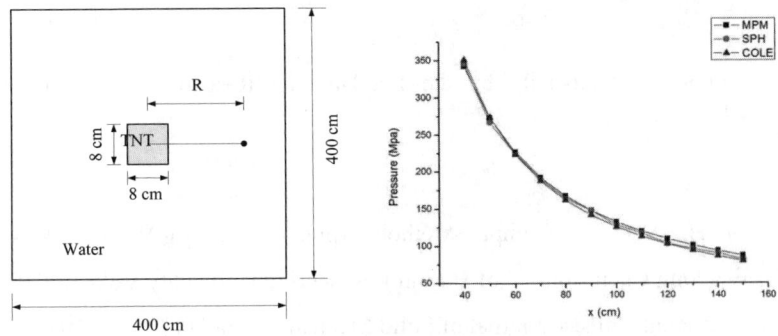

Fig.1 Simulation model   Fig.2 The curve of peak pressure at different location

With the propagation of underwater explosion shock wave, the peak pressure decayed exponentially. It is in good agreement with the attenuation law of shock wave underwater. As can be seen from Fig.2, with increasing distance from the detonation point, the relative errors between the results of MPM and Cole tend to be larger. The MPM is More suitable for simulation of near-field explosion flow .

In MPM the material points carry all material information, the use of Tecplot can draw cloud picture of the particles easily. Fig.2 shows the pressure contour at two typical time. The process of the blast spread can be observed clearly.

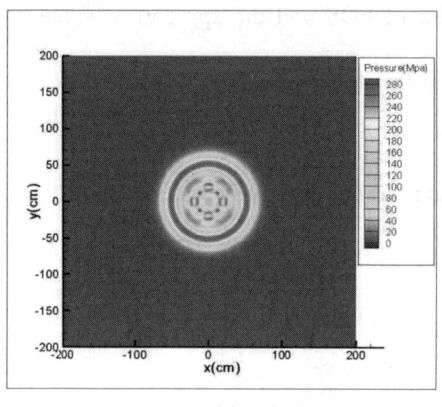 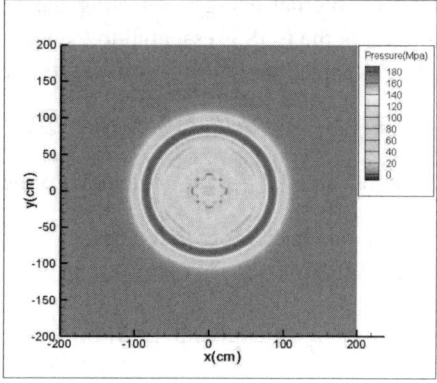

(a) $t=300\mu s$ (b) $t=500\mu s$

Fig. 2 Pressure contour at typical time

Summary

This paper presents the application of the MPM to underwater explosion simulation. The problems encountered by grid-based method such as large deformation, moving material interface and multi-material coupling can be solved effectively by MPM. By numerical simulation the shock wave generation and propagation process in underwater explosion can be well observed. The numerical example demonstrates that the MPM can resolve the simulation of underwater explosion fairly well.

Acknowledgement

The authors were supported financially by the Fundamental Research Funds for the Central Universities (HEUCFZ1126).

References

[1] D.Sulsky, Z.Chen, H.L.Schreyer: Comput. Methods Appl. Mech.Engrg Vol.118(1994),p.179

[2] Zhong Zhang, Weidong Chen: Journal of Harbin Engineering University Vol.31(2010),p.1312

[3] Shang Ma, Xiong Zhang: Chinese Journal of Solid Mechanics Vol.30(2009),p.504

[4] Andersen, S. and L. Andersen: Computers & Structures Vol.88(2010),p. 506

[5] Chiesum J. E. and Shin Y. S: Shock and Vibration Vol.4(1997),p.11

Numerical simulation of fatigue behavior of honeycomb sandwich panels

Jie LU[1, a], G.P. ZOU[b], P.F. YANG[c]

[1]Centre for Mechanics Testing, College of Aerospace and Civil Engineering, Harbin Engineering University, Harbin, 150001, P.R.China

[1,a]lujie@hrbeu.edu.cn, [b]gpzou@hotmail.com, [c]ypfei029@yahoo.com.cn

Keywords: Honeycomb sandwich, high cycle fatigue, residual life, security coefficient

Abstract. In this study, based on Goodman stress correction algorithm, four point bending fatigue behavior of brazed steel honeycomb sandwich panels at room temperature is simulated using Fe-safe emulation module. The cyclic load was sinusoidal with a frequency f= 10 Hz. The residual life and security coefficient are given in the condition of design fatigue lifetime of 10^6 cycles. The results show that, in the local deformation, local maximum deformation occurs in the core wall under the load region, indicate that this region is the failure region of fatigue. And the residual life of the core wall in this region is less because of local stress concentration, the probability of failure becomes high, while the residual life in the region far away from the mid-span is high.

1. Introduction

Honeycomb structures have been received much attention in recent years because of their high strength/weight and stiffness/weight ratios, excellent heat resistance and favorable energy-absorbing capacity. In the past, the problem of the duration of the honeycomb cells and related sandwich panels under cyclic loading remains were solved by many efforts. Most sandwich specimens analyzed in the experimental studies were made of polymer foam cores with composite face sheets or with metal face sheets [1,2]. Relatively many efforts have been made on investigation of the fatigue strength of the sandwich structures with honeycomb cells. Demelio et al. [3] reported their experimental results for the fatigue behavior of the composite sandwich structures fastened with steel plates using blind fasteners. Belingardi et al. [4] investigated the four-point bending fatigue behavior of the undamaged and damaged sandwich beams with aluminum honeycomb cores and carbon-fiber composite face sheets. Belouettar et al. [5] reported their experimental studies of static and fatigue behaviors of honeycomb sandwich composites, made of aramide fibers and aluminum cores through four-point bending tests. In the previous studies of our group, bending fatigue behaviors of steel honeycomb sandwich panels under room temperature and high temperature were investigated in [6] and [7].
In this study, based on Goodman stress correction algorithm, four point bending fatigue behavior of brazed steel honeycomb sandwich panels at room temperature was numerical simulated. The residual life and local deformation were obtained and analyzed. Also the prediction results obtained from the numerical model are compared with the results obtained from the fatigue experiments.

2. Experiments

2.1 Sandwich beam specimens

The mild-steel sheets Q235 were used as the honeycomb core and face sheets materials in the present study. The core structure was folded and brazed together with the upper and lower face layers forming a hexagonal cell structure. The schematic detailed description and geometrical dimensions of the honeycomb sandwich specimens are shown in Fig.1. Parameters in Fig.1 are, l=120mm, b=24mm, t_f=0.50mm, l_c=3.20mm, h_c=8mm and t_c=0.15mm respectively. Both static and fatigue tests were carried out through a four-point bending testing fixture device schematically shown in Fig.1. Such device, designed and built expressly for room temperature conditions, was connected to an electronic servo-hydraulic universal testing machine INSTRON 8801. The bending fixture material was refractory steel and the mid-span was 50mm. Other parameters can be obtained from our previous work [7].

Fig.1 Schematic description of the honeycomb sandwich structure and the four-point bending fatigue tests

2.2 Experimental results

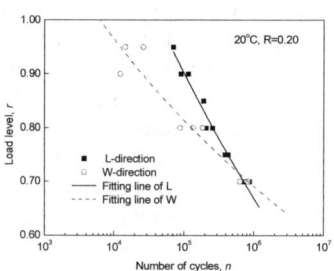

Fig.2 S-N curve at room temperature with load ration R=0.20[7]

The objectives of the investigated tests were to obtain basic knowledge on the fatigue behaviors of steel honeycomb sandwich beams as the standard stress-life, SN diagrams. The test load was sinusoidal with a frequency f=10 Hz. The loading ratio was set at 0.20 and a constant amplitude loading. The maximum applied loads F_{max} in the fatigue tests were selected as 90%, 80%, 70% and 60% of the ultimate applied loads in the static bending tests. The fatigue results were shown in Fig.2.

3. Simulation and discussion

3.1 numerical procedures

According to the experiment and finite element numerical simulation of honeycomb sandwich structure under static loads, it is known that the cycle load is less than the material yield load, so fatigue life was estimated using high cycle fatigue estimation method. The present study was based on the ANSYS and FE-safe[8] module which is widely used in the structure durability analysis of aerospace, machinery, materials, energy and chemical industry.

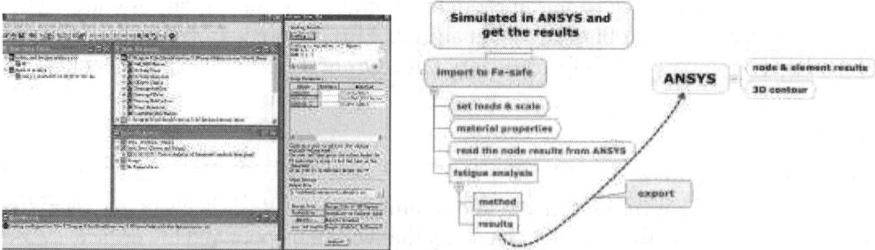

Fig.3 FE-Safe module and simulation procedures

As is shown in Fig.3, the operation interfaces include the user interface, the material database management system, fatigue analysis program and the signal processing program. FE-safe reads the calculated unit load or the actual working load of the elastic stress, and then superpose the results to produce working stress time history based on the actual load condition and alternating load form. At the same time, fatigue results can be calculated and converted to particular type loads of elastic-plastic stress if required. Finally, the results calculated by Fe-safe can be exported to ANSYS and visually obtained.

3.2 Prediction results

In fatigue simulation, the material was chosen as SAE_950C-Manten, elastic modulus and poisson's ratio were the same as the materials used in the previous experiments. The fatigue load frequency was set to 10Hz as shown in Fig.4. Definition design life was 10^6 cyc, thus the residual life and safety coefficient of honeycomb sandwich specimens can be obtained.

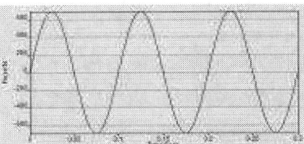

Fig.4 Setup of cyclic load with f=10Hz

Fig.5 Local deformation of core wall near the loading zone and stress concentration, local failure photo

The local stress and deformation initiated between the honeycomb core and face layer by the upper roller loads are simulated. In order to contrast, experimental photographs are also shown in Fig.5. And different color represents different stress and deformation. The results showed that the maximum deformation under fatigue loading appeared in the core wall, which indicated the region might be potential fatigue damage area. The phenomenon is the same as the actual damage failure photo. Because of the applied load, obvious stress concentration initiated. As shown in Fig.5, far away from the loading area, the stress in the core and the face layer are both smaller. In the finite element model, without considering the welding influence between the face and honeycomb core, while the face and core were coupled together to analyze. Thus, the face layers were supported and also restricted by the honeycomb cores, this led to the relatively large stress appeared in the internal area consist of the hexagon.

In the application of the elastic FEA fatigue analysis, strain life or stress life can be obtained. As for stress life, stress range and the average stress should be calculated in every stress cycling. If the maximum stress in cycling is S_{max}, and the smallest stress is S_{min}, then

Stress range: $S_{max} - S_{min}$

Stress amplitude: $(S_{max} - S_{min})/2$

The average stress: $(S_{max} - S_{min})/2$

Usually, the average stress has great influence on the fatigue life of specimens. Goodman and Gerber average stress correction algorithm could be used to avoid this. The stress amplitude and the average stress were used to calculate the equivalent stress amplitude when the average stress in the zero, S_{a0}. Then the parameters reflecting performance of fatigue durability could be calculated by this stress amplitude. The following was calculated based on the Goodman stress correction algorithm.

The residual life and safety coefficient of different parts of structure were listed in the form of contours in Fig.6. For the bottom face, in the design life level, all parts of the residual life value were lower, revealing that when cycle arrived up to 10^6 cyc, the structure tended to be damage. Meanwhile, along the direction of the hexagonal double wall thickness, the residual life is also less. In the real experiments, because of the double wall thickness area of brazing or adhesive is potential crack source area, numerical simulation results proved the defects of the real structure. The joint coupled with the core wall and face sheet has high safety coefficient (Fig. 6a), because of the merge operation in FEM, which reflected the most ideal case. It means that the whole structure of the safety coefficient is standard when welding strength of core and face was high. While in the actual processing, it was difficult to ensure that due to technology and other reasons. That is why the honeycomb sandwich structures tend to appear failure in the welding area in lower fatigue load or relatively low cycles. In figure 6(c) the residual life near support is given. It was clearly that because of the stress concentration, the core wall residual life was less, and the structure was more likely to damage. As shown in 6(d), in mid-span area far from support, due to the lower cycle loads, the residual life of whole area remained higher.

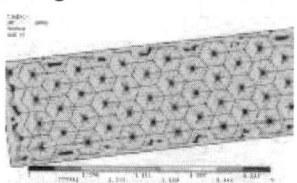 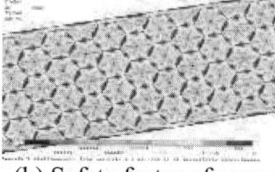

Fig.6(a) Remaining life of lower face layer,   (b) Safety factor of upper face layer

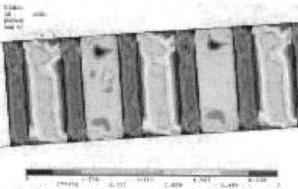 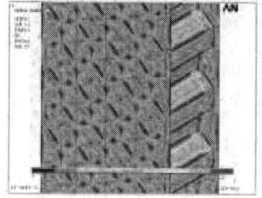

(c) Remaining life near the support, and (d) Remaining life of specimen near the mid-span

4. Conclusion

The fatigue behaviors of steel honeycomb sandwich beams were investigated both in experiments and numerical simulation in this work. The results show that local maximum deformation occured in the core wall under the load region, indicated that this region was the failure region of fatigue. And the residual life of the core wall in this region was less because of local stress concentration, while the residual life in the region far away from the mid-span was high. The simulation results were well proved by the experiments.

Acknowledgement

This work was financially supported by Research Fund for the Doctoral Program of Higher Education of China (Grant No. 20092304110003) and Fund of the central university basic scientific research of Harbin Engineering University (Grant No. HEUCFZ1128).

References

[1] Burman M, Zenkert D: Int. J. Fatigue. Vol. 19(1997), p.551-561.

[2] El Mahi A, Khawar Farooq M, Sahraoui S, et al: Mater. Des. Vol. 25(2004), p.199-208.

[3] Demelio G, Genovese K, Pappalettere C: Composites Part B: Eng. Vol. 32(2001), p. 299-308.

[4] Belingardi G, Martella P, Peroni L. Fatigue analysis of honeycomb-composite sandwich beams. Compo. Part A, Vol. 38(2007), p. 1183-1191.

[5] Belouettar S, Abbadi A, Azari Z, et al: Compos. Struct. Vol. 87(2009), p. 265-273.

[6] Zou G., Lu J, Cao Y, et al: Acta Metall Sinica, Vol.47(2011), p. 1181-1187.

[7] Lu J, Zou G.: PROC. of SPIE, 2009, 7522-141.

[8] Fe-safe user's manual, version 5.00, 2003.

The debond fracture of sandwich plate with corrugated core using cohesive zone element

Guang-ping Zou[a], Peng-fei Yang[b], Jie Lu[c], Yong-gui Li[d]

1Centre for Mechanics Testing, College of Aerospace and Civil Engineering, Harbin Engineering University, Harbin, 150001, P.R.China

[a]gpzou@hotmail.com, [b]ypfei029@yahoo.com.cn, [c]lujie@hrbeu.edu.cn, [d]yblliyonggui@163.com

Keywords: Corrugated core sandwich, interfacial debond, fracture toughness, cohesive zone model

Abstract. In this paper, flatwise tensile test (FWT) and modified double cantilever beam (DCB) experiment were conducted to investigated the debond fracture of sandwich plate with corrugated core. In the experiment, the crack always stays at the face/core interfacial. Tensile bond strength of face core can be given from the flatwise tensile test and we can get the mode I fracture toughness G_{IC} from DCB tests. It is found that the trends of curves change greatly at the beginning, with the propagation of crack, load against open displacement curves change smoothly. In order to simulate the face/core failure of sandwich plate with corrugated core, the cohesive element model is used. Tensile strength and strain energy release rate measured by the experiments presented in this paper are used in as parameters for simulation of the debond fracture. By comparing with the experiment results, the model can express the face/core failure of sandwich plate with corrugated core validly.

1 Introduction

Corrugated-core sandwich structure is comprised of a corrugation core connected to two thin face sheets. The corrugated-core sandwich panels, due to their exceptionally high flexural stiffness-to-weight ratio are commonly used in aviation, aerospace, marine, civil engineering and other applications, where weight is an important design issue. It's unique structure makes it a very light but nevertheless extremely rigid metal sandwich panel which allows for considerable weight savings particularly when used in big formats. It has also got convincing acoustic and dynamic properties [1].

As sandwich structures, a wide range of damages exist during manufacturing and application. The important failure is the debond damages between face and core. The study on the interface fracture of sandwich materials has been found in literature. A modified DCB specimen was proposed and applied to sandwich structures with aluminum face sheet and PVC foam core by Prasad and Carlsson [2,3]. Ural et al[4] carried out two types of experiments to evaluate the adhesive bond in honeycomb sandwich panels and gave fracture toughness values for materials with different face sheet thicknesses and core.

In this paper, Flat wise tensile (FWT) test and double cantilever beam (DCB) test were conducted at room temperature. The cohesive zone model were used to simulate the debond failure of metal corrugated core sandwich structures, and the result is compared with the experimental.

2 Experiment

2.1 Material specimen

The materials of face sheet and core are all Al. The corrugated core is connected to the face sheets by glue. In order to decrease stress concentration, the corrugated core sandwich plate was incise periodic units using wire cutting machine. Geometry dimensions and mechanical parameters of specimens were showed in table 1.

Table 1 Geometry dimensions and mechanical parameters of specimens

specimen	Face thickness/mm	Core height /mm	thickness t/mm	Young's module/GPa	Posson's ration
	1.0	8.00	0.50	75	0.33

2.2 Experimental method

2.2.1 Flatwise tension tests

To obtain the strength of the face-to-core bond, the flatwise tensile (FWT) tests are used. The FWT test is characterized by the application of tensile loading to two opposing faces of a sandwich panel coupon, normal to the plane of the sandwich until ultimate failure occurs. The force is transmitted to the sandwich coupon through thick loading blocks bonded to the face sheets by J-39 adhesive (Fig. 1). The FWT tests were conducted on INSTRON4505 testing machine. The loading was conducted in displacement control with a displacement rate of 0.5 mm/min. Loads and displacements are recorded throughout the test. After installing the specimens into the testing machine, the force was ramped up until failure of the interface occurred. The load-displacement curve is shown in Fig. 2. The dimensions of the FWT test specimens are 60 mm×60 mm×8 mm with four periodic units. Three specimens were tested using this procedure. The ultimate flatwise bond tensile strength was computed by dividing the ultimate failure load by the effective bonded area 720mm^2. Average failure load was found to be 5180 N and average nominal strength was calculated as about 7.2 MPa.

2.2.2 Modified double cantilever beam (DCB) test

A direct approach for determining fracture toughness is the modified DCB test as is shown in Fig. 3. The specimen is loaded symmetrically by tensile forces at both top and bottom faces. The dimensions of the metal corrugated core sandwich specimens are 120 mm×30 mm×8 mm with a 30 mm pre-crack. The specimens include two integrated periodic units. The face sheets were bonded to the hinges using J-39 adhesive. The white paint was used on the surface of specimen over the region where the crack is expected to expand. The fracture tests were carried out by using the INSTRON4505 machine. The loading is conducted in displacement control with a displacement rate of 1 mm/min. Loads and displacement are recorded throughout the test. The CCD photograph technology is used to track the crack tip by recording the picture of crack. The test process was stopped until the crack length reached the scheduled value(~25 mm). The load-displacement curve is shown in Fig. 4.

2.3 Calculation and separation of strain energy release rate

The critical strain energy release rate, G_C, during the crack propagation may be expressed as [5] $G_C = \dfrac{\int_0^\delta p d\delta}{B \Delta a}$, where P is the external load; δ is the displacement; B is the width of specimen and $\triangle a$ is the incremental crack length during the process of test. For the modified DCB test, the energy required for crack growth $\int_0^\delta p d\delta$, may be calculated according to the area enclosed by the loading-unloading path.

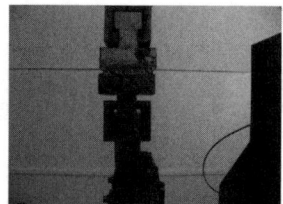

Fig. 1 Photograph of FWT test setup

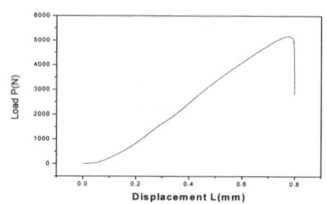

Fig. 2 Load- displacement curves for FWT test

 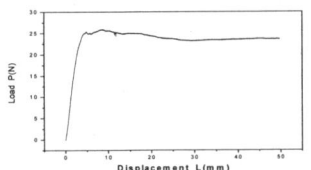

Fig. 3 Photograph of modified DCB Experiment   Fig.4 Load-open displacement curves for modified DCB test

The incremental crack length of specimen, $\triangle a$, may be measured by using CCD photograph. The average of value is about 1200 N/m. Mode I is dominant in the modified DCB test, G_C measured in the experiment can be regard as GI_C.

3. Finite Element Simulation of the modified DCB

The computational model adopted in this paper is based on the cohesive zone model. The cohesive zone model is characterized by a traction-separation law, which is a function of the fracture energy and strength. The interfacial debonding failure of modified DCB specimens were analyzed using the commercial finite element software ABAQUS. The facesheets and the corrugated core were modeled as linear elastic materials. The FE model mesh generation of both face sheet and corrugated core are the same. The facesheets and the corrugated core were both modeled with eight-node solid elements. The zero-thickness cohesive zone element is used to model the interfacial bond between face and core. In the pre-crack region, zero-thickness cohesive zone elements were deleted. Cohesive law is determined experimentally by shear test of structure glue and the result shows that the relation of traction-relative displacement can be described by an approximate power expression.

The size of FE model, clamped and boundary conditions are the same as the test specimen. The total models contain 6990 elements. Only the half of the specimen was modeled using symmetry. Interface elements with a large elastic stiffness should be used with care due to oscillation in computed traction profiles. The two important parameters are interfacial fracture energy and peel strength which can be determined from DCB and FWT tests.

4 Results and Discussion

In the FWT or DCB test, it is found that the propagation of interfacial debonding always stays between the interface of face and core. Because the width of crack propagates is small, actual crack growth length on each side of the specimen indicates that the crack propagates synchronously. In the FWT test, the curve rises steady until the ultimate load then decrease suddenly. The fracture process of corrugated core sandwich structures in the modified DCB test can be seen in Fig. 5. At the beginning, the load increases approximately linearly, then interface debond initiation occurs at the pre-crack front and the curve becomes zigzag, with the propagation of pre-crack front, the load decreases stable.

 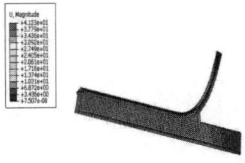

Figure 5. Finite element meshes for DCB simulation   Figure 6 Finite element contour of interface crack growth

In the present simulation, the crack always propagates along the interface between face sheet and core as shown in Fig. 6, we compared the force between simulation and experiment, the simulation agrees well with the observation of modified DCB experiment. The difference between simulation and experiment may because of the processing defects as ununiformity of adhesive of interface in the propagation of crack and the deformation may included other energy which may affect the energy change.

5. Conclusions

In the present paper, the fracture toughness of adhesive bond is determined by using FWT and modified DCB test. The crack curve curves change greatly at the beginning, with the propagation of crack, load against open displacement curves change smoothly. The propagation of interfacial debond always stays on the face/core interface during the FWT or modified DCB tests. Compared with experimental results, it is evident that the model is able to simulate the fracture behavior of modified DCB test validly.

Acknowledgements

This work was financially supported by the Central University Basic Scientific Research HEUCFZ 1128.

References

[1] Wan-Shu Chang, Edward Ventsel. Bending behavior of corrugated-core sandwich plates. Analysis of fracture specimens. Compos Struct 70 (2005) 81–89.

[2] Prasad S, Carlsson LA. Debonding and crack kinking in foam core sandwich beams-I. Analysis of fracturespecimens. Eng Fract Mech 1994;47(6):825–41.

[3] Prasad S, Carlsson LA. Debonding and crack kinking in foam core sandwich beams-II. Experimental investigation. Eng Fract Mech 1994;47(6):825–41.

[4] A. Ural, A.T. Zehnder and A.R. Ingraffea: Eng. Fract.Mech., 2003, 70, 93.

[5] T.L. Anderson: Fracture Mechanics: Fundamentals and Applications, 2nd ed, CRC Press, New York, 1995,4

Study on a numerical calculation method of fatigue behavior in Glare panel

Ma Yu e[1,a], Yun Shuang[1,b]

[1]Box 118, School of Aeronautics, Northwestern Polytechnical University, Xi'An City, P.R.China

[a]ma.yu.e@nwpu.edu.cn, [b]xplane@126.com

Keywords: GLARE, crack growth, delamination, fibre bridging

Abstract. GLARE (Aluminium with glass fibres) has become known since it was successfully and widely used in aircraft structure for its excellent fatigue and damage tolerance behavior. In this study, fatigue testing samples with 3/2 Glare were designed according to the standard. The finite element method was used to build the numerical model of the sample; the fibre bridging was modeled and the bridging stress was calculated. The delamination growth behavior can be predicted based on the energy release rate. This numerical method considered both damage modes (fatigue and delamination) at the same time.

Introduction

Hybrid FMLs take advantages of metal and fiber-reinforced composites, providing superior mechanical properties to the conventional lamina consisting only of fiber-reinforced lamina or monolithic aluminum alloys. Since it was patented at Delft University of Technology, as a family of hybrid materials that consist of bonded thin metal sheets and fibres embedded in epoxy, six grades of Glare have been developed since it was successfully applied in aircraft structure. Fatigue crack propagation in the fibre metal laminate Glare consists of crack growth in the aluminium layers and delamination at the aluminium/prepreg interfaces in the wake of the aluminium crack. The intact fibre layers in the wake of the crack act as a second load path over the crack, which the load around the crack tip in the aluminium layers is reduced significantly. As a result, the crack growth rates in the aluminium layers are significantly lower compared to the crack growth rate in monolithic aluminium under the same applied loading conditions, which was called "fibre bridging" in [1]. R.C. Alderliesten [1] studied fatigue and damage tolerance of Glare and gave the method of fatigue crack growth modeling in Glare [2]; Toi R. [3] investigated and gave an empirical crack growth model for Fiber/Metal laminates; Takamatsu[4] studied the evaluation of fatigue crack growth behavior of Glare3; Vries T.J.[5] carried out research about Glare panel with blunt and sharp notch.

Specimens and damage criteria

The specimen in the study is GLARE 4A-3/2 laminates (2024-T3/0°/90°/0°/ 2024-T3/0°/90°/0°/2024-T3), with 580mm in length and 140mm in width, and a 10 mm centre pre-crack is in the model, as shown in Fig 1. The total thickness of the laminates is 1.662mm, with 0.254mm aluminium per layer and 0.15mm glass fibre per layer. The tensile loading is imposed by pre-scribing a remote normal displacement u∞ at the top and bottom edges of the laminate. As studied in literatures on static and fatigue failure behavior of laminates with pre-crack in the centre, under the applied remote tensile load, the pre-crack in the centre will start to propagate as a mode I crack through the aluminium layers and the delamination damage will occur along the interface between the metal and the fibre layers.

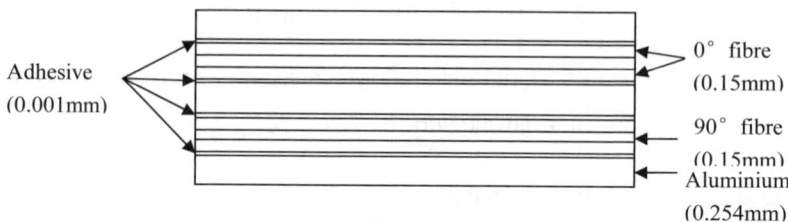

Fig 1 Through –thickness view of the laminate

Aluminium layer was assumed to be isotropic elastic-plastic with isotropic hardening. And the maximal principal stress criterion was used to predict the crack initiation while displacement was used to control the crack propagation with a linear softening mode.

The adhesive bond between the layers was modeled using the cohesive element functionality available in ABAQUS 6.10. A triangular traction-separation cohesive law with linear softening was used to achieve the constitutive mode of the adhesive.

Quadratic nominal strain criterion was used to prediction the initiation of delamination and power law criterion for damage evolution. The XFEM method was used to predict the crack propagation and the cohesive element was introduced to predict the delamination initiation and propagation at the interface. The Hashin damage initiation criteria were used to predict damage for fibre layers.

Finite element model

To predict the crack propagation in aluminium layer and delamination between the aluminium and composite, ABAQUS 6.10 was used. To reduce the degrees of freedom of nodes, only 1/2 of the FML was modeled. The symmetrical boundary condition was applied to the FML along the width. The tensile loading was imposed by pre-scribing a remote normal displacement u∞ at the top and bottom edges of the laminate. Aluminium layers were built with C3D8R elements, composite layers were built with SC8R elements, while adhesive layers were built with COH3D8 elements.

The mesh model in ABAQUS is shown in Fig 2. XFEM crack detail figure is shown in Fig 3. The material properties used in the model were listed in Reference [6].

 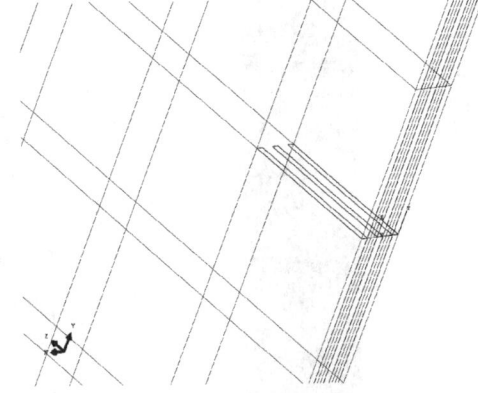

Fig 2 Mesh and load   Fig 3 XFEM crack in detail

Results

Interlayer aluminium stress is shown in Fig 4 and the crack open in this layer is shown in detail in Fig 5. As mentioned before, delamination initial can be seen, and the adhesive layer closed to the mid-layer aluminium is shown in Fig 6, total damage area are shown in Fig 7. Fibre layers are still intact, which indicates the fibre bridging mechanism.

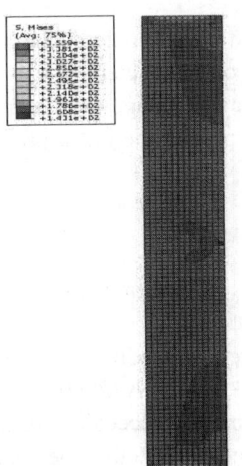 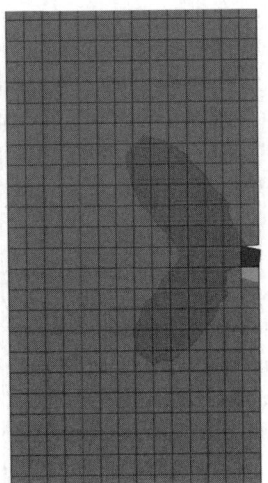

Fig 4 Mid-layer aluminium mises stress   Fig 5 Mid-layer crack open

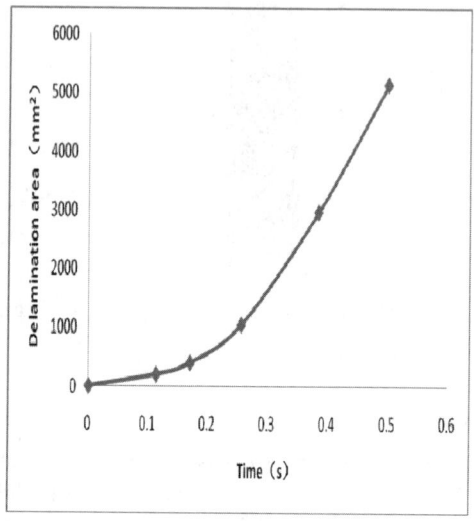

Fig 6 Delamination damage area for the third cohesive

Fig 7 Total delamination damage area as a function of time

Conclusions

A model for predicting crack initiation and delamination behavior was formulated. The maximal principal stress criterion was used to predict the crack initiation, and cohesive element was used to predict the delamination behavior. The result fit well with the laminates damage mechanism. This method can be used to model the response of GLARE panel and to predict the crack propagation and delamination behavior.

References

[1] R.C. Alderliesten, J.J. Homan, *Fatigue and damage tolerance issues of Glare in aircraft structures,* International Journal of Fatigue 28 (2006) 1116–1123.
[2] Alderliesten, R.C., *Fatigue Crack Growth Modelling in Glare,* Proceedings of the USAF Aircraft Structural Integrity Program Conference, San Antonio, Texas, USA (2000).
[3] Toi, R., An Empirical Crack Growth Model for Fiber/Metal Laminates, Proceedings of the 18th symposium of the international Committee on Aeronautical Fatigue, Melbourne, Australia, 899-909(1995).
[4] Takamatsu, T. Shimokawa, T. Matsumura, T. Miyoshi, Y. Tanabe, Y. *Evaluation of fatigue crack growth behaviour of GLARE3 fiber/metal laminates using compliance method,* Engineering Fracture Mechanics 70(2003) 2603-2616.
[5] Vries, T.J. de, *Blunt and sharp notch behavior of Glare laminates,* PhD Thesis, Delft University of Technology, 2001.
[6] Irenensz Lzpczyk, Juan A.Hurtado, *Progressive damage modeling in fiber-reinforced materials,* Composites: Part A 38(2007) 2333-2341

Investigation of residual stress distribution in clad 2024 T3 after laser peening

WANG ZhenHai[1,a], Ma Yu e[2,b]

[1] School of Science, Northwestern Polytechnical University, Xi'An City, P.R.China
[2] Box 118, School of Aeronautics, Northwestern Polytechnical University, Xi'An City, P.R.China

[a]zhwang@nwpu.edu.cn, [b]ma.yu.e@nwpu.edu.cn

Keywords: residual stress, laser peening, K_{res}

Abstract. After laser peening, the residual stress distribution was measured in 2mm thick clad 2024 T3 samples containing surface scribes by the hole drilling method. The profile and tendency of residual stress along sample depth parallel to and perpendicular to the line of the laser spot line was studied. The residual stress perpendicular to the line of the laser spot line is more tensile than the residual stress parallel to it by between 30MPa – 80MPa. The maximum tensile residual stress occurs at the spot surface; while the maximum compressive stress varied between 20-170MPa, these maxima taking place at the depths of 100-400um. The 3D finite element model was built by ABAQUS 6.10. The subroutine program was made and residual stress profiles were input to the numerical model. The effect of residual stress on fatigue parameter was investigated. Stress intensity caused by residual stress--- K_{res} was calculated and studied.

Introduction

Laser shock peening (LSP) is a new surface treatment technique for strengthening metals and improving fatigue lives. It use the laser induced shock wave on the surface of metal, then appropriate residual stress distribution corresponds to specific strengthening effects can be gotten, which can be achieved by the optimized laser peening path. Laser shock peening can produce a deeper compressive residual stress filed near the surface than shock peening. This advantage makes this technique desirable for the manipulation of crack growth rates. The distribution and the effect of residual stress after laser shock peening needs to be studied further.

At present, lots of researchers focus on the effect of residual stress on fatigue life. A. Chahardehi [1] carried out an experimental program to establish this effect in which steel specimens were partially laser peened and subsequently subjected to cyclic loading to grow fatigue cracks and found that compressive stress near surface and tensile stresses in the mid-thickness of the specimens; the crack growth rates more affected by the tensile core than by the compressive surface stresses. Hemanth K. [2] simulated the residual stress induced buy a laser peening process through inverse optimization of material models. Robert A. Brockman [3] predicted the size and distribution of residual stresses from laser shock peening, and give an accurate characterization of the surface treatment and its effect on fatigue behavior. X.C. Zhang [4] studied the improvement of fatigue life of Ti-6Al-4V alloy by laser shock peening. This paper focuses on the character of residual stress in 2.0mm thick clad 2024 T351 panels.

Effects of laser peening on residual stress fields

The samples used for the residual stress peening power trials were approximately 80 mm X 80 mm and were of clad 2024 T3. Laser treatment was performed by Metal Improvements. The laser treatment was conducted with a square spot 5 mm X 5 mm in an overlapping row 80 mm long. The amount of overlap was 10% of the spot diameter. The laser power levels used was 3GW/cm^2. Residual stresses produced by the peening process were investigated using the hole drilling technique. Using this approach it was possible to determine changes in residual stress field with depth beneath the sample surface up to a depth of 1.0 mm.

Residual stress profiles produced were typified by that shown in Figure 1. This shows changes in stress with depth for stress σ_3, perpendicular to the line of the laser spots; and stress σ_1 parallel to the laser spot line. σ_3 was more tensile than σ_1 by about 70MPa. Surface stresses were invariably tensile, by between 20-150MPa in both stress directions; in the majority of samples in both directions moved steadily with increasing depth towards compression, passing through zero into compression at between 100 μm -300 μm depth; the maximum compressive stresses was 120 MPa, occurring at depth of 400 μm.

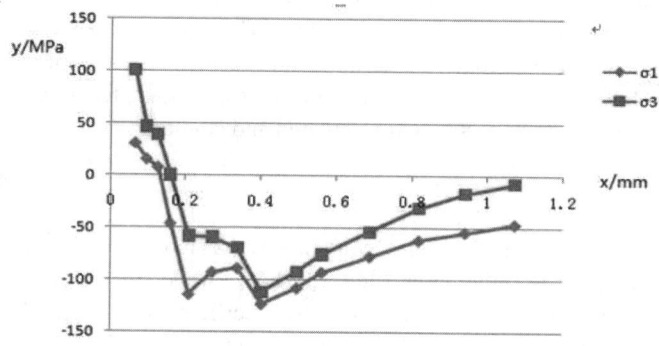

Fig 1 Residual stress profiles after laser peened

K_{res} calculation by finite element method and VCCT

The finite element model was built by ABAQUS 6.10 [5] shown in Fig.2. C3N8 was chosen in this analysis, 24 layer elements were set up along depth 1.0 mm, 8 layer elements were set up along depth from 1.0 mm to 2.0 mm. Total elements are 200438. The subroutine SIGINI was used to make Fortran program to input residual stress profile (shown in Fig.1) to the finite element models. When crack arrives 0.3 mm along depth, profiles of residual stress were shown in Fig.3. It was shown that the maximum of residual stress changes from 100MPa down to 15 MPa. The tendency of residual stress in both directions changed as well.

Finite element analysis was used to calculate stress intensity factor K_{res} from residual stress by using ABAQUS. The virtual crack closure technique (VCCT) method [6] was used for calculating strain energy release rate for unit sample thickness with the formulation:

$$G = \frac{F_j u_i}{2t\Delta c} \qquad (1)$$

where F_j is the reaction force on j node; u_i is the total displacement from i node; t is thickness of samples and Δc is element size.

For plane stress, the relation between the strain energy release rate and stress intensity factor (SIF) is as follows:

$$G = \frac{K^2}{E} \quad \text{(Plane stress)} \qquad (2)$$

If residual stresses were input to this model, K_{res} can be derived from equations (1) and (2). The K_{res} from σ_3 and σ_1 was shown in Fig.4. $K_{res} - \sigma_3$ changes faster than $K_{res} - \sigma_1$.

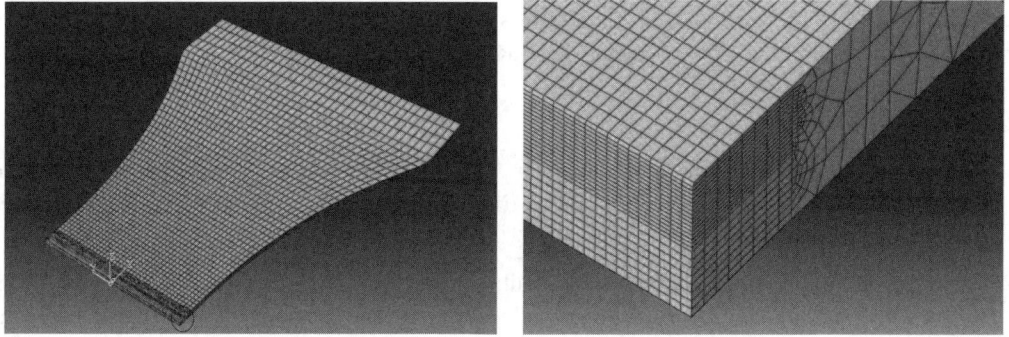

Fig.2 Elements of 2024 T3 panles (a) half panel; (b) elements along depth

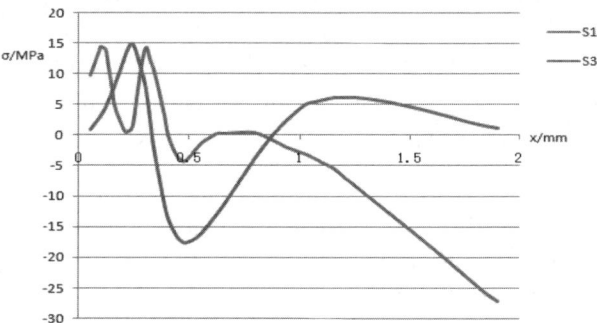

Fig.3 Distributions of residual stress when crack length 0.3mm along depth

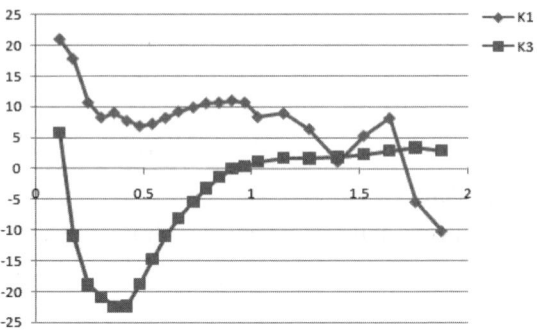

Fig.4 K_{res} profiles along depth

Summary

Residual stresses in both directions were measured in 2024-T3 after laser peening. Stress intensity factors from residual stresses were calculated. The conclusions can be drawn: the redistribution of residual stresses changed with crack through depth and it must be taken into account during studying on effects of residual stress; and stress intensity factors from residual stresses in both directions were different even the tendency residual stresses are similar.

References

[1] A Chahardehi, F.P.Brennan, and A.Steuwer, The effect of residual stresses arising from laser shock peening on fatigue crack growth, Engineering Fracture Mechanics, 77, 2033-2039 (2010).

[2] Hemanth K.Amarchinta, Ramana V.Grandhi, Allan H.Clauer, Kristina Langer,and David S.Stargel, Simulation of residual stress induced by a laser peening process through inverse optimization of material models, Journal of Materials Processing Technology, 210, 1997-2006 (2010).

[3] Robert A.Brockman, William R.Braisted, Steven E.Olson, Richard D.Tenaglia, Allan H.Clauer, Kristina Langer,and Michael J.Shepard, *Prediction and characterization of residual stresses from laser shock peening,* International Journal of Fatigue, 36, 96-108 (2012).

[4] X.C.Zhang, Y.K.Zhang, J.Z.Lu, F.Z.Xuan, Z.D.Wang and S.T.Tu: Improvement of fatigue life of Ti-6Al-4V alloy by laser shock peening, Materials Science and Engineering A 527, 3411-3415(2010).

[5] ABAQUS 6.10 Manual Report.

[6] Ronald Krueger. The virtual crack closure technique (VCCT): history, approach and applications. NASA/CR-2002-211628.

Investigation on the corrosion effect of friction stir welded AA2024 T3 aluminum alloy joints

MA Yu E [1,a], ZHAO ZhenQiang[1,b]

[1]School of Aeronautics, Northwestern Polytechnical University, Xi'An, China

[a]ma.yu.e@nwpu.edu.cn [b]xplane@126.com

Keywords: Friction stir weld; Microstructure of nugget; Corrosion behavior;

Abstract. Before friction stir welded integral panels are used in main aircraft structure, the corrosion behavior of welded joint need to be studied in detail. 2024 T3 samples were designed and welded by friction stir welding; the microstructure crossing the weld zone was observed by scanning electron microscopy (SEM), the feature of different zones (base material, thermo-mechanical affected zone, nugget) was seen; the corrosion testing in NaCl smoking box was carried out, and microstructure was observed after corrosion, localized corrosion predominantly occurs in the thermo-mechanical affected zone.

Introduction

Friction Stir Welding (FSW) is now considered mature for structure applications after patented by TWI in UK. It is shown that even for simple applications like longitudinal fuselage joints weight savings of up to 15% and cost savings of 20% can be achieved. Especially 2024 T3 used widely in aircraft structure can be successfully and easily welded by FSW technology.

Lots of researchers already studied the mechanical properties, fatigue properties, residual stress analysis, the numerical modeling of friction stir welded 2024 joints and so on, but there are a few studies on the corrosion behavior of 2024 FSWed joints. Pao [1] studied the fatigue crack growth rates in friction stir welds in air are slightly higher than those in base metal. Ju Kang [2] studied the surface corrosion behavior of an AA2024-T3 aluminum alloy sheet after friction stir welding. Kalita [3] studied the friction stir welds of some aluminum alloys exhibit relatively poor corrosion resistance. In this study, 2024-T3 plates were joined by friction stir welding. Zhou [4] investigated microstructure and stress corrosion cracking behavior of friction stir welded aluminum 5A06. Dog bone samples were designed and cut from welded plates according to the standard. The microstructures of welded nugget, thermal affected zone and parent material were examined in the scanning electron microscope. Many regular black stringers were observed in the top surface microstructure of etched samples. Pitting corrosions were found. Microstructures of these three zones were compared.

Sample Preparation and Experimental Techniques

2.0 mm thick 2024-T3 was chosen in this study. The samples were designed (shown in Fig.1) to do fatigue testing. The surface of weld seam was polished by use of a mechano polishing method. Etching for metallographic observation was carried out by using Keller's reagent (1ml HF, 2 ml HCl, 2 ml HNO3，50 ml H_2O). The microstructure in the polished surface of the weld seam was observed by SEM.

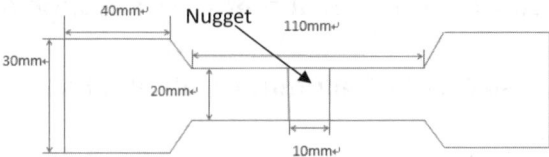

Fig.1 Sample size of 2024-T3

Microstructure of friction stir welded 2024 T3

The microstructure on the surface layer of the friction stir welded joint, as in the cross section of the joints, can also be divided into distinct regions: base material, thermo-mechanical affected zone and nugget (as shown in Fig.2)

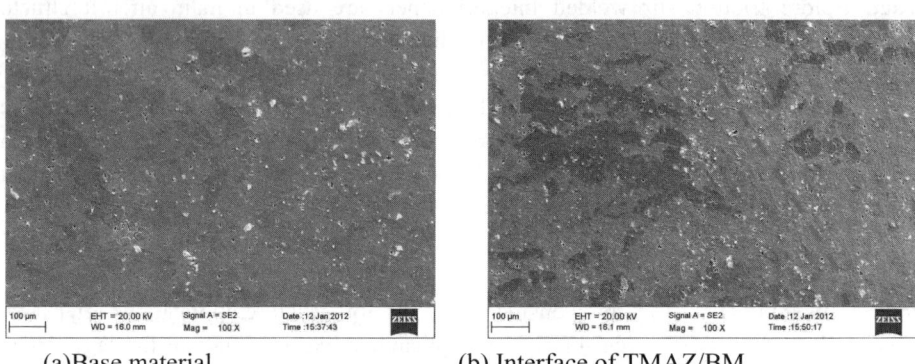

(a) Base material (b) Interface of TMAZ/BM

(c) Nugget

Fig.2 Microstructure on the top of 2024 T3 FSW joints

In Fig.2, the prominent features are obvious difference of the size and shape of the grains among the typical regions. The features of the microstructure are correlated with the particle distribution of the strengthening phase in all zones, while particles are more in the nugget. The grains in the interface of TMAZ/Base material adjacent to the tool shoulder are deformed and elongated as a result of heat and the tool shoulder's friction action, shown in Fig.2 (b). In the nugget (Fig.2 c)--it is seen that the fine, irregular grants have a difference compared with that the in the cross section which is attributed to the stirring action and incomplete dynamic recrystallization. These stringers are dissimilar to the "onion rings" feature in the cross section of the nugget zone. It is noted that they are high density zones of the secondary phase particles.

Microstructure of 2024 T3 after corrosion

After the samples has been input in the smoking box with NaCl for 192 hours, the samples were taken out of box and were observed through the weld regions in SEM, shown in Fig.3.

(a)Base material

(b) Interface of thermal affected zone and base material

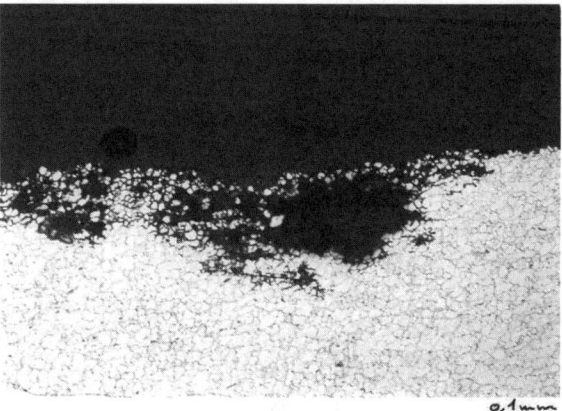
(c) Welded nugget
Fig.3 Microstructure of FSW joints after corrosion

In Fig.3, the area of corrosion in the interface of TMAZ and base material is bigger than base material and nugget. The localized corrosion predominantly occurs nearby the interface of TMAZ and base material.

Summary

In this study, the main conclusions are as follows:
(1) Many regular stringers are seen in the top surface microstructure of the FSW joints after being etched by Killer's reagent.
(2) After being input in the smoking box, the corrosion was observed for all regions of the weld joint. The interface of TMAZ and base material susceptible to corrosion.

References

[1] P.S. Pao, S.J. Gill, C.R. Feng, and K.K. Sankaran, *Corrosion-fatigue crack growth in friction stir welded Al 7075*.Scripta Materialia 45 (2001) 605-612.
[2] Ju Kang, Rui-dong Fu, Guo-hong Luan, Chun-lin Dong, Miao He, *In-situ investigation on the pitting corrosion behavior of friction stir welded joint of AA2024-T3 aluminium alloy*. Corrosion Science 52(2010)620-626.
[3] Samar Jyoti Kalita, *Microstructure and corrosion properties of dilde laser melted friction stir weld of aluminum alloy 2024 T351*. Applide Surface Science 257(2011)3985-3997.
[4] Zhou Yongjie, Sun Dechao, Xing Li, Liu Geping, *Microstructure and Stress Corrosion Cracking Behavior of Friction Stir Welded Aluminum Alloy 5A06*. Journal of International Metal Working 25 (2004) 45-49.

Distortion Induced Fatigue Analysis of Steel Bridges

Chunsheng Wang[1,a], Shenglong Yan[1,b], Jun Cheng[1,c]

[1] Engineering Research Center for Large Highway Structure Safety of Ministry of Education, Chang'an University, Xi'an, Shaanxi Province, China, 710064

[a] wcs2000wcs@163.com, [b] bjysl1987@163.com, [c] cj.jiqimao@yahoo.com.cn

Keywords: curved steel bridges, web gaps, out-of-plane distortional fatigue, finite element model, stress analysis, parameters

Abstract. Induced forces in secondary steel bridge members such as diaphragms and cross-bracing can cause out-of-plane distortion in webs that may lead to fatigue cracking. Such cracking is most likely to occur if the distortion must be accommodated in a short length of the web, such as in the gap between the end of transverse stiffeners and girder flanges, that's because the web gap is subject to double curvature. In this paper, the numerical analysis models of the curved continue steel bridges were established using ANSYS software to calculate the fatigue stress at the web gaps. The numerical analysis results show that both the bottom web gap at middle span area and the top web gap at supported were poor fatigue details, and web thickness and web gap depth give the great affection on the out-of-plane distortion fatigue stress at web gaps of curved steel bridges.

Introduction

In the steel bridge design, lateral connections, including of cross-frames and floor beams were placed between adjacent girders, in order to avoid failures of fatigue details at vertical stiffener and tension flange welds, stiffeners usually didn't weld together with tension flanges, which formed a vertical web gap of centimeters high.

Because of constraints of concrete bridge deck and the main girder flanges, under the action of the vertical wheel loads, the different vertical deflection of the two adjacent longitudinal girders causes notable out-of-plane distortion at the web gaps, thus causes high out-of-plane bending stresses, so fatigue cracks will initiate at the end of welds or weld toes of web gaps in the welded bridges[1-4].

According to the new research results of American scholars, approximately 90% of fatigue cracking is caused by out-of-plane stresses[5].

Case Study Bridge

This paper processes the numerical simulation of a three-span continuous curved steel girder bridge. The span lengths are approximately 25m+35m+25m. The superstructure was composite girder, consisting of five welded plate girders spaced at 2.7m .The concrete deck is 0.22m thick, bolted with steel plate girders. The radius of curvature of the bridge centerline is 80m. Cross-frames, consisting of angles of 150mm width and 12mm thick, forming the shape of an X between adjacent girders, were placed in the radial direction every 5m in the longitudinal direction of the bridge, while groove shape floor beams are placed in the position of abutment. In order to avoid the poor fatigue details, there are 50mm length web gaps in the top flange-to-web connection in the middle supported and bottom flange-to-web connection in the middle span area of the 2nd span, other stiffeners are directly welded to flanges. Fig.1 shows the plane and cross section of this continuous bridge.

Finite Element Model

To analyze the stress and displacement at web gaps, the finite element model (FEM) of the entire real bridge superstructure was created using ANSYS software. The concrete deck was simulated by solid45 element, and steel girders and diaphragms were simulated by shell63 element and beam188 element respectively. The nodes between the bottom surface of concrete deck and the top girder flanges were coupled in X, Y, Z degrees of freedom.

In order to maintain the maximum response of out-plane-distortion in top web gaps, apply one AASHTO fatigue truck to the bridge structure in five load cases. The last axis of AASHTO fatigue truck was located at cross-frame I longitudinal, and changed the horizontal position of truck in turn. Because lack of space, this paper analyzes load case 1 as an example.

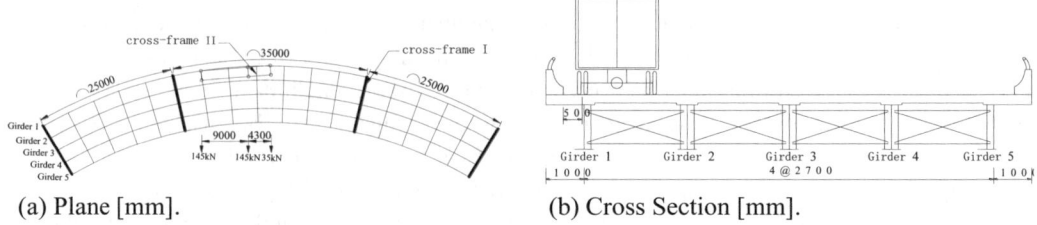

(a) Plane [mm]. (b) Cross Section [mm].

Fig.1 Bridge Plane and Cross Section at the Load Case 1.

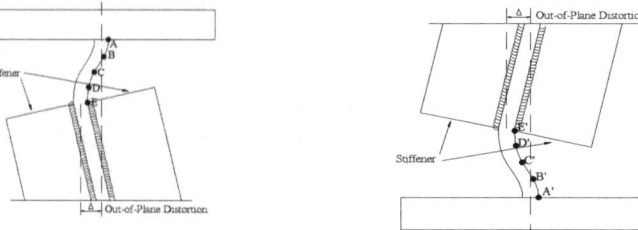

(a) Top Web Gap Analysis Points. (b) Bottom Web Gap Analysis Points.

Fig.2 Analysis Points in the Top and Bottom Web Gap on Girder.

Analysis Results

There were five analysis points in web gaps at cross-frame I, so did cross-frame II, and the distance between adjacent points was a constant. The stresses and displacements of analysis points in web gaps at cross-frame I of the girder 2 and cross-frame II of girder 1 are provided in Table 1 and 2.

Table 1 Analysis Results in Web Gap at Cross-frame I of Girder 2 at the Load Case 1

Points	Web Gap Location	Δ (Relative Deformation) [mm]	Vertical Bending Stress [MPa]
A	Top gap	0.000	222.3
B	Top gap	-0.007	130.0
C	Top gap	-0.028	58.3
D	Top gap	-0.054	-13.8
E	Top gap	-0.080	-105.4

Table 2 Analysis Results in Web Gap at Cross-frame II of Girder 1 at the Load Case 1

Points	Web Gap Location	Δ (Relative Deformation) [mm]	Vertical Bending Stress[MPa]
A'	Bottom gap	0.000	191.7
B'	Bottom gap	-0.031	107.7
C'	Bottom gap	-0.076	42.3
D'	Bottom gap	-0.125	-23.4
E'	Bottom gap	-0.173	-105.2

Notes: Δ - Horizontal displacements of points A,B,C,D,E relative to point A, or points A',B',C',D',E 'relative to point A'.

Table1 shows that each point had occurred horizontal displacement relative to point A. An important finding was that as the distance between concerned points and point A increases, the relative displacement increases. The bending tensile stress of point A was the largest. The largest tensile stress was 222.3MPa at point A. With the increasing distance from point A, the tensile stress gradually decreased, and changed into compressive stress at point D. The compressive stress at point E was the largest. The results show that the point A and point E were suffering reverse moment, and the loading caused the web gaps double curvature. Also, table2 shows that stresses of points in web gap at cross-frame II changed in a similar way to cross-frame I, and the maximum tensional stress was 191.7MPa.

Prametric Analysis

The effects of web gap depth, web thickness, cross-frame type, and curvature radius for out-of-plane stress at web gap were investigated by performing a suite of computer simulations. The web gap depth G evaluated were 10mm, 25mm, 35mm, 50mm, 60mm, 75mm, 90mm, 100mm and 120mm, respectively. Stresses in web gap of girder 2 at the load case 1 were presented in Fig.3. As illustrated in Fig.3, the stresses of point A, B, C were always tension stresses, with G increasing the stresses increased at first and then reduced, the vertical bending stresses were the largest when G was 35mm; the stresses of point D, E, changed from tension stresses to compressive stresses, and compressive stresses increased due to increased G.

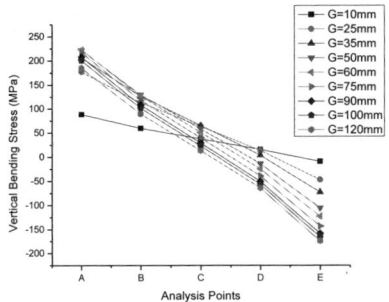

Fig.3 Vertical Bending Stress with Different Web Gap Depth at Cross-frame I of Girder 2 at the Load Case 1

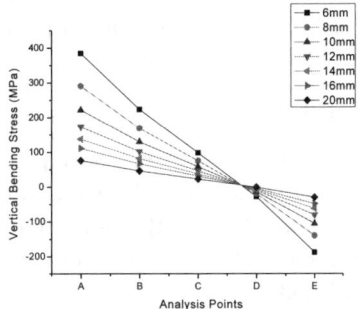

Fig.4 Vertical Bending Stress with Different Web Thickness at Cross-frame I of Girder 2 at the Load Case 1

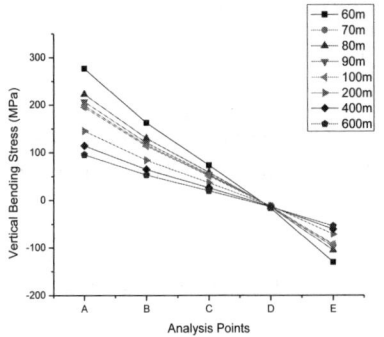

Fig.5 Vertical Bending Stress with Different Curvature Radius at Cross-frame I of Girder 2 at the Load Case 1

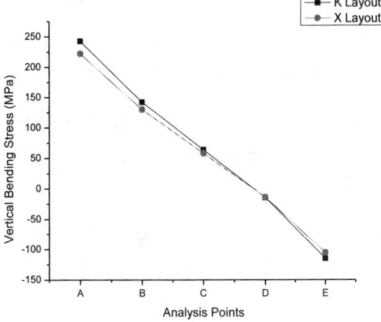

Fig.6 Vertical Bending Stress with X layout Cross-frame and K layout Cross-frame at Cross-frame I of Girder 2 at the Load Case 1

The web thickness evaluated were 6mm, 8mm 10mm, 12mm, 14mm, 16mm, and 20mm, respectively. From Fig.4, we concluded that the vertical bending stresses reduced at web gap due to the increased web thickness. Although this will be beneficial to reduce fatigue damage, the web thickness can't increase unlimitedly. A suitable web thickness should be chosen considering economy and other factors.

In addition, as shown in Fig.5 and 6, increased curvature radius resulted in reduced out-of-plane bending stress, and out-of-plane bending stresses in X layout were a little higher than that in layouts with K cross-frames.

Summary

This paper established a numerical analysis model of a real 3-span continue steel bridge using ANSYS software to calculate the fatigue stress at the web gaps. The following conclusions were drawn by analyzing the results of the numerical simulations: under the action of the vertical wheel loads, different out-of-plane distortion occurred at different positions in web gaps, the web gap was subject to double curvature; both the bottom web gap at middle span area and the top web gap at supported were poor fatigue details, however, only the top web gap at support was poor fatigue detail in straight bridge.

The parametric study showed that web gap depth, web thickness, cross-frame type and curvature radius had an effect on the out-of-plane distortion. Increased web gap depth resulted in increased relative displacement, greater web thickness and curvature radius resulted in lower out-of-plane bending stresses. It was found that out-of-plane bending stresses in X layouts were a little higher than in layouts with K cross-frames.

Acknowledgement

The work described in this paper was partially supported by the National Natural Science Foundation of China (Grant No.51078039), the Foundation for the Author of National Excellent Doctoral Dissertation of the P.R. China (Grant No.2007B49), and the Special Fund for Basic Scientific Research of Central Colleges of the P.R. China, Chang'an University (No.CHD2012ZD008).

References

[1] C.A. Castiglioni, J.W. Fisher and B.T Yen: *Evaluation of fatigue cracking at cross diaphragms of a multi-girder steel bridge*. J. Construct Steel Research, Vol.9 (1988), p. 95.

[2] J. W. Fisher and P. B Keating: *Distortion-induced fatigue cracking of bridge details with web gaps*. Journal of Constructional Steel Research, Vol.12 (1989), p. 215.

[3] J. W. Fisher, Jian Jin, D.C. Wagner and B.T Yen: *Distortion-Induced Fatigue Cracking in Steel Bridges* (NCHRP Rep.336). Transportation Research Board, National Research Council, Washington, D.C. 1990.

[4] T. E. Cousins, J. M. Stallings, T.E. Lower and T. E. Stafford: *Field evaluation of fatigue cracking in diaphragm-girder connections*. Journal of Performance of Constructed Facilities, Vol.12 (1998), p. 25.

[5] R.J. Connor and J. W. Fisher: Identifying effective and ineffective retrofits for distortion fatigue cracking in steel bridges using field instrumentation, Journal of Bridge Engineering, Vol.11(2006), p.745.

Material Properties and Fatigue Safety Evaluation of Old Metal Bridges

Chunsheng Wang[1,a], Haijun Sheng[2,b], Jingyu Hu[1,c], Shenglong Yan[1,d], Lan Duan[1,e]

[1]Engineering Research Center for Large Highway Structure Safety of Ministry of Education, Chang'an University, Xi'an, Shananxi Province, China, 710064

[2]CCCC Road & Bridge Special Engineering Co., Ltd, Wuhan, Hubei Province, China, 430071

[a]wcs2000wcs@163.com, [b]Haijun Sheng@163.com, [c]shuishi1223@163.com, [d]bjysl1987@163.com, [e]dl0310dl@163.com

Keywords: old metal bridge, mechanical property, toughness, fatigue crack growth rate

Abstract. Lanzhou Zhongshan Bridge, which is located in the center of Lanzhou city, is a rivet truss bridge built in 1909. It is the first iron bridge over the Yellow River as well as a national cultural relic. An original steel angel cut off from Lanzhou Zhongshan Bridge was used to carry out a series of tests including material mechanics, fracture and fatigue property. Based on the test results, the fatigue and fracture safety was evaluated during the bridge remaining service life. What's more, remaining service life of Lanzhou Zhongshan Bridge was calculated and found that the trucks weight more than 120 kN should be limited to ensure the safety.

Introduction

In recent years, traffic load and speed on both existing railway and highway steel bridges have increased. Bridge owners are beginning to pay more attention to the actual fatigue remaining life, service safety, maintenance intervals and the ability to increase loads of old metal bridges. To answer such questions, knowledge concerning material properties, fracture toughness and fatigue strength is essential. This paper is regard with material mechanics, fatigue and fracture properties of old metal bridges built about 100 years ago. Lanzhou Zhongshan Bridge, the case study riveted truss structure, is a national cultural relic metal bridge built in 1909. An original steel angel cut off from Lanzhou Zhongshan Bridge was used to carry out series of tests including material, fatigue and fracture property. Based on the tested result, the remaining fatigue life of Zhongshan Bridge was evaluated.

Mechanical strength properties

Tensile test was conducted to study strength properties for old steel including modulus of elasticity, the yield and the ultimate strength etc. The tensile specimens were designed according to the requirements of Chinese GB/T 228-2002 [1]. Three ratio samples with same dimensions were manufactured. According to the requirements of Chinese GB2975-1998 [2], the test specimens were cut off from the leg of the steel angle. The dimensions of tensile test specimens are shown in Figure 1.

During the tensile test, the machine was controlled by displacement and static strain indicator TDS602 was used to collect data. The test results are listed in Table 1. It can be concluded that the strength of the old steel is roughly equal to A3 steel in the old code, but the yield-strength ratio, σ_s/σ_b, is equal to 0.629, which is slightly higher than A3 steel(0.486~0.579).

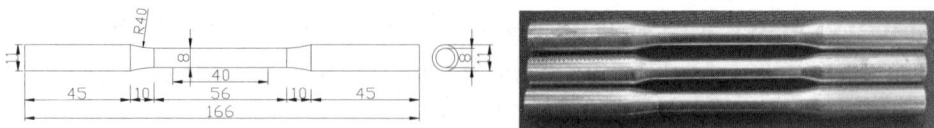

Fig.1 The Tension Specimen (unit:mm).

Table1 Results of Tensile Test.

E(MPa)	σ_s(MPa)	σ_b(MPa)	Elongation (%)	Reduction of Cross Section(%)
1.943E+5	275.7	438	29.1	63.77

Toughness properties

The toughness properties characterize whether steel is brittle or not at different temperature. Old steel is common to tend to be brittle. In order to determine the toughness properties of steel used in Zhongshan Bridge, Charpy-V toughness tests were conducted as specified in Chinese GB/T 229-1994[3].

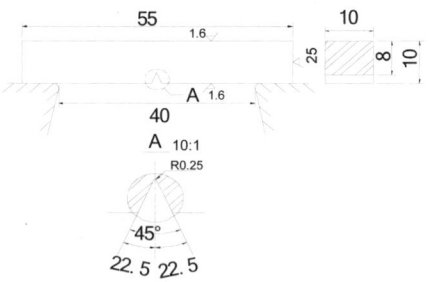

Fig.2 The Charpy-V Test Sample.

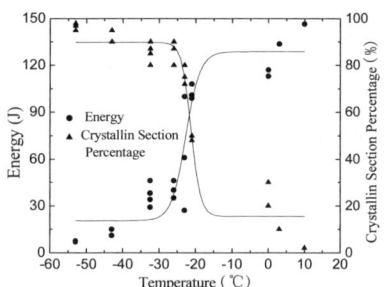

Fig. 3 Ductile -brittle Transition Curve.

Toughness test was conducted for a total of 25 samples at following eight different temperatures from -53°C to 10°C. The cooling medium was ethanol and liquid nitrogen. After the Charpy-V test, crystallinity percentage of fracture surface for each specimen was measured and calculated according to Chinese Code [4]. Ductile-brittle transition curve was drawn from the test, shown in Fig. 3. It is found that the transition region is ranging from -15°C to -28°C, and the impact energy apparently hurried drops in this range. The crystallinity percentage of approximately 50% is at -23°C. The fracture surface can reflect the fracture of the specimen. The fractograph exhibited typical ductile dimple fracture pattern at room temperature. As the temperature drops, fracture mechanism varies from micro-void coalescence fracture to cleavage fracture. At -23°C, the fractograph exhibited dimple as well as a considerable proportion of cleavage. When the temperature drops to -53°C, the fractograph exhibited typical brittle cleavage fracture pattern with barely dimple exhibited.

It can be concluded from the result of Charpy-V tests, the temperature of ductile-brittle of old steel used in Zhongshan Bridge is nearly -23°C. And the transition is very sharp. This means that when the temperature drops to -23°C, the steel will exhibit clear brittleness instantly, and safety of the structure will be out of guaranteed sharply.

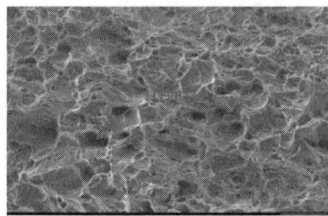 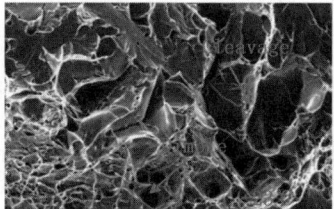 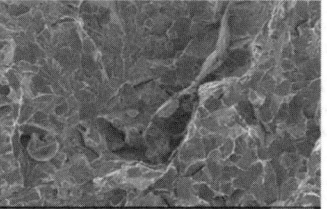

Fig. 4 Fractograph for Specimens at Different Temperatures.

Fatigue Crack Growth Rate

Fatigue crack growth rate is an important parameter which characterizes the material resistance to stable crack extension under cyclic loading. The crack growth can be illustrated with a universal curve shown in Fig. 5. In stage II, the relation between log da/dN and logΔK could be described with a mainly linear expression for the crack growth $da/dN = C(\Delta K)^m$, where C and m are empirically determined material parameters. Six compact [C (T)] specimens were cut from the leg of the old steel angle, shown in Fig. 6. The prefabricated crack for each specimen was oriented perpendicular to the rolling direction of the steel angle.

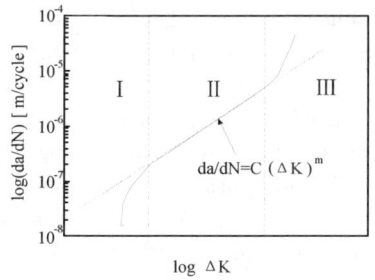 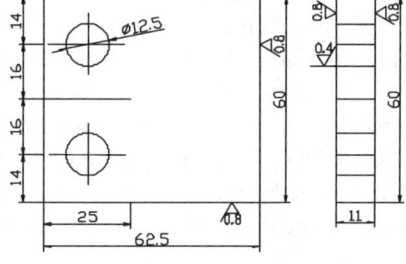

Fig. 5 The da/dN~ΔK Curve.   Fig. 6 Dimension of C (T) Specimen (unit: mm).

The C (T) specimens were tested under load ratio (R) of 0.1, 0.5 and 0.8, respectively. The load ratio, R, and loading frequency are kept constant during test with the loading frequency is 15 Hz. The test is conducted on electro-hydraulic servo-controlled fatigue testing machine and the crack growth is read by the microscope with min scale of 0.01mm. After the test, the seven-point incremental polynomial method is used for crack length and loading cycle. And the stress-intensity factor range ΔK under each crack length is calculated. When ΔK in MPa·m$^{1/2}$ and da/dN in m/cycle, the calculated regression parameters are shown in Table 2. It can be concluded that the fatigue crack growth rate increases with the increase of the load ratio. And the regression parameters of the test are $C=7.88 \times 10^{-12}$, m=3, which is between the test result of Barson($C=6.89 \times 10^{-12}$, m=3) and Sedlacek($C=1.26 \times 10^{-11}$, m=3) [5].

Table 2 Fatigue Crack Growth Rate Test Result.

R	Relgression parameters			
	m	C	logC	Correlation Coefficient
0.1	3.58	1.12×10^{-12}	-11.95	0.992
0.5	3.65	9.55×10^{-13}	-12.02	0.986
0.8	6.14	1.15×10^{-15}	-14.94	0.986

Evaluation of the remaining fatigue life

Based on the analysis of riveted steel bridge fatigue failure mechanism, the assessment model on the basis of S-N curve and fracture mechanics was applied to evaluate the remaining life of Zhongshan Bridge. Detail category D in AASHTO is used during the evaluation, according to the riveted joint fatigue strength analysis of Zhongshan Bridge when it was built, and the fatigue tests results conducted nationally and internationally, as well as the fatigue test result of zhongshan bridge. During the investigating, traffic load such as cars (the weight is 30 kN, and the axle distance is 2.5 m), two axis bus (the weight is 120 kN, and the axle distance is 4 m) and two axis truck (the weight is 150 kN, and the axle distance is 4.5 m) were used. By the method of S-N curve, the remaining fatigue life evaluation of main truss and deck plate in Zhongshan Bridge are all more than 100 years. While fracture mechanics method is used, only the first vertical rod and the first bottom boom tension rod in the fatigue sensetive main truss members can reach a remaining fatigue life more than 100 years under any kind of fatigue load. As a result, the trucks weight more than 120 kN must be limited to ensure the safe service of Zhongshan Bridge.

Summary

The tests in this paper were initiated to investigate the material properties and fatigue crack growth behavior of old steel cut off Lanzhou Zhongshan Bridge. The results have shown that the strength of the old steel is relatively equal to A3 steel in the old code, and the ductile-brittle transition temperature is nearly -23°C, which means that the structure is unsafe in winter due to the cold climate conditions of Lanzhou city. The main regression parameters of the fatigue crack growth rate are $C=7.88\times10^{-12}$ and $m=3$. To some extent, it can represent the old steel bridge built during that period. According to the fatigue remaining life evaluation of main truss and deck plate in Zhongshan Bridge by the S-N curve and fracture mechanics method, it can be concluded that the trucks weight more than 120 kN must be limited to ensure the safe service of Zhongshan Bridge.

Acknowledgement

The work described in this paper was partially supported by the National Natural Science Foundation of China (Grant No.51078039), the Foundation for the Author of National Excellent Doctoral Dissertation of the P.R. China (Grant No.2007B49), the Special Fund for Basic Scientific Research of Central Colleges of the P.R. China, Chang'an University (No.CHD2012ZD008).

References

[1] *Metallic Materials Tensile Testing at Ambient Temperature*(GB/T 228-2002) (China Plan Publications, Beijing 2002).
[2] *Steel and Steel Products—Location and Preparation of Test Pieces for Mechanical Testing*(GB2975-1998) (China Plan Publications, Beijing 1998).
[3] *Metallic Materials-Charpy Notch Impact Test*(GB/T 229-1994) (China Plan Publications, Beijing 1994).
[4] *Determination of Charpy Impact Fracture Surface for Metallic Materials*(GB/T 12778-91) (China Plan Publications, Beijing 1991).
[5] Chunsheng Wang: *Assessment of Remaining Fatigue Life and Service Safety for Riveted Steel Bridges*(The Tongji University Press. Shanghai 2007)

Damage Monitoring and Evaluation Using AE Sensors for Existing Concrete Bridges

Chunsheng Wang[1, a], Musai Zhai[1, b], Lei Tian[1, c]

[1] Engineering Research Center for Large Highway Structure Safety of Ministry of Education, Chang'an University, Xi'an, Shaanxi Province, China, 710064

[a] wcs2000wcs@163.com, [b] zhaimusai@163.com, [c] tl8213310@163.com

Keywords: existing concrete bridge, acoustic emission, fatigue and damage monitoring, safety evaluation, bridge maintenance

Abstract. The existing concrete bridges are often required to carry an increasing volume of traffic, higher speed and heavier trucks, so the bridge owners pay more attention to the actual fatigue damage and service safety of such structures. In this paper, Acoustic Emission (AE) sensors are used to monitor and evaluate the fatigue and damage of Yaoxian Bridge, which is the first field AE monitoring using Physical Acoustics Company equipment and sensors in China existing concrete bridge. This AE sensor has fine frequency bandwidth of interest to bridge monitoring, which is applicable to local cracking and damage positions and can monitor fatigue cracks and damage in close-range. In the field inspection, healthy monitoring was conducted at several locations, including the box concrete girder webs, bottom plates and deck plates using AE sensors. Based on AE monitoring data, the fatigue and damage conditions of box concrete girders was analyzed in order to propose theoretical basis and the rational advices for Yaoxian Bridge maintenance.

Introduction

Yaoxian Bridge, built in 2006, is a key bridge of provincial road 305, from Tongchuan to Jiaopin in Shaanxi, China. It is an 8-span prestressed box girder bridge with span length of 30m. Since the bridge is near a coal mine, most of the vehicles are coal-trucks, so the bridge is under great stress. The bridge owner wants to have a regularly knowledge of the performance of the bridge, and the AE monitoring system is introduced to give a comprehensive results.

Acoustic emission tests structure nondestructively by releasing high-frequency sound waves. AE signal is a transient elastic stress wave produced by the release of material internal local energy. When bridges are under load conditions, the structure of the materials like concrete or steel will change and emit energy in the form of elastic waves [1]. These waves are picked up by the sensors attached to the surface of the structure. After analysis of the collected signals, the evaluating result of the bridge can be given, which is helpful for the future repair and maintenance. The acoustic monitoring was started in 23rd August, 2011, and the total monitoring time was more than 5 weeks. This paper mainly focuses on the AE monitoring of the performance of the tendons in the box girder. From the recorded data, the conditions of the tendons can be obtained and the bridge performance can also be evaluated. It is the first field AE monitoring in existing concrete bridge in China, using Physical Acoustics Company equipment and sensors.

Sensor Locations

The monitoring region is at the NO.2 box girder of the first span of Yaoxian Bridge. The sensor locations are shown in the Fig.1. The basic information of the monitoring positions are described as followings: the NO.1,4,8,12,15 sensors are on the north web, the NO.2,5,9,13 sensors are on the bottom plate, the NO.3,6,10,14,16 are on the south web, and the NO.7,11 sensors are on the deck plate. The distance between NO.1,2,3 sensors to the girder end is 14500mm, the distance between 4,5,6 to the girder end is 12000mm, the NO.7, 8 ,9, 10,11,12,13, 14 is 7500mm, and the NO.15,16 is 1600mm.

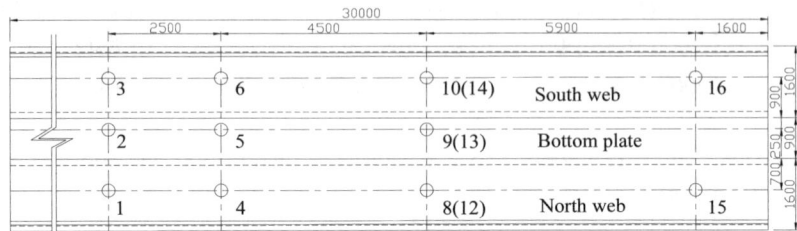

Fig.1 Sensor Locations on the 2# Box Girder (mm).

Monitoring equipment used in Yaoxian Bridge was made by the American Physical Acoustics Company, which includes two kinds of sensors (R3I, R15I) and an independent AE system: Micro-II, which has a 500GB hard disk for saving the data. The equipment is small, light and can be connected to a laptop. And the system can be controlled by remote Internet line, which is easy for site monitoring.

In this AE testing, two kinds of sensors were adopted to give a comparison. One kind is R3I with the 30kHz-resonant-frequency, and the other is R15I with the 150kHz-resonant-frequency. Three locations are chosen to do the comparison.

Recorded Signals and Data Analysis

AE signals can be induced by the plastic deformation and wire break, which are proportional to the release of energy within a material. Most signals are induced by live load, while some scattered signals are not. In general, the number and the energy of the signals induced by wind, noise, insect hit are small. When cars or trucks come across the bridge, the number and the energy of AE signals will increase sharply [2-3]. So AE system can easily filter out the signals induced by the noise and recognize the cars or trucks signals.

Table.1 Recorded Hits of 24 Hours.

Channel	1	2	3	4	5	6	7	8
Recorded Hits	7150	4095	7161	5401	520	40	12900	10278
Channel	9	10	11	12	13	14	15	16
Recorded Hits	30	5913	11755	203	98	64	51	52

Table.1 shows the hits recorded by every channel during 24 hours. It can be seen that Channel.7 and 11 got the most hits, and Channel.8 ranked the second. Since NO.7 and 8 sensors are both on the deck plate, signals of Channel.7 and 8 are picked out to give an analysis.

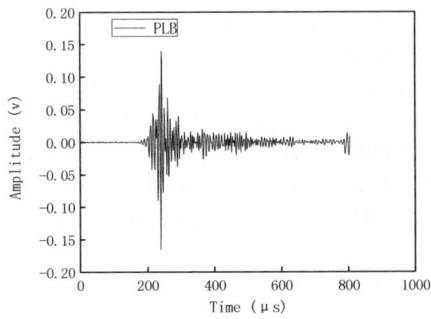 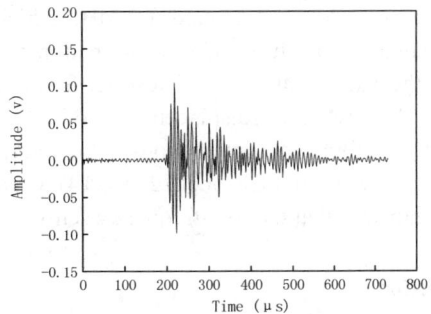

(a)Waveform Induced by PLB (b) Typical Signal Recorded

Fig.2 Typical Signals with Threshold.

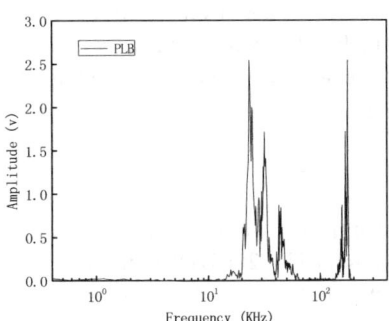 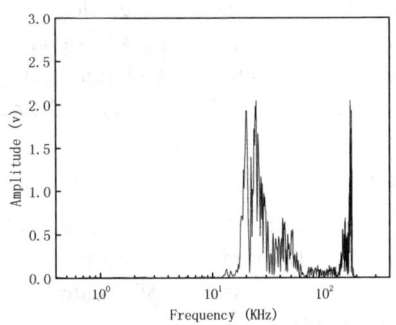

(a) Frequency Spectrogram of PLB (b)Frequency Spectrogramof Typical Signal

Fig.3 Frequency Spectrogram of the Signals.

As shown in Fig.2, they are typical signals recorded in the monitoring. Signal in Fig.2(a) is induced by Pencil Lead Break (PLB), which can be the standard signal for referring to. Fig.2(b) shows the typical signal recorded in site monitoring. Based on the signal information, frequency spectrograms are drawn by MATLAB, shown in Fig.3. Through comparison with the PLB frequency spectrogram, the similar signals to the PLB are found. Because the amplitude of most recorded signals is lower than the PLB, they can not be induced by the crack or damage. These signals may be induced by extruding, friction, grinding of the crack microcosmic and macroscopic surface during the initial periods.

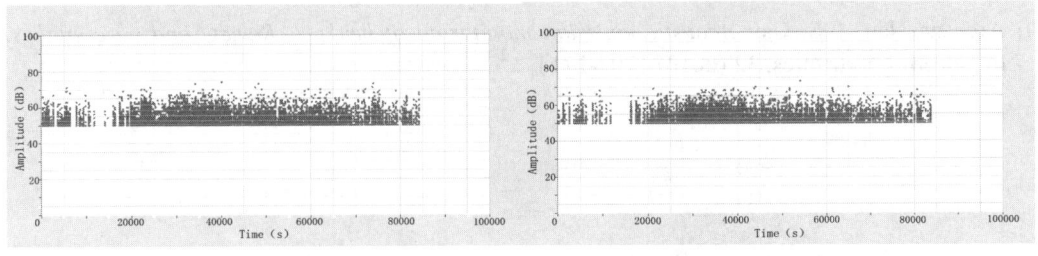

(a)Amplitude of Channel.7 (b) Amplitude of Channel.8

Fig.4 The Amplitude Distribution of Recorded Signals.

Based on the noise level at the site, the threshold level was set at 50dB. Shown in Fig.4, most signal amplitudes are around 60 dB, which indicates the signals are not induced by any crack-damage. And from other examination of it, the bridge is in good conditions, which can help certify the valid results of AE testing. Some high amplitudes were also recorded, which may be caused by relative displacement of the monitored regions due to load effects or concrete-reinforcement interactions at the interface[4]. Through the comparison of two kind sensors (Channel.8, 9, 10 and Channel.12, 13, 14), Channel.8, 9, 10 recorded more hits (shown in Table.1), which indicates that the low-frequency sensor (R3I) is more effective in concrete bridge testing.

Conclusion

This paper examines the validity of using AE monitoring to evaluate the fatigue and damage of concrete bridge. Based on the analysis of the recorded signals, it can be concluded that Yaoxian Bridge is in good condition though there may be some potential damage in the concrete box girder. On the data recorded during the monitoring, it can be concluded that AE monitoring is an effective method for evaluating the performance of the existing concrete bridges. Important information about the concrete bridge can be obtained from the analysis of AE signals, such as whether the wire breaks, whether crack propagate and so on. Using this technique can give an comprehensive diagnosis of bridges. It will be of great value to apply AE technique to the healthy monitoring of structures.

Acknowledgement

The work described in this paper was partially supported by the Foundation for the Author of National Excellent Doctoral Dissertation of the P.R. China (Grant No. 2007B49), the Special Fund for Basic Scientific Research of Central Colleges of the P.R. China, Chang'an University (No.CHD2012ZD008).

References

[1] Ding YL, Deng Y, Li AQ: *Advances in Researches on Application of Acoustic Emission Technique to Health Monitoring for Bridge Structures*. Journal of Disaster Prevention and Mitigation Engineering, 30(2010):341-351. (in Chinses)

[2] Fricker S, Vogel T: *Site installation and testing of a continuous acoustic monitoring*. Construction and Building Materials, 21(2007):501-510.

[3] Yuyama S,Yokoyama K,Niitani K, Ohtsu M, Uomoto T: *Detection and evaluation of failures in high-strength tendon of prestressed concrete bridges by acoustic emission*. Construction and Building Materials, 21(2007): 491-500.

[4] Archana Nair, C.S. Cai: *Acoustic emission monitoring of bridges: Review and case studies*. Engineering Structures, 32 (2010):1704-1714.

Repair and Service Safety Evaluation of a Fire Damaged Concrete Bridge

Chunsheng Wang[1, a], Luping Liu[1, b], Maosen Wei[1, c]

[1] Engineering Research Center for Large Highway Structure Safety of Ministry of Education, Chang'an University, Xi'an, Shananxi Province, China, 710064, P.R. China

[a]wcs2000wcs@163.com, [b]liuluping.1987@163.com, [c]842087513@qq.com

Keywords: steel plate and concrete composite strengthening method, concrete bridge, dynamic strain monitoring, stiffness, ultimate bearing capacity, fatigue life

Abstract. The Yao Zhou downlink Viaduct damaged in a fire accident was repaired using steel plate and concrete composite strengthening (SPCCS) technique. After strengthening and servicing normally for 1.5 years, two loading tests, meanwhile, a three-day traffic information observation and dynamic strain monitoring were carried out. The static loading test and monitoring results all showed that the Yao Zhou downlink Viaduct had good service condition since reinforcement, which satisfying the safety requirement. In order to evaluate the effectiveness of the SPCCS technique in-depth, structural behaviour under the ultimate limit state (ULS) and the serviceability limit state (SLS) were analyzed. The SLS and ULC of the key sections before and after reinforcement were calculated using the simplified formula and the finite element models calibrated by filed test results. The post-evaluation results showed that the behaviour of the spans damaged in fire accident could be improved obviously under the SLS using SPCCS technique, as well as the bridge stiffness enlarged than the original bridge structure. The ultimate bearing capacity (UBC) of the key sections was improved about 75% using SPCCS technique, which was two times as the most disadvantageous load effect combinations. What's more, the fatigue life of fire damaged box girders strengthened using SPCCS technique was evaluated based on monitoring strain data, and the further bridge maintenance suggestions were proposed based on the assessment results.

Introduction

Steel plate and concrete composite strengthening (SPCCS) is a new reinforcement technology on the basis of the steel plate-concrete composite reinforced beam, is an effective combination of the section enlargement method and the steel pasting method. Steel plate-concrete composite reinforcement is that reinforced part and original structure work together as a whole by welding studs on the steel plate, planting steel bars on the surface of the original concrete, casting concrete in the original structure and the reinforcing steel plate, etc [1]. Steel plate-concrete composite reinforcement technique can effectively reinforce existing damaged and unsafe bridge. Some experimental results show that the reinforcement technology can greatly improve the bearing capacity of the concrete beams [1-4]. J.G. Nie in Qinghua University carried out experimental research in the steel -concrete composite reinforced concrete beam [2]. C.S. Wang in Chang'an University once experiment on the flexural performance of the steel-concrete composite reinforced concrete T beam and got some theories [3-4]. On account of the existing research work, steel plate-concrete composite reinforcement technique was applied in the maintenance and strengthening of the Yao Zhou Downlink Viaduct, a prestressed concrete (PC) box girder bridge damaged by fire accident in 2009. All these work in this paper motivate the popularization and application of the strengthening technology in bridge engineering.

Finite Element Model Establishment and Optimization

In order to guarantee finite element model to simulate mechanics behaviors of real bridge effectively, the following assumptions were made. First, it was considered neither the stress redistribution of bridge before and after fire, and assumed the material mechanic of prestressed reinforcement and non-prestressed reinforcement is unchanged. Second, material property of burned concrete should be considered by multiplying reduction factor to consider the effect of fire to stiffness of the whole bridge. Third, it was assumed there was no relative displacement between original girder and new added concrete, between steel plate and new added concrete, and between plate sticking and original girder. In the numerical analysis model, common nodes were used to simulate connection between strengthening steel plate and new added concrete. By generating constraint equation between the interfaces, components were assembled together and work as a whole. Figure 1 shows the whole bridge model. Considering fire effect for bridge integral rigidity, elastic modulus reduction was considered for severe injured concrete. The cross section of severe injury concrete was divided into four layers from outside to inside, shown in Figure 2.

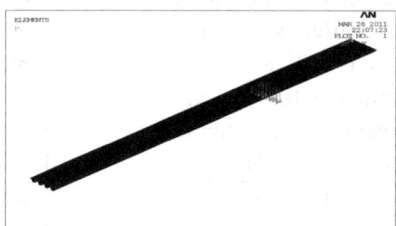
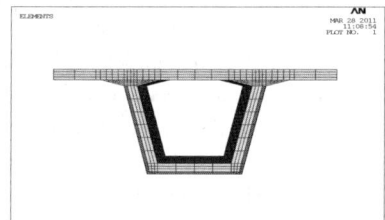

Figure 1. Whole bridge model.   Figure 2. FE details model of burned section.

Load Experimental Study of Real Bridge

The highway from Xi'an to Tong Chuan is the important sections in Shaanxi province of the national highway from Baotou, Inner Mongolia, to Mao Ming, Guangdong. Yao Zhou Viaducts in downlink is a prestressed concrete (PC) continuous box girder bridge. On one hand, the SPCCS method is used for the seriously damaged beams (beam1 in the 82th span). Second, sticking steel board reinforcement method is used for the generally damaged beams (beam 2~4 in the 82th span, beam 1 in the 83th span). For guarantee effectiveness and rationality of loading test, the most disadvantageous loading situation was considered. Static load test of bridge was carried out according to the following loading condition. In load condition 1 and 2, central and eccentric loading were conducted for the most disadvantageous section in positive region in the 82th span. In load condition 3 and 4, central and eccentric loading were conducted for negative region at interior pier in the 82th span. In load condition 5 and 6, central and eccentric loading were conducted at the most disadvantageous section in positive region in the 83th span. Loading position and test point arrangement of all loading conditions are shown in Figure 3 and 4. Dynamic strain monitoring was conducted for three days to evaluate the performance of bridge after reinforcement combined with static-dynamic load test. The dynamic strain monitoring for Yao Zhou Downlink Viaduct is shown in Figure 4. Most measuring dynamic-strains did not exceed the strains produced by corrected design truck load on related sections, which showed the bridge behaviors well in normal operation stage. The result shows that the structural rigidity and vibration met the requirement.

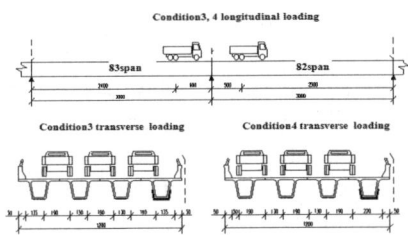
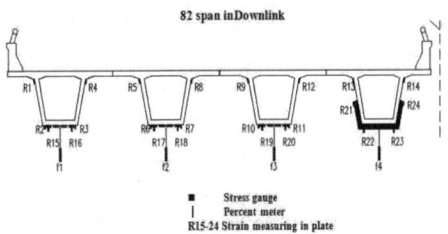

Figure 3. Details of loading position.   Figure 4. Arrangement of measuring points.

Bridge Evaluation under SLS and ULS

For better evaluating the mechanical behavior of strengthened bridge and reflecting stiffness changed in three stages (before fire, after fire and after strengthening), the bridge key-positions using the most adverse load-combination were carried out for stress and deflection comparison before and after reinforcement under SLS. The most adverse load combinations include load combination 1 (dead weight of girder + pre-stressing effect + secondary loads + secondary contraction + secondary creep), load combination2 (dead weight of girder + pre-stressing effect + secondary loads + secondary contraction + secondary creep + the most adverse vehicle load), and load combination 3 (dead weight of girder + pre-stressing effect + secondary loads + dead weight of reinforced concrete etc. + secondary contraction + secondary creep + the most adverse vehicle load). In these three load combinations, the second class vehicle load is used and coefficient of other load effect is 1. From Figure 5, it can be seen that steel plate and concrete composite strengthening can better improve force condition and the integral rigidity of bridge under LSL, as is plate sticking reinforcement. Using software ANSYS, UBC on middle-section of mid-span is 4430kN·m before reinforced in downlink viaduct, and is 10145kN·m after reinforcement. According to results of the UBC of the key sections, it could find that ultimate flexural resistance on middle-section of mid-span was improved by 129% after steel plate and concrete composite strengthening. Using software MIDAS, under ULS and considered partial coefficient of each load, effect of the most adverse load on middle-section of mid-span is 4646kN·m before reinforced in downlink viaduct, and is 5212kN·m after reinforcement. It can be seen that ultimate resistance on middle-section of mid-span is far more than effect of the most adverse load under ULS. Especially, the girders adopted composite reinforced can greatly improve the safety margin of section.

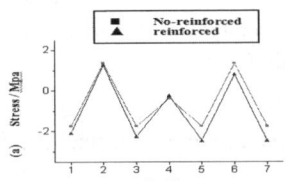
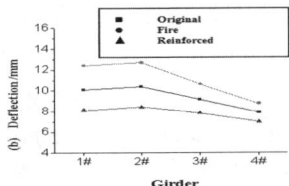

Figure 5. Stress and displacement comparison of the key sections before and after reinforcement.

Conclusions

To effectively simulating mechanical behavior of fire injured Yao Zhou Downlink Viaduct by ANSYS, it was to use respectively different reduction-factor of elastic modulus for concrete according to different burned degree. The load test results from each load condition showed that the measured-deflections of most measuring-points were less than the calculated value in main girder.

Measured time-domain curves of structural vibration conformed vibration regularity of PC continuous girder bridge. It showed that structural stiffness and strength were improved efficiently after reinforcement, which satisfied the demand of design and operation safety. The statistics to monitoring date showed most vehicles load-effects monitored were less than design load-effects, and the partial measuring-points exceeded design value, which coincided with the theory of probability design. After using steel plate and concrete composite reinforcement and plate sticking reinforcement, force condition of girder was to get greatly improved. According to three stages of before fire, after fire, reinforced, it was analyzed deflection response of injury bridge under the most adverse moving-load. It could be seen that integral rigidity of the bridge after composite reinforced met the demand of code. Ultimate flexural resistances got by plastic finite element analysis and theoretical formulae showed that resistance on section after composite reinforced was far more than design-load effect and safety margin of bridge reinforced was obviously improved. The ultimate bearing capacity (UBC) of the key sections was improved about 75% using SPCCS technique, which was two times as the most disadvantageous load effect combinations. What's more, the fatigue life of fire damaged box girders strengthened using SPCCS technique was evaluated based on monitoring strain data, and the further bridge maintenance suggestions were proposed based on the assessment results.

Acknowledgement

The writers would like to acknowledge the financial support provided by the Shaanxi Province Transportation Technology Research Projects (Grant No.07-03k, 07-04k), the Special Fund for Basic Scientific Research of Central Colleges of the P.R. China, Chang'an University (Grant No.CHD2012ZD008), and China West Transportation Development Research Projects (Grant No. 200831849404, 20113185191410).

References

[1] Nie J.G., Zhao J. And Tang L., *"Application of steel plate and concrete composite to strengthenning of reinforced concrete girder"*, Journal of Bridge Construction, 2007, (3):76-79. (in Chinese)

[2] Nie J.G. and Zhao J., *"Experimental study on simply supported RC beams strengthened by steel plate-concerete technique"*, Journal of Building Structures, 2008, 29(5):50-56. (in Chinese)

[3] Wang C. S., Yuan Z. Y., Guo X. Y., Gao S. and Ren T.X., *"Flexural behavior experiment of reinforced concrete T-beams with steel plate-concrete composite strengthening"*, Journal of Traffic and Transportation Engineering, 2010, 9 (6): 21-29. (in Chinese)

[4] Wang C. S., Gao S., Ren T. X. and Xu Y., *"Bending behavior experiment of damaged RC T-beams with steel plate and concrete composite strengthening"*, Journal of Architecture and Civil Engineering, 2010, 27(3):94-101. (in Chinese)

Scaling effects on the tensile strength of fibrous composites

Fang Wang[a], Lu Li[b], and Zhiqian Chen[c]

School of Materials Science and Engineering, Southwest University, Chongqing 400715, China

[a]wfang@swu.edu.cn, [b]neil@swu.edu.cn, [c]chen_zq@swu.edu.cn

Keywords: Composite materials, tensile behavior, Weibull statistics, modelling

Abstract: The primary objective of this paper is to illustrate the effects of weak-link scaling on the tensile behaviour of fiber-reinforced composites. The proposed model takes into account the random nature of fiber strength, which is given by a two-parameter Weibull distribution function. Several hundred Monte-Carlo replications are executed to simulate the statistical strength distributions of the composites. It is shown that probabilistic tensile strength distributions and size scaling is dependent on both the stress redistribution and the fiber strength statistics.

Introduction

Reliability concerns in utilizing advanced fiber-reinforced composites in structural applications have motivated the development of many numerical and analytical statistical strength models. These models are required to identify the dominant failure modes and characterize stress profiles as a result of the interplay between fiber stress concentrations and the considerable variation in fiber strength due to statistic distribution of defects [1,2].

The catastrophic failure of such composites is largely dominated by the nucleation and interaction of fiber breakage. Under simple tension, failure in composite materials usually begins with random fiber breaks that develop at flaws under increasing load. Consequently failure involves a complex statistical progression of random fiber failure, local stress transfer from broken to surviving fibers though shear stresses, local matrix deformation and interfacial debonding and slipping around the fiber breaks. Therefore, the strength of fibrous composites depends on the stochastic damage evolution, which is a combination of these complex mechanisms [3].

The focus of the present work is on investigating scaling effects involved in the prediction of ultimate tensile strength of fiber-reinforced composite materials. Several hundred Monte-Carlo simulations are executed to determine the statistical strength distributions of the composite for two values of the fiber Weibull modulus, $\kappa = 10$ and 20. Material sizes of 20, 25 and 30 fibers are taken to determine the dependence of strength on the composite size. These are analyzed according to the two-parameter Weibull distribution. The accuracy of using weak link scaling statistics for composite strength is examined as well.

Flaw statistics

For fibrous composites of interest here, the tensile strength distribution of fibers is a key constitutive property of composite materials. The statistical nature of brittle fibers is due to the varying severity of material defects that are randomly distributed along their lengths. As is observed experimentally for such fibers [4], it is assumed that the strength of high performance fiber under tension can be well described by means of the two-parameter Weibull function which is of the form,

$$P(\sigma_f) = 1 - \exp\left\{-\left(\frac{L}{L_0}\right)\left(\frac{\sigma_f}{\sigma_0}\right)^\kappa\right\}. \qquad (1)$$

where P in the range of [0,1] is the probability of failure of each segment at an applied stress less than or equal to σ_f. σ_0 and L_0 is the reference strength and length parameters, respectively. L is the fiber gauge length. κ is the Weibull modulus or shape parameter, which controls the degree of disorder in the distribution. Fig. 1 illustrates that smaller κ have larger strength variability and vice versa.

Owing to the statistical nature of the fiber strengths, fibers in a composite do not necessarily fail in region of high stress concentrations. It is essential to know about where and at what stress the fiber breaks will occur for comprehensively understanding the failure behavior. In this section, each fiber with total length L_T is divided into N_T segments of equal length Δx such that $\Delta x = L_T / N_T$. Therefore, the strengths of the potential failure locations in a single fiber are assumed to follow a Weibull distribution,

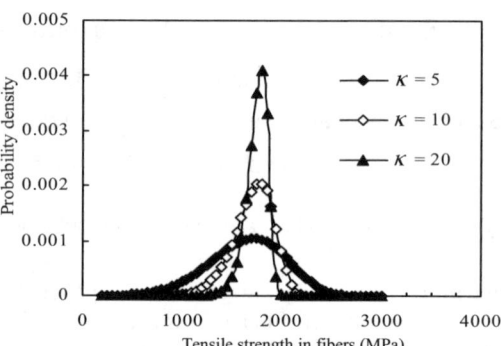

Fig. 1 Influence of Weibull modulus on the probability density distribution of tensile strength in fibers

$$\sigma_f = \sigma_0 \left\{ \left(\frac{L_0}{\Delta x}\right) \ln\left(\frac{1}{1-P}\right) \right\}^{1/\kappa}. \qquad (2)$$

Since in reality many defects are distributed in fibers randomly, Monte-Carlo method is needed to carry out the numerical analysis of the proposed model. At the beginning of each simulation, all fiber segments are assumed to be intact. Fiber breakage at some location takes place when the axial stress in fiber becomes equal to the strength, σ_f, at this position.

Weak-link scaling

Weak link theory, which could be applied to many brittle materials, is based on the assumption that the material is made up of smaller elements linked together and that failure of the materials as a whole occurs when any one of these elements or 'links' fail [5,6]. The probability of failure of each link subjected to a stress increase from 0 to σ is described by the distribution function $P_f(\sigma)$. The probability of survival of that link is then given by,

$$P_s(\sigma) = 1 - P_f(\sigma). \qquad (3)$$

It is also assumed that $P_f(\sigma)$ describes the strength distribution for every element and each $P_f(\sigma)$ is an independent randomly distributed variable. The probability of survival of n elements in series is then given by,

$$P_{s,n}(\sigma) = [1 - P_f(\sigma)]^n. \qquad (4)$$

Hence the probability of failure of a chain of n elements is given by,

$$P_{f,n}(\sigma) = 1 - [1 - P_f(\sigma)]^n. \qquad (5)$$

This equation is known as the basis of statistical weak-link scaling.

It is claimed that a 'large enough' composite length and number of fibers can be used along with weak-link scaling to predict the strengths of larger composites [5]. If the strength of a material is governed by weak-link statistics, then the cumulative probability of failure, P_{f,V_2}, of a composite with volume V_2 loaded to stress σ can be related to the cumulative probability of failure, P_{f,V_1}, of a composite with volume V_1 by

$$P_{f,V_2}(\sigma_c) = 1 - [1 - P_{f,V_1}(\sigma_c)]^{\frac{V_2}{V_1}}. \qquad (6)$$

where the normalized volume of the composite in our model is given by $V = N \cdot (W + D) \cdot L_T$. N represents number of fibers. If the diameter, D, of all fibers and the width, W, of all matrix are assumed to be the same, then Eq. (6) can be written as at a specified length, L_T,

$$P_{f,V_2}(\sigma_c) = 1 - [1 - P_{f,V_1}(\sigma_c)]^{\frac{N_2}{N_1}}. \qquad (7)$$

Results and Discussions

The analytical method under the framework of shear-lag argument developed recently in Ref. [2,7] can derive stress profiles around multiple damage events. Monte-Carlo simulation technique [8] coupled with the methodology is used to predict the tensile strength distributions. 200 simulations are run for each distribution owing to computation time constraints. Meanwhile, an element length of 5×10^{-2} mm is available for the sake of efficiency.

Fig. 2 is a plot of the cumulative probability of failure on Weibull coordinates for composite sizes of 10, 20 and 30 fibers. The figure also shows that the strength distribution depends on the physical volume of the composite (N fibers). The variability of the composite strength decreases steadily as the number of fibers N in the composite increases. It should be mentioned that the size effect in composite length is not included in this model.

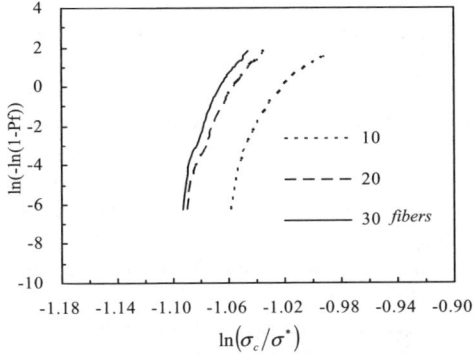
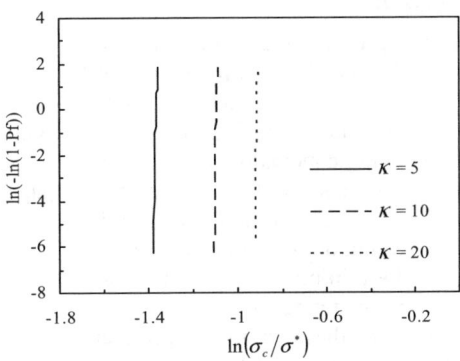

Fig. 2 Strength distributions for composites with $\kappa = 10$ for specimen sizes of 10, 20 and 30 fibers

Fig. 3 Strength distributions for composites with 50 fibers for shape parameter β including 5, 10, 20

Fig. 3 presents Weibull plots for the three different scale parameters of 5, 10 and 20. Here, $\sigma_0 = 1800$ MPa is used for the simulation. It indicates that the effect of varying the Weibull modulus κ on the distribution of composite tensile strength is significant, showing a weakening trend as κ decreases from 20 to 5. As the scatter of the fiber tensile strength increase the tensile strength for composites will dramatically decrease.

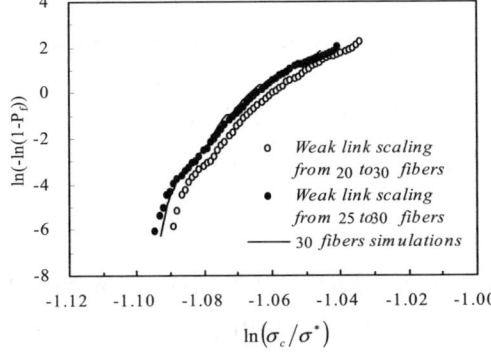
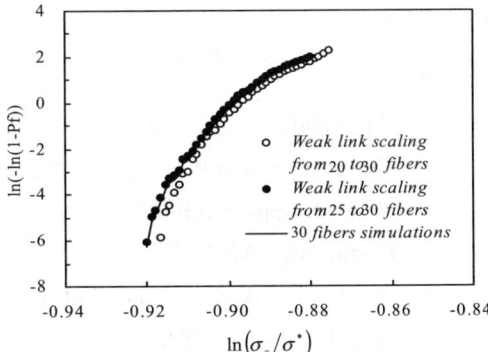

Fig. 4 Comparison of the simulated results for 30 fibers with the two results from 20 and 25 fibers scaled to 30 fibers for $\kappa = 10$

Fig. 5 Comparison of the simulated results for 30 fibers with the two results from 20 and 25 fibers scaled to 30 fibers for $\kappa = 20$

In order to assess the accuracy of weak-link scaling, predictions pf tensile strengths of larger composite are made by the size scaling from smaller one. Fig. 4 contains the distribution for specimen sizes of 30 fibers with $\kappa = 10$ along with the distributions for specimen sizes of 20 fibers and 25 fibers, weak link scaled to 30 fibers, i.e. with $V_2/V_1 = 30/20$ and $V_2/V_1 = 30/25$. The results show that weak link scaling works well for the 25 to 30 scalings. This is due to the composite strength gets to be convergent when $N = 25$. Therefore, for a given κ, it is possible to find a large enough fiber size for which weak link scaling is valid [1,9].

$\kappa = 20$ is present on Fig. 5 for comparison, plotting the strength distribution for 30 fibers. Also shown are results of two weak linked distributions. The open and solid circles represent the $N = 20$ and $N = 25$ distribution weak link scaled to $N = 30$, respectively. For $\kappa = 20$ we find that smaller composite sizes of 20 fibers can be scaled to a volume of 30 fibers. There is much more distributed damage in the composites with $\kappa = 10$ fibers than in composites with $\kappa = 20$ fibers. This explanation is supported by the results shown in Fig. 1. Specifically, the critical composite size necessary to produce weak link scaling increases with decreasing the fiber Weibull modulus.

Conclusions

In the present study, we have attempted to predict the strength distributions using weak link scaling. Monte-Carlo simulation has been implemented to investigate the scaling effects in the tensile properties of fiber-reinforced composites. The approach involves the determination of the critical size and comparing the prediction with strength values determined by numerical simulation. The following conclusions were obtained:

1. Materials that fail by unstable propagation of localized damage must, for sufficiently large sizes, have strength distributions that obey weak-link scaling. In other words, weak-link scaling does set in at an appropriate composite size
1. The minimum size needed for weak link scaling depends on the fiber Weibull modulus. As the scatter of the fiber strength increases the critical size will be increased.

Together, these results lend confidence to the suggested approach of obtaining Weibull parameters from tests of specimens and using the parameters for strength distribution predictions of larger structures.

Acknowledgements

The financial support for this subject was provided by the National Science Foundation of China (Grant No. 11102169), and the Fundamental Research Funds of the Central Universities (Grant No. XDJK2011C070 and XDJK2010C011) for which the authors are grateful.

References

[1] C.M. Landis, I.J. Beyerlein and R.M. McMeeking: J. Mech. Phys. Solids Vol.48 (2000), p. 621-648

[2] J.Q. Zhang and F. Wang: Mech. Adv. Mater. Struct Vol.16 (2009), p. 522-535

[3] L. Mishnaevsky Jr and P. Brøndsted: Mater. Sci. Eng. A Vol. 498 (2008), p. 81-86

[4] W. Weibull: J. Appl. Mech Vol.18 (1951), p. 293

[5] W.A. Curtin: Adv. Appl. Mech Vol.36 (1999), p. 163-253

[6] L.S. Sutherland, R.A. Shenoi and S.M. Lewis: Compos. Sci. Technol Vol. 59 (1999), p. 209-220

[7] F. Wang, Z.Q. Chen and Y.Q. Wei et al: J. Compos. Mater Vol.44 (2010), p.2325-2340

[8] J.Q. Zhang and F. Wang: Int. J. Damage Mech Vol.19 (2010), p. 851-875

[9] S.L. Phoenix, M. Ibnabdeljalil and C.Y. Hui: Int. J. Solids Struct Vol.34 (1997), p. 545-568

Analysis of fatigue and cumulative damage characteristics for concrete in freezing-thawing environment

XIAO Qian-hui

(School of architecture and civil engineering, XI'AN University of science and Technology, XI'AN 710054, China)

Keywords: concrete;fatigue;cumulative damage;freezing-thawing environment

Abstract: Through analyzing the characteristic of concrete freezing-thawing cumulative damage, concrete freezing-thawing damage can be with fatigue problems with similar methods to analysis. Coagulation Earth material property discrete and cause damage in a given amount of state natural freeze-thaw cycle times is a random variable, analysis found that it is better to obey three- parameter Weibull distribution. Based on three parameters Weibull distribution model, the test data in had established, on the basis of freezing-thawing environment concrete damage and freeze-thaw cycles of the probability of the relation curve, and regression to obtain different guarantee rate of freezing-thawing damage cumulative model for freezing-thawing environment,which could be provides reference for concrete service life prediction and health diagnosis.

At present, reinforced concrete is the most prevalent construction material used in building structure .Freezing-thawing cycle can be caused destruction of concrete are generally exhibit surface of osteoporosis, spalling, aggregate exposure, and even exposed steel bar etc. The concrete buildings in the cold area, especially those in the alternating wet and dry environment structure, such as dams, bridge, culvert, port engineering, often due to alternation of freezing-thawing cycles occur durability damage, can not meet the design service life.The concrete building in the environment of alternating positive and negative temperaturea, and internal containing more water, will occur freezing-thawing damage, so the freezng- thawing damage is the most representative indicator of concrete durability [1]. In this paper, based on the existing experiment data and three parameters of the Weibull distribution model, established the curves of concrete in freezing-thawing circumstance are defined to describe probabilistic relationships between damage and freezing-thawing cycle number. It can quantitative analysis the concrete damage and the freezing-thawing cycle number in freezing-thawing environment, study of concrete fatigue and damage accumulation rule in freezing-thawing environment, for concrete health diagnosis and life prediction offer reference and basis.

1 Mechanism of freezing-thawing damage of concrete

About the concrete freezing-thawing damage mechanism, scholars at home and abroad for a lot of the theoretical and experimental research, many theories put forward, in which the Powers T.C. hydrostatic pressure theory and osmotic pressure is the most classical theory[2-3].

For hydrostatic pressure hypothesis, in the freezing process, because the pores in the concrete part of pore solution expands when it freezes around 9%, forcing an ice hole solution from icing area outward migration, pore solution in permeable cement slurry structure in mobile, must overcome resistance, resulting in hydrostatic pressure, formation stress.

Osmotic pressure hypothesis that, due to cement slurry in pore solution was alkaline, ice crystal formation make these pores without freezing hole concentration rises, and other smaller pores formed between non freezing hole solution concentration difference. In this concentration difference, the smaller pore without freezing hole solution to oneself appear ice crystals larger pores in migration, osmotic pressure generated in the pore solution. The migration to freeze ice and solution pore in increasing the volume, osmolality is increasing. The role of osmotic pressure in cement paste, leading to internal cracking of cement paste.

2 Freezing-thawing damage of concrete analysis

LiJinYu etc[1] research that concrete freezing-thawing damage similar to the fatigue damage of concrete,concrete damage can be understood as hydrostatic pressure and osmotic pressure in the role of concrete and cause and internal damage accumulation of concrete results. To assess structure of the degree of damage in frost environment, how to life prediction of concrete structure, all the countries researchers interested in question, and to date have not yet been unfathomed on international major problem.

In the freezing-thawing cycle process, the hydrostatic pressure and osmotic pressure produced by the inner stress is cycle. A natural freezing-thawing cycle is every time the freeze-thaw cycle of damage produced gradually accumulated a development. The freezing-thawing destruction is a natural pure physical process, is the internal have complex stress function results. In the freezing-thawing cycle, the temperature reaches the highest point minimum stress, low stress when most. With the natural freeze-thaw cycle, stress the loading, unloading constantly, round and round, which causes the disintegration of concrete micro, strength is lost, and form a tiny cracks, even development to that concrete destruction. Fatigue problems is similar to this, also because alternating load of tiny cracks in the materials developed and expanded the cause. Therefore, concrete freezing-thawing damage can be with similar methods to analysis.

3 Concrete under freezing-thawing environment fatigue damage probability model

3.1 Model selection

The biggest characteristic of concrete material is its multiphase and not evenness, in material internal there are a lot of micro cracks, space, or defects such as impurity. After a load of the repeated and discharge, near the defects in damage, and accumulation, lead to fatigue crack extension, link up and eventually destroying. Concrete can be used the fatigue of the normal distribution, logarithmic distribution or Weibull distribution model describes three parameters. But normal theory only when $\varepsilon = -\infty$, its survival rate to equal 100%, obviously this is not accord with the actual; And three parameters of Weibull distribution advantage is that existing parameters can be minimum concrete survival rate to 100%, the actual life span character with material[4]-[5]. Therefore, this paper choose three-parameters of the Weibull distribution model.

3.2 Model validation

According to three parameters Weibull probability model[6], can get concrete freeze-thaw cycles of life can be expressed as density function:

$$f(N) = \frac{\beta}{\eta}\left(\frac{N-\gamma}{\eta}\right)^{\beta-1} \exp\left[-\left(\frac{N-\gamma}{\eta}\right)^{\beta}\right] \tag{1}$$

The corresponding distribution function is

$$F(N) = 1 - \exp\left[-\left(\frac{N-\gamma}{\eta}\right)^{\beta}\right] \tag{2}$$

In which, γ is minimum life parameter, η is characteristic parameter for life, β is weibull shape parameter.

Transform equation (2), $X = \ln(N-\gamma)$, $Y = \ln[-\ln(1-F(N))]$, type (2) can be written as a linear function:

$$Y = a + bX \tag{3}$$

The determination of γ use precision of maximum likelihood method, make the scatterplot (Xi, Yi) to achieve best linear state, of which the estimation of failure rate adopted an average rank. Description of the failure of concrete frost environment of damage can be expressed as D[7]:

$$D = 1 - \frac{E_t}{E_0} \tag{4}$$

In which, E_0 is dynamic elastic modulus before damage; E_t is dynamic elastic modulus after damage.

Based on the existing test data (quick freezing method) selected 25 concrete test data as sample[8]-[11], the sample includes a concrete dynamic elastic modulus decreased to 60% (the D is 0.4) , the number of freeze-thaw cycles were analyzed, the results as shown in Figure 1. Basic data points fall in a straight line, correlation coefficient R = 0.9911, to illustrate the application of three parameter Weibull distribution as to the amount of loss damage D under freezing-thawing cycles N probability distribution model is reasonable.

Then you can get the damage of distribution function:

$$F(N) = 1 - \exp\left[-\left(\frac{N-91}{170.12}\right)^{1.24}\right] \qquad (5)$$

4 The probabilistic relationship curves of concrete damage and freezing-thawing cycles number in freezing-thawing environment

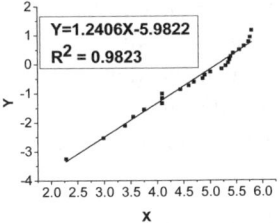 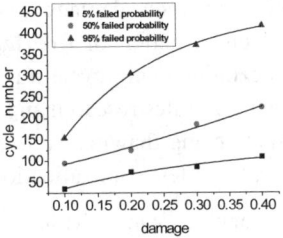

Fig. 1. Three-parameter Weibull distribution   Fig. 2. Probabilistic relationships between damage and freezing-thawing cycle number

Use reliability life data analysis software Weibull + + 7 is obtained from table 1 freeze-thaw cycle distribution parameters and different guaranteed rate under the condition of freeze-thaw cycles, the table 1 damage quantity with different guaranteed rate under freeze-thaw cycles in cartesian coordinates can be drawn, concrete under freeze-thaw environment damage and frost cycles the probability curve, as shown in figure 2.

Table 1 Distribution parameter in different damage and cycle number in variousre liability probabilities

damage D		0.1	0.2	0.3	0.4
Least life parameter γ		25	51	74	91
Characteristic life parameter η		1.585 1	1.080 1	1.153 2	1.240 8
Weibull Shape Parameter β		80.590 2	107.297 6	144.757 2	170.115 6
Correlation coefficient		0.985 8	0.993 7	0.991 5	0.991 1
Cycle number in different liability probabilities	95%	35	74	86	109
	50%	95	125	185	225
	5%	154	305	362	402

On Figure 2 data regression analysis, we can get different guaranteed rate damage quantity of cycle of freezing and thawing relations function.

When the guarantee rate is 95%, the correlation coefficient is 0.9680,

$$N = 135.38 - 149.86 \exp^{-4.12D} \qquad (6)$$

When the guarantee rate is 50%, the correlation coefficient is 0.9812,

$$N = -239.72 + 349.20 \exp^{D} \qquad (7)$$

When the guarantee rate is 5%, the correlation coefficient is 0.9986,
$$N = 453.63 - 589.71\exp^{-6.79D} \tag{8}$$

5 Conclusion

(1)Through analyzing the characteristic of concrete freezing-thawing damage, concrete freezing-thawing damage can be with fatigue problems with similar methods to analysis. Because of concrete material properties of discrete and lead to a given amount of damage state natural freezing-thawing cycle times is a random variable, analysis found that it is better to obey three parameter Weibull distribution model.

(2)Based on three parameter Weibull distribution model, concludes that the typical damage of the number of freezing-thawing cycle and the corresponding distribution parameters of the number of freezing-thawing assurance, further give the probabilistic relationship curve of freeze-thaw ordinary concrete damage and freeze-thaw cycles number under the environment.

(3)Analysis the curves of concrete in freezing-thawing environment are defined to describe probabilistic relationships between damage and freezing-thawing cycle number, defined the cumulate damage model under action of freezing-thawing cycle in various reliability. We found concrete damage and freezing-thawing cycle number accord with exponential function. The result provides a reference to damage rules research and life prediction in freezing-thawing environment. The results show that this freezing-thawing cycle cumulative damage model has higher fitting precision, but still needs a lot of test data validation.

Biography: XIAO Qianhui, Dr., Xi'an 710054, P.R.China,Tel:13669189037, E-mail: xiaohui_99@126.com

References

[1] LI Jin-yu, CAO Jian-guo, XU Wen-yu, etal. Stydy on the mechanism of concrete destruction under frost action [J].Journal of hydraulic engineering shuili xue bao, 1999（1）:41-49.

[2] Powers T. C. A working hypothesis for further studies of frost resistance [J]. Journal of the ACI. 1945,16(4):245-272.

[3] Powers T .C , Helmuth R .A . Theory of Volume Change in Hardened Portland Cement Paste During Freezing, Proc., Highway Research Board, Vol.32 , 1953: 85-21.

[4] Wang Fengwu, Zhang Lixiang, Xu Reiping, eta1. Areliability calculation method for concrete strength by means of confidence limits[J]. Journal of kunming university of science and technology，2000, 25(4): 72-75.

[5] H Matsushita, Y Toskumitsu. A study on compressive fatigue strength of concrete considered survival probability[J]. Proc.of JSCE,1979,198:127-138.

[6] JIANG Ren-yan. Characteristic, parameter estimate and application of Weibull model [M].Bei jing: Science Press, 1998 .

[7] SONG Yu-pu, JI Xiao-dong. Analysis on reliability of concrete under freezing-thawing action and evaluation of residual life[J].Journal of hydraulic engineering shuili xue bao, 2006，37（3）：259－263.

[8] SHI Shi-sheng. Effect of freezing-thawing cycles on mechanical properties of concrete. [J]. China Civil Engineering Journal，1997, 30(4): 35-42.

[9] SHANG Huai-shuai, SONG Yu-pu, QIN Li-kun. Experimental study on properties of concrete after freezing and thawing cycles[J]. China Concrete and Cement Products, 2005,{2).

[10] TANG Guang-pu,LIU Xi-la. A study on concrete frost model in view of phenomenological damage theory[J]. Sichuan Building Science,2007,33(3):138-143.

[11] XU Hui, XIE You-jun, LONG Guang-cheng, MA Kun lin. Research on Freezing-Resistance Durability of Air-Entraining Fly Ash Concrete[J]. Coal Ash China, 2004，6：3－5.

Residual Fatigue Life Analysis of Submarine Pressure Structure Based on Probabilistic Fracture Mechanics

Zhiyong Zhang[1, a], Tianshu Song[1, b], Yang He[1, c]

[1]College of Aerospace Engineering and Civil Engineering, Harbin Engineering University, Harbin, Heilongjiang, 150001, China

[a]zzyhunan@163.com, [b]songts@126.com, [c]woshiheyang888@163.com

Keywords: submarine pressure structure, residual fatigue life, Monte Carlo method, probabilistic fracture mechanics, combination method

Abstract. The residual fatigue life of a submarine pressure structure is investigated, based on the combination between the methods of conventional Monte Carlo and classical probabilistic fracture mechanics. Firstly, Monte Carlo method is employed to obtain the reliability of given initial fatigue life. Secondly, the two induced factors M_{A1} and M_{A2} in the paper are estimated according to the initial fatigue life and the reliability. Thirdly, based on the two factors, the residual fatigue life based on other reliability is obtained by using classical probabilistic fracture mechanics method. Numerical examples show that the proposed method is more efficient without accuracy loss for residual fatigue life compared with Monte Carlo method. This method can also be employed to predict the residual fatigue life on other analogue structures.

Introduction

A submarine pressure structure shown in Fig. 1 is frequently used in submarine structure design [1]. Fatigue crack is always found in the weld toe where subjected to high stress gradient and complex load. Several submarine failures were reported to be caused by the fatigue crack initiated in the weld toe [2-4]. Strictly speaking, the fatigue life is affected by several parameters which present random distribution features. Therefore, it is reasonable to apply residual fatigue life analysis of submarine pressure structure based on probabilistic fracture mechanics. In general, the Monte Carlo method and probabilistic fracture mechanics method are frequently implemented [5, 6]. However, both the computation cost of Monte Carlo method and the accuracy of probabilistic fracture mechanics showed to be unacceptable. In the current research, the author combines these two methods and obtains a highly precise result by a small amount of sampling data.

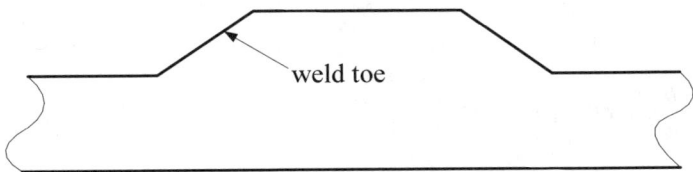

Fig. 1 Sketch map of a submarine pressure structure

Stress intensity factor for a semi-elliptical surface crack

In most cases, a fatigue crack is initiated on the surface of a weld toe of the submarine submarine pressure structure shown in Fig. 1. The fatigue crack is usually treated as a semi-elliptical surface crack in a finite-thickness plate from an engineering consideration. The author consulted correlative references [7], present a stress intensity factor as follows:

$$K_I = (\sigma_t + H\sigma_b)\sqrt{\frac{\pi a}{Q}} F(\frac{a}{t}, a/b, \frac{2b}{w}, \theta). \tag{1}$$

where, $F = \left[M_1 + M_2 (a/t)^2 + M_3 (a/t)^4 \right] f_\theta g f_w$;

$M_1 = 1.13 - 0.09 a/b$;

$M_2 = -0.54 + \dfrac{0.89}{0.2 + a/b}$;

$M_3 = 0.5 - \dfrac{1}{0.65 + a/b} + 14(1 - a/b)^{24}$;

$f_\theta = \left[(a/b)^2 \cos^2 \theta + \sin^2 \theta \right]^{1/4}$;

$f_w = \left[\sec\left(\dfrac{\pi b}{2w} \cdot \sqrt{a/t} \right) \right]^{1/2}$;

$g = 1 + \left[0.1 + 0.35(a/t)^2 \right] (1 - \sin \theta)^2$;

$H = H_1 + (H_2 - H_1) \sin^p \theta$;

$p = 0.2 + a/b + 0.6 a/t$;

$H_1 = 1 - 0.34 a/t - 0.11(a/b)(a/t)$;

$H_2 = 1 + G_1 a/t + G_2 (a/t)^2$;

$G_1 = -1.22 - 0.12 a/b$;

$G_2 = 0.55 - 1.05 (a/b)^{0.75} + 0.47 (a/b)^{1.5}$;

$Q = 1 + 1.464 (a/b)^{1.65}$;

σ_t — remote uniform tension stress, MPa ;

σ_b — remote bending stress on outer fiber, MPa ;

a — depth of crack, mm ;

b — half-length of crack, mm ;

t — plate thickness, mm ;

w — width of cracked plate, mm ;

θ — parametric angle of ellipse.

For the deepest crack tip $\theta = 0$, the corresponding stress intensity factor is given as follows:

$$K_{IA} = M F_A (\sigma_t + H_A \sigma_b) \sqrt{\dfrac{\pi a}{Q}} . \tag{2}$$

where, $F_A = [\sec(\dfrac{\pi b}{2w} \sqrt{a/t})]^{1/2}$; $H_A = H_2$.

Fatigue life reliability analyses by combination method

It is complicated to estimate the stress intensity factor with Eq. 1 and Eq. 2 using classical probabilistic fracture mechanics. In current study, Monte Carlo method is combined with classical probabilistic fracture mechanics (combination method) to solve the above problem. The stress amplitude follows:

$$\Delta K_{IA} = M F_A \Delta \sigma_t \sqrt{\dfrac{\pi a}{Q}} + M F_A H_A \Delta \sigma_b \sqrt{\dfrac{\pi a}{Q}} . \tag{3}$$

Suppose $M_1 = M\sqrt{1/Q}$, $M_2 = MH_2\sqrt{1/Q}$, then

$$\Delta K_{IA} = (M_{A1}\Delta\sigma_t + M_{A2}\Delta\sigma_b)F_A\sqrt{\pi}\sqrt{a}. \quad (4)$$

Classical probabilistic fracture mechanics can be applied here. Because the value of M_{A1} and M_{A2} would effect the accuracy of the results, reasonable M_{A1} and M_{A2} value become the critical factor to the final results.

The procedures of "combination method" in residual fatigue life analysis is as follows:

(1) Choose a small reliability and obtain the residual life under this reliability with Monte Carlo method;

(2) Estimate M_{A1} and M_{A2} according to the reliability and residual life results in step (1);

(3) Based on M_{A1} and M_{A2}, obtain the residual life under other reliability by using classical probabilistic fracture mechanics method.

Calculation Examples

The parameters of one submarine submarine pressure structure are listed in table 1. Assuming the required reliability is 0.999, the number of submerges and floating that the submarine can experience safely can be estimated followed procedures like:

(1) Firstly set a relative small reliability and it can be choose as 0.995, the residual life evaluated by Monte Carlo method is 89048;

(2) program the computer to calculate M_{A1} and M_{A2}. Here, the results are $M_{A1} \approx 0.963647$ and $M_{A2} \approx 0.784851$;

(3) Based on $M_{A1} \approx 0.963647$ and $M_{A2} \approx 0.784851$, use classical probabilistic fracture mechanics method to estimate the residual life under the reliability 0.999, and the result is 65787.

The comparisons of Monte Carlo method and combination method are listed in table 2. It can be observed that the sampling size of the combination method is much smaller than that of the Monte Carlo method. As a result, the computational time of the combination method is less than that of Monte Carlo method. Assuming the result of Monte Carlo method as exact value, the results of combination method is highly accurate and conservative.

Table 1 Parameters and their distribution type

parameters	initial crack depth a_0 [mm]	critical crack length a_c [mm]	half-length of Initial crack b_0 [mm]	shell thickness t [mm]	stress amplitude due to bending $\Delta\sigma_b$ [MPa]	stress amplitude due to tension $\Delta\sigma_t$ [MPa]	material yield strength σ_s [MPa]
mean value	0.5	4	6	20	258.6558	-103.3373	630
standard deviations	0.018	0.04	0.063	0.5	17.24	6.88	41.9
distribution type	normal						

Table 2 Comparisons of the results

	residual life	minimum sampling size
combination method	69878	20000
Monte Carlo Method	73036	100000
difference	4.32%	—

Conclusions

The sampling size and computing time of the proposed combination method are much more efficient than the conventional Monte Carlo method. Furthermore, the residual life results of combination method are highly accurate and conservative. The combination method can provide a new numerical method in estimating the residual fatigue life of submarine structures, and it can be extended to predict the residual fatigue life of analogue structures.

Acknowledgements

This work was financially supported by Harbin Engineering University Groundwork Foundation (HEUFT05001).

References

[1] Binghan Xu, Bangjun Zhu, Lvwei Ouyang, Junhou Pei. *Theory and experiments on modern submarine structure strength*. Beijing: National Defense Industry Press, 2007. (in Chinese)

[2] Hongbin Cui, Dexin Shi, Han Li, Xianqian Qu. Evaluation the size of initial crack in welding toes of joint of cylinder and cone. *Journal of Harbin Engineering University*, 2006, 27(4), pp. 492~500 (in Chinese)

[3] Liangbi Li, Zili Wang. Summarization about fatigue strength research of submarine structures. *Journal of East China Shipbuilding Institute (Natural Science Edition)*, 2004, 18(3), pp. 15~20 (in Chinese)

[4] L. B. R. Robles, M. A. Buelta, E. Goncalves. A method for the evaluation of the fatigue operational life of submarine pressure hulls. *International Journal of Fatigue*, 2000, 22(1), pp. 41~52

[5] Genki Yagawa, Shinobu Yoshimura. A study on probabilistic fracture mechanics for nuclear pressure vessels and piping. *International Journal of Pressure Vessels and Piping*, 1997, 73(1), pp. 97~107

[6] G. Yagawa. Probabilistic fracture mechanics analysis of nuclear structural components. *Nuclear Engineering and Design*, 2001, 207(3), pp. 269~286

[7] Zuoshui Xie, Zili Wang, Jianguo Wu. *Stucture analysis of submarine*. Wuhan: Huazhong University of Science and Technology Press, 2004. (in Chinese)

Geometrical parameters influencing a hybrid mechanical coupling

G. Lamanna[a], F. Caputo[b], A. Soprano[c]

Dept. of Aerospace and Mechanical Engineering, Second University of Naples,
via Roma 29, 81031 Aversa, Italy

[a]giuseppe.lamanna@unina2.it, [b]francesco.caputo@unina2.it, [c]alessandro.soprano@unina2.it

Keywords: Hybrid joint, shank-hole, clearance.

Abstract. Coupling techniques for components of different materials is spreading in mechanical industry; the test case studied in this work deals with the connection of an aluminium alloy component with a carbon fibre composite one. In particular, the first component is made of an aluminium-zinc alloy and exhibits an isotropic behaviour, while the second is made of a carbon fibre reinforced polymer (CFRP) and shows a strongly anisotropic behaviour; both materials are widely used in engineering applications. A titanium bolt connects the parts. This work is focused on the influence of the geometrical parameters which characterize the coupling between the components. In particular, a study has been carried out on the influence of the shank-hole clearance, the bolt head size, the bolt preload and the shape of the bolt head. A numerical model has been built and statically tested; the results have been compared with the experimental ones from literature. Once validated, the same numerical model has been used to evaluate the performance of the joint in presence of a change of the above mentioned characteristic parameters. The required numerical analyses have been performed using Abaqus/Standard® numerical code.

Introduction

Carbon fibre reinforced polymers composite materials are widely used in aerospace and automotive industry due to their high specific strength, stiffness and high resistance to fatigue and corrosion. The joints between components made of those materials are often the critical point of the whole structure as a result of the physical discontinuity between the parts, the stress concentrations, which develop around the joint area, and the corrosive attack from the environment. The above aspects also reflect upon the joint design of metal parts. The manufacturing of a reliable structural joint with composite materials is more challenging than those where metals are used, because of their reduced shear and bearing strengths. All these factors can exert a really high influence over the performance of the joint. In particular, their effects can be controlled by acting on both the technical and technological factors related to the geometry and adopted materials.

Some machine parts designed and built with composite materials are often innovative technological solutions, which are widely disseminated in mechanical industry. In many cases these components have to be coupled with parts made of metallic material since, evidently, at the current state of the art, not all components of such machines can be manufactured by composite materials or, at least, the use of an hybrid technology makes it more convenient than a technological solution based exclusively on composite materials.

As mentioned above the main feature of the CFRP is certainly the high strength/weight ratio which makes it suitable for many structural uses. The resistance to corrosion and to fatigue loads represents further relevant features. With reference to the last aspect, it must be stressed that difficulties are met when evaluating the birth and the propagation of a damage in the neighborhood of the bolt and predicting the residual life of a CFRP damaged component. Many scientific works deal with such issues from a numerical point of view and all of them point out the difficulties of FE modeling [1].

The performances of hybrid mechanical joints are strongly influenced by several variables of the manufacturing process. The shape of bolt head plays an important role on the performance of the joint; in particular two different titanium bolt configurations (both of them with a 6 mm shank diameter) have been investigated in [2], by considering a protruding head for the first one and a countersunk type for the second one (maximum pin diameter: 12 mm; height of conical portion: 3 mm, with reference to Fig. 1). The said paper shows how the countersunk-head bolt generally provides higher performance in comparison with the protruding-head bolt. However, it must be considered that the countersunk head requires a more expensive manufacturing process and that therefore the protruding-head solution should not be given up.

In the present work the performance of a hybrid joint has been studied, varying the clearance between the bolt shank and the hole drilled in the two plates; the same performance has been also evaluated varying the tightening torque on the bolt.

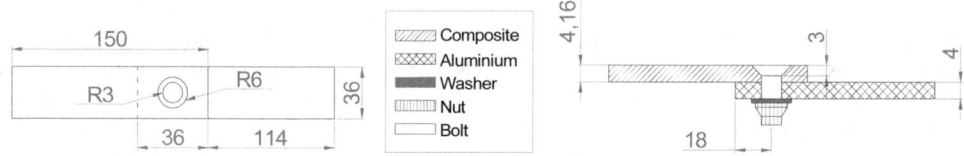

Fig. 1 – Specimen dimensions (mm).

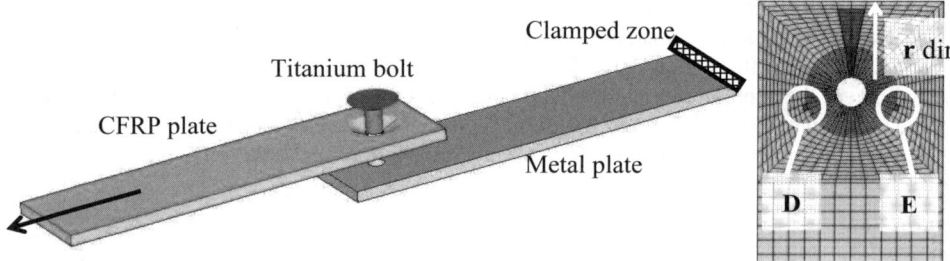

Fig. 2 – The numerical model and strain gauges positioning.

The test case refers to the joining of an aluminium-zinc alloy (AA7475-T76) component, which exhibits an isotropic behaviour, with a carbon fibre reinforced polymer (CFRP) composite, [(±45,0,90)4]s, which shows a high anisotropic behaviour; a titanium bolt (Ti6Al14VSTA) connects the parts. The mechanical properties of the different materials are shown in Table 1. Both plates measure 150x36 mm, have slightly different thickness (4.16 mm CFRP plate; 4.00 mm metal plate) and are drilled (radius 3 mm). That hybrid mechanical coupling is fixed at the end of the metal plate, faraway from the bolt, and is subject to a longitudinal loading at the end of composite plate according to Fig. 2 scheme. The testing conditions and the manufacturing process of plates are described in [3] where results of the experimental tests have been reported and compared with the numerical ones obtained through Abaqus/Standard® code. The validated model has been used to evaluate the influence of the hole-shank clearance and of the tightening torque on the joint strength.

Description of the numerical model

The numerical model has been built by Abaqus® template of the HyperMesh® software. It consists of three main components: the two plates and the bolt. The last has been considered as a single component together with the washer and the nut in order to simplify the modelling phase. The whole model has been entirely discretized through tridimensional solid elements, C3D8 of Abaqus/Standard® element library, with about 44,264 elements and 47,215 nodes; that option can be justified by considering that the selected element can be constituted of a single homogeneous material, which has been used for the metal plate and the bolt, or can include several layers of different materials, which is required to analyse the behaviour of the CFRP plate. The free end of the metal sheet has been fixed while a displacement has been imposed to the free end of the composite sheet in order to generate a traction load of 4 kN upon the specimen.

Table 1 – Material mechanical properties.

Ply properties	E_{11} [GPa]	E_{22} [GPa]	E_{33} [GPa]	E_{12} [GPa]	E_{13} [GPa]	E_{23} [GPa]	ν_{12}	ν_{13}	ν_{23}
	141	10	11	5.2	5.2	3.9	0.3	0.5	0.5
Aluminium	E [GPa]		ν		Titanium		E [GPa]		ν
AA7475-T76	69		0.28		Ti6A114VSTA		110		0.29

The modelling of the interface elements on the contact surfaces completed the numerical model: five different interfaces have been defined to avoid undesired stress concentrations around the contact areas [4]. The contact pairs can be used to define interactions between bodies in mechanical simulations. A particular contact formulation has been assigned to each contact pair and has been related to a given interaction property. A "small-sliding" technique has been applied as a tracking contact approach and the "node-to-surface" algorithm has been used to discretize contacts.

Different friction coefficients have been used according to the type of material involved in the contact, while the other boundary conditions are the same above described. The hole-bolt clearance and the axial preload over the bolt shank have been simulated by using the contact interference option of Abaqus/Standard® [5].

Results and discussion

Results providing for the validation of the numerical model are shown in Figg. 3 and 4. The strains have been compared at a load value of 4kN and the relative displacements between the plates have been compared up to this load value. The agreement between numerical and experimental computed strains is fairly good. For the case of relative displacements the numerical results are slightly higher than the experimentally measured ones at low load values but in any case the numerical results are enclosed within error ranges of experimental ones. Comparison between measured and computed strains at gauge locations D and E is shown in Fig. 4. It is possible to observe from the plots in the figure that the agreement between measured and computed strains in generally good.

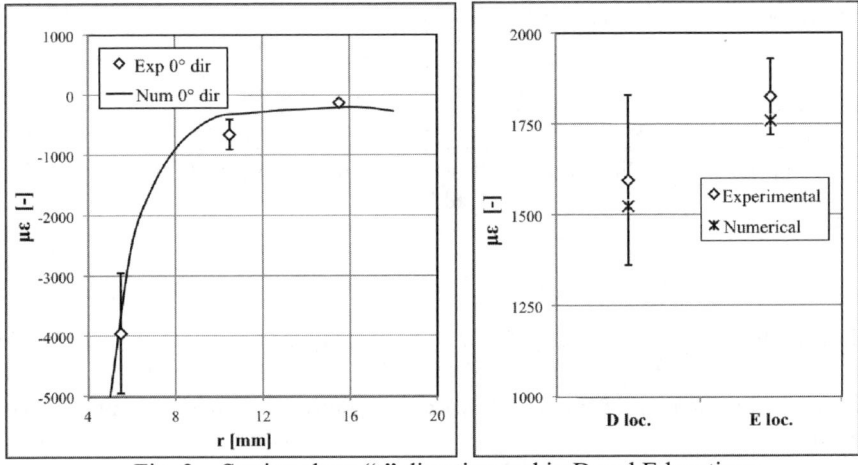

Fig. 3 – Strains along "r" direction and in D and E locations.

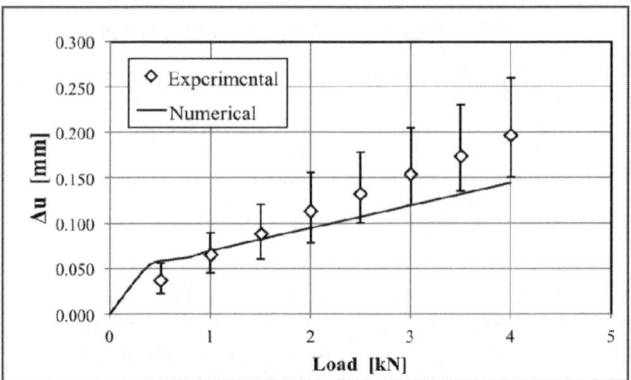

Fig. 4 – Relative displacements between the plates.

The validated model technique has been applied to the joint with countersunk bolt and the effects of the tightening torque and of the bolt-hole clearance on the strength of the joint have been investigated (Fig. 5). Strains have been calculated along "r" direction (see Fig. 2) at 4kN load value. It is clear from the figures that the strain increases for small values of bolt-hole clearance; the strains begin to decrease once the clearance achieves values over 1% of hole diameter. For high values of bolt-hole clearance the joint loses its capacity to transfer loads between the plates.

The tightening torque in general helps to reduce the values of the computed strains. Under quasi finger-tightening torque values the strains values decrease quickly; starting from 5 Nm tightening torque value, the strains values continue to decrease with a lower slope. Higher values of tightening torque cannot be taken into account because the stress concentrations in the neighborhood of the bolt become too high.

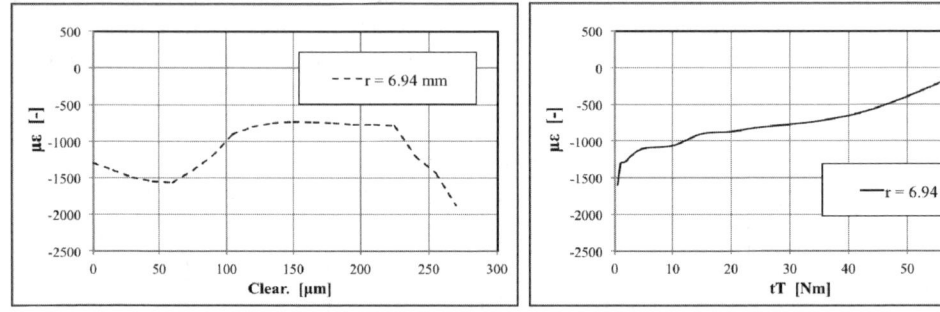

Fig. 5 – Strains along "r" direction vs bolt-hole clearance and vs tightening torque (tT) at 4 kN load level.

References

[1] W. Hufenbach, L.A. Dobrzański, M. Gude, J. Konieczny, A. Czulak: Journal of Achievements in Materials and Manufacturing Engineering Vol. 20, Iss. 1-2 (2007), p. 119.

[2] F. Caputo, G. Lamanna, A. Soprano: Structural Durability and Health Monitoring Vol.7, No.4 (2011), p. 283.

[3] T. Ireman: Compos Struct Vol. 41 (1998), p. 195.

[4] T.J. Whitney, E.V. Iarve and R.A. Brockman: Int J Solid Struct Vol. 41 (2004), p. 1893.

[5] Abaqus 6.10, Analysis User's Manual, Dassault Systèmes Simulia Corp., Providence, RI, USA (2010).

Fatigue test of small sized AZ31 magnesium alloy using micropillar specimen

Satoshi Mizuno[1,a], Toshifumi Kakiuchi[1,b], Yoshihiko Uematsu[1,c]

[1]Gifu University, 1-1 Yanagido, Gifu 501-1193, Japan

[a]q3122040@edu.gifu-u.ac.jp, [b]kakiuchi@gifu-u.ac.jp, [c]yuematsu@gifu-u.ac.jp

Keywords: Micro material, Micro mechanical testing, Fatigue, Size effect, Magnesium alloy

Abstract. The bending fatigue tests were performed using the small sized specimens of AZ31 magnesium alloy fabricated by the focused ion beam (FIB) processing to investigate the scale effect on the fatigue behavior. For the fatigue test of the small sized specimens, the fatigue testing apparatus was constructed by the piezo actuator and the high-resolution microscope for controlling the very small displacement. The specimen was micropillar-shaped with the rectangular cross section of 3μm x 8μm and the height of 40μm. The fatigue strengths of the small sized specimens were higher than those of the bulk sized specimens. The fracture surfaces were also investigated carefully compared with those of the bulk specimen.

Introduction

The microminiaturization of various devices is enabled by the recent rapid innovation, and the micromachine is developing with it in various fields; for example the field of the medicine. For the use of the micromachine, the evaluation of the mechanical properties of micro sized materials (micro material) constituting the parts of the micromachine is important. However, the problems such as the handling of the specimen or the measurement of the stress exist in the mechanical testing of micro materials because the micro material is extremely small. Therefore the universal testing methods have not been established up to the present and the reports of mechanical strengths of micro materials are limited. In recent years the material processing using the focused ion beam (FIB) has been developed remarkably and came to be used for the fabrication of the micro sized specimens [1, 2]. The FIB processing is flexible, and with the FIB processing it is easy to fabricate micro sized specimens in the bulk material, which could be handled easily. In this study, pillar-shaped specimens of micro dimensions, so called micropillars, were fabricated by the FIB processing. Using the specimens, the bending fatigue tests were performed by the fatigue testing apparatus for the micro material which was designed for the present study using the piezo actuator and the high-resolution microscope, and the fatigue behavior of the micro material was investigated.

Experimental Details

Material and Specimen. The material used is the magnesium alloy AZ31 of which the FIB processing is easy since the resistance to the processing is small. Fig. 1 shows the schematic configuration of the specimen fabrication. In the FIB processing, positive gallium ions (G^+) are sputtered to the object to be processed. The specimens were micropillar-shaped with the rectangular cross section and they were fabricated by the FIB processing in a side of the cube-shaped bulk block of the material so that they were easily loaded to be pushed perpendicularly to the side surface of the bulk block. The dimensions of the micropillar are 8μm in the width, 3μm in the thickness and 40μm in the height. The average grain size of the material is 29μm, so one grain is included in the width and the thickness directions, and 2 or 3 grains are included in the height direction. Fig. 2 shows an example of the scanning ion microscope (SIM) micrograph of the micropillar. It was observed that the micropillar was successfully fabricated with the precise dimension of the micro-order by the FIB processing.

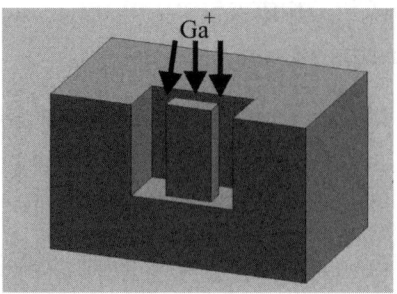 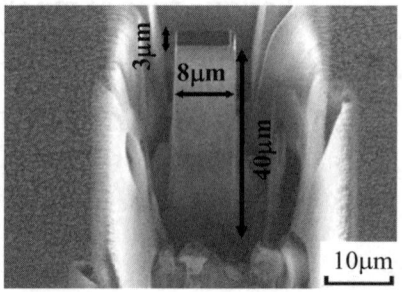

Fig.1 Schematic configuration of micropillar fabrication by FIB processing.

Fig.2 SIM micrograph showing a micropillar.

Fatigue Test. Fig. 3 shows the appearance of the fatigue testing apparatus for the micro material designed for the present study. The micropillar-shaped specimen is fabricated at the side of the bulk block which is set on the xyz stage. The specimen is pushed and loaded by the glass probe with the end tip radius of approximately 3.5μm. The glass probe is attached to the piezo actuator to be controlled by the very small displacement. For controlling the displacement, the signal voltage is input to the piezo actuator from the PC and through the piezo actuator where the input voltage is converted into the displacement so that the fatigue test is displacement-controlled. During the fatigue test, the specimen is observed by the high-resolution digital microscope which is set above the specimen. The displacement of the micropillar is measured by the microscopy observation. The fatigue testing conditions were set to be the load sinusoidal frequency $f=15$Hz and the stress ratio $R=0$ in the laboratory room atmosphere. After the fatigue tests, the fracture surfaces were observed by a scanning electron microscope (SEM).

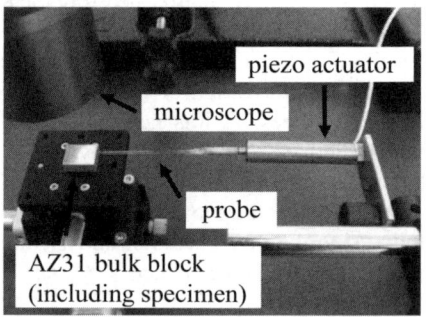

Fig.3 Fatigue testing apparatus.

Results and Discussion

Cyclic Loading. Fig. 4 shows the magnified view of the micropillar-shaped specimen and the settings around the specimen before the onset of the fatigue test by the digital microscope above the specimen. The probe is fixed and the position of the bulk block is adjusted by the xyz stage so that the specimen is moved to the initial position of the fatigue test to be contacted with the probe end. During the adjustment of the position, the specimen is being observed by the digital microscope. Fig. 5 shows a series of time-resolved divided photographs during the loading cycles with the loading of 2μm displacement. As shown in Fig. 5, the testing apparatus was able to give the cyclic loading of displacement to the specimen continuously.

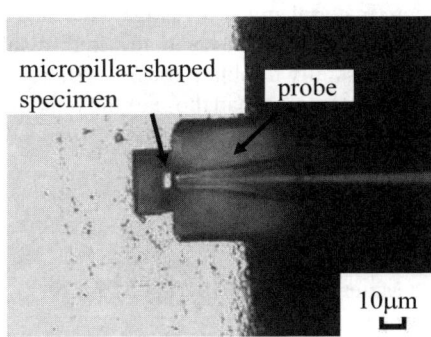
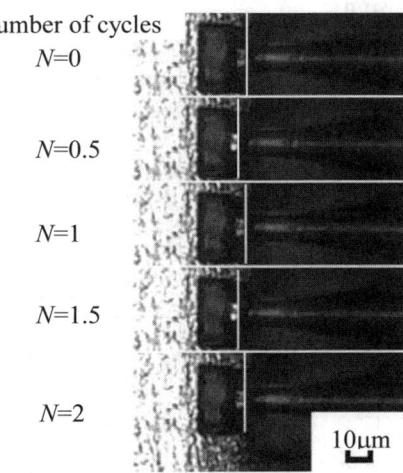

Fig.4 Magnified view of specimen and settings befor fatigue test.

Fig.5 Time-resolved photography during fatigue test with loading of 2μm displacement.

Since the micropillar in this test is assumed to be a cantilever, the maximum stress occurs at the root of the micropillar and it is estimated by the following equation (1) known as the formula of a cantilever,

$$\sigma = \frac{3Eh\delta}{2l^2}, \quad (1)$$

where E is the elastic modulus, h is the thickness (3μm) of the micropillar, l is the length (40μm) and δ is the displacement of the micropillar at the loaded point. In this paper, it is supposed that the elastic modulus of the micropillar is the same as that of the bulk material and that it is 45GPa. On this supposition, the cyclic applied stress in the test shown in Fig. 5, where the cyclic displacement is 2μm, is calculated as 253MPa, which is higher than the yield stress of the bulk material AZ31, 170MPa. However the micropillar did not yield in this test.

Fatigue Strength. Fig. 6 shows the S-N curves of the micropillar specimens and the bulk ones. The applied stress of the micropillar is calculated by substituting the displacement of the micropillar for Eq. 1. where the elastic modulus is 45GPa and the displacement of the micropillar is determined by the microscopy observation as mentioned above. The S-N data of the bulk material were obtained by the plane bending fatigue tests performed at the stress ratio $R=-1$. The equivalent stress amplitude of bulk material at $R=0$, σ_{eq}, was calculated using modified Goodman's diagram as follows.

$$\sigma_{eq} = \frac{\sigma_a}{1-(\sigma_{mean}/\sigma_B)}, \quad (2)$$

where σ_a is stress amplitude, σ_{mean} is mean stress and σ_B is tensile strength. The fatigue limits are 120MPa and 75MPa for the micropillar and bulk material at $R=0$, respectively. The higher fatigue limit of micropillar could be attributed to the size effect.

Fracture Surface Observation. All the micropillars were fractured at the root of the pillar where the applied stress was the highest. Fig. 7 shows an example of the fatigue fracture surface micrographs. In the bulk material, dimples are usually observed in the unstable fracture region of the fatigue fracture, but as seen in Fig. 7, no dimples were observed in the micro material. The fracture surface was flat and the crack initiation site was not identified.

Conclusion

In the present study, the micropillar-shaped specimens of magnesium alloy AZ31 were fabricated by the FIB processing to perform the fatigue tests of the micro material and to investigate the size effect. The fatigue testing apparatus was constructed using the piezo actuator and the high-resolution microscope. The bending fatigue tests were performed successfully to obtain the S-N relation of the micro material. The micro material showed the higher fatigue strengths than those of the bulk material due to the size effect. The fracture surfaces were flat and no dimples were observed.

References

[1] E. Lilleodden: Scripta Materialia 62 (2010), pp. 532-535

[2] C.M. Byer, B. Li, B. Cao and K.T. Ramesh: Scripta Materialia 62 (2010), pp. 536-539

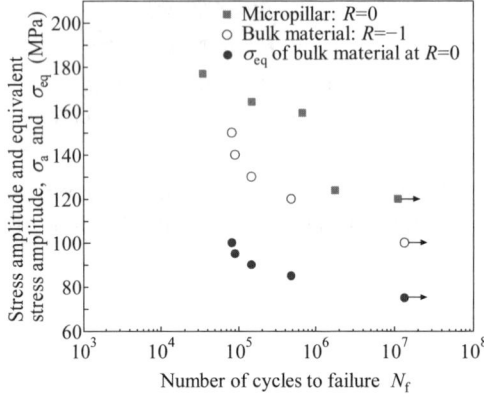

Fig.6 S-N curves of AZ31.

Fig.7 SEM micrograph showing fracture surface. (σ_a=124MPa, N_f=1.8 × 10^6)

Fatigue Behavior of A356 Cast Aluminum Alloy Microstructurally Modified by Friction Stir Processing under Low Strain Rate Condition

Toshifumi Kakiuchi[1,a], Yoshihiko Uematsu[1,b] and Yasunari Tozaki[2,c]

[1]Gifu University 1-1 Yanagido, Gifu 501-1193, Japan

[2]Gifu Prefectural Research Institute for Machinery and Materials, 1288 Oze, Seki 501-3265, Japan

[a]kakiuchi@gifu-u.ac.jp, [b]yuematsu@gifu-u.ac.jp, [c]tozaki-yasunari@pref.gifu.lg.jp

Keywords: cast aluminum alloy, friction stir processing, microstructural modification, fatigue behavior.

Abstract The fatigue behavior of cast aluminum alloy, A356-T6, microstructurally modified by the friction stir processing (FSP) was investigated. The FSP conditions were set to be the tool rotational speed of 500 rpm and traveling speed of 200 mm/min, in which the strain rate was relatively low. Plane bending fatigue tests have been performed using the as-cast and friction stir processed (FSPed) specimens. Fatigue strengths in the finite life region and the fatigue limit of the FSPed specimens were highly improved compared with the as-cast ones resulting from the elimination of casting defects by the FSP. However, the crack growth rates of the FSPed specimens were faster than those of the as-cast ones due to the softening of the material by heat input during the FSP. The effects of FSP with low stain rate were discussed based on the microstructural consideration.

Introduction

The cast aluminum alloy is widely used as the light structural material. In the cast alloy, the fatigue crack is initiated from the casting defect. It is difficult to eliminate casting defects only by the casting process but the microstructural modification after casting process is effective. The friction stir processing (FSP), an applied technique of friction stir welding (FSW), is utilized as one of the techniques of microstructural modification of cast alloys to eliminate casting defects [1]. In the FSP, the material is stirred strongly with the plastic flow and the microstructures are modified. By the FSP, the elliptical banded structure, or so-called the onion structure, is formed. The onion structure could be crack initiation sites and crack growth paths resulting in the harmful affect on the fatigue behavior. In this paper, the heat-treated cast aluminum alloy A356-T6 was modified by the FSP at the low strain rate to reduce the onion structure and the fatigue behavior of the FSPed material was investigated.

Material and Experimental Procedures

Material and FSP Condition. The material used is heat-treated cast aluminum alloy A356-T6. The chemical composition and the mechanical properties are shown in Table 1 and Table 2. The FSP tool is 14 mm in the shoulder diameter and 4.7 mm in the probe length. The probe is M6-threaded in a left-handed screw. In the FSP, the tool was rotated clockwise and tilted 3° opposite to the processing direction. The FSP conditions were set to be the tool rotational speed of 500 rpm and the traveling speed of 200 mm/min. The strain rate of the FSP set in the present study was relatively low compared with our previous study in which the conditions were set to be the tool rotational speed of 1000 rpm

Table 1 Chemical composition of A356-T6 aluminum alloy (wt%).

Material	Si	Mg	Fe	Ti	Ni	Cr	Sr	Al
A356-T6	7.1	0.39	0.06	0.13	0.003	0.001	0.0042	Bal.

Table 2 Mechanical properties of A356-T6 aluminum alloy.

Material	0.2% proof stress, $\sigma_{0.2}$ (MPa)	Tensile strength, σ_B (MPa)	Elongation, δ (%)	Young's modulus, E (GPa)
A356-T6	264	324	10.9	75.1

and the traveling speed of 150 mm/min [2, 3]. The plate for the FSP was taken from the boat-shaped cast aluminum ingots and after the FSP it was machined to the fatigue test specimen, denoted as FSPed. The longitudinal direction of the specimen was set parallel to the processing direction. The fatigue specimen machined from the plate without the FSP, denoted as as-cast, was also prepared for the comparison with the FSPed specimen. The configuration of fatigue test specimen is; the thickness is 4mm, the length is 120mm and the minimal width of the gauge section is 5mm for the FSPed specimen and 8 mm for the as-cast specimen. Before the fatigue test, the surface of the gauge section of the specimen was polished using #2000 grade emery paper and buff-finished.

Experimental Procedures. The plane bending fatigue tests were performed using the resonance type bending fatigue testing machine with the frequency f=33.3 Hz and the stress ratio R=-1 in the laboratory air. The crack growth behavior was observed by a plastic replication technique.

Experimental Results

Microstructures. Fig. 1 shows the microstructures of the as-cast and the FSPed materials. In the as-cast material (Fig. 1(a)), the dendrite structure where eutectic silicon is net-like distributed is observed. In the microstructures, casting defects are also observed. In the FSPed material (Fig. 1(b)), the eutectic silicon is distributed uniformly. The dendrite structure or the casting defects are not observed after the FSP; that is the FSP is effective to eliminate the casting defects. Fig. 2 shows the macroscopic views of the microstructures of FSPed material. The material which was FSPed at higher strain rate [2, 3] is also shown. The elliptical banded onion structures are observed in both materials;

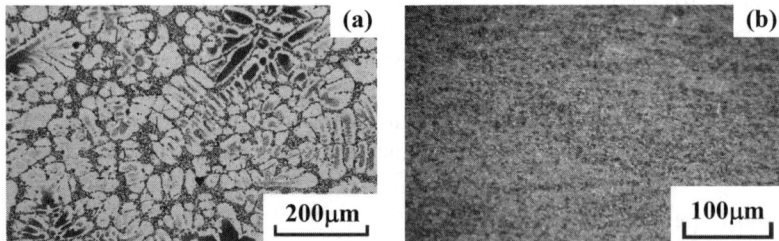

Fig.1 Microstructures of the materials: (a) As-cast, (b) FSPed.

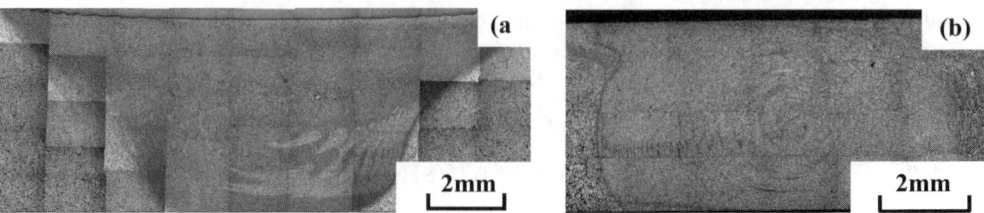

Fig.2 Macroscopic views of microstructures of FSPed materials:
(a) at low strain rate, (b) at high strain rate [2, 3].

that FSPed at lower strain rate (Fig. 2(a)) and that FSPed at higher strain rate (Fig. 2(a)). However, the onion structure is reduced in the material FSPed at lower strain rate; that is the low strain rate is effective to restrain forming onion structures. The Vickers hardness of the as-cast material is 118 HV and that of the FSPed material is 59 HV which becomes considerably low. It is attributed to the resolution of the hardened precipitates by the heat input during the FSP.

Fatigue Behavior. Fig. 3 shows the S-N diagram. The fatigue limit of the as-cast specimen is 130 MPa and that of the FSPed specimen is 200 MPa. The fatigue strength in the finite life region is also highly improved by the FSP. Fig. 4 shows the crack initiation sites in the as-cast and the FSPed specimens. In the as-cast specimen, the crack initiation site is the casting defect. On the other hand, in the FSPed specimen, there is no casting defect observed in the microstructure as in Fig. 1(b) and the crack initiation site is not the casting defect.

Discussion

For the further discussion of the fatigue behavior, the crack growth behavior was investigated by the plastic replication technique, of which the results are shown in Fig. 5. Fig. 5(a) shows the relationship between the surface crack length $2c$ and the fatigue cycle ratio N/N_f. In the as-cast specimen, the crack is non-propagating in the early stage after the crack initiation. However, the crack initiation of the as-cast specimen is earlier than the FSPed specimen; that is the crack initiation resistance becomes higher by the FSP, which is attributed to the transition of the crack initiation mechanism as seen in Fig. 4. Fig. 5(b) shows the relationship between the crack growth rate dc/dN and the maximum stress

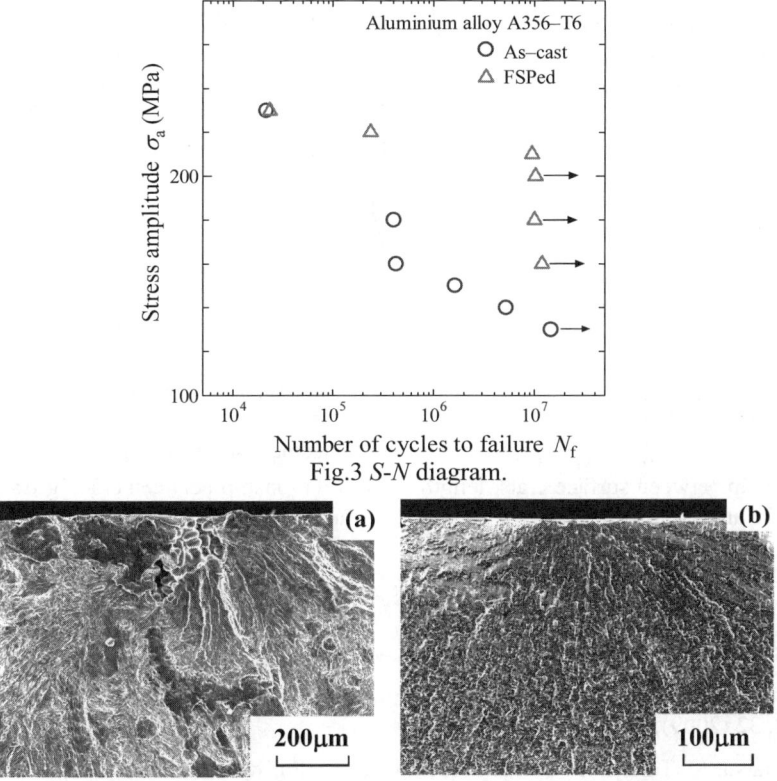

Fig.3 S-N diagram.

Fig.4 Fatigue fracture surfaces near crack initiation sites:
(a) as-cast (σ=180MPa), (b) FSPed (σ_a=220MPa).

intensity factor K_{max}. On the contrary to the crack initiation behavior, the crack growth rate becomes faster in the FSPed specimen than in the as-cast specimen; that is the crack growth resistance becomes lower by the FSP, which is attributed to the softening of the microstructures due to the heat input. Although the crack growth resistance in the FSPed specimen is lowered, the improvement of the crack initiation resistance results in the improvement of the fatigue strength. The specimen FSPed at the higher strain rate [2, 3] had 40MPa higher fatigue limit than the as-cast specimen under axial loading condition, while the FSP reduced the fatigue strengths in the finite life region. As shown in Fig.3, the specimen FSPed at the lower strain rate has 70MPa higher fatigue limit, and the fatigue strengths in the finite life region were also improved. The lower strain rate condition reduced the softening of the T6-treated A356 due to the lower heat input during FSP. Furthermore, the formation of onion rings, which had detrimental effect on the crack initiation, was suppressed under low strain rate. Consequently, the FSP at low strain rate could improve fatigue strength better than the FSP at high strain rate.

Conclusion

In this study, the heat-treated cast aluminum alloy A356-T6 was modified by the FSP at the low strain rate and the fatigue behavior was investigated. By the FSP, casting defects were eliminated and the fatigue strength was highly improved. Compared with the FSP at the high strain rate, the formation of the onion structure was reduced in the FSP at the low strain rate and it is considered to be more effective to improve the fatigue strength.

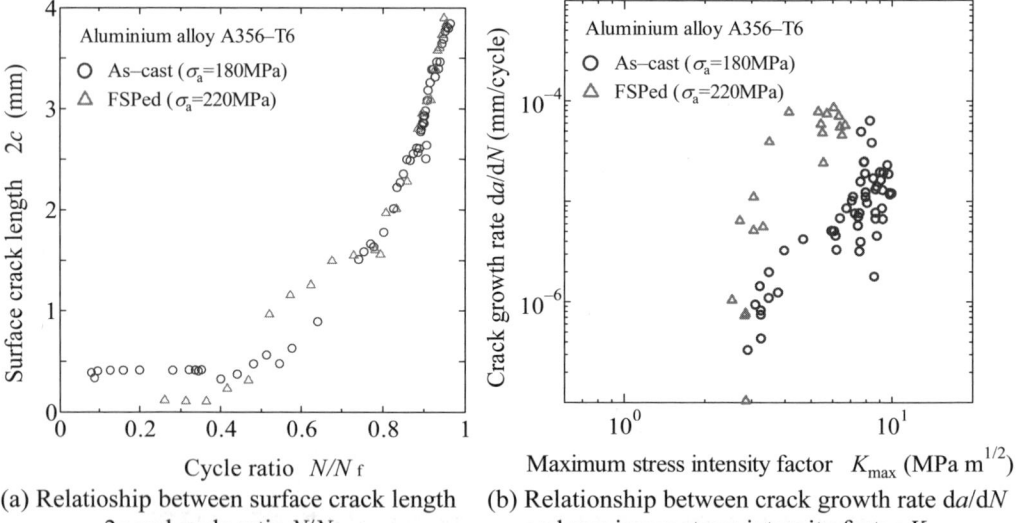

(a) Relatioship between surface crack length $2c$ and cycle ratio N/N_f.

(b) Relationship between crack growth rate da/dN and maximum stress intensity factor K_{max}.

Fig.5 Crack growth behavior.

References

[1] Y. Uematsu, K. Tokaji, K. Fujiwara, Y. Tozaki and H. Shibata: Fatigue Fract. Engng. Mater. Struct., Vol. 32 (2009), pp. 541 – 551

[2] Y. Uematsu, K. Tokaji, Y. Tozaki, H. Shibata, K. Fujiwara and T. Murayama: Journal of the Society of Materials Science, Japan, Vol. 58 (2009), pp. 69 – 75 (in Japanese)

[3] Y. Uematsu and K. Tokaji: Materials Science Forum, Vol. 638-642 (2010), pp. 3727–3730

Load Influence on the Behavior of Micro-crack in the Particulate Composite

Zdeněk Majer[1,a]

[1]Faculty of Mechanical Engineering, Brno University of Technology, Technická 2896/2,

616 69 Brno, Czech Republic

[a]majer@fme.vutbr.cz

Keywords: polymer particulate composites, interphase, fracture mechanics, non-linear matrix FEM, shielding effect.

Abstract. Particulate composite with soft polymer matrix and rigid mineral fillers are one of most frequently used construction and engineering materials. The main focus of a present paper is an estimation of the load influence on behavior of micro-crack placed in close proximity to the particle with interphase in soft matrix. The particulate composite with polymer matrix filled by magnesium-based mineral fillers is investigated by means of the finite element method. A non-linear material behavior of the matrix was considered. Numerical model on the base of representative plane element (RPE) was developed. The conclusions of this paper can contribute to a better understanding of the behavior of micro-crack in particulate composites with soft polymer matrix.

Introduction

Polymer particulate composites are frequently used in many engineering applications and they are of great practical importance due to the possibility of modifying the mechanical properties of the resulting composite, see e.g. [1]. The composite was modeled as three-phase continuum. Together with particles and matrix is considered an interphase which is created at the matrix-particle interface due chemical interaction. In fact, the presence of interphase changes adhesion between the matrix and the particle and can play a significant role in micro-crack initiation and behavior in composite, see e.g. [2-4]. Moreover, the mineral particles are usually chemically treated to reach better dispersion in the polymer matrix and this process positively influences formation of interphase.

The interphase properties can be estimated only indirectly from the macroscopic composite properties (modulus, yield stress, tensile strength). The interphase thickness varies in range 0.012 μm to 0.160 μm according to used method, see e.g. [4,5]. In the paper [4] the interphase thickness is correlated with the work on adhesion and for the uncoated particles is estimated as 0.1 μm.

Material model and experimental data

Several various experiments focused on estimation of polymer particulate composite macroscopic properties were performed at the Institute of Material Science and Engineering, Faculty of Mechanical Engineering (Brno University of Technology) and at the Polymer Institute Brno, spol. s r.o [6].

The particle size was determined from experiments which were done at the Polymer Institute Brno, spol. s r. o. From these measurements the typical size of the particles was estimated as 1 μm. The influence of particle size on macroscopic composite behavior considering linear elastic properties of matrix was formerly studied in [7].

Numerical model

Micro-crack behavior in the three phase composite was numerically simulated on a microscopic scale using the finite element program ANSYS. The numerical model was developed assuming that only the particles located close to the micro-crack tip significantly influence micro-crack behaviour.

The geometry of the model is shown in Fig. 1b. For given particle size, distance between particles *2b* corresponds to the volume filler fraction (VFF) of the composite which in our case is 25%. For reason of symmetry only one half of the representative plane element was modelled and symmetrical boundary conditions were used, see Fig. 1c.

Preconditions. During creation of the numerical model it is very important keep the following general terms [8]: (i) particles should be of small size (less than 5 μm), (ii) the aspect ratio must be close to unity to avoid high stress concentration, (iii) the particles must debond prior the polymer matrix reaches the yield strain in order to change the stress state of the matrix material and (iv) particles must be dispersed homogeneously in the polymer matrix. The homogenously distributed spherical particles in the matrix are considered, see Fig. 1a. The influence of non-spherical particles was discussed i.e. in [9] (only linear behavior of all composite components was considered).

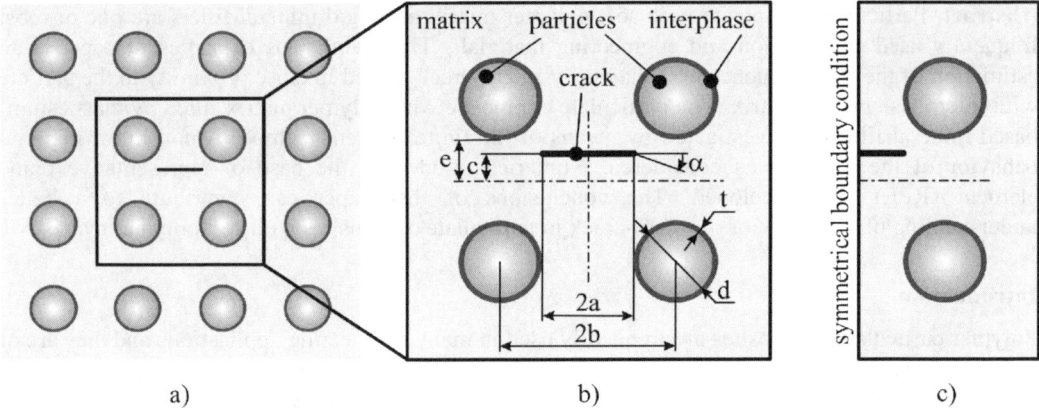

Fig. 1 a) Homogenous and regular distribution of the particles inside polymer matrix with micro-crack location, b) geometry of the model used for micro-crack behavior estimation in the polymer composite and c) simplified computational model

Material properties. The Young modulus of the particles E_p was considered as 72 GPa, and the value of Poisson's ratio $v = 0.29$. The corresponding parameters of the soft polymer matrix were estimated from the experimental measurements. The thickness of the interphase t is considered here as 0.1 μm [4]. The value of interphase Young modulus E_i is considered between 0.05 and 3.3 GPa. Value 3.3 GPa corresponds to Young modulus of pure matrix measured at -10°C. Constant value of Young modulus E_i through the interphase thickness was assumed in numerical models.

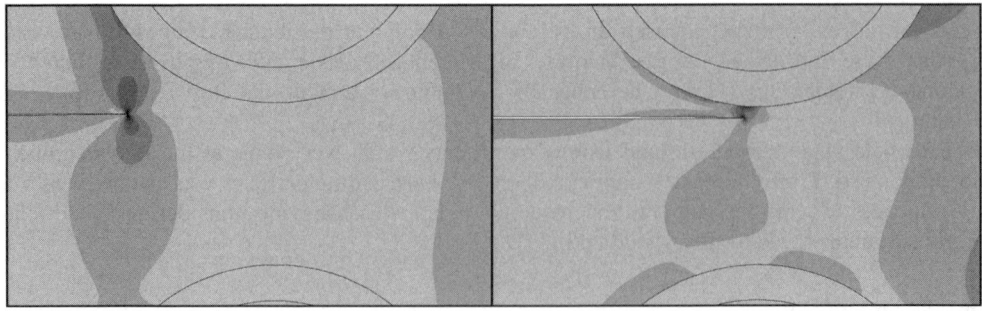

Fig. 2 The range of the analyzed micro-crack sizes *a* (from 45% to 95% of *b*)

Results and discussion

Basic geometry of numerical model of polymer particulate composite mentioned above was used in FE simulations for estimation of load. The same size of particles was considered in all calculations. Results obtained are shown in Fig. 3. The results describe patterns of local mechanical stress (von Mises). On the results is shown that the maximum value of von Mises stress is calculated approximately 49 MPa. This value was achieved already for load 3 MPa. The increasing of load leads only to change of patterns of local mechanical stress but the maximum stress value remains the same.

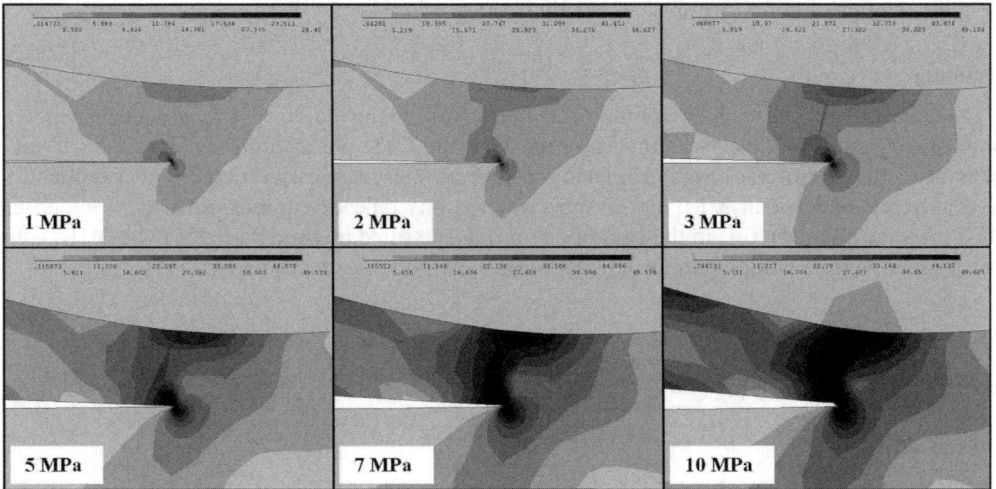

Fig. 3 Patterns of local mechanical stress (von Mises) for different values of load

On the base of previous results the value of used load for next calculations was determined as 3 MPa. For this load value the estimation of micro-crack behaviour was provided.

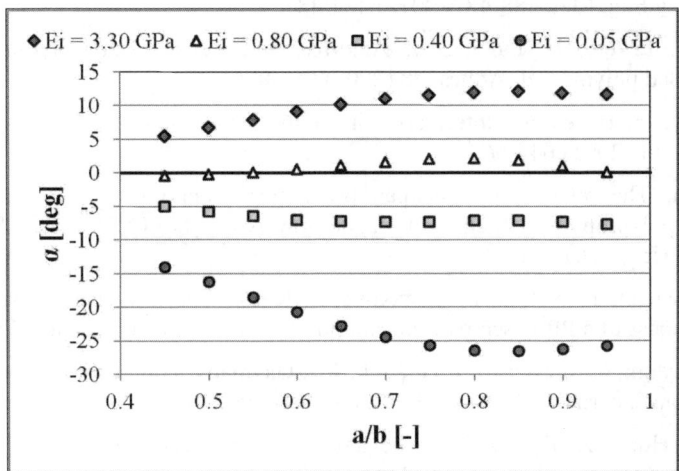

Fig. 4 Dependence of micro-crack propagation direction α on micro-crack length for ratio $c/e = 0.9$ and for different values of interphase Young modulus (matrix properties at -10°C are considered)

Assuming that particles are regularly distributed in the matrix, the behavior of the micro-crack was studied for particle size 1μm, see Fig. 4. The interphase Young modulus was changed from 0.05 GPa to 3.3 GPa (special case the configuration corresponding to perfect adhesion between particles and matrix). In the case of $E_i = 3.3$ GPa the angle of micro-crack propagation direction α is always positive. It means that the micro-crack avoids regions with rigid particles and grows only in the soft matrix. For smaller values of the interphase Young modulus ($E_i = 0.05$ GPa) rigid particle is completely shielded by softer interphase and the calculated angle of micro-crack propagation direction α was always negative. The limit value of interphase Young modulus $E_i = 0.8$ GPa was calculated. This is the highest value even when rigid particle is at least partially shielded by a softer interphase and the micro-crack does not avoid regions with rigid particles.

Summary

The obtained results can be summarized by the following points: (i) the toughened polymer composite by rigid particles was investigated using the three-phase finite element model and the behavior of the micro-crack was estimated for several interphase properties; (ii) it was found that the numerical model load 3 MPa was sufficient for this type of composite (matrix properties at -10°C were considered); (iii) in the case of polypropylene filled by rigid particles (Mg(OH)$_2$) is limit value of interphase Young modulus 0.8 GPa for matrix properties measured at -10°C.

Acknowledgement

This contribution was supported by Grant No. P107/10/P503 of the Czech Science Foundation.

References

[1] H.G. Elias, An Introduction to Plastics, second ed., Wiley-VCH GmbH & Co.KGaA, Weinheim, 2003.

[2] A. Ayyar, N. Chawla, Microstructure-based modeling of crack growth in particle reinforced composites, Comp. Sci. Tech. 66 (2006) 1980-1994.

[3] Z. Majer, P. Hutař, The effect of nonlinear matrix on crack propagation in the particulate composite, Key Eng. Mat. 488-489 (2012) 484-487.

[4] J. Moczó, F. Fekete, B. Pukánszky, Acid-base interactions and interphase formation in particulate-filled polymers, J. Adhes. 78 (2002) 861-876.

[5] B. Pukánszky, Interfaces and interphases in multicomponent materials: past, present, future, Eur. Polym. J. 41 (2005) 645-662.

[6] E. Molliková, The relationship between production technology, structure and mechanical properties of polypropylene filled with magnesium hydroxide, Ph.D. thesis, FME BUT, Brno, 2003, pp. 103 (in czech).

[7] P. Hutař, Z. Majer, L. Náhlík, L. Šestáková, Z. Knésl, The influence of particle size on the fracture toughness of a PP-based particle composite, Mech. of Comp. Mat. 45 (2009) 281-286.

[8] W.C.J. Zuiderduin, C. Westzaan, J. Huétink, R.J. Gaymans, Toughening of polypropylene with calcium carbonate particles, Polymer 44 (2003) 261-275.

[9] Z. Majer, P. Hutař, Z. Knésl, Crack behaviour in polymeric composites: The influence of particle shape, Key Eng. Mat. 465 (2011) 564-567.

Pressure pipe damage: Numerical estimation of point load effect

Michal Zouhar[1,2,a], Pavel Hutař[1,b], Martin Ševčík[3,c] and Luboš Náhlík[3,d]

[1] Institute of Physics of Materials, Žižkova 22, 616 62 Brno, Czech Republic

[2] Brno University of Technology, Technická 2, 616 69 Brno, Czech Republic

[3] CEITEC IPM, Institute of Physics of Materials, Žižkova 22, 616 62 Brno, Czech Republic

[a]zouhar@ipm.cz, [b]hutar@ipm.cz, [c]sevcik@ipm.cz, [d]nahlik@ipm.cz

Keywords: polymer pressure pipes, point load, stress intensity factor, crack shape prediction

Abstract. The most relevant loading conditions for real polymer pipe systems are not only internal pressure, but also loading caused by sand embedding including bending or different kinds of point loads. It has been shown that service lifetime of buried pipes can be reduced especially due to stress concentration caused by external point loads. If the pipe is loaded locally the stress is concentrated here and a crack can initiate at this position or the existing crack can be affected by corresponding stress redistribution. In the paper the effect of the hard indenter, Poisson's ratio, hoop stress level and pipe wall thickness on the crack shape was estimated using numerical simulations of the creep crack propagation based on finite element method. Relation between crack length and crack width was found and expressed by simple relationship. A deeper understanding of the point load effect in order to prevent unexpected failure of the pipelines is of paramount importance for pipeline design.

Introduction

The main advantages of polymer pipe materials in the comparison with traditional one can be summarized as follows: better corrosion, chemical and microorganism resistance, material properties can vary for different kinds of requirements and lifetime of the modern polymer pipes is predicted up to 100 years. The traditional method of lifetime estimation of these systems is based on hydrostatic pressure tests [1]. Hydrostatic pressure tests are conducted in specific environments, at various pressure levels and at different temperatures. Failure time is usually expressed by a log-log diagram hoop stress (burst stress) versus failure time, see e.g. [2,3]. The most relevant for real service failure of PE pipes is area of slow crack growth (in the region of quasi-brittle fracture) which can be described by linear elastic fracture mechanics concepts [2,4]. Numerical prediction of the pipe lifetime based on approach of linear elastic fracture mechanics approach is describes in details in [5]. The most relevant loading conditions for real polymer pipe systems are not only internal and external pressure, but also loading caused by sand embedding including bending or different kinds of point loads [6]. It has been shown that service lifetime of buried pipes can be reduced especially due to stress concentration caused by external point loads [7]. This aspect corresponds to real conditions of soil embedding.

Presented article is focused on the assessment of point load effect on the damage of pressurized polymer pipe. Using numerical simulations (based on finite element method) the effect of the material properties, different pipe and indenter geometry on the shape of propagating crack and fracture parameters are estimated.

Numerical model

The parametric finite element model of the pipe with indenter was created in software Ansys, see Fig.1. Dimensions of the pipe are given by particular standard dimension ratio (SDR). The SDR is the ratio of the nominal outside diameter D of a pipe to its nominal wall thickness s. The usual pipe SDRs used in practice are 9, 11, 13.6, 17. Numerical simulations presented in the work follow those geometries for $D = 40$ mm. It was shown experimentally, that the failure of the polymer pipe occurs by slow crack growth from inner surface defect [8]. A stress intensity factor K was used ss a relevant fracture parameter. Than the stress field around the crack tip for loading mode I can be described by following equation [9].

$$\sigma_{ij} = \frac{K_I}{\sqrt{2\pi r}} \cdot f_{ij}(\theta), \tag{1}$$

where K_I is stress intensity factor, r and θ are polar coordinates with the origin at the crack tip and f_{ij} is known function, see [9] for details. Stress intensity factor was estimated by direct method. For better accuracy of the direct method fine finite element mesh near the crack tip is necessary, see Fig.1. With the assumption that the crack growths with constant increments [10] the crack shape was determined using an assumptions of constant value of the stress intensity factor along the crack front and elliptical crack geometry is characterized by ratio $b/2a$.

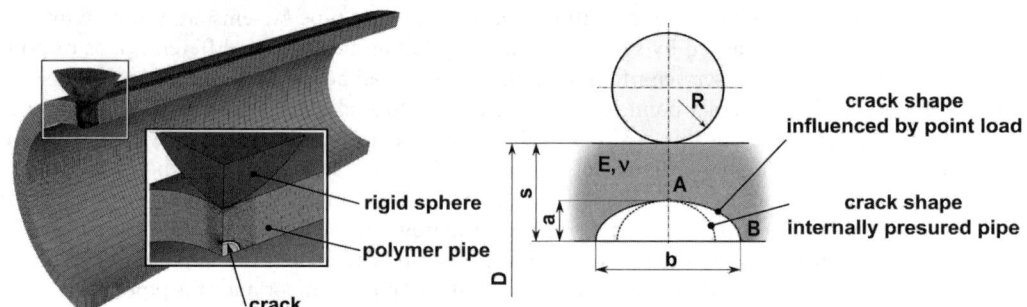

Fig. 1 Numerical model of the pipe used for calculations (left), geometrical configuration of the crack bellow the indenter (right)

In this study the linear elastic material properties of the polymer pipe were used, Young's modulus was $E = 940$ MPa corresponding to 20°C, Poisson's ratio $v = 0.33$. Numerical model of the pipe with initial crack was loaded by inner pressure p which corresponds to hoop stress σ_{hoop} in the range of 6-10 MPa. The point load was modeled by contact of the outer pipe wall with a rigid spherical indenter. This numerical model also corresponds to the usual experimental set up [11] used for point load effect evaluation. Spherical indenter geometry with radius R from 2.5 to 25 mm simulate contact of small sharp stone or big blunt stone respectively.

Results and conclusions

Crack geometry has significant influence on the resulting stress intensity factor value and lifetime of the whole system. The shape of the crack was numerically estimated using a special algorithm, which ensures constant stress intensity factor (SIF) along the crack front. The parametric study of the effects which can occur in polymer pipe application was performed by finite element method. The influence of the crack shape expressed by aspect ratio $b/2a$ on hoop stress level is shown in Fig. 2. The effect of the hoop stress level on the crack shape increases with increase of the crack depth (for the crack depth ratio $a/s = 0.6$ is around 4%). The influence of the wall thickness s characterized by standard dimension ratio SDR on crack shape is shown in Fig. 3. It can be seen that the influence of the SDR on the crack aspect ratio is stronger than in previous case and maximum difference for $a/s = 0.6$ is approximately 13%. The effect of Poisson's ratio which is varied from 0.25 to 0.4 on crack shape is shown in Fig. 4. Discrepancies in this case are about 5%, but for real polymer materials, where Poisson ratio varies usually in the range 0.29-0.35, this effect can be neglected. The effect of the size of the rigid indenter on crack shape in dependency on crack depth was studied as well, see Fig. 5. This effect was smaller than 10% for the mentioned polymer pipe configuration. Increase of the sphere radius leads to increase of the crack width for the same loading conditions.

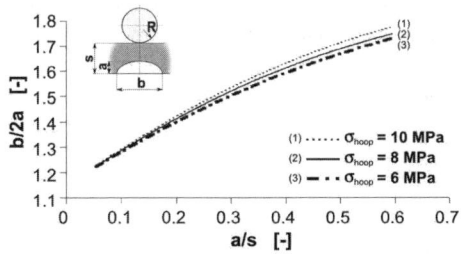

Fig. 2. Influence of the hoop stress σ_{hoop} on crack shape $b/2a$ in dependency on the crack depth, SDR 11, $R = 10$ mm

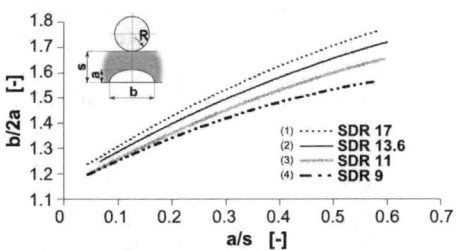

Fig. 3. Influence of the SDR ratio on crack shape $b/2a$ in dependency on the crack depth, $\sigma_{hoop} = 6$ MPa, $R = 2.5$ mm

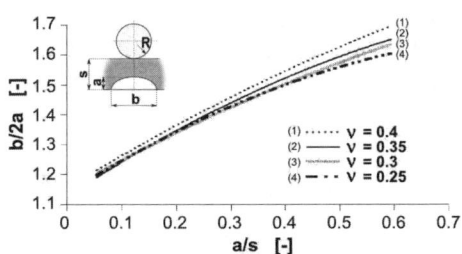

Fig. 4. Influence of the Poisson's ratio v on the crack shape $b/2a$ in dependency on the crack depth, SDR 11, $\sigma_{hoop} = 6$ MPa, $R = 2.5$ mm

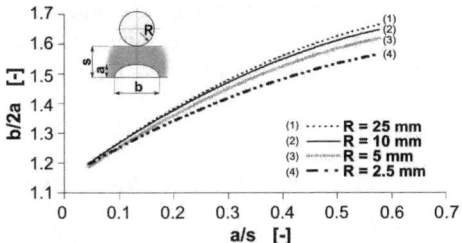

Fig. 5. Influence of the rigid sphere radius R on the crack shape $b/2a$ in dependency on the crack depth, SDR 9, $\sigma_{hoop} = 10$ MPa

The results show that the main influence on the crack shape has the SDR ratio and the size of rigid sphere but for all pipe configurations discrepancy is not higher than 13%. Final summary of nonlinear numerical simulation results is shown in Fig. 6. Plotted data are fitted by second degree polynomial function and compared with same kind of relation found for the pipe loaded by inner pressure only. It is shown, that crack influenced by point load effect has in all cases higher aspect ratio than only internally pressured pipe. The stress intensity factor is higher due to point loads and consequently it leads to the shortening of the residual lifetime of the pipe system. Crack shape of the pipe loaded by inner pressure and influenced by external point load could be described by following approximate equation:

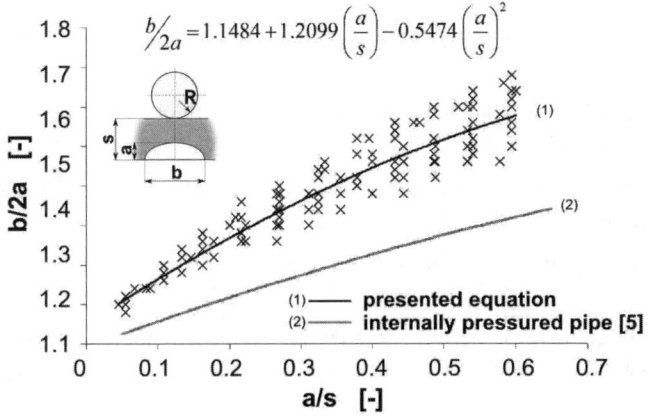

Fig. 6. The summary of the crack aspect ratios influenced by point load effect.

$$b/2a = 1.1484 + 1.2099\left(\frac{a}{s}\right) - 05474\left(\frac{a}{s}\right)^2, \qquad (2)$$

where a is a crack depth, b is a crack width and s is a pipe wall thickness, see Fig. 1. By the presented equation (2) it is possible to predict the crack shape with less than 10% error and its validity is limited for presented range of parameters. Results presented can helpful for deeper understanding of the point load effects and prevent unexpected failures of the pipelines.

Acknowledgement

The investigation has been supported by grants P108/12/1560 and 106/09/H035 of the Czech Science Foundation and by the Specific academic research grant No. FSI-S-11-11/1190 of the Ministry of Education, Youth and Sports of the Czech Republic provided to Brno University of Technology. The research work of this paper was partially performed at the Polymer Competence Center Leoben GmbH (PCCL, Austria) within the framework of the COMET program of the Austrian Ministry of Traffic, Innovation and Technology with contributions by the University of Leoben, AGRU Kunststofftechnik GmbH (Bad Hall, A), Dow Europe GmbH (Horgen, CH) and ÖVGW - Österreichische Vereinigung für das Gas- und Wasserfach (Vienna, A). The PCCL is funded by the Austrian Government and the State Governments of Styria and Upper Austria.

References

[1] Janson L.E., Plastic Pipes for Water Supply and Sewage Disposal, Borealis, Stockholm; 1999.

[2] Pinter, G., Haager, M., Lang, W.G., Influence of nonylphenol–polyglycol–ether environments on the results of the full notch creep test, Polymer Testing, Vol. 26, Issue 6, p. 700-710, 2007.

[3] Farshad, M., Determination of the long-term hydrostatic strength of multilayer pipes, Polymer Testing, Vol. 24, Issue 8, p. 1041-1048, 2005.

[4] X. Lu, N. Brown, A test for slow crack growth failure in polyethylene under a constant load, Polymer Testing, Vol. 11, Issue 4, p. 309-319, 1992.

[5] Hutař, P., Ševčík, M., Náhlík, L., Pinter, G., Frank, A., Mitev, I., A numerical methodology for lifetime estimation of HDPE pressure pipes, Engineering Fracture Mechanics, Vol. 78, Issue 17, p. 3049-3058, 2011.

[6] Watkins R.K., Anderson L. R., Structural Mechanics of Buried Pipes, CRC Press, Boca Raton, 2000.

[7] Hessel, J., Minimum service-life of buried polyethylene pipes without sand-embedding. 3R international 40, Special Plastic Pipes, p. 4-12, 2001

[8] Schouwenaars R., et.al, Slow crack growth and failure induced by manufacturing defects in HDPE-tubes, Engineering Failure Analysis, vol. 14, p. 1124-1134, 2007

[9] Anderson, T. L. Fracture mechanics: fundamentals and applications. 2nd ed. London: CRC Press, 2000

[10] Branco, R., Antunes, F.,V., Finite element modeling and analysis of crack shape evolution in mode-I fatigue Middle Cracked Tension specimens, Engineering Fracture Mechanics, 75, p. 3020-3037, 2008

[11] Blümich, A., et al.: Mobile NMR for Analysis of Polyethylene Pipes. Acta Physica Polonica A. Vol. 108, p.13-23. 2005.

Quasi-static simulation of crack growth in elastic materials considering internal boundaries and interfaces

P. Judt[1,a], A. Ricoeur[1,b]

University of Kassel, Institute of Mechanics, 34125 Kassel, Germany

[a] judt@uni-kassel.de, [b] ricoeur@uni-kassel.de

Keywords: J-integral, crack growth, cohesive zone, path independence

Abstract. This work presents numerical methods used for predicting crack paths in technical structures based on the theory of linear elastic fracture mechanics. The FE-method is used in combination with an efficient remeshing algorithm to simulate crack growth. A post processor providing loading parameters such as the J-integral and stress intensity factors (SIF) is presented. Path-independent contour integrals are used to avoid special requirements concerning crack tip meshing and to enable efficient calculations for domains including interfaces and internal boundaries. In particular, the interaction of cracks and internal boundaries and interfaces is investigated. The simulation combines crack propagation within elastic bodies and at bi-material interfaces. The latter is based on a cohesive zone model. The presented numerical results of crack paths are verified by experiments.

Introduction

The J-Integral based on the formulation by Eshelby [1] and interpreted by Rice [2] for strain concentration problems is a path independet conservation integral. In LEFM J is a crack loading quantity equivalent to SIF and the energy release rate. Exploiting its path independence, the J-integral provides very good results for integration contours far from the crack tip. Especially for simulation of crack propagation within boundary value problems in conjunction with internal boundaries or interfaces, a contour close to the geometrical outer boundary is beneficial with regard to automatic calculation of the J-integral.

Path independent contour integral

The general representation of the J-integral can be given as a line integral of the Energy-Momentum-Tensor Q_{kj} multiplied by the local unit normal vector n_j

$$J_k = \lim_{\epsilon \to 0} \int_{\Gamma_\epsilon} Q_{kj} n_j \mathrm{d}s = \lim_{\epsilon \to 0} \int_{\Gamma_\epsilon} \left(u \delta_{kj} - \sigma_{ij} u_{i,k} \right) n_j \mathrm{d}s \tag{1}$$

where the crack tip is enclosed by the contour Γ at a distance ϵ. Integration along finite contours Γ in general leads to path dependence. The strain energy density $u\left(\epsilon_{ij}\right)$ is calculated from the stress tensor σ_{ij} and strain tensor ϵ_{ij} assuming linear constitutive behaviour:

$$u = \frac{1}{2} \sigma_{ij} \epsilon_{ij} \tag{2}$$

Coherent crack loading quantities. The coordinates of the two-dimensional J-integral vector J_1 and J_2 can be related to mixed mode stress intensity factors K_I and K_{II} and the energy release rate G. From substituting the near tip stress and displacement fields into Eq. (1), Bergez [3] obtained the following expressions:

$$J_1 = \frac{K_I^2 + K_{II}^2}{E'} \tag{3a}$$

$$J_2 = -2\frac{K_I K_{II}}{E'} \tag{3b}$$

For plain stress $E' = E$ and for plain strain $E' = E/(1-\nu^2)$. The total value of the energy release rate is given by the projection of J_k on the unit vector of crack propagation z_k:

$$G = J_k z_k \tag{4}$$

Path independence. To supply path independence for the J-Integral considering curved cracks and internal boundaries or interfaces, Eq. (1) must be spezified for finite contours Γ and thus be extended by additional integrals:

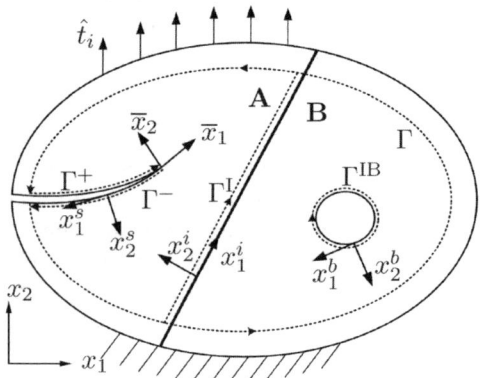

Fig. 1: Boundary value problem involving a curved crack, internal boundary and interface

$$J_k = \int_\Gamma Q_{kj} n_j \mathrm{d}s + \int_{\Gamma^+ + \Gamma^-} Q_{kj} n_j \mathrm{d}s + \int_{\Gamma^{IB}} Q_{kj} n_j \mathrm{d}s + \int_{\Gamma^I} \left(Q_{kj}^{(A)} - Q_{kj}^{(B)} \right) n_j \mathrm{d}s \tag{5}$$

When integrating along crack faces or internal boundaries free of tractions, perpendicular normal and shear stresses do not exist leading to a simplified integrand in its local coordinate systems x_i^s and x_i^b:

$$Q_{kj}^s n_j^s = \frac{1}{2}\sigma_{11}^s \epsilon_{11}^s n_k \quad \text{respectively} \quad Q_{kj}^b n_j^b = \frac{1}{2}\sigma_{11}^b \epsilon_{11}^b n_k \tag{6}$$

When bimaterial interfaces are considered, path independence is realized by integrals along the interfaces including the jump of Eshelby's tensor Q_{kj}. Normal and shear stresses as well as tangential derivatives of displacements are continuous leading to the following simplified integrand of the interface integral in its local coordinate system x_i^i:

$$\left(Q_{2j}^{(A)} - Q_{2j}^{(B)} \right)^i n_j^i = \left(u^{(A)} - u^{(B)} \right) n_2^i - t_i^i \left(u_{i,2}^{(A)} - u_{i,2}^{(B)} \right)^i \tag{7}$$

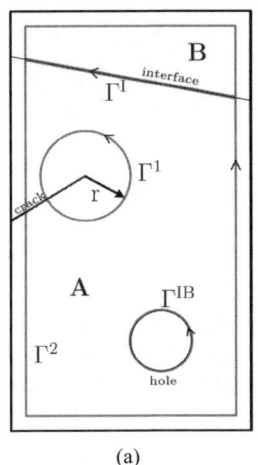 

Contour	J_1	J_2
Γ^1 (r=10mm)	0.0572	-0.0208
Γ^2	0.0269	-0.0503
Γ^{IB}	-0.0019	-0.0007
Γ^I	0.0322	0.0303
$\Gamma^2 + \Gamma^{IB} + \Gamma^I$	0.0573	-0.0207

(a) (b)

Fig. 2: Model for the numerical verification of path independence of the J-Integral and results obtained for illustrated integration contours.

Verification of path independence. The considered boundary value problem includes a slanted crack and a hole. Two regions with unequal elasticity moduli $E_A = 210000$MPa and $E_B = 70000$MPa are connected by a perfect interface. To show path independence the J-integral is calculated for two different contours, i.e. one circular integration contour Γ^1 close to the crack tip and another integration contour Γ^2 close to the outer boundary. In Fig. 2(a) and 2(b) the model and mesh as well as integration contours are illustrated. Path independence is obvious comparing results from first and last rows of the table in Fig. 2.

Crack propagation simulation

Crack growth can be simulated by continuous incremental crack extensions. This leads to a continuous modification of the geometry and therefore to the necessity of intelligent re-meshing.

Re-meshing algorithm. The element size at the crack tip region and along the crack faces is finer compared to the remaining mesh. In the case of a crack growing towards an internal boundary, an interface or any other edge, the related element size of the edge is refined, if the distance between crack tip and edge falls below a critical value.

Crack propagation. Different crack growth and deflection criteria have been formulated by Erdogan & Sih [4], Nuismer [5] and Sih [6]. Our work is based on the criteria of critical energy release rate G_C for threshold of initiation and the maximum energy release rate for the angle of deflection θ. The energy release rate G is calculated from Eq. (4). Here, it is obvious, that the condition $dG/d\theta = 0$ is equivalent to the fact, that the crack will propagate into the direction of J_k.

Experimental validation. An aluminium alloy specimen including a hole is imposed to a subcritical cyclic loading with a frequency of 3Hz. The boundary conditions are similar to those shown in Fig. 2(b). A controlled increase of displacement ensures stable crack growth.

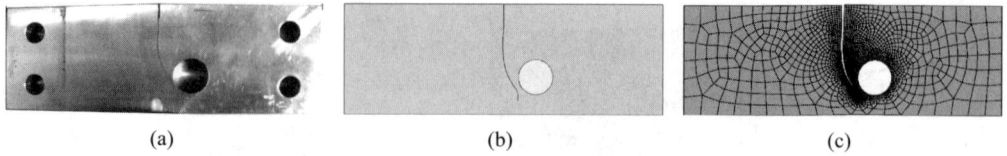

Fig. 3: Comparison of experimental and numerical results of crack paths.

In Fig. 3(a) and 3(b) experimental and numerical crack paths are compared. The adapted mesh of the numerical calculation is shown in Fig. 3(c). The numerical result coincides in principle with the experimental findings. First, the crack slightly departs from the hole and is then strongly attracted by it. However, the experiment shows a slightly stronger deflection of the crack path finally ending at the hole. This might be due to the choice of the criterion for crack propagation possibly depending on the material.

Interface crack growth model

Different from the perfect interface, two bodies may be connected by an adhesive layer having a finite compliance. Avoiding the common construction of cohesive zone elements, a cohesive zone model is implemented introducing restraining stresses at edges of finite elements at the interface. An iterative procedure seeks the equilibrium of stress and crack opening displacement.

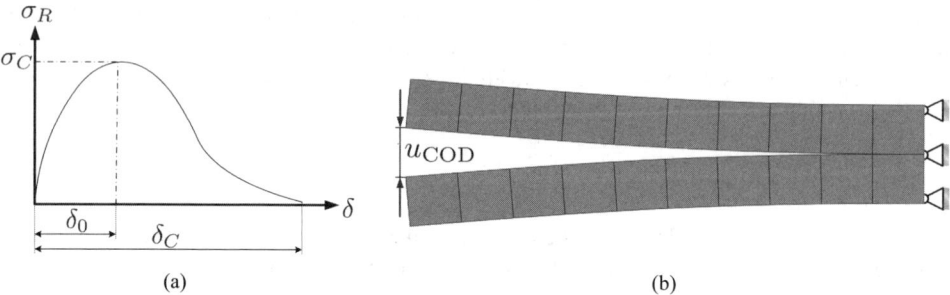

Fig. 4: (a) Typical cohesive law, i.e. restraining stress as function of displacement. (b) Delamination at a DCB specimen with crack opening displacemen u_{COD}.

Fig. 4(b) shows a delamination process based on numerical simulation with the boundary condition $u_{COD} = $ const.

References

[1] J. D. Eshelby: Solid State Physics Vol. 3 (1956), p. 79-144
[2] J. R. Rice: Journal of Applied Mechanics Vol. 35 (1968), p. 379-386
[3] D. Bergez: Mechanics Research Communications Vol. 1 (1974), p. 179-180
[4] F. Erdogan, G. C. Sih: Journal of Basic Engineering Vol. 85 (1963), p. 519-527
[5] R. J. Nuismer: International Journal of Fracture Vol. 11 (1975), p. 245-250
[6] G. C. Sih: International Journal of Fracture Vol. 10 (1974), p. 305-321

Deterioration of FRC plate due to explosion and change of temperature

Petr P. Prochazka

Association of Czech Civil Engineers, Prague 6, Komornická 15, Czech Republic

petr.proch@gmail.com

Keywords: Extreme temperature, load due to explosion, fiber reinforced concrete, gas dynamics, surface cracks, finite volume method.

Abstract: Recently it was shown that the temperature plays very important role in coupling with the impact of shock wave caused by explosion in a free space. The deterioration of fiber reinforced concrete structures, FRC, at the face to which the main impact of the coupled load is directed is not easy to describe. In couple of previous papers of the author the problem was solved under condition that the structure behaves elastically. It appeared that a numerical tool known as free hexagon method, which is based on soft contact problem in combination with boundary elements, seems a powerful approach solving any disconnecting media. The soft contact is identified by nonlinear spring rules. Similar idea is used in the suggested paper for solving the coupled problem envisaged, but the mesh of free hexagons is split into the air and the structure. If the impact of explosion and the subsequent shock wave cause is viewed separately from the changes in material due to temperature a very fast changes in stresses and movements in material are observed in the first case while the changes due to temperature develop much slower. On the other hand in combination of both influences the temperature increase the activity inside of structures considerably. In the paper the theory is briefly described and selected examples illustrate the mobility of the application of the method to a problem of FRC plate.

Introduction

It has recently been shown that the behavior of fiber reinforced concrete exposed to high temperature and explosion effects is influenced by many factors, including both environmental factors and constituent materials. Extreme temperature is accompanied by a volume change of air and water in the voids created nearby the heated face within the trial material, see [1], for example. Consequently, the superheated vapor causes spalling at the heated face of structure, [2]. If also the influence of explosion is present the gasdynamical waves coexist and interact with each others and the solution of such a problem, as strongly nonlinear, has to be obtained by numerical means. The gasdynamical behavior of the air is described in [3], for example, where application to shock wave impact on civil engineering structures is studied. It appears that free hexagon method, [4-6], is an appropriate numerical tool for the formulation and solution of problems of this kind, if the propagation in the air is not considered. Simplified three-phase medium is emerged in paper [7], for instance. For describing the influence of the shock wave and temperature the free hexagons fail in describing the gasdynamics.

Under fire conditions, materials are subjected to transient processes and therefore there is a big need to quantify these properties being determined under transient conditions. Hereinafter the combustion (free hexagons) together with gasdynamics (discontinuous finite elements) is studied for underground concrete tunnel linings of a special shape (model structure), in order for free hexagons (perfectly describing possible damage of material) to apply to damage processes. Construction of theoretical models of solid materials that burn and release most of their energy in the gas-phase is carried out. A matter of validation of free hexagons is concerned for acquisition of a sufficiently accurate model of the complex processes that are involved in two-phase material, which consists of energetic solid part and the gases that, among others evolve from it.

The free hexagon method, which serves here as one of the numerical tools, starts with division of the domain (the material being confined within it) into non-overlapping hexagonal particles (elements). The particles create the solid phase while the influence of gas is concentrated in the surrounding of them. During the heating of a concrete face the tensile strength cannot be exceeded and if so, during the cooling process the tensile strength descends to zero and the admissible set of displacements turns to a cone.

"Finite volume", [8], for example, refers to a small volume surrounding each node point on a mesh. The basic idea of the finite volume method consists in converting volume integrals in a partial differential equation that contain a divergence terms to surface integrals using the divergence theorem. These terms are then evaluated as fluxes at the interfaces of each finite volume. If no internal resources of the heat are present in a volume element the flux entering a given volume is identical to that leaving the adjacent volume, so that this method is conservative. Another advantage of the finite volume method is that it is easily formulated to allow for unstructured meshes. In our case the hexagonal meshes are created for the application of the finite volume method in the air in order to hold the continuity with the free hexagon method.

The change of temperature is described by the augmented Poisson equation in both phases and the interaction of temperature and pressure is there involved, see, e.g., [1].

Conservation equations for the gas

First the notation of the quantities under study is as follows: ρ is the density, $\boldsymbol{u} = \{u_1, u_2, u_3\}$ is the flow velocity vector, $E = e + \frac{1}{2} u_j u_j$ is the total energy density, e is the specific internal energy, which is a function of the pressure p and the specific volume $v = \frac{m}{\rho}$, m is the mass. Generally also the presence of sources of mass M, momentum F, and energy H can be taken into account. The complete set of basic conservation equations, which is linked up with gasdynamics, is found in [3]. Recall that they can be stored in the column vector notation as:

$$\frac{d}{dt}\int_{\Omega} \boldsymbol{U} \, d\Omega + \int_{\Gamma} \boldsymbol{F}_j n_j \, d\Gamma = \int_{\Omega} \boldsymbol{S} \, d\Omega \tag{1}$$

where

$$\boldsymbol{U} = \begin{pmatrix} \rho \\ \rho u_i \\ \rho E \end{pmatrix}, \quad \boldsymbol{F}_j = \begin{pmatrix} \rho u_j \\ \rho u_j u_i + p \\ \rho u_j (E + pv) \end{pmatrix}, \quad \boldsymbol{S} = \begin{pmatrix} M \\ F \\ H \end{pmatrix}. \tag{2}$$

The column vector $\boldsymbol{S}$ specifies the fact that point sources are in the field. Note that the effects of these quantities are of the principal importance in case the Sedov explosion is studied.

Suppose now that Ω is independent of time and is covered by a sum of $\Omega_g + \sum_{k=1}^{n} \Omega_k$, where volume Ω_g is occupied solely by gaseous material, and Ω_k with their boundaries Γ_k are n subdomains (particles) of the hexagonal shape. Since (1) is applied to the gaseous part only in the new set-up it is written as:

$$\frac{d}{dt}\int_{\Omega_g} \boldsymbol{U} \, d\Omega + \int_{\Gamma - \sum_k \Gamma_k} \boldsymbol{F}_j n_j \, d\Gamma = \int_{\Omega_g} \boldsymbol{S} \, d\Omega,$$

$$\int_{\Omega_g} \frac{d\boldsymbol{U}}{dt} \, d\Omega + \int_{\Gamma} \boldsymbol{F}_j n_j \, d\Gamma - \sum_{k=1}^{n} \int_{\Gamma_k} (\boldsymbol{F}_j - \boldsymbol{U} u_j^k) n_j \, d\Gamma = \int_{\Omega_g} \boldsymbol{S} \, d\Omega \tag{3}$$

where u_j^k is the velocity vector along the surface of the particle k and appears as a consequence of time-dependent volume Ω_k. The third term of the left hand side describes the mutual effect of gas and solid particle at the interfaces. The idea will be made clear from the adjacent examples.

Example and conclusions

Number of hexagons is 3064 and number of finite elements is 19320, Mirror symmetry is applied along x_2-axis, the other boundaries are defined by rollers. Movements of particles in cuts of the cross-section of the tunnel are depicted in Fig. 4 to Fig. 7 at different time instant. The fissures appear on the surface because of various interactions of vapor and gas states inside of the concrete and rock. The combination of free hexagons (admitting cracks in natural way as disturbed contacts) and finite elements solving one dimensional gasdynamics seems a promising tool for solving problems of this sort.

One mesh of regular hexagonal elements (particles) is created, involving both the air and the solid plate, which cover a cross-section of $40 \times 50 \, \text{m}^2$. Maximum applied temperature is 1200 ^{0}C, the ignition of which is located at the origin (located at the left down position of the mesh), the igniter mass flux is 40 kg/(m^{3}s), which appears in H as a source term at $t = 0$, the molar gas constant $R = 8.314472$ J·K^{-1}·mol^{-1}. A simplification is made here that the material properties of the solid phase remain unchanged inside of the particles and are given in a standard way by modules taken for linear elasticity: $E = 208 \times 10^9$ N/m^2, $v = 0.29$. In Fig. 1 the mesh with movements of particles after 10 msec is depicted. The solid plate is shaded in this picture. Number of particles is 2000 and the ratio of an inscribed circle to a particle is 0.5 m long.

Two basic cases of the shape of plate are reflected. In Fig. 2 the most unfavorable case of a rectangle plate concerning stresses is cut out the mesh (first undeformed and then the moved particles are shown in one picture). The second case emerges the situation when an inclination of the plate is assumed in the lower part of the rectangle. The undeformed and moved particles are seen in Fig. 3.

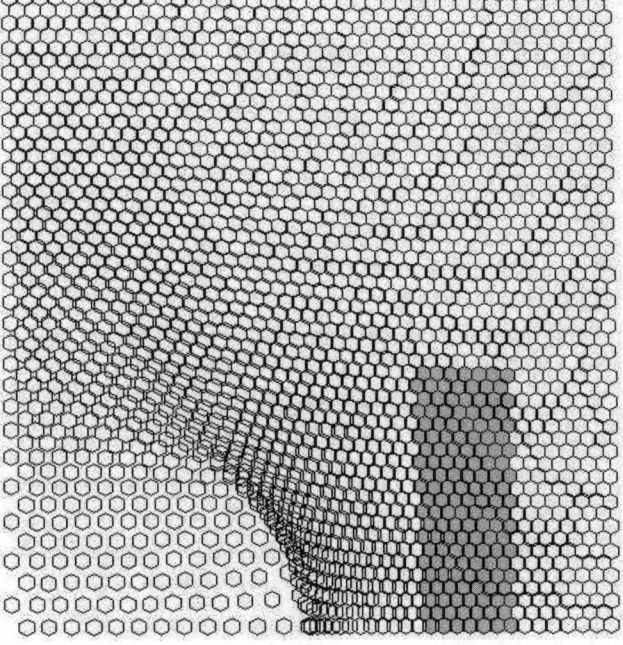

Fig. 1: Symmetric mesh for discontinuous finite elements

Along the whole face of the pure rectangle tensile stresses appear while in case of the lower inclination the most exposed regions (circled) are located at the lower part (this will be easily overcome by strengthening by rebars or denser fiber reinforcement) and in the most upper part, where damage can really occur.

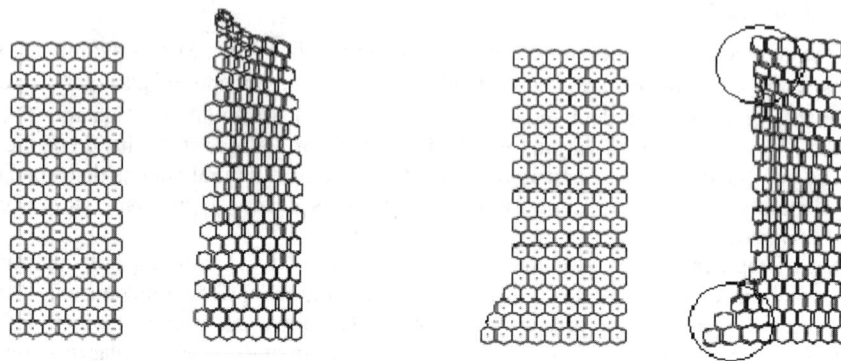

Fig. 2: Rectangular plate   Fig. 3: Case with lower inclination

Conclusions

Combination of finite volume and hexagonal particles solves a complicated non-linear problem of interaction of shock wave and solid phase. In order to hold a compatibility of both phases (air and solid), hexagonal mesh is created. Inside the particles describing the air finite volume element method is applied and in the particles describing the solid structure boundary element method is applied. The compatibility of movements, pressures (stresses) and temperature along the interface are ensured. The change of temperature is described by the augmented Poisson equation in both phases and the interaction of temperature and pressure is there involved, see, e.g., [1]. The results prove a capability of the method provided.

Acknowledgment: This paper was financially supported by GAČR, project No. P104/10/1021.

References

[1] S. Peskova and P.P. Prochazka: CTU Reports 1, Vol. 12, (2008), pp. 150

[2] P.P. Prochazka, M.V. Valek, S. Peskova: Study on spalling of extremely heated FRC by free hexagons. Key Eng. Mater. 452-453: (2011), p. 693-696

[3] P. Prochazka, A. Kravtsov: Shock Waves as a Main Destructive Dynamical Loading of Structures. Czech Technical University in Prague (2010), pp. 132

[4] P.P. Prochazka: Application of discrete element methods to fracture mechanics of rock bursts. Engineering Fracture Mechanics 71 (2004) p. 601–618

[5] P.P. Prochazka, T-S Lok: Hereditary problems in long-wall minimg by free hexagons. Int. J. Comp. Meths 8(2) (2011) p. 293-313, DOI: 10.1142/S0219876211002587

[6] P.P. Prochazka, T-S Lok: Explosion and temperature resistance of underground structures by free hexagons. To appear in Key Engineering Materials (2012)

[7] G.W. Ma, J.C. Li, J. Zhao: Three-phase medium model for filled rock joint and interaction with stress waves. Int. J. Numer. Anal. Meth. in Geomech 35(1) (2011) p. 97-110

[8] H-J. Kim, J.W. Lee, K-S. Yoon, et al: Numerical Analysis of Flood Risk Change due to Obstruction. KSCE J. of Civil Engng 16(2) (2012), pp. 207-214

Monitoring fatigue cracks and stress intensity factors based on the distributed dislocation technique

Ramdane Boukellif [1,a] and Andreas Ricoeur [2,b]

[1,2] University of Kassel, Institute of Mechanics, 34125 Kassel, Germany

[a] ramdane.boukellif@uni-kassel.de, [b] ricoeur@uni-kassel.de

Keywords: Monitoring fatigue cracks, dislocation technique, stress intensity factors, inverse problems

Abstract. We present a method for crack detection and stress intensity factor measurement in plate structures by using strain gauges and applying the dislocation method. The presented approach is based on the strain measured at different locations on the surface of the structure. This allows both the identification of crack position parameters, such as length, location and angles with respect to a reference coordinate system and the calculation of stress intensity factors (SIF). The method solving the direct problem is based on the idea of representing the crack by a line of point dislocations. The latter are formed by applying a constant displacement between adjacent points located at either side of the crack. Thus, the approach is based on the weighted superposition of elastic Green's functions representing the strain field due to the presence of a crack, where the weights are being identified by inverse problem solution. Since the strain fields are controlled by both external loads and the crack growth the unknown parameters are crack length, position and inclination as well as loading quantities. The particle swarm algorithm (PSO) came out to be most suitable for parameter identification in a high dimensional space.

Introduction

Engineering structures are in general exposed to cyclic or stochastic mechanical loading. Exhibiting incipient cracks, particularly light-weight shell and plate structures suffer from fatigue crack growth, limiting the life time of the structure and supplying the risk of a fatal failure. Due to the uncertainty of loading boundary conditions and the geometrical complexity of many engineering structures, numerical predictions of fatigue crack growth rates and residual strength are not reliable. Most experimental monitoring techniques, nowadays, are based on the principle of wave scattering at the free surfaces of cracks. Many of them are working well, supplying information about the position of cracks. One disadvantage is, that those methods do not yield any information on the loading of the crack tip. On the other hand, there are techniques to measure the stress intensity factors e.g. applying a strain gauge in front of the crack tip. However, this method does not allow for crack growth and implies the knowledge of crack position. Goal of our work is the development of a monitoring concept supplying both the information on the actual crack position and the stress intensity factors. This enables a more comprehensive and reliable survey of structures, based on the actual crack configuration in connection with a numerical prediction of further crack development from crack tip loading quantities.

Theoretical Background

The dislocation method was proposed by [1,2]. Assume that the crack is filled with the same material layer by layer as shown in Fig. 1 (left). Each inserted layer of material will generate a displacement jump which can be interpreted as an alignment of point dislocations, see Fig. 1 (right). Using Green's functions, the stress field due to the inserted strip is obtained. In this manner, each strip is regarded as a dislocation-loop. In 2D crack applications, the crack is assumed to be infinitely long in one direction. Thus, the material strip must also be infinite in one direction, and the corresponding dislocation loop can be assumed to be a dislocation dipole with two infinitely long and straight dislocations of opposite line directions.

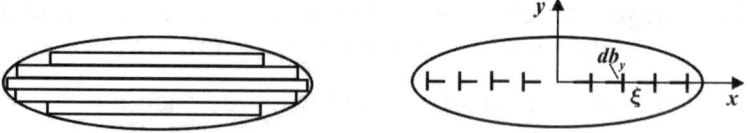

Fig. 1, Eshelby's interpretation of crack dislocations by material insertion [3]

Stress at an arbitrary point according to the dislocation method

We consider an infinite plate containing a crack with a length $2a$ under remote stresses $\sigma_{ij}^\infty(x,y)$. The total stress field $\sigma_{ij}(x,y)$ by using the superposition principle is as follows [4]:

$$\sigma_{ij}(x,y) = \sigma_{ij}^A(x,y) + \sigma_{ij}^D(x,y), \qquad (1)$$

where $\sigma_{ij}^A(x,y)$ is the stress field induced in the plate without a crack.

The stresses $\sigma_{ij}^D(x,y)$ ($ij = xx, yy, xy$) induced at a point (x,y) due to a single dislocation located at the origin of the coordinate system ($\xi = 0$) with Burgers vector $\boldsymbol{b}$ and its components b_x and b_y may be found from the corresponding Airy stress functions, given by [4]:

$$\sigma_{xx}^D(x,y) = \frac{2\mu}{\pi(\kappa+1)}\left\{b_x\left[-\frac{y}{r^4}(3x^2+y^2)\right] + b_y\left[\frac{x}{r^4}(x^2-y^2)\right]\right\} \qquad (2)$$

$$\sigma_{yy}^D(x,y) = \frac{2\mu}{\pi(\kappa+1)}\left\{b_x\left[\frac{y}{r^4}(x^2-y^2)\right] + b_y\left[\frac{x}{r^4}(x^2+3y^2)\right]\right\} \qquad (3)$$

$$\sigma_{xy}^D(x,y) = \frac{2\mu}{\pi(\kappa+1)}\left\{b_x\left[\frac{x}{r^4}(x^2-y^2)\right] + b_y\left[\frac{y}{r^4}(x^2-y^2)\right]\right\}, \qquad (4)$$

where $r^2 = x^2 + y^2$. κ is a constant related to Poisson's ratio ν as $\kappa = (3-\nu)/(1+\nu)$ for plane stress and $\kappa = 3 - 4\nu$ for plane stain, μ is the shear modulus. Strains $\varepsilon_{ij}(x,y)$ are calculated applying Hooke's law.

The calculation of crack length

To obtain an expression for the crack length $2a$, we first consider a single concentrated dislocation within $[\xi, \xi + d\xi]$ with the respective Burgers vector:

$$db_y = B_y(\xi)d\xi, \qquad (5)$$

where $B_y(\xi)$ is the density of the dislocation at point ξ, see Fig. 1 (right). Assuming pure mode-I loading, the normal stresses arising along the crack faces due to a single dislocation are given by setting $b_x = 0$ and $y = 0$ so that

$$\sigma_{yy}^D(x, y=0) = \frac{2\mu}{\pi(\kappa+1)}\frac{db_y(\xi)}{x-\xi} = \frac{2\mu}{\pi(\kappa+1)}\frac{B_y(\xi)}{x-\xi}d\xi. \qquad (6)$$

The stresses due to a continuous distribution of dislocations along the crack line are then given by:

$$\sigma_{yy}^D(x, y=0) = \frac{2\mu}{\pi(\kappa+1)}\int_{-a}^{+a}\frac{B_y(\xi)}{x-\xi}d\xi. \qquad (7)$$

On the other hand, the crack surface must be traction free. In an infinite domain, e.g., the boundary conditions can be written as follows:

$$\sigma_{yy}(x, y = 0) = \sigma_{yy}^{\infty}(x, y = 0) + \sigma_{yy}^{D}(x, y = 0) = 0, \qquad |x| < a \tag{8}$$

$$\sigma_{xy}(x, y = 0) = \sigma_{xy}^{D}(x, y = 0) = 0, \qquad |x| < a \tag{9}$$

Based on these boundary conditions, the Eq. (7) can be written as follows [4]:

$$\sigma_{yy}^{\infty}(x, y = 0) = -\frac{2\mu}{\pi(\kappa + 1)} \int_{-a}^{+a} \frac{B_y(\xi)}{x - \xi} d\xi. \tag{10}$$

Solving Eq. (10) numerically, the half crack length a is found as follows:

$$a = \frac{b_y 2\mu}{\sigma_{yy}^{\infty}(\kappa + 1)} \tag{11}$$

Solving the inverse problem

The solution of the inverse problem is carried out applying the PSO algorithm by minimizing a *fitness* function, defined as [5,6,7]:

$$fitness = \sum_{m=1}^{M} \sum_{ij=xx,yy,xy} \{\bar{\varepsilon}_{ij}(P_m) - \varepsilon_{ij}(P_m)\}^2 \tag{12}$$

The fitness function is defined by a square sum of residuals between measured $\bar{\varepsilon}_{ij}(P_m)$ and computed $\varepsilon_{ij}(P_m)$ strains distributions for an assumed crack. The unknowns are the component b_y of Burgers vector b and the loading σ_{yy}^{∞}. The crack length $2a$ is calculated using Eq. (11).

Verification of the concept

Whereas experiments are in progress, first verifications of the monitoring concept have been carried out numerically. We consider a center crack located at (x_0, y_0). The strain fields are taken from FEM analyses. The strain $\bar{\varepsilon}_{ij}(P_m)$ "measured" at points $P_m(m = 1, ...,5)$ for $a = 1$mm, $a = 2$mm and $a = 4$mm subjected, respectively, to the loading stresses of $\bar{\sigma}_{yy} = 50$MPa, $\bar{\sigma}_{yy} = 70$MPa and $\bar{\sigma}_{yy} = 30$MPa are used as testing parameters, as shown in Fig. 2.
Table 1 shows the obtained results. The SIFs of the "given problem" represent the exact results from numerical calculation. The "obtained results" emanate from the solution of the inverse problem.

Table 1: Results of the inverse problems for the center crack.

given problem	obtained results
$\bar{\sigma}_{yy} = 50$MPa	$\bar{\sigma}_{yy} = 49,29$MPa
$a = 1$mm	$a = 0,755$mm
$K_I = 88,6$MPa$\sqrt{mm}$	$K_I = 75,91$MPa$\sqrt{mm}$
$\bar{\sigma}_{yy} = 70$MPa	$\bar{\sigma}_{yy} = 67,25$MPa
$a = 2$mm	$a = 1,53$mm
$K_I = 175,46$MPa$\sqrt{mm}$	$K_I = 147,44$MPa$\sqrt{mm}$
$\bar{\sigma}_{yy} = 30$MPa	$\bar{\sigma}_{yy} = 28,33$MPa
$a = 4$mm	$a = 3,43$mm
$K_I = 106,35$MPa$\sqrt{mm}$	$K_I = 92,99$MPa$\sqrt{mm}$

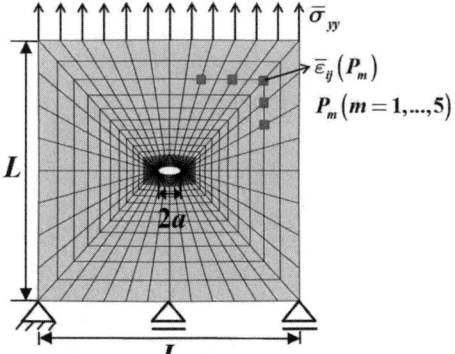

Fig. 2, FE-model for analysis, where $L = 200$mm.

Conclusion

In this work, the dislocation method is used for the detection of cracks and the calculation of SIFs. We verified this method numerically for a center crack in an infinite domain under uniform mode-I loading. The solution of the inverse problem is in fairly good agreement with the model data. The method introduced in the present work can be extended to determine cracks with random orientations.

References

[1] J. Friedel, Fracture, chapter 24 (Wiley, New York, 1959), pp. 498–523.

[2] B.A. Bibly and J.D. Eshelby, Fracture (Wiley, New York, 1968).

[3] Ghoneim, N.M. and Huang, J. "The elastic field of general-schaped 3-D cracks". Phil. Mag., Vol. 86 (2006), 4195-4212.

[4] D.A. Hills, P.A. Kelly, D.N. Dai and A.M. Korsunsky, "Solution of Crack Problems, the Distributed Dislocation Technique", (Kluwer Academic Publisher, 1996).

[5] Chen, D. H. and Nisitani, H. (1993), "Detection of a Crack by Body Force Method", Engineering Fracture Mechanics, Vol. **45**, 671-685.

[6] J. Kennedy and R. Eberhart, Proc. of IEEE international Conf. on Neural Networks, Vol. **IV** (1995), 1942-1948.

[7] K. Oda, N. Yoshimatsu and T. Morisaki,. "Detection of Plural Cracks in an Infinite Plate by PSO Algorithm", Engineering Materials, Vol. **452,453** (2011), 393-396.

Computational Characterization of Micro- to Macroscopic Deformation Behavior of Double Network Hydrogel

Isamu Riku[1,a] and Koji Mimura[1,b]

[1]Department of Mechanical Engineering, Osaka Prefecture University,
1-1, Gakuen-cho, Naka-ku, Sakai-shi, Osaka 599-8231, Japan

[a]riku@me.osakafu-u.ac.jp, [b]mimura@me.osakafu-u.ac.jp

Keywords: Double Network Hydrogel, Necking, 3D Simulation, Molecular Chain Network Model

Abstract. To take advantage of the toughness mechanism of DN gels and explore the possibility for engineering application as the structural member, the information on the mechanical behaviour of DN gels under various loading conditions is indispensable. Therefore, in this paper, we at first constitute a model of DN gel by paralleling a slider element with a nonlinear rubber elasticity spring element based on the nonaffine molecular chain network model, where each element represents the first and the second network of DN gel respectively. The theoretical stress-strain relation of this model shows a strain softening and subsequent strain hardening response, which has been considered as an agent of the propagation of the necking during the simple tension of glassy polymer. Continuously, based on this model, we propose a constitutive equation for DN gel and a three-dimensional simple tension simulation is performed. The computational results show that the propagation of the necking together with the macroscopic mechanical response of DN gel can be reproduced by the proposed model very well.

Introduction

Double network (DN) hydrogels have drawn much attention as an innovative material having both high water content and high mechanical strength and toughness. DN gel are characterized by a special network structure consisting of two types of polymer components with opposite physical natures: the minor component is abundantly cross-linked polyelectrolytes (rigid skeleton) and the major component comprises of poorly cross-linked neutral polymers (ductile substance)[1]. In this paper, the former and the latter components are referred to as the first network and the second network respectively. Extensive experimental studies have shown that the rigid, brittle first network serves as a sacrificial bond that fractures at a relatively low stress, while the soft, ductile second network serves as hidden length that sustains stress by large extension afterwards[1]. Furthermore, based on the experimental results of DN gel under cyclic loading condition, it can be understood that DN gel behaves like a glassy polymer and the onset and propagation of necking occurs in the initial cycle and behaves like a rubber in the subsequent cycles.

In order to investigate the mechanical behaviour of DN gel under various loading conditions, we at first constitute a molecular chain network model of DN gel, in which the first and the second network are represented by a slider element and a nonlinear rubber elasticity spring element respectively. Continuously, based this model, the constitutive equation of DN gel is proposed. And then, three-dimensional simple tension simulation is performed to reproduce the propagation of the necking as well as the macroscopic mechanical response of DN gel.

Constitutive Model

Model Constituents. In this paper, two parallelized elements are empolyed to account for both networks' resistance to deformation observed for DN gel. These are illustrated schematically in Figure 1(b) for a one-dimensional analog to the DN gel deformation model shown in Figure 1(a). The two elements are: a slider used to monitor the fracture of the first network at a relatively early deformation stage and a nonlinear rubber elasticity spring element that accounts for an anisotropic resistance to chain alignment at later deformation stage. Moreover, for the three-dimensional representation of the deformation of DN gel, the one-dimensional model is introduced to a network of eight non-Gaussian chains[2], as sketched in Figure 1(c), where each chain is described by statistical mechanics. The total applied stress, σ, is given by $\sigma = \sigma^{1st} + \sigma^{2nd}$. On the other hand, the stretch on any individual chain in the network is given[2] as a function of the applied principal stretches λ_1, λ_2, λ_3

$$\lambda_{chain} = \frac{1}{\sqrt{3}}\left(\lambda_1^2 + \lambda_2^2 + \lambda_3^2\right)^{1/2}. \tag{1}$$

The individual chain stress-stretch relationship has the form

$$\sigma_{chain} = \lambda_{chain} k_B T \mathsf{L}^{-1}\left\{\frac{\lambda_{chain}}{\sqrt{N}}\right\}, \tag{2}$$

where k_B is Boltzmann's constant, T is temperature, $\mathsf{L}^{-1}\{\lambda_{chain}/\sqrt{N}\}$ is the inverse Langevin function, N is the number of segments per single chain and $\sqrt{N}$ is a statistical parameter related to the limiting value of chain stretch. The Langevin function will result in asymptotically increasing stress as the chain stretch reaches $\sqrt{N}$. The three-dimensional network's response to the principal stretches λ_i, is given in terms of the difference in two principal stresses to eliminate an arbitrary pressure term that arises from the condition of incompressibility

$$\sigma_1 - \sigma_2 = \frac{nk_B T}{3}\sqrt{N}\mathsf{L}^{-1}\left\{\frac{\lambda_{chain}}{\sqrt{N}}\right\}\frac{\lambda_1^2 - \lambda_2^2}{\lambda_{chain}}, \tag{3}$$

where n is the number of chains per unit volume. Furthermore, to account for the re-division of the cluster of the first network after fracture, which playing a role of cross-linker of the second network, the number of segments per single chain of the first network is proposed to increase whereas that of the second network to decrease according to the value of local principal stretch.

Figure 2 shows the theoretical calculation result of the normalized stress-stretch relation of the network components of DN gel. In the case of the first network, after a nonlinear elastic increase of stress, the local principal stretch reaches the limiting value of chain stretch $\sqrt{N}$ and the fracture of the molecular chain occurs, which leads to a dramatic decrease of stress at early deformation stage. In the case of the second network, the stress increases dramatically at the later deformation stage, which is due to increase of the cross-linker (re-divided cluster of the first network). With regard to the total mechanical response, it is quite similar to that of the glassy polymer, i.e. a strain softening and subsequent strain hardening deformation behavior occurs.

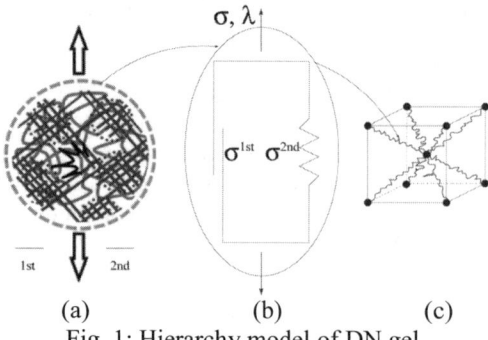
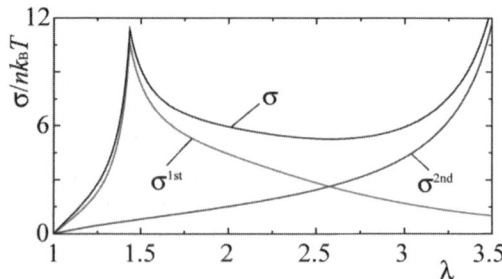

(a)　　　　　　(b)　　　　　(c)　　　　　Fig. 2: Normalized stress-strain relation of
Fig. 1: Hierarchy model of DN gel.　　　　　　the network components of DN gel.

Constitutive Equation. The rate-type expression of the constitutive equation the proposed model of DN gel, which relates the rate of Kirchhoff stress $\dot{S}_{ij}$ to strain rate $\dot{\varepsilon}_{ij}$, i.e. the symmetric component of the velocity gradient tensor, can be given as[3]

$$\dot{S}_{ij} = (R_{ijkl} - F_{ijkl})\dot{\varepsilon}_{kl} - \dot{p}\delta_{ij},\qquad(4)$$

$$R_{ijkl} = \frac{1}{3}nk_B T\sqrt{N}\left\{\left(\frac{\xi}{\sqrt{N}} - \frac{\beta_c}{\lambda_c}\right)\frac{A_{ij}A_{kl}}{A_{mm}} + \frac{\beta_c}{\lambda_c}\left(\delta_{ik}A_{jl} + A_{ik}\delta_{jl}\right)\right\},$$

$$F_{ijkl} = \frac{1}{2}\left(\sigma_{lj}\delta_{ki} + \sigma_{kj}\delta_{li} + \sigma_{li}\delta_{kj} + \sigma_{ki}\delta_{lj}\right),$$

$$\xi_c = \left.\frac{d}{dx}L^{-1}(x)\right|_{x=\beta_c} = \frac{\beta_c^2}{1-\beta_c^2\operatorname{csch}^2\beta_c},\quad \beta_c = \frac{\lambda_c}{\sqrt{N}},\quad \dot{p} = -K\dot{\varepsilon}_{mm},$$

where σ_{ij}, δ_{ij} and A_{ij} are the Cauchy stress tensor, Kronecker's delta and the left stretch tensor, respectively. To introduce the volume constant constraint to DN gel, we use the penalty method with the penalty parameter $K = 5000\,\mathrm{MPa}$.

Computational Model

To clarify the three-dimensional neck propagation behavior, the computational model of a specimen subjected to simple tension at both ends under shear-free conditions, as shown in Figure 3, is employed. In order to initiate necking at the center of the specimen, an initial geometrical imperfection of the width is introduced. Due to the symmetry of the deformation, the finite element discretization with quadratic block elements is employed for 1/8 of the specimen. The computation is performed at room temperature ($T = 296K$) and the applied tensile strain rate is fixed to the value of $1.0\times 10^{-5}/s$. The material parameters for the first network is $nk_B T = 0.4\,\mathrm{MPa}$, $N = 1.2$ and that for the second network is $nk_B T = 0.4\,\mathrm{MPa}$, $N = 6$. Furthermore, with the consideration of computational stability, lower limit value of the number of segments per single chain of the second network is set to half of its initial value.

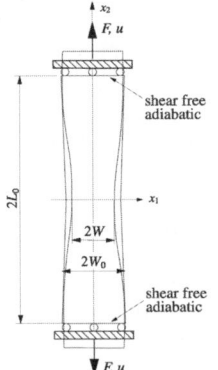
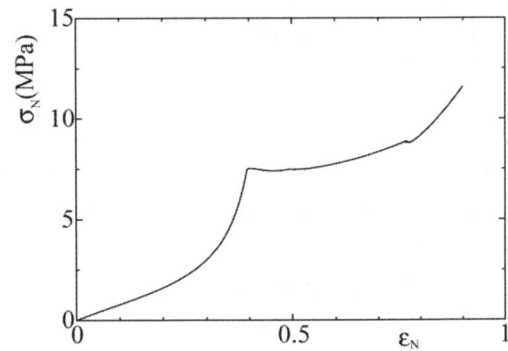

Fig. 3: Computational model.   Fig. 4: Nominal stress-strain relation under simple tension.

Results

Figure 4 shows the nominal stress-strain relation of DN gel under simple tension, where nominal stress and strain are calculated by $\sigma_N = F/A_0$ and $\varepsilon_N = (L-L_0)/L_0$. F is the force applied at the upper end and A_0 is the initial area of the upper end. Figure 5 shows the distribution of tensile stress at different deformation stages. At the early deformation stage, the deformation resistance arises mainly from the first network and increases continuously until the onset of necking. At the deformation stage $\varepsilon_N = 0.4$, the first network begins to fracture at the lower end near to the side face. As a result, strain softening occurs there, which leads to a temporal decrease of the resistance and the onset of necking. And then, the resistance remains almost constant and the necking propagates along the tensile direction due to the dramatically increased strain and the consequent strain hardening of the second network. At the deformation stage $\varepsilon_N = 0.5$, the propagation of necking finishes and the high stress region begins to expand to the whole specimen. Finally, a homogeneous distribution of stress is observed at the deformation stage $\varepsilon_N = 0.8$ and the resistance increases quickly again due to the approach of the chain stretch of the second network to its lower limit value.

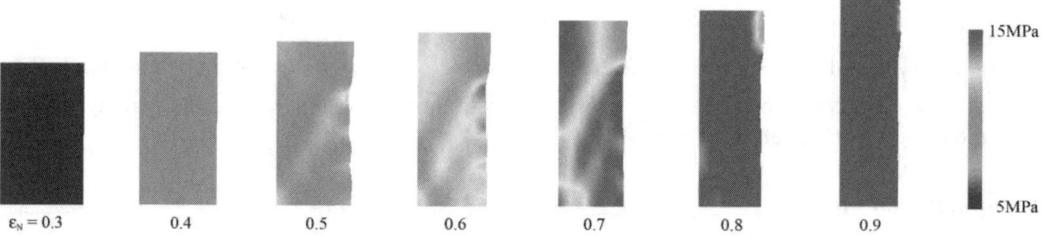

Fig. 5: Distribution of tensile stress at different deformation stages.

Conclusions

In this paper, to investigate the mechanical behaviour of DN gel under various loading conditions, two-element model, in which a slider is parallelized with a nonlinear rubber elasticity spring element, is introduced to a network of eight non-Gaussian chains. Employing this proposed model, three-dimensional simple tension simulation is performed. Comparing with the experimental results shown in Ref. [1], we found that the proposed model can reproduce both the propagation of the necking and the macroscopic mechanical response of DN gel very well.

Acknowledgements

Grant-in-Aid for Scientific Research (B) from Japan Society for the Promotion of Science(JSPS) is acknowledged with thanks.

References

[1] J.P. Gong: *Soft Matter*, Vol. 6, pp. 2583-2590, (2010).

[2] E.M. Arruda and M.C. Boyce: *Journal of the Mechanics and Physics of Solids*, Vol. 41, pp. 389-412, (1993).

[3] I. Riku, K. Mimura and Y. Tomita: *Journal of Engineering Materials and Technology*, Vol. 130, pp. 021017(1-9), (2008).

Multilevel/Multiobjective Design of Composite Structures

O. Bacarreza[1, a], M. H. Aliabadi[2, b], A. Apicella[3, c]

[1]Imperial Consultants, 58 Prince's Gate, SW7 2PG, London, UK

[2]Department of Aeronautics, Imperial College London, SW7 2AZ, London, UK

[3]Alenia, Viale dell'Aeronautica, Pomigliano d'arco, 80038, Naples, Italy

[a]o.bacarreza-nogales@imperial.ac.uk, [b]m.h.aliabadi@imperial.ac.uk, [c]aapicella@alenia.it

Keywords: Optimisation, Multilevel, Multiobjective, Composites.

Abstract. A multilevel multiobjective platform for structural sizing reproducing the sequence of actions taken during design and structural sizing in industry is presented in this paper. This platform is integrated at two design levels labeled as Preliminary Design Level and Detailed Design Level. The set of design variables can be divided into a group of variables describing the main conceptual layout that affect the dimensions and architecture of the model and a second group of variables influencing the material and mechanical behavior. This kind of approach can be effective if it is possible to separate the constraints that are strongly dependent on the design variables of different design levels.

Introduction

The computational resources required for the solution of an optimization problem, especially if a multiobjective optimization problem is faced, typically increase with dimensionality of the problem at a rate that is more than linear. Then, massive computer resources are required for the design of realistic structures carrying a large number of load cases and having many components with several parameters describing the detailed geometry, like for direct one-level optimization.

One obvious solution is to break up large optimization problems into smaller sub-problems and a coordination problem to preserve the couplings among these sub-problems.

In this paper an innovative multilevel approach is presented which simulates the sequence of actions taken during design and structural sizing in contemporary aeronautical departments in industry. A multilevel multiobjective approach for structural design of composite structures is presented in this paper, integrated at two design levels. The two levels are labeled as preliminary design and detailed design. The set of design variables can be divided into a group of variables describing the main conceptual layout that affect the dimensions of the model and a second group of variables influencing the material behavior.

Proposed methodology

In Fig. 1 the multilevel optimization platform is schematically represented. The solutions at Level 1, Preliminary Design, are found using elastic analysis and trying to define the optimum geometry of the stiffened panel for the given boundary conditions. At Level 2, Detailed Design, material nonlinearities have to be included, keeping the geometry obtained in the previous level constant.

The separation of design variables for different levels makes this approach very efficient, since a lower number of solutions have to be tried because there are less design variables to mix. Another point is that the more complex geometry optimization is done elastically, taking lower time to calculate than a full material nonlinear analysis, so that more runs can be done in a reasonable amount of time. The material properties, including stacking sequence of the composite layup, are then optimized in the next level.

The iteration between levels can be finished at any point where the decision maker believes she got a reasonable or the optimum solution. There is no need for the iteration to always finish after both levels have been optimized. The only requirement is that any level is optimized at least once.

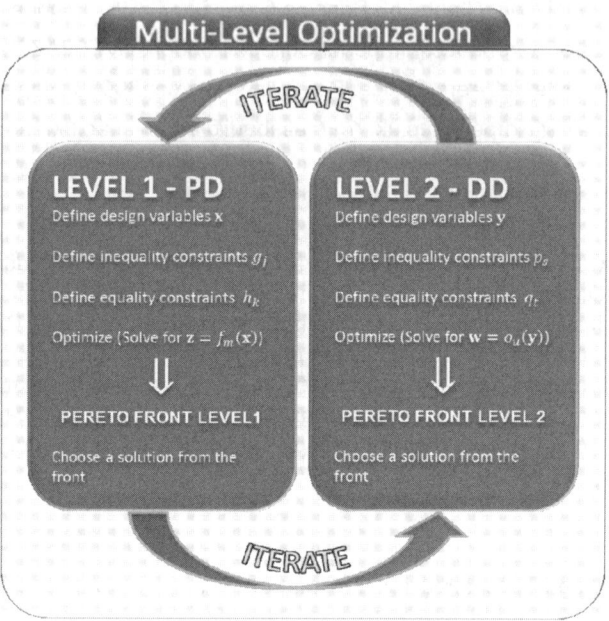

Fig. 1 - Multilevel Multiobjective design

Multilevel/multiobjective design of a composite stiffened panel under post-buckling

A composite stiffened panel is to be optimized; the only fixed dimensions are the overall dimensions of the panel W and L as shown in Fig. 2. The designer wishes to obtain a panel with minimum mass while being able to carry as much load as possible in post-buckling.

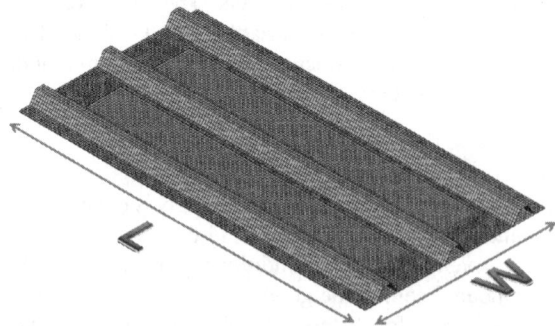

Fig. 2 - Stiffened Panel Geometry

Level 1 – Preliminary Design. The design space for level 1 takes into account all the parameters that define the geometry of the panel, including thickness of the skin (by taking into account the number of layers), the thickness and cross section of the stringers and the number of stringers. The design variables only consider panels that can be manufactured, so it implicitly takes into account the manufacturing constraints, e.g. a panel with 9 layers in the skin would not be manufactured.

The design objectives at this Level are to minimize the total mass while maximizing the reaction force that can the panel can take and maximize the Tsai-Wu index in order to obtain panels that will not fail at the given shortening of 6mm. The objective space for level 1 is constrained by the following constraints: the mass of the panel should be less than 5kg and the panel should vary a load of at least 0.5 MN.

The optimization at Level 1 is done using the AMGA [1] (Archive based Micro Genetic Algorithm), which is an evolutionary optimization algorithm and relies on genetic variation operators for creating new solutions. It uses a generational scheme, however, it generates a small number of new solutions at every iteration, therefore it can also be classified as an almost steady-state genetic algorithm. The algorithm works with a small population size and maintains an external archive of good solutions obtained. The Pareto front, shown in Fig. 3, is constructed from all the feasible solutions that do not violate the constraints using non-dominating sorting.

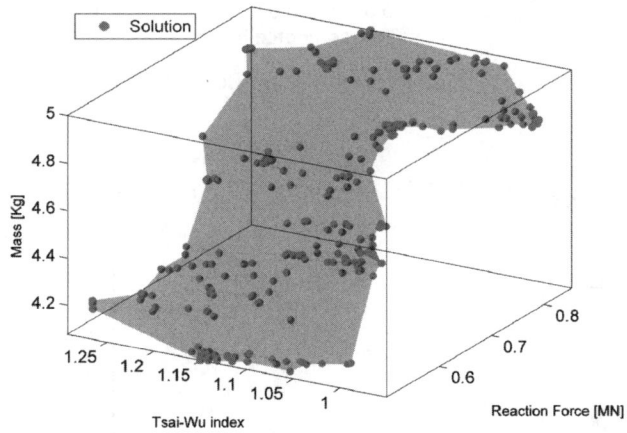

Fig. 3 - Level 1 Pareto Front

Level 2 – Detailed Design. During the Preliminary Design Level optimization, the geometry of the panel was established, leading to the use of 8 layers in the skin and 10 in the stringers. From the manufacturing point of view, only four orientations are possible (-45°,0°,45°,90°), and the layup has to be symmetric. The design variables for this Level are the layups of the skin and stringers. The internal energy of the model and the total reaction force are to be maximized at Level 2 while obtaining a reaction force at least as high as the one from the design obtained in the previous Level 1.

The optimization at this level is done using the Non-dominated Sorting Genetic Algorithm (NSGA-II) [2] which is a multi-objective technique that deals with the high computational complexity of non-dominated sorting, lack of elitism and need of a sharing parameter specification by using a fast non-dominated sorting, an elitist Pareto dominance selection and a crowding distance method.

In this level there are only two objectives present, so that they can be presented in a simple way to the decision maker on a table or on a scatter plot of the solutions. Fig. 4 shows all the solutions that were obtained this level, it can be seen that there are several solutions violating the constraints, and an overall optimum solution maximizing both of the objectives, the reaction force and the internal energy. In this case, it can be said that the Pareto front converges to a single point.

A comparison of the solutions obtained in Level 1 and Level 2 is shown in Fig. 5, it can be seen that the reaction force was dramatically improved from 0.658 MN to 0.826 MN, the solution is also stiffer, but it fails at a lower shortening.

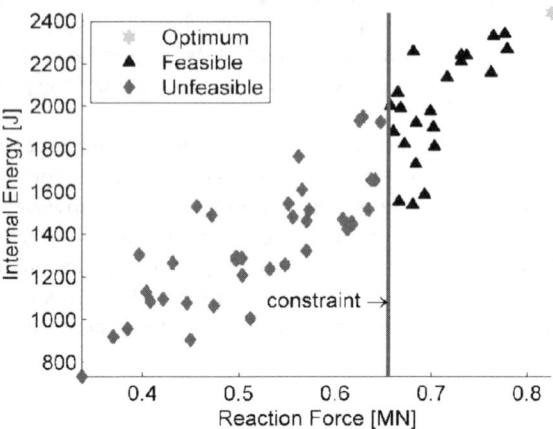

Fig. 4 - Solutions for Level 2

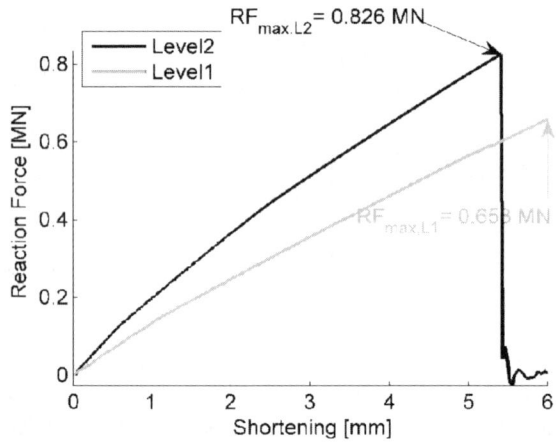

Fig. 5 - Response of the optimum solutions at different levels

Conclusions

A combination of optimization methods is the best solution when dealing with a multilevel optimization, and the use of a multilevel iteration scheme can be integrated into a Pareto front search algorithm, which can use genetic/evolutionary algorithms to explore the design space at different levels and for different purposes. In order to find the optimum design of a real component, subjected to different type of load combinations, a more realistic design should include more load cases generating more objectives and constraints and increasing complexity to the problem.

References

[1] S. Tiwari, P. Koch, G. Fadel, and K. Deb, "AMGA: an archive-based micro genetic algorithm for multi-objective optimization," 2008, pp. 729-736.

[2] K. Deb, A. Pratap, S. Agarwal, and T. Meyarivan, "A fast and elitist multiobjective genetic algorithm: NSGA-II," *Evolutionary Computation, IEEE Transactions on,* vol. 6, pp. 182-197, 2002.

The Role of Grain Size on Deformation of 316H Austenitic Stainless Steel

S Mahalingam[1] PEJ Flewitt [1,2] A Shterenlikht[3]

[1]Interface Analysis Centre, University of Bristol, BS2 8BS, UK

[2]School of Physics, H H Wills Physics Laboratory, University of Bristol, BS8 1TL, UK

[3]Deaprtment of Mechanical Engineering University of Bristol, BS2 8BS, UK

Email: Sunthar.Mahalingam@bristol.ac.uk;Peter.Flewitt@bristol.ac.uk;Mexas@bristol.ac.uk

Keywords: Austenitic stainless steel, grain size, deformation, FE analysis, three point bend

Abstract
The polycrystalline high purity 316H austenitic stainless steel has been thermo-mechanically treated to produce material with two layers of grain size, one of coarser and the other of finer grains. Small three point bend specimens containing a notch positioned in either the coarser or finer layer have been tested at a constant strain rate and a temperature of -196°C. The results are discussed with respect to the effect of grain size on the underlying deformation between the two layers of different grain size.

Introduction
The austenitic stainless steels (SS) have excellent mechanical and physical properties including resistance to corrosion and oxidation, high strength and high ductility over a temperature range from cryogenic to elevated temperatures [1]. The effect of grain size on deformation is well established for various metals and alloys. The yield stress/flow stress varies with grain size based on the Hall-Petch inverse relationship [2,3]. The grain refinement induces superior mechanical properties but ductility and work hardening are compromised [4]. A bimodal grain size distribution is a way to overcome this obstacle [5]. In addition, the deformation mechanisms are also reported to vary with grain size over a range of test conditions; both slip and twinning increase with increasing grain size. In this paper we consider the effect of grain size on deformation of high purity 316H austenitic SS at low temperature when the deformation initiates in a relatively thin layer of finer grain size material and propagates to the coarser and vice versa.

Experimental
The material employed is a high purity 316H austenitic stainless steel (SS) with a chemical composition given in Table 1. The initial block of material 20 mm×10 mm×100 mm thick was subjected to a solution heat treatment at 1150°C and water quenched. The thickness of the block was reduced by 20% by cold rolling. Then it was heat treated at 1100°C for 3 hours followed by water quenching. Layered grains were produced through the thickness with the finer grains confined to the surface and coarser grains in the interior, Figure 1(a). This was followed by a heat treatment at 600°C for 100 hours with furnace cooling. Table 2 summaries the thermo-mechanical history of the each specimen. Three point bend geometry specimens with a dimension of 4 mm ×

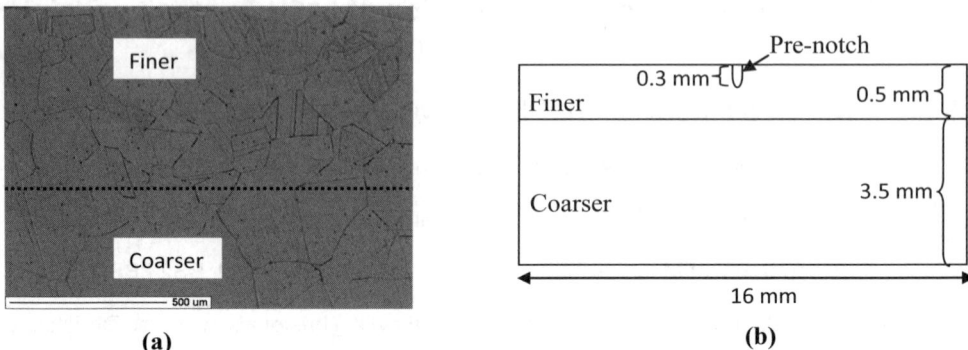

Figure 1 (a) Optical micrograph of layered grain size material (b) schematic of three point bend geometry specimen, S3.

4 mm × 16 mm containing a notch of depth about 0.3 mm were prepared using EDM discharge machining. For specimen S3 the notch was positioned in the finer grain size material, Figure 1(b) and for specimen S4 it is vice versa. The tests were conducted using a Zwick tensometer (load cell range 0-10 kN), equipped with an environmental chamber as described elsewhere [6]. The specimens were deformed under displacement control with a strain rate of 1.40 ×10^{-3} s^{-1} at a temperature of -196°C.

The finite element analysis (FEA) was carried using the ABACUS code with a 30μm cell size and some of the elements were uniformly distributed around the notch root. The stress-strain behaviour for the austenitic stainless steel was considered to obey a power law hardening relationship $\sigma = K\varepsilon^n$, where K is the strength factor and n is the work hardening coefficient. To undertake the analysis values for K and n taken as 1400 MPa and 0.44 respectively [7] and the yield stress is consider to vary as grain size $^{-1/2}$.

Results and Discussion

Specimen S1 contained an appreciable amount of carbide ($M_{23}C_6$) precipitate at the grain boundaries and evidence within the grains of slip and narrow deformation twins. After ageing at 600°C for 100 hours the proportion of both inter- and intra- carbide precipitation increased. Specimen S3 and S4 with the layered microstructure had an average grain size, mean linear intercept, for the finer grains of 121±27 μm compared with that for the coarser grain region of 215±48 μm, Figure 1(a). The amount of carbide precipitate for S1, S2, S3 and S4 was measured by percentage linear coverage at the grain boundary, Table 3.

The stress-strain curves for specimens S1, S2, S3 and S4 tested at -196°C are presented in Figure 2. Following initial bedding-in there is a linear displacement. Departure from the linearity is used to provide a measure of the yield stress, σ_y, for the material and the linear gradient is a measure of the elastic modulus. These results are summarized in Table 3. The yield stress is highest for S1, but significantly the value for S3 exceeds that for specimen S4. Also given in Table 3 is a measure of overall ductility where the values are similar for specimens S3 and S4. Here specimens S1 and S2 are shown for completeness but it is noteworthy that S2 failed by brittle fracture. The 40% of brittle intergranular fracture is promoted by the greater percentage of carbide precipitates at the grain boundaries [8].

Table 1 Chemical composition (wt%)

C	Si	Mn	P	S	Cr	Mo	Ni	Co	N	Fe
0.1	0.51	1.61	<0.002	<0.002	16.9	2.25	11.9	<0.02	0.003	Bal

Table 2 Thermo-mechanical history for specimens

Specimen ID	Thermo-mechanical History ST1=1150°C for 0.5 h ST2=1100°C for 3 h	Heat treatment 600°C for 100 h	Grain size (μm)		
			Fine	Coarse	Average
S1	ST1+cold work 20%	x	-	-	125±27
S2	ST1+cold work 20%	√	-	-	103±33
S3	ST1+cold work 20%+ST2	√	121±27	215±48	-
S4	ST1+cold work 20%+ST2	√	121±27	215±48	-

Table 3

Specimen	E (MPa)	σ_y(MPa)	σ_f(MPa)	Ductility or ε_f (%)	Failure mode	% Carbide
S1	1000±50	1500±55		45	Ductile bending	35
S2	1000±100	800±80	2200±125	50	Brittle	58
S3	1000±64	673±48		70	Ductile bending	44
S4	1000±82	489±35		70	Ductile bending	44

It is clear from the differences in the yield stress between specimens S3 (finer to coarser) and S4 (coarser to finer) that the surface layer of finer grain size material containing the notch has a significant contribution to the overall deformation behavior of these bend geometry specimens at a temperature of -196°C. Since this three point bend specimen, S3, has a relatively thin surface layer containing the notch, Figure1, a FE analysis has been undertaken to consider the distribution of stress and model the stress-strain curves associated with specimens S3 (finer to coarser grains) and S4 (coarser to finer grains). In the case of the latter specimen the finer grain size region represents a layer approximately 0.5 mm thick on the back face of the total 4.0 mm thickness, whereas in the former this notch is contained within the finer grains, Figure 1(b).

FE model shows distinct deformation in the layered grain structure where plasticity propagates from the finer to coarser grain region (S3) and coarser to finer grain region (S4). A localised deformation is observed for S3, Figure 3(a) as opposed to S4 where the contribution from the finer grain layer is negligible when plasticity propagates across the specimen. The FE analysis also shows the maximum stress is at the notch root region for both cases. The predicted stress-strain curves from the FE analysis are shown in Figure 3. These data are in agreement with the experimentally observed curves and shows the finer layer of grains contribution to the overall stress-strain curve. For the prediction, the K value is adjusted from 1400 MPa to 800MPa to get the best fit to the experimental observation.

Acknowledgement: Authors acknowledge to EPSRC for providing financial support, Grant No: EP/H006729/1. They also thank the high performance computing facility provided by the University of Bristol.

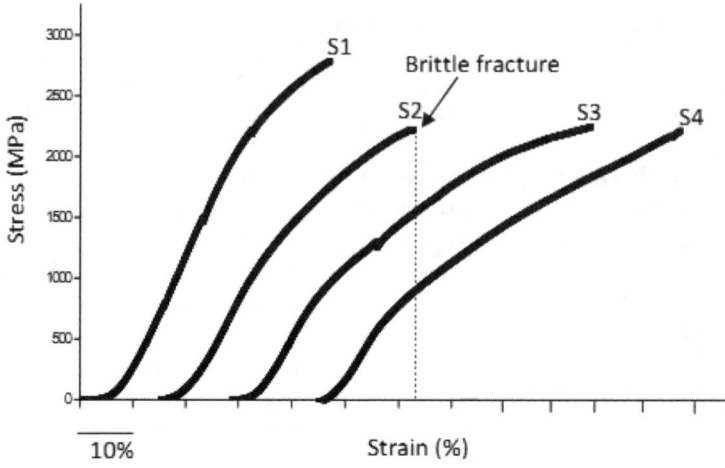

Figure 2 Stress-strain curves for specimens S1, S2, S3 and S4

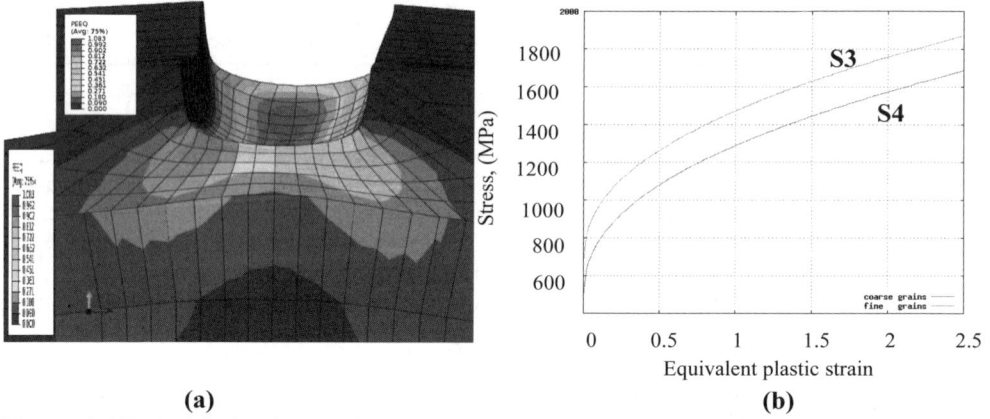

(a) (b)

Figure 3 FE model showing (a) localized deformation at the notch root in the finer grain region (b) stress-strain curves for specimens S3 and S4

References
[1] P.D. Harvey: *Engineering Properties of Steels* (1982) American Society For Metals, Ohio.
[2] E.O. Hall: Proceedings of the Physical Society of London Section B 64 (1951), p747.
[3] N.J. Petch: Journal of the Iron and Steel Institute 174(1953), p25.
[4] M.J.N.V Prasad, S.Suwas, and A.H.Chokshi: Materials Science and Engineering a-Structural Materials Properties, Microstructure and Processing 503(2009), p86.
[5] W.C. Xu, P. Q. Dai, and X.L.Wu (2010). Materials Science and Technology 26(2010), p591.
[6] S. Mahalingam and P. E. J. Flewitt: *Influence of Grain Size on Briitle Crack Propagation* Conference Series of Electron Microscopy (2011), IOP Publishing, Bristol.
[7] E. I. Samuel, B.K. Choudary and K. Bhanu Shakara Rao:Scripta Materialia 46(2002), p507
[8] B. Chen, P. E. J. Flewitt, and D.J.Smith: Materials Science and Engineering A-Structural Materials Properties, Microstructure and Processing 527(2010), p7387.

Experimental investigation of the influence of the bond conditions on the shear bond strength between steel and self-compacting concrete using push-out tests

Peter Helincks[1,2,a], Wouter De Corte[1,2,b], Jan Klusák[3,c], Veerle Boel[1,2] and Geert De Schutter[2]

[1]Department of Construction, Faculty of Applied Engineering Sciences, University College Ghent, Valentin Vaerwyckweg 1, 9000 Ghent, Belgium

[2]Department of Structural Engineering, Faculty of Engineering Sciences and Architecture, Ghent University, Technologiepark-Zwijnaarde 904, 9000 Ghent, Belgium

[3]Institute of Physics of Materials, Academy of Sciences of the Czech Republic, Žižkova 22, 616 62 Brno, Czech Republic

[a]peter.helincks@hogent.be, [b]wouter.decorte@hogent.be, [c]klusak@ipm.cz

Keywords: steel-concrete joint, epoxy adhesive layer, push-out test, shear bond strength

Abstract. Steel-concrete joints are often provided with welded shear studs. However, stress concentrations are induced in the structure due to the welding. Moreover, a reduction in toughness and ductility of the steel and a decreased fatigue endurance of the construction is observed. In this paper the shear bond strength between steel and ultra-high performance concrete (UHPC) without mechanical shear connectors is evaluated through push-out tests. The test samples consist of two sandblasted steel plates with a thickness of 10 mm and a concrete core. The connection between steel and concrete is obtained by a 2-component epoxy resin. Test samples with a smooth adhesive layer are compared with those with an epoxy layer, which is applied with a toothed paddle and/or gritted with small aggregates. In this research, specimens prepared with river gravel, crushed stone, and steel grit are compared and also two different epoxy resins are used. During the tests, the ultimate shear force is recorded as well as the slip between steel and concrete. All test specimens exhibited a concrete-adhesive or concrete failure. Furthermore, test results show that the use of a more fluid epoxy resin improves the anchorage of the gritted aggregates in the adhesive layer, resulting in higher shear bond stresses. No significant difference is found between specimens, gritted with river gravel or crushed stone. Applying the adhesive layer with the toothed paddle in horizontal direction slightly improves the bond behaviour. Finally, the experimental results of the test members with a smooth epoxy layer without gritted aggregates, provide test data for a fracture mechanics approach, which uses a 2D numerical model of the test specimen, composed of steel, epoxy resin, and concrete.

Introduction

In the construction industry, concrete and steel are often combined in structural elements. The tensile strength of steel and the compressive strength of concrete co-operate to allow the member to sustain the service-load stresses. However, a good connection between both materials is required to obtain this good structural performance. Mechanical shear connectors, welded on the steel surface, are often applied to ensure this connection [1]. This paper investigates the shear bond strength between steel plates and ultra-high performance concrete (UHPC), connected by using a two-component epoxy adhesive, avoiding the welding and consequent stress concentrations and decrease of toughness, ductility, and fatigue endurance of the steel [2]. In this research, the influence of different bond conditions on the shear bond strength is investigated. Two types of adhesive are applied, which are gritted with three different aggregates. Based on earlier research, steel plates with a thickness of 10 mm are used in the push-out test specimens to avoid secondary stresses prevailing during tests with thinner steel plates [3]. During the push-out tests, the maximum force and relative slip between the concrete core and the steel plates are recorded and evaluated.

Experimental program

Materials. Sandblasted steel plates (S235) with a thickness of 10 mm are used. They are cleaned with acetone before application of the adhesive layer to remove grease, oil, and dust, in order to achieve a better bonding performance. The adhesives used in these experiments are two-component epoxy resins. The first one, *PC® 5800/BL*, which has a good bonding at concrete, metal, and carbon, is applied with a toothed paddle. Hence, this adhesive layer (epoxy 1) has vertical (direction of loading) or horizontal ridges. After applying the epoxy resin on the steel plates, the layer is gritted with river gravel 2/4 or crushed stone 2/4 (Fig. 1a and 1b). The second adhesive is *Tecnoepo 400*, supplied by Tecnochem Italiana S.p.A. (epoxy 2). Because of its higher fluidity, it is applied on the steel plates with a paint roller. Afterwards, the adhesive layer is gritted with steel grit 1/2 (Fig. 1c).

Figure 1. River gravel 2/4 (a), crushed stone 2/4 (b), and steel grit 1/2 (c)

The test samples are cast with an ultra-high performance concrete (UHPC) premix, produced by Tecnochem Italiana S.p.A. The REFOR-tec GF5 / ST-HS is a tri-component cement based product (powder, liquid, and fibres), which combines the self-levelling rheology with exceptional physico-mechanical properties and ductility. Table 1 lists some important characteristics of this concrete, according to the technical data sheet.

Table 1. Concrete properties

	UHPC mixture
Density [kg/m³]	2 450
f_{ck} [N/mm²]	130
f_{ct} [N/mm²]	8.5
$f_{ct,fl}$ [N/mm²]	32
E_c [N/mm²]	38 000

Test specimens. The dimensions of the push-out specimens, which are based on these used in push-out tests executed by Aboobucker [4], are shown in Fig. 2. All different specimens (4 of each type) are listed in Table 2. Test member 0-0-0 is prepared without adhesive layer and specimen 1-S-0 is made by attaching the hardened concrete prism with epoxy 1 to the steel plates under pressure in order to obtain the best adhesion between both smooth layers. For the other specimens, the adhesive layer is applied and gritted with aggregates as described before. According to the specifications, the cure time of the epoxy resins is 24h. After the curing of the adhesive layer, the UHPC mixture is poured into the mould, between the prepared steel plates. The test samples were demoulded after 24h, sealed and stored at 20 ± 2 °C until the age of testing.

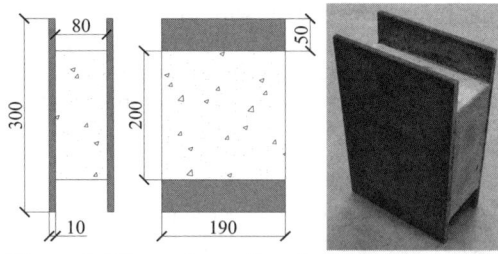

Figure 2. Dimensions of push-out test members [mm]

Table 2. Preparation of the different push-out specimens

Name	Epoxy 1	Epoxy 2	Smooth	Ridge Horizontal	Ridge Vertical	Aggr (a)	Aggr (b)	Aggr (c)
0-0-0								
1-S-0	x		x					
1-R-H-a	x			x		x		
1-R-V-a	x				x	x		
1-R-H-b	x			x			x	
1-R-V-b	x				x		x	
2-S-c		x	x					x

Test procedure. The samples were tested at the age of 7 days by means of push-out tests at a constant rate of 1 kN/s as shown in Fig. 3. Load spreaders with a width of 75 mm are used to convey the applied load to the concrete top surface. Two dial gauges are connected to the steel plates to measure the slip between concrete and steel. In order to exclude unevenness of the test specimen, one fixed and one roller support are used.

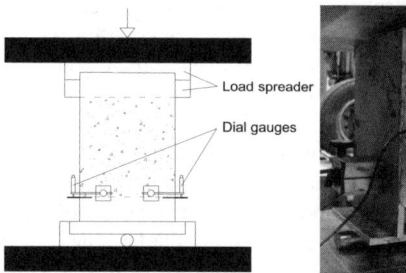

Figure 3. Test set-up push-out test

Experimental test results and discussion

All test specimens exhibited a concrete-adhesive or concrete failure (Fig. 4). For all 1-R specimens, most of the gritted aggregates are pulled out of the adhesive layer (Fig. 4a), causing a loss of bonding between epoxy and concrete. Only for specimens 2-S-c, a concrete failure is observed (Fig. 4b). Here, an improved gritting is obtained due to the higher fluidity of the adhesive. To investigate this failure mechanism of push-out tests in detail, a fracture mechanics approach is recommended. The test results of specimens 1-S-0 are suitable to compare with theoretical fracture mechanics calculations, where perfect interfacial conditions are supposed and simulated.

(a)   (b)

Figure 4. Concrete surface after concrete-adhesive failure (a) and concrete failure (b)

The shear bond stress τ_n [N/mm²] is calculated from the failure load F [N] by Eq. 1, where A [mm²] is the area of the contact surface between steel and concrete.

$$\tau_n = \frac{F}{A} = \frac{F}{2 \cdot 200 \text{ mm} \cdot 190 \text{ mm}} \tag{1}$$

Table 3 summarizes the mean values and standard deviations (stdev) of the failure load F, the shear bond stress τ_n, and the according maximum relative slip s for the different push-out specimens. The mean shear bond stresses τ_n are also visualized in Fig. 5.

Table 3. Results push-out tests

Test member	Failure load F [kN]		Shear bond stress τ_n [N/mm²]		Maximum relative slip s [mm]	
	mean value	*stdev*	*mean value*	*stdev*	*mean value*	*stdev*
0-0-0	115	21	1.51	0.28	0.02	0.01
1-S-0	222	45	2.92	0.59	0.05	0.03
1-R-V-a	214	44	2.82	0.57	0.01	0.01
1-R-H-a	227	29	2.99	0.38	0.01	0.01
1-R-V-b	177	49	2.33	0.64	0.02	0.01
1-R-H-b	207	53	2.73	0.70	0.02	0.02
2-S-c	353	85	4.65	1.12	0.03	0.01

First, similar relative slip values at failure load are found for all specimens, varying from 0.01 to 0.05 mm. However, in case of a ridged adhesive layer (samples 1-R), the lowest slip values between the steel plates and concrete core are measured.

Then, as expected, bond performance between steel and concrete is improved by using an adhesive layer. When interfacial conditions are gained by connecting the hardened concrete prism to the steel plates with an adhesive layer, a mean bond stress of 2.92 N/mm² is measured (1-S-0). These results can be used as test data for the fracture mechanics approach. When epoxy 1 is applied, lower shear bond strengths are recorded (mean value of 2.72 N/mm²). Further, a decrease in bond strength (14.8 %) is observed when crushed stone 2/4 instead of river gravel 2/4 is used as gritted aggregates. Also, when applying the adhesive layer with the toothed paddle in horizontal direction, an improved bond shear behaviour is noticed (increase of 9.7 %). Only for 2-S-c, significant higher bond shear stresses are recorded than the other specimens, made with epoxy 1 (1.7 times higher). This increase can be explained by the use of an epoxy resin with higher fluidity and smaller aggregates. Due to these conditions, a stronger adhesive interlayer is obtained, resulting in a concrete failure.

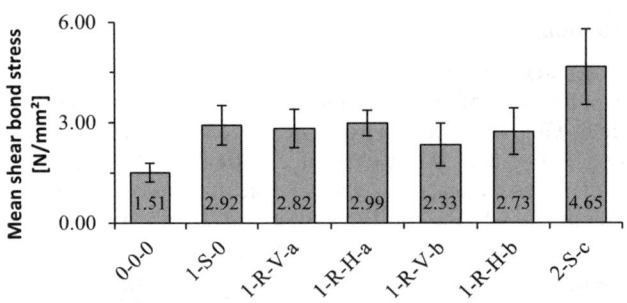

Figure 5. Mean experimental shear bond stresses [N/mm²]

Conclusions

The aim of this research was to investigate the bond strength between steel and concrete using different bond conditions. The results of the experiments have led to the following conclusions: and

- In relation to the smooth epoxy interlayer (1-S-0), the use of a thick adhesive with low fluidity, from which the gritted granules are pulled out during failure, leads to worse bond behaviour (decrease in mean value of 7.1 %). Using a more fluid adhesive and smaller aggregates results in higher shear bond stresses (increase of 59.1 %).
- Gritting the epoxy adhesive layer with crushed stone instead of river gravel decreases the bond strength with 14.8 %.
- Applying the adhesive layer with the toothed paddle horizontally slightly increases the maximum shear bond stress. An increase of 9.7 % is observed.

Acknowledgements

The authors thankfully acknowledge the Research Fund University College Ghent and Czech Science Foundation (grants P108/10/2049 and P105/11/1551) for financial support, Tecnochem Italiana S.p.A. for providing the epoxy resin Tecnoepo 400, and students Frederic Verplancken and Bart Van Gyseghem for their assistance in the preparation and execution of the push-out tests.

References

[1] Richard Liew J.Y., Sohel K.M.A. (2009), *Lightweight steel-concrete-steel sandwich system with J-hook connectors*, Engineering Structures 31, pp. 1166-1178

[2] Johnson R.P. (2000), *Resistance of stud shear connector to fatigue*, Journal of Constructional Steel Research 56, pp. 101-116

[3] Helincks P., De Corte W., Klusák J., Seitl S., Boel V., De Schutter G. (2012), *Failure conditions from push-out tests of a steel-concrete joint: experimental results*, Key Engineering Materials Vols. 488-489, 2012, pp. 714-717

[4] Aboobucker M.A.M., Wang T.Y., and Richard Liew J.Y. (2009), *An experimental investigation on shear bond strength between steel and fresh cast concrete using epoxy*, The IES Journal Part A: Civil & Structural Engineering Vol. 2 (No. 2, May), pp. 107-115

Experimental study of the influence of the initial notch length in cubical concrete wedge-splitting test specimens

Sara Korte[1,2,a], Veerle Boel[1,2,b], Wouter De Corte[1,2,c], Geert De Schutter[2,d], Stanislav Seitl[3,e]

[1]Department of Construction, Faculty of Applied Engineering Sciences, University College Ghent, Valentin Vaerwyckweg 1, 9000 Ghent, Belgium

[2]Department of Structural Engineering, Faculty of Engineering Sciences and Architecture, Ghent University, Technologiepark-Zwijnaarde 904, 9000 Ghent, Belgium

[3]Institute of Physics of Materials, Academy of Sciences of the Czech Republic, Žižkova 22, 616 62 Brno, Czech Republic

[a]Sara.Korte@hogent, [b]Veerle.Boel@hogent.be, [c]Wouter.DeCorte@hogent.be, [d]Geert.DeSchutter@UGent.be, [e]Seitl@ipm.cz

Keywords: concrete, wedge-splitting test, fracture mechanics, crack growth

Abstract. The wedge-splitting test (WST) is a frequently used test configuration for performing stable crack fracture experiments on concrete specimens, thus allowing to determine the fracture process and crack propagation in the heterogeneous material. However, there are no standard rules regarding the wedge-splitting specimen's geometry, groove dimensions or notch length. This paper concentrates on the influence of the initial notch length in geometrically identical, cubical specimens, cast from vibrated concrete. The experimental results of nine WSTs under monotonic loading, including F_{sp}-CMOD curves - splitting force versus crack mouth opening displacement - and fracture energy G_f, are presented. An important effect of the starting notch length on the fracture properties is observed.

Introduction

The design of concrete structures is typically based on the strength properties of the material, rather than considering fracture mechanics. One of the reasons is the heterogeneous nature of the material and its inherent flaws (such as pores, water inclusions, microscopic cracks due to shrinkage…), which prohibit the use of LEFM-theory [1-3]. Strain softening, caused by distributed cracking, transition of micro-cracks to macro-cracks prior to failure, and bridging stresses at the fracture front, requires a non-linear approach (NLFM) because of the large and more complicated fracture process zone (FPZ) at the crack tip [1]. Only during the last few decades researchers have established NLFM-models for concrete and cementitious materials, as well as test configurations to determine its fracture parameters. For instance, the wedge-splitting test on specimens with a groove and an initial notch was originally introduced by Linsbauer & Tschegg [4] in the 1980s and further developed by Brühwiler & Wittman [5]. Gradually, it has become a common method to investigate the fracture behaviour of concrete. However, it is no standardized method and there are no exact rules regarding the wedge-splitting specimen's geometry, groove dimensions or notch length. Therefore, this study examines the effect of the latter parameter on the fracture properties of several WST samples.

Experimental program

Mixture. The concrete's composition, along with its compressive strength, is listed in Table 1. In order to determine the strength, standardized compressive tests [6] were carried out on cubical control specimens (side 100mm) at an age of 28 days. The resulting value of $f_{c,cub100,m}$ is 54.68MPa.

Table 1. Concrete composition and properties

COMPOSITION	Quantity [kg/m³]
CEM I 52.5 N	365
Water	126
Sand 0/4	759
Crushed sea gravel 4/16	1101
Glenium 27 (superplasticizer)	2.9
STRENGTH	**[MPa]**
$f_{c,\,cub100,m}$	54.68

Specimens. The cubical wedge-splitting specimens in this experiment (Fig. 1) were obtained by placing a wooden rod with rectangular section (30x22mm) at the side of a standard cube mould (150x150x150mm), thus creating a plain top surface with guiding groove. These dimensions are based on the findings of Löfgren et al. [7]. After demoulding, the rod was removed and the specimens were stored under water at 20 ± 2°C. Approximately two days before testing the 3mm wide starter notch was made, by wet diamond sawing.

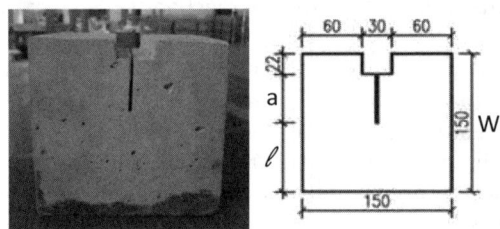

Fig. 1. Wedge-splitting specimen geometry and dimensions

In order to investigate the influence of the notch length (a) in relation to the specimens height (W) on the concrete's fracture properties, the ratio a/W was chosen 0.2, 0.3 and 0.37 (noted as 0.4), corresponding to notch lengths of 30mm, 45mm and 55mm, respectively. Per ratio three wedge-splitting specimens (A, B and C) were tested.

Test procedure. A 25kN capacity compression test device was used to apply a vertical, static load onto a transfer beam with two metal wedges. Each of these wedges (total angle 30°) move between two roller bearings, mounted on two metal caps, which rest on the edges of the specimen's guiding groove. This way, the applied vertical displacement (F_v) is converted to horizontal splitting forces (F_{sp}), thus causing the specimen to start cracking at the notch tip. Underneath the concrete cube a steel plate with two linear supports is placed (Fig. 2).

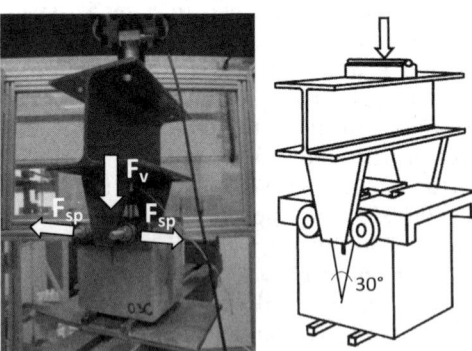

Fig. 2. Wedge splitting test setup

The testing machine was set in displacement mode, making the wedges move downward with an initial speed of 0.48mm/min. When the peak load was nearly reached, the speed was lowered to 0.01mm/min. During the entire test, the vertical compressive load (F_v) was continuously registered with a computer controlled data acquisition system. Likewise, the CMOD (crack mouth opening displacement) was measured by a clip gauge, mounted at the top of the guiding groove.

Results and discussion

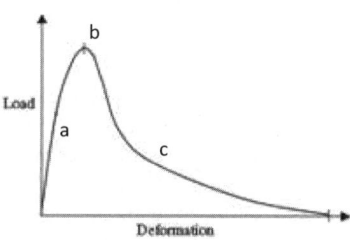

Fig. 3. Strain softening curve [6]

A general load-deformation response of a pre-cracked concrete specimen, characterized by progressive softening, is depicted in Fig. 3. During the ascending branch of the curve (a), only micro-cracks occur. Subsequently, after the peak load (b) is reached, the micro-cracks coalesce in the fracture process zone, but still, the load continues to decrease with increasing strain (c), due to aggregate bridging [8]. This phenomenon is called strain softening and is also present in the results of the conducted wedge-splitting tests.

The graphs below show the CMOD evolution for the three series of WSTs as a function of the horizontal splitting force (F_{sp}), which is calculated by Eq. 1.

$$F_{sp} = \frac{F_v}{2 \tan \alpha} \quad (1)$$

where α is the wedge angle, equal to 30°. The roller bearings' frictional effect on F_{sp} is not taken into account, since the introduced error by doing so is negligible [7,8].

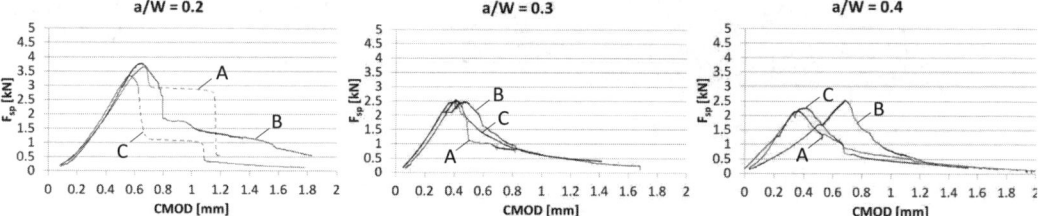

Fig. 4. F_{sp}-CMOD curves for different a/W ratios

From the curves, it can be seen that the ultimate load decreases from ca. 3.75kN over 2.50kN to ca. 2.25kN, as the a/W ratio (or the notch length a) increases (Fig. 4). Indeed, the longer the initial crack (notch), the less force is needed to split the specimen. The CMOD, corresponding to the peak load, measures 0.6mm, 0.4mm, and 0.4mm in case of an a/W ratio of 0.2, 0.3, and 0.4, respectively (specimen 0.4B excluded). As a result of the varying peak load, the fracture energy (G_f), which is basically derived from the surface under the curve, differs, as well. Table 2 lists the values of G_f as calculated by Eq. 2. Note that, for specimens 0.2A and 0.2C, no representative fracture energy could be obtained due to an incomplete descending branch of the F_{sp}-CMOD curve.

$$G_f = \frac{surface\_under\_curve}{W.\ell} \quad (2)$$

Table 2. Fracture energy G_f [N/m]

	a/W=0.2	a/W=0.3	a/W=0.4
Specimen A	-	112	116
Specimen B	189	108	•⁻
Specimen C	-	114	119
AVERAGE	**189**	**111**	**118**

In case of a/W=0.2 more energy is dissipated during the cracking of the specimen, compared to the larger ratios, which is not astonishing, given the larger fracture surface (W x ℓ). Surprisingly, little distinction in G_f is noticed between the 0.3 and the 0.4 ratio. Moreover, in the latter case, the fracture energy is slightly larger.

Finally, it is noticed that cracks, initiating at the corners of the guiding groove, as theoretically found by Vesely et al. [9], do not occur during these experiments, not even in case of a small a/W ratio (Fig. 5).

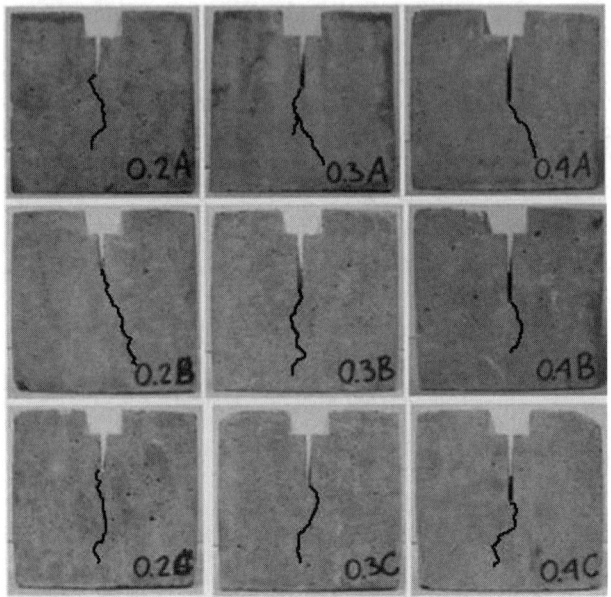

Fig. 5. Cracked wedge-splitting specimens

Conclusions

As could be expected, the shorter the initial notch, the larger the ultimate load. Fracture energy calculation consequently points out that the smallest a/W ratio, investigated in this experiment (0.2), produces the largest dissipated energy during fracture. Considering the other ratios (0.3 and 0.4), however, little difference is present.

Acknowledgments

The authors wish to express their gratitude to Vanholle T. and Mingneau S. for their assistance in the preparation of the tests and also to the research fund University College Ghent for the financial support.

References

[1] Bazant, Zdenek P. 2002. Concrete fracture models: testing and practice, *Engineering Fracture Mechanics 69:* 165-205.

[2] Moes, N., Belytschko, T. 2002. Extended finite element method for cohesive crack growth, *Engineering Fracture Mechanics* 69: 813-33.

[3] Cedolin, L., Cusatis, G. 2008. Identification of concrete fracture parameters through size effect experiments, *Cement and Concrete Composites 30 (9)*: 788-797.

[4] Linsbauer, H.N., Tschegg, E.K. 1986. Fracture energy determination of concrete with cube shaped specimens, *Zement und Beton 31*: 38-40.

[5] Brühwiler, E., Wittmann, F.H. 1990. The wedge splitting test, a new method of performing stable fracture mechanics test, *Engineering Fracture Mechanics 35 (1-3)*: 117-125.

[6] CEN (2001). EN 12390-3: Testing hardened concrete – Part 3: Compressive strength of test specimens.

[7] Löfgren, I., Olesen, J. F., Flansbjer, M. 2005. The WST-method for fracture testing of fibre-reinforced concrete, *Nordic Concrete Research 34 (2)*: 15-33.

[8] Hanjari, K. Z. 2006. Evaluation of WST Method as a Fatigue Test for Plain and Fiber-reinforced Concrete. Göteborg, Sweden: Chalmers University of Technology.

[9] Vesely, V., Routil, L., Seitl, S. 2011. Wedge-Splitting Test – Determination of Minimal Starting Notch Length for Various Cement Based Composites: Part I: Cohesive Crack Modelling, *Key Engineering Materials 452-453*: 77-80.

Micro-analyses of small cracks in 6061-T6 aluminium alloy subjected to high-cycle fatigue

Yoshimasa TAKAHASHI[1,a], Takahiro SHIKAMA[2,b] and Hiroshi NOGUCHI[3,c]

[1]Department of Mechanical Engineering, Kansai University, 3-3-35 Yamate-cho, Suita-shi, Osaka, 564-8680 Japan

[2]Kobe Steel Ltd., Aluminum & Copper Company (Chofu Works), 14-1 Chofu Minato-machi, Shimonoseki, Yamaguchi 752-0953 Japan

[3]Department of Mechanical Engineering, Kyushu University, 744 Motooka, Nishi-ku, Fukuoka-shi, Fukuoka 819-0395 Japan

[a]yoshim-t@kansai-u.ac.jp, [b]shikama.takahiro@kobelco.com, [c]nogu@mech.kyushu-u.ac.jp

Keywords: aluminium alloy, high-cycle fatigue, small crack, persistent slip band

Abstract. The growth of a small crack controlling the high-cycle fatigue life of a precipitation-strengthened 6061-T6 aluminium alloy was critically investigated. As the applied stress lowered, the small crack was arrested for a long period (over 10^6 cycles) at grain boundaries before regrowth, which resulted in a significantly slow growth process. The morphological and crystallographic details of the small crack were then analyzed with focused ion beam and transmission electron microscopy. It was revealed that the small crack was formed along fine persistent slip bands (PSBs) whose structure was fairly different from that reported for other metals. The concept of PSB-limited fatigue strength may be extended to include the present material type.

Introduction

Steels are known to exhibit distinct fatigue limit at high cycle fatigue (HCF) regime; a *knee point*, where fatigue life abruptly increases from 10^5-10^6 orders to over 10^7 orders, exists in the *S-N* diagram [1]. Aluminium (Al) alloys, on the other hand, do not exhibit a clear knee point in the HCF regime (see e.g. Fig. 2). This still makes it essentially difficult to conduct reliable fatigue design of machine components made of Al alloys, which needs to be improved through rational approaches.

The above difference in the *S-N* diagram is attributed to the different growth properties of initial small cracks (typically less than ca. 1 mm in dimension): the small cracks in steels at low stress level become non-propagating ones soon after their initiation while those in Al alloys continue to grow irrespective of stress level. From a phenomenological viewpoint, such growth properties have been well characterized e.g. by surface replica observations [2]. On the other hand, the underlying micro-mechanism of small crack growth, particularly its metallographic aspects, has mostly been discarded due to the analytical difficulties stemming from the smallness of the target crack.

The aim of this study is to critically investigate the micro-mechanism of small crack growth in a typical precipitation-strengthened Al alloy (JIS 6061-T6) through precise metallographic analyses. The results are then discussed in relation to the HCF life property.

Materials and methods

The chemical composition of the 6061 alloy used in this study is listed in Table 1. The ingot was extruded at 773 K and water-quenched at 293 K. The material was then age-hardened by the T6 heat treatment (450 K×28.8 ks). The 0.2% proof strength ($\sigma_{0.2}$) and absolute strength (σ_B) were 358 MPa and 379 MPa, respectively. Fig. 1 shows the shape and dimension of the smooth round bar specimen machined from the extruded bar. The gauge portion of the specimen was electropolished to remove damage layer. The specimen was then fatigued by using a rotating bending machine (frequency: 55 Hz, stress ratio: -1). The tests were conducted in air at room temperature. Fatigue crack growth was observed by the replica method.

Table 1 Chemical composition of 6061 alloy [wt.%]

Si	Fe	Cu	Mn	Mg	Cr	Zn	Ti	Al
0. 65	0.08	0.22	0.10	0.92	0. 23	0.01	0.02	Bal.

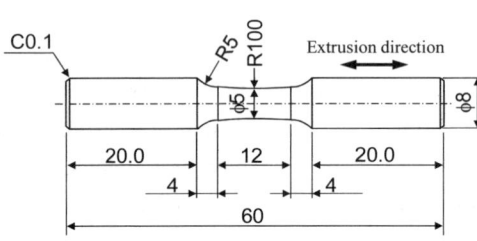

Fig. 1 Shape and dimension of specimen

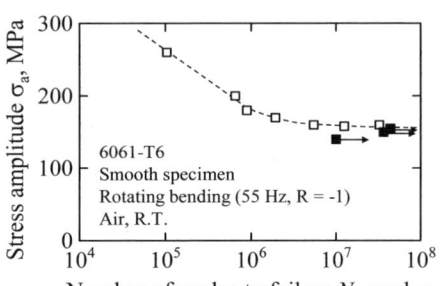

Fig. 2 S-N curve of 6061-T6 alloy

Results and discussion

Fig. 2 shows the obtained S-N curve of 6061-T6. The curve has no clear knee point around 10^5-10^6 cycles, which is in contrast to those for steels that exhibit clear knee points [1]. Fig. 3 shows the initial small crack observed in a specimen subjected to relatively low stress amplitude. The crack grew oblique to the principal stress plane. The crack tip was frequently arrested for more than 10^6 cycles and then began to grow at a relatively large rate. An electron backscatter diffraction (EBSD) analysis revealed that the crack tip was arrested at grain boundaries. Note that the regrowth of crack did not necessarily occur from the previous crack tip; a microcrack was sometimes formed near the crack tip and it coalesced with the main crack. The intermittent growth of the initial oblique crack occupied over 90% of the total life in this specimen. This explains why the slope of the S-N curve shown in Fig. 2 is reduced after 10^6 cycles: as the stress level lowers, the number of fatigue cycles consumed during the crack arrest increases.

Fig. 4 shows the secondary electron images of a small crack (left; plan view, right; cross section) observed by a focused ion beam (FIB). The crack extends from specimen surface and reaches only a few tens of micrometers inside; the aspect ratio of the crack (depth from surface/surface length) is less than 0.1, which is smaller than a normal semi-circular surface crack (aspect ratio: ≈ 0.5). The sub-surface crack plane is significantly tilted away from the principal stress plane, indicating that

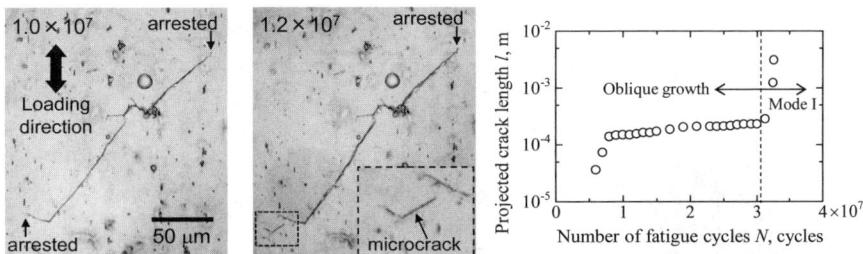

Fig. 3 Example of crack growth observed by replica and crack growth curve (σ_a = 160 MPa)

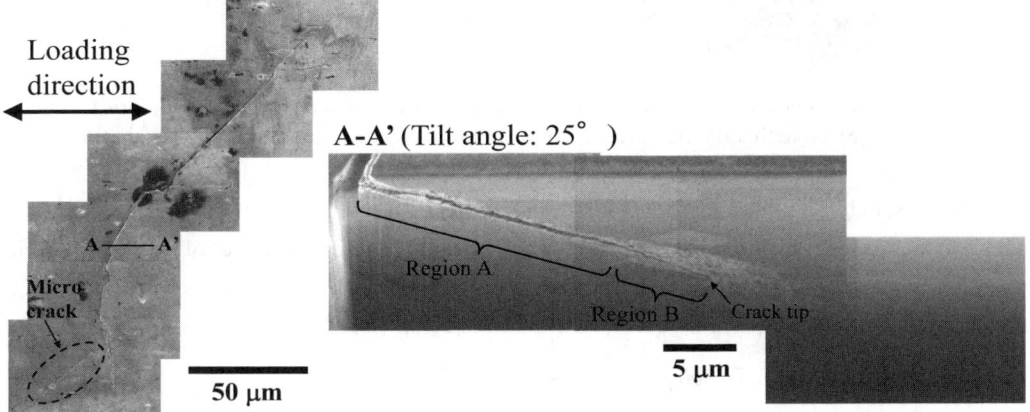

Fig. 4 A small crack observed by FIB (σ_a = 155 MPa, N = 4.3 × 10^7 cycles)

the crack tip was subjected to strong shear-mode stress intensity. The crack path of Region A is rather straight and accompanies no eminent plastic zone, while that of Region B is undulated and accompanies a severely deformed plastic zone. The three-dimensionally *anisotropic* crack growth is probably caused by the different growth resistance between Regions A and B.

Fig. 5 shows the TEM image of crack tip region located near the specimen surface. The beam direction was set parallel to the <110> direction of G1 so that the {111} planes in G1 are edge-on to the printed surface. The crack grows within a significantly localized slip band structure (i.e. persistent slip band; PSB) whose width is not more than 200-300 nm. The fine PSB and the associated crack are very different from the reported ones in pure FCC metals that accompany the formation of dislocation walls (e.g. vein, labyrinth or ladder structure [3]). The characteristic appearance of the fine PSB/cracking probably stems from the *inhomogeneity* (i.e. precipitation-reinforced matrix structure) of the present material type; once the precipitates on a certain slip plane are sheared, further slip along the same plane is facilitated [4].

The results suggest that the observed small crack growth in HCF regime can essentially be regarded as the initiation and extension of PSB cracking that exclusively occur on specimen surface. This process, the unit of which is grain or a group of sub-grains, is entirely crystallographic; it is easily impeded by grain boundaries, and the subsequent process depends primarily on the position of active dislocation sources in the adjacent grain. The micro-mechanism shown here is believed to be applicable to other precipitation-strengthened alloys such as SUH660 stainless steel [5] in which similar small crack growth behavior is reported.

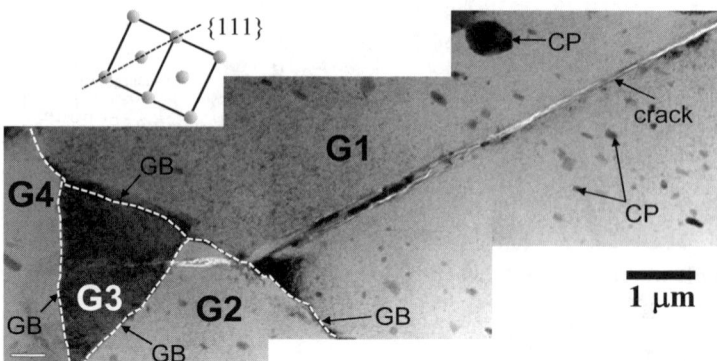

Fig. 5 TEM image of crack wake corresponding to Region A in Fig. 4 (GB; grain boundary, CP; coarse precipitate). Note that the fine precipitates reinforcing the matrix are not resolved in this image.

Finally, it should be added that the PSB-limited fatigue strength concept [6] still appears to be hold for the present material at very high-cycle fatigue (VHCF) regime of over 10^9 cycles according to our recent investigation using ultrasonic fatigue tester (see Fig. 6). The results will be further compared with the present data and discussed in detail elsewhere.

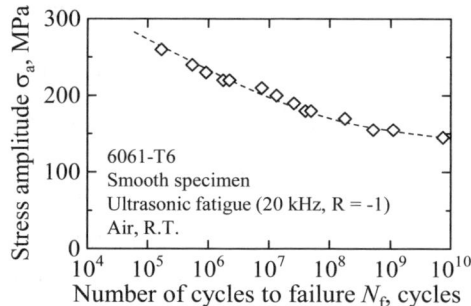

Fig. 6 S-N curve of 6061-T6 alloy including VHCF regime

Summary

The initial small fatigue crack in 6061-T6 alloy was critically analyzed by FIB and TEM. It was revealed that the small crack growth was attributed to the initiation and extension of PSBs whose structure was very different from those in pure FCC metals. The concept of PSB-limited fatigue strength may generally apply to the same material type (i.e. precipitation-strengthened alloys).

References

[1] M. Goto: Fatigue Fract. Eng. Mater. Struct. Vol. 14 (1991), p. 833
[2] H. Nisitani, K. Takao: Eng. Fract. Mech. Vol. 15 (1981), p.445
[3] C. Laird, P. Charsley, H. Mughrabi: Mater. Sci. Eng. Vol. 81 (1986), p. 433
[4] E. Hornbogen, K.H. Zum Gahr: Acta Metall. Vol. 24 (1976), p. 581
[5] H. Wu, Y. Oshida, S. Hamada, H. Noguchi: Procedia Eng. Vol. 10 (2011), p. 1973
[6] H. Mughrabi: Int. J. Fatigue Vol. 28 (2006), p. 1501

Fatigue Properties of Solution-treated Type 304 Stainless Steel after Nitriding

Yuki Nakamura[1, a], Masaki Nakajima[1, b], Toshihiro Shimizu[1, c], Kentaro Suzuki[2, d], Yu Bai[3,e] and Yoshihiko Uematsu[4,f]

[1]Toyota National College of Technology, 2-1 Eisei-cho, Toyota 471-8525, JAPAN

[2]Advanced Course Student, Toyota National College of Technology, JAPAN

[3]Graduate Student, Gifu University, JAPAN

[4]Gifu University, 1-1 Yanagido, Gifu 501-1193, JAPAN

[a]nakamura@toyota-ct.ac.jp, [b]nakajima@toyota-ct.ac.jp, [c]shimizu@toyota-ct.ac.jp, [d]kentaro_america@gmail.com, [e]p3122033@edu.gifu-u.ac.jp, [f]yuematsu@gifu-u.ac.jp

Keywords: Fatigue, Solution treatment, Nitriding, Austenitic stainless steel, Strain-induced martensitic transformation

Abstract. The solution treatment after nitriding (STAN) was performed to stabilize the γ- phase in a metastable austenitic stainless steel, type 304, and to improve the strength of type 304 by the solid solution of nitrogen. Plasma nitriding was conducted at 500°C for 8.5h, and then solution treatment was performed at 1200°C for 1h. As a result, the static strength and the hardness were improved by the STAN. Rotary bending fatigue tests were performed on the specimens with STAN (solid solution strengthened) together with the untreated and the nitrided ones in laboratory air and in 3%NaCl solution. In laboratory air, the fatigue strength of the solid solution strengthened specimen increased compared to that of the untreated specimens, where fatigue limits were 340MPa and 290MPa for the solid solution strengthened and the untreated, respectively. However, the fatigue limit of the solid solution strengthened specimen was lower than that of the nitrided specimen, that is, 380MPa. On the other hand, in 3%NaCl solution, the fatigue strengths of the nitrided specimens and the solid solution strengthened specimens decreased significantly compared to those in laboratory air. After the fatigue tests at the stress level of fatigue limit in laboratory air, the strain-induced martensitic transformation was examined by XRD. In the solid solution strengthened specimens, the strain-induced martensitic transformation was not detected during fatigue tests until 3×10^7 cycles, indicating that the γ- phase was stabilized by the solid solution of nitrogen.

Introduction

Austenitic stainless steel, type 304, is extensively used as a structural material due to its good ductility, high corrosion resistance and so on. However, since the nickel equivalent is less as seen in the Schaeffler diagram [1], it is known to develop the strain-induced martensitic transformation because of the lower stability of austenitic phase. Therefore, it is expected that the fatigue behavior of type 304 could be influenced by the strain-induced martensitic transformation, because the martensitic phase has higher hardness than austenitic phase and has lower corrosion resistance than austenitic phase [2-4].

In this study, the solution treatment after nitriding (STAN) was performed to stabilize the γ-phase in type 304 and to improve the strength of type 304 by the solid solution of nitrogen. Then, the rotary bending fatigue tests until 10^8 cycles were performed in order to clarify the effects STAN on fatigue properties. In addition, the results obtained were compared to those of high temperature gas nitriding (HTGN) specimens.

Experimental Procedures

Materials and Treatment Conditions. The material used was an austenitic stainless steel, type 304, whose chemical composition (wt.%) is as follows; C: 0.06, Si: 0.29, Mn: 1.65, P: 0.04, S: 0.03, Ni: 8.03, Cr:18.62, Fe: bal. This material was solution treated at 1080°C for 30min followed by water quenching. After the solution treatment, hourglass-shape fatigue specimens with a reduced section of 4.5mm diameter were machined as shown in Fig.1 and the reduced section was polished before experiment. After polishing, plasma nitriding was conducted at 500°C for 8.5h, and then solution treatment was performed at 1200°C for 1h followed by oil quenching.

Procedures. Fatigue tests were conducted using cantilever-type rotary bending fatigue testing machines on the specimens with the STAN (solid solution strengthened), the nitrided and the untreated operating at a frequency of 53 Hz in laboratory air and in 3%NaCl solution. The strain-induced martensitic transformation was detected by X-ray diffraction method (XRD).

Experimental Results and Discussions

Microstructures. Figure 2 shows microstructures of the untreated, the nitrided and the solid solution strengthened specimens. As can be seen in Fig.2 (b), the nitrided layer of 50μm in thickness was formed on the surface of specimen. The average grain sizes of the respective specimens were 65μm, 73μm, 69μm, indicating that the grain size was not influenced by nitriding and STAN.

Mechanical Properties. Mechanical properties of each specimen are shown in Table 1. The solid solution strengthened specimen exhibits a little higher 0.2% proof stress and tensile strength than those of the other specimens, where STAN leads to the higher static strength.

Hardness Distributions. The Vickers hardness distributions of the untreated and the solid solution strengthened specimens are shown in Fig.3. In the figure, the hardness distributions of the specimen with STAN for 45min and 90min are also plotted. Furthermore, the hardness distributions of high temperature gas nitrided (HTGN) specimens, which were annealed from 1100°C and 1200°C in nitrogen gas, are also plotted in the figure for comparison [5]. All the specimens except for the untreated specimen exhibit higher hardness at the surface than that of untreated specimen, indicating the solid solution strengthening by nitrogen. The hardness layers of the solid solution strengthened specimens for 60min and 90min are thicker than that of other specimens, suggesting nitrogen was diffused deeper.

Fatigue Strength. Fig.4 indicates the S-N curves in laboratory air. The fatigue test data of the HTGN specimens are also plotted in the figure. It is clear that the fatigue strength of the solid solution

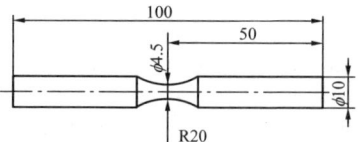

Fig.1 Specimen configuration.

Table 1 Mechanical properties.

Material	0.2% proof stress $\sigma_{0.2}$ (MPa)	Tensile strength σ_B (MPa)	Elongation δ (%)	Reduction of area φ (%)
Untreated	268	606	62	83
Nitrided	263	616	64.7	76.8
Solid soulution strengthened	282	640	62.2	77.7

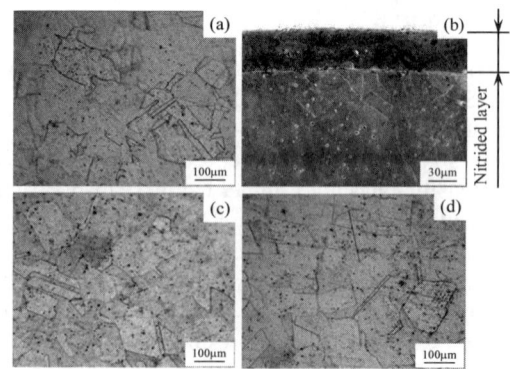

Fig.2 Microstructures: (a) untrteated, (b) nitrided layer, (c) nitrided and (d) solid solution strengthned.

strengthened specimens increased compared to those of the untreated and the HTGN at 1100°C specimens, where fatigue limits were 340MPa, 290MPa and 290MPa for the solid solution strengthened, the untreated and the HTGN at 1100°C specimens, respectively. The improvement of fatigue strength was due to the solid solution of nitrogen. However the fatigue limit of the solid solution strengthened specimens was lower than those of the nitrided and the HTGN at 1200°C specimens, 380MPa, because the nitrides (CrN), which have high hardness, were formed on the surface for the nitrided and the HTGN at 1200°C specimens. In addition, the subsurface fracture due to fish-eye was found in the nitrided at σ_a=390MPa, N_f=9.4x10^4 cycles.

The *S-N* curves in 3%NaCl solution are represented in Fig.5, together with the data in laboratory air. In the untreated specimens, the fatigue strength in 3%NaCl solution is nearly the same as that in laboratory air. However, it is clear that the fatigue strength of the nitrided specimens and the solid solution strengthened specimens in 3%NaCl solution decreased significantly compared to those in laboratory air. In addition, fatigue failure occurred at applied stresses below the fatigue limits in laboratory air, indicating the reduction of fatigue strength in the corrosive environment. The fatigue limits for the nitrided and the solid solution strengthened specimen disappeared under the corrosion fatigue condition. This behavior may be attributed to the sensitization caused by nitriding and STAN. The *S-N* characteristics of the HTGN specimens in 3%NaCl solution also showed the similar tendency.

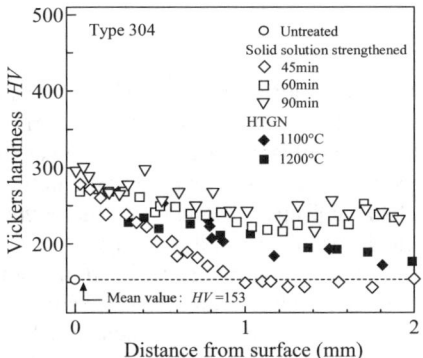

Fig.3 Vickers hardness distribution.

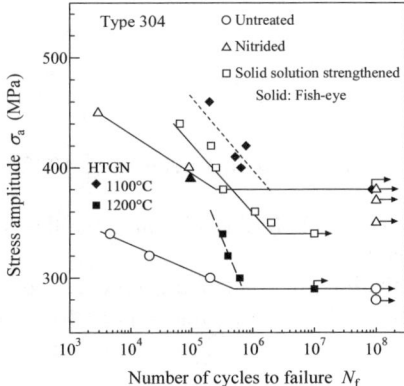

Fig.4 *S-N* curves in laboratory air.

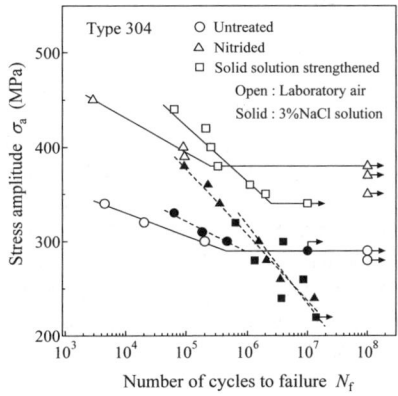

Fig.5 *S-N* curves in 3%NaCl solution and in laboratory air.

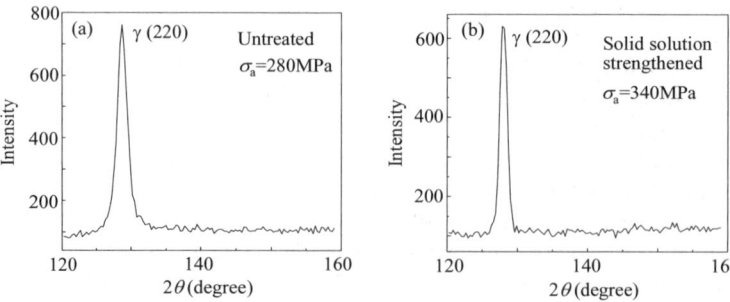

Fig.6 X-ray diffraction patterns: (a) untreated and (b) solid solution strengthened.

Martensitic Phase Transformation. After the fatigue tests at the stress level of fatigue limit, the strain-induced martensitic transformation was examined by XRD for the untreated and the solid solution strengthened specimens. The results are shown in Fig.6. The strain-induced martensitic transformation was not detected in both materials. In the solid solution strengthened specimens, though the fatigue limit was 60MPa higher than that of the untreated specimens, the strain-induced martensitic transformation did not take place during fatigue tests until 3×10^7 cycles. Thus it implies that the γ-phase was stabilized by the solid solution of nitrogen.

Conclusions

In the present study, the solution treatment after nitriding (STAN) was performed on the austenitic stainless steel, type 304, and rotary bending fatigue tests were performed to investigate the effect of STAN on the fatigue property. The results obtained are summarized as follows;
(1) The grain size was not changed by STAN, as compared to the untreated specimen.
(2) The STAN enhanced the mechanical properties and hardness in type 304.
(3) The solid solution strengthened specimen exhibited higher fatigue limit than the untreated specimen, but lower fatigue limit than the nitrided specimen in laboratory air.
(4) The fatigue strengths of the nitrided specimens and the solid solution strengthened specimens in 3%NaCl solution decreased significantly compared to those in laboratory air.
(5) The strain-induced martensitic transformation was not detected after the fatigue tests at the stress level of fatigue limit for the untreated and the solid solution strengthened specimens. Thus, it was considered that the γ-phase of type 304 was stabilized by STAN.

References

[1] A. L. Schaeffler: Metal Progress, 56 (1949) p. 680
[2] L. J. Qiao, J. L. Luo: Corrosion, 54 (1998) p. 281
[3] M. Nakajima, Y. Uematsu, T. Kakiuchi, M. Akita, K. Tokaji: Procedia Engineering, 10 (2011) p. 299
[4] M. Nakajima, Y. Uematsu, T. Kakiuchi, M. Akita, K. Tokaji, T. Murasaki: Journal of the Sosiety of Materials Science, Japan, 60 (2011) p. 796 (in Japanese)
[5] Y. Bai, Y. Uematsu, M. Akita, T. Kakiuchi, M. Nakajima, Y. Nakamura: Proceedings of the Second China-Japan Joint Symposium on Fatigue of Engineering Materials and Structures, (2011) p. 109

Fatigue Crack Growth of Alloy 7050-T7451 Plate in L-S Orientation

Bao Rui[1,a], Zhao Xiaochen[1,b], Zhang Ting[1,c] and Zhang Jianyu[1,d]

[1] Institute of Solid Mechanics, School of Aeronautic Science and Engineering, Beihang University 100191, PR China

[a]rbao@buaa.edu.cn, [b]melo_striving@126.com, [c]13466380139@163.com, [d]jyzhang@buaa.edu.cn

Keywords: fatigue crack growth; crack deflection; aluminium alloys.

Abstract. Experiments have been conducted to investigate the crack growth characteristics of 7050-T7451 aluminium plate in L-S orientation. Two loading conditions are selected, i.e. constant amplitude and constant stress intensity factor range (ΔK). The effects of ΔK-levels and stress ratios (R) on crack splitting are studied. Test data shows that crack splitting could result in the reverse of crack growth rate trend with the increasing R ratio at high ΔK-level. The appearance of crack splitting depends on both ΔK and R.

Introduction

Alloy 7050 is the premier choice for aerospace applications requiring the best combination of strength, stress corrosion cracking (SCC) resistance and toughness [1]. Typical applications for alloy 7050 plate include fuselage frames and bulkheads. The 7050-T7451 alloy is one of the highly anisotropic materials, consequently, crack growth behavior along the short transverse direction (L-S) is found to be quite different from that along the long transverse direction (L-T). Most strength and fatigue properties available are in L-T orientation [1]. Nevertheless, the L-S orientated plates have found some applications in the spar caps and stringer webs of machined integral skin-stringer panels, therefore, full characterization of crack growth in different orientations are needed for use. Progress has been made in understanding the crack growth behavior in L-S orientated AA 7050-T7451 plates under constant amplitude loading of different stress ratios [2], which shows that crack turning, kinking, splitting and branching appeared when the specimens were loaded under tension load only. Crack growth rates in AA 7050-T7451 subjected to simple load sequences containing underloads are reported in [3]. Ref. [4, 5] present the crack growth behavior of 7050-T7451 under truncated spectrum loading, in which comparisons of the failure modes between L-S and T-L plates are also presented.

Lots of evidence of crack path changes have been found in the experiments on AA 7050-T7451 L-S orientation specimens under tension dominated loading condition as mentioned above, which are quite different from the typical Mode I crack growth morphology. To understand the factors leading to crack branching in 7050-T7451, further experimental investigations are conducted and reported in this paper.

Experiments

Material and specimens. The following material data is provided for reference from the guaranteed ALCOA material spec [1]: (WT%), Si 0.12, Fe 0.15, Cu 2.0 – 2.6, Mn 0.10, Mg 1.9 – 2.6, Zn 5.7 – 6.7, Zr 0.08 – 0.115, Ti 0.06, others each 0.05, others total 0.15, Balance Aluminum. The mechanical properties are also available in [1], i.e. tensile Strength 510 MPa, yield strength 441MPa, elongation 9%, K_{IC} values for thin plate (25-51 mm) L-T 31.9 MPa$\sqrt{m}$, T-L 27.5 MPa$\sqrt{m}$, S-L N/A and for thick plate (127-152 mm) L-T 26.4 MPa$\sqrt{m}$, T-L 24.2 MPa$\sqrt{m}$, S-L 23.1 MPa$\sqrt{m}$.

Standard compact tension (CT) specimens in L-S orientation are used in this study. The configuration of the specimens is shown in Fig.1.

Experimental procedure. All the fatigue crack growth tests were accomplished using MTS880 fatigue test system, and an observation system consisting of digital microscope, servo motor, and raster ruler is used to register the position of crack tip.

Two kinds of loading conditions are selected, one is constant amplitude (CA) loading, and the other is constant stress intensity factor range ΔK. There are three stress ratios (R), i.e. R= 0.1, 0.3 and 0.5, and three ΔK levels, i.e. ΔK = 15, 18 and 20 MPa$\sqrt{m}$, adopted in the experiments. All the tests conducted in this study are listed in Table 1.

Table 1 Loading conditions involved in this study

Loading conditions		stress ratio		
		R=0.1	R=0.3	R=0.5
CA		✓	✓	✓
Constant ΔK (MPa$\sqrt{m}$)	ΔK=15	---	---	✓
	ΔK=18	✓	✓	✓
	ΔK=20	---	---	✓

Notes:
    ✓ represents that the tests under this loading condition are conducted.
    --- represents that the tests under this loading condition are not conducted.

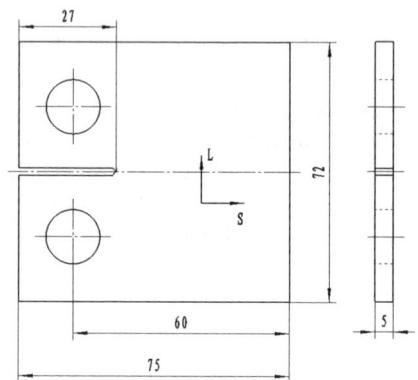

Fig. 1 CT specimens principal dimensions (Unit: mm)

All the specimens were fatigue precracked up to 1 mm from the notch tip under constant amplitude load. The final ΔK during precracking is kept less than 7 MPa$\sqrt{m}$ for the tests under CA loading condition, or reaches the level of crack growth test for the constant ΔK loading condition, which is ensured by choosing a reasonable applied stress level.

Constant amplitude crack propagation tests were performed in accordance with ASTM E647 (2011). Tests were run at 8 Hz, in laboratory air. Constant ΔK loading condition was achieved by decreasing the applied stress range. ΔK was calculated when the crack size was measured during the test. If the difference between the current ΔK level and the required ΔK level is greater than 3% of the required level, the applied stress is decreased.

Results and discussion

Results under constant amplitude loading condition. The original data obtained from the tests were crack size versus elapsed cycles data (a versus N). The (a,N) data were then converted to crack growth rate da/dN vs. ΔK. da/dN was computed by incremental polynomial method, and the corresponding stress intensity factors were calculated using the expression recommended by ASTM E647 (2011). The test results under CA loading conditions are shown in Fig. 2. It is amazing that the da/dN vs. ΔK curves under the three R ratios crossed each other. It is can be seen that (1) in the lower ΔK range (ΔK<10 MPa$\sqrt{m}$), da/dN increases with the increase of R at the same ΔK level; (2) when ΔK increased above 10 MPa$\sqrt{m}$, the increasing speed of da/dN for R =0.5 dropped remarkably, while the da/dN for R = 0.1 and 0.3 increased continuously, which results in the cross of the three da/dN~ΔK curves; (3) when ΔK

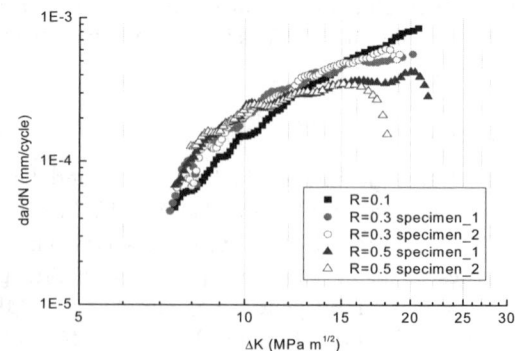

Fig. 2 da/dN~ΔK curve under CA loading condition

increased over 16 MPa$\sqrt{m}$, the da/dN~ΔK curve of $R = 0.3$ lies below that of $R = 0.1$, that is to say, under the same ΔK level, the da/dN decreases with the increase of R which is contrary to the basic idea of linear elastic fracture mechanics; (4) The sudden drop of da/dN at the end of the curve for $R = 0.5$ is due to crack splitting, the crack morphology can be find in the figures in the section of discussion.

Results under constant stress intensity factor range loading condition. Since the da/dN~ΔK curve for $R = 0.5$ shows uncommon trend, further tests were conducted under constant ΔK loading condition of three ΔK levels, i.e. ΔK =15, 18 and 20 MPa$\sqrt{m}$. Tests under $\Delta K = 18$ MPa$\sqrt{m}$, $R = 0.1$ and 0.3 were also done. Since ΔK was fixed in the test, the results are given in terms of da/dN vs. a in Fig. 3. For comparison, the da/dN data for 7050-T74511 L-T specimen from Ref. [6], the changing trend of which is accordance with the common knowledge, are plotted together with the test data in Fig.3. The figure indicates that (1) da/dN at ΔK =15 and 18 MPa$\sqrt{m}$, $R = 0.5$ show little difference; (2) at the given ΔK level of 18 MPa$\sqrt{m}$, da/dN of R =0.5 is the lower than that of R =0.1 and 0.3 regardless of the crack length. This is in accordance with the results of the test under CA loading condition. There is no da/dN vs. a data of $\Delta K = 20$ MPa$\sqrt{m}$ in Fig. 3(a) because crack splitting appeared in the very early stage in the test, when the length of the main crack is about 1mm from the notch tip as shown in Fig.4(a). The propagation of the "main" crack, which is perpendicular to the loading direction, stopped, and more splits were observed as the increasing of loading cycles, see Fig. 4(b) and (c).

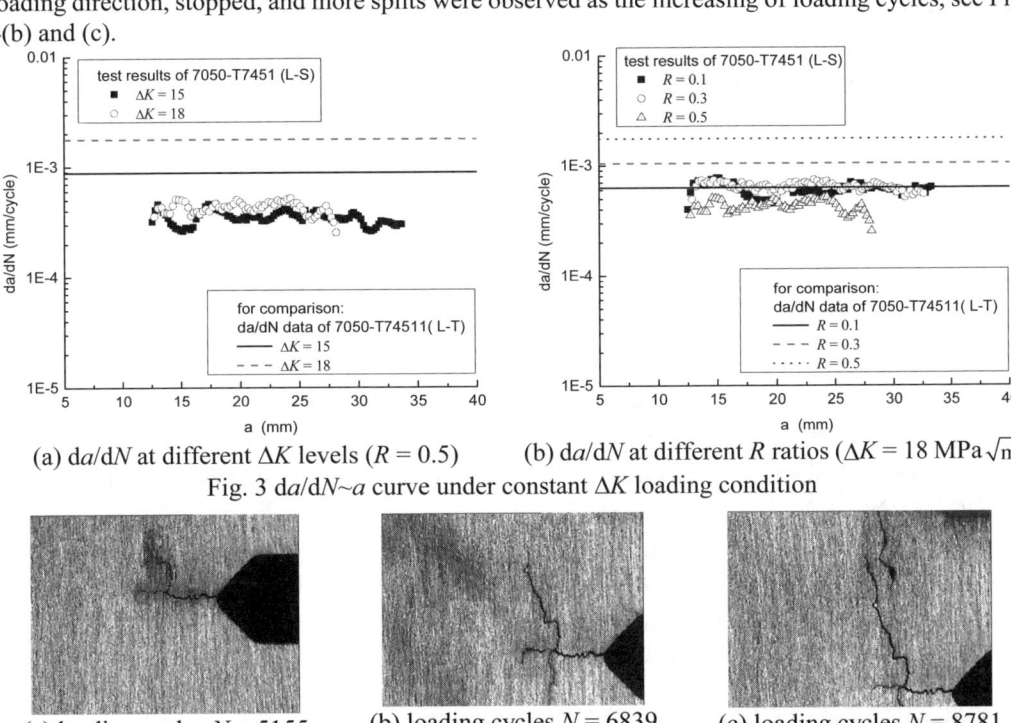

(a) da/dN at different ΔK levels ($R = 0.5$) (b) da/dN at different R ratios ($\Delta K = 18$ MPa$\sqrt{m}$)

Fig. 3 da/dN~a curve under constant ΔK loading condition

(a) loading cycles $N = 5155$ (b) loading cycles $N = 6839$ (c) loading cycles $N = 8781$

Fig. 4 Crack patterns under constant ΔK= 20 MPa$\sqrt{m}$, $R = 0.5$

Discussion. According to linear elastic fracture mechanics, da/dN is a function of ΔK, and this function depends on the stress ratio R. In the Paris-ΔK-region, da/dN increases with the increase of ΔK at a fixed R ratio, and da/dN increases with the increase of R ratio at a given ΔK-level [7]. However, the tested da/dN trend is contrary to the common idea, the reason for which is the crack branching or splitting observed in the experiments. This phenomenon was also reported by JJ Schubbe in Ref. [2]. Schubbe points out that significant forward growth retardation or splitting is evident at threshold ΔK values in the range of 10-15 MPa$\sqrt{m}$, and a distinct arrest point where vertical growth is dominating is found between 18 and 20 MPa$\sqrt{m}$. The tests in this study show that (1) under CA loading conditions, the ΔK-levels that dominant the changing points of the da/dN~ΔK curve, see Fig.2, agree quite well with the result of Schubbe; (2) ΔK is not the only factor which affected the crack splitting, see Fig. 5, macro-noticeable crack splitting appears only in the high R ratio condition; (3) the tests under constant ΔK of R =0.5 indicates that R is not the only factor which affected the crack splitting either, see Fig. 6 and 4, remarkable macro-level crack splitting was not observed in the high R ratio test of ΔK= 15 MPa$\sqrt{m}$. Besides the final splitting in the specimen tested under ΔK= 15 MPa$\sqrt{m}$, R = 0.5, early branching (marked by circle in Fig.6) was also observed.

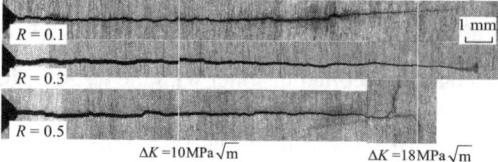

Fig. 5 Crack patterns under CA loading conditions

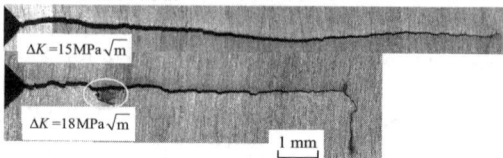

Fig. 6 Crack patterns under constant ΔK loading conditions (R = 0.5)

Conclusions

Fatigue crack growth test results for this study of 7050-T7451 aluminum alloy in L-S orientation show that (1) for CA loading condition, da/dN~ΔK curves of R = 0.1, 0.3 and 0.5 crossed each other, the reason for which is the crack branching or splitting; (2) ΔK and R have combined effect on the redirection of crack growth.

Acknowledgement

The National Natural Science Foundation of China is acknowledged for supporting the project (10802003).

References

[1] Alcoa Mill Products: 7050 Aluminium Alloy Plate and Sheet; Website (accessed Aug. 2009): http://www.alcoa.com/mill_products/catalog/pdf/alloy7050techsheetrev.pdf.

[2] J J Schubbe. Fatigue crack propagation in 7050-T7451 plate alloy, Eng Frac Mech, 76(2009): 1037-1048.

[3] P White, S A Barter, C Wright. Small crack growth rates from simple sequences containing underloads in AA7050-T7451, Int J Fatigue, 31 (2009) 1865–1874.

[4] J J Schubbe. Evaluation of fatigue life and crack growth rates in 7050-T7451 aluminum plate for T-L and L-S oriented failure under truncated spectra loading, Eng Fail Anal, 16(2009): 340-349.

[5] R Bao, X Zhang. Fatigue crack growth behaviour and life prediction for 2324-T39 and 7050-T7451 aluminium alloys under truncated load spectra, Int J Fatigue 32 (2010) 1180–1189.

[6] AFGROW – US Air Force Crack Growth Software; V4.12.15.0, 2009.

[7] K Walker. The Effect of Stress Ratio During Crack Propagation and Fatigue for 2024-T3 and 7075-T6 Aluminum, ASTM STP 462, American Society for Testing and Materials, 1970.

Structural Design and Finite Element Analysis of Composite Wind Turbine Blade

Shifan Zhu[1, 2, a] and Ibrohim Rustamov[1, 3 b],

[1]College of Mechanical and Electrical Engineering, Harbin Engineering University, Harbin, China

[2]Heilongjiang Modern Manufacturing Engineering Research Center, Harbin, China

[3]Department of Mechanical Engineering, Tashkent State Technical University, Tashkent, Uzbekistan

[a]zhushifan@hrbeu.edu.cn, [b] ibrohim105@yahoo.com

Keywords: wind turbine blade, structural design, composite materials, finite element method

Abstract. This paper presents structural studies of a medium scale composite wind turbine blade construction made of epoxy glass fiber for a 750kW rated power stall regulated horizontal axis wind turbine system. The complex geometry of the blade with a skin-spar foam sandwich structure was generated by utilizing commercial code ANSYS finite element package. Dimensions of twist, chord and thickness were developed by computer program. NREL S-series airfoils with different chord thickness are used along current blade cross-sections. The current design method uses blade element momentum (BEM) theory to complete satisfactory blade design and can be carried out using a spreadsheet, lift and drag curves for the chosen aerofoil. According to composite laminate theory and finite element method, optimal blade design was obtained. The focus is on the structural static strength of wind turbine blades loaded in flap-wise direction and methods for optimizing the blade cross-section to improve structural reliability. Moreover, the natural frequencies and modal shapes of the rotor blade were calculated for defining dynamic characteristics. Structural analysis was performed by using the finite element method in order to evaluate and confirm the blade to be sound and stable under various load conditions.

Introduction

Blades of horizontal axis wind turbines are being manufactured using polymer matrix composite materials, in a combination of monolithic (single skin) and sandwich structures. Composite materials satisfy complex design constraints such as lower weight and proper stiffness, while providing good resistance to the static and dynamic loading [1].

Many researchers investigated on different techniques such as design, testing, structural and fatigue strength analysis of wind turbine blades. A computerized method has been developed by Bir [2] to aid preliminary design of composite wind turbine blades. The method in his work allows for arbitrary specification of the chord, twist, and airfoil geometry along the blade and an arbitrary number of shear webs. Minimization of the blade weight was the main purpose in the study of Veers at al. [3] who described the currently dominant glass fiber technology and the potential use of carbon fibers. Kong and Bang [4] investigated structural design of medium scale composite wind turbine blade by determining design loads from various load cases and fatigue life. A full-scale 34m composite wind turbine blade was tested to failure under flap-wise loading in the work of Jensen [5]. He observed the ovalization of the load carrying box girder in the full scale test and simulated in non-linear finite element calculations.

In this paper, a structural design procedure of a medium scale E-glass/epoxy composite wind turbine blade was proposed. With a layered shell element, the geometrical blade model was created by ANSYS commercial finite element package. The finite element model is validated through tip deflection, modal analysis and comparison of the structural results of the blade with the measured ones.

Properties of the Blade

The blade geometry studied in this work is 24m long skin-spar-foam structure, maximum chord of 2.2 m, tip chord of 0.45, with a low pressure (LP) shell on the downwind side, a high pressure (HP) shell on the downwind side, and two shear webs bonded between the two shells. The exterior blade shape uses NREL S814 airfoil at the root to keep good structural performance, S809 airfoil was applied for the outboard section. For good aerodynamic performance, S810 airfoil shape was used at the tip of the blade. Thickness of the skin was reduced along the blade length and the shell also twisted near to 15^0 because of aerodynamic reasons while the shape is tapering from root to tip. The FE model of the blade is as shown in Fig. 1. Spar caps make a use of 80% unidirectional fiber (0^0) in combination with 20% of ($\pm 45^0$) double bias, while the two shear webs and outer skin uses ± 45 fabric with a foam core as filler in sandwich type layups. Table 1 shows the mechanical properties of the blade materials [6].

Table 1 Mechanical properties of involved materials in the blade

Material, property	Unidirectional GFRP	$\pm 45^0$ Fabric GFRP	Foam
E_x (GPa)	35.7	22.14	0.060
E_y (Gpa)	10.6	2.65	0.059
G_{xy} (Gpa)	2.8	1.6	0.019
v_{xy}	0.32	0.3	0.2
ρ (kg/m^3)	1800	1870	119

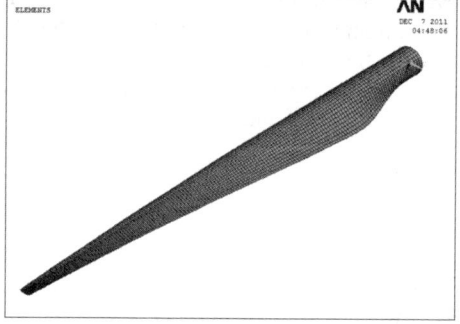 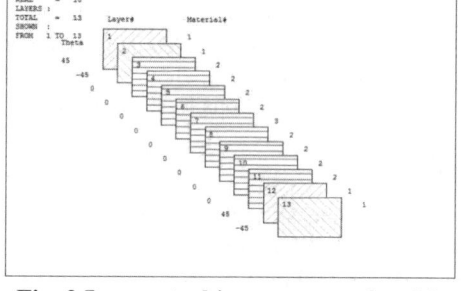

Fig. 1 FE mesh model   Fig. 2 Layer stacking sequence for skin

Finite Element Analysis

Basically, the blade model consists of shell elements. The surface geometry is defined by connecting a number of predefined blade cross sections with radial splines. The element type used for the static stress analysis is the 3D linear layered structural shell element (SHELL99), which is used for up to 250 layered applications of a structural model. It has six degrees of freedom at each node x, y, and z axes. Composite materials are regarded as orthotropic having material properties of Young`s modulus $E_x$, $E_y$, $E_z$, Poisson`s ratio v_{xy}, and shear modulus G_x, G_y, G_z for each material as shown above. Fiber stacking layers and angles was determined in the interface program with the thickness of 0.6mm and 1.2mm. Fig. 2 shows lay-up sequence for the skin compressive upper blade part. The final model consists of 16881 elements and 49596 nodes.

In the static loading analysis, the blade structure was treated as a cantilever beam by fixing all six degrees of freedom for nodes in the root end. Three concentrated loads were applied throughout the blade to extract all displacement, stiffness-strength data. Geometry shape, lay-up sequence and the thickness of the plies were modified iteratively until convergence is obtained when additional deformation is negligible.

In order to avoid the resonance, the natural frequency and modal shape of the rotor blade must be calculated for defining dynamic characteristics. The first four steps of flapping and lagging natural frequencies and modal shapes should be focused on, because the natural frequency and modal shape has high influence on dynamic characteristics of a wind turbine rotor blade.

Results and Discussion

In the static analysis, the blade tip deflection and the Von Misses stresses were analyzed. The blade bends in the direction of the lift forces causing the low pressure shell to be in compression, while the low pressure shell is in tension. In these conditions the largest deflection occurs in the blade tip with 1.94m, which is acceptable value because the tip-to-tower distance of the blade is 4.5m. Fig. 3 depicts the deformation of the blade. As shown in Fig. 4, the maximum stresses are located at the blade root and decrease towards the tip. According to the results of this analysis, the maximum tensile and compressive stresses developed in the blade due to cyclic bending were 55.45 MPa and -47.28 MPa, respectively. Additionally, some other values of the blade were extracted such as the mass of the blade and span-wise center of mass. Table 2 illustrates the comparison of the current blade values with these obtained from the analytical work of [4]. The results are found to be in good agreement with this experiment.

Table 2 Comparison of the blade values with the measured one

	Tip deflection(mm)	Total mass(kg)	Center of Mass(m)
Measured	1978.6	2951	8.451
FEA model	1938	2947	8.214
Accuracy(%)	97.94	99.86	97.19

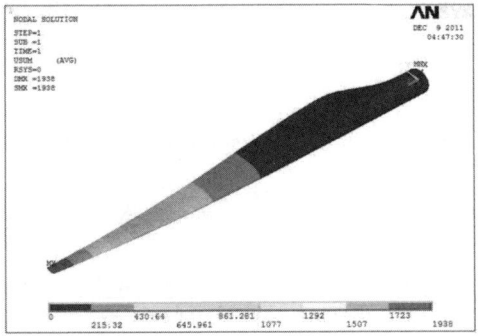

Fig. 3 Total deformation of the blade   **Fig.4 Von Misses stresses**

The mode shape results, associated with the lowest blade natural frequencies, are listed in Table 3. The dynamic analysis takes into account the blade natural modes of vibration. Those must be uncoupled with the natural frequencies of the other components of the wind turbine in order to avoid mechanical resonance. In Fig. 6, the calculated natural frequencies have been compared with the corresponding natural frequencies obtained from FE model of the investigated blade in [4], in order to evaluate the state-of-the-art blade modeling capacity. According to the results of free vibration analysis, it can be say that the model is in accordance with real structures from different aspects such as dimension, material properties and lay-up stacking sequence.

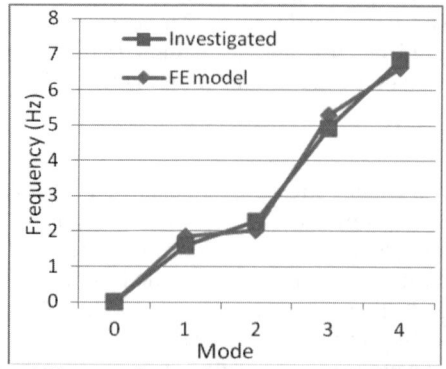

Fig. 6 Comparison of the natural frequencies

Table 3 Modal analysis results

Mode description	Frequency (Hz)
1st flap-wise bending	1.814
1st chord-wise bending	2.16
2nd flap-wise bending	5.19
2nd chord-wise bending	6.75
1st torsional	18.58

Conclusion

A medium scale composite wind turbine blade was designed, optimized, and structural analysis was carried out by using finite element method. E-glass/epoxy demonstrated excellent strength and weight reduction with few lay-up changes. This study also provides a potentially dominant method for designing the blade structure to minimize ply drops and simplify manufacturing. The relative stacking of the blade elements to form the final blade shape was manipulated to decrease the blade tip deflection and stress within the blade during design operating condition.

The predicted mass, span-wise center of gravity, blade tip deflection and natural frequencies have a good agreement with the corresponding measured values. Through linear static stress, modal, and maximum blade tip deflection analyses, it was confirmed that the blade structure was safe and stable to operate at various load conditions.

References

[1] M. S. Mahmood, R. Rafiee: Simulation of fatigue failure in a full composite wind turbine blade. Composite Structures 74 (2006) 332–342.

[2] Bir GS: Computerized method for preliminary structural design of composite wind turbine blades. J Sol Energy Eng 2001; 123:372–81.

[3] P.S. Veers, T.D. Ashwill, H.J. Sutherland, D.L. Laird, D.W. Lobitz, et al: Trends in the design, manufacture and evaluation of wind turbine blades. Wind Energy 6, 245–259 (2003).

[4] C. Kong, J. Bang, Y. Sugiyama: Structural Investigation of Composite Wind Turbine Blade Considering Various Load Cases and Fatigue Life. Energy, 2005, (30):2101-2114.

[5] F.M. Jensen, B.G. Falzon, J. Ankersen, H. Stang: Structural testing and numerical simulation of a 34 m composite wind turbine blade. Composite Structures. 76, 52–61 (2006).

[6] C. Kong, H. Kim, J. Kim: A study on structural and aerodynamic design of composite blade for large scale HAWT system. Final report, Hankuk Fiber Ltd; 2000.

Determination of Intensity Factors for Interfacial Cracks in TIP materials

Hongjun Zhong[1, a] and Jun Lei[1, b]

[1] Department of Engineering Mechanics, Beijing University of Technology, Beijing 100124, China

[a]zhj2010@emails.bjut.edu.cn, [b]leijun@bjut.edu.cn

Keywords: interfacial crack, intensity factors, transversely isotropic piezoelectric

Abstract. An explicit extrapolating formula for a general case of the interfacial crack plane lying with an angle to the poling axis in transversely isotropic piezoelectric (TIP) materials is derived, which is very feasible to determine the intensity factors for numerical methods such as FEM or BEM. Additionally, a more concise extrapolating formula for the typical state of the interfacial crack plane lying perpendicular to the poling axis is also presented in this paper.

Introduction

In contrast to cracks in homogeneous, isotropic/anisotropic and linear elastic materials, the local asymptotic displacement and stress fields near the tips of interfacial cracks in dissimilar, piecewise homogeneous and anisotropic piezoelectric materials are much more complicated. After determining the eigenvectors and following the Stroh's method, Suo et al [1] derived a relationship between the intensity factors and the crack opening displacements, electric potential jump at a distance behind the interfacial crack tip for an interface crack in dissimilar piezoelectric materials. But the application of this representation to determine the intensity factors for numerical methods is not directly feasible because of the numerical complexity of solving a complex eigen equation $\bar{\mathbf{H}}\mathbf{w} = e^{2\pi\gamma}\mathbf{H}\mathbf{w}$. To avoid solving this eigen equation, a displacement extrapolation technique for an interface crack in dissimilar anisotropic piezoelectric materials is proposed in this paper. Furthermore, an explicit extrapolating formula for the general case of the interfacial crack plane lying with an arbitrary angle to the poling axis in transversely isotropic piezoelectric (TIP) materials is presented, which is very feasible to determine the intensity factors for numerical methods such as FEM or BEM. Additionally, the displacement extrapolating formula for a typical state of the interfacial crack plane lying perpendicular to the poling axis, which has been studied by Nishioka et al [2], is also considered and amended in this paper.

Classification of TIP Bi-materials

Considering a generalized plane strain interfacial crack problem for linear piezoelectric bimaterials, Suo et al [1] obtained the following solution

$$\mathbf{h}(z) = \mathbf{w} z^{-1/2+i\gamma}, \qquad (1)$$

where γ is the singularity parameter; $\mathbf{w}$ is the eigenvector of the following eigenvalue problem

$$\bar{\mathbf{H}}\mathbf{w} = e^{2\pi\gamma}\mathbf{H}\mathbf{w}, \qquad (2)$$

in which $\mathbf{H} = \mathbf{Y}^{(1)} + \bar{\mathbf{Y}}^{(2)}$ and $\mathbf{Y} = i\mathbf{P}\mathbf{L}^{-1}$ is a Hermitian matrix determined by the material properties, which can be referred to [1].

Let $\mathbf{H}=\mathbf{V}+i\mathbf{F}$ and $\mathbf{E}=\mathbf{V}^{-1}\mathbf{F}$. Then Eq. 2 becomes $(\mathbf{E}+i\beta\mathbf{I})\mathbf{w}=0$, and its characteristic equation can be written in the form

$$\|\mathbf{E} + i\beta\mathbf{I}\| = \beta^4 + 2b\beta^2 + c = 0, \qquad (3)$$

where $\gamma = -\frac{1}{\pi}\tanh^{-1}\beta$, $b = \frac{1}{4}\text{tr}[\mathbf{E}^2]$, $c = \|\mathbf{E}\|$, with tr[ ] and $\|\ \|$ standing for the trace and the module of a matrix, respectively.

Then Eqs. 3 yield $\gamma_{1,2} = 0, \gamma_{3,4} = \pm\frac{1}{\pi}\tanh^{-1}\left[\sqrt{-\text{tr}[\mathbf{E}^2]/2}\right]$. Let ε and κ be the real and imaginary roots, respectively, then

$$\varepsilon = \frac{1}{\pi}\tanh^{-1}[(b^2 - c)^{1/2} - b]^{1/2}, \kappa = \frac{1}{\pi}\tan^{-1}[(b^2 - c)^{1/2} + b]^{1/2}. \tag{4}$$

For all TIP bi-materials, according to Ou's research [3], $c = \|\mathbf{E}\| = 0$ always exists when the poling direction of the materials are perpendicular to the crack face. In fact, the equation $\|\mathbf{E}\| = 0$ is always valid after any in-plane coordinate rotations. Supposing that $\mathbf{R}$ is the transformation matrix, the rotated $\mathbf{H}$ is obtained as $\mathbf{H}' = \mathbf{R}\mathbf{H}\mathbf{R}^T$. With substitution of $\mathbf{V}$ and $\mathbf{F}$, $\mathbf{H}' = \mathbf{R}(\mathbf{V} + i\mathbf{F})\mathbf{R}^T$, so the transformed $\mathbf{E}$ can be expressed as

$$\mathbf{E}' = (\mathbf{V}')^{-1}\mathbf{F}' = (\mathbf{R}^T)^{-1}\mathbf{E}\mathbf{R}^T. \tag{5}$$

The module of $\mathbf{E}'$ can be easily obtained as

$$c' = \|\mathbf{E}'\| = \|\mathbf{E}\| = 0. \tag{6}$$

So Eq. 6 is valid when the materials have the same poling direction.

For any TIP bi-material with the same poling direction for both materials, one of the parameters ε and κ always vanishes but the other one remains non-zero. Consequently, TIP bi-materials can be classified into two classes: one with vanishing κ (termed as ε-class TIP bimaterials) and the other one with vanishing ε (κ-class). When the materials have different poling directions, TIP bi-materials don't have the classification.

Explicit Extrapolating Formula for Intensity Factors

Let us consider in-plane deformation and in-plane electric potential for piecewise homogeneous piezoelectric bimaterial systems. The anti-plane u_3 decouples from u_1, u_2 and ϕ. With u_3 ignored, the bimaterial matrix $\mathbf{H}$ is a 3×3 Hermitian matrix. $\mathbf{V}$ is real and symmetric and $\mathbf{F}$ is real and antisymmetric. They take the following component forms

$$\mathbf{V} = \begin{bmatrix} v_{11} & v_{12} & v_{13} \\ v_{12} & v_{22} & v_{23} \\ v_{13} & v_{23} & v_{33} \end{bmatrix}, \mathbf{F} = \begin{bmatrix} 0 & f_{12} & -f_{13} \\ -f_{12} & 0 & f_{23} \\ f_{13} & -f_{23} & 0 \end{bmatrix}. \tag{7}$$

Then Eq. 3 can be expressed as

$$\beta^3 + b\beta + ic = 0. \tag{8}$$

Since $c = 0$, the roots of Eq. 8 is $-\sqrt{-b}$, $\sqrt{-b}$ and 0. According to the relationship between ε and β, the corresponding eigenvalues can be expressed as ε, $-\varepsilon$ and 0 for $b \leq 0$ and κi, $-\kappa i$ and 0 for $b > 0$.

For interfacial crack, the oscillatory singularity of the stress and electric field can be imaginary or real. When $b \leq 0$, the oscillatory singularity is imaginary. When $b > 0$, the oscillatory singularity is real.

Unified Form of the Explicit Formula. Suppose $\beta = -\sqrt{-b}$ when $b \leq 0$. According to the relationship $\varepsilon = \frac{1}{2\pi} \ln \frac{1-\beta}{1+\beta}$ between ε and β, the corresponding eigenvalues and eigenvectors can be expressed as (ε, w), $(-\varepsilon, \overline{w})$ and $(0, w_3)$. While $b > 0$, $\beta = -i\sqrt{b}$, $\kappa = \frac{\tan^{-1}(-\beta) - \tan^{-1}\beta}{2\pi}$, the eigenvalues and eigenvectors are $(\kappa i, w)$, $(-\kappa i, \overline{w})$ and $(0, w_3)$. The components of w and $\overline{w}$ are conjugate complex numbers. Suppose $\Lambda = [w \ \overline{w} \ w_3]$, $\Lambda \mathrm{diag}[\mu_\alpha] \Lambda^{-1}$ can be represented as the linear combination of three matrices,

$$\Lambda \mathrm{diag}[\mu_\alpha] \Lambda^{-1} = \mathbf{G}_1 \mu_1 + \mathbf{G}_2 \mu_2 + \mathbf{G}_3 \mu_3, \qquad (9)$$

where $\mathrm{diag}[\mu_\alpha]$ is a diagonal matrix. Considering the properties of Λ, the following matrix equation system of $\mathbf{G}_1$, $\mathbf{G}_2$ and $\mathbf{G}_3$ can be obtained after some algebraic operations as

$$\mathbf{G}_1 + \mathbf{G}_2 + \mathbf{G}_3 = \mathbf{I}, \quad -i\beta \mathbf{G}_1 + i\beta \mathbf{G}_2 = \mathbf{E}, \quad -\beta^2 \mathbf{G}_1 - \beta^2 \mathbf{G}_2 = \mathbf{E}^2. \qquad (10)$$

With the method of elimination, Eq. 10 can be solved and the roots are

$$\mathbf{G}_1 = -\frac{1}{2\beta^2}(\mathbf{E}^2 - i\beta \mathbf{E}), \quad \mathbf{G}_2 = -\frac{1}{2\beta^2}(\mathbf{E}^2 + i\beta \mathbf{E}) = \overline{\mathbf{G}}_1, \quad \mathbf{G}_3 = \frac{1}{\beta^2}(\mathbf{E}^2 + \beta^2 \mathbf{I}), \qquad (11)$$

where $\mathbf{I}$ is a unit matrix.

Due to the oscillatory singularity in the near-tip field, the individual stress and electric displacement intensity factors K_1, K_2 and K_4 for the interfacial crack cannot be uniquely associated with mode I, mode II and mode IV fracture as defined in homogeneous piezoelectric materials. However, they still represent three different modes of fracture behavior. To coincide with the traditional definition in homogeneous piezoelectric materials, we introduce $\mathbf{K} = [K_2, K_1, K_4]^T$. Then, the crack opening displacements and electric potential jump at distance r behind the crack tip can be expressed as

$$[\delta_1 \ \delta_2 \ \delta_4]^T = \frac{4\sqrt{r}}{\sqrt{2\pi}} \mathbf{H} \Lambda \mathrm{diag}[\varsigma_1 \ \varsigma_2 \ 1] \Lambda^{-1} (\mathbf{I} + \overline{\mathbf{H}}^{-1} \mathbf{H})^{-1} [K_2 \ K_1 \ K_4]^T, \qquad (12)$$

where $\varsigma_1 = r^{i\varepsilon} \frac{\cosh \pi \varepsilon}{1 + 2i\varepsilon}$, $\varsigma_2 = \overline{\varsigma}_1$ when $b \leq 0$; $\varsigma_1 = r^\kappa \frac{\cos \pi \kappa}{1 + 2\kappa}$, $\varsigma_2 = r^{-\kappa} \frac{\cos \pi \kappa}{1 - 2\kappa}$ when $b > 0$. With the substitution of Eq. 11 into Eq. 9, $\Lambda \mathrm{diag}[\varsigma \ \overline{\varsigma}_1 \ 1]\Lambda^{-1}$ can be expressed as the linear combination of $\mathbf{I}$, $\mathbf{E}$ and $\mathbf{E}^2$. The corresponding matrices $\mathbf{T}_1 = 2\mathbf{H}(\mathbf{I} + \overline{\mathbf{H}}^{-1}\mathbf{H})^{-1}$, $\mathbf{T}_2 = 2\mathbf{H}\mathbf{E}(\mathbf{I} + \overline{\mathbf{H}}^{-1}\mathbf{H})^{-1}$ and $\mathbf{T}_3 = 2\mathbf{H}\mathbf{E}^2(\mathbf{I} + \overline{\mathbf{H}}^{-1}\mathbf{H})^{-1}$ have the following explicit form

$$\mathbf{T}_1 = \frac{1}{d_1}\mathbf{D} + \mathbf{V}, \quad \mathbf{T}_2 = \frac{d_1 + d_2}{d_1}\mathbf{F}, \quad \mathbf{T}_3 = \frac{d_1 + d_2}{d_1^2}\mathbf{D}, \quad \mathbf{D} = \begin{bmatrix} d_3 & d_4 & d_5 \\ d_4 & d_6 & d_7 \\ d_5 & d_7 & d_8 \end{bmatrix}, \qquad (13)$$

where $d_1 = v_{11}c_1 + v_{12}c_2 + v_{13}c_3$,

$d_2 = f_{12}(f_{12}v_{33} + f_{13}v_{23}) + f_{13}(f_{13}v_{22} + f_{12}v_{23}) + 2f_{23}(f_{13}v_{12} + f_{12}v_{13}) + f_{23}^2 v_{11}$,

$d_3 = f_{12}^2 c_4 + 2f_{12}f_{13}c_5 - f_{13}^2 c_7$, $d_4 = f_{12}^2 c_2 - f_{12}f_{13}c_3 - f_{12}f_{23}c_5 + f_{13}f_{23}c_7$,

$d_5 = -f_{12}f_{13}c_2 + f_{13}^2 c_3 - f_{12}f_{23}c_4 - f_{13}f_{23}c_5$, $d_6 = -f_{12}^2 c_1 + 2f_{12}f_{23}c_3 - f_{23}^2 c_7$,

$d_7 = f_{12}f_{13}c_1 - f_{12}f_{23}c_2 - f_{13}f_{23}c_3 + f_{23}^2 c_5$, $d_8 = -f_{13}^2 c_1 + 2f_{13}f_{23}c_2 + f_{23}^2 c_7$,

with $c_1 = v_{23}^2 - v_{22}v_{33}$, $c_2 = v_{12}v_{33} - v_{13}v_{23}$, $c_3 = v_{13}v_{22} - v_{12}v_{23}$, $c_4 = v_{11}v_{33} - v_{13}^2$,

$c_5 = v_{11}v_{23} - v_{12}v_{13}$, $c_6 = v_{13}v_{23} - v_{11}v_{13}$, $c_7 = v_{12}^2 - v_{11}v_{22}$.

After the substitution of $\mathbf{T}_1$, $\mathbf{T}_2$ and $\mathbf{T}_3$ into Eq. 12, the explicit extrapolating formula for intensity factors can be obtained as

$$[K_2 \ K_1 \ K_4]^T = \sqrt{\frac{\pi}{2r}}[(1 - \frac{\varsigma_1 + \varsigma_2}{2})\frac{1}{\beta^2}\mathbf{T}_3 + \frac{i(\varsigma_1 - \varsigma_2)}{2\beta}\mathbf{T}_2 + \mathbf{T}_1]^{-1}[\delta_1 \ \delta_2 \ \delta_4]^T. \quad (14)$$

With the variation of ς_1 and ς_2, Eq. 14 can be applied to TIP materials of both ε-class and κ-class, so the uified form of the explicit formula is obtained.
Consider the typical state of the interfacial crack plane lying perpendicular to the poling axis. In this case, the components $v_{12} = v_{13} = f_{23} = 0$ and the expressions of $\mathbf{T}_1$, $\mathbf{T}_2$ and $\mathbf{T}_3$ are

$$\mathbf{T}_1 = \frac{1}{v_{11}}\begin{bmatrix} v_{11}^2 b_1 & 0 & 0 \\ 0 & a_2 & a_3 \\ 0 & a_3 & a_4 \end{bmatrix}, \mathbf{T}_2 = b_1 \mathbf{F}, \mathbf{T}_3 = \begin{bmatrix} \frac{v_{11}(f_{12}a_5 + f_{13}a_6)}{a_1} & 0 & 0 \\ 0 & -f_{12}^2 & f_{12}f_{13} \\ 0 & f_{12}f_{13} & -f_{13}^2 \end{bmatrix}, \quad (15)$$

where $a_1 = v_{23}^2 - v_{22}v_{33}$, $a_2 = v_{22}v_{11} - f_{12}^2$, $a_3 = v_{23}v_{11} + f_{12}f_{13}$, $a_4 = v_{33}v_{11} - f_{13}^2$, $a_5 = v_{33}f_{12} + v_{23}f_{13}$, $a_6 = v_{23}f_{12} + v_{22}f_{13}$; $b_0 = v_{11}a_1 + f_{12}a_5 + f_{13}a_6$, $b_1 = b_0/(v_{11}a_1)$. They are well agreed with Nishioka's results [2] but a minus $\mathbf{T}_2$ which may be caused by the literal errors.

Acknowledgements

This work is supported by the Natural Science Foundation of China under Grant No. 11002006, which is gratefully acknowledged.

References

[1] Suo Z, Kuo C M, Barnett D M and Wills J R. Fracture mechanics for piezoelectric ceramics. Journal of the Mechanics and Physics of Solids. 1992, 40(4): 739-765.

[2] Nishioka T, Shen S, Yu J. Dynamic J integral, separated dynamic J integral and component separation method for dynamic interfacial cracks in piezoelectric bimaterials. International Journal of Fracture. 2003, 122(3-4): 101-130.

[3] Ou Z C. Singularity parameters ε and κ for interface cracks in transversely isotropic piezoelectric bimaterials. International Journal of Fracture. 2003, 119(2): L41-L46.

Failure of composite T-joints in bending with through-the-thickness reinforcement: stitching vs Z-pinning

H Cui[a], Y-L Li[b]

School of Aeronautics, Northwestern Polytechnical University, Youyi Xilu 127, Xian, China

[a] cuihao_nwpu@hotmail.com, [b] liyulong@nwpu.edu.cn

Keywords: Z-pin; Stitch; Delamination; Fracture toughness;

Abstract. The stitched composite T-joints and Z-pinned ones subject to bending load were investigated in this paper. A simple theoretical model characterizing the failure process of through the thickness reinforcement (TTR) during mode I delamination was presented. The experimental results showed that the initial damage load and maximum load of stitched specimens are higher than that of Z-pinned ones, while the energy absorption of stitched specimens during delamination is lower than that of Z-pinned ones. The energy absorption values predicted by the present model meet the experiments reasonably well. High friction force at the interface between TTR tow and matrix, with a long pull-out displacement of the tow, helps to improve the delamination resistance.

1. Introduction

Delamination resistance has become a major concern for composite laminates in the last twenty years. Z-pinning and stitching are two of the most common ways to reinforce composite laminates in the thickness direction. Z-pinning has been reported to be able to improve the delamination toughness of composite laminates significantly [1, 2], which is concluded to be not effective at resisting the crack initiation, but effective at resisting the propagation of long cracks [3]. Stitching is capable of resisting damage growth, arresting crack propagation and retarding final fracture [4]. The mode I fracture toughness can be improved by 47 times [5] with Kevlar stitches, and increasing the cross-sectional area of the stitch has a more profound influence on fracture toughness than increasing the stitch density [6]. Most of existing work focuses on comparison between TTR reinforced specimens and unreinforced specimens, while less comparative investigations between stitched laminates and Z-pinned laminates have been conducted. Besides, traditional data reduction methods for evaluating delamination toughness based on linear elastic fracture mechanics may be not applicable, due to the large bridging zone during the delamination process [7, 8]. The

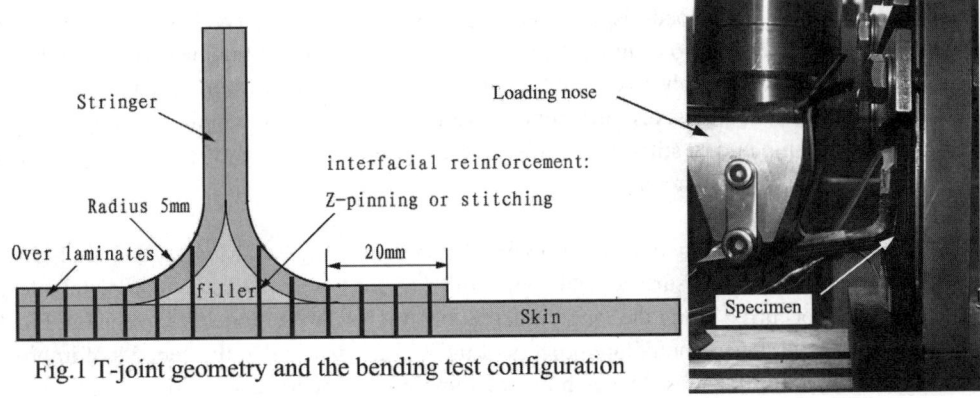

Fig.1 T-joint geometry and the bending test configuration

delamination resistance is strongly dependent upon specimen geometry [7], and expensive tests need to be carried out for every single laminate with different geometry. A theory that can predict the reinforced delamination toughness is strongly desired by the industry.

Bending tests of composite laminated T-joints reinforced respectively by Z-pinning and stitching have been carried out in this paper, in purpose of comparing their crack resistance performance. An analytical model is developed to highlight the difference between Z-pinning and stitching, and predict the mode I fracture toughness of TTR strengthened composite laminates.

2. Specimen introduction and experiment configuration

Fig.1 illustrates the T-joint specimens used in this investigation. The Z-pins were made of T300/BMI, their diameter was 0.3 mm, and the pin to pin spacing was 3 mm. The stitch yarns were Kevlar-29 fiber and the stitch to stitch spacing was 5 mm. The specimen was placed on a steel support during the experiment as shown in Fig.1. The bending test was conducted by a static load frame with a built-in load monitor cell. The displacement of and load on the indenter have been recorded.

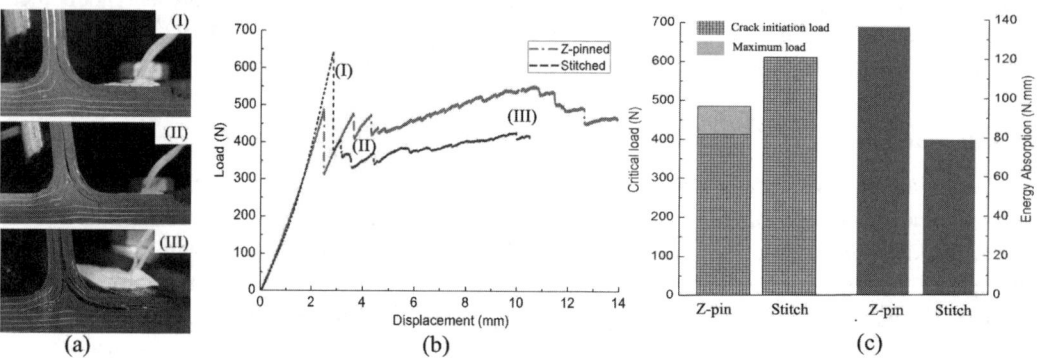

Fig.2. Failure process of T-joint in bending

3. Results and discussion

3.1 Comparison between Z-pinning and stitching

The failure process of a Z-pinned specimen-joint is shown in Fig.2-a, which is similar to that of stitched specimen. The load-displacement curves (Fig.2-b) are almost linear at the initial part of the curve until the onset of matrix cracking at the intersection radius. At the instance of damage initiation, a sharp drop of the load-displacement curve can be observed. The load increases linearly again with a reduced slope. Two or more "zigzag like" distinct loops of the load increasing linearly and dropping sharply can be observed until catastrophic failure comes out. It may be concluded from Fig.2-c that Z-pinning is less effective at resisting damage initiation and improving the maximum load as compared to stitching. However, the Z-pinning is more efficient at improving the delamination toughness than stitching.

3.2 Generalized TTR model

The failure modes of Z-pins and stitch fibers are shown in Fig.3-a. Most of the stitch threads ruptured near the delamination surface, while most of the Z-pins were pulled out completely despite some splitting that occurred along the longitudinal direction. The TTR model is sketched in Fig.3-b. The energy absorption during the failure process of the TTR, in particular, the benefits of improving interfacial delamination toughness by stitching or Z-pinning, is discussed as following.

Fig.3. (a) Failure modes and (b) schematics of through thickness reinforcement

For stitching reinforcement, the energy was mainly dissipated by rupture of the stitch fibers, the interfacial debonding between stitch fiber and matrix, and the pull out process of the ruptured stitch fibers. Hence, the total energy absorption of the stitch fiber is

$$E_s = E_r + E_d + E_f = \frac{1}{2}\sigma_s\varepsilon_s LA_f + G_{IIC}^D \pi D\hat{L} + \int_0^{\hat{L}} f\pi Dl dl \tag{1}$$

Where E_r is the fracture energy per unit Kevlar fiber cross section area, E_d is the energy dissipation caused by interfacial shear debonding, E_f is the energy absorption during the pulling out process; h and D represents the length and diameter of the TTR fiber that has experienced elastic deformation during the fracture process [9], $\hat{L}$ is the length of the ruptured stitch that will be pulled out of the matrix, for Z-pin tow, $\hat{L}$ is equal to L. G_{IIC}^D is the interfacial mode II fracture toughness, σ_s, ε_s is the fracture stress and fracture strain of TTR fibers.

The energy dissipation of Z-pins was mainly caused by interfacial debonding and the pull out process of the Z-pin fibers. Therefore, the total energy absorption for Z-pins is:

$$E_z = G_{IIC}^D \pi DL + \int_0^L f\pi D l\, dl \tag{2}$$

Finally, the energy absorption of the TTR per unit laminate area can be obtained as:

$$G_S = \frac{1}{2}\sigma_s\varepsilon_s L\eta + \frac{G_{IIC}^D \hat{L} + \frac{1}{2}\overline{f}\hat{L}^2}{\sqrt{A_f}}\sqrt{\frac{4\pi}{60\%}}\eta \quad \text{(Stitching)} \tag{3}$$

$$G_Z = \frac{4G_{IIC}^D L + 2\overline{f}L^2}{D}\eta \quad \text{(Z-pinning)} \tag{4}$$

3.3 Model Validation

The G_{IIC}^D of Z-pinning and stitching are assumed to be the same, with a typical mode II fracture toughness of 1.0 N/mm. The frictional stress between the TTR rod and matrix is chosen as 15 Mpa [10]. The reinforced area of the specimens tested here is about 50 mm long. Hence, the difference of energy absorbed between stitched joints and Z-pined counterparts is 58.5 N. mm when the parameter $\hat{L}$ is 0.5 mm for stitching. When $\hat{L}$ is 1 mm, the difference is 38.5 N. mm. According to the test results, the average energy absorption of Z-pinned specimens is 58.7 N. mm higher than the stitched ones. The presented model meets experimental results reasonably well.

The sensitivity of the energy absorption on these uncertain parameters has been evaluated. It was found that the frictional stress has significant influence on the energy absorption performance, while the influence of interfacial bonding is negligible. Keeping the TTR tow area fraction constant as 0.8 % for both stitching and Z-pinning, the energy absorption is also calculated for both Z-pinning and stitching of different diameters. The efficiency of both Z-pins and stitches decrease with the increase of diameter. For thick laminates where the pull-out length is longer, the effect of tow diameter is more significant.

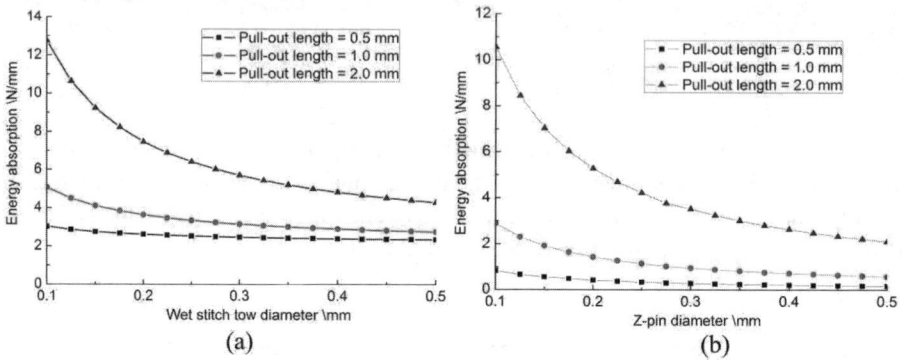

Fig. 4. Energy absorption for stitching and Z-pinning with different diameters.

4. Summary

Experiments show that Z-pinning is not effective at resisting the crack initiation, but is able to resist long cracks efficiently. Stitching is more effective at raising the crack initiation load when compared with Z-pinning. A simple mechanical model characterizing is presented to predict the efficiency of Z-pinning and stitching. The model presented here supplies a convenient method to evaluate the efficiency of TTR on improving the delamination toughness.

Acknowledgement

This work is funded by National Natural Science Foundation of China (No.10932008) and the 111 project (No.B07050) in China.

References

[1] G. Freitas, C. Magee, Dardzinski, T. Fusco. J Adv Mater Vol. 25 (1994), p. 36
[2] K.L.Rugg, B.N.Cox, M.Massabo. Composites Part A Vol. 33 (2002), p.177
[3] A.P. Mouritz. Composites: Part A Vol. 38 (2007), p. 2383
[4] K. Dransfield, C. Baillie, Y.W. Mai. Composites Science and Technology Vol. 50 (1994), p. 305
[5] L.Chen, P.G. Ifju , B. V. Sankar. Journal of Composite Materials Vol.35 (2001), p.1137
[6] L.Chen, B.V. Sankar, P.G. Ifju. 44th AIAA/ASME/ASCE/AHS Structures, Structural Dynamics, and Materials Conference 7-10 April 2003, Norfolk, Virginia
[7] I.K. Partridge, D.D.R. Cartie. Composites: Part A Vol. 36 (2005), p. 55
[8] D.D.R. Cartié, J.M. Laffaille, I.K. Partridge, A.J. Brunner. Engineering Fracture Mechanics Vol. 76 (2009), p. 2834
[9] K. P. Plain, L. Tong. Composite Structures Vol. 88 (2009), p. 558
[10] M. Meo, F. Achard, M. Grassi. Composite Structures Vol. 71 (2005), p. 383

Investigation on T_{11}-stress for semi-elliptical surface cracks in finite thickness plates under remote tension

Wei Xie

School of Aeronautics, Northwestern Polytechnical University, Xi'an 710072, Shaanxi, China

nwpuxiewei@ nwpu.edu.cn

Keywords: T_{11}-stress; Constraint effect; Surface crack; Three-dimensional finite element analysis

Abstract. In the present work, three-dimensional finite element analyses have been conducted to calculate the T_{11}-stress for semi-elliptical surface cracks in finite thickness plates under remote tension. The T_{11}-stress solutions are presented along the crack front for cracks with a/t values of 0.2, 0.4, 0.6 or 0.8 and a/c values of 0.2, 0.4, 0.6 or 1.0. The current T_{11}-stress solutions are suitable to be used as the constraint parameter for the fracture analysis.

Introduction

The limited ability of a single parameter such as the stress intensity factor (SIF) K for linear elastic fracture mechanics to characterize crack tip conditions irrespective of geometry and load level has been recognized for years. To overcome this problem, two-parameter of the crack-tip stress-strain state have been studied over the past two decades [1]. Williams [2] proposed a two-parameter $K-T$ approach for the isotropic linear elastic materials. Bilby et al. [3] and Larsson and Carlsson [4] showed that the T-stress can strongly affect the magnitude of hydrostatic triaxiality in the near crack tip elastic-plastic fields and influence crack tip constraint. Positive T-stress strengthens the level of crack tip stress triaxiality and leads to high crack tip constraint; while negative T-stress reduces the level of crack tip stress triaxiality and leads to the loss the crack tip constraint [5]. Subsequently, many authors made great efforts [6 - 9] and studied the in-plane constraint of the crack border [10,11]. These works were focused on 2D (in-plane) crack tip constraint issues, and thus, the methodology was effective in regard to explaining the effect of crack depth on the fracture toughness testing. Hereafter, in-plane T-stress will be denoted as T_{11} [1].

On the other hand, out-of-plane crack tip constraint is also known to have a significant influence on the fracture behavior of materials, and work have been done to express this constraint along the 3D crack front by the out-of-plane T-stress T_{33} [1]. The variation of T_{11} and T_{33} along 3D crack front were analyzed by Toshiyuki Meshii [1] and E. Giner [12].

In order to apply the two parameter fracture mechanics methodology, it is important to provide T-stress solutions for various crack configurations. So in this paper, three dimensional finite element analyses were conducted to calculate the T_{11}-stress along the crack front for a surface crack in a finite thickness plates under tension.

Extraction of T_{11}-stress

In an isotropic linear elastic body containing a crack subjected to symmetric (mode I) loading, the leading terms in a series expansion of the stress field very near the crack front are

$$\sigma_{11} = \frac{K_I}{\sqrt{2\pi r}}\cos\frac{\theta}{2}(1-\sin\frac{\theta}{2}\sin\frac{3\theta}{2})+T_{11}, \quad \sigma_{22} = \frac{K_I}{\sqrt{2\pi r}}\cos\frac{\theta}{2}(1+\sin\frac{\theta}{2}\sin\frac{3\theta}{2}),$$
$$\sigma_{33} = \frac{K_I}{\sqrt{2\pi r}}2\upsilon\cos\frac{\theta}{2}+T_{33}, \quad \sigma_{12} = \frac{K_I}{\sqrt{2\pi r}}\sin\frac{\theta}{2}\cos\frac{\theta}{2}\cos\frac{3\theta}{2}, \quad \sigma_{13} = \sigma_{23} = 0$$

(1)

Where the subscripts 1, 2 and 3 suggests a local Cartesian coordinate system formed by the plane normal to the crack front and the plane tangential to the crack front at point s; r and θ are the local polar coordinates; K_I is the mode I local K; E is the Young's modulus and υ is the Poisson's ratio. The terms T_{11} and T_{33} are the amplitudes of the second orders term in the three dimensional series expansion of the crack front stress field. In the current analyses, the interaction integral method introduced by Nakamura and Parks [13] is used to exact the extract T_{11}-stress.

Three-dimensional finite element model

3D finite elements are used to model the symmetric quarter of an elastic plate containing a semi-elliptical surface crack. The geometry and coordinate are shown in Fig. 1(a) and (b). The finite element analyses are made using ANSYS, version11.0, with 20-noded isoparametric three dimensional solid elements and reduced integration. The 3D singular elements with four mid-side nodes at the quarter points are used around the crack front to simulate the inverse square root singularity at the crack tip [14].

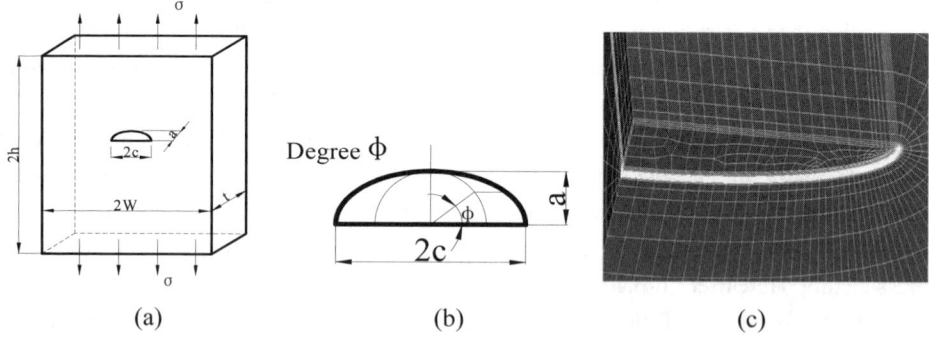

Fig. 1 Geometry and typical finite element mesh

In order to model the 3D stress field accurately, the thicknesses of the successive element layers are gradually reduced toward the free surface with respect to the crack front line because of the strong variations of the stress gradients. A typical finite element mesh is illustrated in Fig. 1(c). In the finite element models, h/w ratio is fixed to 1, and w/c is fixed to 10 for all analyses in the current calculation. Total of 18 elements were used along the crack front. Because of the symmetry of the model and boundary condition, only the uncracked planes in the xoz plane and yoz plane are symmetrically constrained. The uniform tension stress is set on the top plane.

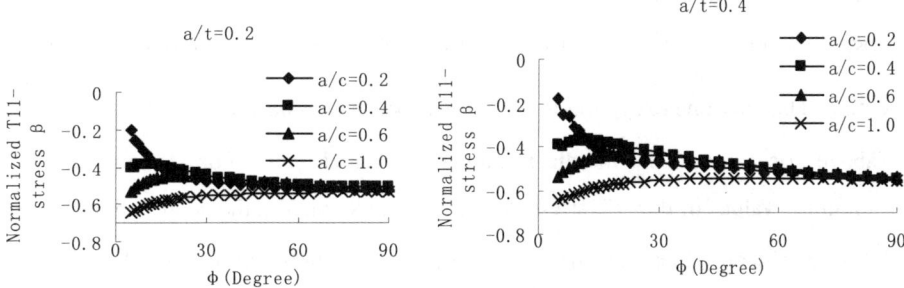

Fig.2 Results for variation a/c ($a/t = 0.2$)   Fig. 3 Results for variation a/c ($a/t = 0.4$)

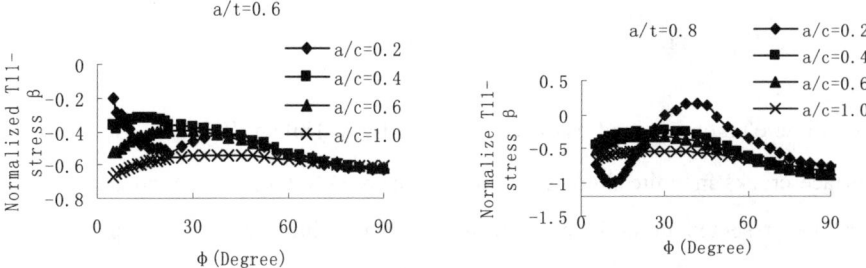

Fig.4 Results for variation a/c ($a/t = 0.6$)   Fig. 5 Results for variation a/c ($a/t = 0.8$)

Results and Discussion

The T_{11}-stress distributions along the crack front for semi-elliptical surface cracks (a/t =0.2, 0.24, 0.6 or 0.8) in a finite thickness plate with aspect ratios, a/c, of 0.2, 0.4, 0.6 or 1 under remote tension loads have been calculated. The Poisson's ratio used in the current analysis is 0.3.

The T_{11}-stress results have been normalized as follows:

$$\beta(\phi) = \frac{T_{11}(\phi)}{\sigma_0} \tag{2}$$

Where $\beta(\phi)$ is the normalized T_{11}-stress at specific angle ϕ, and σ_0 is the normal tensile stress.

It is well known that at the intersection of the crack front and the free surface, the $r^{-1/2}$ singularity vanishes [5]. That is, the $r^{-1/2}$ singularity as described in Eq. (1) occurs only near crack front points embedded entirely in material. Values of the T_{11}-stress for $\phi < 5°$ are therefore neglected in this paper. "Surface point" in this point is referring to the point at $\phi = 5°$.

The results are plotted in Fig. 2-5. From these results, the following are observed. The T_{11}-stress are negative along the crack front for all crack aspect ratios a/c and relative depth a/t with one exception. The only one exception is for a low aspect ration deep crack, T_{11}-stress for this case reaches above zone around the middle portion of the crack front. For shallow cracks, the maximum or minimum values of the T_{11}-stress generally occurs either at the deepest point or at the surface point; while critical values may occur in the middle portion of the crack front for deep cracks. The variation near the deepest point ($\phi = 90°$) is small for all depth a/t; while the variation near the surface point ($\phi = 5°$) is apparent for shallow and mid-deep crack.

Conclusions

Three-dimensional finite element method has been conducted to calculate the elastic T_{11}-stress for semi-elliptical surface cracks in finite thickness plates under remote tension was considered. The T_{11}-stress solutions are presented along the crack front for cracks with a/t values of 0.2, 0.4, 0.6, and 0.8, a/c values of 0.2, 0.4, 0.6, and 1. The current T_{11}-stress solutions are suitable to be used as the constraint parameter for the fracture analysis.

Acknowledgments

This paper was supported by National Natural Science Foundation of China and by the 111 Project Nos. B07050.

References:

[1] Toshiyuki Meshii, Kai Lu: Engineering Fracture Mechanics. Vol.77 (2010), p. 2467-2478.
[2] Williams ML: crack. J Appl Mech.Vol. 24 (1957), p.109 – 114.
[3] Bilby BA, Cardew GE: London: Mechanical Engineering Publications Limited. 1986. p.37 – 46.
[4] Larsson SG, Carlsson AJ: J Mech Phys Solids. Vol. 21 (1973), p. 447 – 473.
[5] Wang X: Eng Fract Mech. Vol. 70 (2003), p.731 – 756.
[6] Hutchinson JW: J Mech Phys Solids. Vol. 16 (1968), p. 13 – 31.
[7] Rice JR, Rosengren GF: J Mech Phys Solids. Vol. 16 (1968) p. 1 – 12.
[8] Sharma SM, Aravas N: J Mech Phys Solids. Vol. 39 (1991), p. 1043 – 1072.
[9] O'Dowd NP, Shih CF: J Mech Phys Solids. Vol. 39 (1991), p. 989 – 1015.
[10] Xia L, Wang TC, Shih CF: J Mech Phys Solids. Vol. 41 (1993), p.665 – 687.
[11] Yang S, Chao YJ, Sutton MA: Eng Fract Mech. Vol. 45 (1993), p. 1 – 20.
[12] E Giner: Engineering Fracture mechanics. Vol. 78 (2011), p. 412-427
[13] Nakamura T, Parks DM: Int J Solids Struct. Vol. 29 (1992), p.1597-1611
[14] Junhua Zhao, Wanlin Guo: International Journal of Fatigue. Vol. 29 (2007), p.435 – 443.

Investigation of thermal damage and the consequent fatigue property of laser surface heat treated component

W. Tian[1,a], C. Zeng[1,b] and L. Hua[1,c]

[1]College of Mechanical and Electrical Engineering, Nanjing University of Aeronautics and Astronautics, Nanjing 210016, China

[2]Nanjing Institute of Railway Technology, Nanjing 210031, China

[a]tw_nj@nuaa.edu.cn, [b]poetal@126.com, [c]hua672@163.com

Keywords: Thermal Damage, Fatigue Property, Laser Surface Treatment, Finite Element Analysis.

Abstract. In order to signify the thermal damage and the influence it has on the fatigue property, a 3D finite element method, as well as experiment investigation, was employed for this research. By the study, it suggests that laser surface scanning can bring about some tensile residual stress and a heat affected zone in the workpiece, so as to lead a fall in fatigue life cycles. Since the fatigue character is affected by the residual stress and the microstructure change, which relate to the laser process parameters, it is expected to draw a scalar variable from stress feature and microstructure parameter to stand the level of thermal damage influence.

Introduction

Laser technology has been widely used in the machinery industry by means of laser machining, laser welding, laser peening, laser surface heat treatment, laser cladding, etc. The high energy of laser and its fast speed of heating and cooling directly contribute to the change of the material microstructure and surface state which, in part, act as the thermal damage and is responsible for the change of fatigue properties of the applied component. For the prediction of the fatigue life of those products which laser was used during their manufacturing process, researches have carried out with a mass of studies focused on the fatigue properties of how they go on with the working load of those components based on experiment [1-4]. It has been reached the common point that it is, mostly, the residual stress introduced by the laser method that devotes to the influence of the fatigue endurance [5-7]. Thermal damage has been often studied qualitatively and not clearly defined, which turns up in literatures as the describing of the metallographic structure, residual stress state and the hardness state as well [8-10]. Quantitative study of thermal damage is very rare.

In this paper, thermal damage and fatigue property of laser surface heat treatment of steel Q235 is studied through finite element analysis (FEA) while supported by experiment. By the simulation, the temperature field during the process and the residual stress state of the cooled test piece is studied, as well as the subsequent fatigue property.

Finite Element Modeling

The FEA was done in the circumstance of Ansys software (version 12.1). Indirect analysis method was engaged, which the stress simulation was based on the previous temperature simulation. A schematic illustration of laser heating trial is shown in Fig. 1, where the workpiece and the coordinate system are fixed and the laser beam moves at velocity v transversely. When the finite element model was created, a finer mapped mesh was used in the higher temperature area directly below the incident Gaussian laser heat flux. This fine mesh method enables the calculation of the steep temperature gradients with precision. The movement of Gaussian heat source was realized by writing an APDL program. The thermal stress was derived for the latter fatigue simulation, while the external working load was applied with maximum stress 160 MPa and stress ratio -1.

The thermo-physical property of the material is shown in Table 1. In this work, the treated workpiece material is assumed to be homogeneous and the density of the material is supposed to keep the same value 7860 kg/m³. In the simulation, the ambient temperature is 20 ℃, while air convection coefficient is 100 [W/(m²·°C)]. Besides, in the centre of the laser beam, when the temperature exceeds the melting point (1468 °C) the node is supposed to remain in the mesh, and the latent heat of the fusion is simulated by artificially increasing the liquid specific heat. The imposed laser heat source was handled as Gaussian plane distribution model of absorbed laser heat flux q(x, y) which is given as:

$$q(x,y) = \frac{2P\eta}{\pi r^2} \exp(-\frac{2(x^2+y^2)}{r^2}) \quad (1)$$

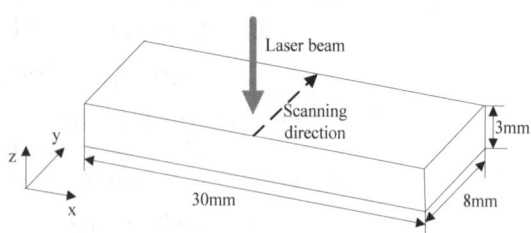

Fig. 1. A schematic illustration of the laser heating experiment.

Where η, P and r refers to the average absorptivity of the Q235 plate workpiece which is assumed to be 0.408, laser power P=1800W and the radius of laser spot r=0.35 mm, respectively. Besides, the scanning speed v=0.3 mm/s.

Table 1 Thermo-physical properties of Q235

Temperature [°C]	Specific Heat [J/(kg·°C)]	Thermal Conductivity [W/(m·°C)]	Thermal Expansion [°C^{-1}]
20	460	50	1.10E-5
250	480	47	1.22 E-5
500	530	40	1.39 E-5
750	675	27	1.48 E-5
1000	670	30	1.34 E-5
1500	660	35	1.33 E-5
1700	780	140	1.32 E-5

Discussion of FEA Result

Features of Temperature Field. A 3D transient temperature field analysis was run taking advantage of the APDL of Ansys software. The model, containing 7259 nodes and 5760 elements, was meshed with the 8-node thermal element SOLID 70. The numerical simulation result is shown in Fig.2. We can have a direct viewing of the temperature contour distribution of the workpiece form Fig.2 (a), meanwhile the picture (b), (c) and (d) clearly exhibit a magnitude relationship in corresponding with the geometric dimensioning. In general, Fig.2 delivered the information that, in the center of laser beam the temperature achieves maximum value on the workpiece and decreases around this point. The temperature profile is symmetric about the laser center on the longitude direction, as is shown in Fig.2 (c). Since the austenite begin to grow above 727°C, on both sides of the heat source along x orientation the heat affected zone (HAZ) has achieved about 1 mm, while through thickness direction under about 1.2 mm they are all HAZ, that is approximately 1.8 mm depth.

Features of Residual Stress. The residual stress was derived by structure simulation based on the former temperature result while switching the element type SOLID 70 to SOLID 185, the structure element. Since the fatigue life of a component is much more affected by the surface residual stress compares with those in the inner part, the study is focused on surface stress state of the treated face, illustrated in Fig.3. In this figure, (a) shows the contour plot of the nodal von Mises stress while (b) signifies the stress profile in corresponding with it. Similarly, the profile is symmetric about the laser center on the longitude direction. Furthermore, it is obvious that on both sides of the laser scanned track, the most areas stand a state of tensile stress. However, the stress is not that notable just with the maximum value of 9.9 MPa or so (in Fig.3 (b)) in x orientation which is in accordance with the latter fatigue load direction. Besides, it is seen that the maximum value of von Mises stress is 28.0 MPa.

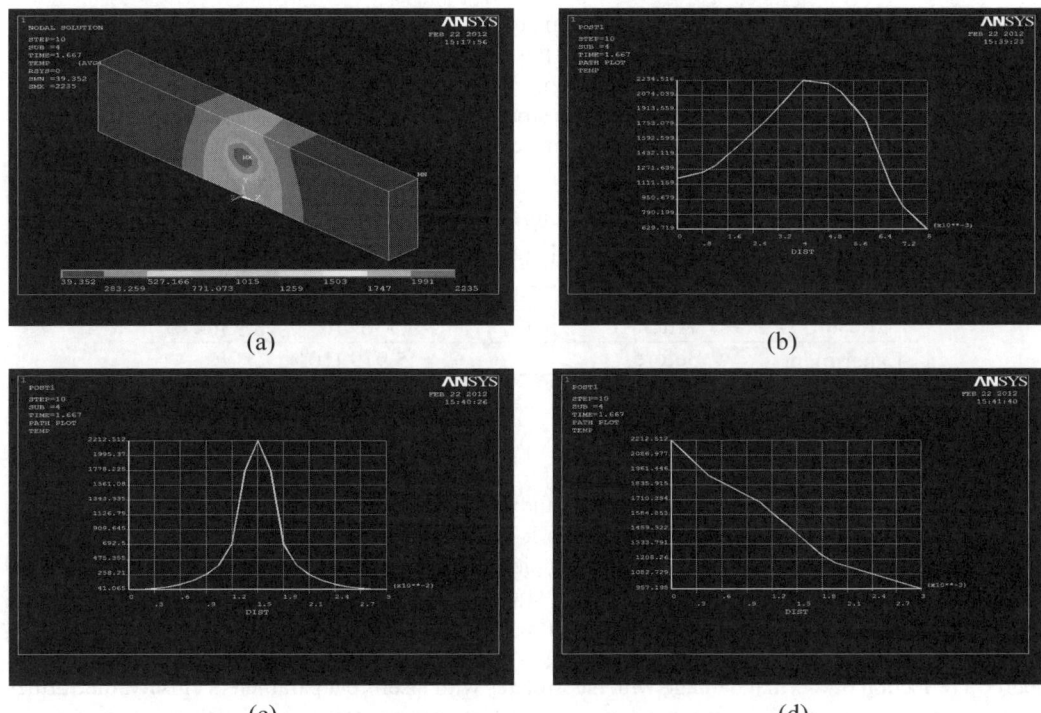

Fig.2. Transient temperature field at a time point during laser scanning
(a) Contour plot of the nodal temperature; (b) temperature profile of y direction across heat source center; (c) temperature profile of x direction across heat source center; (d) temperature profile of z direction across heat source center

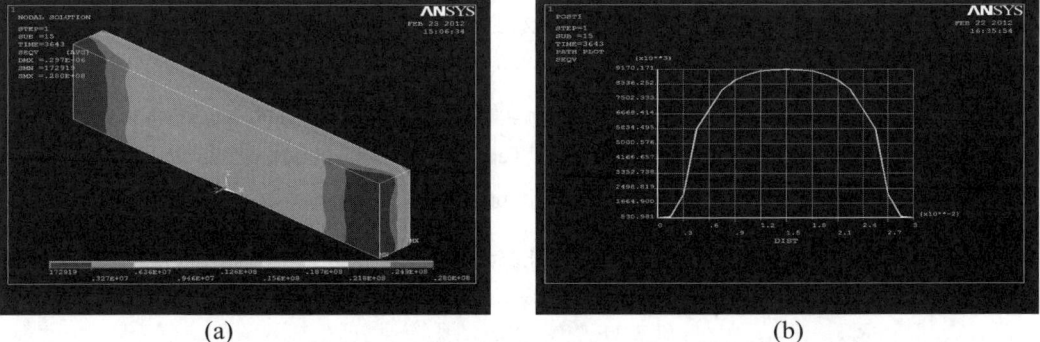

Fig.3. Residual stress distribution caused by laser heat treatment
(a) Contour plot of the nodal von Mises stress; (b) von Mises stress profile of longitude direction

Result of Fatigue Properties

From the former section, where the tensile surface stress achieves maximum value the location on the component is more likely to break down. Therefore, such points are chosen as the locations to store stress to analysis their fatigue properties. In this model three locations were chosen with coordinate values (0, 0, 0), (0, 0, -0.0021) and (0, 0, -0.003), respectively. Results are listed in Table 2. It is seen that, when the specimen was treated by laser it suffered a slightly fall in fatigue life cycles. However, this change is not that remarkable just by about 11000 cycles, that is approximately 0.19 %

of the normal life. Similarly, this fall in fatigue life cycles happened in experiment case. It is clearly to see that, in the experiment, after laser burning the fatigue life of the specimen falls remarkably by about 65.2%. This is mainly caused by the change of the section area, which would result in stress concentration, and the change of metallographic structure. Therefore, it suggests that the ansys model need to be modified. Besides, the fatigue life before laser treatment agrees with the simulation result very well with error of 3.0%.

Table 2 The fatigue life results of simulation and experiment

	Laser Treatment	Location 1	Location 2	Location 3
Simulation Result	Before	5.748×10^6	5.748×10^6	5.748×10^6
	After	5.737×10^6	5.736×10^6	5.736×10^6
Experiment Result	Before		5.919×10^6	
	After		2.06×10^6	

Conclusion

By finite element analysis and experiment study of thermal damage and the subsequent fatigue character of laser surface treated component, it is believed that the laser burning can caused performance deterioration of the treated working piece. However, due to the smaller magnitude of the residual stress this influence is limited to a relatively lower level. As is known, it is the residual tensile stresses that lead to the shorter fatigue endurance. Therefore, it is expected to build a relationship between the residual stress character and the fatigue life decrease. While the residual stresses are in a degree a reflection of thermal damage which correlates with laser treat parameters closely, our further work is to find this rule and extract the mathematical scalar parameter to signify thermal damage.

References

[1] D. Boronski: Int. J. Fatigue. Vol.28(2006), p. 346-54

[2] F.J. Carpio, D. Araújo, F.J. Pacheco, D. Méndez, A.J. García, M.P. Villar, R. García, D. Jiménez and L. Rubio: Appl. Surf. Sci. Vol.208-209(2003), p. 194-8

[3] M. Heitkemper, C. Bohne, A. Pyzalla and A. Fischer: Int. J. Fatigue. Vol.25(2003), p. 101-6

[4] C.M. Sonsino, M. Kueppers, M. Eibl and G. Zhang: Int. J. Fatigue. Vol.28(2006), p. 657-62

[5] T.-L.Teng and P.-H. Chang: J. Mat. Process. Technol. Vol.145(2004), p. 325-35

[6] C.D.M. Liljedahl, O. Zanellato, M.E. Fitzpatrick, J. Lin, L. Edwards: Int. J. Fatigue. Vol.32(2010), p. 735-43

[7] G. Bussu and P.E. Irving: Int. J. Fatigue. Vol.25(2003), p. 77-88

[8] N. Farabi, D.L. Chen, J. Li, Y. Zhou, S.J. Dong: Mat. Sci. Eng. A. Vol.527(2010), p. 1215-22

[9] A. Di Ilio and V. Tagliaferri: Composites. Vol.20(1989), p. 115-9

[10] B. Valsecchi, B. Previtalia, E. Gariboldia, A. Liu: Procedia Eng. Vol.10(2011), p. 2851-6.

A 3-Dimensional Model Related to Stress-Strain-Time of Rock Salts

ZHANG Huabin[1,a], WANG Zhiyin[1,b] and MA Jiwei[1,c]

[1] Key Laboratory of Urban Oil and Gas Distribution Technology, China University of Petroleum, Beijing 102249, China

[a] upc_zhb@163.com, [b] wzy3360@163.com, [c] jiweimada@163.com

Key words: rock salts, creep failure, resultant stress-strain curve, stress-strain-time.

Abstract. To build a stress-strain-time relationship of salt rock with unified instantaneous law and creep law, we combine the stress-strain curve with whole process of creep curve, looking for inner relation between softening curve and creep deformation. Finally, the stress-strain-time space surface is obtained. The researching results demonstrated that time and space development law of rock salts can be well reflected by it.

Introduction

During the extended operation of underground salt caverns, the operation pressure is lower than the rock salts yield strength. The failure is mainly aroused by the accumulated creep deformation during operation stage. During the long-term creep process, the creep law, failure deformation and the failure time of surrounding rock are closely related to the stress-strain curve. Pan Juncheng [1] obtains the stress-strain-time 3D surface relationship through his study on the rheological soils. Goodman R.E. [2] puts forward the concept of creep termination trajectory to predict creep failure by analyzing his test data. Through a systematic research on the whole process of rock creep deformation under different stress states, Wang ZhiYin etc [3] carry out the corresponding relationship between the conventional tri-axial compression stress strain curve and the creep deformation with start and end phases as well as that of different creep failures attributes. Many achievements have been obtained until now in the field of researching on the mechanical features of rock salts. For example, Liu Jiang etc[4] analyze and compare the short-term strength and deformation characteristics by conducting the conventional mechanical tests on the rock salts in Hubei Yingcheng and Jiangsu Jintan; Yang ChunHe [5] analyses the rock salts steady creep law and proposes the steady creep 3D analysis equation through a lot of rock salts creep experiments on site; Chen Weizhong etc [6] raise the rock salts Nonlinear Damage Constitutive Model based on Burgers Model according to the tri-axial creep results of Jintan gas storage reservoir. But there are rarely documents which unify rock salts stress-strain-time law. Based on the previous studies, the authors research on the rock salts stress-strain-time relationship according to rock salts indoor test results and rheological theory and more intuitive stress-strain-time space stretch surface is given.

Modeling Basis

The authors have finished a lot of statistics analysis and researches on the data of numerous rock salts indoor instantaneous tests and creep tests. The typical stress-strain relationships in the instantaneous test are shown in Fig.1. Rock salts stress-strain curve has obvious strain hardening-softening feature. With the increase of confining pressure, peak stress increases simultaneously and so as the peak strain and residual strength, presenting a comparative strong ductility and the simple 6 order polynomial can be used to fit the stress-strain curve. The typical strain-time relationships under different stress states in creep test are shown in Fig.2. Thanks to the soft property of rock salt, creep phenomenon can take place, in accordance with Burgers rheological model even when stress level is very low. While the stress level is above a certain threshold, creep will stay constant for a period of time at first before increasing gradually, i.e. presenting the accelerated creep stage until the failure. The test results show that rock salts can bear larger strain when the stress is lower than the peak stress level. And the strain magnitude is related to the creep duration. Therefore, creep failure is not only related to stress but also to strain and time.

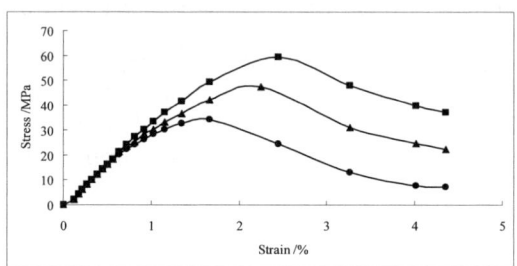

Fig. 1   The stress-strain relation in the instantaneous test under different confining pressures

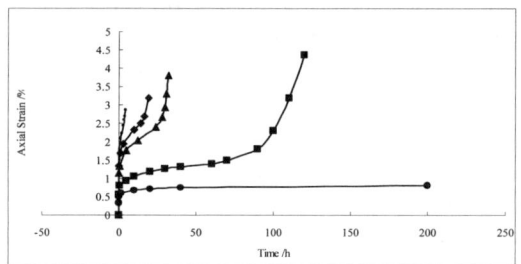

Fig. 2   Rock salts creep curve under different stress levels

CHEN Zongji [7] considers there exists a critical value for creep strength. If the strength is lower than the value, the deformation will be steady creep, otherwise there will be accelerated creep. This value is called the long-term strength of the rock. Sun Jun and Hu Yuyin[8] believe that rock strength is related to time and will decrease in exponential function forms with time and approach to the long-term strength gradually. According to the creep test, the tautochrone curves have obvious inflection points and the change gradients of stress-strain before and after inflection point are apparently different, i.e. the inflection point is the long-term strength value. Through the analysis of test material, the ratio between the long-term strength and the instantaneous peak is about 0.65 under the same circumstances and will increase with the confining pressure. The stress threshold, at which rock salts accelerated creep failure appears, can be deemed as the long-term strength value.

Model Analysis

According to the creep deformation and softening curve relationship in the stress-strain surface, which is put forward by Wang Zhiyin etc[3, 9] with combination with the rock salts test law, the stress-strain-time 3-dimensional graphics of rock salts in certain conditions are formed (Fig. 3(a),Fig. 3(b)). σ_p represents for instantaneous peak strength (deviatoric stress), ε_p for the corresponding creep peak strain, σ_{CL} for creep long-term strength (deviatoric stress), ε_{CL} for the strain under the creep long-term strength, σ_{FR} for the stress level in a certain accelerated creep failure phase (deviatoric stress), ε_{FR} for the creep failure strain value for the stress level, t_{FR} for the creep failure time for the stress level, OK for the terminal trajectory of creep and the stress at K points is the creep long-term strength σ_{FR}. The Fig. 3(a) shows that when stress level is below σ_{CL}, creep strain will incline to be steady while the creep strain increases to the intersection point with OK before when the strain no longer increases with time. It can be regarded as unbounded surface within human's scope (time infinite). In this state, the stress range is $(0, \sigma_{CL})$, the strain range $[0, \varepsilon_{CL})$ and the time range $[0, \infty)$. When the stress exceeds σ_{CL}, accelerated creep stage will present and surface curvature change. The variation is matched with creep rate change law. Curved surface will bent between the I and II creep stage as well as II and III creep stage with the bending rate more and more obvious as the increase of stress. The creep failure will occur when creep strain equals to the strain value of the after peak strength under the same stress state correspondingly. The higher the level of stress, the smaller creep failure strain, the shorter creep failure time will be and more close to the instantaneous stress-strain curve. Finally, they all join together at the peak point of whole stress-strain curve. The projection of the surface at three planes of coordinate axis is also in conformity with the existed experiment law (Fig. 3(b))

(a) If t=0, it is the whole stress-strain curve in instantaneous test. If $t = l_1, l_2...l_n$, it is the isochron under different stresses. Its $\sigma - \varepsilon$ plane projection is the surrounded area with coordinate axis within the stress-strain curve boundary.

(b) If $\sigma = \sigma_1, \sigma_2...\sigma_n$, it is the creep curve under the given stress. And its projection in $\varepsilon - t$ are several different creep rate curves as shown in Fig.2.

(c) If $\varepsilon = \varepsilon_1, \varepsilon_2,...\varepsilon_n$, it is the stress-time curve, which is in accordance with the conclusion by Sun Jun and Hu Yuyin. And its projection in $\sigma - t$ is a family of radiation straight line. Stress level decreases in the form of index along with the time, which demonstrates as stress relaxation phenomena.

Based on mathematical regression equations, the whole stress-strain curve can approximately satisfy 6 polynomial relationships and the creep curves are described by piecewise function which can conform to Burgers model and accelerated Burgers model respectively. The relationship is shown in formula (1), among which a, b, c, d, r, M are all constant coefficients. The range of α is (0, 1), which reflects the characteristics of accelerating period. The greater the stress level, the higher the value of α will be.

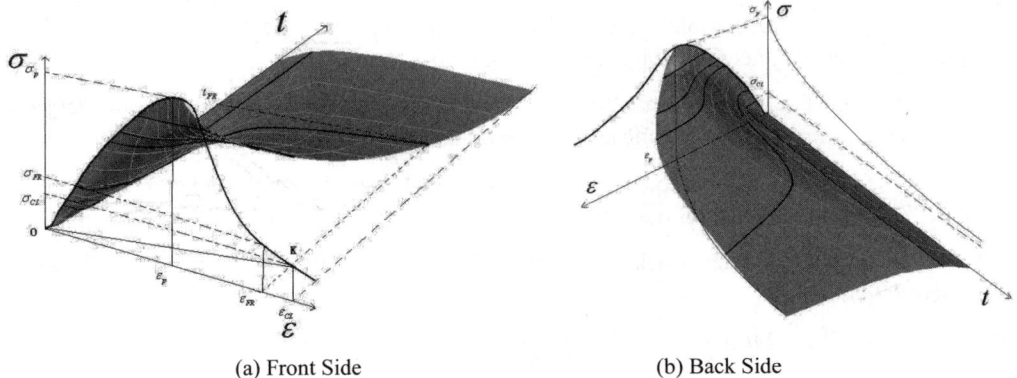

(a) Front Side   (b) Back Side

Fig.3 Rock Salts Stress-Strain-Time Graph

$$\begin{cases} \varepsilon = a + b(1-e^{-rt}) + ct & \sigma < \sigma_{CL} \\ \varepsilon = a + b(1-e^{-rt}) + c[1-(d-t)^{\alpha}] & \sigma \geq \sigma_{CL} \\ \sigma = \sum_{i=0}^{6} M_i \varepsilon^i \end{cases} \quad (1)$$

Conclusions

Combined with the test data and existing theory, the stress and strain-time 3-Dimensional rheological surface is drawn and the change law of the space stretches surface is given. The inner relationship between rock salts creep failure and instantaneous test can be very intuitively described by this surface, whose function equations are also carried out. But the conclusion of this paper still needs further validation considering the discontinuity of the test data. In addition, the influences of confining pressure and temperature should be taken into consideration while researching on the loading path effect to the surface The larger the confining pressure, the higher the instantaneous peak strength and the lower the creep rate. The surface gently expands with the surface change in the normal direction along with the change of whole stress-strain curves. And the influence of the temperature is equivalent to a strong contraction of the surface.

References

[1] Pan Junzheng.Rheological Soils Stress-Strain-Time Graph and its Application Journal of Zhenjiang Agricultural Machinery College, 1982, 3:1-4.

[2] Googman R.E.Introduction to Rock Mechanics. New York:John wiley and sons,1980.

[3] Wang Zhiyin, Tang Mingming, Sun Yili, Wang Yi. Study on the Relationship between Rock Salts Creep Whole Process and Triaxial Stress-strain Curves [A]. The Paper Collection of the 9th CCS&CSTAM rheological academic conference[C], 2008.

[4] Liu Jiang, Yang Chunhe, Wu Wen, Li Yinping. Experiment Study On Short-Term Strength and Deformation Properties of Rock Salts[J]. Chinese Journal of Rock Mechanics and Engineering, 2006, 25(s1):3104-3109.

[5] Yang Chunhe, Chen Feng, Zeng Yijin. Investigation on Creep Damage Constitutive Theory of Salt Rock [J]. Chinese Journal of Rock Mechanics and Engineering, 2002, 21(11):1602-1604.

[6] Chen Weizhong, Wang Zhechao, Wu Guojun, Yang Jianping, Zhang Baoping. Nonlinear Creep Damage Model for Rock Salts and its Engineering Application [J]. Chinese Journal of Rock Mechanics and Engineering, 2002, 21(11):1602-1604.

[7] CHEN Zongji.The Mechanical Problems for the Long-Term Stability of Underground Galleries[J]. Chinese Journal of Rock Mechanics and Engineering, 1982, 1(1):1-20.

[8] Sun Jun, Hu Yuyin. Time-Dependent Effects on the Tensile Strength of Saturated Granite at Three Gorges Project [J]. Journal of Tongji University, 1997, 25 (2):127~134.

[9] Wang Zhiyin, Li Yunpeng. Rock Rheological Theory and Numerical Simulation[M]. Beijing: Science Press, 2007.

[10] Cai Meifeng. Rock Mechanics and Engineering [M]. Science Press, 2002.

Experimental research on interlaminar shear strength of GFRP bridge decks under simulated concrete environment

XUE Weichen[a], FU Kai[b]

College of Civil Engineering, Tongji University, Shanghai, 200092, China

[a]xuewc@tongji.edu.cn, [b]eric_fk@sohu.com

Keywords: Simulated concrete environment; GFRP bridge deck; Interlaminar shear strength; SEM analysis

Abstract. Fiber reinforced plastic (FRP) composite which has high strength, high fatigue resistance, low density, and better corrosion resistances is desirable characteristics for bridge applications, especially decks. According to the ACI 440.3R–04, Glass fiber reinforced plastic (GFRP) bridge deck samples were immersed into the simulated concrete environment at 60°C for 92d (corresponds to the natural environment 25 years). The results show that, with the time increased, the interlaminar shear strength of GFRP bridge decks decreased significantly. After being exposed to the simulated concrete environment for 3.65d, 18d, 36.5d and 92d, the interlaminar shear strength degradation of GFRP bridge decks were 18.69%, 25.90%, 50.93% and 53.74%, respectively. The micro-formation of the GFRP bridge deck sample surface was surveyed under scanning electron microscopy (SEM), which indicated that with the aging time increased, corrosion pits in the surface of GFRP bridge decks became more obviously and the interface between fiber and resin was severely damaged. Therefore, the degradation of FRP under the simulated concrete environment should be considered in the design of FRP bridge decks.

Introduction

Bridge decks are subjected to severe environmental conditions and heavy traffic load and sometimes account for a major percentage of a bridge structure's dead load, hence transportation infrastructure systems deteriorate at an alarming rate in many countries ([1][2]). For example, highway bridge repair each year cost $ 50 billion in the United States.

In response to the call for more durable bridge construction, a great deal of interest has been given to GFRP composite. Its high strength, high fatigue resistance, low density, and better corrosion resistance are the desirable characteristics for bridge applications, especially decks ([3]).

The existing study of GFRP bridge deck at home and abroad related to aspects of design theory, carrying capacity, and production technology, etc., and there is a lack of information related to the durability of GFRP bridge decks ([4] [5]).

According to the ACI 440.3R–04([6]), GFRP bridge deck samples were immersed in 60°C of the simulated concrete environment for 92d (Corresponds to the natural environment 25 years) ([7] [8]), the mechanical behavior of GFRP bridge decks was evaluated by focusing on effects of corrosion time on the interlaminar shear strength.

1 Materials and Test Methods

In this paper, GFRP bridge decks made up of E-glass fibers and vinyl ester resin were selected due to their wide applications in civil engineering.

Groups of GFRP bridge deck samples were tested before and after immersing into the simulated concrete environment at 60°C for 3.65d, 18d, 36.5d and 92d, respectively. The simulated concrete environment (pH 12.6 to 13.0) for this study was prepared from a mixture of 118.5 g of Ca (OH)$_2$, 4.2 g of KOH and 0.9 g of NaOH in 1 L of deionized water, which is made to simulate the pore solutions of concrete according to the ACI 440.3R–04 (Table 1).

Table 1 Composition of simulated concrete environment

Solution	Grams of solute contained in 1 liter of water (g/L)		
	Ca(OH)$_2$	KOH	NaOH
Alkaline solution	118.5	4.2	0.9

The interlaminal shear tests were conducted to measure the interlaminal shear strength of the GFRP bridge deck samples by three-point bending a short beam method in accordance with ASTM D 2344. The tests were carried out in a hydraulic testing machine, and the duration of the loading was 2 to 4 minutes for each sample.

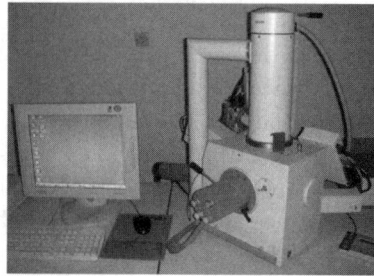

Fig.1 Scanning electron microscopy

In addition, scanning electron microscopy (SEM) observation and image analysis were performed to analyze sample microstructure before and after immersing into the simulated concrete environment, such as matrix cracking, fiber damage, and fiber-matrix interface damage (Fig. 1).

2 Results and discussion

2.1 Microstructure analysis

The surface conditions of the GFRP bridge decks before and after corrosion for 92 days were shown in Fig. 2 and Fig. 3, respectively.

Fig. 2 Surface conditions of the GFRP bridge deck samples before corrosion

Fig. 3 Surface conditions of the GFRP bridge deck samples after corrosion for 92 days

The surface of samples was analyzed before and after corrosion for 92 days using the SEM, and it could be seen from the control samples (Fig.2) that fibers and matrix exhibited a strong bond before corrosion. Corrosion pits appeared in the surface of GFRP bridge deck samples when the sample immersed in simulated concrete environment at 60°C for 92 days. And it was obviously to see the resin deterioration after corrosion for 92 days in Figs. 3. For the samples immersed in simulated concrete environment, the failure resulted from the matrix cracking, the interface loosing, and matrix-fiber deboning.

2.2 Interlaminal shear strength

The interlaminar shear strength values are plotted against corrosion time and are shown in Fig.4.

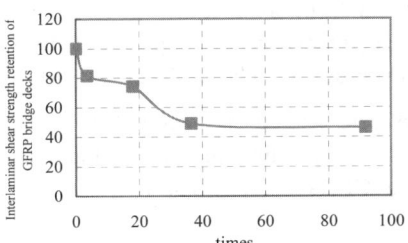

Fig.4 Degradation of the interlaminal shear strength of GFRP bridge decks

Fig.4 shows that the interlaminar shear strength of GFRP bridge decks decreased significantly as the corrosion times increasing. After being exposed to the simulated concrete environment at 60°C for 3.65d, 18d, 36.5d and 92d, the interlaminar shear strength degradation of GFRP bridge decks were 18.69%, 25.90%, 50.93% and 53.74%, respectively.

3 Conclusions

(1) The micro-formation of the GFRP bridge decks was surveyed under SEM, and it indicated that after being immersed in the simulated concrete environment at 60°C for 92 days, GFRP bridge deck samples deterioration was serious, the failure resulted from the matrix cracking, the interface loosing, and matrix-fiber deboning.

(2) The interlaminar shear strength of GFRP bridge decks decreased significantly as the corrosion times increasing. After being exposed to the simulated concrete environment at 60°C for 3.65d, 18d, 36.5d and 92d, the interlaminar shear strength degradation of GFRP bridge decks were 18.69%, 25.90%, 50.93% and 53.74%, respectively.

Acknowledgments

The authors gratefully acknowledge the financial support provided by the Chinese National Natural Science Foundation (Project No. 50978193), and the Project of Shanghai Science and Technology Commission (Project No. 10dz1202200, 10dz0583700 and 09231200400).

References

[1] Wan Shui, Hu Hong. Development and application of FRP bridge decks [J]. Journal of Highway and Transportation Research and Development, 2004, 21(8): 59-63.

[2] Bakis C. Fiber-reinforced polymer composite for construction state-of-art-review [J]. Journal of Composite for Construction, ASCE, 2002, 6(2): 73-87.

[3] Wu, H. C., G. Fu, et al. Durability of FRP composite bridge deck materials under freeze-thaw and low temperature conditions [J]. Journal of Bridge Engineering, ASCE, 2006, 11(4): 443-451.

[4] Karbhari, V. and S. Zhang. E-glass/vinyl ester composites in aqueous environments–I: experimental results [J]. Applied Composite Materials, 2003, 10(1): 19-48.

[5] Park, K. T., Y. K. Hwang, et al. Experimental study on durability comparison of the GFRP decks by resin types [J]. KSCE, Journal of Civil Engineering, 2007, 11(5): 261-267.

[6] American Concrete Institute. Guide test methods for fiber-reinforced polymers (FRPs) for reinforcing or strengthening concrete structures. ACI Report 440.3R~04, Farmington Hills, 2004.

[7] Valter Dejke. Durability of FRP Reinforcement in Concrete–Literature Review and Experiments [D]. Thesis for the Degree of Licentiate of Engineering. Department of Building Materials, Chalmers University of Technology. Göteborg, Sweden, 2001.

[8] Porter M L, Barnes B A. Accelerated Aging Degradation of Glass Fiber Composites[C]. Fiber Composites in Infrastructure: Proceeding of the Second International Conference on Fiber Composites in Infrastructure ICCI'98, Vol. 2, Tucson, 1998, pp: 446-459.

Vibration Fatigue Behavior of 2024-T62 Aluminum Alloy Cantilever Beam Under Different Vibration State

Haitao Hu[1,a], Yulong Li[1,b], Jinli Wang[1,c]

[1]School of aeronautics, northwestern polytechnical university, Xi'an, 710072, China

[a]huhaitao@mail.nwpu.edu.cn, [b]liyulong@nwpu.edu.cn, [c]wangjinli.nwpu@163.com

Keywords: fatigue; natural frequency ;resonance; damage; vibration state

Abstract: The vibration fatigue experiments of cantilever beam structures were performed to investigate the fatigue behavior of 2024-T62 aluminum alloy. Two types of cantilever beams with various natural frequencies under the sinusoidal excitation were investigated. The initial stress of two types of specimens were set in the same amplitude by adjusting the acceleration of electrodynamic shaker. Based on the stress history recorded by the strain gauge in fatigue test and the Miner's liner cumulative damage rule, the fatigue damage of the cantilever beam was calculated. The effect of vibration state on the vibration fatigue behavior of the cantilever beam was discussed. The experiment results show that the fatigue life of the cantilever beam, of which the initial vibration state is resonance, is longer than that of non-resonance. The calculated damage results were in accord with the reduction of the natural frequency measured in experiment. The reduction of natural frequency could be used to evaluate the fatigue damage of structures.

Introduction

Some structures of military aircraft often experience various dynamic loads during high speed flight and maneuvers. These loads can lead to premature vibration fatigue in metallic structures. The fatigue which is related to dynamic property of structure is called vibration fatigue.

The difference between vibration fatigue and conventional fatigue is that the former need to consider the dynamic property of structure, such as natural frequency and damping. When the dynamic response plays a major role leading to fatigue failure, the structure would occur in the vibration fatigue problem. Previous works looked into the effect of loading frequency on structural fatigue behavior [1].In 1999, Morrissey [2] has studied the loading frequency and stress ratio effects in high cycle fatigue of Ti-6Al-4V, but the structural dynamic response has not been considered. Zhanfei Shi and Yulong Li [3] reviewed the frequency effect in fatigue characteristics of some alloy steel, titanium alloy, aluminum alloy. Ruijie Wang and Deguang Shang [4-6] compared the changes of specimen's natural frequency in the weld crack propagation process, and predicted its fatigue damage parameter. Whether the vibration state (resonance or non-resonance) will influence the fatigue life of structure under dynamic loading? Investigations about this issue has rarely been reported in literature.

The main goal of this paper is to evaluated the effect of vibration state on the vibration fatigue behavior of 2024-T62 cantilever beam. Sinusoidal loads with the same frequency were applied on two types of cantilever beams with different natural frequency The value of excited frequency is equal to the natural frequency of one type of cantilever beam. The beams of two types had the same initial stress under different vibration state. Then the fatigue damage was calculated by stress records.

The calculated results were compared with the reduction of natural frequency measured by experiment (ie, the percentage drop of specimen's natural frequency since the test began, the same as following).

Materials and Test Method

The specimen is made of 2024-T62 aluminum alloy. The mechanical properties of 2024-T62 aluminum alloy are E=72.4 GPa, ρ=2780 kg/m^3. All the specimens were fabricated in the same rolling direction without surface scratch. Fig. 1 shows the geometry and dimensions of cantilever beam specimen with thickness 4mm. The first-order natural frequencies of two types of specimens were 146Hz and 150Hz when the mass mounted at different locations. The strain history of the cantilever beam was measured by the strain gauge mounted near the clamped end.

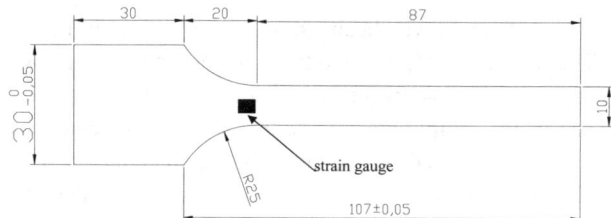 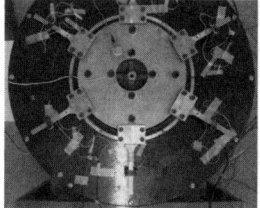

Fig.1 Dimensions of the specimen(unit: mm).   Fig.2 The photo of clamped specimens.

Figure 2 shows the setup of the experiment system including Y51150 electrodynamic shaker, specimens and fixtures, and data acquisition system. The frequency of sinusoidal excitation was 150Hz. The maximum stress amplitude of the specimen under different vibration state was 145MPa. by controlling the acceleration of shaker. The dynamic responses of two types of specimens are shown in Fig.3. The natural frequencies of the specimens were tested after each 500000 loading cycles. The 5% reduction of the natural frequency would be supposed as fatigue failure criterion[7].

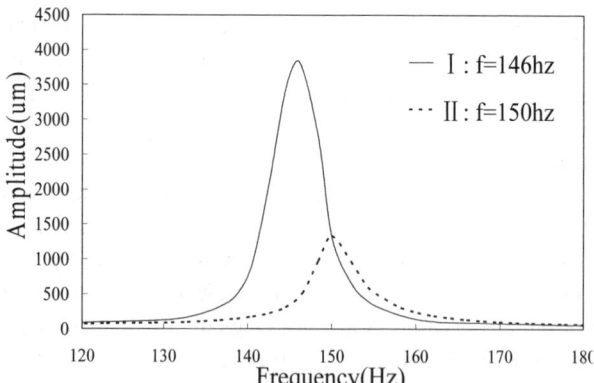

Fig.3 Frequency response curves of two types of specimens.

Results and Discussion

The fatigue damage of specimens was calculated based on its stress history. The Miner linear damage accumulation model was selected to characterize the fatigue life of the 2024-T62 aluminum alloy specimens and was expressed by:

$$D = \sum_i \frac{n_i}{N_i} \tag{1}$$

where D is the cumulative damage at i load cases with n_i cycles at the stress level s_i, N_i was the allowable cycles at the stress level s_i. The accumulated results which were used to evaluate the damage failure of specimen. were shown in Table 1.

Table1 The damage of specimens corresponding to cycles.

Type No.	type I							type II						
Specimen No.	CH1	CH2	CH3	CH4	CH5	CH6	CH7	CH1	CH2	CH3	CH4	CH5	CH6	CH7
Cycles*10⁴	210	210	210	210	210	210	210	200	320	300	160	310	310	300
cumulative damage	1.262	1.399	1.587	1.434	1.463	1.473	1.416	0.609	0.482	0.356	0.320	0.749	0.738	0.536

Table1 shows that the cumulative damage of the specimens of type I is significantly greater than that of type II. The initial vibration state of the specimens in type I and II are non-resonant and resonant, respectively. The cumulative damage of structure is related to cracks in structure. The cracks generate and expand as the damage increase. The calculated results show that the vibration fatigue property of specimens in type II are better than that of type I under this test condition.

Fig.4(a) shows stress history of two types of specimen in experiment by relevant statistical approach[8] and Fig.4(b) shows the reduction of natural frequency. Both the stress amplitudes of two types are reduced slowly within the first 500 000 cycles. Then the stress amplitudes of type II is reduced rapidly than that of type II until up to 1500000 cycles. The stress amplitudes of type I is reduced sharp near 1500000 cycles and at the same time the reduction of natural frequency reach the default threshold of 5%.

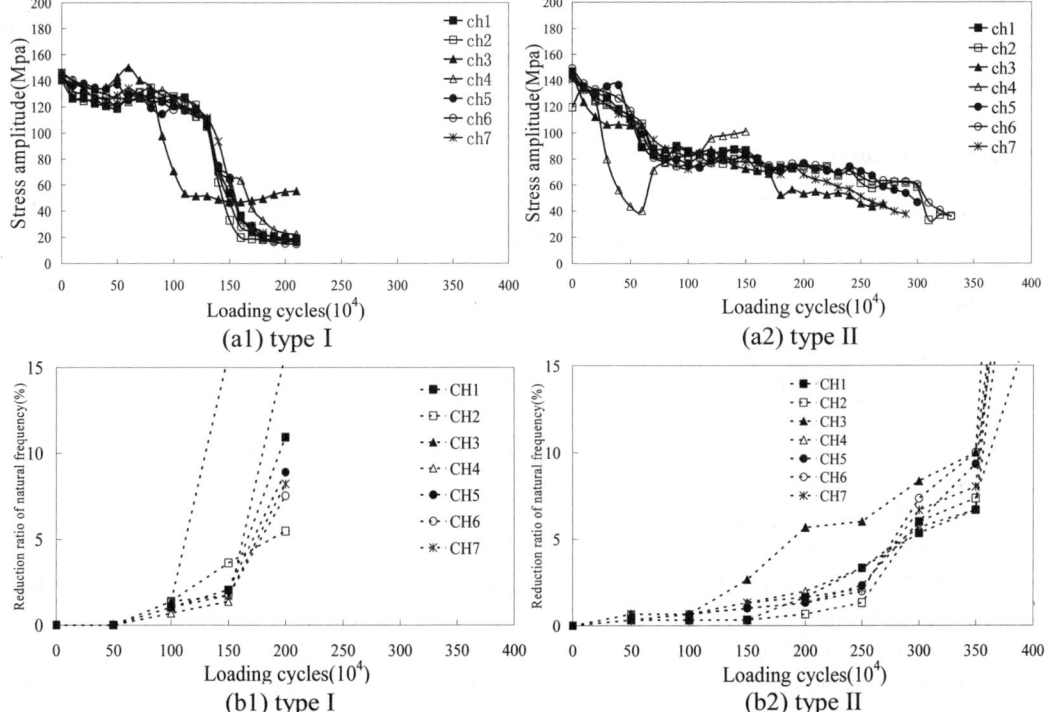

Fig.4 The stress history (a) and reduction of natural frequency (b) in two types of specimens during experimental process.

Fig.4(b) shows that the reduction of the natural frequencies is different when the initial vibration state is different. The reduction of natural frequencies of type II begin earlier than that of type I, but the rate of reduction of type I is faster and achieved 5% default failure threshold firstly. The faster natural frequencies reduce, the faster internal cracks expand and greater damage is generated. Throughout the test, cumulative damage in specimens of type I is greater than that of type II. This results also shows that the specimens of type II has better vibration fatigue property than that of type I. So the results obtained from the reduction of natural frequencies are agree with the calculated results.

The resonance state is dangerous in conventional vibration [9], the response of structure was amplified several times in resonance state, so it was more prone to generated structural damage. But in this paper the initial stress amplitude of two types of specimens are the same. The damage within structure would reduce structural rigidity and natural frequency. Once the natural frequency is deflected from the resonant frequency, the stress amplitude of the specimen reduces quickly. That means the increase of cumulative damage will slow-down.

4. Conclusion

The vibration fatigue behavior of 2024-T62 cantilever beam under different vibration states has been investigated using experimental study. The effect of the vibration state can not be neglected for the estimation of fatigue damage. In the fatigue testing, the initial vibration states of the two types of the cantilever beams are resonance and non-resonant, and initial stress amplitude of them are same. The stress history and the reduction of natural frequency of the two types specimens are different. The vibration fatigue performance of the specimen, of which the initial vibration sate is resonance, is better than that of the other one because the stress amplitude of structure is related to the dynamic response of structure. The reduction of natural frequency can be used to evaluate the structural fatigue damage.

Acknowledgements

This work was supported by the National Natural Science Foundation of China (contract No.10932008) and the 111 Project of China (contract No.B07050).

References

[1] Jianghua Liu. M.S.Thesis. (Northwestern Polytechnical University, China,2008).
[2] R.J.Morrissey, D.L.McDowell, T.Nicholas. *International Journal of Fatigue*, Vol.21 (1999), p.679–685.
[3] Zhanfei Shi, Yulong Li, etc. *Journal of Materials Science and Engineering*, Vol.27,No.3 (2009), p.488-492.
[4] Rui-Jie Wang, De-Guang Shang, Li-Sen Li, Cheng-Shan Li. *International Journal of Fatigue*. Vol.30 (2008), p.1047–1055.
[5] Rui-Jie Wang, De-Guang Shang. *International Journal of Fatigue*, Vol.31 (2009), p.361 – 366.
[6] De-Guang Shang. *Materials and Design*, Vol.30 (2009), p.1008 – 1013.
[7] Xiao Shouting, Du Xiude. *Journal of Mechanical Strength*. Vol.17 (1995), p.22-24.
[8] Gao Zhentong: *Fatigue Test Design and Data Treatment*. (Beijing University of Aeronautics and Astronautics Press , China,1999).
[9] Qihang Yao, Jun Yao. *Chinese journal of applied mechanics*, Vol.23 (2006), p.12-15.

Simulation of pylon emergency break-away of large commercial aircraft

Zhang Xin[1, a], Li Yulong[1, b]

[1]School of Aeronautics, Northwestern Polytechnical University, Xi'an, China

[a]wozhangxin@gmail.com, [b]liyulong@nwpu.edu.cn

Keywords: engine, pylon, emergency break-away, SPH.

Abstract. Most large commercial aircraft engines are hanged below wings. When the aircraft makes an emergent landing process, at the same time, the landing gear cannot work or the centrifugal force becomes unbalanced as the fan blades fly away. In order to ensure safe landing, overall tank crack and fuel leak should be avoid at the crack area.

Access to a large number of emergent landing accidents, emergency break-away is essential. In this paper, the break-away position, which locates between the pylon and the wing, is mainly considered. We choose true size in the models of wing, pylon and engine. High strength steel is employed for the bolts which connect pylon and wing. Earth and lake are employed concrete and pure water respectively. SPH method is applied in the case that the aircraft lands on the lake. What's more, different landing cases have been analyzed. By constantly adjusting the size of pins, a set of conclusions of the emergent landing problem are obtained in the simulation process. The locus of centroid of engine and pylon is obtained, and then the condition which may achieve safe flight and avoid the secondary damage to wings can be chosen, from which we can provide reasonable designing strength of the wing box and accordingly provide reference for the design of aircraft structure.

Introduction

Large aircraft project is one of Chinese long-term projects in the last two decades, the connection of the engine and the wing is the key technology in aircraft designing. The pylon, which connects the engine and the wing, plays a significant role in supporting engine and transforming engine trust.

Many works have been done perfectly at crashworthiness. Karen E. Jackson et al. [1] established the aircraft and earth model using LS-DYNA software, good coherence have been obtained from the compare between simulation results and experiment outcomes in every second, Qiuling Qu et al. [2] investigate the mechanic behaviors of aircraft which has an emergency landing on water by using numerical method, structural failure will be more serious when landing with angle of attack of 6 degree and oppositely 14 degree at least. Edwin L. Fasanella and Karen E. Jackson et al. [3] carried out an experiment on the B737 fuselage that made by Boeing, normal landing, over weighted landing and landing with the auxiliary tank have been investigated, they also simulated all these conditions, from which, design proposals have been supplied to Boeing.

Few works have been done at the pylon emergency break-away. There should be two kinds of emergency break-away, the pylon separate from the wing and the engine separate from the pylon, whatever which style we choose, the overall tank should not crack and leak fuel under impact [4].

In this paper, the break-away position which happens between the pylon and the wing is mainly considered. We choose true model of wing, pylon and engine, and the simulations divided into two conditions: the aircraft lands on the airport runway and lands on the lake.

Statistics of emergency landing accidents

Emergency landing accidents have been investigated on internet [5]. According to these accidents as shown in Table 1, it is found that most commercial aircraft involving Boeing and Airbus series' airplanes have the pylon emergency break-away designing, when the aircraft lands emergently on airport runway, sea or marsh. Accordingly, the designing of emergency break-away of pylon is of great important..

Table1 Emergency landing cases

	Emergency landing cases	Engine separate cases	Engine unseparated cases	Indistinct cases
number	25	14	9	2

Finite elemental model

Models of the pylon, the engine and the wing are found by CATIA, and then we mesh models using HyperWorks, assemble and process them by using PAM-CRASH as shown in Fig. 1. The bolts are made of high strength steel, concrete for earth and pure water for lake, in order to be more authentic and effective in emergency landing on the lake, we choose SPH method. The aircraft emergency landing contains only the nose landing gear can work, only the main landing gear can work and both of them cannot work, by giving a different impact angle to achieve the different conditions.

Air friction and engine trust have been ignored in simulations. At first we suppose that the airplane approach with constant velocity, in which condition friction between the fuselage and the airport runway or lake is ignored. If the pylon obtains a reasonable separation and does not impact the wing, a conclusion of successful emergency break-away can be drawn. Oppositely, if mesh elements of the pylon penetrate into the wing, deceleration of the airplane that produced by friction should be considered. It has been prescribed in CCAR that the coefficient of front utmost inertial tolerant load of passengers is 9.0. Accordingly, 9.0 times acceleration of gravity is selected as the deceleration of the airplane. Forward and downward velocities have been evaluated 70.0m/s and 10.7m/s separately according to CCAR.

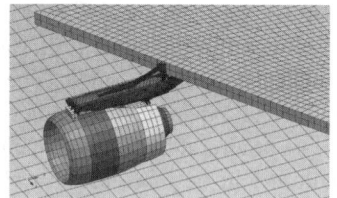 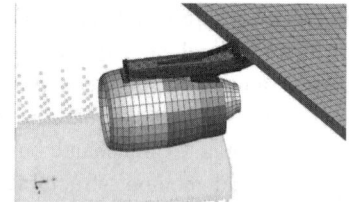

(a) The aircraft impacts the runway (b) The aircraft impacts the lake
Fig. 1 The whole model of simulation

Simulation results

From Airframe Structural Design [5], we can learn that when an airplane takes an emergency landing, for more stable pneumatic layout, the upper beam of the pylon trends to retain connection with the wing, and the nether beam connect with the pylon. We also know that in order to make a safe separation, the pylon and the engine trend to throw away outside and in front of the airplane.

All pins have been numbered as shown in Fig. 2 in order to facilitate the analyses. The aircraft have angles of attack of 0 degree, 5 degree and 10 degree under both runway and lake landing conditions.

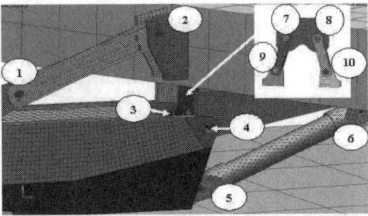

Fig. 2 The whole model of simulation

1) The aircraft impacts the runway of airport with 0 degree angle of attack.

We simulate in PAM-CRASH, and some outcomes are obtained. It is found that number 1, 3, 4, 6, 7, 8, 9, 10 pins have failed, and the failure order is 1-6-3-4-7-8-10-9. The failure of pin No.1 and non-failure of pin No.2 mean that the upper beam is connected with the wing and separated from the pylon. Similarly, the failure of No.6 and non-failure of No.5 mean that the down beam connects with the pylon and separates from the wing, as shown in Fig. 3. The measured load-time curves of pins are shown in Fig. 4. By examining the time at which the load goes back to zero, we can obtain the failure order of the pins, which agrees well with the conclusion above.

The maximum contact load is 664KN; accordingly the tolerant load of the wing box which connects with the lug should be 764KN, 15% abundant of the contact load. From the simulation results, we can also find that the pylon is threw away to the front and outside of the fuselage, a successful pylon break-away has been achieved.

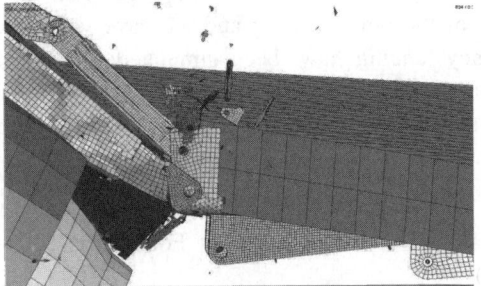 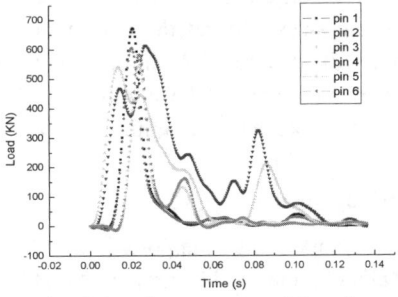

Fig. 3 The pylon separate from the wing   Fig. 4 Load responses of the pins

2) The aircraft impacts the lake of airport with 0 degree angle of attack.

The lake model was established SPH model. From the results, the failure order of the pin is 1-3-4-7-8-6-10-9. The measured load-time curves of pins are shown in Fig. 5, from researching the time that the load goes back to zero, we can obtain the failure order of the pins, and we find it agrees with the conclusion ahead. The tolerant load of the wing box which connects with the lug should be 989KN.

The scene can be discerned in Fig.6 that mesh elements of the pylon penetrate into the wing, we evaluate deceleration with 9.0 times acceleration of gravity, and the wing will have 0.75 meter displacement backward compare to the condition without the deceleration in 0.13 second that the pylon separated from the wing. In that case, the pylon and engine will not impact the wing. From the simulation results, we can find that the impact load and the friction between the fuselage and the water work together to lead a reasonable pylon break-away, and the pylon is threw away in front of the fuselage.

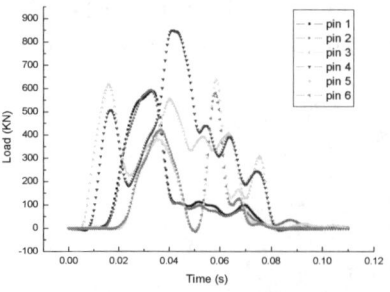 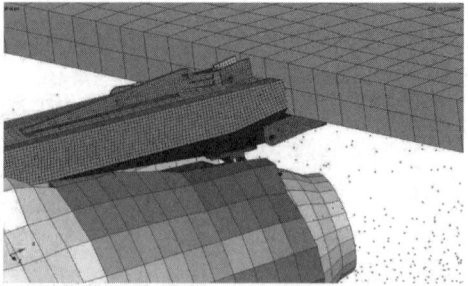

Fig. 5 Load responses of the pins   Fig. 6 The pylon separate from the wing

When the aircraft takes an emergency landing on the runway at 5 and 10 degree angle of attack, the reasonable pylon break-away should be accomplished by both friction and impact load, and the pylons are threw away both outside and in front of the fuselages. When the aircraft takes an emergency landing on the lake at 5 and 10 degree angle of attack, the reasonable pylon break-away can be accomplished only by impact load, pylons are casted in front of the fuselages, but an outside cast cannot be fulfilled.

We also check the security of these pins when the engine supplies maximum trust and a conclusion of safe flight have been obtained.

Summary

The connection of the engine and the wing is the key technology in aircraft designing. In this paper, we collect large amounts of emergency landing accident reports and learn that pylon emergency break-away is essential. According to the simulation of emergency landing under different pin sizes, we get the reasonable sizes of the pin structure; and we have validated their securities. Six different conditions of emergency landing have been simulated; the reasonable break-away of pylon is the combined infection of impact and friction produced by the airport runway or the lake. The maximum load of the wing box that should be carried has also been lay out.

Acknowledgements

This work was Supported by the National Natural Science Foundation of China (No.10932008) and the 111 project (No.B07050).

References

[1] Jackson K E, Fasanella E L: *Development of an LS-DYNA model of an ATR42-300 aircraft for crash simulation* (Dearborn MI 2004).

[2] Qiuling Qu et al. Numerical research of mechanical behaviors of a civil aircraft lands on the water. Civil Aircraft Design and Research, 2009, S1, p. 64-69.

[3] Fasanella E L, Jackson K E, Jones Y T, et al: Crash Simulation of a Boeing 737 Fuselage Section. Langley research center (2004)

[4] Michael C. Y. Niu. *AIRFRAME STRUCTURAL DESIGN*. (Publications, Beijing 2008)

[5] Information on http://www.faa.gov.com

Dynamic Mechanical Behavior of Two Fiber-reinforced Composites

Y.Z. Guo[1,a], X. Chen[1], X.Y. Wang[2], S.G. Tan[2], Z. Zeng[1], Y.L. Li[1,b]

[1] School of Aeronautics, Northwestern Polytechnical University, Xi'an, Shaanxi 710072, China
[2] Department of Strength, The First Aircraft Insitute, Xi'an, Shaanxi 710072, China
[a]email: guoyazhou@nwpu.edu.cn, [b]email: liyulong@nwpu.edu.cn

Keywords: composites, dynamic, mechanical, fracture, deformation modes

Abstract: The mechanical behavior of two composites, i.e., CF3031/QY8911 (CQ, hereafter in this paper) and EW100A/BA9916 (EB, hereafter in this paper), under dynamic loadings were carefully studied by using split Hopkinson pressure bar (SHPB) system. The results show that compressive strength of CQ increases with increasing strain-rates, while for EB the compressive strength at strain-rate 1500/s is lower then that at 800/s or 400/s. More interestingly, most of the stress strain curves of both of the two composites are not monotonous but exhibit double-peak shape. To identify this unusual phenominon, a high speed photographic system is introduced. The deformation as well as fracture characteristics of the composites under dynamic loadings were captured. The photoes indicate that two different failure mechanisms work during dynamic fracture process. The first one is axial splitting between the fiber and the matrix and the second one is overall shear. The interficial strength between the fiber and matrix, which is also strain rate dependent, determines the fracture modes and the shape of the stress/strain curves.

Introduction

The deformation and fracture behavior of fiber-reinforced composites have drawn tremendous attention of researchers due to their potential engineering applications. Their light weight, high specific strength and stiffness, superior corrosion resistance and good designability make fiber-reinforced composites ideal structural matierials in aircrafts, automobiles, missles, armors, buildings and so on. Most of these structures are inevitably subjected impact loading within their service lifetime. In that case, the mechanical properties of fiber-reinforced composites at high strain rates are of special interest. However, the mechanical behavior of these materials under dynamic loading are much less studied comparing to those under quasi-static loadings. The limited literature indicates that mechanical behavior at high strain rates differ a lot from quasi-static case[1-7]. The dynamic fracture mechanisms, including crack intiation and propagation and their relationship with the overall mechanical properties of the composites, are still poorly understood. A major limitation is the experimental technique associated with dynamic tests.

The split Hopkinson pressure bar (SHPB) apparatus, originally developed by Kolsky[8], has been widely used for the determination of the dynamic mechanical properties of metals and of composites[2-4, 6-7, 9-11] in recent years. Depending on the specific design of a SHPB system, it allows mechanical testing at strain rates from $\sim 10^2 s^{-1}$ to $\sim 10^4 s^{-1}$. The histories of stress, strain and strain rate could be calculated based on theory of one dimensional wave propagation. However, due to the transient nature of dyanmic loading(usually dozens to hundreds of micro-seconds), the deformation and fracture process are difficult to observe. The previous studies on fracure mechanism of fiber-reinforced composites are limmited to post-test obsevation of the fractured pieces of the materials. The failure mode, as well as crack initiation and propagation are rarely observed.

In this paper, two fiber-reinforced resin matrix composites will be studied by SHPB apparatus. The mechanical properties at strain rates ranging from 400/s to 1500/s will be obtained. A high speed photographic system is used to capture the deformation and fracture process of the composites. The failure modes and their relationship to the mechanical response will also be discussed.

Experiment

The SHPB setup locates at Northwestern Polytechnical Universtiy. The lengths of the input bar and output bar are 1200mm and 1000mm respecitvely. Their diameters, together with the striker bar, are all 12.7mm. The length of the striker bar varies from 100mm to 260m to get appropriate strain rate and deformation. A copper disc with diameter of 3mm is used as pulse shaper. The thickness of the disc varies according to the loading rates. The input pulse shape is controlled to ensure stress equilibration within the specimen. Another function of the pulse shaper is to achieve constant strain rate during loading.

Two kinds of 2D braided composites, i.e. CQ and EB, are selected as the example material. CQ is a carbon fiber reiforced resein matrix composite, while EB is glass fiber reiforced. The ply orientation for both of the composites is 0/90. The loading direction is along one of the fiber direction for all tests. At least four tests are repeated for each experimental condition to make sure the validity of the tests.

Results and Disscusion

The stress-strain curves of the two kinds of composites are acquired by using traditional SHPB theory. Fig. 1 presents the typical curves at different strain rates. It could be noticed that both of the composites exhibit strain rate dependent characteristics in terms of compressive strength. For CQ, compressive strength increases with increment of strain rate, while the compressive strength of EB at 1500/s is lower than that at 800/s and 400/s. The variation of compressive strength and failure strain with respect to strain rates for the two composites are illustrated in fig.2. Clearly, the compressive strength as well as failure strain of CQ increases with increasing strain rate from 400/s to 1400/s. The synchronous enhancement of strength and strain indicates that this material may absorb more energy at higher strain rates and could be potentially used in dynamic environments. Different from CQ, the compressive strength of EB gets maximum at 800/s, corresponding a minimum failure strain. The variation of the strength and strain may be related to the transition of failure modes at different strain rates, which will be discussed later.

Another unusual feature for most of the curves is that the stress does not increase monotonicly with the strain, but decreases slightly after its initial elevation, except for the case of EB at strain rate of 800/s. To identify the reason of this special phenominon, a high speed photographic system is introduced. The frequency of the high speed camera could reach 200,000/s with reasonably good quality of the picture. Since the typical duration of deformation process of the materials in this paper is about 200μs, this frequency is high enough to capture deformation and fracture characteristics of the samples.

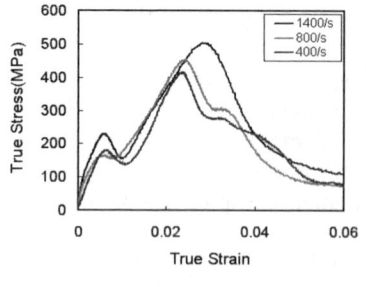

(a)

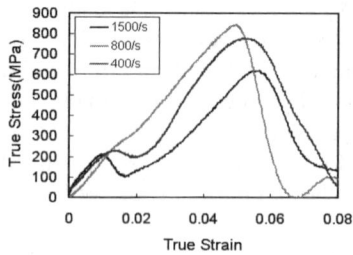

(b)

Fig.1. Compressive Stress/strain curves of CF3031/QY8911 (a) and EW100A/BA9916 (b) at different loading rates.

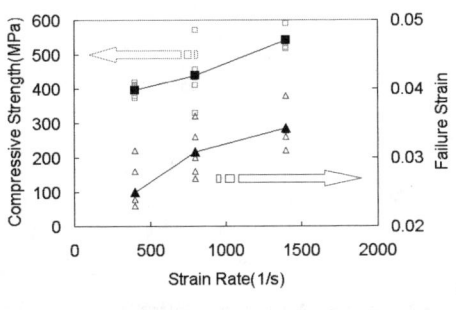 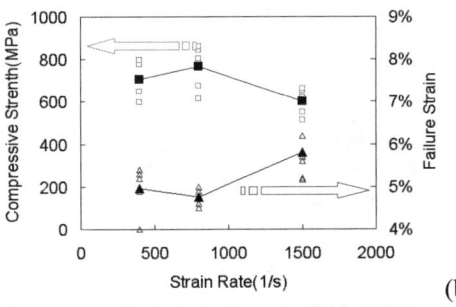

Fig.2. Compressive strength and failure strain of CF3031/QY8911 (a) and EW100A/BA9916(b) composites under different loading rates. Square and triangle represent compressive strength and failure strain, respectively. The hollow icons represent experimental points and solid icons represent their average values.

Two kinds of fracture modes, i.e. axial splitting and overall shear, are observed during our tests, listed in fig.3 and fig.4. The stress/strain curves and stress/loading-time curves are also presented. The present experiments indicate that the double-peak shape stress/strain curves are always corresponding to the fracture mode of axial splitting. From the pictures in fig.3, the specimen appears uniform deformation at the initial stage of loading until axial cracks initiates at 119μs. Splitting occurs right after the crack initiation and the stress drops sharply at the same time. More parallel cracks are developed and the specimen loses its load carrying capasicity completely at 156μs. The stress/loading-time curve indicates that there is a stress drop about 50μs after loading, but no evident overall deformation is observed in the pictures(C and D in fig.3.b). This stress decline could be induced by the microbuckling of the fibers. Different from tension, where fibers carry most of the tensile load, a majority of compressive load is undertaken by the matrix. Part of the fibers may carry load at the initial stage of loading. However, when the load is large enough, and if the interficial strength between these fibers and the matrix is not so strong, the fibers may be buckled and the stress drops. The microbuckling of these fibers will soon be supported by the matrix and the stress will clamb up. The stress will not fall down untill macrocrack and splitting form, which indicates an overall buckling of the specimen.

Fig.4 presents us another deformation mode. The stress goes up monotonicly with increasing strain till final fracture. The photos show that the specimen fractures 138μs after loading in the form of overall shear. Our experiments indicate that this fracture mode is always corresponding to the monotonic stress/strain curves. It is well known that the interficial strength between the fibers and the matrix is a key factor to the mechanical behavior of fiber-reinforced composites. To some extent, the inerficial strength determines the compressive strenth of the composites. A large interficial strength ties the fiber and matrix together firmly and makes them into a whole, which gives birth to a maximum compressive strength(see fig.1.b and 2.b).

Conclusion Remarks

Our experimental work indicate that the interficial strengths of CQ and EB fiber-reiforced composites are both strain rate sensitive. The higher strain rate is, the larger interficial strength is, which leads to the increase of the compressive strength. For EB, the variation of interficial strength even induces a transformation of the fracture modes, i.e. from axial splitting to overall shear. Further increase of strain rate could bring both of the two mechanism into the deformation process of EB and results in a decline in compressive strength(see Fig.2(b)).

Ackownledgement

The authors would like to thank the financial support by NSFC(No.10932008 and No.11102166) and by the 111project(No.B07050).

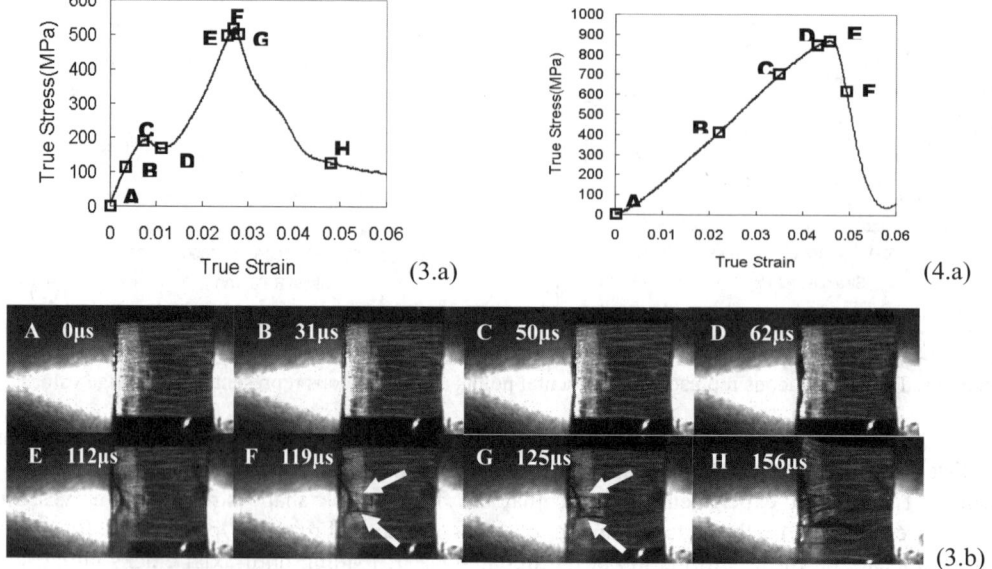

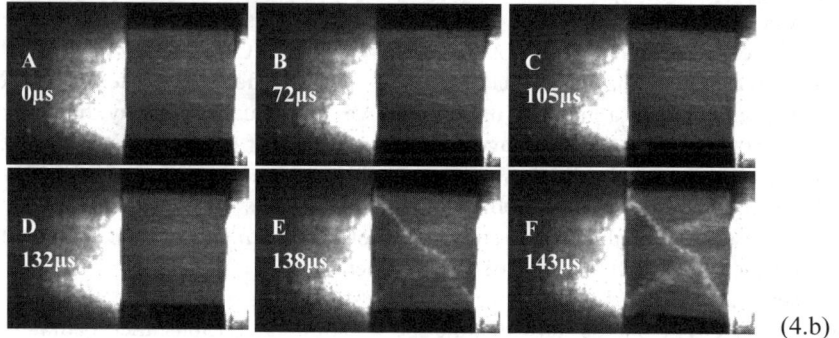

Fig. 3. Stress/Strain curve(3.a) of CF3031/QY8911 composites. The letters in (3.a) are corresponding to photos in (3.b). The arrows in the figure marked the cracks. Strain rate 400/s.

Fig. 4. Stress/Strain curve(4.a) of EW100A/BA9916 composites. The letters in (4.a) are corresponding to photos in (4.b). Strain rate 800/s.

References

[1] Kim, J.K. and Y.W. Mai, Composites Science and Technology, 1991. **41**(4): p. 333-378.
[2] Oguni, K. and G. Ravichandran, Journal of Materials Science, 2001. **36**(4): p. 831-838.
[3] Song, B., W.N. Chen, and T. Weerasooriya, Journal of Composite Materials, 2003. **37**(19): p. 1723-1743.
[4] Tsai, J.L. and C.T. Sun, International Journal of Solids and Structures, 2004. **41**(11-12): p. 3211-3224.
[5] Liu, M.S., et al., Materials Science and Engineering: A, 2008. **489**(1-2): p. 120-126.
[6] Yuan, Q., et al., Carbon, 2008. **46**(4): p. 699-703.
[7] Xuan, C. and L. Yulong, Materials Science and Engineering: A, 2011. **528**(22–23): p. 6998-7004.
[8] Kolsky H. *Proceedings of the Physical Society B*. 1949. London.
[9] Zhao, H., Computers and Structures, 2003. **81**: p. 1301-1310.
[10] Ninan, L., J. Tsai, and C.T. Sun, International Journal of Impact Engineering, 2001. **25**(3): p. 291-313.
[11] Guo, Y.Z. and Y.L. Li, Materials Science and Engineering A, 2007. **458**: p. 330-335.

Numerical Simulation of Compression-after-Impact Process of Composite Laminates

Li Biao, Li Yazhi, Li Xi, Yao Zhenhua

School of Aeronautics, Northwestern Polytechnical University, Xi'an 710072, China

libiao1109@126.com, yazhi.li@nwpu.edu.cn, shannonlixi@hotmail.com, yao_21g@163.com

Key words: composite laminate, low-velocity impact, damage, residual compressive strength.

Abstract: The residual compressive strength of composite laminates subjected to low-velocity impact (CAI) was analyzed using the ABAQUS/Explicit package through a two-step calculation. The finite element model was composed of solid elements and interfacial cohesive elements. The out of plane low-velocity impact process was simulated in the first step and the results of which were taken as the input for the second step of the in-plane compression, until the collapse of the laminate. The usefulness of the explicit solution algorithm in dealing with the quasi-static procedure of the in-plane compression was investigated by examining the effect of different initial velocities of the compression loading on CAI values. The simulation results agree well with the experimental results.

Introduction

Laminated composites are highly sensitive to impact from foreign bodies. The damage induced to such materials by low-velocity impact, even hardly visible, will reduce their compression strength remarkably[1-3]. Study on the resistance the material to low-velocity impact and residual compressive strength after impact damage is of great importance.

1. Specimen and the finite element model

1.1 Specimen

The specimen was made of CCF300/QY8911 fiber-reinforced composite laminate with the stacking sequence as [45/0/-45/90/0/0/45/0/-45/-45]$_s$. The basic properties are listed in Tab. 1. The specimen size is 150mm×100mm×2.4mm according to the ASTM test standard D7136/7137.

Tab. 1 Mechanical properties of the unidirectional laminate

E_{11}/GPa	E_{22}/GPa	E_{33}/GPa	G_{12}/GPa	G_{13}/GPa	G_{23}/GPa	v_{12}	v_{13}
134	7.9	7.9	4.62	4.62	3.2	0.33	0.33
v_{23}	X_T/MPa	X_C/MPa	Y_T/MPa	Y_C/MPa	S_{12}/MPa	S_{13}/MPa	S_{23}/MPa
0.48	1590	1250	49	192	98.2	98.2	98.2

1.2 Finite element modeling

The ABAQUS/Explicit package was employed for the simulation of impact and compression. Every lamina of the laminate is meshed with 8-node solid elements (C3D8R), and the interfacial thin layers among the laminas were meshed with cohesive elements. The minimum element size in the central area of the specimen is 1mm.

2. Damages and failure criteria

Hou[4] proposed a series of stress based criteria for different intralaminar failure modes. Huang [5] argued that the local degradation of stiffness coefficients to denote certain kind of failure usually causes the sharp alteration of stress level at that location. The change of strains as the result of such degradation is continuous and smoother than that of the stresses. So the strain levels should be the better indication to assess laminate failure. In this paper, Hoe's criteria [4] are modified as:

Fibre failure in tension:
$$e_{ft} = \left(\frac{\varepsilon_{11}}{X_T^\varepsilon}\right)^2 + \left(\frac{\varepsilon_{12}}{S_{12}^\varepsilon}\right)^2 + \left(\frac{\varepsilon_{13}}{S_{13}^\varepsilon}\right)^2 \geq 1 \quad (1)$$

Fibre failure in compression:
$$e_{fc} = \left(\frac{\varepsilon_{11}}{X_C^\varepsilon}\right)^2 + \left(\frac{\varepsilon_{12}}{S_{12}^\varepsilon}\right)^2 + \left(\frac{\varepsilon_{13}}{S_{13}^\varepsilon}\right)^2 \geq 1 \quad (2)$$

Matrix cracking in tension:
$$e_{mt} = \left(\frac{\varepsilon_{22}}{Y_T^\varepsilon}\right)^2 + \left(\frac{\varepsilon_{12}}{S_{12}^\varepsilon}\right)^2 + \left(\frac{\varepsilon_{23}}{S_{23}^\varepsilon}\right)^2 \geq 1 \quad (3)$$

Matrix cracking in compression:
$$e_{mc} = \frac{1}{4}\left(\frac{-\varepsilon_{22}E_{22}}{S_{12}^\varepsilon G_{12}}\right)^2 + \left[\frac{\varepsilon_{22}}{Y_C^\varepsilon}\left(\left(\frac{E_{22}}{4G_{12}}\right)^2 \frac{Y_C^{\varepsilon 2}}{S_{12}^{\varepsilon 2}} - 1\right)\right] + \left(\frac{\varepsilon_{12}}{S_{12}^\varepsilon}\right)^2 \geq 1 \quad (4)$$

where, $\sigma_{11} = E_{11}\varepsilon_{11}$, $\sigma_{22} = E_{22}\varepsilon_{22}$, $\tau_{12} = G_{12}\gamma_{12}$, $\tau_{13} = G_{13}\gamma_{13}$, $\tau_{23} = G_{23}\gamma_{23}$, $X_T = E_{11}X_T^\varepsilon$, $X_C = E_{11}X_C^\varepsilon$, $Y_T = E_{22}Y_T^\varepsilon$, $Y_C = E_{22}Y_C^\varepsilon$, $S_{12} = G_{12}S_{12}^\varepsilon$, $S_{13} = G_{13}S_{13}^\varepsilon$, $S_{23} = G_{23}S_{23}^\varepsilon$.

The interfacial delamination was simulated by the failure of cohesive elements in two stages, damage initiation and evolution, as shown in Fig. 1. Damage is assumed to initiate when a quadratic criterion

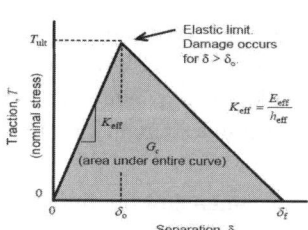

$$e_t = \left(\frac{\langle\sigma_n\rangle}{\sigma_{nc}}\right)^2 + \left(\frac{\sigma_t}{\sigma_{tc}}\right)^2 + \left(\frac{\sigma_s}{\sigma_{sc}}\right)^2 \geq 1; \quad \langle\sigma_n\rangle = \begin{cases} \sigma_n, & \text{when } \sigma_n > 0 \\ 0, & \text{when } \sigma_n \leq 0 \end{cases} \quad (5)$$

is met, where σ_n, σ_t and σ_s are the interfacial normal stress and

Fig. 1 Delaminating process

two shear stresses. σ_{nc}, σ_{tc} and σ_{sc} are the ultimate values of those individual stress components respectively and approximately replaced by the intralaminar strength parameters.

Damage evolution is based on the energy dissipated as the result of damage growth, which can be defined by:

$$e_t = \left(\frac{G_I}{G_{IC}}\right)^2 + \left(\frac{G_{II}}{G_{IIC}}\right)^2 + \left(\frac{G_{III}}{G_{IIIC}}\right)^2 \geq 1 \quad (6)$$

where G_I, G_{II} and G_{III} are the energy release rates of the cohesive elements in the normal and the both parallel directions to the interface, and G_{IC}, G_{IIC} and G_{IIIC} are their critical values which are 0.227J/mm^2, 1.105 J/mm^2 and 1.105J/mm^2 respectively.

Fig. 2 The area of impact damage
(a) Simulation (b) C-scan

3. The results of low-velocity impact

The impact energy was 16J. Fig. 2 shows that the magnitude of the simulated delamination area is 523.3mm^2, which is quite close to the actual damage area of 536.0mm^2 measured by C-scanning. The error between them is only 2.4%. So the simulation results of the impact step could be imported as the initial condition for the simulation of compression-after-impact.

4. The results of compression after impact

The simulation based on implicit solution algorithm might cause convergence problem if material degenerates during the simulation. Such problem may be avoided by the use of explicit solution algorithm. In order to implement the quasi-static analysis by explicit solution algorithm and save the computation time as much as possible, we desire that the velocity of displacement-controlled compressive loading be as quick as possible, and in the mean while the effect of inertia force is not significant.

Fig. 3 is the nominal compressive stress and displacement curves in different loading velocities. The curves are linear while the load is small. When the load is raised up to a certain extent, curves begin to deviate from linear and the slopes decrease, indicating that damage is growing in the laminate and the stiffness is declining. The highest points of the curves determine the CAI values under different loading velocities. Loading velocity has little effect on compressive failure displacement, but has marked effect on the CAI levels. Table 2 shows the CAI values for different loading velocities. When the velocity is less than 150mm/s, the CAI value tends to be stable.

Tab. 2 CAI values for different loading velocities

Loading velocity /(mm/s)	800	400	250	150	100	50
CAI /MPa	299.5	292.5	266.5	245.5	240.6	240.2

The influence of inertia force can be ignored if the ratio of kinetic energy to internal energy is below the level of 10%[6]. As shown in Fig. 4, if the loading velocity is slower than 100mm/s, the ratio will be below 10%, and the corresponding loading can be considered as quasi-static.

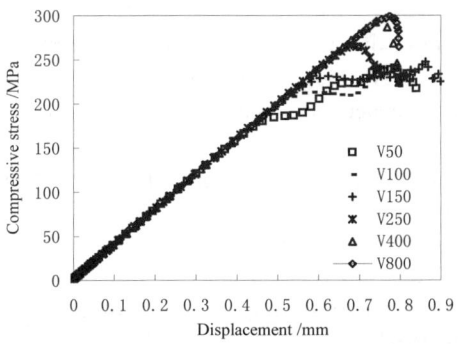
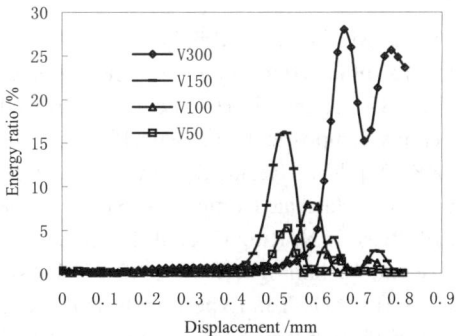

Fig. 3 CAI value in different loading velocity   Fig. 4 Ratios of kinetic energy to internal energy

The compressive load is taken mainly by the plies of 0°. Fig. 5 illustrates the growth of fiber compressive damages of 0° plies as well as the corresponding compressive stress levels at different moments of the compression, where Fig. 5(a) gives the impact damage, Fig 5(b) shows the damage pattern when the compressive loading reaches the maximum and Fig 5(c) demonstrates that the damage has spread to both sides of the laminate as the load descends from its peak value. In Fig. 5

the plies are numbered from upper (impact) surface to the bottom surface. We can see that the growth of fiber failure damage is faster in the plies close to the impact surface than those close to the bottom surface, because of the more severe in-plane impact damages and the deformation of the laminate along the direction of the impact.

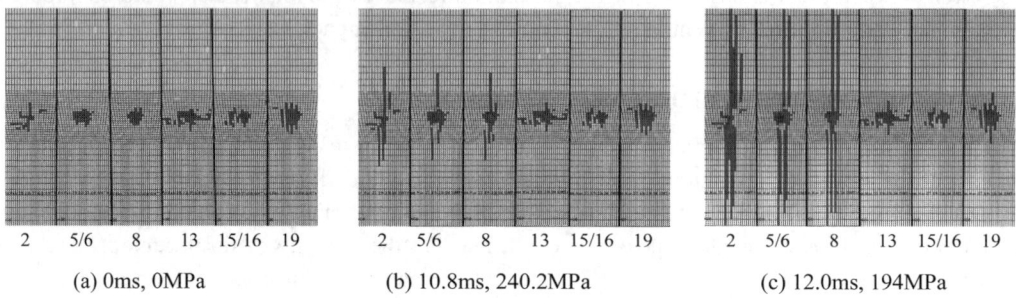

(a) 0ms, 0MPa (b) 10.8ms, 240.2MPa (c) 12.0ms, 194MPa

Fig. 5 Fiber compression damages in 0° plies

5. Conclusions

A whole-process analysis was conducted to predict the residual compressive strength of composite laminates subjected to low-velocity impact. Conclusions are drawn as follows:

(1) The cohesive elements are able to simulate the delamination damage induced during the low-velocity impact of the laminate.

(2) The predicted CAI value agrees well with the test result. The explicit finite element solver can be used to simulate the quasi-static compression-after-impact process, as long as the loading velocity is set slow enough.

(3) The final collapse of the laminate during the compression-after-impact process is mainly due to the compressive failure of 0° fibres and the growth of it from impact damage area and perpendicular to loading direction.

References

[1] de Freitas M, Silva A, Reis L. Numerical evaluation of failure mechanisms on composite specimens subjected to impact loading. Composites: Part B 2000; 31: 199–207

[2] Hitchen SA, Kemp RMJ. The effect of staking sequence on impact damage in a carbon fibre/epoxy composite. Composite 1995; 26: 207–14

[3] Lin ZY, Xu XW. Residual compressive strength of composite laminates after low-velocity impact. Acta Materiae Compositae Sinica, 2008, 25(1): 140-146

[4] Hou JP, Petrinic N, Ruiz C, et al. Prediction of impact damage in composite plates. Composites Science and Technology, 2000, 60(2): 273-281

[5] Huang CH, Lee YJ. Experiments and simulation of the static contact crush of composite laminated plates. Composite Structures, 2003, 61(3): 265-270

[6] ABAQUS Analysis User's Manual, Version6.8, Pawtucket, RI, USA: Karlsonand Sorensen, Inc. 2008

Deposition Process and Interface Properties of Electro-thermal Explosion Sprayed WC/Co Coating

Cui Xiu-fang[1,a], Jin Guo[1], Li Qing-fen[2,b]

[1] College of Material Science and Chemical Engineering, Harbin Engineering University,

[2] College of Mechanical and Electrical Engineering, Harbin Engineering University,

Harbin 150001, China

[a]cuixf97721@yahoo.com.cn, [b]qingfli@yahoo.com.cn

Keywords: deposition process; interface properties; electro-thermal explosion spraying; WC/Co coating

Abstract. The deposition process and interface properties of electro-thermal explosion sprayed WC/Co coating were studied by numerical simulation and experimental observation in this paper. The variety rule of the deposition particles / metal substrates interface temperature, the critical remelting condition of the substrates surface, and the interface removment speed of deposition particles were numerical simulated by finite element method. Results show that the remelting depth of the substrate increased with increasing spraying particle temperature and particle size. Structures and interface properties of the electro-thermal explosion sprayed WC/Co coating was analysed based on the numerical results. The morphologies, AES patterns, element distributions, TEM micrographs were experimentally observed and compared with the simulatin results. Which show that the finite element analysis is in good agreement with the experimental results, suggesting that the analysis in the present work is reliable.

Introduction

Cermet thermal spraying coatings are widely used in wear situations because they combine several advantages such as resistance to abrasion, erosion, corrosive atmospheres and high temperature [1]. WC/Co is one of the most useful cermet thermal spraying materials. But for the traditional thermal spraying technology, it is difficult to gain the high quality cermet coatings for high porosity and poor bonding of the coatings. Electro-thermal explosion directional spraying (EEDS) is a new technique [2] based on the explosion mechanism of the impulsive discharge for high voltage and quickly solidification mechanism of the droplets with high temperature. However, It is very difficult to observe the process of electro-thermal explosion spraying, where, the particles deposit on the surface of the metal substrates with extremely high speed in high-temperature and high-pressure condition[3]. Structures and properties of the coating mainly lie on the deposition process [4, 5]. It is therefore necessary to study the deposition process and interface properties of electro-thermal explosion sprayed coating.

In this paper, the deposition process and interface properties of electro-thermal explosion sprayed WC/Co coating were studied by numerical simulation and experimental observation.

Numerical Simulation

Model Description. The physical model of particles deposition process of electro-thermal explosion spraying was firstly built. Where, the substrate thickness (a) is 30 mm, and the deposition particles (b) are 2 μm and 20 μm respectively. The substrate is in the room temperature (20℃), and the deposition particles are supposed in 3000℃ and 4000℃ respectively.

The interface temperature of substrate / particles, T_i, is calculated according to the formula in reference [6] as follow:

$$\frac{T_{sub,l} - T_i}{T_i - T_{p,l}} = \left(\frac{\beta_p \rho_p C_{pp}}{\beta_{sub} \rho_{sub} C_{psub}}\right)^{0.5} \tag{1}$$

and $\quad T_i = \dfrac{T_{sub,I} + \alpha T_{P,I}}{1+\alpha}$ (2)

where, $\alpha = \left(\dfrac{\beta_p \rho_p C_{pp}}{\beta_{sub} \rho_{sub} C_{psub}} \right)^{0.5}$ (3)

In which, β is the thermal conductivity, ρ is the density and C_p is the specific heat. The physical property parameters used in calculation are shown in Table 1.

Table 1 Physical property parameters used in calculation

Material	P (kg/m³)	Melting point T_m(°C)	specific heat C_p(J/kg·k)	Heat conductivity B (W/m·°C)	Initial temperature T_I(°C)
WC/Co	14230	2870	295	45	4000 / 3000
substrate	7850	1510	461	52.3	20

From formula (3), the α for the WC/Co coating is calculated as:
$$\alpha = 1.834$$
Then, according to formula (2), we have

$T_i = 1948.5$°C (for $T_{WC,I} = 3000$°C) and $T_i = 2595.6$°C (for $T_{WC,I} = 4000$°C)

And that, the critical interface temperature is defined as $T_{WC,I,C}$, when the substrate temperature is 20°C, we have

$T_{WC,I,C} = 2322$°C (for $T_{45,I} = 20$°C)

Which shows that the critical interface temperature (2322°C) is lower than the melting point (2870°C) of the WC/Co coating material, and that the substrate surface may be melt easily when electro-thermal explosion directional spraying WC/Co impact the substrate surface.

Simulation of EEDS WC/Co Coatings Process. A finite element model was developed by using finite element code ANSYS (revision 8.0). The interface region of EEDS WC/Co coatings was modeled using the thermal-structural element PLANE 55 and the element size of 4 × 4 mm was adopted and found to produce suitable results.

Simulation results of EEDS WC/Co coatings are shown in Table 2. Temperature of the bottom surface of particle and upper surface of substrate versus time during EEDS WC/Co process is given in Fig. 1. The interface location of particle and substrate versus time during EEDS WC/Co process is given in Fig.2. Results show that the remelting depth of the substrate increased with increasing spraying particle temperature and particle size. And that remelting occured on the substrate surface in all the four situations, and the remelting phenomena is more obvious when the particle thickness is 20μm. We can also see that, the liquid state of the bottom surface of particle and upper surface of substrate coexisted, which will induce metallurgy combination when temperture reached 4000°C.

Table 2 Simulation results of EEDS WC/Co coatings

b (μm)	$T_{sub,I}$ (°C)	$t_{sub, melt initial}$ (μs)	$t_{sub, max. depth}$ (μs)	$t_{sub, hard set}$ (μs)	$t_{p, initial solidify}$ (μs)	$t_{p, hard set}$ (μs)	h_{max} (μm)
2	3000	0.02	0.11	0.27	0.01	0.20	0.26
	4000	0.01	0.33	0.83	0.09	2.0	0.82
20	3000	0.02	20.9	30.1	0.01	10.9	5.67
	4000	0.01	100.9	160.3	0.26	51.14	14.80

Cooling rates of the bottom surface of over-heated particles during EEDS process is given in Fig. 3. It can be seen that the cooling rate of the deposition particles was very high (about $10^{10} K/s$) at the beginning. In that case, the super-cooling degree of deposition particles will reach the critical value ΔT_N and nucleation takes place soon.

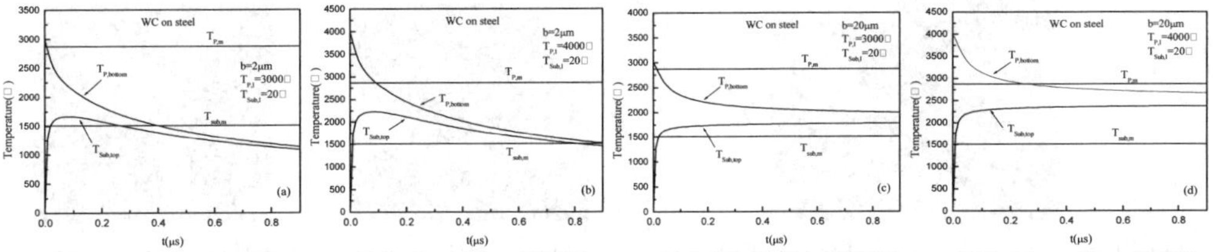

(a) b=2um, Tp,₁=3000°C (b) b=2um, Tp,₁=4000°C (c) b=20um, Tp,₁=3000°C (d) b=20um, Tp,₁=4000°C
Fig. 1 Surface temperature of particle and substrate versus time during EEDS WC/Co process

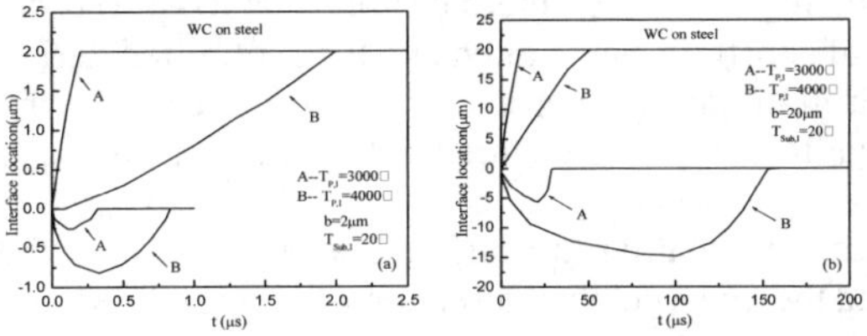

(a) b=2um   (b) b=20um   Fig. 3 The cooling rates of particles' bottom surface
Fig. 2 Interface location versus time during EEDS WC/Co process   during EEDS process

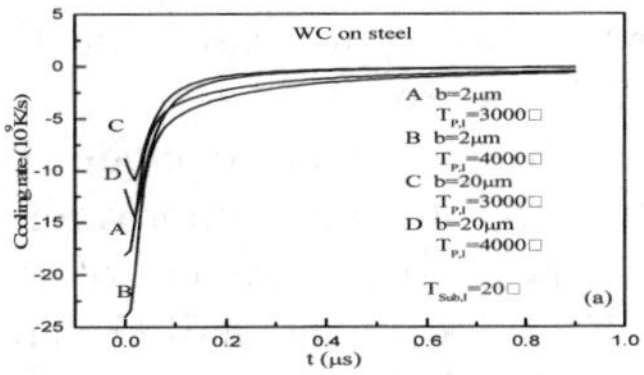

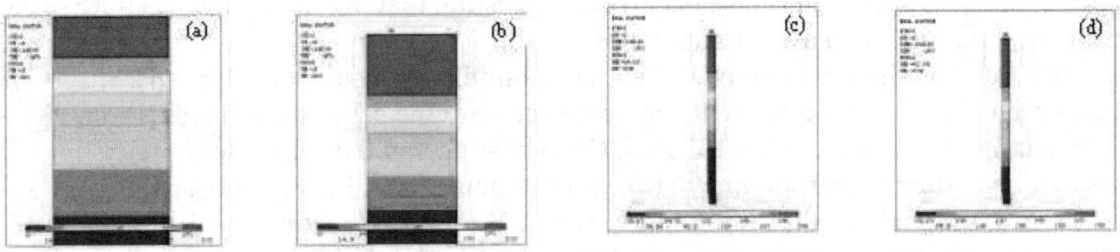

(a) b=2um Tp,₁=3000°C (b) b=2um, Tp,₁=4000°C (c) b=20um, Tp,₁=3000°C (d) b=20um, Tp,₁=4000°C
Fig. 4 The temperature distribution at the largest melt depth during EEDS WC/Co process

Fig. 4 gives the temperature distribution during EEDS WC/Co process when the melt depth of substrate reached the maxmum value, where, the temperature of deposition particles are still very high. However, the transfer heat from the particles to the molten substrate is lower than the one from the molten substrate to the inner part. In that case, the substrate no longer melts down and will solidify reversely.

Experimental Observation

The morphologies of EEDS WC/Co coatings in the bond area, AES patterns in the transition area of EEDS WC/Co coatings, element distributions of the EEDS WC/Co coating, TEM micrographs of the interface and substrate were experimentally observed and given in Fig. 5 - 8 respectively.

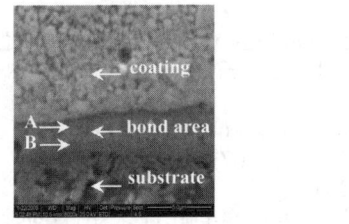

Fig.5 Bond area morphologies of EEDS WC/Co coatings

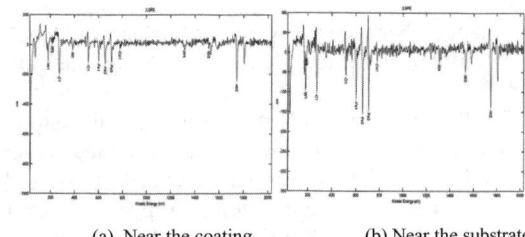

(a) Near the coating   (b) Near the substrate

Fig. 6 AES patterns in the transition area of EEDS WC/Co coatings

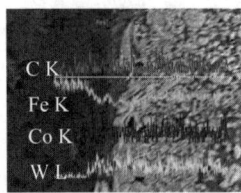 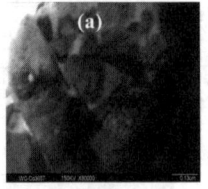 

Fig.7 Element distributions of the coating   (a) Bright field image   (b) Diffraction pattern
Fig. 8 TEM micrographs of the interface

From Fig. 5, it can be obviously seen that the coating / substrate interface area (about 4 μm) is the remelting bond area (a), indicating that the surface of substrate was molten and then solidified during the EEDS WC/Co process. And that, the columnar grain and isometric crystal are presented in the transition area (b). It may be concluded that the metallurgical bond occured in the interface area, and the bond strength is therefore improved.

Fig. 6 shows that the content of W and C near the coating are higher than the one near the substrate, whereas, the content of Fe is reverse. And there is a coexisted section of W, C and F where the content of W and C decreased slowly from the coating to the substrate (Fig. 7). It indicates that the elements mutual diffusion appered in the transition area during the EEDS WC/Co coating process. Besides, a peak value of C at the interface occurred, suggesting that the remelting substrate surface in the solidification period induced a segregation of C.

The above results are in good agreement with the simulation results (Table 1 and Fig. 1-2).

TEM was used to further investigate the microstructure of the coating bond area. From Fig. 8 (a), it can be seen that the bond between the EEDS WC/Co coating and the substrate is pretty fine. Fig. 8 (b) is the diffraction pattern corresponding to (a), where a new phase, Fe_2W_4C phase ([110]), has been detected. It evidently approves that the bond between the EEDS coatings and the substrate is a metallurgical bond. Which is also in good agreement with the simulation results.

Due to space limitation, more detail analysis is omitted.

Summary

The deposition process and interface properties of electro-thermal explosion sprayed WC/Co coating were studied by numerical simulation and experimental investigation. The finite element analysis is in good agreement with the experimental results,

Acknowledgement

This work is financially supported by the National Basic Research Program of China (973 Program) (No. 2011CB013404) and National Natural Science Foundation of China (Nos. 50905038, 50875053).

References

[1] H. Liao, B. Normand, C.Coddet: Surface and Coatings Technology Vol. 124 (2000) , p. 235
[2] F.Mizusako, H.Tamura, K.Horioka, Y.Harada: Surface and Coatings Technology Vol. 187(2004) , p. 257
[3] S.X. Hou, Z.D. Liu, D.Y. Liu: Surface and Coatings Technology Vol.205(2011), P.4562
[4] P. Fauchais, G. Montavon: Advances in Heat Transfer Vol. 40(2007), p. 205
[5] I.P. Jain, G. Agarwal: Surface Science Reports Vol. 66(2011), P. 77
[6] E.R.G. Eckert, R.J.Goldstein, W.E.Ibele, S.V.Patankar, T.W.Simon, P.J.Strykowski, K.K. Tamma,T.H.Kuehn,A.B.Cohen,J.V.R.Heberlein, J.H.Davidson, J.Bischof, F.Kulacki, U. Kortshagen: International Journal of Heat and Mass Transfer Vol. 42(1999), P. 2717

Grain-boundary Segregation of Phosphorus and Inter-granular Fracture Behavior under Low Tensile Stresses

Fu Yu-dong [1,a], Li Qing-fen [2,b], Sun Wei-xin

[1] College of Material Science and Chemical Engineering, Harbin Engineering University,

[2] College of Mechanical and Electrical Engineering, Harbin Engineering University,

Harbin 150001, China

[a]fuyudong@hrbeu.edu.cn, [b]qingfli@yahoo.com.cn

Keywords: non-equilibrium grain-boundary segregation (NGS); phosphorus; intergranular fracture; low tensile stress

Abstract. The present work is an effort to provide experimental results focusing on segregation behavior of phosphorus at grain boundary and the intergranular fracture behavio under low tensile stresses. AES (Auger electron spectroscopy) experiments and dynamic analyses on the non-equilibrium grain-boundary segregation (NGS) of phosphorus and the SEM photos of intergranular fracture in Auger specimens in 12Cr1MoV steel were carried out in this paper. The variation of phosphorus segregation level in grain boundary under different low tensile stresses and at different temperature were obtained. Results show that NGS of phosphorus occurred in the experimental steel while subjected to low tensile stresses. Maximum values of phosphorus segregation level were obtained at the critical times. SEM photos of intergranular fracture in Auger specimens of the test steel show that the intergranular fracture rate increased with increasing concentration of phosphorus. The intergranular fracture behavior is accordant with the segregation behavior of phosphorus at grain boundary.

Introduction

Fracture failure along grain boundaries occur often by the micro-segregation of embrittling impurity to the grain boundaries, which are envolved under the influence of heat and stress. Therefore, for a long time, the study of grain boundary micro-segregation has been being one of the most attractive fields in material science and engineering. The segregation of solute atoms in grain boundaries is classified into equilibrium and non-equilibrium segregation. McLean first proposed models for the thermodynamics and isothermal kinetics of equilibrium segregation [1]. Guttmann worked out the ternary segregation theory in 1975 [2]. Xu developed it further and proposed the non-equilibrium grain boundary segregation and co-segregation theory [3-6]. Non-equilibrium grain-boundary segregation （NGS） is a complex dynamic process and can be classified in three general categories: (1) thermally induced segregation; (2) stress-driven segregation; (3) neutron irradiation-induced segregation. In the three categories, oversaturated vacancies are all produced and then induced NGS.

The first studies on the effects of applied stress on intergranular segregation is Shinoda and Nakamura [7]. However, only a few studies about the stress-driven NGS on the condition of low stress were reported since then. A further study about this problem is therefore needed.

The present work is an effort to provide experimental results focusing on segregation behavior of phosphorus at grain boundary and the inter-granular fracture behavior under low tensile stresses.

Experimental

Experimental studies were carried out on a low alloy industrial 12Cr1MoV steel of composition listed in Table.1. Specimens of 16 mm x16 mm x250 mm were heat treated in a vacuum condition. The heat treatment includeds: solution treated at 1050°C/1h, water quenching to 20°C, tempering at 200°C/2h, air cooling, isothermal holding at 540℃ for 1800h.

Tension test specimens were made according to the standard of GB/T2039—1997. Tension tests were carried out on a RD2—3 creep testing machine at 540°C or 500°C and under constant loading of 20MPa, 30MPa, and 40 MPa respectively.

The NGS concentrations of phosphorus were carefully measured with Auger Electron Spectroscopy (AES). Cylindrical samples of Φ3.68mm with sharp notch were fractured by impact at a temperature of liquid nitrogen. On the freshly prepared fracture surfaces, 20 inter-granular facets were observed and subsequently analyzed. The phosphorus grain boundary concentration was calculated according to [8] and [9].

Table 1 Chemical Compositions of Steel 12Cr1MoV (wt %)

C	P	Mn	Si	Cr	Mo	V	S	Ni	Cu
0.14	0.019	0.62	0.22	1.05	0.27	0.17	0.015	0.02	0.008

Results and discussion

Peaks of AES patterns of steel 12Cr1MoV in 30Mpa tensile stress held at different temperature for different time are given in Fig. 1. The Auger electron peaks for P was measured and peak to peak heights of the element was normalized with respect to Fe.

The variation of phosphorus segregation levels in grain boundary held at 540°C and 500°C in 30Mpa tensile stressed condition were given in Fig. 2 and 3 respectively. The variation of phosphorus segregation levels in grain boundary under different tensile stress (20MPa, 30MPa, and 40 MPa respectively) were given in Fig. 4.

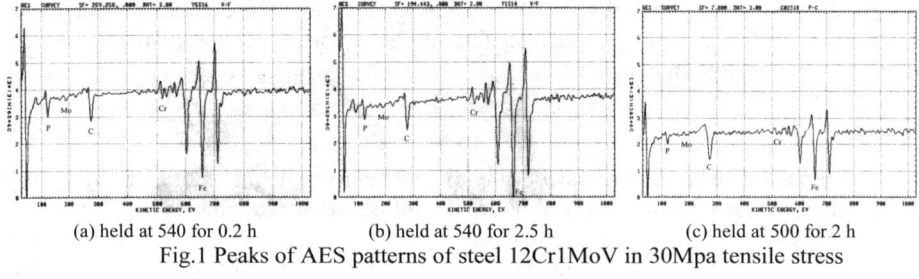

(a) held at 540 for 0.2 h   (b) held at 540 for 2.5 h   (c) held at 500 for 2 h
Fig.1 Peaks of AES patterns of steel 12Cr1MoV in 30Mpa tensile stress

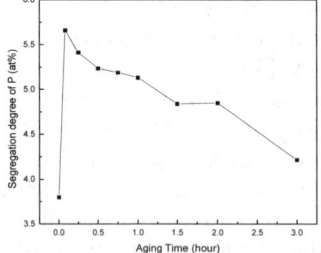 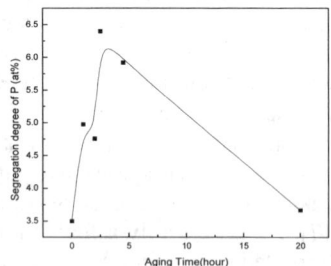

Fig.2 Variation of phosphorus segregation level   Fig.3 Variation of phosphorus segregation level
(held at 540°C in 30Mpa tensile stress)   (held at 500°C in 30Mpa tensile stress)

SEM photos of intergranular fracture in Auger specimens of 12Cr1MoV steel held at 540°C for different time in 30MPa tensile stress were shown in Fig. 5. SEM photos of intergranular fracture in Auger specimens under different tensile stress (held at 500°C for 2 hours) were shown in Fig. 6.

From Fig. 2, it is seen that from the start to about 0.1h, the degree of phosphorus segregation at grain boundaries increases rapidly with increasing isothermal holding time. When the holding time is longer than 0.1h, the degree of phosphorus segregation decreases continuously with increasing holding time, and the maximum value of segregation degree appears at about 0.1h. This phenomenon can be easily explained by the mechanism of diffusion of vacancy-solute atom complex for non-equilibrium grain-boundary segregation proposed by Xu in [4-6]. It is clear that the process of segregation to the grain-boundaries is dominant when the isothermal holding time is shorter than the critical time corresponding to the maximum value of segregation degree. When the holding time exceeds 0.1h, the process of de-segregation is dominant. We may therefore conclude that phosphorus does have the characteristic of non-equilibrium grain-boundary segregation, and the critical time for phosphorus non-equilibrium grain-boundary segregation is about 0.1h at 540°C in 30Mpa tensile stressed condition for this experimental steel.

It is also seen from Fig.2 that the NGS is a kinetic process. In the segregation process ($t < t_c$) complex diffusion to the grain boundary is dominant, a high gradient of complex concentration drives the complexes to diffuse to the grain boundary and the rate of phosphorus segregation is high due to a large diffusivity of the complex. In the desegregation process($t > t_c$), the diffusion of phosphorus atoms from grain boundaries to grain centers is dominant and the level of phosphorus segregation decreases with increasing stress aging time t. It is at the critical time a kinetic process in which the reverse phosphorus atoms diffusion from the grain boundaries to the center balances the complex diffusion to grain boundaries, and the concentration of phosphorus atoms in grain boundaries reaches a maximum.

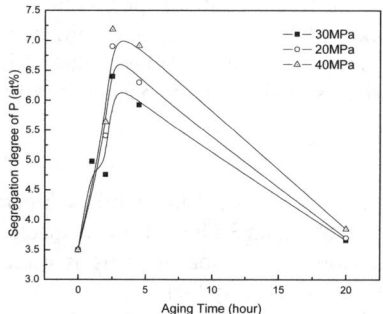

Fig.4 Variation of phosphorus segregation level under different tensile stress（held at 500°C）

Fig. 3 shows that the critical time is about 3.5h when specimens held at 500°C in 30Mpa tensile stressed condition for this steel.

From Fig.4, we see that the the degrees of phosphorus segregation at grain boundaries under different stresses are different, whereas, the critical time almost the same. Suggesting that the grain-boundary segregation level of phosphorus is affected by the tensile stress values, whereas, the critical time is not affected by them. We can also see that the grain boundary segregation level of phosphorus in 30Mpa tensile stress is lower than the ones in 20Mpa and 40Mpa.

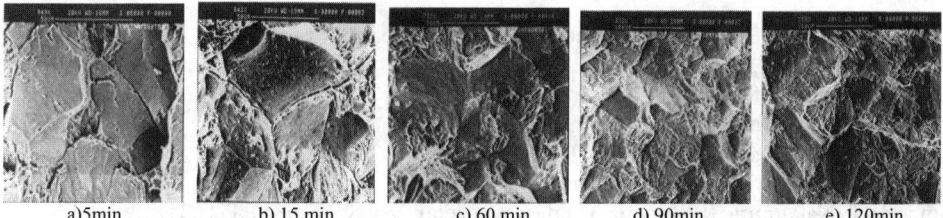

a)5min   b) 15 min   c) 60 min   d) 90min   e) 120min

Fig.5 SEM photos of intergranular fracture in Auger samples held at 540°C for different time in 30MPa tensile stress

From Fig.5, the SEM photos of intergranular fracture in Auger samples held at 540°C for different time in 30MPa tensile stress, it is seen that the intergranular fracture rate is high when the holding time near the critical time (5 min and 15 min), and that the intergranular fracture rate obviously decreased with increasing holding time then. It is accordant with the segregation behavior of phosphorus at grain boundary shown in Fig. 2. Where, the phosphorus atoms at grain boundarie reached the maximum value of segregation level at the rcitical time, and then decreased with incresing holding time. It is clear that the intergranular fracture rate increased with increasing concentration of phosphorus.

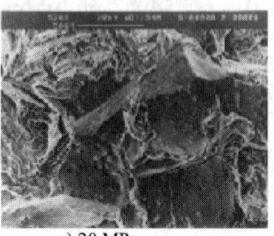 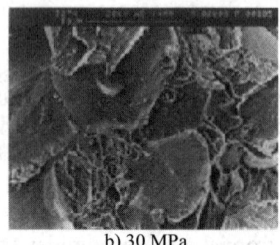 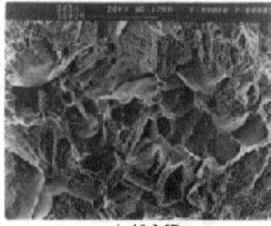

a) 20 MPa   b) 30 MPa   c) 40 MPa

Fig.6 SEM photos of intergranular fracture in Auger samples under different tensile stress (held at 500°C for 2 hours)

Fig. 6 shows that the intergranular fracture rate of specimens under 30 MPa tensile stress is lower than the ones under 20 MPa and 40 Mpa. It is also accordant with the segregation behavior of phosphorus at grain boundary shown in Fig. 3. Where, the grain boundary segregation level of phosphorus in 30Mpa tensile stress is lower than the ones in 20Mpa and 40Mpa.

Summary

1) NGS of phosphorus occurred in steel 12Cr1MoV while subjected to low tensile stresses, and the critical time is about 0.1h at 540°C, and 3.5h at 500°C respectively in 30Mpa tensile stresse.
2) The grain-boundary segregation level of phosphorus is affected by the tensile stress values, whereas, the critical time is not affected by them.
3) SEM photos of intergranular fracture in Auger specimens of the test steel show that the intergranular fracture rate increased with increasing concentration of phosphorus.
4) The intergranular fracture behavior is in accordance with the segregation behavior of phosphorus at grain boundary.

Acknowledgement

The work is financially supported by the National Basic Research Program of China (973 Program, No. 2011CB013404).

References

[1] D. McLean, Grain Boundaries in Metals (Oxford University Press, Amen House, London; 1957).
[2] M. Guttmann, Surf. Sci., **53** (1975) 213
[3] T. D. Xu, J. Mater Sci., **22** (1987) 337.
[4] T.D. Xu, Acta Metall. 1989: 37: 2499.
[5] T. D. Xu, Scripta Materialia, **37** (1997) 1643
[6] T. D. Xu, Progress in Materials Science, **49** (2004) 109
[7] Shinoda T, Nakamura T. Shinoda T, ed. Acta Metal. Vol. 29 (1981), P 1631
[8] L.E. Davis, N.C. McDonald, P.W. Palmberg, G.E. Riach and R.E. Weber. Handbook of Auger Electron Spectroscopy. 2nd edn. Minnesota: Phys. Electronics Industries Press, 1976.
[9] Eds. D. Briggs and M.P. Seah, Practical Surface Analysis. Chichester: Wiley Press, 1990.

Effect of the Neodymium Content on Mechanical Properties of the Electro-brush Plated Nano-Al$_2$O$_3$/Ni Composite Coating

Jin Guo[1,a], Cui Xiu-fang[1], Liu Er-bao[1], Li Qing-fen[2,b]

[1] College of Material Science and Chemical Engineering, Harbin Engineering University,

[2] College of Mechanical and Electrical Engineering, Harbin Engineering University,

Harbin 150001, China

[a]jg97721@yahoo.com.cn, [b]qingfli@yahoo.com.cn

Keywords: Neodymium; mechanical properties; electro-brush plated nano-Al$_2$O$_3$/Ni composite coating

Abstract. The effect of the neodymium content on mechanical properties of the electro-brush plated nano-Al$_2$O$_3$/Ni composite coating was investigated in this paper. The microstructure and phase structure were studied with scanning electron microscope (SEM) and X-ray diffraction (XRD). The hardness and abrasion properties of several coatings with different neodymium content were studied by nano-indentation test and friction / wear experiment. Results show that the coatings are much finer and more compact when the neodymium was added, and the hardness and abrasion property of the coatings with neodymium were improved obviously. Besides, the small cracks conduced by the upgrowth stress in the coatings were ameliorated when the rare earth neodymium was added. The improvement mechanism was further discussed.

Introduction

The composite plating coatings are widely used in many industrial fields due to their excellent properties, and nano-particle materials began to be used for many composite plating coatings [1, 2]. The rare earth elements have been used in manufacture of function materials for their special 4f electron structure and excellent physical and chemical properties [3]. Previous work pointed out that it is effective to advance the coating properties by adding the rare earths into the plating electrolyte [4, 5]. However, reports about the application of neodymium are not available yet so far.

To the author's knowledge it is the first time the rare earth, neodymium, was used to prepare the electro-brush plated nano-Al$_2$O$_3$/Ni composite coating, and the effect of different content of neodymium on mechanical properties of the coating was investigated in this paper.

Experimental

The chemical composition of steel GCr15 used in the present investigation is shown in Table 1. Test specimen (Φ18 × 8 mm) surface were successively polished by using waterproof abrasive paper 80#, 320#, 1000# and 2000#, and then immersed in acetone solution and cleaned in the ultrasonic cleaner for 5 minutes. After that, specimens were electr-cleaned and activated. Electro-brush was then carried out at ambient temperature (about 25 °C) with a velocity of 6 m/min~12 m/min for 5 min. The nano-Al$_2$O$_3$/Ni composite coatings with different neodymium content were made by adding different content of NdCl$_3$ into the plating electrolyte.

Surface micrographs and microstructures of the test specimens were observed by scanning electron microscopy (SEM) FEI Quant200 and X-ray diffraction (XRD) X-Pert-Pro. The microhardness of coating was examined by HVS 1000 sclerometer with load fixed at 25 g and load-time fixed at 10 s and the average value of five test data was adopted. Nano-hardness of coating was examined by the nano-mechanical testing system from Hysitron. Friction/wear test was carried out in a ball-on-disk friction/wear tester to investigate the wear resistance of the coating in dry friction condition at room temperature. The load was kept at 1000 g with rotate speed at 240 r/min during the test. The wear time was 30 min.

Table 1 Chemical Composition of steel GCr15 [wt %]

C	Si	Mn	S	P	Cr
0.95-1.05	0.15-0.35	0.2-0.4	≤0.02	≤0.027	1.30-1.65

Results and Discussion

The SEM photographs of specimens with different coatings are given in Fig. 1, where, the nano-Al_2O_3/Ni composite coatings with different neodymium content are shown in (a-d), and the common quick Ni-base (QN) plating without $NdCl_3$ is in (e). It can be seen that coatings are all showing the tipical cluster crystal with cauliflower shape, where, every cluster crystal is composed of many small crystal-cell, and that the crystal grains are finer, boundings are more compact and the delamination of plating coatings are improved when different content of $NdCl_3$ was added.

Fig. 1 (a) shows that the cluster crystal is small and the crystal grain bounding is compact when the content of $NdCl_3$ is 0.1g/l, and Fig. 1 (b) shows that the cluster crystal is smaller and the crystal grain bounding is more compact when the content of $NdCl_3$ is 0.3g/l. However, when the content of $NdCl_3$ is 0.5g/l, the cluster crystal is bigger and the delamination of plating coating can be seen (Fig. 1 c). When the content of $NdCl_3$ is 0.7g/l, the cluster crystal is bigger, the crystal grain bounding is incompact and the delamination of plating coating is more obvious (Fig. 1 d).

Besides, it can be seen that small cracks conduced by the upgrowth stress in the coating reduced much more when the neodymium was added. It is concluded that coating with neodymium were improved obviously

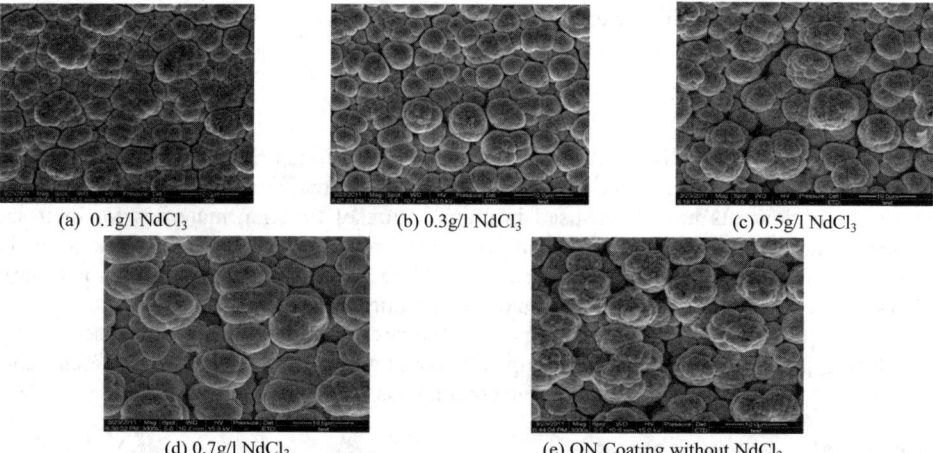

(a) 0.1g/l $NdCl_3$ (b) 0.3g/l $NdCl_3$ (c) 0.5g/l $NdCl_3$

(d) 0.7g/l $NdCl_3$ (e) QN Coating without $NdCl_3$

Fig. 1 SEM photographs of coatings with different $NdCl_3$ content in electrolyte

Fig.2 gives the XRD patterns of specimens with different coating. It can be seen that the diffracton peaks of the four nano-Al_2O_3/Ni composite coatings with different neodymium content are (111), (200) and (220), the three most intensive peaks of Ni, indicating that the substrate is completely covered with the nano-Al_2O_3/Ni composite coatings.

The grain size can be calculated according to the Scherrer formula:

$$D = 0.89\lambda/(\beta cos\theta) \qquad (1)$$

Where, D is the grain size, λ is the wavelength of X-ray, β is the full width at half maximum (FWHM) of the differaction peak, θ is the differaction angle.

The average grain size of the common quick Ni-base plating is 9.6 nm, the nano-Al_2O_3/Ni composite coatings with different neodymium content are 10.57 *nm*, 8.87 *nm*, 11.7 *nm, and* 9.77 *nm*, respectively. It can be seen from the calculation results that the minimum grain size was obtained when the content of $NdCl_3$ in electrolyte is 0.3g/l. It is in agreement with the the results of Fig. 1.

The microhardness of specimens with different coating is shown in Fig. 3. It is seen that the microhardness of specimens with different neodymium content are all higher than the one of specimen without neodymium content. The microhardness incresed at first, and then decreased with increasing neodymium content, and that, the maximum microhardness value, 771.9 HV, was obtained when the content of $NdCl_3$ is 0.3g/l.

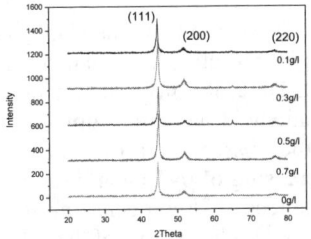 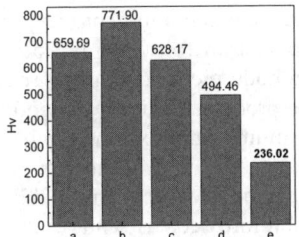

(a) 0.1g/l $NdCl_3$ (b) 0.3g/l $NdCl_3$ (c) 0.5g/l $NdCl_3$
(d) 0.7g/l $NdCl_3$ (e) QN Coating without $NdCl_3$

Fig. 2 XRD of specimens with different coating   Fig. 3 Microhardness of specimens with different coating

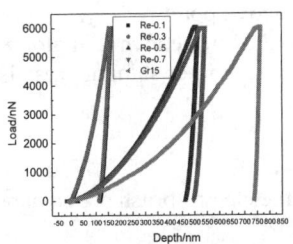 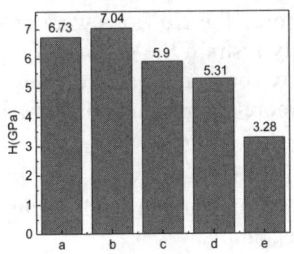

(a) 0.1g/l $NdCl_3$ (b) 0.3g/l $NdCl_3$ (c) 0.5g/l $NdCl_3$ (d) 0.7g/l $NdCl_3$ (e) QN

Fig.4 Nano-hardness of specimens with different coating

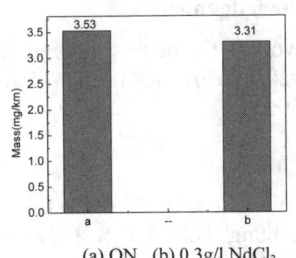

(a) QN   (b) 0.3g/l $NdCl_3$ (a) QN (b) 0.3g/l $NdCl_3$
Fig. 5 SEM of coatings section   Fig. 6 Wear resistance of different coating

Fig.4 gives the nano-hardness of specimens with different coatings. These results are in agreement with the ones in Fig. 3. Further indicating that the coating hardness can be effectively improved by adding the neodymium into the Ni-base plating coatings, and when the content of $NdCl_3$ is 0.3g/l, the hardness reached the maximum value.

Fig. 5 gives the SEM of quick Ni-base (QN) plating section (a) and the nano-Al_2O_3/Ni composite coatings with 0.3g/l $NdCl_3$ (b). It is seen that the thickness of QN plating layer is about 20 μm and the one of 0.3g/l $NdCl_3$ coating is 27μm. Indicating that both the plating layer thickness and the plating aggradation rate increased obviously when rare earth was added. Besides, there are some microcracks in both coatings, however, the situation was ameliorate more when the neodymium was added.

The wear resistances of QN plating and nano-Al_2O_3/Ni composite coating with 0.3g/l $NdCl_3$ are given in Fig. 6. Which shows that the wear resistance property is improved when the neodymium was added. The wear loss of nano-Al_2O_3/Ni composite coating with 0.3g/l $NdCl_3$ decreased about 8% compare with the one of QN plating without $NdCl_3$.

From the above results, it is clear that the mechanical properties of the electro-brush plated nano-Al_2O_3/Ni composite coating were remarkably improved by adding different content of neodymium. And that the best result was obtained when 0.3g/l $NdCl_3$ was added into the electrolyte.

The improvement mechanism is analyzed as follow:

The rare earth, neodymium, has its particularity that once it was aded into the plating electrolyte, it may decrease the transform energy from the metal complex ion to active intermediate, and thereby increase the cathode electrochemistry reaction revert speed. Here, the neodymium play a role of activator in the process. Besides, the neodymium and organic acid metal chelator may come into being neodymium-complex ions, which may impede the dissociation of organic acid and consequently decrease the denseness of H^+, result in the decreasing of hydrogen dissolving out and increasing of cathod electric current efficiency. The phenomena of hydrogen embrittlement and porosity will therefore decrease and result in the low porocity and high compact of the coating. In that case, the coating hardness can be effectively improved.

Besides, a proper concent of neodymium in the plating electrolyte may induce proper electrodeposit rate, which may refine the crystal grain and make the coating microstructure more compact, bring on a better coating hardness.

Furthermore, the Nd^+ ions were prone to adsorb on the active points of crystal growth, which may effectively restrain the growth of crystal, and the compounds of neodymium along the crystal boundaries may come into being the new nucleating particle, where finer shining crystals prodused. Conducing a more compact and higher hardness coating.

Summary

1) It is the first time that the neodymium was used to prepare the electro-brush plated nano-Al_2O_3/Ni composite coating.
2) The mechanical properties of the electro-brush plated nano-Al_2O_3/Ni composite coating were remarkably improved by adding different content of neodymium. And the best result was obtained when 0.3g/l $NdCl_3$ was added into the plating electrolyte.

Acknowledgement

This work is financially supported by the National Basic Research Program of China (973 Program) (No. 2011CB013404) and National Natural Science Foundation of China (Nos. 50905038, 50875053).

References

[1] W.W. Chen, W. Gao, Y.D. He: Surface and Coatings Technology Vol. 204(2010), P. 2493
[2] B. Jiang, B.S. Xu, S.Y. Dong, Y. Yi, P.D. Ding: Surface and Coatings Technology Vol. 202 (2007), P. 447
[3] R. L. Satet, M. J. Hoffmann, R.M. Cannon: Materials Science and Engineering: A Vol. 422 (2006), P. 66
[4] J.L. Tang, Z.Z. Han, Y.Zuo, Y.M. Tang: Applied Surface Science Vol. 257(2011), P.2806
[5] B.L. Han, X.C. Lu: Surface and Coatings Technology Vol. 202(2008), P. 3251

FE Simulation Methods to Predict Welding Residual Stresses

LiLi [1, a], Khurram Asifa [2, b], Hong Li [1, c], Shehzad Khurram [3, d]

[1] College of Materials Science Engineering, Harbin Engineering University, Harbin 150001, China

[2] College of Aerospace and Civil Engineering, Harbin Engineering University, Harbin 150001, China

[3] College of Shipbuilding Engineering, Harbin Engineering University, Harbin 150001, China

[a] lili_heu@hrbeu.edu.cn, [b] Asifa_khuram@hotmail.com, [c] lihong@hrbeu.edu.cn, [d] Khurram_1977@hotmail.com

Keywords: FE simulation, welding, residual stresses

Abstract: The purpose of this review paper is to summarize the novel and advanced finite element simulation techniques devised in recent years to predict the welding residual stresses. The finite-element (FE) simulation methods are commonly used to predict the thermal, material, and mechanical effects of welding because of their efficiency and flexibility. This study presents an overview of the research, conducted to more accurately simulate the welding process. Some recommendations and simplification techniques presented by researchers are also discussed in this study which provides a foundation for further development in this field.

Introduction

Welding is a complex process includes thermal mechanical and metallurgical coupling. Therefore, computational support is necessary for prediction of distortions and residual stresses. The soft computing methods provide a substitute for modelling; optimization and control of weld quality without any mathematical model [1]. The primary challenge with computational techniques is to model the complex and at times indeterminate nature of the welding process, in a simple and transparent manner [2]. FE simulation model for welding generally involves a moving heat source, material deposition, non-linear temperature dependent material behaviour, metal plasticity and elasticity, calculations of the temperature fields and the residual stress distributions. More authentic simulation models include other factors such as; boundary conditions, filler material and pass sequence in multi-pass welding are complex and require greater computational power and CPU time. This study reviews some advanced FE simulation methods and simplification techniques.

Simulation Models

Thermo-Elastic-Plastic FE Model. Numerical simulation has been playing a significant role in manufacturing recently. The coupled and un-coupled thermo-elastic-plastic FE models are analysed using various software, in the mechanical process as during welding [3].

Coupled FE Simulation Model. A coupled thermo-mechanical simulation model comprises a moving heat source, material deposition, temperature-dependent material properties, phase-change phenomena defined in terms of latent heat release and shrinkage effects, metal plasticity and elasticity, transient heat transfer and mechanical analysis. ANSYS, ABAQUS, SYSWELD, WELD PLANNER, SOLIDWORKS, SIMWELD, VRWELD and ADINA system is some modern welding simulation software [4, 13].

Un-Coupled FE Simulation Model. In un-coupled FE simulation model, the transient thermal analysis independently performed to estimate temperature distribution at each time step. The mechanical analysis is based on the temperature history of the thermic analysis [14]. Uncoupled FE methods evaluate effects of heat transfer during welding, heat treatment and modeling of residual stress and distortion [15, 16].

Other Simulation Model. Other simulation models practiced in recent years as an efficient predictive tool include genetic algorithm method, inherent strain method, cellular automata method, finite volume method, volumetric method, improved energy-based method, mesh less local Petrov-Galerkin (MLPG) method, control volume approach, smoothed particle hydrodynamics (SPH) method and sensitivity analysis methodology [17,25].

Heat Source Model

The distribution of heat input can be classified as superficial or surface and volumetric [26]. Welding heat source is modeled as surface heat source following Gaussian distribution heat flux [27,28]. Goldak's et al. proposed two semi-ellipsoidal model, Nguyen et al. proposed an analytical solution of a single-ellipsoidal model and V.D. Fachinotti's numerically computed solutions for double-ellipsoidal and double-elliptical source models as an extension of Nguyen et al. numerical model [29,31]. A. Ghosh et al. proposed double central coinciding heat source to simulate submerged arc welding [31]. Y.H. Choi investigated temperature distributions of the plates using surface heat source suggested by Rykalin [32]. An analytical approach of the discretely distributed point heat source model is an intermediate stage between the experiments and the numerical model [33].

Temperature Dependent Material Properties Model

Using material mechanical properties at room temperature, except for the yield stress, gives reasonable predictions for the transient temperature fields, residual stress and distortion [34]. Continuous cooling transformation (CCT) diagrams established from transformation data contained within the SYSWELD, and Kirkaldy's reaction rate equations are used for welding simulations [35,37]. Welding volumetric changes and material thermal properties, mainly enthalpy, which are very important due to significant material phase changes, are incorporated into the FE model [38, 39].

Restrains and Boundary Conditions Model

The geometry of weld pool and distribution of residual stresses are affected by heating temperature prior to welding, and application of fixture and constraints [40, 42].

Material Deposition Model

Metal Deposition and Pass Sequence. Welding speed and geometric shape of weld groove affect the weld bead and penetration. Element birth and death technique is used to simulate mass addition from the filler metal into the weld pool with the heat source movement [43,45]. Accurately predicting welding residual stresses and developing a convenient welding sequence for a weld system is an appropriate task [46]. The simulation model consists of sequentially coupled thermal and structural analyses. In multi-pass welding simulation, using element birth and death technique, effect of welding lag, inter-pass temperature and welding order on welding residual stresses are studied [47,48].

Simplification Techniques

Some suppositions are used in the simulation models to reduce computation, i.e. block dumping technique, reducing the model from three-dimensional to two dimensional, indirect method to calculate the constructive field, technique of sub modeling and substructure, replacing 3D solid elements with 3D shell elements and macro weld deposit methodology [51, 57]. Other methods to improve efficiency of the simulation process include segmented moving heat source, adaptive mesh technique, similitude principles, special transition element, material modeling at high temperature and parallel computation, etc. [58].

Conclusion

The summary presented is a broad representation of different finite element models. It provides ample information to the readers seeking novel research ideas in analysis and simulation of welding.

Refferences

[1] K. Pal and K. Surjya: Int. J. of Manf. Research., Vol. 6(2011), p.15

[2] P. Mollicone, et al.: J. of Mat. Proc. Tech. Vol. 176 (2006), p.77

[3] W. Yang, L. Yu and Z. Xueyuan: Trans. of JWRI, Vol.39 (2010), p.82

[4] D. Stamenković and I. Vasović: Sci. Tech. Review, Vol. LIX, No.1 (2009), p.57

[5] G.M. Newaz, R. Patwa and H. Herfurth: J. Eng. Mater. Tech. Vol. 132 (2010), p.1

[6] A.S. Aloraiera and S. Joshib: Mat. Sc. and Engg. Vol. 534 (2012), p. 13

[7] M.S. Sulaiman, et al.: J. of Mech. Sci. and Tech. Vol. 25 (2011), p.2641

[8] G.S. Brar and R. Kumar: Int. J. of Engg. Vol. 2 (2010), p. 271

[9] M. Chiumenti, et al.: Com. Meth. in App. Mech. and Engg., Vol. 199 (2010), P. 2343

[10] M. Perić, D. Stamenković and V. Milković: Sci. Techn. Rev. Vol. 60 (2010), p. 22

[11] W. Bleck, et al.: Adv. Engg. Mat. Vol. 12 (2010), p. 147

[12] J. Goldak: Comparison of Residual Stresses, Goldak Technology Publisher (2010).

[13] P. Ferro and F. Bonollo: Int. J. of Com. Mat. Sci. & Surf. Engg., Vol. 3 (2010), p. 114

[14] A. Anca, A. Cardona and J. Risso: App. Math. Mod., Vol 35 (2011), p. 688

[15] A.A. Deshpande, et al.: J. of Mat. Dgn. and App., Vol. 225 (2011), p. 1

[16] A.R. Michaleris: Sci. & Tech. of Weld. & Joining, Vol. 16 (2011), p. 215

[17] O.E. Canyurt: Int. J. of Mech. Sci. Vol. 47 (2005), p.1249

[18] K. Asifa, H. Li, L. Li and S. Khurram, in proc. 4th IEEE ICCSIT. 2011, p.635

[19] J. Xu, L. Chen, J. Wang and C. Ni: Int. J. Adv. Manuf. Technol., Vol. 35 (2008), p. 987

[20] W. Zhong-Hui and W. Yu: Elec. Weld. Mach. Vol. 40 (2010), p. 24

[21] Y. Susumu, et. al.: Trans. of JWRI, Vol.39 (2010), p. 37

[22] M. Shibahara and S.N. Atluri: Int. J. of Therm. Sci. Vol. 225 (2011), p.1

[23] M. Sunar, B.S. Yilbas and K. Boran: J. of Mat. Proc.Tech.Vol.172 (2006), p.123

[24] R. Das and P.W. Cleary: 16th Australasian Fluid Mech. Conf. (2007), p. 253

[25] O. Asserin, et al.: F.E in Anal. & Dgn. Vol. 47 (2011), p.1004

[26] Fazal and M. Arif: Adv. Mat. Research Vol. 83 (2009), p.858

[27] K. Asifa, H. Li and L. Li: Adv. Mat. Res. Vol. 328 (2011), p. 492

[28] K.H. Tseng and K.L. Chen: key. Engg. Mat. Vol 480-481(2011), p.459

[29] D. Gery, H. Long and P. Maropoulos: J. of Mat. Proc. Tech. Vol.167 (2005), p. 393

[30] V.D. Fachinotti, et al.: Int. J. for Num. Meth. in Biomed. Engg. Vol. 27 (2011), p. 595

[31] A. Ghosh and S. Chattopadhyaya: 2nd Int. Conf. on Mech. & Elect. Tech. (2010), p. 733

[32] Y.H. Choi, Y.W. Lee, K. Choi and D.H. Doh: J. of Therm. Sci. Vol. 21 (2012), p.82
[33] A.S. Azar, K. Sigmund and M. Akselsen: Comp. Mat. Sci. Vol. 54 (2012), p. 176
[34] X.K. Zhu and Y.J. Chao: Comp. and Struct. Vol. 80 (2002), p. 967
[35] C. Heinze, C. Schwenk, M. Rethmeier and J. Caron: Fron. Mat. Sci. Vol. 5 (2011), p.168
[36] J. Caron et al.: Weld. J. Vol. 8 (2010), p.151
[37] H. Dai, Francis and J. A. Withers and P. J.: Mat. Sci. & Tech. Vol. 26 (2010), p. 940
[38] C. Lee and K. Chang: Comp. & Struct. Vol. 89 (2011), p. 256
[39] G.A. Moraitis and G.N. Labeas: Int. J. of Press. Vess. and Piping Vol. 86 (2009), p. 133
[40] Y. Danisc, E. Lacoste and C. Arvieu: J. of Mat. Process. Tech. Vol. 210 (2010), p. 2053
[41] R. Kohandehghan and S. Serajzadeh: J. of Mat. Engg. & Perf. (2011), p. 1
[42] R. Kohandehghan and S. Serajzadeh: J. of Manuf. Proc. Vol. 13 (2011), p. 96
[43] A. Traidia, F. Roger and E. Guyot: Int. J. of Ther. Sci. Vol. 49 (2010), p.1197
[44] F. Kong and R. Kovacevic: J. of Mat. Proc. Tech. Vol. 210 (2010), p. 941
[45] J.J. Díaz, P.M. Rodríguez and P. J. Nieto: App. Therm. Eng. Vol. 30(2010), p. 2448
[46] S. Ziaee, M.H. Kadivar and K. Jafarpur: Ira. J. of Sci. & Tech. Vol. 32, p. 367
[47] E. Armentani, et al.: J. of Ach. in Mat. & Manu. Vol. 20 (2007), p. 319
[48] L. Gannon, Y. Liu, N. Pegg and M. Smith: Mar. Struct. Vol. 23 (2010), p. 385
[49] A. Pahkamaa, L. Karlsson and J. Pavasson: Int. Dsgn. Engg. Tech. Vol. 3 (2010), p. 81
[50] I.F.Z. Fanous, M.Y.A. Younan and A.S. Wifi: J. of Press. Vess. Tech. Vol. (2005), p.487
[51] H.S. Bang, H.S. Bang, Y. C. Kim and I. H. Oh: Mat. & Dsgn. Vol. 32 (2011), p. 2328
[52] M. Iranmanesh and A.R. Darvazi: Asian J. of App. Sc., Vol. 1 (2008), p. 70
[53] Z. Xueyuan, L. Yu and W. Yang: Trans. of JWRI Vol. 39 (2010), p.73
[54] Z. Zeng, L. Wang, P. Du and X. Li: Comp. Mat. Sci. Vol. 49 (2010), p. 535
[55] Y. Gao and F. Zhang: 2nd Inter. Conf. on AIMSEC (2011), p. 3988
[56] P. Mrvar, J. Medved, and S. Kastelic: Weld. J. Vol. 90 (2011), p. 148
[57] Y.J. SUN and Y. Zang: 2nd Int. Conf. on Info. & Comp. Sci. (2009), p.92
[58] A. Barroso, et al.: Mat. & Dgn. Vol. 31 (2010), p.1338

Study on the Pin-load Distribution of Multiple-bolted Composite to Metal Joints

Liu Xiangdong, Li Yazhi, Yao zhenhua, Shu Huai

School of Aeronautics, Northwestern Polytechnical University, Xi'an 710072, China

18991143745@163.com, yazhi.li@nwpu.edu.cn, yao_21g@163.com, shu2035153@yahoo.com.cn

Key words: multi-bolted joint; fastener load distribution; experiment; numerical analysis

Abstract: The experiment and finite element analysis were made to determine the pin-load distribution of multiple countersunk bolted single-lap joints. In the experiment, the pin-load fractions were evaluated indirectly by the lap-sheet surface strains collected from a few rows of strain gages. The joint strains and pin-load distribution were also obtained directly in the finite element analysis. The calculated strains correlated well with the experiment. Nevertheless, the pin-load fraction results of the both techniques are quite different. The further analysis revealed that the procedure of transforming the measured strains into pin loads is not reliable, since the intrinsic additional bending had not been taken into account. Therefore the appropriate way to determine the pin-load distribution should be the numerical analysis validated by the strain measurement. The another attempt showed that the pin-load distribution can be evaluated by the finite element modeling of two-dimensional shells and beams as well with satisfied accuracy.

Introduction

Fastener joints are extensively used in engineering structures. The significant input in predicting the load-carry capacity of multiple-bolted joints is the pin-load distribution which could be evaluated by the calculations (analytically [1,2] and numerically [5,6]) as well as the experiments[3,4]. The usual way of measuring the pin-load distribution is to measure the lap-sheet surface strain distribution and transform it into the pin loads. In the present work, such means of measurement is to be evaluated with our experiments and numerical analysis.

1. Experiment

1.1. Specimen

The test specimens are the multiple-bolted metal to composite single-lap joints being compose of the TC4 titanium alloy sheet and CCF300/QY8911 composite laminate connected by 3 (row)×3 (column) TC4 countersunk bolts. The specimen geometry is show in Fig. 1. The countersinks are located on the laminate sheet. The metallic sheet has the same nominal thickness of 6mm as the laminate, and the latter has the ply stacking-sequence as $[45/0/-45/90/45/0/-45/0]_{3s}$. The glass/epoxy doublers are bonded to both griping ends of the specimens to eliminate the loading eccentricity. There are three specimens which are numbered from the S-01 to S-03.

The metallic material constants: $E = 115 \text{GPa}$, Poisson's ratio $v = 0.31$. The composite constants:

$E_1 = 133 \text{GPa}$, $E_2 = E_3 = 9.9 \text{GPa}$, $G_{12} = G_{13} = 6.67 \text{GPa}$, $G_{23} = 3.9 \text{GPa}$, $v_{12} = v_{23} = v_{13} = 0.27$.

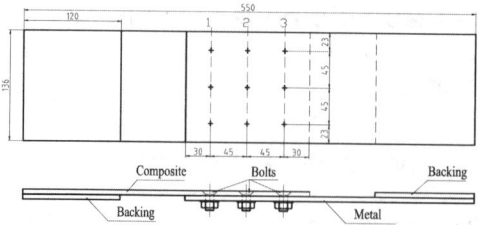
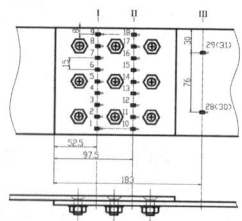

Fig. 1 Test specimens

Fig. 2 The layout of the strain gages layout

1.2. Test method

The pin-load distribution was determined by using strain gage measurement. Three rows of strain gages were bonded to the three sections on the metallic sheet. Two rows are in the outer surface of overlap area, and another row is in both faces of the sheet closing to the gripping end, as shown in Fig. 2. The specimen was gripped at both ends, ensuring the loading-axis through the centre line of the specimen, and the anti-bending fixture, shown in Fig. 3, was attached to the specimen to minimize the overall bending deflection.

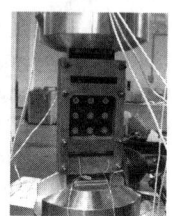

Fig. 3 The testing machine griping and the anti-bending fixture of the specimen

1.3. Calculation of pin-load proportion

For a given load, denoting the mean values of strains measured by the strain gages at rows I, II and III of the metallic plates as $\overline{\varepsilon}_I$, $\overline{\varepsilon}_{II}$ and $\overline{\varepsilon}_{III}$ respectively, we can determine the pin-load fractions of each row of bolts.

$$r_1 = \overline{\varepsilon}_I / \overline{\varepsilon}_{III} \times 100\% \quad , \quad r_2 = (\overline{\varepsilon}_{II} - \overline{\varepsilon}_I) / \overline{\varepsilon}_{III} \times 100\% \quad , \quad r_3 = (\overline{\varepsilon}_{III} - \overline{\varepsilon}_{II}) / \overline{\varepsilon}_{III} \times 100\% \quad (1)$$

2. Finite element modeling

With the ANSYS software, strains and pin-load distribution of the specimens were calculated by two-dimensional (2-D) and three-dimensional (3-D) finite element models respectively.

The 2-D finite element model is shown in Fig. 4. Elements beam188, shell63 and shell99 were used to simulate bolts, metallic sheet and laminate sheet respectively.

In 3-D model, the elements Solid45 were used to mesh metallic sheet and bolts, and Solid46 for laminate sheets. The global mesh and its detail of bolt assembly are shown in Fig. 5 and 6 respectively. Conta173 and Target170 were used to define contact pairs. The pre-tighten forces of the bolts were applied by the elements Prets179.

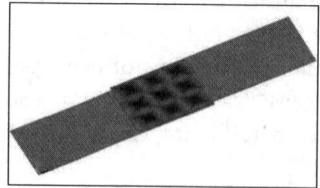

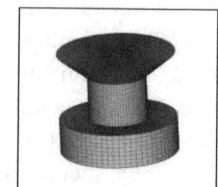

Fig. 4 2-D finite element model   Fig.5 3-D finite element model   Fig. 6 Mesh Detail of the bolt

3. Results of the pin-load distribution

3.1. Test results

Fig. 7 shows the test results of pin-load distribution of specimen S-01. It is found that the fractions of pin-load shear vary disorderly with the tensile load.

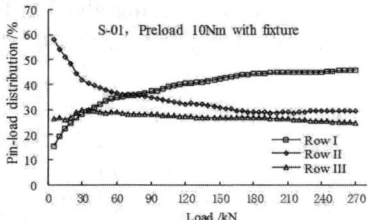

Fig. 7   Test results of pin-load distribution

3.2. FE Results

Pin-load distributions calculated with two meshes are shown in Fig. 8. In 2-D model, pin loads came from the nodal forces of the beam elements. In 3-D model, pin loads were obtained by collecting the contact forces at each bolt-hole interfaces. We can see from Fig. 8 that: i) In linear elastic condition, the pin-load distribution is independent of load level; ii) The difference between the results of 2-D model and those of 3-D model is negligible; iii) The load share by the bolts of row I or row III are higher than the row in middle; iv) For each row of the bolts, the load share is a little higher at both ends than at the middle.

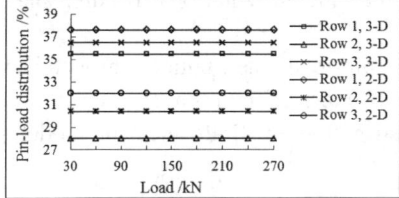 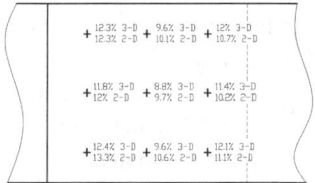

Fig. 8   Pin-load distribution of specimen

4. Discussion

Pin-load distribution given in section 3 reveals a significant difference between the test results and calculations. The reason of the deviation should be sought and the solution should be given for reliable results.

Liu etc. [7] have studied the pin-load distribution for the single-lap joints with an array of three bolts using the finite element technique. They found that the bolt preload and the friction between the contact pairs do not play important roles in pin-load distribution, but the uneven bolt-hole clearance does make the load shares change with external load. This conclusion, however, is not the major reason for the observed deviation between test and calculation.

In reality, the additional bending is the distinct feature of single-lap joints. Although the anti-bending fixture can to some extent reduces the overall bending deflection, the local bending can not be avoid due to the load eccentricity to each joint half, the fastener rotation and the non-uniform contact between fastener and hole. The results of 3-D FE analysis given in Fig. 9 demonstrate that the additional bending moments in metallic sheet increase rapidly with tensile load.

Fig. 10 plots the calculated mean strain-load curves of the specimen for cross-section I, also shown is the relevant test results of specimen S-01. The correlation among the curves indicates that the FE analysis has been verified by the strain measurement. Similar results are observed in other specimens.

In the case that a practically effective way of determining the pin-load distribution of multiple bolted lap-joints is not available, the authors suggest that the 3-D FE model validated by the strain measurement test should be a better option. 2-D FE analysis has satisfied accuracy if only the pin-load distribution is needed.

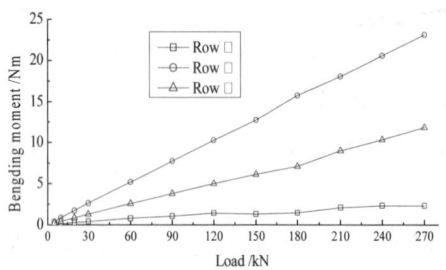
Fig. 9 Additional bending moments

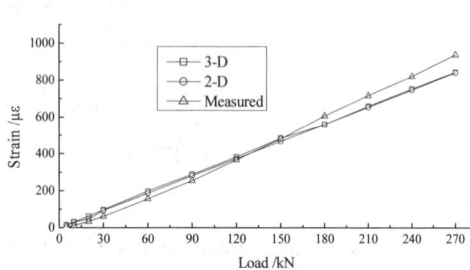
Fig. 10 Strain-load curves

5. Conclusions

In this paper, pin-load distribution is measured and analyzed by using strain electrometric measurements and finite element method. Conclusions are drawn as follows:

(1) In the multiple bolted single-lap joints, the inherent additional bending plays significant role affecting the pin-load distribution;

(2) The traditional way of determining pin-load distribution through surface strain measurement is questioned, because the additional bending effect has not been considered;

(3) It is feasible to determine pin-load distribution by finite element model validated by experimental strain measurements.

References

[1] Xu Yao-ling, Zhang Zhen-xian. Pin-load computations for composite laminate joints with multi-fasteners. Journal of Yanshan University, 2005, 29(2): 178-181

[2] Zhang chi-ping, Zhang-xing. An analytical method of pin load computations for joints with multi-fasteners. Acta Aeronautica Et Astronautica Sinica, 1994, 15(3): 310-317

[3] Lawlor V. P, McCarthy M. A, Stanley W. F.. An experimental study of bolt-hole clearance effects in double-lap, multi-bolt composite joints. Composite Structures, 2005, 71: 176-190

[4] Stanley W. F., McCarthy M. A., Lawlor V. P.. Measurement of load distribution in multi-bolt, composite joints, in the presence of varying clearance. Journal of Plastics, Rubber and Composites, 31(9): 412-418

[5] Xiong Y., Bedair O. K.. Analytical and finite element modeling of riveted lap joints in aircraft structure. AIAA Journal, 1999, 37(1): 93-99

[6] Madenci E, Shkarayev S, Sergeev B, et al. Analysis of composite laminates with multiple fasteners. Solids and Structures, 1998, 15(35): 1793-1811

[7] Liu Xing-ke, Li Ya-zhi, Liu Xiang-dong, etc. Study on load distribution of multiple-bolted metal to composite joints. Advance in Aeronautical Science and Engineering, 2011, 2(2): 193-198

The sensitivity of the foam-core parameters under impact loading

Xu Fei[1,a*], Duan Min-ge[1,b]

[1]School of Aeronautics, Northwestern Polytechnical University, Xi'an, China, 710072

[a]xufei@nwpu.edu.cn, [b]duanminge@sina.com

*Corresponding author, Email:xufei@nwpu.edu.cn; Tel:029-88493705

Keywords: Sensibility; Foam; Material parameters; FEA; Impact

Abstract. This study presents the numerical investigation of the low-velocity impact for the foam-cored sandwich composites. Firstly, the proposed FEA model is validated by comparing the results between simulation and test. The user subroutine VUMAT and the crushable foam model are chosen to describe the damage of the face sheets and the characteristics of the foam material, respectively. The detailed damage process of the sheets and the foam is clearly shown. The sensitivity of seven parameters related to foam-core material are studied. It is shown that the yield strength, the fracture strain and the fracture displacement have significant effects on the impact-resistance of the foam-cored sandwich composites.

Introduction

Composite laminates and composite sandwich structures are widely used in the field of aeronautic, astronautic and automotive for their low density, large stiffness, obvious advantages in the flexural shear performance and outstanding design ability. However, these structures are possibly subjected to impacts, such as the tool drops, hail, bird strikes and runway debris, which have great influence on the bearing capacity of the structure. Therefore, increasing experimental and numerical studies are performed in the impact performance of the composite sandwich structures[1-8]. Unfortunately, there are few researches on the effect and sensibility of the foam-core parameters.

Two mathematical model to predict the impact force history and the structure's overall impact response[5] have been used for composite laminates[6] and composite sandwich structures[7], which are the spring-mass and the energy-balance models. The spring-mass model sometimes overestimates the impact load for the plates after the onset of damage and the energy-balance model is inaccurate when damage initiates at higher impact energies.

The crushable foam model is intended for the analysis of foams that are typically used as energy absorption structures and the foam material, which deforms in compression due to cell wall buckling[9]. Companied with ductile and shear damage, the material parameters required in the material mode, are the Young's modulus, the yield stress, the fracture strain, the fracture displacement, the compression yield stress ratio, hydrostatic yield stress ratio and etc. Unfortunately, it is very difficult to get the true value of these parameters. So the research on the parameters would supply some advises to the related tests and the simulations.

The FEA model

Three types of foam-cored sandwich composite structures were investigated with experimental methods by Anderson et. al [1]. The impact energy in the tests ranges from 3J to 25J. The face sheets with the material of LTM45EL/CF0111, are comprised of either 0/90/0(0.792mm thickness) or $0_2/90_2/0_2$(1.584mm thickness). The size of specimens is 76.2mm×76.2mm, with the 12.7mm thick core. Two kinds of Rohacell polymethacryimide foam, low density 51WF and high density 110WF, are chosen for the core materials. Three kinds of structures, structure I, structure II and structure III are comprised of thin panel with low-density core, thin panel with high-density core and thick panel with high-density core, respectively.

As shown in Fig.1, the FEA model is consists of composite face sheets, foam core and the impactor, based on the Anderson et. al's tests. The element type of C3D8R is chosen with the mesh size of 1mm.

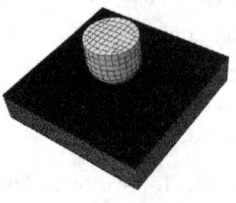
Fig. 1 The FEA model

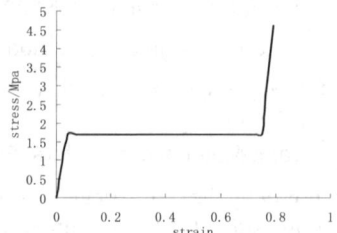
Fig. 2 The typical stress-strain curve for the foam

For the face sheets, the damage constitutive material model is carried out by the user's subroutine VUMAT. By comparing different damage criteria[3], Hou's criteria show its capability of dealing with woven composites. The stiffness of the woven composites degrades linearly after the damage initiation.

For the foam, the typical stress-strain curve is shown in Fig. 2. Material parameters are listed in Table.1, where σ_c, σ_τ, ε_{cf} and $\varepsilon_{\tau f}$ refer to the yield stress and the fracture strain of ductile damage and shear damage, respectively. K and K_t are the compression yield stress ratio and hydrostatic yield stress ratio.

Table 1 Material parameters for Rohacell-51/110 WF

	E[Mpa]	G[Mpa]	ρ [kg·m⁻³]	σ_c [Mpa]	ε_{cf} [%]	σ_τ [Mpa]	$\varepsilon_{\tau f}$ [%]	K	Kt
51	75	24	51	0.8	68.0	0.8	34.0	0.85	0.82
110(Y)	180	70	110	3.6	68.5	2.4	38.5	0.85	0.82

The validation of the model

The time history of load (contact force between impactor and composite sheet), impactor velocity, deflection (displacement of impactor), indentation (displacement of composite sheet), KE (the kinetic energy of the impactor) and IE (the inner energy of the foam) can be obtained through numerical simulations. The results of structure III are shown in Fig. 3. Fig.3(a) shows that the peak load comes a little earlier than the rebounding of the impactor. During the impactor rebounding process, the load decreases to zero while the velocity increases to a stable value. The deflection and the indentation history are shown in Fig.3(b). After the rebounding process, the deflection of the impactor changes to the opposite direction, while deflection of the composite sheet keeps a stable value, which is called indentation in this paper. The kinetic energy of the impactor and the inner energy of the foam core are shown in Fig.3(c). It is clear that the minimum KE correspond to the maximum IE. After the fully transfer of these energies at this point, the impactor rebound to regain the kinetic energy, while the inner energy of the foam decrease a little due to the elastic deformation.

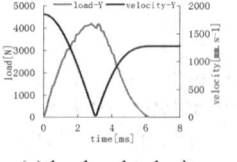
(a) load and velocity

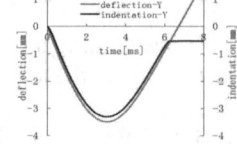
(b) deflection and indentation

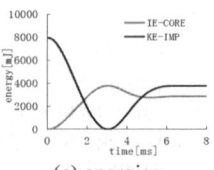
(c) energies

Fig. 3 The results of structure III

In Fig.4 the black color indicates the damaged element while the white color refers to undamaged parts. Fig.4(a) shows that the damage occurs on the edge of the specimen for structure I, while the damage is found below the impactor site for structure II and III. Fig.5(a) and (b) show the Tresca contour of the cross-section view. It is found that the 45° shear bands locates near the periphery of the structure I, while, it positions around the impact site in structure II. All the damage modes and impacting responds of the three structures in simulation agree well with those in the tests.

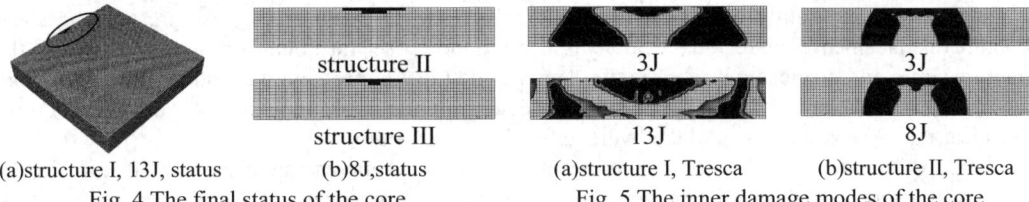

(a)structure I, 13J, status   (b)8J,status   (a)structure I, Tresca   (b)structure II, Tresca
Fig. 4 The final status of the core   Fig. 5 The inner damage modes of the core

The sensibility of the foam parameters

This section is focused on the parameter analysis of foam-core materials based on structure III under the impact energy of 8J. The differences of Young's modulus from 60Mpa, 120Mpa to 180Mpa are shown in Fig. 6, where Y refers to the result from the base parameters listed in Table 1. Fig.6 (a), (b), (c) and (d) show the history of the load, velocity, deflection and indentation, respectively. With the increasing of the Young's modulus, the time to the peak load and to the rebounding of the impactor comes earlier, the maximum deflection gets lower, and the indentation get deeper, which may be due to the earlier damage initiation caused by higher stresses in foam and higher stiffness of the whole structure. The differences between 60Mpa and 120Mpa are bigger than those between 120Mpa and 180Mpa.

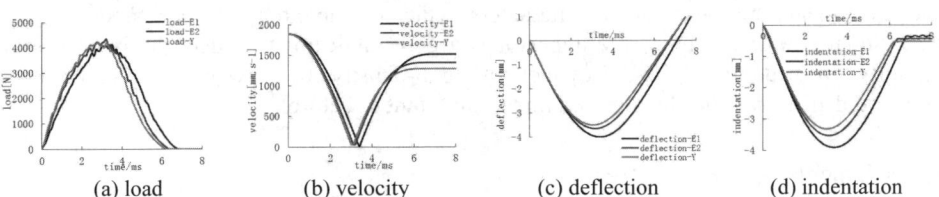

(a) load   (b) velocity   (c) deflection   (d) indentation

Fig. 6 The influence of the Young's modulus

Table 2 The sensibility of different foam parameters

		Pmax [N]	velocity [mm·s^{-1}]	Deflection [mm]	Indentation [mm]	Damage-area [mm*mm]	IE [J]	KE [J]
Young's modulus	E1(60Mpa)	4122	1514	4.00	0.34	11.03*11.03	1.42	5.34
	E2(120Mpa)	4344	1378	3.65	0.44	16.04*16.04	2.36	4.42
	Y(180Mpa)	4194	1278	3.50	0.53	18.05*18.05	2.91	3.81
Yield strength	S1(1.8Mpa)	3462	1170	4.06	1.07	26.07*26.07	3.72	3.19
	Y(3.6Mpa)	4194	1278	3.50	0.53	18.05*18.05	2.91	3.81
	S2(5.4Mpa)	4142	1228	3.79	0.52	18.05*16.04	2.36	3.52
Fracture strain	D1(0.358)	4214	1287	3.50	0.52	18.05*18.05	2.85	3.86
	Y(0.685)	4194	1278	3.50	0.53	18.05*18.05	2.91	3.81
	D2(0.985)	4214	1287	3.50	0.52	18.05*18.05	2.84	3.86
Fracture displacement	DP1(0.685mm)	4214	1287	3.50	0.52	18.05*18.05	2.84	3.86
	Y(1.37mm)	4194	1278	3.50	0.53	18.05*18.05	2.91	3.81
	DP2(2.055mm)	4214	1287	3.50	0.52	14.04*14.04	2.84	3.86
Shear fracture strain	SD1(0.185)	3648	1304	3.85	0.27	37.10*36.09	2.92	3.96
	Y(0.385)	4194	1278	3.50	0.53	18.05*18.05	2.91	3.81
	SD2(0.585)	4304	1222	3.49	0.83	8.02*8.02	3.12	3.48
Shear fracture displacement	SDP1(0.385mm)	3892	1309	3.65	0.36	28.08*28.08	2.81	3.93
	Y(0.685mm)	4194	1278	3.50	0.53	18.05*18.05	2.91	3.81
	SDP2(1.155mm)	4258	1250	3.49	0.71	10.03*10.03	3.09	3.64

	Y(0.82)	4194	1278	3.50	0.53	18.05*18.05	2.91	3.81
Compression yield stress ratio	K1(0.75)	4254	1288	3.50	0.53	18.05*18.05	2.93	3.87
	K2(0.45)	4167	1280	3.51	0.56	19.05*18.05	2.91	3.82
Hydrostatic yield stress ratio	KT1(0.973)	4277	1140	3.50	0.56	18.05*18.05	3.05	3.03
	Y(0.85)	4194	1278	3.50	0.53	18.05*18.05	2.91	3.81
	KT2(0.700)	4172	1170	3.48	0.58	18.05*18.05	3.02	3.19

More detailed data for other parameters are listed in Table 2. The peak load, the damage area and the indentation are apparently influenced by the yield stress, the shear fracture strain, and the shear fracture displacement. The residual velocity is affected mostly by the Young's modulus. However, the fracture strain and displacement for ductile-damage and the compression yield stress ratio show very few differences. So it is of great importance to determine the accurate values of Young's modulus, the shear-damage parameters and the hydrostatic yield stress ratio.

The effect of the shear fracture parameters is analyzed. With the increasing of shear fracture strain, the indentation increases while the damage area decreases, which means that the impact-resistance of the structure cannot be reflected reasonably by only one or two physical variables. The energy transfer between different parts of the structure is probably a good indicator to reflect the impact-resistance capability. And the residual kinetic energy of the impactor is found to be one of the indicators. The smaller Young's modulus, the smaller shear fracture parameters and the larger yield stress would reflect better impact-resistance capability.

Summary

The FEA model for foam-cored sandwich composites structure under impact is established, and the model is proved to be reasonable and effective. Sensitivity of the foam parameters on the impact-resistance capability is obtained.
1) The impact-resistance capability of the structure is greatly influenced by the Young's modulus, the yield strength and the shear fracture parameters of the crushable foam damage model.
2) The residual kinetic energy of the impactor is a good indicator to reflect the impact-resistance capability. For structure III the thick panel with high-density core, lower Young's modulus of the foam would create a better foam-cored impact-resistant structure.

Acknowledgments

The work is supported by the 111 Project (B07050) and the Aeronautical Science Foundation of China (2010ZF53071).

References

[1] T.Anderson, E.Madenci. Composite Structures. 2000,50:239-247
[2] Xia Long, Xu Fei, Li Qiao, Yao Xionghua. Advances in Aeronautical Science and Engineering. 2011,4:425-431(in Chinese)
[3] J.P.Hou, N.Petrinic, C.Ruiz, S.R. Hallett. Composites Science and Technology. 2000,60:273-281
[4] Levent Aktay, Alastair F. Johnson, Martin Holzapfel. Computational Materials Science. 2005, 32:252–260
[5] Serge Abrate. Composite Structures. 2001,51:129-138
[6] P Feraboli, KT Kedward. Composites Science and Technology. 2006, 66:1336–47.
[7] T Anderson. Composites Part B. 2005,36:135–42.
[8] T.Anderson, E.Madenci. Engineering Fracture Mechanics. 2000, 67: 329-344
[9] ABAQUS Analysis User's Manual
[10]   Z. Hashin and A. Rotem. Journal of Composite Materials. 1973,7: 448-464
[11]   Z. Hashin. Journal of Applied Mechanics. 1980, 47(2):329-335

Dynamic Anti-plane Behaviors on Two Dissimilar Piezoelectric Media with an Interfacial Non-circular Cavity

Tianshu Song[1,a], Dong Li[2,b], Mingju Zhang[1,c] and Yuefa Zhou[1,d]

[1] School of Aerospace and Civil Engineering, Harbin Engineering University, Harbin, 150001, China
[2] Department of Civil Engineering, Hebei Jiaotong Vocational and Technical College, Shijiazhuang, 050091, China

[a]songts@126.com, [b]lidong242@163.com, [c]mingju861102@163.com, [d]zhouyuefa@163.com

Keywords: two dissimilar piezoelectric media; dynamic stress concentration factor (DSCF); Green's function; non-circular cavity; conformal mapping

Abstract. Dynamic anti-plane behaviors are studied on two dissimilar piezoelectric media with an interfacial non-circular cavity subjected to time harmonic incident anti-plane shearing. Based on Green's function and conformal mapping method, the dynamic stress concentration factors at the edge of the non-circular cavity are obtained by applying the orthogonal function expansion technique. Numerical cases about two dissimilar piezoelectric media with an elliptic cavity are provided with different elliptic axial length ratio, different wave number and different piezoelectric characteristic parameter. The calculating results show that dynamic analyses are of importance at lower frequencies and larger piezoelectric characteristic parameters.

Introduction

Due to the instinct electromechanical coupling behavior, piezoelectric materials have been widely used in smart components or structures, such as piezoelectric actuators and transducers. However, various types of defects occur in newly developed piezoelectric materials during their manufacturing and polling process. Great efforts were made and a lot documents were published in the past thirty years It should be noted that most of published documents on fracture mechanics of piezoelectricity were static. In 21st century, dynamic stress concentrations have been paid more attention than static ones. Wang et al investigated scattering of anti-plane shear wave by a piezoelectric circular cylinder with an imperfect interface [1]. Narita et al: studied circular inclusion problem in dynamic anti-plane piezoelectricity [2]. Wang analyzed dynamic behavior of interacting interfacial cracks in piezoelectric media [3]. Bian and Wang researched dual equations and solutions of I-type crack of dynamic problems in piezoelectric materials [4].

The objective of the present work is to provide a theoretical method for dynamic anti-plane behaviors on two dissimilar piezoelectric media with an interfacial non-circular cavity. The emphasis is focused on dynamic stress concentration factors at the cavity's edge. The boundary conditions on the cavity are assumed to be traction free and electrically permeable. While the interface is assumed to be well-bonded excluding the cavity

Governing Equations

Consider two dissimilar piezoelectric media with an interfacial non-circular cavity subjected to a remote uniform time-harmonic anti-plane shearing directed at an incident orientation α_0 with the positive direction of the x-axis, as shown in Fig. 1. The piezoelectric media are assumed to have been poled along the z-axis which is perpendicular to the xy-plane. In the absence of body forces and free charges, the governing equations of linear piezoelectricity for dynamic anti-plane shearing with exponential time-harmonic factor $Exp(-i\omega t)$ are given by

$$c_{44}\nabla^2 w + e_{15}\nabla^2 \phi + \rho\omega^2 w = 0, \quad e_{15}\nabla^2 w - \kappa_{11}\nabla^2 \phi = 0 \tag{1}$$

in which, c_{44}, e_{15} and κ_{11} are shear elastic modulus, piezoelectric and dielectric constants of the media, respectively; while w, ϕ, ρ and ω are elastic displacement, electric potential, mass density and frequency of the incident wave, respectively. The time factor $Exp(-i\omega t)$ will be omitted hereafter. Introducing a complex argument $x + iy = \omega(\eta)$ and its conjugate $x - iy = \overline{\omega(\eta)}$, if only $\omega'(\eta) \neq 0$, the external domain of a non-circular cavity in the xy-plane can be transformed into one of a unit circle in mapping plane η based on conformal mapping technique. In mapping plane η expressed by polar coordinates (r, θ), the constitutive relations can be written as

$$\tau_{rz} = \frac{c_{44}}{|\omega'(\eta)|}\left(\frac{\partial w}{\partial \eta}e^{i\theta} + \frac{\partial w}{\partial \overline{\eta}}e^{-i\theta}\right) + \frac{e_{15}}{|\omega'(\eta)|}\left(\frac{\partial \phi}{\partial \eta}e^{i\theta} + \frac{\partial \phi}{\partial \overline{\eta}}e^{-i\theta}\right)$$

$$\tau_{\theta z} = \frac{ic_{44}}{|\omega'(\eta)|}\left(\frac{\partial w}{\partial \eta}e^{i\theta} - \frac{\partial w}{\partial \overline{\eta}}e^{-i\theta}\right) + \frac{ie_{15}}{|\omega'(\eta)|}\left(\frac{\partial \phi}{\partial \eta}e^{i\theta} - \frac{\partial \phi}{\partial \overline{\eta}}e^{-i\theta}\right)$$

$$D_r = \frac{e_{15}}{|\omega'(\eta)|}\left(\frac{\partial w}{\partial \eta}e^{i\theta} + \frac{\partial w}{\partial \overline{\eta}}e^{-i\theta}\right) - \frac{\kappa_{11}}{|\omega'(\eta)|}\left(\frac{\partial \phi}{\partial \eta}e^{i\theta} + \frac{\partial \phi}{\partial \overline{\eta}}e^{-i\theta}\right) \tag{2}$$

$$D_\theta = \frac{ie_{15}}{|\omega'(\eta)|}\left(\frac{\partial w}{\partial \eta}e^{i\theta} - \frac{\partial w}{\partial \overline{\eta}}e^{-i\theta}\right) - \frac{i\kappa_{11}}{|\omega'(\eta)|}\left(\frac{\partial \phi}{\partial \eta}e^{i\theta} - \frac{\partial \phi}{\partial \overline{\eta}}e^{-i\theta}\right)$$

where τ_{rz}, $\tau_{\theta z}$, D_r and D_θ are two shear stress components and two electric displacement components, respectively. Eq. 1 can be simplified further in mapping domain η as

$$\frac{\partial^2 w}{\partial \eta \partial \overline{\eta}} = \left(\frac{ik}{2}\right)^2 \omega'(\eta)\overline{\omega'(\eta)}w(\eta,\overline{\eta}), \quad 4\frac{\partial^2 g}{\partial \eta \partial \overline{\eta}} = 0, \quad \phi = \frac{e_{15}}{\kappa_{11}} + g \tag{3}$$

where $k^2 = \rho\omega^2/c^*$, $c^* = c_{44}(1+\lambda)$, $\lambda = e_{15}^2/c_{44}\kappa_{11}$. While all the formulae will be expressed in the plane η.

Boundary Value Problem and DSCF

Firstly, in the plane η, take apart the whole infinite media into two semi-infinite media along the interface and construct a couple of Green's functions G_w and G_ϕ, which are fundamental solutions of elastic displacement and electric potential for a semi-infinite piezoelectric medium with a semi-circular notch under a pair of time harmonic anti-plane line force and in-plane line charge at an arbitrary point r_0 on the interface, respectively. The boundary conditions at the surface are considered to be traction free and electrically impermeable, while the notch is considered to be traction free but electrically permeable. G_w^I and G_ϕ^I in medium I can be solved as the following forms

$$G_w^I = \frac{iH_0^{(1)}\left(k|\omega(\eta)-\omega(\eta_o)|\right)}{2c_{44}(1+\lambda)} + \frac{i}{2c_{44}(1+\lambda)}\sum_{n=0}^{\infty} A_n H_n^{(1)}\left(k|\omega(\eta)|\right)\cdot\left\{\left[\frac{\omega(\eta)}{|\omega(\eta)|}\right]^n + \left[\frac{\omega(\eta)}{|\omega(\eta)|}\right]^{-n}\right\}$$

$$G_\phi^I = \frac{e_{15}}{\kappa_{11}}G_w + B_0 + \sum_{n=1}^{\infty}\left(B_n \eta^{-n} + C_n \overline{\eta}^{-n}\right) \tag{4}$$

$$G_\phi^C = \frac{i}{2c_{44}(1+\lambda)}[D_0 + \sum_{n=1}^{\infty}\left(D_n \overline{\eta}^n + E_n \eta^n\right)]$$

in which, $H_n^{(1)}$ and G_ϕ^C indicate Hankel's function of the first kind and Green's function in the notch; respectively. A_n, B_n, C_n, D_n and E_n are coefficients which can be obtained by solving boundary value problem. The difference of Green's functions between media I and II is only on materials' parameters.

Secondly, a pair of additional unknown shear stresses f^I and f^{II} are applied on the two media's surfaces excluding the notch respectively, so as to satisfy the continuity conditions of the interface excluding the notch. The following relation should be satisfied so as to fulfill the conjunction

$$\tau_{\theta z}^I \cos\theta_0 + f^I(r_0,\theta_0) = \tau_{\theta z}^{II} \cos\theta_0 + f^{II}(r_0,\theta_0), \quad \theta_0 = 0, \pi \tag{5}$$

The fuctions f^I and f^{II} can be obtained by solving the following integral equations

$$\int_{R_0}^{\infty} \{f_1(r_0,\pi)[G^I(r,\theta,r_0,\pi) - G^{II}(r,0,r_0,\pi)] + f_1(r_0,0)[G^I(r,\theta,r_0,0) - G^{II}(r,\theta,r_0,0)]\}dr_0$$
$$= w(r,\theta) + \int_{R_0}^{\infty} [\tau_{\theta z}^I(r_0,\theta) - \tau_{\theta z}^{II}(r_0,\theta)]G^{II}(r,\theta,r_0,0)dr_0, \quad \theta = 0, \pi \tag{6}$$

Finally, DSCF at the non-circular cavity's edge can be defined as

$$\tau_{\theta z}^* = \left|\frac{\tau_{\theta z}}{\tau_0}\right| \tag{7}$$

in which, τ_0 refers to shear stress magnitude of the incident wave; $\tau_{\theta z}$ is dynamic stress in the cavity's edge.

Calculating Cases

As numerical examples, DSCFs around an interfacial elliptical cavity's edge in piezoelectric bimaterials at the situation of $\lambda^I/\lambda^{II} = 2$ are calculated and plotted based on Eq. 7. The mapping function has the following form:

$$\omega(\eta) = \frac{a+b}{2}\eta + \frac{a-b}{2\eta} \tag{8}$$

in which, a and b indicate the two semi-axial lengths of the elliptic cavity, respectively. Figs. 2~4 plot DSCFs at the cavity's edge so as to show how the structural geometry, the two materials' parameters and the incident wave's frequency influence on the DSCFs. Fig. 2 displays DSCFs around the cavity under horizontal incidence at a quasi-static situation, from which the asymmetric phenomenon between the upside and the underside can be obviously found. This is due to the different physical parameters of the two dissimilar media. DSCFs for piezoelectric bimaterials are larger than those for a uniform medium by comparison with document [5]. Fig. 3 shows DSCFs with different wave numbers under vertical incidence which implies that DSCFs are larger at lower frequencies (Fig. 3a) than those at higher frequencies (Fig. 3b). The influence of the incident frequency on DSCFs can be clearly seen in Fig.4 that the maximal DSCFs always appear at the range of lower frequencies $0.3 < ka < 0.5$.

Summary

Dynamic analysis for dynamic stress concentration factor for two dissimilar piezoelectric media is more important than that on one medium. Dynamic analyses are very important to piezoelectric bimaterials with an interfacial non-circular cavity at lower frequencies and larger piezoelectric characteristic parameters. The method employed in present paper can serve as a useful tool for modeling of a piezoelectric medium with complex defects and for the design of piezoelectric structures.

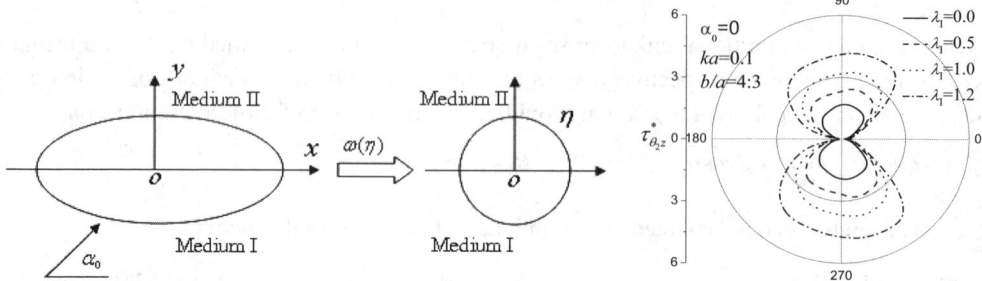

Fig. 1 A model of conformal mapping   Fig. 2 DSCFs under horizontal incidence

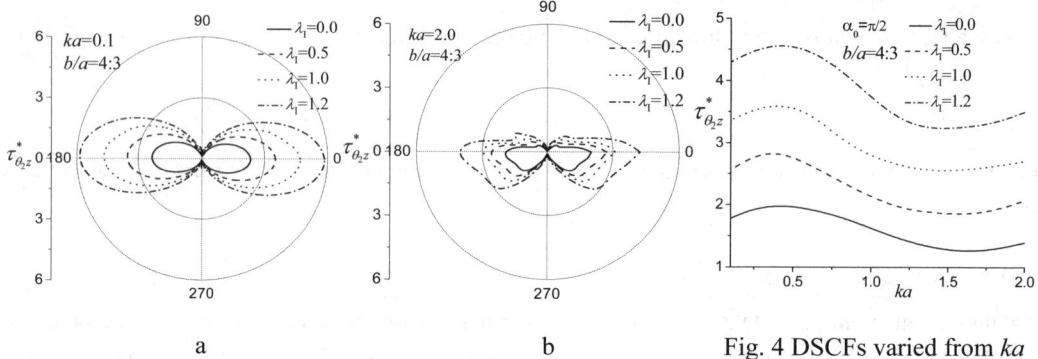

a   b   Fig. 4 DSCFs varied from ka
Fig. 3 DSCFs under vertical incidence   under vertical incidence

Acknowledgments

The present work is supported by the Fundamental Research Funds of China Central Universities (HEUCFZ1125).

References

[1] X. Wang, E. Pan and A.K. Roy: Scattering of antiplane shear wave by a piezoelectric circular cylinder with an imperfect interface. Acta Mechanica, 193(3-4): 177-195(2007).
[2] F. Narita, Y. Shindo and H. Moribayashi: Circular inclusion problem in dynamic antiplane piezoelectricity. The International Society for Optical Engineering, 4235: 69-78(2001).
[3] X.D. Wang: On the dynamic behavior of interacting interfacial cracks in piezoelectric media. Int. J. Solids and Structures, 38: 815-831(2001)
[4] W.F. Bian and B. Wang: Dual equations and solutions of I-type crack of dynamic problems in piezoelectric materials. Applied Mathematics and Mechanics, 28(6): 731-739(2007).
[5] T.S. Song, L.L. Sun and D.K. Liu: Dynamic anti-plane characteristic of an infinite piezoelectric medium with a non-circular cavity. Proceedings of IDETC/CIE ASME 2005(2005)

Analysis of crack tip field in materials with creep behavior

Qinghua Meng[1,a], Wenyan Liang[1,b], Zhenqing Wang[1,c]

[1]College of Aerospace and Civil Engineering, Harbin Engineering University, Harbin 150001, China

[a]mengqinghua@hrbeu.edu.cn, [b]liangwenyan@hrbeu.edu.cn, [c]wangzhenqing@hrbeu.edu.cn

Keywords: Crack tip field; Creep conditions; Effective stress; Mode I

Abstract. The stress and strain field near the tip of Mode I growing crack in materials under creep conditions is examined. The case of the effective stress equal to zero is considered. The asymptotic equations for the crack of the crack tip field are derived and solved numerically. It is concluded that the rates of stress and strain posses the $r^{\delta-1}$ singularity near the tip crack, and the stresses remain finite at the crack tip.

Introduction

Creep failure is an important consideration in the design of engineering structures. Character of the crack tip stress and strain field depends on the mathematical description of the constitutive relation. Using a new constitutive relation that contains all of the three stages of creep for materials with creep behavior, the stress and strain distribution near the tip of a Mode III growing crack is examined by Gao Yuchen[5]. Moreover, what was shown by Gao Yuchen[6] is that the deformation fields near a Mode I growing crack tip for materials with creep behavior. Unfortunately, the analysis only considers one case that creep stress equal to finite value.

The purpose of the present paper is thus to completely examine the stress and strain field near tip field of Mode I growing crack in materials under creep conditions. The case of the effective stress equal to zero is considered.

Basic equations

Considered is a semi-infinite Mode I crack as shown in Fig.1. A coordinates (x, y) is fixed to the moving crack tip. The velocity of crack tip V is constant. Assume that the material is incompressible, and then we can introduce a potential Ω such that the displacement u_x and u_y can be expressed as

$$u_x = -\frac{\partial \Omega}{\partial y}, \quad u_y = \frac{\partial \Omega}{\partial x} \quad (1)$$

The strains become

$$\varepsilon_x = -\varepsilon_y = -\frac{\partial^2 \Omega}{\partial x \partial y}, \quad \varepsilon_{xy} = \frac{1}{2}(\frac{\partial^2 \Omega}{\partial x^2} - \frac{\partial^2 \Omega}{\partial y^2}) \quad (2)$$

The velocity can be expressed as

$$v_x = -\frac{\partial \dot{\Omega}}{\partial y}, \quad v_y = \frac{\partial \dot{\Omega}}{\partial x} \quad (3)$$

The superscript dot denotes the time derivative

$$\dot{(\ )} = \frac{d}{dt}(\ ) = -V\frac{\partial}{\partial x}(\ ) \quad (4)$$

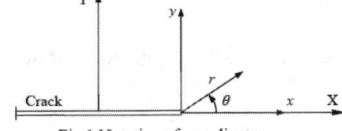

Fig.1 Notation of coordinates

The strain rate is

$$\dot{\varepsilon}_x = -\dot{\varepsilon}_y = -\frac{\partial^2 \dot{\Omega}}{\partial x \partial y}, \quad \dot{\varepsilon}_{xy} = \frac{1}{2}(\frac{\partial^2 \dot{\Omega}}{\partial x^2} - \frac{\partial^2 \dot{\Omega}}{\partial y^2}) \quad (5)$$

For a quasi-static problem, a stress function φ can be introduced such that

$$\sigma_x = \frac{\partial^2 \varphi}{\partial y^2}, \quad \sigma_y = \frac{\partial^2 \varphi}{\partial x^2}, \quad \sigma_{xy} = -\frac{\partial^2 \varphi}{\partial x \partial y} \quad (6)$$

The stress rates is

$$\dot{\sigma}_x = \frac{\partial^2 \dot{\varphi}}{\partial y^2}, \quad \dot{\sigma}_y = \frac{\partial^2 \dot{\varphi}}{\partial x^2}, \quad \dot{\sigma}_{xy} = -\frac{\partial^2 \dot{\varphi}}{\partial x \partial y} \quad (7)$$

For general stress state under creep condition takes the form

$$\dot{\varepsilon}_c = \frac{3}{2} c e_c^{-\alpha}(e_{cf} - e_c)^{-\beta} s^{n-1} \mathbf{S} \quad (8)$$

The constitutive relation in materials with creep behavior can be written as

$$\begin{cases} \dot{\varepsilon}_x - \dot{\varepsilon}_y = \frac{1}{2G}(\dot{\sigma}_x - \dot{\sigma}_y) + \frac{3c}{2} e_c^{-\alpha}(e_{cf} - e_c)^{-\beta} s^{n-1}(\sigma_x - \sigma_y) \\ \dot{\varepsilon}_{xy} = \frac{1}{2G} \dot{\sigma}_{xy} + \frac{3c}{2} e_c^{-\alpha}(e_{cf} - e_c)^{-\beta} s^{n-1} \sigma_{xy} \end{cases} \quad (9)$$

where G is elastic shear modules, c, α, β, n the material constants that may depends on the temperature, e_c the effective creep strain, e_{cf} the failure effective creep strain and s the effective stress

$$s = \sqrt{3}[\frac{1}{4}(\sigma_x - \sigma_y)^2 + \sigma_{xy}^2]^{1/2} \quad (10)$$

Asymptotic equations

Far from the crack tip, the stress and strain are relatively small. When the crack tip is approached, the effective creep strain would increase monotonically up to the value e_{cf}, but the effective stress may only increase up to a distance and then decrease because of creep relaxation. More specifically, e_c reaches the value e_{cf} at the crack tip while s may equal to a finite value s^* which may also become zero. The case that s vanishes at the crack tip will be analyzed subsequently. The asymptotic equation for the case of $s^* = 0$ will be considered.

It is expedient to introduce the polar coordinates

$$r = (x^2 + y^2)^{1/2}/a, \quad \theta = \arctan\frac{y}{x}$$

where a is a typical length. Let σ denote the normal stress in the far field. Near the crack tip, let

$$\Omega = a^2 r^2 \left[\frac{\sqrt{3}}{4} e_{cf} \sin 2\theta + \frac{\sigma}{2G} r^\delta g(\theta)\right] \quad (11)$$

where δ is an exponent to be determined. Using Eq. (4), then

$$\dot{\Omega} = Var\left[-\frac{\sqrt{3}}{2} e_{cf} \sin\theta + \frac{\sigma}{2G} r^\delta \eta\right] \quad (12)$$

in which

$$\eta = g'\sin\theta - (2+\delta)g\cos\theta \quad (13)$$

According to Eqs. (1) and (2), the displacement and strain is obtained

$$u_r = ar[\frac{\sqrt{3}}{2} e_{cf} \cos 2\theta + \frac{\sigma}{2G} r^\delta g'], \quad u_\theta = ar[\frac{\sqrt{3}}{2} e_{cf} \sin 2\theta + \frac{\sigma}{2G}(2+\delta) r^\delta g] \quad (14)$$

$$\varepsilon_r = -\varepsilon_\theta = -[\frac{\sqrt{3}}{2}e_{cf}\cos 2\theta + \frac{\sigma}{2G}(\delta+1)r^\delta g'], \quad \varepsilon_{r\theta} = \frac{\sqrt{3}}{2}e_{cf}\sin 2\theta + \frac{\sigma}{2G}r^\delta[\delta(2+\delta)g - g''] \quad (15)$$

According to Eqs. (3) and (5), the velocity are

$$v_r = V[\frac{\sqrt{3}}{2}e_{cf}\cos\theta - \frac{\sigma}{2G}r^\delta \eta'], \quad v_\theta = V[-\frac{\sqrt{3}}{2}e_{cf}\sin\theta + \frac{\sigma}{2G}(1+\delta)r^\delta \eta] \quad (16)$$

The strain rates are

$$\dot{\varepsilon}_r = -\dot{\varepsilon}_\theta = -V\frac{\delta\sigma}{2Ga}r^{\delta-1}\eta', \quad \dot{\varepsilon}_{r\theta} = -V\frac{\sigma}{4Ga}r^{\delta-1}[(1-\delta^2)\eta + \eta''] \quad (17)$$

Further, considering that the basic stress states at crack tip is uniaxial tension, the stress function is written as

$$\varphi = a^2\sigma r^{\delta+2}f(\theta) \quad (18)$$

And hence

$$\dot{\varphi} = aV\sigma r^{\delta+1}\zeta \quad (19)$$

where

$$\zeta = f'\sin\theta - (2+\delta)f\cos\theta \quad (20)$$

Using Eqs. (6) and (7), the stress can be obtained

$$\sigma_r - \sigma_\theta = \sigma r^\delta[f'' - \delta(2+\delta)f], \quad \sigma_{r\theta} = -\sigma r^\delta(1+\delta)f' \quad (21)$$

The stress rates are

$$\dot{\sigma}_r - \dot{\sigma}_\theta = V\frac{\sigma}{a}r^{\delta-1}[\zeta'' + (1-\delta^2)\zeta], \quad \dot{\sigma}_{r\theta} = -V\frac{\sigma}{a}\delta r^{\delta-1}\zeta' \quad (22)$$

The effective creep strain can be written as

$$e_c = e_{cf} - \frac{\sigma}{G}r^\delta \psi \quad (23)$$

Substituting Eq. (23) into (8) and considering (4), matching the order of r and taking the dominant terms only, it follows that

$$\psi'\sin\theta - \delta\psi\cos\theta + K\psi^{-\beta} = 0, \quad \delta = \frac{1}{\beta+1-n} \quad (24)$$

in which

$$K = \frac{ac}{V}\sigma^{n-\beta-1}e_{cf}G^{\beta+1} \quad (25)$$

Substituting (29) and (33) to (35) into (9), considering (10), and matching the singularity, there results

$$\zeta'' + (1-\delta^2)\zeta + 2\delta\eta' + 3K\psi^{-\beta}F^{n-1}[f'' - \delta(2+\delta)f] = 0$$
$$\eta'' + (1-\delta^2)\eta - 2\delta\zeta' - 6K\psi^{-\beta}F^{n-1}(1+\delta)f' = 0 \quad (26)$$

where

$$F = \sqrt{3}\left\{\frac{1}{4}[f'' - \delta(2+\delta)f]^2 + [(1+\delta)f']^2\right\}^{1/2} \quad (27)$$

Boundary Conditions and Numerical Results

For Eq. (24), the regularity at $\theta = 0$ requires that

$$\psi(0) = (\frac{K}{\delta})^\delta \quad (28)$$

For Eq. (20), the regularity at $\theta = 0$ requires that

$$\zeta(0) = -(2+\delta)f(0) \quad (29)$$

For Eq. (13), since $g(\theta)$ and $\eta(\theta)$ are odd function. Further the regularity at $\theta = 0$ requires that

$$g(0) = 0, \quad \eta'(0) = -(1+\delta)g'(0) \tag{30}$$

where $g'(0)$ and $f(0)$ is to be determined by the boundary conditions at the point of failure. For Eq. (26), the following condition should be supplied

$$\eta(0) = 0, \quad \zeta'(0) = 0 \tag{31}$$

The location of failure $\theta = \theta_0$ should be determined by the conditions that the tractions become free at the failure site, i.e.,

$$f(\theta_0) = f'(\theta_0) = 0, \quad \psi(\theta_0) = 0 \tag{32}$$

According to analyze (25) under conditions (28) and (32), the function $\psi(\theta)$ can be express

$$\psi(\theta) = (\frac{K}{\delta})^\delta (\cos\theta)^\delta$$

Eq. (26) under conditions (29) to (32) can be solved numerically. The solutions are obtained for large β and small n, i.e. $\beta + 1 - n > 0$. The curves of function $\psi(\theta)$ and $f(\theta)$ are shown in Fig. 2 and 3 for $K = 1$.

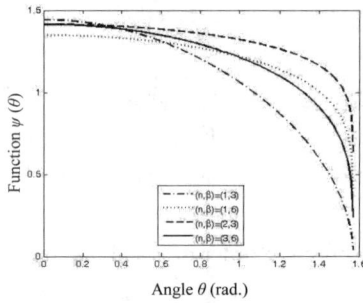

Fig.2 Angular variation of $\psi(\theta)$

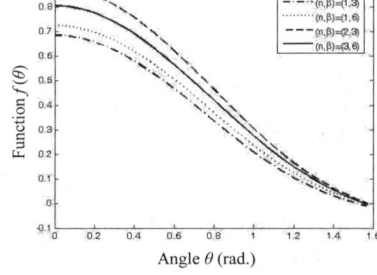

Fig.3 Angular variation of $f(\theta)$

Conclusion

According to the constitutive relation for creep used in this work, the near tip field of a Mode I growing crack is obtained. The possibility of $s^* = 0$ remains open, when $\beta + 1 - n > 0$. The rates of stress and strain are singular of the order $r^{\delta-1}$ near the crack tip, and the stresses remain finite at the crack tip.

Acknowledgements

This work is supported by the Research Fund for the Doctoral Program of Higher Education of China (No. 20112304110015).

References

[1] C. Y. Hui and H. Riedel:Int. J. Fract. Vol.17 (1981), p.409.
[2] W. Yang and L. B. Freund:Int. J. Fract. Vol.30(1986), p.157.
[3] T. J. Delph:Int. J. Fract. Vol.68(1994), p.183.
[4] Y. C. Gao:J. Theor. Appl. Fract. Mech. Vol.14(1990), p.233.
[5] Y. C. Gao and Z. Q. Wang:J. Theor. Appl. Fract. Mech. Vol.25(1996), p.113.
[6] Y. C. Gao:Solid Mechanics and Its Applications. Vol.64(2002), p.47.

The Conservation Laws and Path-independent Integrals for Piezo-magnetic Media

SONG Haiyan[1, a], ZHOU Jiansheng[1, b], LIU Zongmin[1, c]

[1] College of Aerospace and Civil Engineering, Harbin Engineering University, Harbin 150001, China

[a] shyivg@yahoo.com.cn, [b] zhoujiansheng@hrbeu.edu.cn, [c] lzm1976@yahoo.com.cn

Keywords: conservation laws, path-independent integrals, piezo-magnetic media

Abstract. Path-independent integrals have important application value in dislocation, fracture mechanics and other defects theories. Motivated by concepts of Jacobi integral and cyclic integral in analytical mechanics and energy-momentum tensor in electro-magntic field, the conservation laws and path-independent integrals for piezo-magnetic media are derived in this paper.

Introduction

Starting from Noether's theorem [1] on variational principles invariant under a group of infinitesimal transformations, Knowles and Sternberg [2], Fletcher [3], Li [4] obtained some conservation laws in elasticity. In fact, Rice's J-integral [5] is a special case of one of these conservation laws derived in the refs. above. Path-independent integrals have important application value in dislocation, fracture mechanics and other defects theories. There have been some researches on computing methods and applications of these path-independent integrals.

Conservation laws that lead to path-independent integrals can be derived in various ways. A familiar one, involving use of the Lagrangian for the system under consideration, consists of allowing a large class of variations and imposing the invariance conditions on the corresponding action integral. An alternate direct procedure was shown by Eshelby [6] for a static elastic continuum. It involves the Lagrangian density only and consists of differentiating it with respect to the independent variables.

It is well known that piezo-electric/piezo-magnetic media produce electric/magnetic field when deformed and undergo deformation when subjected to electric/magnetic field. Due to the intrinsic coupling phenomenon, piezo-electric/piezo-magnetic materials are applied widely in science and technology field. In ref.[7], the conservation laws and path-independent integrals for piezo-electric media were obtained. Motivated by concepts of Jacobi integral and cyclic integral in analytical mechanics [8] and energy-momentum tensor [9] in electro-magntic field, the conservation laws and path-independent integrals for piezo-magnetic media are derived in this paper.

Basic equations of piezo-magnetic media

It is assumed that there is a magneto-elasticity, which the volume is expressed as V, the boundary is expressed as S, $S = S_\sigma + S_u = S_B + S_\psi$. It is assumed that magneto-elasticity is linear elasticity. The basic equations of magneto-elasticity are expressed as follows:

Governing equations

$$\sigma_{ij,j} + f_i = 0 \qquad \text{in } V \qquad (1)$$

$$\varepsilon_{ij} - \frac{1}{2}u_{i,j} - \frac{1}{2}u_{j,i} = 0 \qquad \text{in } V \qquad (2)$$

$$B_{i,i} = 0 \qquad \text{in } V \qquad (3)$$

$$H_i + \psi_{,i} = 0 \qquad \text{in } V \qquad (4)$$

boundary conditions

$$\sigma_{ij}n_j - \overline{P}_i = 0 \qquad \text{on } S_\sigma \qquad (5)$$

$$u_i - \overline{u}_i = 0 \qquad \text{on } S_u \qquad (6)$$

$$B_i n_i - \overline{b} = 0 \qquad \text{on } S_B \qquad (7)$$

$$\psi - \overline{\psi} = 0 \qquad \text{on } S_\psi \qquad (8)$$

constitutive equations [10-12]

$$\sigma_{ij} = \frac{\partial h^m}{\partial \varepsilon_{ij}}, \quad B_i = -\frac{\partial h^m}{\partial H_i} \qquad (9)$$

$$\varepsilon_{ij} = -\frac{\partial h^{el}}{\partial \sigma_{ij}}, \quad H_i = \frac{\partial h^{el}}{\partial B_i} \qquad (10)$$

For linear elasticity, magnetic enthalpy and elastic enthalpy can be expressed as follows

$$h^m = \frac{1}{2}a_{ijkl}\varepsilon_{ij}\varepsilon_{kl} - \frac{1}{2}\mu_{ij}H_iH_j - h_{kij}H_k\varepsilon_{ij} \qquad (11)$$

$$h^{el} = -\frac{1}{2}c_{ijkl}\sigma_{ij}\sigma_{kl} + \frac{1}{2}\beta_{ij}B_iB_j - b_{kij}B_k\sigma_{ij} \qquad (12)$$

where, σ_{ij} is Cauchy stress, f_i is body force, ε_{ij} is Cauchy strain, u_i is displacement, B_i is magnetic induction intensity, H_i is magnetic field intensity, ψ is magnetic potential, $\overline{P}_i$ is surface force, n_i is surface direction cosine, h^m is magnetic enthalpy, h^{el} is elastic enthalpy, $a_{ijkl}, c_{ijkl}, \mu_{ij}, \beta_{ij}, h_{kij}, b_{kij}$ are constitutive constants, "," is partial derivative of space coordinate, the other variables with overline are known variables.

Conservation equations

Conservation equatons can be expressed with magnetic enthalpy. Let functional π_1

$$\pi_1 = h^m - f_i u_i \qquad (13)$$

Considering Eqs.(2) and (4), we obtain

$$\pi_{1,k} = \frac{\partial \pi_1}{\partial u_i}u_{i,k} + \frac{\partial \pi_1}{\partial \varepsilon_{ij}}u_{i,jk} + \frac{\partial \pi_1}{\partial \psi}\psi_{,k} - \frac{\partial \pi_1}{\partial H_i}\psi_{,ik} \qquad (14)$$

Substituting constitutive equation (9) into Eq.(14), we have

$$\pi_{1,k} = (\sigma_{ij}u_{i,k} + B_j\psi_{,k})_{,j} \tag{15}$$

According to Eq.(15), the first conservation equaton on magnetic enthalpy can be expressed as follow

$$(\pi_1\delta_{kj} - \sigma_{ij}u_{i,k} - B_j\psi_{,k})_{,j} = 0 \tag{16}$$

The integral form of the conservation equaton on magnetic enthalpy is as follow

$$\oint_s (\pi_1\delta_{kj} - \sigma_{ij}u_{i,k} - B_j\psi_{,k})n_j ds = 0 \tag{17}$$

Because linear hypothesis is not used in the derivation above, Eqs. (16) and (17) are also suitable for nonlinear materials. Similar to energy-momentum tensor defined by Eshelby, and considering Eqs. (16) and (17), energy-momentum tensor of piezo-magnetic material with magnetic enthalpy is defined as follow

$$P_{1kj} = \pi_1\delta_{kj} - \sigma_{ij}u_{i,k} - B_j\psi_{,k} \tag{18}$$

Considering Eq.(15) and Eq.(18), $x_j P_{1jk}$ can be expressed as follow

$$(x_j P_{1jk})_{,k} = 3\pi_1 - (\sigma_{ij}u_{i,j} + B_j\psi_{,j}) \tag{19}$$

For linear materials, and considering constitutive equations and governing equations (2) and (4), π_1 can further be expressed as follow

$$\pi_1 = \frac{1}{2}\sigma_{ij}u_{i,j} + \frac{1}{2}B_j\psi_{,j} - f_i u_i \tag{20}$$

Substituting Eq.(20) into Eq.(19), and considering Eq.(1) and Eq.(3), we obtain

$$(x_j P_{1jk})_{,k} = \frac{1}{2}(\sigma_{ik}u_i + B_k\psi)_{,k} - \frac{5}{2}f_i u_i \tag{21}$$

When body force $f_i = 0$, the second conservation equaton on magnetic enthalpy can be derived from Eq.(21)

$$[x_j P_{1jk} - \frac{1}{2}(\sigma_{ik}u_i + B_k\psi)]_{,k} = 0 \tag{22}$$

Substituting Eq.(18) into Eq.(22), we obtain

$$[x_k\pi_1 - x_j(\sigma_{ij}u_{i,j} + B_k\psi_{,j}) - \frac{1}{2}(\sigma_{ik}u_i + B_k\psi)]_{,k} = 0 \tag{23}$$

The integral form of the second conservation equaton on magnetic enthalpy is as follow

$$\oint_s [x_k\pi_1 - x_j(\sigma_{ij}u_{i,j} + B_k\psi_{,j}) - \frac{1}{2}(\sigma_{ik}u_i + B_k\psi)]n_k ds = 0 \tag{24}$$

According to the Eqs. (17) and (24), the path-independent integrals on magnetic enthalpy are as follows

$$J_{1k} = \int(\pi_1\delta_{kj} - \sigma_{ij}u_{i,k} - B_j\psi_{,k})n_j ds \tag{25}$$

$$M_1 = \int[x_k\pi_1 - x_j(\sigma_{ij}u_{i,j} + B_k\psi_{,j}) - \frac{1}{2}(\sigma_{ik}u_i + B_k\psi)]n_k ds \tag{26}$$

Let functional $\pi_2 = h^{el}$. According to the same way, the conservation equatons on elastic enthalpy can also be derived. Due to the limit of space, they are not described in more details. The path-independent integrals on elastic enthalpy can be expressed as follows

$$J_{2k} = \int (\pi_2 \delta_{kj} + u_i \sigma_{ij,k} + \psi B_{j,k}) n_l ds \tag{27}$$

$$M_2 = \int [x_k \pi_2 + x_j (u_i \sigma_{ik,j} + \psi B_{k,j}) + \frac{3}{2}(u_i \sigma_{ik} + \psi B_k)] n_k ds \tag{28}$$

Summary

Motivated by concepts of Jacobi integral and cyclic integral in analytical mechanics and energy-momentum tensor in electro-magntic field, the conservation laws and path-independent integrals for piezo-magnetic media are derived in this paper.

Acknowledgement

This research is supported by the Fundamental Research Funds for the Central Universities of China through Grant No.HEUCF100205 and the Heilongjiang Postdoctoral Funds for Scientific Research Initiation.

References

[1] Noether E.: submitted to Math. Phys. Klasse, 1918, 2, pp.235-257.

[2] Knowles J.K., Sternberg E.: submitted to Arch. Rational Mech. Anal., 1972, 44, pp.187-211.

[3] Fletcher D.C.: submitted to Arch. Rational Mech. Anal., 1976, 60, pp.329-353.

[4] Xu Li: submitted to Engng. Fract. Mech., 1988, 29, pp.233-241.

[5] Rice J.R.: submitted to J. Appl. Mech., 1968, 35, pp.379-386.

[6] J.D. Eshelby, in: *Inelastic Behavior of Solids*, edtied by M.F. Kanninen, McGraw Hill, New York, 1970, pp.77-113.

[7] X.M. WANG, Y.P. SHEN: submitted to Acta Mech. Sin., 1995, 27(1), pp.86-92.

[8] Chen Bin: *Analytical Dynamics* (Peking University Press, Beijing 1987).

[9] L.D. Landau, E.M. Lifshitz: *The Classical Theory of Fields* (Pergamon Press, Oxford 1975).

[10] Song Haiyan, Li Haibo, Liang Lifu and Zhou Zhengong Structure & Environment Engineering, 2010, 37(2), pp.8-16.

[11] Song Hai-Yan, Zhou Zhen-Gong, Liang Li-Fu and Liu Zong-Min: Key Engineering Materials, 2010, Vols.419-420, pp:153-156.

[12] Song Haiyan, Liang Lifu and Zhou Zhengong: Journal of Harbin Engineering University, 2011, 32(1), pp:33-37.

Scattering of Anti-plane SH-wave by Multiple Cylindrical Inclusions in Elastic Semi-space

HongLiang Li[1,a], Yong Yang[1,b]

[1]Department of Engineering Mechanics, College of Aerospace Engineering and Civil Engineering, Harbin Engineering University, Harbin, 150001, China

[a]leehl@sina.com, [b]yangharbin@163.com

Keywords: semi-space; cylindrical inclusion; SH-wave scattering; dynamic stress concentration factor (DSCF)

Abstract. Multiple circular inclusions exists widely in natural media, engineering materials and modern municipal construction. The scattering field produced by multiple circular inclusions determines the dynamic stress concentration factor around the circular inclusions, and therefore determines whether the material is damaged or not. These problems are complicated, because there are many factors influenced. Researchers solved these problems by analysis and numerical methods. It is hard to obtain analytic solutions except for several simple conditions. In this paper, the solution of displacement field for elastic semi-space with multiple cylindrical inclusions by anti-plane SH-wave is constructed. In complex plane, considering the symmetry of SH-wave scattering, the displacement field aroused by the anti-plane SH-wave and the scattering displacement field impacted by the cylindrical inclusions comprised of Fourier-Bessel series with undetermined coefficients which satisfies the stress-free condition on the ground surface are constructed. Through applying the method of multi-polar coordinate system, the equations with unknown coefficients can be obtained by using the displacement and stress condition around the edge of cylindrical inclusions. According to orthogonality condition for trigonometric function, these equations can be reduced to a series of algebraic equations. Then the value of the unknown coefficients can be obtained by solving these algebraic equations. The total wave displacement field is the superposition of the displacement field aroused by the anti-plane SH-wave and the scattering displacement field. By using the expressions, an example is provided to show the effect of the change of relative location of the cylindrical inclusions. Based on this solution, the problem of interaction of multiple cylindrical inclusions and a linear crack in semi-space can be investigated further.

Introduction

Multiple circular inclusions exists widely in natural media, engineering materials and modern municipal construction. The problem of scattering of SH waves by multiple circular inclusions is one of the important and interesting questions in mechanical engineering and civil engineering for the latest decade. There are lots of materials obtained by theoretical research and earthquake damage investigation. The scattering field produced by multiple circular inclusions determines the dynamic stress concentration factor around the circular inclusions, and therefore determines whether the material is damaged or not. These problems are complicated, because there are many factors influenced. Researchers solved these problems by analysis and numerical methods[1-4]. It is hard to obtain analytic solutions except for several simple conditions.

In this paper, the solution of displacement field for elastic semi-space with multiple cylindrical inclusions by anti-plane SH-wave is constructed. By using the solution, an example is provided to show the effect of the change of relative location of the cylindrical inclusions. Based on this solution, the problem of interaction of multiple cylindrical inclusions and a linear crack in semi-space can be investigated further.

Model and Governing Equation

The model is shown as Fig.1, an elastic semi-space containing multiple shallow-buried inclusions. In this paper, the anti-plane shear wave model is studied.

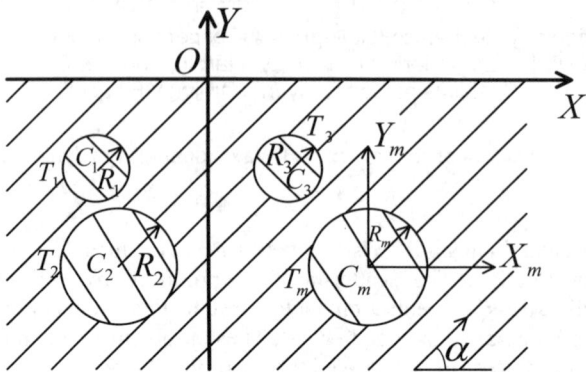

Fig.1 Model of the problem

In this paper, the anti-plane shear wave model is studied. The displacement in elastic semi-space is expressed as $W(x,y,t)$. The displacement in inclusions are expressed as $W_m(x,y,t)$ (m=1,2,...,N). In complex plane, the displacement function W and W_m satisfies the following governing equations:

$$\frac{\partial^2 W}{\partial z \partial \bar{z}} + \frac{1}{4}k^2 W = 0, \tag{1a}$$

$$\frac{\partial^2 W_m}{\partial z \partial \bar{z}} + \frac{1}{4}k_m^2 W_m = 0 \quad (m=1,2,...,N). \tag{1b}$$

where $k_i = \frac{\omega}{C_{Si}}$, $C_{Si} = \sqrt{\mu_i/\rho_i}$, ω is the circular frequency of the displacement, C_{Si} stands for the shear wave velocity, ρ_i and μ_i are the mass density and the shear modulus of elasticity respectively.

The boundary conditions can be expressed as below:

$$\tau_{\theta z} = 0 (where\ \theta = 0\ and\ \theta = \pi) \tag{2a}$$

$$W = W_m\ (|z-c_m|=R_m, m=1,2,...,N), \tau_{rz} = \tau_{rzm}\ (|z-c_m|=R_m, m=1,2,...,N). \tag{2b}$$

The solution in elastic semi-space which satisfies the control equation (1a) and the boundary conditions (2b) should include two parts of motion: the disturbance of incident SH-wave and the scattering wave incited by multiple cylindrical inclusions. To satisfy the boundary condition (2a), the wave displacement of the elastic semi-space due to the incident wave can be given as[5,6]:

$$W^{(i)} = W_0 (e^{\frac{ik}{2}(ze^{-i\alpha}+\bar{z}e^{i\alpha})} + e^{\frac{ik}{2}(ze^{i\alpha}+\bar{z}e^{-i\alpha})})$$

The scattering wave incited by multiple cylindrical inclusions and the displacement in the inclusions can be written as:

$$W^{(s)} = \sum_{j=1}^{N} \sum_{n=-\infty}^{\infty} A_n^j [H_n^{(1)}(k|z-c_j|)(\frac{z-c_j}{|z-c_j|})^n + H_n^{(1)}(k|z-\bar{c}_j|)(\frac{z-\bar{c}_j}{|z-\bar{c}_j|})^{-n}],$$

$$W_m = \sum_{n=-\infty}^{\infty} B_n^m J_n(k_m|z-c_m|)(\frac{z-c_m}{|z-c_m|})^n \quad (m=1,2,...,N)$$

where $H_n^{(1)}(*)$ is the first kind of Hankel function and n order, A_n^j and B_n^m are unknown coefficients.

Therefore, the total wave field in elastic semi-space can be written as:

$$W = W^{(i)} + W^{(s)} \tag{3}$$

The wave field must satisfy the conditions on inclusions, so by using the method of transferred coordinate, W and W_m can be written as:

$$W = W_0(e^{\frac{ik}{2}((z_m+c_m)e^{-i\alpha}+(\bar{z}_m+\bar{c}_m)e^{i\alpha})} + e^{\frac{ik}{2}((z_m+c_m)e^{i\alpha}+(\bar{z}_m+\bar{c}_m)e^{-i\alpha})})$$

$$+ \sum_{j=1}^{N}\sum_{n=-\infty}^{\infty} A_n^j [H_n^{(1)}(k|z_m-d_{mj}|)(\frac{z_m-d_{mj}}{|z_m-d_{mj}|})^n + H_n^{(1)}(k|z_m-d'_{mj}|)(\frac{z_m-d'_{mj}}{|z_m-d'_{mj}|})^{-n}]$$

$$W_m = \sum_{n=-\infty}^{\infty} B_n^m J_n(k_m|z_m|)(\frac{z_m}{|z_m|})^n .$$

where $d_{mj} = c_j - c_m$, $d'_{mj} = \bar{c}_j - c_m$.

And the boundary conditions could be changed to
$$W = W_m,$$
$$\tau_{rz} = \tau_{r_m z}^{(i)} + \tau_{r_m z}^{(is)} = \tau_{r_m z} \quad (|z_m| = R_m)..$$

Substituting the wave filed W and W_m to the boundary conditions, it can be obtained that

$$\sum_{j=1}^{N}\sum_{n=-\infty}^{\infty} A_n^j \phi_{mn}^j + \sum_{n=-\infty}^{\infty} B_n^m \phi_{mn} = \phi_m , \tag{4a}$$

$$\sum_{j=1}^{N}\sum_{n=-\infty}^{\infty} A_n^j \varphi_{mn}^j + \sum_{n=-\infty}^{\infty} B_n^m \varphi_{mn} = \varphi_m . \tag{4b}$$

where

$$\phi_{mn}^j = H_n^{(1)}(k|z_m-d_{jm}|)(\frac{z_m-d_{jm}}{|z_m-d_{jm}|})^n + H_n^{(1)}(k|z_m-d'_{jm}|)(\frac{z_m-d'_{jm}}{|z_m-d'_{jm}|})^{-n},$$

$$\phi_{mn} = -J_n(k_m|z_m|)(\frac{z_m}{|z_m|})^n,$$

$$\phi_m = W_0(e^{\frac{ik}{2}[(z_m+c_m)e^{-i\alpha}+(\bar{z}_m+\bar{c}_m)e^{i\alpha}]} + e^{\frac{ik}{2}[(z_m+c_m)e^{i\alpha}+(\bar{z}_m+\bar{c}_m)e^{-i\alpha}]}),$$

$$\varphi_{mn}^j = \frac{1}{2}\mu k A_n^j \{[H_{n-1}^{(1)}(k|z_m-d_{jm}|)(\frac{z_m-d_{jm}}{|z_m-d_{jm}|})^{n-1} - H_{n+1}^{(1)}(k|z_m-d'_{jm}|)(\frac{z_m-d'_{jm}}{|z_m-d'_{jm}|})^{-(n+1)}]e^{-i\theta_m}$$

$$-[H_{n+1}^{(1)}(k|z_m-d_{jm}|)(\frac{z_m-d_{jm}}{|z_m-d_{jm}|})^{n+1} - H_{n-1}^{(1)}(k|z_m-d'_{jm}|)(\frac{z_m-d'_{jm}}{|z_m-d'_{jm}|})^{-(n-1)}]e^{-i\theta_m}\},$$

$$\varphi_{mn} = \frac{1}{2}\mu_m k_m B_n^m [J_{n-1}(k_m|z_m|)(\frac{z_m}{|z_m|})^{n-1}e^{i\theta_m} - J_{n+1}(k_m|z_m|)(\frac{z_m}{|z_m|})^{n+1}e^{-i\theta_m}],$$

$$\varphi_m = -\frac{1}{2}i\mu k[e^{i(\theta_m-\alpha)} + e^{-i(\theta_m-\alpha)}]W_0 e^{\frac{ik}{2}[(z_m+c_m)e^{-i\alpha}+(\bar{z}_m+\bar{c}_m)e^{i\alpha}]}$$
$$-\frac{1}{2}i\mu k[e^{i(\theta_m+\alpha)} + e^{-i(\theta_m+\alpha)}]W_0 e^{\frac{ik}{2}[(z_m+c_m)e^{i\alpha}+(\bar{z}_m+\bar{c}_m)e^{-i\alpha}]}$$

By multiplying both sides of (4) with $e^{-im\theta_m}$ and integrating in interval $[-\pi,\pi]$, it can be obtained that

$$\sum_{j=1}^{N}\sum_{n=-\infty}^{\infty} A_n^j \Phi_{mn}^j + \sum_{n=-\infty}^{\infty} B_n^m \Phi_{mn} = \Phi_m \tag{5a}$$

$$\sum_{j=1}^{N}\sum_{n=-\infty}^{\infty} A_n^j \Psi_{mn}^j + \sum_{n=-\infty}^{\infty} B_n^m \Psi_{mn} = \Psi_m \tag{5b}$$

Where
$$\Phi_{mn}^j = \frac{1}{2\pi}\int_{-\pi}^{\pi} \phi_{mn}^j e^{-im\theta_m} d\theta_m, \quad \Phi_{mn} = \frac{1}{2\pi}\int_{-\pi}^{\pi} \phi_{mn} e^{-im\theta_m} d\theta_m, \quad \Phi_m = \frac{1}{2\pi}\int_{-\pi}^{\pi} \phi_m e^{-im\theta_m} d\theta_m.$$

$$\Psi_{mn}^j = \frac{1}{2\pi}\int_{-\pi}^{\pi} \varphi_{mn}^j e^{-im\theta_m} d\theta_m, \quad \Psi_{mn} = \frac{1}{2\pi}\int_{-\pi}^{\pi} \varphi_{mn} e^{-im\theta_m} d\theta_m, \quad \Psi_m = \frac{1}{2\pi}\int_{-\pi}^{\pi} \varphi_m e^{-im\theta_m} d\theta_m.$$

Equation (5) is a set of infinite algebraic equation for determining the coefficients A_n^j.

Substituting the coefficients A_n^j to Expression (3), the total wave field W of this problem can be obtained.

Example

In this paper, we pay attention to a representative kind of models. There are two cylindrical inclusions in elastic semi-space. The radius of the inclusions equals 1. The other parameters are shown in the figures. In Fig.2, it can be found that with the increasing of the burying depth of the inclusions, the influence of the ground to DSCF at the left inclusion edge is decreasing. In Fig.3, the distance between the centre of the left inclusion and the centre of the right inclusion is changed from 3 to 9. It can be found that the influence of the right inclusion is decreasing.

Summary

In this paper, by using the method of multi-polar coordinate system, the solution of displacement field for elastic semi-space with multiple cylindrical inclusions by anti-plane SH-wave is solved. Then an example is given. Based on this solution, the problem of interaction of multiple cylindrical inclusions and a linear crack in semi-space can be investigated further.

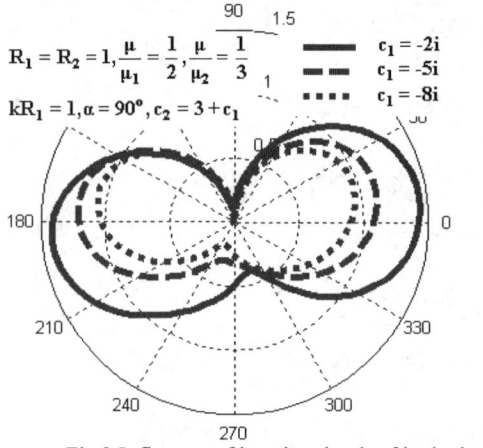
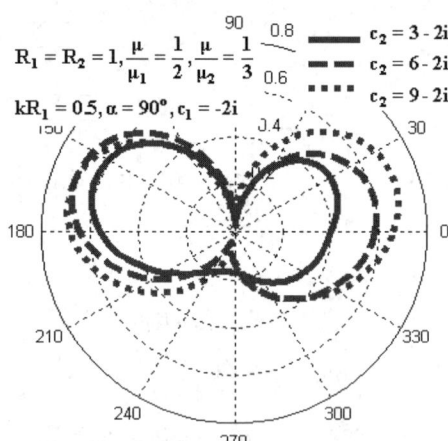

Fig.2 Influence of burying depth of inclusions Fig.3 Influence of right inclusion

Acknowledgments

Thanks for the support by the Fundamental Research Funds for the Central Universities HEUCF110201.

References

[1] Liu Diankui, Lin Hong. Scattering of SH-wave by an Interface Linear Crack and a Circular Cavity near Biomaterial Interface. Acta Mechanica Sinica, Vol.20(2004), No.3, p.317-326
[2] Wang Ren, Huang Kezhi, Yang Wei, Zheng Quanshui, He Yousheng, An Introduction to the 19th ICTAM. Mechanics and Practice, Vol.19(1997), No.1, p. 57-64
[3] Y. Yang and A. Norris: Shear Wave Scattering from a Debonded Fiber. Mech. Phys. Solids, Vol.39(1991), p.273-280
[4] J.Tian, U.Gabbert, H.Berger, and X.Y.Su: Lamb Wave Interaction with Delamination in CFRP Laminates. Computers, Materials and Continua, Vol.1(2004), p.327-336
[5] Li Hongliang, Zhang Cun: Green's Function Solution of the Semi-space with Double Shallow-buried Cavities. Key Engineering Materials, Vol.417-418(2010), p.145-148
[6] Li Hongliang, Li Hong, Yang Yong: Dynamic Stress Intensity Problem of SH-Wave by Double Linear Cracks near A Circular Hole. Key Engineering Materials, Vol.385-387(2008), p.105-108

Multiscale simulation of the size effect of nano-indentation

Zhenqing Wang[1,a], Qinghua Meng[1,b], Zengjie Yang[1,c]

[1]College of Aerospace and Civil Engineering, Harbin Engineering University, Harbin, China

[a]wangzhenqing@hrbeu.edu.cn, [b]mengqinghua@hrbeu.edu.cn, [c]yangzengjie@hrbeu.edu.cn

Keywords: Multiscale simulation; Nano-indentation; Size effect

Abstract. In this paper, the size effect of copper, aluminum and silver nanoindentation has been studied by adopting the new multiscale method. According to the analysis and comparison of load-displacement cures for the three materials under the four specimens, the effects of various physical parameters on the dislocation nucleation have been discussed. For different specimen lengths, simulation results show the apparent size effect, especially in the initial elastic deformation stage, the difference of the curve slope is very obvious. Comparing these three kinds of metal materials of the load displacement curve, it can be found that the metal copper thin film and the metal aluminum thin film in the load process, only experience one time reduced the process, but the metallic silver thin film reveals a fluctuation.

Introduction

In recent decades, large-scale integrated circuit technology and nanotechnology have development rapidly, especially when the characteristic width of the circuit reduces to the nanometer level, the study on indentation, which reveals the nano-scale mechanical properties in various types of electronic components, has become one focus of application researches. For ultra-thin coating materials, the size of the test objects is small, traditional tensile test can not meet mechanical strength and other physical quantities measurement. But for a few precious cultural relics and crafts, due to concerns about its mechanics measurement is easy to cause irreparable damage, the impact or pressure law can not meet the determination of their material hardness. Therefore nano-indentation measurement method, which is weak to the test specimen destructiveness and has a small the following influence, becomes an effective way. Nano-indentation test has received the test specimen surface defect, the size effect, the crystal orientation etc. complex factor influence. Especially when the measurement of the sample material within the grain size reduced to nano-scale, the Young's modulus, hardness etc. physical quantities have a certain correlation to shape and size of the indenter.

Yet the solving of these problems is not separable in the development and perfection of multiscale methods. In the last years, researchers have made the model discrimination in continuum scale, replaced the finite element system with particle system, and developed a new method, which is called as atomic/particle multiscale method. Based on previous atomic/particle multiscale analysis methods, a separate Quasi-particle method has been proposed [5].

In the paper, the initial plastic deformation process of aluminum, copper and silver single crystal films nano-indentation was simulated by adopting the new multiscale method, which reveals the elastic and plastic transition mechanism of the thin film under the different indenter size, and discusses the influence of the factor on nano-indentation elastic-plastic deformation process of the sample.

Simulation procedures

Separate quasi-particle simulations are performed to study the processes of the initial plastic deformation under the action of a rigid indenter on three FCC metals thin film (Al, Cu and Ag) which has a rectangle surface defect. The Lennard-Jones potential is adopted in a rectangular region, whose parameters as shown in Table1 [4]. The size of the simulated thin film is L (1000Å, 2000Å, 4000Å, 8000Å) in x-direction, H (600 Å) in y-direction, and W (8.064 Å) in z-direction, the width of indenter is $4a$. as shown in Fig.1.

Assume that the indenter and the contact surface of the thin film use frictionless contact condition and the number of atoms in both interface remains the same. The whole model adopts a periodic boundary condition in z-direction. Separate quasi-particle mode will be divided into the atomic system region and simulated particle region (as shown in Fig.2). The area size of the atomic system: length is $40a$, height is $80a$. The perimeter of the atom region is a particle system.

In model computational process, the direction of the indenter parallels to the close-packed FCC crystal, and the dislocation occurs on the close-packed surface, which is advantage to observe the dislocation nucleation and emission process. The simulation is conducted by controlling the displacement of the indenter, and the indenter dropped about 0.1Å to the inferior of the films at each step. In addition, total 81 steps are performed in this modeling until the depth of the indentation reaches 8.1Å. The energy minimization is executed in the end of each step so that these quasi-particles or atoms could be able to reach their equilibrium positions. The maximum displacement of the indenter contrasted to the thickness of the film, 8.1Å, is still very small, thus the far-field boundary conditions will not affect the result of the computation.

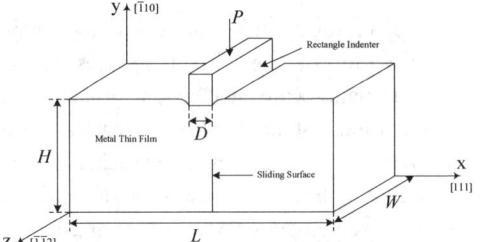
Fig. 1 Geometric model of nano-indentation model for the metal film

Fig. 2 Separate quasi-particle model for nano-indentation simulation

Table 1 Lattice constant and parameters of Lennard-Jones potential function in different metal materials

Metal Elements	e_0(eV)	r_0(Å)	a(Å)
Cu	0.004096	2.338	3.615
Al	0.005000	2.850	4.032
Ag	0.003450	2.644	4.085

Results and discussion

With the same the width of the indenter, different specimen sizes of Al, Cu and Ag film are loaded, load-displacement response cures of three metal films under the conditions of different sample sizes are shown in Fig.3(a), Fig.3(b) and Fig.3(c), respectively. Intuitively, the load value is equal to pressure divided by the width of the indenter, unit is N/m.

Load-displacement response cures of the copper films under the conditions of different sample sizes are shown in Fig.3(a), L is the specimen length. Due to this model take displacement increments of the indenter as the control quantity, the load - displacement response curve presents discontinuous and fluctuation characteristics in the whole load process, which is in good agreement with Ref.1 and 2. In the initial stage of the indenter load is applied, the load-displacement curve in the four cases are approximate monotone linear rise, which indicates that this stage is the material elastic deformation stage. With the load increase, when the load reaches its maximum value, the defects are activated below the indenter and the dislocation starts the shape nucleus, which began to transition from elastic deformation to plastic deformation. At this time the maximum load under various conditions is in the range of 30.17N/m ~ 31.46N/m, which shows change of the specimen size did not significantly affect to the maximum indenter load. But in four cases the indenter displacement corresponding to the load maximum are difference, and its value as 4Å, 3.5Å, 3.3Å, 3.2Å, respectively, which present that when the specimen is light-sized specimen, the material elastic deformation has obvious impediment to the indenter. But as the sample size increases, this effect will gradually be diminished. As the indenter displacement increases, the load value is gradually reduced, the material enters the plastic deformation stage, and the region below indenter presents dislocation nucleation. After the curve dropped to its lowest level, started to rise, which shows that internal dislocation of the material below indenter has reached a new equilibrium, and the material is strengthened. Combining the above curve, the following conclusions can be drawn: the specimen length of the copper crystal film is greater, first inflection point of the curve is the closer to the left, namely the corresponding indenter displacement is also smaller, and the dislocation is more approaches to the lower part of the thin film.

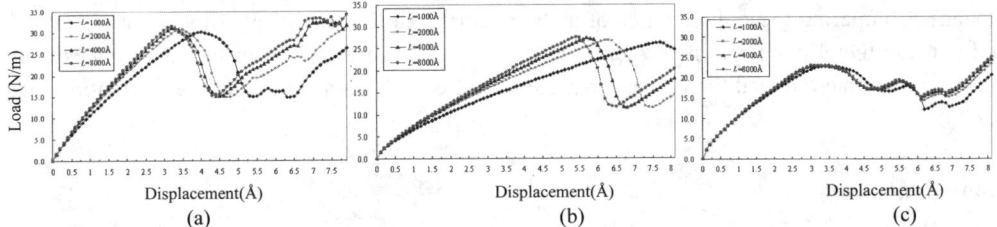

Fig. 3 Load-displacement response cures of the thin films under the conditions of different sample sizes

Load-displacement response cures of the aluminum films are shown in Fig.3(b). At the initial stages of loading, the curve is monotonously increase, which shows that at this stage material is perfect elasticity. The maximum load under various conditions is in the range of 26.23N/m ~ 27.56N/m, which corresponding to the indenter displacement are 4Å, 3.5Å, 3.3Å, 3.2Å, respectively. Comparison of the four curves, it can be found that the sample size is smaller, and the elastic deformation process is longer. With the increasing of the indenter displacement, load curve began to decline, which indicates that material begins as a state of plastic deformation. And the load from a maximum value to a minimum value of the amplitude is basically the same, which shows that material to resist external load capacity does not change with the work piece size. When the curve reaches a low point, internal dislocation have been completed and reached a new equilibrium state. After the curve dropped to its lowest level, the dislocation nucleation has occurred. As the indenter displacement increases, the curve starts to rise. It is not difficult to find the minimum load value of the materials are basically the same and the slope of the curve also remained unchanged, which shows that the dislocation nucleation for different scales of the specimen has only a delayed effect, it will not change the material constitutive characteristics.

Load-displacement response cures of the silver films under the conditions of different sample sizes are shown in Fig.3(c). At the initial stage of the indenter loading, load-displacement cures under the four conditions are monotone rise, which shows material is perfect elasticity. The maximum approximation of the indenter load is about 23N/m, the range of variation is very small, which corresponding to the indenter displacement is in the range of 3.2 Å -3.6 Å. As the indenter displacement increases, the load value is gradually reduced. With the loading displacement increases, the metal silver crystal material began to show plastic characteristics. Internal dislocation below the indenter material will constantly form new structure through the nucleation, accumulation, cross diffusion process. As the completion of the dislocation nucleation and the continuous release of energy, the crystal material will form a stable microscopic structure. Overall speaking, with the loading displacement increases, the metal silver thin film material in the four case of load variation is consistent, which illustrates the specimen scale change to silver metal film material effect is relatively small.

Conclusions

According to the analysis and comparison of load-displacement cures for the three materials under the four specimens, the effects of various physical parameters on the dislocation nucleation have been discussed. For different specimen lengths, simulation results show the apparent size effect, especially in the initial elastic deformation stage, the curve slope difference is very obvious. When the material enters the plastic deformation stage, the load will suddenly decrease. This process corresponds to the material internal dislocation nucleation. With the loading displacement increases, the material will be strengthened, the load-displacement response curve will enter the next stage. Comparing these three kinds of metal materials of the load displacement curve, it can be found that the metal copper thin film and the metal aluminum thin film in the load process, only experience one time reduced the process, but the metallic silver thin film reveals a fluctuation many times.

Acknowledgements

This work is supported by the Research Fund for the Doctoral Program of Higher Education of China (No. 20112304110015).

References

[1] J. R. Rice: J. Mech. Phys. Solids. Vol.40(1992), p.239.
[2] W. W. Gerberich and J. C. Nelson: Acta Materialia. Vol.44(1996), p.3585.
[3] H. T. Wang, Z. D. Qin, Y. S. Ni and W. Zhang: Trans Nonferrous Met Soc China. Vol.18(2008), p.1164.
[4] J. Fan: John Wiley & Sons, Ltd. 2010.
[5] Z. Q. Wang, Z. J. Yang and Q. H. Meng (to be published).

J-Integral and Its Dual Form Based on Finite Deformation Theory

LIU Zongmin[1, a], MAO Jize[1, b], SONG Haiyan[1, c]

[1] College of Aerospace and Civil Engineering, Harbin Engineering University, Harbin 150001, China

[a] lzm1976@yahoo.com.cn, [b] maojize@hrbeu.edu.cn, [c] shyivg@yahoo.com.cn

Keywords: J-integral, dual form, finite deformation

Abstract. In general, there are large strains and displacements on the crack tip. So it is necessary to study J-integral and its dual form based on finite deformation theory. Base forces theory is a new theory for describing finite deformation. J-integral and its dual form based on base forces theory are presented in the paper. This work provides theoretical foundation for studying crack propagation.

Introduction

J-integral is a very important concept in fracture mechanics[1]. Within the confines of small deformation, J-integral, energy release rate and srress intensity factor are all equivalent fracture parameters[2,3]. However for a finite deformation body, there are large strains and displacements on the crack tip[4]. So it is necessary to study J-integral and its dual form based on finite deformation theory.

Base forces theory, which was proposed by Y.C. Gao[5], is a new theory for describing finite deformation. When base forces are adopted, a series of problems of large strain elasticity with singular points can be easily solved[6]. According to the dual relationship between potential energy and complementary energy in general mechanics[7], J-integral and its dual form based on base forces theory are presented in the paper. This work provides a new theoretical foundation for studying crack propagation.

Basic equations of base forces theory

It is assumed that there is a finite deformation body, which the volume is expressed as V, the boundary is expressed as S, $S = S_\sigma + S_u$. Basic equations of base forces theory are as follows

equilibrium equation $\quad\quad \dfrac{1}{V_P}\dfrac{\partial \boldsymbol{T}^i}{\partial x^i} + \rho_0 \boldsymbol{f} = 0 \quad\quad$ in V $\quad\quad$ (1)

geometric equation $\quad\quad \boldsymbol{g}_i = \dfrac{\partial \boldsymbol{u}}{\partial x^i} \quad\quad$ in V $\quad\quad$ (2)

constitutive relations $\quad\quad \boldsymbol{T}^i = \rho_0 V_P \dfrac{\partial A}{\partial \boldsymbol{g}_i} \quad\quad$ in V $\quad\quad$ (3)

$$g_i = \rho_0 V_P \frac{\partial B}{\partial T^i} \quad \text{in } V \tag{4}$$

boundary conditions
$$\frac{1}{V_P} T^i m_i = \bar{\tau}_0 \quad \text{on } S_\sigma \tag{5}$$

$$\boldsymbol{u} = \bar{\boldsymbol{u}} \quad \text{on } S_u \tag{6}$$

where, T^i is base forces, $\boldsymbol{f}$ is body forces per unit mass, ρ_0 is original mass density, g_i is displacement gradient, $\boldsymbol{u}$ is displacement, V_P is original base volume, A is strain energy per unit, B is strain complementary energy per unit, m_i are components of unit normal vector of the surface, $\bar{\tau}_0$ is traction per unit original area of a given boundary, $\bar{\boldsymbol{u}}$ is given displacement.

J-integral based on base forces theory

Rice's J-integral is a potential energy type integral. For a given two-dimensional crack system (shown in Fig.1), J-integral can be expressed as potential energy release rate, i.e.

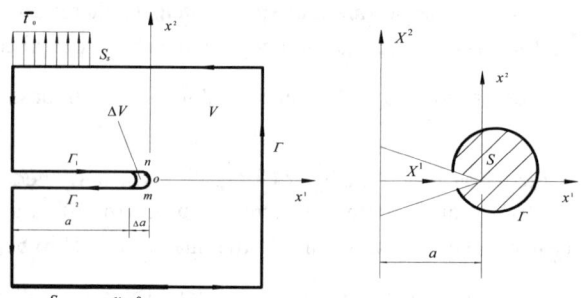

Fig.1 Alternative integral contours for 2D crack system and local coordinate system

$$J = -\frac{d\Pi_P}{da} \tag{7}$$

where, Π_P is functional of potential energy, a is crack length. When it is free of body forces in V, $\Pi_P = \iiint_V \rho_0 A dV - \iint_{S_\sigma} \bar{\tau}_0 \cdot \boldsymbol{u} dS$ [8]. It is assumed that crack was propagated along X_1 axis. Introducting local coordinate (x^1, x^2), and considering $x^1 = X^1 - a$, $x^2 = X^2$ (shown in Fig.1), we get

$$\frac{d}{da} = \frac{\partial}{\partial a} + \frac{\partial}{\partial x^1} \frac{\partial x^1}{\partial a} = \frac{\partial}{\partial a} - \frac{\partial}{\partial x^1} \tag{8}$$

Considering Eq.(8), the right side of Eq.(7) is rewritten as

$$\frac{d\Pi_P}{da} = \iint_S \rho_0 [\frac{\partial A}{\partial a} - \frac{\partial A}{\partial x^1}] dS - \int_\Gamma \bar{\tau}_0 \cdot [\frac{\partial \boldsymbol{u}}{\partial a} - \frac{\partial \boldsymbol{u}}{\partial x^1}] d\Gamma \tag{9}$$

According to virtual work principle and Green theorem, and considering Eq.(5), the first integral on the right side of Eq.(9) is rewritten as

$$\iint_S \rho_0 \frac{\partial A}{\partial a} dS = \iint_S \rho_0 \frac{\partial A}{\partial g_i} \cdot \frac{\partial g_i}{\partial a} dS = \iint_S \frac{1}{V_P} T^i \cdot \frac{\partial g_i}{\partial a} dS = \int_\Gamma \bar{\tau}_0 \cdot \frac{\partial u}{\partial a} d\Gamma \qquad (10)$$

$$\iint_S \rho_0 \frac{\partial A}{\partial x^1} dS = \int_\Gamma \rho_0 A dx^2 \qquad (11)$$

Substituting Eqs.(9) ~ (11) into Eq.(7), we get

$$J = -\frac{d\Pi_p}{da} = \int_\Gamma [\rho_0 A dx^2 - \bar{\tau}_0 \cdot \frac{\partial u}{\partial x^1} d\Gamma] \qquad (12)$$

Eq.(12) is J-integral based on base forces theory.

I-integral based on base forces theory

According to the dual relationship between potential energy and complementary energy in general mechanics [7], J-integral dual form, which is named I-integral, can be expressed as the complementary energy release rate, i.e.

$$I = \frac{d\Pi_c}{da} \qquad (13)$$

where, Π_c is functional of complementary energy, $\Pi_c = \iiint_V \rho_0 B dV - \iint_{S_u} \frac{1}{V_P} \bar{u} \cdot T^i m_i dS$ [8].

In this case (shown in Fig.1), $\partial V = \Gamma + \Gamma_1 + \Gamma_2 + nm$ and $nm = -mn$ indicated the front of crack, i.e., the tip of a smooth ended notch, as Rice [1] termed. The right side of Eq.(13) is rewritten as

$$\frac{d\Pi_c}{da} = \frac{\partial \Pi_c}{\partial a} - \frac{\partial \Pi_c}{\partial x^1} - \int_m^n \rho_0 B dx^2 \qquad (14)$$

It is convenient to use the complementary virtual work principle, which states that for the stress field satisfying the equilibrium equation and the prescribed traction boundary condition, there exsits the energy identity

$$\iiint_V \frac{1}{V_P} g_i \cdot T^i dV = \iint_{S_u} \frac{1}{V_P} \bar{u} \cdot T^i m_i dS \qquad (15)$$

Considering Eqs.(13) and (15), the first and second terms on the right side of Eq.(14) is rewritten as

$$\frac{\partial \Pi_c}{\partial a} = \iint_S \rho_0 \frac{\partial B}{\partial T^i} \cdot \frac{\partial T^i}{\partial a} dS - \int_\Gamma \frac{1}{V_P} \bar{u} \cdot \frac{\partial T^i}{\partial a} m_i d\Gamma = \iint_S \frac{1}{V_P} g_i \cdot \frac{\partial T^i}{\partial a} dS - \int_\Gamma \frac{1}{V_P} \bar{u} \cdot \frac{\partial T^i}{\partial a} m_i d\Gamma = 0 \qquad (16)$$

$$\frac{\partial \Pi_c}{\partial x^1} = \int_\Gamma \rho_0 B dx^2 - \int_m^n \rho_0 B dx^2 - \int_\Gamma \frac{1}{V_P} u \cdot \frac{\partial T^i}{\partial x^1} m_i d\Gamma + \int_m^n \frac{1}{V_P} u \cdot \frac{\partial T^i}{\partial x^1} m_i d\Gamma \qquad (17)$$

Regardless of stress distributions near the crack tip are symmetric or anti-symmetric [9], we obtained

$$\int_m^n \frac{1}{V_P} u \cdot \frac{\partial (t^{ij} Q_j)}{\partial x^1} m_i d\Gamma = \int_m^n \frac{\partial}{\partial x_j} (\frac{1}{V_P} u \cdot t^{i2} Q_2) dx_j \qquad (18)$$

$$\oint \frac{\partial}{\partial x_j}(\frac{1}{V_P}\boldsymbol{u}\cdot t^{i2}\boldsymbol{Q}_2)dx_j = \int_\Gamma \frac{\partial}{\partial x_j}(\frac{1}{V_P}\boldsymbol{u}\cdot t^{i2}\boldsymbol{Q}_2)dx_j - \int_m^n \frac{\partial}{\partial x_j}(\frac{1}{V_P}\boldsymbol{u}\cdot t^{i2}\boldsymbol{Q}_2)dx_j = 0 \qquad (19)$$

where, t^{ij} are components of $\boldsymbol{T}^i$ on current base vectors $\boldsymbol{Q}_j$. Substituting Eqs.(16) ~ (19) into Eq.(14), we get

$$\frac{d\Pi_c}{da} = -\int_\Gamma \rho_0 B dx^2 + \int_\Gamma \frac{1}{V_P}\boldsymbol{u}\cdot\frac{\partial(t^{ij}\boldsymbol{Q}_j)}{\partial x^1}m_i d\Gamma + \int_\Gamma \frac{\partial}{\partial x_j}(\frac{1}{V_P}\boldsymbol{u}\cdot t^{i2}\boldsymbol{Q}_2)dx_j = I \qquad (20)$$

Eq.(20) is I-integral based on base forces theory. When the admissible variables $\boldsymbol{u}, \boldsymbol{T}^i, \boldsymbol{g}_i$ are actual solutions, $\Pi_c = -\Pi_p$. According to Eqs.(12) and (20), we derived

$$J = -\frac{d\Pi_p}{da} = \frac{d\Pi_c}{da} = I \qquad (21)$$

From Eq.(21), it can be verified that J-integral and I-integral are dual form.

Summary

Base forces theory is a new theory for describing finite deformation. J-integral and I-integral based on base forces theory are presented in the paper. This work provides a new theoretical foundation for studying crack propagation.

Acknowledgement

This research is supported by Fundamental Research Funds for the Central Universities of China (HEUCF100205, HEUCFZ1102), National Natural Science Foundation of China (50908059), and Heilongjiang Postdoctoral Funds for Scientific Research Initiation.

References

[1] Rice J.R.: submitted to J. Appl. Mech., 1968, 35, pp.379-386.

[2] M.F. Kanninen: submitted to J. Int. Fracture, 1974, 10, pp.415-430.

[3] SONG Haiyan, FAN Tao: Energy Principle in Crack Propagation, Key Engineering Materials, 2012, Vols.488-489, pp: 593-596.

[4] Wu X.F., Fan T.Y. and Liu C.H.: submitted to Appl. Math. Mech., 1999, 20(3), pp.301-304.

[5] Gao Y.C.: submitted to Arch. Appl. Mech., 2003, 73, pp.171-183.

[6] Gao Y.C., Jin M., Dui G.S.: submitted to Appl. Mech. Rev., 2008, 61, pp. 030801-1-16.

[7] Song Hai-yan, Liang Li-fu, Zhou Yi-lei : Dual Form of Generalized Variational Principles in General Mechanics, Journal of Harbin Engineering University, 2004, 25(6), pp: 740-742, 760.

[8] LIU Zongmin: *Quasi-variational principles for large elastic deformation based on base forces*, Harbin Engineering University Doctor Degree Dissertation, (2008).

[9] C.C.Wu, Q.Z.Xiao and Genki Yagawa: submitted to Int. J. Solids Struct., 1998, 35(14), pp.1635-1652.

Study on the Diffusion Coefficient in NGS under Low Tensile Stress

Qiao Ying-jie[a], Fu Hong-bo[b], Li Chun-kai

College of Material Science and Chemical Engineering, Harbin Engineering University,
Harbin 150001, China

[a]qiaoyingjie@hrbeu.edu.cn, [b]fuhongbo@hrbeu.edu.cn,

Keywords: Diffusion coefficient; non-equilibrium grain-boundary segregation (NGS); phosphorus; low tensile stress

Abstract. In this paper, the diffusion coefficient of vacancy- solute complex and diffusion coefficient of solute atoms, P, in non-equilibrium grain-boundary segregation under low tensile stress in steel 12Cr1MoV are calculated based on the model developed by Xu. Both the simulation and experimental kinetics curves are achieved. Results show that the calculated result with the kinetic equations perfectly fits with the experimental observations for phosphorus in steel 12Cr1MoV under low tensile stress. In the segregation process, the rate of phosphorus segregation is high due to a large diffusivity of the complex. In the de-segregation process, the level of phosphorus segregation decreases with increasing stress aging time. But the rate of phosphorus desegregation from grain boundaries to centre is slower compared with that in segregation for the diffusion coefficient of phosphorus atoms is far lower than the complexes.

Introduction

The subject of the grain-boundary segregation of solute has attracted considerable attention, since it is widely recognized that such solute segregation not only control many of the mechanical, chemical and electronic properties of technologically important materials, but also induce grain boundary structural transformations. The solute segregation to grain boundaries may be classified into equilibrium and non-equilibrium segregation. Non-equilibrium grain-boundary segregation（NGS）is a complex dynamic process and can be classified in three general categories: (1) thermally induced segregation; (2) stress-driven segregation; (3) neutron irradiation-induced segregation. In the three categories, oversaturated vacancies are all produced and then induced NGS.

Generally, steel structures are subjected to the working stresses which are far lower than the yield stresses, therefore, the investigations on the NGS induced by low stresses are helpful to predict the service life of the steel structures, especially for the boiler steel. This is because boiler usually works at temperatures of 450°C-540°C and under low tensile stresses about 15MPa-30MPa, where, it can be easily caused the NGS of embrittling impurity in steel structures. Study the diffusion coefficient of complex and solute atom in NGS, is the basic work to analyse the NGS behaviour of the steel.

In this paper, the diffusion coefficient of vacancy- solute complex and diffusion coefficient of solute atom, P, in NGS under low tensile stress in steel 12Cr1MoV are calculated based on the model developed by Xu.

NGS Model under Tensile Stress

A model of non-equilibrium segregation in grain boundary under low stress was developed by Xu in references [1-3], in which the non-equilibrium segregation kinetics process under low tensile stress can be characterized on experimental foundation. The NGS was simulated by Xu based on the data of Shinoda [4] and Misra [5, 6].

The maximum vacancy concentration in the grain-boundary region, $C_{V(\sigma=\sigma)}$, induced by the tensile stress is given by equation (1) [1, 2]:

$$C_{V(\sigma=\sigma)} = C_{V(\sigma=0)} + (K_0/2)\sigma^2/(EF_V) \qquad (1)$$

Where $C_{V(\sigma=0)}$ is the equilibrium vacancy concentration in the grain-boundary region under no stress. F_V is the formation energy of a vacancy in the grain boundary region, E is the elastic or Young's modulus of grain boundary region, K_o is a constant coefficient as a geometric factor.

It is assumed that one vacancy and one solute atom combine to form one complex, which diffuses from the grain-boundary to the interior. The maximum concentration of solute, induced by the applied tensile stress σ in the boundary region, $C_{b(\sigma=\sigma)}$, is given by equation (2) [1, 2]:

$$C_{b(\sigma=\sigma)} - C_{b(\sigma=0)} = (K_0/2)\sigma^2/(EF_V) \qquad (2)$$

Where $C_{b(\sigma=\sigma)}$ is the equilibrium grain boundary concentration of solute under no stress.

The model in [1, 2] is valid only when the applied stress and aging temperature does not reach a threshold value to cause any plastic flow (no any diffusional creep).

Kinetics Equations of NGS

The fact that a critical time exists is the most characteristic aspect of non-equilibrium segregation. The critical time at a stress-aging temperature T, t_c (T), is given by [1]

$$t_c(T) = d^2 \ln(D_c/D_i)/[4\delta(D_c - D_i)] \qquad (3)$$

where d is the average grain size, δ is the critical time constant, D_c is the diffusion coefficient of vacancy-solute complex in the matrix and D_i is the diffusion coefficient of solute atoms in the matrix.

The isothermal kinetic relationship for non-equilibrium segregation in the segregation processes($t < t_c$), is given in [6] as

$$[C_b(t) - C_{b(\sigma=0)}]/[C_{b(\sigma=\sigma)} - C_{b(\sigma=0)}] = 1 - \exp(4D_c t/\alpha_{i+1}^2 d_n^2)\mathrm{erfc}[2(D_c t)^{1/2}/\alpha_{i+1} d_n] \qquad (4)$$

where $C_b(t)$ is the concentration of the concentrated layer in the grain-boundary region when stress aging time is equal to t, changing with the stress aging time. $D = D_c$ is the diffusion coefficient of complexes, d is the average grain size, t is the isothermal holding time at the constant temperature.

Equation (4) describes the non-equilibrium segregation concentration, $C_b(t)$, of the solute atoms at the grain boundary as a function of the isothermal holding time t at the constant temperature when t is shorter than the critical time t_c ($t < t_c$), i.e. the segregation process of the complexes to the grain boundaries from the center is dominant.

In the de-segregation processes($t > t_c$), the equation is given as:

$$[C_b(t) - C_g]/[C_{b(\sigma=\sigma)} - C_g] = 1/2\{\mathrm{erf}[(d_n/2)/[4D_i(t-t_c)]^{1/2}$$
$$- \mathrm{erf}[(-d_n/2)/[4D_i(t-t_c)]^{1/2}\} \qquad (5)$$

Where D_i is the diffusion coefficient of solute, C_g is the concentration of solute atoms within the grain.

Equation (5) describes the non-equilibrium segregation concentration, $C_b(t)$, of the solute atoms at grain boundary as a function of the holding time t at constant temperature when $t > t_c$, i.e. the diffusion process of solute atoms from grain boundaries to center is dominant.

Kinetic Simulation on Experimental Data

Experimental studies were carried out on a low alloy industrial 12Cr1MoV steel of composition listed in Table.1 [8]. In the experiment, all specimens were firstly solution treated at 1050°C and maintained for 1 hour. After quenching into water of 293K, they were tempered at 473K for two hours. And then isothermal holding at 813K for 1800h.

Table 1 Chemical Compositions of Steel 12Cr1MoV

Element.	C	Si	Mn	P	S	Cr	Ni	Mo	V
Wt.%	0.14	0.22	0.62	0.019	0.015	1.05	0.02	0.27	0.17

The experimental non-equilibrium grain-boundary segregation kinetics curves of phosphorus for 12Cr1MoV steel (30MPa, 540°C) is given in Fig.1.

Basing on the experimental data and Xu's model, the simulation was carried out by the calculation program. The simulation result of 12Cr1MoV steel (30MPa, 540°C) was achieved and given in Table 2. The NGS kinetic curve obtained with equations (4) and (5) was given in Fig. 2. Where, the critical time is 5 minute (0.0833h), parameters d (the grain size) is 0.01 μm, and C_g (the concentration of phosphorus atoms within the grain) used in the simulation is 0.001790209. (According Table 1, chemical composition of 12Cr1MoV steel, the at% of P element can be calculated. It is 0.1790209%).

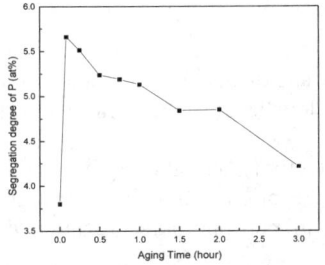 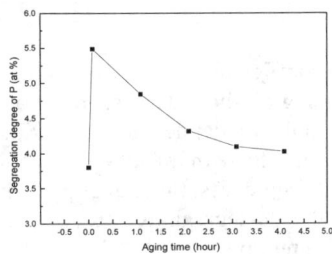

Fig. 1 Experimental kinetic NGS curve (30MPa, 540°C)   Fig. 2 Simulation kinetic NGS curve

Table 2 Simulation data of 12Cr1MoV steel

Time (h)	D_c (10^-12 m2s-1)	D_i (10^-23 m2s-1)
0	0.8583774	
0.0833	0.8583774	
0.0833		178.6611
1.0833		141.5012
2.0833		112.0702
3.0833		88.76058
4.0833		70.29916
8.0833		27.66115

It is clear that there are two processes, one is the diffusion of complex (P-Vacancy) in segregation process and another is the diffusion of P atom in de-segregation process. For the segregation process, the diffusion coefficient of complexes, D_c, can be calculated according to the equation (4) and the segregation concentration of phosphorus. The relation of D_c vs stress aging time t is as follow:

$$D_c(t)=10^{\wedge}(0.8053175\times t-12.06632)m^2s^{-1} \qquad (6)$$

For the de-segregation process, the diffusion coefficient of P atom, D_i, can be calculated according to the equation (5) and the segregation concentration of phosphorus. The relation of D_i vs stress aging time t is as follow:

$$D_i(t)=10^{\wedge}(-0.10127\times t-20.73953)m^2s^{-1} \qquad (7)$$

It is obvious that the concentration of phosphorus atoms in grain boundaries increased first, and then decreased with increasing aging time. The diffusion velocity of complex did not change in the segregation processes, whereas the diffusion velocity of P atom decreased with aging time in the de-segregation processes.

From Fig. 1 and 2, it is seen that the simulation result with the kinetic equations (4) and (5) perfectly fits with the experimental observations for phosphorus in steel 12Cr1MoV. It also shows that in the segregation process, the rate of phosphorus segregation is high due to a large diffusivity of the complex, D_c. In the de-segregation process, the level of phosphorus segregation decreases with increasing stress aging time t. But the rate of phosphorus desegregation from grain boundaries to center is slower compared with that in segregation for the diffusion coefficient of phosphorus atoms in the matrix, D_i, is far lower than the complexes, D_c. It is at the critical time that the reverse phosphorus atoms diffusion from the grain boundaries to the center balances the complex diffusion to grain boundaries, and the concentration of phosphorus atoms in grain boundaries reaches a maximum.

Conclusions

1) The diffusion coefficient of vacancy- solute complex and diffusion coefficient of solute atoms, P, in non-equilibrium grain-boundary segregation under low tensile stress in steel 12Cr1MoV are calculated based on the model developed by Xu.
2) The calculated results with the kinetic equations perfectly fit with the experimental observations.
3) In the segregation process, the rate of phosphorus segregation is high due to a large diffusivity of the complex. In the de-segregation process, the level of phosphorus segregation decreases with increasing stress aging time.

Acknowledgement

The work is financially supported by the Fundamental Research Foundation of HEU.

References

[1] T. D. Xu, Progress in Materials Science, **49** (2004) 109

[2] T. D. Xu, Scipta Materialia. Vol. 46 (2002), P. 759-763

[3] T. D. Xu, Journal of Materials Science. Vol. 35 (2000), P. 5621- 5628

[4] Shinoda T, Nakamura T. Shinoda T, ed. Acta Metal. Vol. 29 (1981), P 1631

[5] Misra. R. D. K. Acta Mater. Vol. 44 (1996), P. 885-890

[6] Misra. R. D. K. Scripta Mater. Vol. 35 (1996), P. 755-760

[7] T. D. Xu, S. H. SongActa Matall. Vol. 37 (1989), P. 2499

[8] Y. D. Fu, Key Engineering Materials, Vols. 353-358 (2007) pp. 396-399

Pressureless Sintering and Properties of Boron Carbide-Titanium Diboride Composites by In-situ Reaction

Liu Ai-dong[a], Qiao Ying-jie[b], Liu Ying-ying

College of Material Science and Chemical Engineering, Harbin Engineering University,

Harbin 150001, China

[a]liuaidong@hrbeu.edu.cn, [b] qiaoyingjie@hrbeu.edu.cn

Keywords: Pressureless sintering; boron carbide; titanium diboride; mechanical properties

Abstract. Pressureless sintering to obtain high density boron carbide-titanium diboride composites by in-situ reaction was studied. Pressureless sintering behavior of this material was investigated between 1800-2150 °C .The effects of composition, sintering temperature and tine were examined. Density up to 98.5% T.D. was reached at 2150°C. Maximum values of flexural strength (502 MPa), hardness (33 Gpa) and fracture toughnes (4.6 MPa·m$^{1/2}$) were observed in the specimens containing 15 vol.% TiB$_2$.

Introduction

Boron carbide (B$_4$C) is a strategic material, finding applications in nuclear industry, armor for personnel and vehicle safety, rocket propellant, etc [1]. Boron carbide is also known as black diamond and is the third hardest material after diamond and cubic boron nitride[2-4]. Boron carbide possesses high hardness and low density (2.52g/cm^3) [5]. Moreover the low density of B$_4$C and its high Young's modulus recommend this material for the construction of light-weight armor such as in the military helicopter and similar aero-applications [6]. However, the application range of boron carbide has been greatly limited by its brittleness, low strength, poor sinterability, and the need of very high sintering temperatures [7-8]. In order to promote densification at relatively low temperatures, different sintering additives have been proposed such as Si, Al, Al$_2$O$_3$, SiC, ZrB$_2$, BN [9–11]. Among the studied systems, the TiB$_2$ phase is introduced through reactive synthesis routes [12]. Through these reactions, B4C and TiB$_2$ with active surfaces are formed without introduction of surface B$_2$O$_3$ impurity [13]. In the present study, in order to obtain high density boron carbide-titanium diboride composites, B$_4$C–TiB$_2$ composites were in-situ synthesized using pressureless sintering. Pressureless sintering behavior and mechanical of this material were investigated.

Experimental

Commercially available high-purity B$_4$C (Chengdu Greenworld Chemical, China, B:C = 3.8–3.9, average particle size = 1.3μm), TiO$_2$ (Chengdu Greenworld Chemical, China, 99.99% pure and particle size <1μm) powders , and glucose (99.99% pure, Tianda, Co., China)were used as starting materials. The compositions of each batch along the expected sintered compositions are listed in Table1.The powder was dispersed by ZrO$_2$ balls with diameters of 2 and 5 mm were used as milling media. The blended powders were dried at 75°C for 24 hours. The powder blends were compacted by cold uniaxial pressing in a tool-steel die at a pressure of 10 MPa into parallelepiped shaped green compacts with dimensions of 50x50x10 mm. Samples were then heated for 1 h at 900–1000°C to burn out the phenolic resin binder. After reaching 1800°C, the heating schedule proceeded as follows:(i) The specimensi ntended for the constant heating rate studies were heated up to a temperature1850°C and 2150 °C (with an increment of 50°C) with a rate of 1 to10°C/min, and then furnace cooled to room temperature; (ii) The specimens intended 0C/min to a sintering temperature in range of 2050—2150°C at which were held for 1-60 minutes and then cooled with a rate of 20°C/min.

Bulk and apparent density of the samples were measured by Archimedes's method. Crystalline phases were characterized by X-ray diffraction (XRD) (Philips, model Xpert).The hardness (HV) was evaluated by micro-Vickers hardness tester (401MVD, Wowei scientific and technical Co. Ltd., Beijing, China). The flexural strength and the fracture toughness were evaluated using hydraulic universal testing machine (WE-10A, Mts Systems (China) Co. Ltd.,China), respectively. The flexural strength was measured by three point bending tests (specimen size=40mm×4mm×3mm, bend span=20mm, load speed=0.05mm/min). The fracture toughness was measured by single edge notched beam method (specimen size=30mm×5mm×5mm, notch width=0.2mm, notch depth=3mm, bend span=20mm, load speed=0.05mm/min).

Table 1 Compositions of specimens and crystalline phases detected by XRD analysis

Specimen	Green			Sintered	
	B_4C wt.%	TiO_2 wt.%	glucose wt.%	TiB_2 vol.%	Phases
BC	100	0	0	0	B_4C
BT5	88.69	9.23	5.2	5	TiB_2/B_4C
BT10	79.42	16.79	9.48	10	TiB_2/B_4C
BT15	71.72	23.08	13	15	TiB_2/B_4C
BT20	65.19	28.40	16.02	20	TiB_2/B_4C
BT25	59.60	32.97	18.58	25	TiB_2/B_4C

Results and discussion

Table1 gives the crystalline phases detected by XRD in the B_4C–TiB_2 composites. The TiB_2 was obtained by the reaction of B_4C, TiO_2 and glucose shown as the following:

$$B_4C + 2TiO_2 + 3C \rightarrow 2TiB_2 + 4CO \uparrow \qquad (1)$$

Fig.1 shows that, at 1800°C, the relative densities of all specimens are lower than the corresponding green densities. At temperatures higher than 1800°C, the difference in relative density between samples with different compositions diminishes. At a sintering temperature of 2150°C, the sintered densities of all compositions, except of pure B_4C, fall in the range of 95 to 96%, while the density of the pure B_4C specimen remains at the level of 88%.This demonstrates the poor sinterability of undoped boron carbide and the significance of the use of TiO_2 and glucose sintering aids.

Fig. 2 shows the isothermal evolution of the relative density of the B_4C-TiB_2 specimens, and sintered by heating with a rate of 10°C/min followed by holding at 2100°C for 1-60 min. Inspection of Figure 2 shows that, at 2100°C, between 20 and 60 minutes, the density increases at a moderately low rate and reaches a level of 95-98% at 60 minutes of sintering.

Fig.3 shows the relative densities of the B_4C-TiB_2 specimens, and sintered by 60 minute holding at sintering temperatures 2150°C.The presence of TiB_2 obtained by the in-situ conversion of TiO_2 is responsible inenhancing the driving force for sintering, and therefore, the increase in the level of additives will increase the sinterability. However, it is worth pointing out that the addition of TiO_2 is accompanied by the outburst of the CO gas (Eq. 1).The large volumes of CO may slow down the densification process and lead to some reduction in density. This behaviour is depicted in Figure 2 which shows a slight decrease in density when the level of TiO_2 exceeds 23 wt.%.

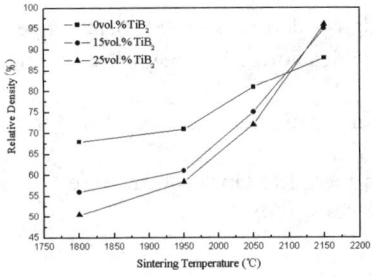
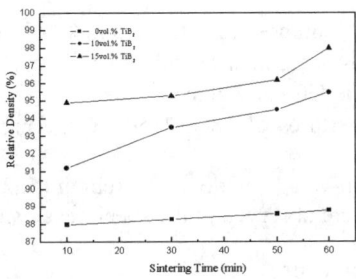

Fig. 1 Effect of sintering temperature on relative density of B$_4$C-TiB$_2$ specimens

Fig. 2 Effect of sintering time on relative density of B$_4$C-TiB$_2$ specimens

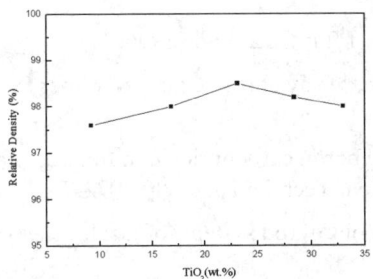
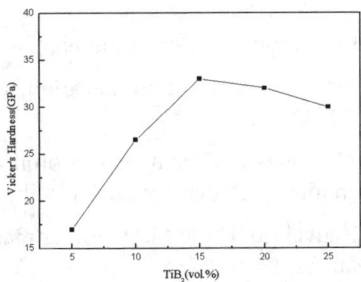

Fig. 3 Effect of wt.% TiO$_2$ on relative density of B$_4$C-TiB$_2$ specimens

Fig. 4 Effect of vol.% TiB$_2$ on Vicker's hardness of B$_4$C-TiB$_2$ specimens

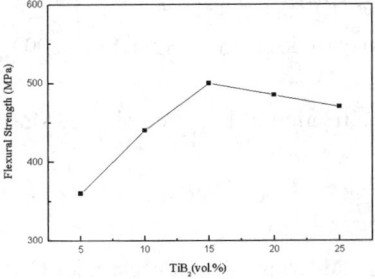
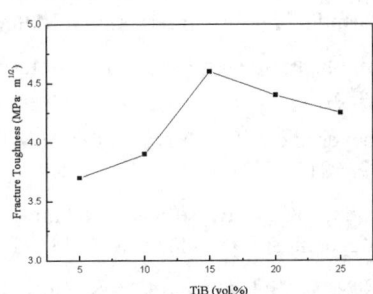

Fig. 5 Effect of vol.% TiB$_2$ on flexural strength of B$_4$C-TiB$_2$ specimens

Fig. 6 Effect of vol.% TiB$_2$ on fracture toughnes of B$_4$C-TiB$_2$ specimens

Fig.4 shows that the hardness increases from 17 GPa for 5 vol.%TiB$_2$ up to its of 33 GPa for 15 vol.%TiB$_2$, and then gradually decreases to 30Gpa for 25 vol.%TiB$_2$. Fig.5 shows the flexural strength increases from 363 MPa for 5 vol.%TiB$_2$ to its value of over 502MPa for 15 vol.% TiB$_2$, and then gradually decreases to the level of about 470 MPa. Fig.6 shows the fracture toughness increases from 3.6 MPa·m$^{1/2}$ at 5vol.%TiB$_2$ to its maximum value of 4.6 MPa·m$^{1/2}$ at15vol.%TiB$_2$ and then drops to 4.2 MPa·m$^{1/2}$ at 25 vol.%TiB$_2$. It should be noted that the high hardness, high flexural strength and high fracture toughness obtained in the present ceramic system are caused by the presence of TiB$_2$ particles .

Conclusions

1) B_4C–TiB_2 composites was pressureless sintered to relative densities of 98.5% at temperatures of 2150° C. The ceramics contained 5-25 vol.%TiB_2 particles formed through the in-situ reaction between B_4C, TiO_2 and glucose.

2) At temperatures of 1800-2150° C, the presence of fine TiB_2 particles promotes the sintering process.

3) Maximum values of flexural strength (502 MPa), hardness (33 Gpa) and fracture toughnes (4.6 MPa·$m^{1/2}$) were observed in the specimens containing 15 vol.% TiB_2.

Acknowledgement

The work is financially supported by the Science and Technique Foundation of Heilongjiang province of China (GB08A202).

References

[1] Thevenot F. Boron carbide – a comprehensive review. J Eur Ceram Soc 1990; 6:205-25.

[2] Jianxin D. Erosion wear of boron carbide ceramic nozzles by abrasive air jets. Mater Sci Eng A 2005; A408:227-33.

[3] Yanfeng C, Yip-Wah C, Shuyou L. Boron carbide and boron carbonitride thin films as protective coatings in ultra-high density hard disk drives. Surf Coat Technol 2006;200:4072-7

[4] Dunner, Heuvel HJ, Horle M. Absorber materials for control rod systems of fast breeder reactors. J Nucl Mater 1984;124:185-94

[5] Jianxin D, Jun Z, Yihua F, Zeliang D. Microstructure and mechanical properties of hot-pressed B4C/(W,Ti)C ceramic composites. Ceram Int 2002;28:425-30.

[6] A. K. Suri, C. Subramanian, J. K. Sonber and T. S. R. Ch. Murthy. Synthesis and consolidation of boron carbide:a review. International Materials Reviews.2010, 5, (1),4-40.

[7] H. Lee and R. F. Speyer.Pressureless sintering of boron carbide.J. Am. Ceram. Soc., 2003, 86, (9), 1468–1473.

[8] R. F. Speyer and J. Lee.Advances in pressureless densification of boron carbide. J. Mater. Sci. , 2004, 39, 6017–6021.

[9] Y. Kanno, K. Kawase, K. Nakano, Additive effect on sintering of boron carbide, Yogyo-Kyokai-Shi 95 (11) (1987) 1137–1140.

[10] T.K. Roy, C. Subramanian, A.K. Suri, Pressureless sintering of boron carbide, Ceram. Int. 32 (2006) 227–233.

[11] A. Goldstein, Y. Geffen, A. Goldenberg, Boron carbide–zirconium boride in situ composites by the reactive pressureless sintering of boron carbide zirconia mixtures, J. Am. Ceram. Soc. 84 (3) (2001) 642–644.

[12] Levin L, Frange N, Dariel MP. A novel approach for the preparation of B4C-basedcermets. Int J Refract Met Hard Mater 2000;18:131–135.

[13] Changming Xu, Yanbing Cai , KatarinaFlodstrom ,ZheshenLi , Saeid Esmaeilzadeh, Guo-Jun Zhang. Spark plasma sintering of B4C ceramics: The effects of milling medium and TiB_2 addition. 2004;24:2303-11. Int. Journal of Refractory Metals and Hard Materials .2012, 30:139–144.

Galvanic Corrosion of Titanium/Cu-Ni Alloy/High Strength Steel Multiphase Material System in Seawater

Wang Chun-li[1,a], Li Qing-fen[2,b], Wu Jian-hua[3,c]

[1,3]College of Materials Science and Chemical Engineering, Harbin Engineering University of China
[2]College of Mechanical and Electrical Engineering, Harbin Engineering University of China
[1,3]State Key Laboratory for Marine Corrosion and Protection, Luoyang Ship Materials Research Institute of China
[a]spring555555@126.com, [b]qingfli@yahoo.com.cn, [c]wujh@sunrui.net

Keywords: Galvanic corrosion, titanium, Cu-Ni Alloy, high strength steel

Abstract. The galvanic corrosion behavior of titanium/ Cu-Ni Alloy / high strength steel in multiphase material system has been studied under open circuit conditions using a zero-resistance ammeter (ZRA) in seawater. After the tests, the surface morphologies of the samples were detected by SEM. These results have been compared with results estimated from the polarization curves according to the mixed potential theory (imposed potential measurements). Results showed that Galvanic corrosion behavior of titanium (TA2)/Cu-Ni Alloy(B10) / high strength steel(921A) fulfill the mixed potential theory, 921A acts as the anode and bothTA2and B10 act as the cathodes. The overall galvanic reaction is mainly governed by the anodic oxidation of the 921A.

1. Introduction

Galvanic corrosion is originally defined as the enhanced corrosion between two or more electrically connected "dissimilar metals"[1-3]. The galvanic corrosion is stimulated because of the potential difference existed among the different metals in ocean engineering structures that a lot of metal materials are widely used. The effect of coupling different metals/alloys together, either directly or through an external path, increases the corrosion rate of the anodic alloy and reduces or suppresses the corrosion rate of the cathodic alloy. When two different metals or alloys are coupled together, the metal or alloy with the more negative potential undergoes anodic dissolution. On the surface of the more positive alloy, these excess electrons are consumed by a cathodic reaction that, in seawater, is usually oxygen reduction. The driving force for galvanic corrosion is the potential difference between two or more metals or alloys in a conductive medium that generates current flow between the anodic and cathodic members. The extent of galvanic corrosion between two or more coupled dissimilar alloys depends on other factors such as the effective area ratio of the anodic vs. cathodic members, solution conductivity, temperature and the stability of passive films, the magnitude of the potential difference between the dissimilar alloys, oxygen content of the solution, and the cathodic efficiency and polarization characteristics of the more noble metal or alloy[3,4].

Several authors have studied galvanic corrosion in seawater in the fields of bimetal material coupling system. However, studies the galvanic corrosion of multiphase material system in seawater are scarce in the literature. The objective of this work was to study the galvanic corrosion between commercially pure titanium (TA2), Cu-Ni 90/10(B10) Alloy and high strength steel (921A) in seawater (three alloys electrically connected is possible in the real-world engineering structures).

2. Experimental details

The materials tested were Grade 2 commercially pure titanium (0.024% N, 0.024% C, 0.16% O, 0.111% Fe, 0.001% H, Bal. Ti), Cu-Ni 90/10Alloy (10.13% Ni, 0.0059% C, 0.834% Mn, 1.71% Fe, 0.026% Zn, 0.01% Pb, 0.005% S, Bal. Cu) and the high strength steel 921A (0.12% C, 0.33% Si, 0.37% Mn, 0.08% Pb, 0.04% S, 2.27% Ni, 1.05% Cr, 0.24% Mo, 0.08% V, Bal. Fe). The test samples with a size of 60mm×20mm×3mm were made from the coupled materials whose surfaces were wet abraded from 100 SiC (silicon carbide) grit to 600 SiC grit then degreased with acetone and rinsed with absolute alcohol. And then measuring the sizes of specimen covered partially by use of Model AB glue in order to determine the working areas of anodic material and cathodic material of galvanic couples to 10cm^2. The specimens were stored in desiccators before use. Measurements were carried out in natural seawater collected from the North Yellow Sea and off the Qingdao Coast.

Potentiodynamic polarization measurement was conducted using electrochemical workstation IM6. Potential sweep rate was 1 mV s^{-1}. Rectangle test specimens, sized 10mm×10mm×4mm, were made from the coupled materials. Before the tests, the sample surfaces were pre-treated with the same method as above. And then measuring the sizes of specimen covered partially by use of Model AB glue in order to determine the working areas to 1cm^2. The specimens were stored in desiccators before use.

To carry out the galvanic corrosion experiment, the three dissimilar metals were immersed face to face at each vertex of equilateral triangle with sides of 8cm in stagnant seawater programmable temperature and humidity chamber Model ESS-SDJ201. In experiment the galvanic corrosion currents, the galvanic potential and self-corrosion potential were measured over a zero-resistance ammeter (ZRA). A detailed description of this type of measurement is given elsewhere [5,6]. By inserting a zero-resistance amperometer in the electrical circuit that connects three members, it was possible to measure the galvanic current between members and simultaneously, the galvanic potential of the triune was registered against a saturated calomel electrode (SCE) as a reference electrode. The galvanic current and potential was measured at sampling rate of 1 point h^{-1}. The ZRA was programmed to record 240 points potential and current plots for each hour of total 240 h measurement duration. Corrosion experiments were carried out at 25°C. After 240 h, the corrosion samples removed from the chamber and rinsed with deionized water. The surface morphologies of the samples were detected by SEM in order to estimate possible microstructural variations. Energy dispersive X-ray analysis (EDX) was adopted to trace the galvanic corrosion production.

3 Results and discussions

3.1. The individual behavior of coupled alloys

Fig.1 shows the self-corrosion potential data plots of TA2, B10 and 921A. It can be seen that the self-corrosion potential of TA2, B10 and 921A is 123 mV, -72mV and -780 mV respectively. In general, at the mixed potential (E_{couple}) of multiphase material systems the conditions $\sum i_{anode} = \sum i_{cathode}$ has to be fulfilled to maintain charge neutrality[7]. For TA2, B10 and 921A this condition can be fulfilled if the high strength steel (921A) acts as the anode and both titanium (TA2) and Cu-Ni Alloy (B10) act as the cathodes. Accordingly, the mixed potential (E_{couple}) for three coupled phases is a consequence of $i_{921A} = i_{B10} + i_{TA2}$. Under this condition and as shown in Figure 2, the estimated value of (E_{couple}) for the three alloys composition is -0.695V vs SCE.

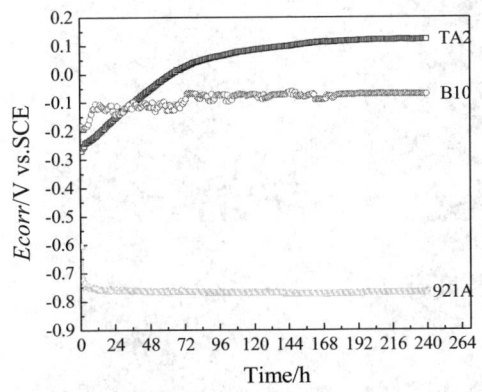
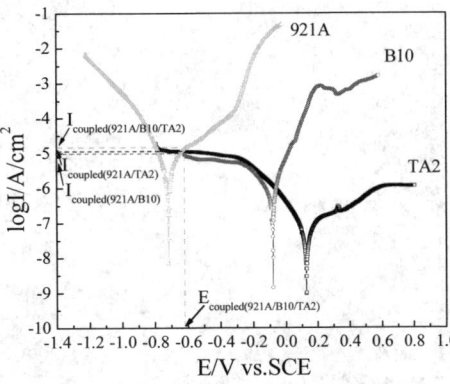

Fig.1. Potential data plot at 25°C of TA2, B10 and 921A

Fig.2. The plots of potentiodynamic polarization for TA2, B10 and 921A

3.2. The behavior of galvanic corrosion

For time-resolved characterization of the galvanic corrosion behaviour of all members, ZRA measurements were carried out under the experimental conditions previously described. Fig.3 shows, as an example, the galvanic potential and the galvanic current density data of 921A, B10 and TA2 members immersed in natural seawater at 25°C during 240h.

The result shows the mixed potential (E_{couple}) is approximately -0.730V vs SCE, the galvanic current density tends to stabilise from the sixth day, i_{921A} is around − 0.016 mA/cm^2, i_{B10} and i_{TA2} are around +0.007mA/cm^2 and +0.009 mA/cm^2. This value of mixed potential is nearby the intersection of polarization curves of 921A, B10 and TA2 as shown in Figure 2. Total galvanic current density $i_{921A} = i_{B10} + i_{TA2}$. The negative sign of the galvanic current density baseline indicated that 921A was the anode, so that it was corroding. Both TA2 and B10 act as the cathodes, so that they were protecting.

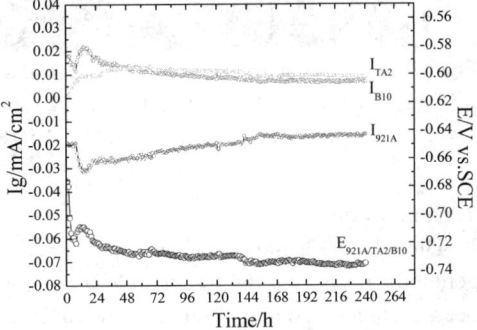

Fig.3. Galvanic current density and galvanic potential of TA2, B10 and 921A

3.3. Microstructural examination

Fig.4 presents SEM micrographs of the three alloy members surface obtained after self-corrosion and electrically connected for 240h immersed in natural seawater at 25°C. It can be seen that 921A as anode is suffered from corrosion greater than that of self-corrosion(Fig. 4a(2),(3)), both TA2 and B10 as cathodes is suffered from corrosion lighter than that of self-corrosion.(Fig. 4b(2),(3) and 4c(2),(3)). Yellow thick corrosion product is formed on the surface of the 921A (Figure 4a(1)), which when analysed in the SEM with EDX revealed the presence of significant amounts of Fe, O and Cl (iron oxide). White thick corrosion product is formed on the surface of the B10 and TA2, the presence of significant amounts of Ca, O and C(calcium carbonate). According to this result, anodic dissolution of iron takes place on the surface of 921A as anode, oxygen reduction reaction occurs on the surface of the B10 and TA2. Hence, the overall galvanic reaction in the multiphase material system seems to be mainly governed by the anodic oxidation of the 921A, which is the highest active and least polarizable.

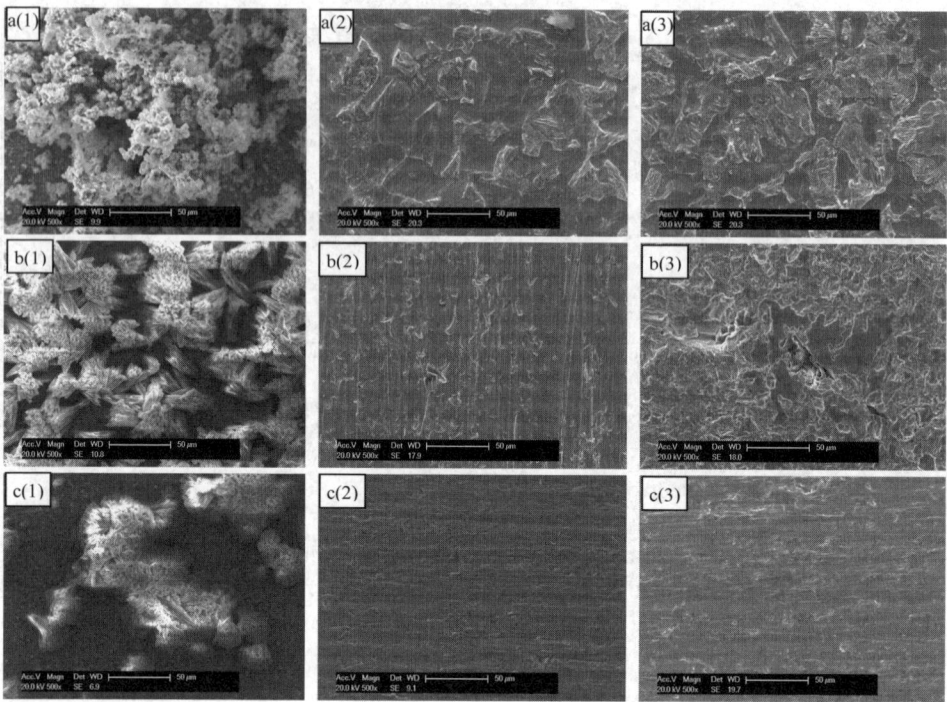

Fig.4.The surfaces morphology of 921A, B10 and TA2 for 240h immersed in natural seawater at 25°C(500x): a(1),(2) 921A, b(1),(2)B10 and c(1), (2)TA2 show corrosion products and surfaces rinsed respectively after electrically connected; a(3) 921A, b(3)B10 and c(3)TA2 show surfaces rinsed of self-corrosion for 240h.

4. Conclusions

The electrochemical behavior of uncoupled and three galvanically coupled single phases generally occurring in titanium/Cu-Ni Alloy/high strength steel multiphase material system has been characterized in natural seawater at 25°C for 240h. The following conclusions can be drawn:

(i) The order of corrosion sensitivity of the uncoupled single phases is: titanium < Cu-Ni Alloy < high strength steel; (ii) Galvanic corrosion behavior of titanium (TA2)/Cu-Ni Alloy(B10)/high strength steel(921A) fulfill the mixed potential theory, 921A acts as the anode and both TA2 and B10 act as the cathodes. The overall galvanic reaction is mainly governed by the anodic oxidation of the 921A, which is the highest active and least polarizable.

References

[1] J.W. Oldfield, ASTM STP 978, ASTM, West Conshohocken, PA, (1988), p. 5.

[2] R. Baboian, G.S. Harness, ASTM STP 558, ASTM, Philadelphia, PA, (1974), p. 171.

[3] R. Baboian, ASTM STP 516, ASTM, Philadelphia, PA,(1972), p. 145.

[4] R. Francis, Br. Corros. J. 29 (1994),p.53.

[5] G.W. Warren, G. Gao and Q. Li, J. Appl. Phys. 70 (1991),p.6609.

[6] T.S. Chin, R.T. Chang, W.T. Tsai and M.P. Hung, IEEE Trans. Magn. 24 (1988),p.1927.

[7] 25. G. Barkleit, A.M. El-Aziz, F. Schneider and K. Mummert, Mater. Corros. 52 (2001),p.193.

Wear Failure Behavior of Steel Surface with Palygorskite Powders as Lubricant Additives

Wang Li-min[1,a], Xu Bin-shi[1,2], Xu Yi[2], Zhang Bo[2], Yu He-long[2], Li Qing-fen[3,b]

[1]Engineering Training Center, Harbin Engineering University, Harbin 150001, China
[2]National Key Laboratory for Remanufacturing,
Academy of Armored Force Engineering, Beijing 100072, China
[3] College of Mechanical and Electrical Engineering,
Harbin Engineering University, Harbin 150001, China

[a]sunrise_hit@163.com，[b]qingfli@yahoo.com.cn

Keywords: Wear failure; steel surface; palygorskite powders; lubricant additives

Abstract. Wear failure behavior of steel surface with palygorskite powders (Palys) as lubricant additives was investigated in this paper. Different content of Palys was added in different lubricating oils. The wear failure behavior of the steel–steel contact surfaces was studied by using an optimal SRV oscillating friction/wear tester. The elemental component and morphology on worn surface were analyzed by SEM and EDS. From the results of the friction/wear tests and analysis of the worn surfaces, we can see that specimens with Palys as lubricant additives exhibit excellent wear reducing properties and introduce a smooth and compact oxide layer on the worn surface. This may be attributed to the layer-chain fibrous crystal structure of palygorskite and the complex physicochemical nature in friction process.

Introduction

The tribological properties of ultrafine powders as lubricant additives have been studied over the last few years [1-3]. It was found that the tribological properties were significantly improved when nano-sized particles were used [4, 5]. Palygorskite is a clay mineral widely used in different industrial fields. It is possible that the Palys which have nano-sized needle shape fibrous crystals may be used as lubricant additives and the tribological properties may be improved. However, few reports about this work are available so far.

In the present work, the improvement of wear failure behavior of steel surface with Palys as lubricant additives was investigated.

Experimental

The Palys were prepared by using mechanical ball-grinding method with the natural palygorskite stick from Xuyi (Jiangsu province, China). The microstructure of the powder was observed by using a JEM-1011 transmission electron microcopy.

Different content of Palys were added and ultrasonic-dispersed in three different lubricating oils (150SN, CD15W-40 and PAO40). The physicochemical properties of the three experimental oils are listed in Table 1.

The wear failure behavior of the steel–steel contact surfaces was investigated by using a standard Optimal SRV-IV oscillating friction/wear tester in ball-on-disk contact configuration. The GCr15 bearing steel ball (Φ10 mm, HRC61-63) was used as a rider, which was pressed against the stationary 45 steel disk (Φ24×8mm, HRC42-45) to reciprocate by a horizontal oscillating rod. The tests were carried out in 50℃ environment under load of 20 N, frequency 10 Hz, oscillating amplitude of 1mm, and duration of 60 min. All specimens were cleaned ultrasonically in ethanol and dried in oven before and after tests.

The elemental component and morphology on worn surface were analyzed by Quanta200 scanning electron microscope (SEM) equipped with energy dispersive spectrometry (EDS).

Table 1 Physicochemical property of the experimental lubricating oils

	Item value(SAE)				
	Density (gcm^{-3})	Kinematic viscosity (mm^{2}s^{-1})	Viscosity index	Pour point(℃)	Flashing point(℃)
150SN	0.863	5.43, 100℃ / 32.28, 40℃	101	-12	202
CD15W-40	0.794	15.02, 100℃ / 110.6, 40℃	141	-27	228
PAO40	0.842	39, 100℃ / 396, 40℃	147	-30	280

Results and discussion

The TEM images of Palys are given in Fig. 1, where, the layer-chain fibrous crystal structure of palygorskite can be clearly seen.

Fig. 2 shows the elemental distribution patterns of worn surfaces lubricated with 150SN base oil and 150SN+0.4wt% Palys oils respectively. It is seen that both the surfaces have the Fe, C, O elements, however the worn surface lubricated with 150SN+0.4% Palys shows a significant increasing in the content of O, and denier Mg, Al and Si can be observed there. This indicates that the Palys may facilitate the enrichment of O element and cause the formation of oxide film in the friction surfaces.

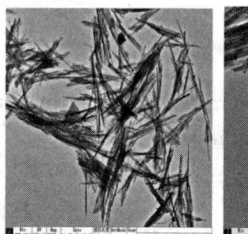  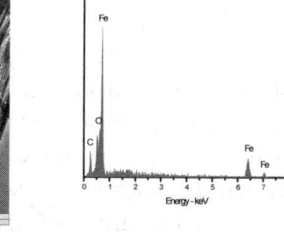 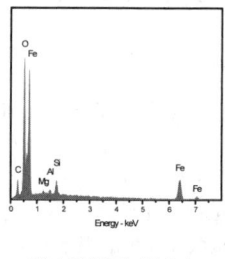

                (a) 150SN   (b) 150SN+0.4% Palys

  Fig.1 TEM image of palygorskite   Fig.2 EDS patterns of worn surfaces

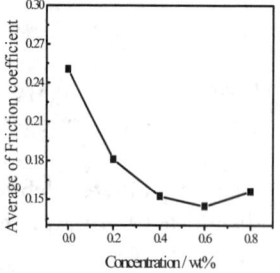 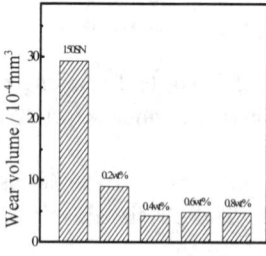 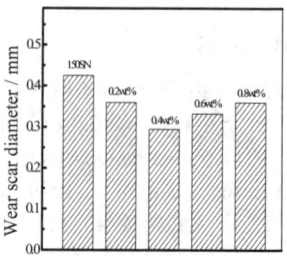

(a) Average friction coefficient (b) Wear volume of the disk (c) Wear scar diameter of the ball.

Fig.3 Effect of the palygorskite concentration on the tribological properties (in 150SN oil)

The average friction coefficient, wear volume of the disk and wear scar diameter of the ball investigated in the base 150SN oil and oil with different content of Palys are given in Fig.3. From Fig.3 (a), it is seen that the average friction coefficient of specimens in the oil with Palys as lubricant additives are much smaller than that in the base oil, and that it reached to the minimum value of 0.1446 (decreased about 42.32%) when the content of Palys is 0.6 wt%, where, the best anti-wear property was obtained. Fig.3 (b) shows that the wear volume of the disk in the oil with Palys decreased obviously compare with the one in base oil. The minimum wear volume value, namely the best anti-wear property was obtained at 0.4 wt% content of Palys. Results of wear scar diameter of the ball, Fig.3 (c), gave the same conclusion of (b).

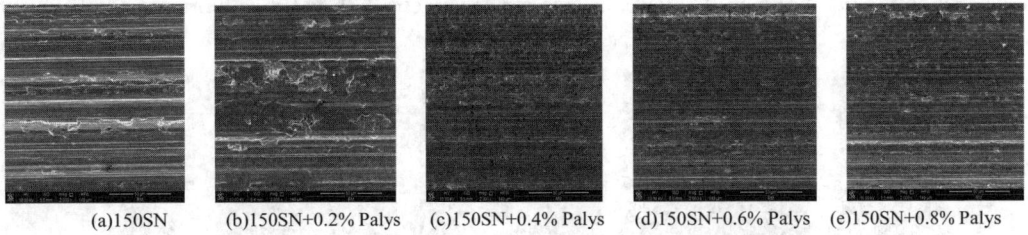

(a)150SN (b)150SN+0.2% Palys (c)150SN+0.4% Palys (d)150SN+0.6% Palys (e)150SN+0.8% Palys

Fig.4 SEM morphologies of worn surfaces (in 150SN oil)

The SEM morphologies of worn surfaces in 150SN oil are given in Fig.4. From (a), it is seen that the obvious furrows and grooves in sliding direction, formed by the wear debris, are wide and deep for the disk in the base 150SN oil, showing severe abrasive wear. The furrows become shallow and narrow when adding 0.2 wt% Palys to oil (b). The worn surfaces lubricated with 0.4 – 0.8 wt% Palys oils are much smoother and few furrows can be found (c, d, e) and the best appearances are the ones with 0.4 and 0.6 wt% (c, d). This result is in accordance with the above results (Fig.3) that the best tribological properties were obtained at 0.4 - 0.6 wt% content of Palys. Besides, we can see that a smooth and compact oxide layer on the worn surface for specimens with Palys as lubricant additives (b-e). It is concluded that proper content of Palys added in 150SN oil may significantly improve the anti-wear property of the steels.

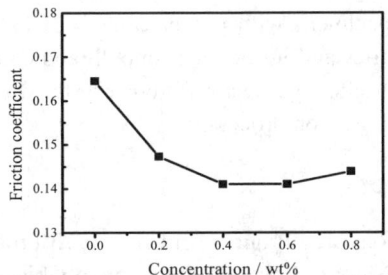 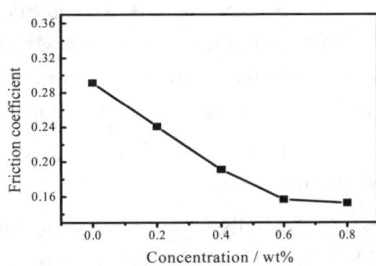

Fig. 5 Effect of the palygorskite concentration on friction coefficient (in CD15W-40 oil)

Fig.6 Effect of the palygorskite concentration on friction coefficient (in PAO40 oil)

Effect of the palygorskite concentration on friction coefficient in oil CD15W-40 and PAO40 are given in Fig. 5 and 6 respectively. Fig. 5 shows that the friction coefficient of specimens in the oil with Palys as lubricant additives are much smaller than that in the base oil, and that it reached to the minimum value of 0.141 (decreased about 14.3%) when the content of palygorskite is 0.4 and 0.6 wt% in the CD15W-40 oil. Fig. 6 shows that the friction coefficient decreased with increasing palygorskite concentration and reached to the minimum value of 0.152 (decreased about 47.7%)

when the content of palygorskite is 0.8 wt% in the PAO40 oil. It is known that the viscidity of PAO40 oil is much larger than the ones of 150SN and CD15W-40 (see Table 1). The activation of nano-sized particles of palygorskite will be restrained and therefore the formation of lubricating film needs more palygorskite particles added there. That is why the best result was obtained at 0.8 wt% in the PAO40 oil.

The SEM morphologies of worn surfaces in oil CD15W-40 and PAO40 are given in Fig. 7 and 8 respectively. Where, for CD15W-40 oil, the worn surfaces lubricated with 0.4 and 0.6 wt% Palys are smoother and few furrows can be found (Fig.7 c, d). For PAO40 oil, the best appearance is the one with 0.8 wt% Palys (Fig.8 e). These results are in accordance with the ones given in Fig. 5 and 6. The smooth oxide layer on the worn surface for specimens with Palys as lubricant additives can also be observed.

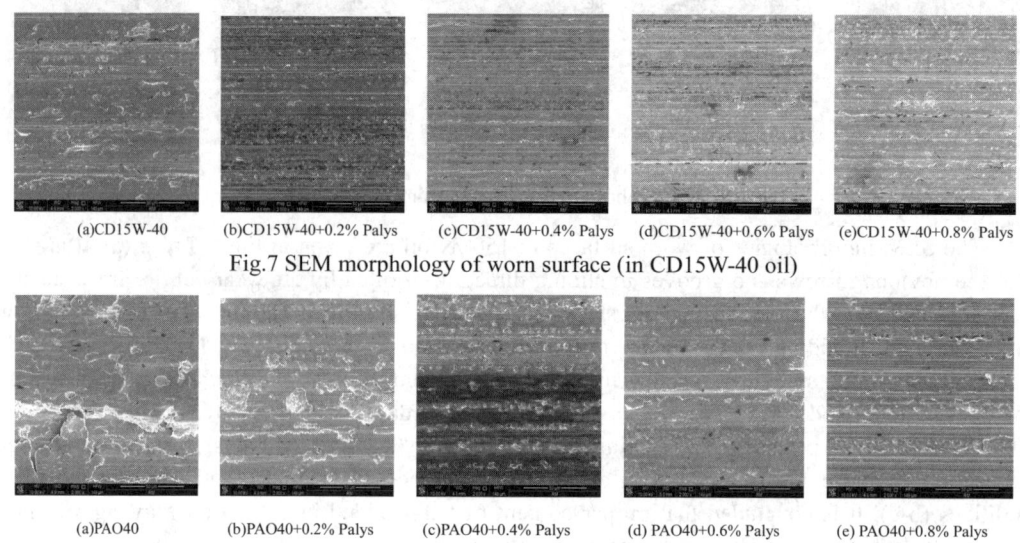

(a)CD15W-40 (b)CD15W-40+0.2% Palys (c)CD15W-40+0.4% Palys (d)CD15W-40+0.6% Palys (e)CD15W-40+0.8% Palys

Fig.7 SEM morphology of worn surface (in CD15W-40 oil)

(a)PAO40 (b)PAO40+0.2% Palys (c)PAO40+0.4% Palys (d) PAO40+0.6% Palys (e) PAO40+0.8% Palys

Fig. 8 SEM morphology of worn surface (in PAO40 oil)

From all the above results we may conclude that specimens with proper content of Palys as lubricant additives exhibit excellent wear reducing properties and introduce a smooth and compact oxide layer on the worn surface. This may be attributed to the layer-chain fibrous crystal structure of palygorskite and the complex physicochemical nature in friction process.

Acknowledgement

The work is financially supported by the National Basic Research Program of China (973 Program, Nos. 2011CB013405 and 2007CB607601), and National Natural Science Foundation of China (Nos. 5073506, 50805146, 50904072, 51005243).

References

[1] Gao Y X, Zhang H C, Xu X L, Lubrication Engineering, 2006 (10): 39- 42
[2] Guo Y B, Xu B S, Ma S N, et al.. Tribology, 2004, 24 (6): 512-515
[3] Tian B, Wang C B, Yue W, et al.. Tribology, 2006 (6): 574-578
[4] Lu Y L, Li Z, Yu Z Z,et al. Composites Science and Technology, 2007,67: 2903-2913.
[5] Chen C H..Journal of Physics and Chemistry of Solids, 2008, 69: 1411-1414.

FATIGUE ASSESSMENT OF TRIMARAN STRUCTURE BASED ON SIMPLIFIED PROCEDURE

Ren Huilong[1, a], Khurram Shehzad[1, b], Zhen Chunbo[1, c], Asifa Khurram[2, d]

[1] College of Shipbuilding Engineering, Harbin Engineering University, Harbin 150001, China

[2] College of Materials Science Engineering, Harbin Engineering University, Harbin 150001, China

[a] Renhuilong@263.net, [b] Khurram_1977@hotmail.com, [c] Zhenchunbo@yahoo.com.cn, [d] Asifa_Khuram@hotmail.com

Keywords: Trimaran, Fatigue, Simplified Procedure

Abstract: In recent years, Trimaran platform design has got the attention of naval architects owing to its superior seagoing performance. Trimaran structure experiences severe loads due to its unique configuration and high speed, causing stress concentration, especially in cross deck region and accelerate fatigue damage. This paper presents fatigue strength assessment of Trimaran structure by simplified procedure. A methodology is proposed to evaluate fatigue loads and loading conditions by load combinations of direct calculation procedure of Lloyd's Register Rules for Classification of Trimaran (LR Rules). Global FE analysis, in ANSYS, is performed to investigate the stress response. The stress range is computed by hot-spot stress approach, and its long term distribution is specified by Weibull distribution. Fatigue damage of selected critical details is calculated using mathematical formulation of simplified fatigue assessment procedure of Common Structure Rules (CSR).

Introduction

The trimaran platform design has gained enormous attention in recent years owing to its superior seagoing performance. The trimaran offers significant advantages in terms of low resistance at high speed, excellent sea-keeping characteristics, massive deck space and stealth. Due to its unique configuration and high operating speed, trimaran experiences severe structural loads, which include splitting moment, wet deck slamming and stress concentration in the cross deck region. These loads accelerate fatigue damage; hence, evaluation of fatigue strength is vital for trimaran design.

The research work focusing on sea keeping aspects of the novel trimaran platform emerged after launching of RV Triton in 2000, being the world's largest trimaran of that time [1]. However, very little material is available on fatigue strength assessment of trimaran structure. The fatigue strength assessment can be carried out either by *Simplified method* or *Spectral method* [2]. A few researchers have investigated the fatigue strength of trimaran cross structure by the spectral approach [3]. However, no literature has been published on simplified method. This study is focused on fatigue strength assessment of trimaran by the simplified method.

Fatigue assessment of ship structure by simplified method is based on the guidelines of classification societies. Application of this method for trimaran is complex since guidelines for fatigue loads, load cases and loading conditions are not available. Lloyd's Register Rules for Classification of Trimaran (LR Rules) include formulation of design loads and strength analysis but do not address the fatigue issue [4]. Load combinations of direct calculation procedure of LR rules are summarized in Table 1. This study manipulates load combinations of LR Rules to determine fatigue loads, which are systematically employed for fatigue assessment of trimaran platform.

Table 1. LR Rule load combinations

Wave Direction	Load Cases	M_{swh}	M_{sws}	M_{wh}	M_{ws}	M_h	M_{sph}	M_{sps}	M_{lt}	M_{tt}
Head Seas	LC1	1.0	-	1.0	-	-	0.3	-	-	0.2
	LC2	-	1.0	-	1.0	-	-	0.3	-	0.2
Beam Seas	LC3	1.0	-	0.1	-	-	1.0	-	0.2	-
	LC4	-	1.0	-	0.1	-	-	1.0	0.2	-
Oblique Seas	LC5	-	-	-	-	-0.3	0.4	-	1.0	0.3
	LC6	-	-	-	-	1.0	0.4	-	-	0.2
	LC7	1.0	-	-	0.2	-0.2	0.6	-	-	1.0

In Table 1, M_{swh}, M_{sws} are still water bending moments, M_{wh}, M_{ws} are vertical wave bending moments, M_{sph}, M_{sps} are splitting moments for corresponding hog and sag condition respectively. M_h is horizontal bending moment and M_{lt}, M_{tt} corresponds to longitudinal and transverse torsional moments.

Method

FE Modelling. A 3D FE model of trimaran is generated in ANSYS software as shown in Fig.1. Global FE analysis is performed as per direct calculation procedure of *LR Rules* [5]. Based on the results, following details are selected at transverse bulkhead location (Frame -33) for fatigue assessment (Fig. 2):
- Connection of main hull with wet deck structure (Hot Spot 1)
- Intersection of longitudinal and transverse bulkhead near side hull ((Hot Spot 2)
- Connection of side hull with wet deck structure (Hot Spot 3)

Local fine mesh of size t x t is used at selected hot spot locations to take into account stress increase due to change in component geometry as shown in (Fig. 3).

Loading Conditions & Fatigue Loads. Fatigue analyses are carried out for the representative loading conditions according to the intended operation of the ship. Table 1 reveals three loading scenarios of trimaran corresponding to heading directions, i.e. head sea, beam sea and oblique sea. Same are used as fatigue loading conditions. Since, only fluctuating loads contribute to fatigue damage, therefore, fatigue loads are acquired by exempting hydrostatic pressure.

Stress Range Evaluation. A global FE analysis of trimaran structure is performed for fatigue load cases using inertia relief boundary condition. Hot spots stresses are extracted by extrapolation method for each load case according to CSR guidelines [6].

Fig 1. Global FE model of Trimaran

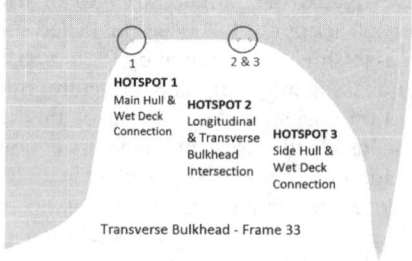

Fig 2. Hot spot location

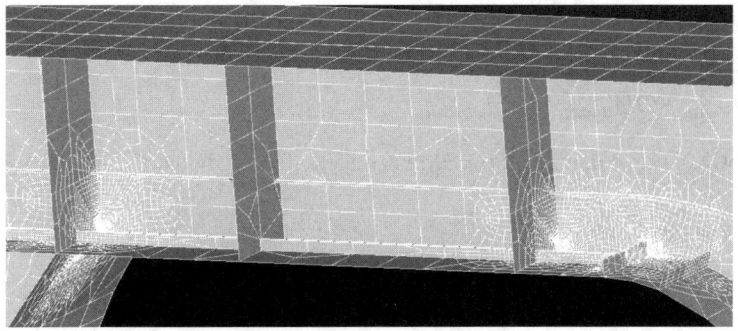

Fig 3. Fine mesh at hot spots

Stress Range Evaluation. A global FE analysis of trimaran structure is performed for fatigue load cases using inertia relief boundary condition. Hot spots stresses are extracted by extrapolation method for each load case according to *CSR* guidelines.

The difference of hot spot stress values of load case LC1 and LC2 corresponds to stress range for head sea condition. Similarly, the difference of hot spot stress values of the load cases LC3 and LC4 represent stress range for beam sea condition. Stress range for oblique sea condition is calculated by using Eq. 1.

$$\Delta \sigma_{oblique} = \frac{1}{3}[2\sigma_{LC5} + 2\sigma_{LC6} + 2\sigma_{LC7}] \quad (1)$$

Simplified Fatigue Calculation. In simplified fatigue analysis, long term distribution of stress range is represented by a two parameter Weibull distribution. According to LR rule, long term response predictions for trimaran are based on a probability of exceedance of 10^{-8}. Assuming Weibull long term distribution of stress and using a two slope SN curve of Common Structure Rules (CSR) and Palmgren-Miner's rule, cumulative fatigue damage DM_i for i-th loading condition is calculated by Eq. 2.

$$DM_i = \frac{\alpha_i N_L}{K_2} \frac{S_{Ri}^m}{(\ln N_R)^{m/\xi}} \mu_i \Gamma\left(1 + \frac{m}{\xi}\right) \quad (2)$$

Where:
- α_i Proportion of ship's life in different loading conditions
- N_L Number of cycles for expected design life
- m, K_2 SN curve parameters
- S_{Ri} Stress range at the representative probability level of 10^{-8}, in N/mm^2
- N_R 10^8 number of cycles corresponding to probability level of 10^{-8}
- ξ $f_{Weibull}\left(1.1 - 0.35\dfrac{L-100}{300}\right)$
- L   Rule Length
- $f_{weibull}$ Area dependent modification factor
- Γ Gamma function
- μ_i $1 - \dfrac{\left[\gamma\{1+\frac{m}{\xi},v_i\} - v_i^{-\Delta m/\xi}\gamma\{1+\frac{m+\Delta m}{\xi},v_i\}\right]}{\Gamma\left[1+\frac{m}{\xi}\right]}$
- v_i $\left[\dfrac{S_q}{S_{Ri}}\right]^\xi \ln N_R$
- S_q Stress Range at the intersection of two segments of S-N curve
- Δm Slope change of upper to lower segment of SN curve
- (a,x)   Incomplete Gamma function, Legendre form

Assuming equal probability of occurrence for each loading condition, Cumulative fatigue damage is computed using Eq. 3:

$$DM = DM_{Head} + DM_{Beam} + DM_{Oblique} \qquad (3)$$

Finally, cumulative fatigue damage ratio DM is converted to predicted fatigue life by Eq. 4:

$$\text{Fatigue Life} = \frac{\text{Design Life}}{DM} \qquad (4)$$

RESULTS

Cumulative fatigue damage and corresponding fatigue life of selected locations are presented in Table 2. Predicted fatigue life of hot spot 3 is less than the design value, which specifies the need to improve structure design at particular location. A design life of 25 years is used in this study.

Table 2. Cumulative Damage and Fatigue Life

	DM_{Head}	DM_{Beam}	$DM_{Oblique}$	DM	Fatigue Life [Years]
Hot Spot 1	0.0061	0.6550	0.3730	1.0341	19.34
Hot Spot 2	0.00002	0.0021	0.0002	0.0023	< 300
Hot Spot 3	0.0003	0.1174	0.0027	0.1204	166.04

Conclusion

This study proposes a methodology for fatigue strength assessment of Trimaran structure. Fatigue loads, load cases and loading conditions are determined by manipulating load combinations of direct calculation method of LR rules. Stress response is investigated by FE analysis based on hot spot approach. Following the guidelines of CSR, fatigue life of selected details is computed. The method provides a simplified and efficient tool for fatigue life prediction of trimaran structure.

References

[1] T. Blanchard and C. Ge: J. Nav. Eng. Vol. 44-1 (2007), p. 1

[2] Y. Wang: Int. J. Fatigue Vol. 32 (2010), p. 310

[3] Z. Chunbo, R. Huilong, F. Guoqing and L. Chenfeng: App. Mech. Mat. Vol. 148-149 (2012), p. 393

[4] Rules for the Classification of Trimarans, Lloyd's Register Rules and Regulations (2006)

[5] K. Shehzad, R. Huilong, A. Khurram: submitted to AEMT (2012)

[6] Common Structure Rules for Double Hull Oil Tankers, IACS (2010)

A Comparative Study between Steel Grillage and SPS Stiffening Plate Based on FEA Eigen Value Analysis[1]

[a]XUE Qichao, [b]ZOU Guangping, [c]WU Ye, [d]LI Jia, [e]SHANG Lei

(College of Aerospace and Civil Engineering, Harbin Engineering University, Harbin, China)

[a]xue1736@163.com, [a]zouguangping@hrbeu.edu.cn, [c]wuye@hrbeu.edu.cn, [d]Lijia@hrbeu.edu.cn, [e]shanglei@hrbeu.edu.cn

Keywords: Sandwich plate system; Composite; Buckling; Stiffening plate; Ship structure; FEA

Abstract. Sandwich plate system (SPS) with polyurethane elastomer is more and more used on ship buildings as grillage. A Comparative Study on critical buckling loads between steel grillage and SPS stiffening plate are carried out in this paper. Plate elements and beam elements are used in numerical simulation and a rectangular model is taken as a calculating example. By using of FEA software ANSYS, ultimate buckling loads are calculated with different influential factors between steel grillage and SPS stiffening plate, including the number of stiffening ribs, inertia moment of stiffening ribs, and the thickness of plates. the analysis results of comparison shows that SPS stiffening plates have better performance than steel grillage under in plane compressive loadings.

Introductions

Sandwich plate system (SPS) is composed of two pieces of steel faceplates and one soft polyurethane elastomer core. It is developed by Intelligent Engineering Limited and is more widely used in civil engineering and shipbuilding fields[1]-[3] for weight reduction, rapid reparation, vibration isolation, noise reduction and resisting impact. Fig. 1 is a typical companion between SPS structure and conventional structure. Some research show that SPS construction has a number of benefits over conventional steel structures, including simplified structure, reduced weight, increased fatigue resistance and reduced susceptibility to corrosion, leading to less maintenance and easier inspection[4].

Global buckling theories of steel grillage and sandwich plates

Buckle calculating of single sandwich plate was firstly proposed by Reissener in about 1947[5], Global buckling of plates with stiffeners were also studied before 1950s by S. P. Timoshenko and J. M. Gere[6]. Base on Reissner's theory, elastic global buckling of sandwich plates with ribs are calculated by energy method, in which assumed that the core is thick and softer relatively to faceplates. An established coordinate system is showed in Fig.2 and the critical buckling loads of stiffened plates σ_{cr} with single ribs in x direction is:

$$\sigma_{cr}=\frac{\pi^2 D}{b^2(h+2t)}\frac{(1+\beta^2)^2+\varphi+2\gamma}{\beta^2(1+2\delta)}, \text{ in which } \varphi=\frac{\pi^2 D}{G_c\beta^2 b^2(h+t)^2}(1+\beta^2)^3 \text{ ; } \frac{EI}{bD}=\gamma \text{ ; } \frac{P}{bN}=\delta \quad (1)$$

Where parameter D is the bending rigidity of SPS; β =a/b; h and t are thickness of core and faceplate respectively, E and G_c are modulus and shear modulus of elasticity. I is inertia moment of stiffeners. P is force endured by stiffener.

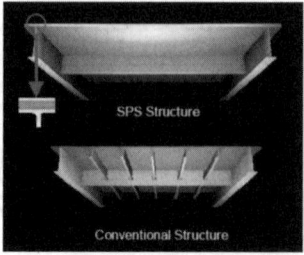 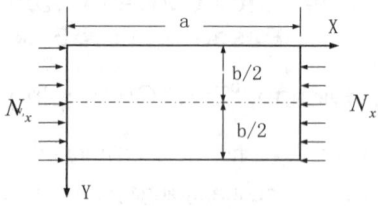

Fig.1 A typical SPS structure and conventional structure   Fig.2 Coordinate system of stiffened sandwich plate.

FEM analysis of buckling for stiffened steel grillage and SPS

FEM Eigen value buckling analysis are carried out to study the influence of different factors of conventional plates and SPS with stiffeners, including the number of stiffeners, ratio of length and wide, thickness of plates and inertia moments of stiffeners. Without loss of generality, only rectangular plates with simple supported boundaries are taken as examples to calculated in this paper. Table.1 is some parameters and material properties in FEM analysis. Finite element model is showed in Fig. 3 with β=2 and four ribs distributed along each side. Fig.3 and Fig. 4 are the first order buckling of conventional plate with $\beta = 2$ and $\beta = 5$.

Table.1 Is some parameters and material properties in FEM analysis.

Plate form	Stiffener materials	modulus of elasticity (faceplates)	Poisson's ratio (faceplates)	modulus of elasticity of core	Poisson's ratio of core	Element Type in FEM simulation
Plate with stiffeners	steel	2.1e5 MPa	0.25			Shell43 and beam188
SPS with stiffeners	steel	2.1e5 MPa	0.25	800MPa	0.3	Shell181 and beam188

Fig.3 Finite element model of plates with stiffeners   Fig.4 First order of buckling model for stiffened plates with β =2

Firstly influence of β is studied, in which other parameters are fixed, including the wide of conventional plate b=2000mm, thickness h=4mm; SPS thickness is equal to 16mm with faceplate thickness 2mm; four ribs in each side with rectangular sections stiffened to the plate whose inertia moment is 22500mm^4. Fig.5 is the graph of critical buckling loads with different value of β.

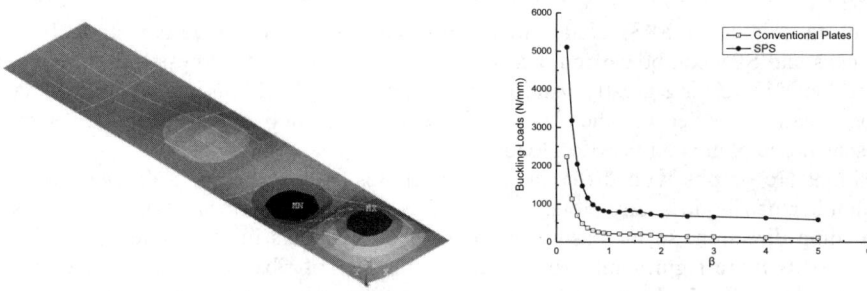

Fig.4 First order of buckling model for stiffened plates with β=5

Fig.5 Critical loads with different ratio of length and wide

Fig.5 shows when the thickness of conventional plate equal to thickness of the sum of two faceplates of SPS, critical buckling loads of SPS in different β is about 3 times larger than conventional plates. This is because the existing of elastomer core in SPS raise the value of bending rigidity. Along with the increase of β, two curves has similar variation trend. And at the location about β =1.4, each curves has a little heave which means buckling shape is changing from one half wave to two half waves.

Fig.6 is the graph of critical buckling loads with different inertia moment of stiffeners. It shows critical buckling loads is increasing as the moment of inertia rises. There are approximate liner relationship between critical buckling loads and inertia moments of ribs. But for conventional plates, if the inertia moment reaches a certain threshold, buckling loads tend to be unchanged. At this time, first order buckling is local buckling in single grillage instead of global buckling. So when the number of ribs in plates is fixed, using bigger area of stiffeners can improve stability significantly until local buckling appears in some single grillages.

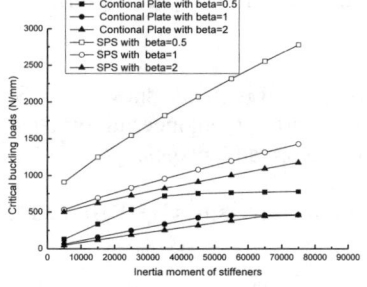
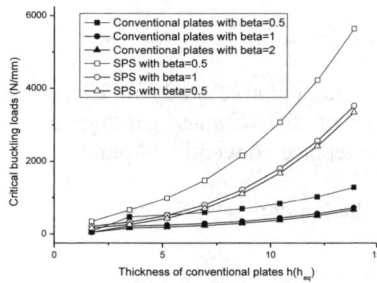

Fig.6 Graph of critical buckling loads with different inertia moment of stiffeners

Fig.7 Graph of critical buckling loads with different thickness

Fig.7 is critical buckling load with different h. Because SPS is composed of three parts with different thickness, an equivalent thickness can be calculated by using of sandwich plate theory [5] based on equal bending rigidity:

$$h_{eq} = \sqrt[3]{6(h_c + t)^2 t} \qquad (2)$$

Here h_{eq} is an equivalent thickness; h_c is the thickness of polyurethane elastomer core, and t is the thickness of faceplates of SPS. Under this equivalent case, a comparative study between conventional plates and SPS can be carried out. The result shows when thickness h is growing, the critical buckling loads increasing greatly. And in same value of h, buckling loads of SPS is larger than conventional plates. It is because the elastomer core has a certain bending rigidity and shearing rigidity that resists the in plane compression loads.

Fig. 8 and Fig.9 are graphs with different number of ribs, in which Fig. 8 draw curves with changing the number of ribs in x directions and Fig. 9 is y directions. Because x direction is the compression loading direction, so increasing the number of stiffeners in x direction can raise the critical buckling loads more significant than increasing number of ribs in y directions. Curves in Fig.8 are more flat than Fig. 7. That means transverse stiffeners in plane cannot increase critical buckling loads efficiently when β is greater than 0.5. in additional to above, the result also shows when use SPS structures, using fewer stiffeners can get even more larger critical loads than conventional stiffening plates.

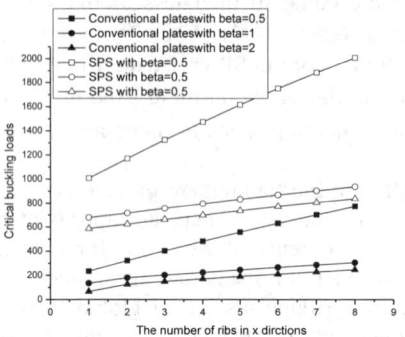

Fig.8 Critical buckling loads with different number of ribs in x directions

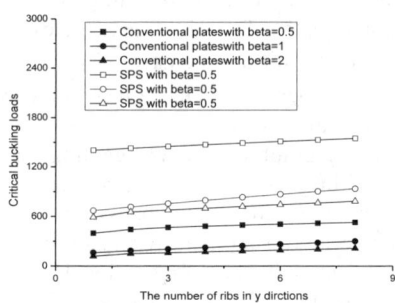

Fig.8 Critical buckling loads with different number of ribs in y directions

Conclusion

The result of the comparative study of Eigen value buckling in this paper shows that SPS can reduce the number of stiffeners and get lager critical buckling loads. In engineering practice, using sandwich plate to replace conventional plates can benefit much more advantages.

[1] Supported by "the Fundamental Research Funds for the Central Universities" (**HEUCF100211**)

References

[1] SHAN Chenglin, Research on mechanical properties of orthotropic sandwich bridge deck., China journal of highway and transport. 2012.1
[2] Stephen J. Kennedy. An Innovative "No Hot Work" Approach to Hull Repair on In-Service FPSOs Using Sandwich Plate System Overlay. Offshore Technology Conference. 2003.
[3] Devin K. Harris; Tommy Cousins; Thomas M. Murray; and Elisa D. Sotelino Performce of Constructed Facilities 2008 / 305
[4] M A Brooking; Dr S J Kennedy. The Performance, Safety and Production Benefits of SPS Structures for Double Hull Tankers
[5] The Chinese academy of sciences Beijing solid mechanics laboratory, plate and shell group; The bending, stability and vibration of sandwich plate and shell. Science Press, 1977.
[6] S. P. Timoshenko, J. M. Gere. Stability theory. Science Press, 1965.

Yield Criterion and Crack Tip Plastic Zone of Nickel-Based Single Crystal

YANG Lihong [1,2, a], ZOU Guangping[1, b] and QU Jia [1, c]

[1]College of Aerospace and Civil Engineering, Harbin Engineering University, Harbin 150001, China
[2]Center for Composite Materials, Harbin Institute of Technology, Harbin 150001, China

[a]yanglihong@hrbeu.edu.cn, [b]zouguangping@hrbeu.edu.cn, [c]qujia@hrbeu.edu.cn

Key words: Nickel-based single crystal; Modified Hill yield criterion; Mixed mode I, II crack; Plastic zone; Temperature

Abstract. A modified Hill yield criterion considering tension-shear stress coupling was given and material parameters of nickel-based single crystal in this criterion was determined by the Least Square Method. Crack tip plastic zones of nickel-based single crystal plate were analyzed by using this modified yield criterion and the plastic zones corresponding to mixed mode I, II crack, were derived. The crack tip plastic zones of nickel-based single crystal were compared to that of isotropic material. The influences of tension-shear stress coupling, compound ratio of mixed mode I, II crack and temperature on crack tip plastic zones were discussed. The results obtained in this paper indicate that the increasing rate of crack tip plastic zones with compound ratio in the plane stress condition is greater than that in the plane strain condition for nickel-based single crystal and the temperature has greater influence on the size of crack tip plastic zones.

Introduction

Because of excellent anti-fatigue and anti-creep in high temperature condition, nickel-base single crystal has become the key materials in manufacturing blades of aircraft engines and gas turbines. In the past few years, many states have pay much more attention to the development and research of the nickel-base single crystal and have obtained a series of meaningful results.

Fracture and crack problem is one of research focus in the field of mechanics. The shape and size of crack tip plastic zone are important to analyzing crack problem, and there is close contact between crack propagation and crack tip plastic zone[1]. By using the theory of the macroscopic fracture mechanics, Gao Xin[2] analyzed crack tip plastic zone of oblique crack of orthotropic composites, and the results obtained show that the angle of crack had a greater effect on the scope and shape of crack tip plastic zone.

Hill yield criterion is most widely used in the elastic-plastic problems of orthotropic composites. But it doesn't consider the tensile-shear stress coupling of the composite, which may affect the accuracy of results while analyzing the off-axis loaded problem. Ding Zhiping[3] proposed a new yield criterion of nickel-based single crystal by adding a fourth-order correction term to the Hill yield criterion. Yield stresses forecasted by this criterion gave the significance of considering tensile-shear coupling. In this paper, by the Hill yield criterion and a modified Hill yield criterion, the crack tip plastic zones of nickel-based single crystal are analyzed.

Yield Criterion of Nickel-Base Single Crystal

Hill Yield Criterion. For nickel-base single crystal, ignoring the Bauschinger effect and considering the mechanical properties in three crystal axis [100], [010] and [001] are the same, Hill yield criterion becomes

$$\frac{F}{2}\left[(\sigma_{22}-\sigma_{33})^2+(\sigma_{33}-\sigma_{11})^2+(\sigma_{11}-\sigma_{22})^2\right]+L\left(\sigma_{23}^2+\sigma_{31}^2+\sigma_{12}^2\right)=\sigma_s^2. \quad (1)$$

where σ_{ij} ($i, j=1, 2, 3$) are stresses along the main coordinate axis, σ_s is the reference yield stress and F, L may be obtained based on yield stresses along [001] and [111], as shown in Table 1.

For Eq. 1, $\sigma_{33} = \sigma_{13} = \sigma_{23} = 0$ in the plane stress condition and $\sigma_{33} = \gamma(\sigma_{11} + \sigma_{22})$, $\sigma_{13} = \sigma_{23} = 0$ in the plane strain condition. Here γ is the Poisson's ratio of material.

Modified Hill Yield Criteria with Tensile-Shear Coupling. Because tensile-shear coupling in off-axis direction is not considered in Hill yield criterion, its accuracy is not sufficient enough to describe the yield properties of single crystal material. A new yield criterion for nickel-based single crystal is given in this paper by adding a second-order correction term in Hill yield criterion, that is

$$\frac{f}{2}\left[(\sigma_{22}-\sigma_{33})^2+(\sigma_{33}-\sigma_{11})^2+(\sigma_{11}-\sigma_{22})^2\right]+l(\sigma_{23}^2+\sigma_{31}^2+\sigma_{12}^2)+r\left[\sigma_{23}(2\sigma_{11}-\sigma_{22}-\sigma_{33})\right.$$
$$\left.+\sigma_{31}(2\sigma_{22}-\sigma_{33}-\sigma_{11})+\sigma_{12}(2\sigma_{33}-\sigma_{11}-\sigma_{22})\right]=\sigma_s^2 \quad (2)$$

Here, f, l and r are determined based on yield stresses along several crystal axis orientations by using the Least Square Method, as shown in Table 1. In the plane stress condition, this yield criterion becomes

$$f\left(\sigma_{22}^2+\sigma_{11}^2-\sigma_{11}\sigma_{22}\right)+l\sigma_{12}^2-r\left(\sigma_{12}\sigma_{11}+\sigma_{12}\sigma_{22}\right)]=\sigma_s^2. \quad (3)$$

and in the plane stress condition, it becomes

$$f_1\left(\sigma_{22}^2+\sigma_{11}^2\right)-f_2\sigma_{11}\sigma_{22}+l\sigma_{12}^2+r_1\left(\sigma_{12}\sigma_{11}+\sigma_{12}\sigma_{22}\right)=\sigma_s^2. \quad (4)$$

where, $f_1 = f(1-\gamma+\gamma^2)$, $f_2 = f(1+2\gamma-2\gamma^2)$ and $r_1 = r(2\gamma-1)$.

Table 1 Material parameters of DD3 nickel-based single crystal

	950/°C	850/°C	760/°C		950/°C	850/°C	760/°C
F(=f)	1	1	1	σ_s/MPa	525	870	915
L	2.8310	2.7305	3.8015	E/GPa	93.5	101	106
l	2.9305	2.6557	3.5846	G/GPa	96	104	109
r	0.1788	-0.0034	0.4582	γ	0.330	0.330	0.322

Crack Tip Plastic Zone of Nickel-Based Single Crystal Plate

Stress Field near the Crack Tip. The infinite nickel-base single crystal plate with a center crack is shown in Fig.1, and the crack is along the crystalline axis.

Assuming the coordinates coincide with the crystalline axis of material, flexibility coefficients of nickel-based single crystal alloy are, respectively

$$S_{11} = S_{22} = S_{33} = \frac{1}{E},\ S_{12} = S_{13} = S_{23} = -\frac{\gamma}{E},\ S_{44} = S_{55} = S_{66} = \frac{1}{G} \quad (5)$$

where E, G are, respectively, elastic modulus and shear modulus of material along the crystalline axis and γ is the Poisson's ratio of material.

The characteristic equation of nickel-base single crystal alloy for the plane problem is represented by[4]

$$a_{11}\lambda^4+(2a_{12}+a_{66})\lambda^2+a_{22}=0. \quad (6)$$

In the plane stress condition, we can get $a_{11} = a_{22} = S_{11}$, $a_{66} = S_{66}$, $a_{12} = S_{12}$, and in the plane strain condition, we can get $a_{11} = a_{22} = (1-\gamma^2)S_{11}$, $a_{66} = S_{66}$, $a_{12} = (1+\gamma)S_{12}$.

The crack tip stress field of orthotropic plate with mixed mode I, II crack is given in reference [4], as follows

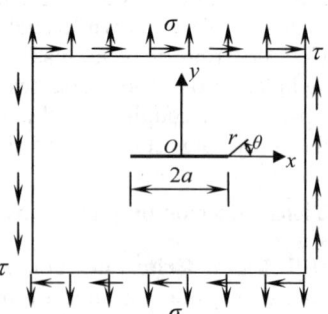

Fig. 1 Mixed mode I, II crack of nickel-based single crystal plate

$$\sigma_{11} = \frac{K_I}{\sqrt{2\pi r}}P_{11}^I + \frac{K_{II}}{\sqrt{2\pi r}}P_{11}^{II},\ \sigma_{22} = \frac{K_I}{\sqrt{2\pi r}}P_{22}^I + \frac{K_{II}}{\sqrt{2\pi r}}P_{22}^{II},\ \sigma_{12} = \frac{K_I}{\sqrt{2\pi r}}P_{12}^I + \frac{K_{II}}{\sqrt{2\pi r}}P_{12}^{II}. \quad (7)$$

where K_I, K_{II} are, respectively, the stress intensity factor of mode I, mode II crack and

$$P_{11}^I = \text{Re}\left\{\frac{\lambda_1\lambda_2}{\lambda_1-\lambda_2}\left(\frac{\lambda_2}{k_2}-\frac{\lambda_1}{k_1}\right)\right\}, \; P_{11}^{II} = \text{Re}\left\{\frac{1}{\lambda_1-\lambda_2}\left(\frac{\lambda_2^2}{k_2}-\frac{\lambda_1^2}{k_1}\right)\right\}, \; P_{22}^I = \text{Re}\left\{\frac{1}{\lambda_1-\lambda_2}\left(\frac{\lambda_1}{k_2}-\frac{\lambda_2}{k_1}\right)\right\}$$
$$P_{22}^{II} = \text{Re}\left\{\frac{1}{\lambda_1-\lambda_2}\left(\frac{1}{k_2}-\frac{1}{k_1}\right)\right\}, \; P_{12}^I = \text{Re}\left\{\frac{\lambda_1\lambda_2}{\lambda_1-\lambda_2}\left(\frac{1}{k_1}-\frac{1}{k_2}\right)\right\}, \; P_{12}^{II} = \text{Re}\left\{\frac{1}{\lambda_1-\lambda_2}\left(\frac{\lambda_1}{k_1}-\frac{\lambda_2}{k_2}\right)\right\} \quad (8)$$

where, λ_1, λ_2 are roots of the Eq. 6, $k_1 = \sqrt{\cos\theta + \lambda_1\sin\theta}$ and $k_2 = \sqrt{\cos\theta + \lambda_2\sin\theta}$.

Crack Tip Plastic Zone. For Hill yield criterion, substituting Eq. 7 into Eq. 1, we can give the crack tip plastic zone of nickel-base single crystal plate in the following form in plane stress condition:

$$r_P = \frac{1}{2\pi}\left(\frac{K_I}{\sigma_s}\right)^2\left\{F\left[\left(P_{11}^I+P_{11}^{II}\right)^2+\left(P_{22}^I+P_{22}^{II}\right)^2-\left(P_{11}^I+P_{11}^{II}\right)\left(P_{22}^I+P_{22}^{II}\right)\right]+L\left(P_{12}^I+P_{12}^{II}\right)^2\right\}. \quad (9)$$

in plane strain condition:

$$r_P = \frac{1}{2\pi}\left(\frac{K_I}{\sigma_s}\right)^2\left\{F_1\left[\left(P_{11}^I+P_{11}^{II}\right)^2+\left(P_{22}^I+P_{22}^{II}\right)^2\right]-F_2\left(P_{11}^I+P_{11}^{II}\right)\left(P_{22}^I+P_{22}^{II}\right)+L\left(P_{12}^I+P_{12}^{II}\right)^2\right\}. \quad (10)$$

For the modified Hill yield criterion with considering tension-shear coupling, substituting Eq. 7 into Eq. 3, we can give the crack tip plastic zone of nickel-base single crystal plate in plane stress condition in the following form

$$r_P = \frac{1}{2\pi}\left(\frac{K_I}{\sigma_s}\right)^2 \left\{F\left[\left(P_{11}^I+P_{11}^{II}\right)^2+\left(P_{22}^I+P_{22}^{II}\right)^2-\left(P_{11}^I+P_{11}^{II}\right)\left(P_{22}^I+P_{22}^{II}\right)\right]+L\left(P_{12}^I+P_{12}^{II}\right)^2\right.$$
$$\left. -R\left[\left(P_{12}^I+P_{12}^{II}\right)\left(P_{11}^I+P_{11}^{II}\right)+\left(P_{12}^I+P_{12}^{II}\right)\left(P_{22}^I+P_{22}^{II}\right)\right]\right\} \quad (11)$$

In plane strain condition, substituting Eq. 7 into Eq. 4, we obtain

$$r_P = \frac{1}{2\pi}\left(\frac{K_I}{\sigma_s}\right)^2\left\{F_1\left[\left(P_{11}^I+P_{11}^{II}\right)^2+\left(P_{22}^I+P_{22}^{II}\right)^2\right]-F_2\left(P_{11}^I+P_{11}^{II}\right)\left(P_{22}^I+P_{22}^{II}\right)\right.$$
$$\left. +L\left(P_{12}^I+P_{12}^{II}\right)^2+R_1\left[\left(P_{12}^I+P_{12}^{II}\right)\left(P_{11}^I+P_{11}^{II}\right)+\left(P_{12}^I+P_{12}^{II}\right)\left(P_{22}^I+P_{22}^{II}\right)\right]\right\} \quad (12)$$

Analysis on Examples. The material parameters of nickel-based single crystal DD3 are given in reference [5], as shown in Table 1. The crack tip plastic zones $r_P \Big/ \dfrac{2\pi}{(K_I/\sigma_s)^2}$ are shown in Fig. 2~Fig. 5.

From Fig.2, the crack tip plastic zones symmetrically distribute along the crack based on the Hill yield criterion. In the plane stress condition, the scope of crack tip plastic zone of nickel-base single crystal alloy is smaller than that of the homogeneous material; in the plane strain condition, it is larger than that of the homogeneous material.

The sizes of crack tip plastic zones of nickel-base single crystal alloy are given, respectively, based on Hill yield criterion and the modified Hill yield criterion under $k=1$ and 760°C, shown in Fig. 3. From Fig. 3, we find that the size of crack tip plastic zone becomes smaller in the plane stress condition and larger in the plane strain condition when considering the tensile-shear stress coupling.

Based on the modified Hill yield criterion, according to the different compound ratio $k = K_{II}/K_I$ of mixed mode I, II, we get the crack tip plastic zones of nickel-base single crystal alloy under 760°C, shown in Fig.4. In the plane stress condition, the increasing rate of the size of crack tip plastic zone with compound ratio k is greater than the increasing rate in the plane strain condition.

Based on the modified Hill yield criterion, crack tip plastic zones of nickel-base single crystal alloys under different temperature and $k=1$, are shown in Fig. 5. The results which can be seen from Fig. 5 are that the temperature has greater effect on size of the nickel-base single crystal alloy crack tip plastic zone. For the nickel-based single crystal alloys DD3, the size of plastic zone changes greatly when the temperature changes between 760°C and 850°C, but slightly between 850°C and 950°C.

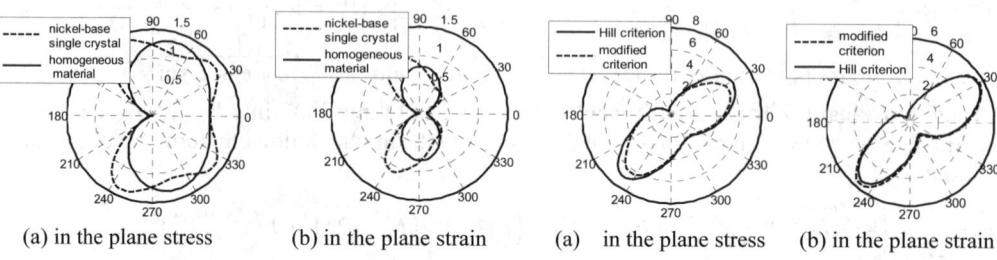

(a) in the plane stress (b) in the plane strain   (a) in the plane stress (b) in the plane strain

Fig. 2 Crack tip plastic zone of nickel-based single crystal and homogeneous materials (760°C, $k=0$)

Fig. 3 Crack tip plastic zone of nickel-based single crystal (760°C, $k=1$)

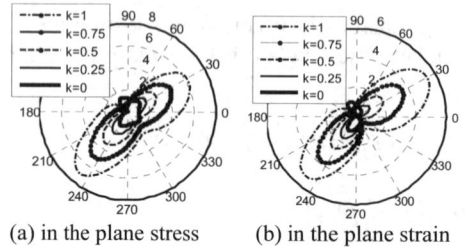 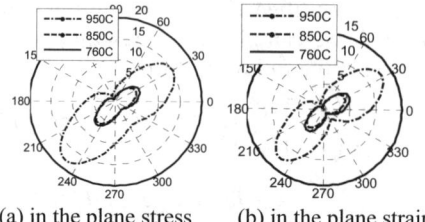

(a) in the plane stress (b) in the plane strain   (a) in the plane stress (b) in the plane strain

Fig. 4 Mixed mode I, II crack tip plastic zone of nickel-based single crystal for different k (760°C)

Fig. 5 Mixed mode I, II crack tip plastic zone of nickel-based single crystal for different temperatures ($k=1$)

Conclusions

Based on Hill yield criterion and a modified Hill yield criterion of considering tensile-shear coupling, we derive the expressions of crack tip plastic zone of nickel-base single crystal alloys plate with mixed mode I, II crack, respectively, in the plane stress condition and in plane strain condition. We can conclude from the analysis in the paper that the tensile-shear stress coupling should be considered in the yield criterion for nickel-base single crystal alloys. The influence of the temperature on sizes of crack tip plastic zones of the nickel-base single crystal alloy is greater at 760°C to 850°C.

References:

[1] Liu C, Huang Y, Stout M G. On the Asymmetric Yield Surface of Plastically Orthotropic Materials: A Phenomenological Study. Acta Mater. Vol. 45(1997), p. 2397-2406.
[2] Gao Xin, Kang Xinwu, Wang Hangong. Solution to the Crack Tip Plastic Zone of Orthotropic Unidirectional Composites. Journal of Solid Rocket Technology. Vol. 32(2009), p. 554-559.
[3] Ding Zhiping, Liu Yilun, Yin Zeyong. Study of Elastic-Plastic Constitutive Model for Single Crystal Nickel-Based Superalloy. Journal of Aerospace Power. Vol. 19(2004), p. 755-760.
[4] Zhang Shaoqin. The Special Method for Fracture Analysis of Composite Materials. Chapter, 3, China's Weapon Industry Press (2003).
[5] Editorial Board of China's Handbook of Aeronautical Materials. China Aeronautical Materials Handbook (Volume II) (Edition 2). Standards Press of China (2002).

Interaction of elliptical inclusion and crack in half-space under SH-waves

Zailin Yang[1,2a], Huanan Xu[1,b], Baoping Hei[1,c] and Yong Yang[1,d]

[1]College of Aerospace and Civil Engineering, Harbin Engineering University, Harbin, 150001, China

[a]yangzailin00@163.com, [b]hntiger_86@126.com, [c]hbp0627@126. com,[d]yanghrb@hrbeu.edu.cn

Keywords: scattering of SH-wave, multiple cracks, circular lining structure, Green's function, surface displacement

Abstract. The methods of Green's function, complex function and multi-polar coordinates are applied here to report interaction of an elliptical inclusion and a crack in half-space under incident SH-waves. Based on the symmetry of SH-waves scattering, the "conformal mapping" technology was developed to construct a suitable Green's function, a fundamental solution to the displacement field for the elastic half space containing elliptical inclusion while bearing out-plane line source load at arbitrary point, for creating a beeline crack with arbitrary length at any position combined with "crack-division" technology. The displacement field and stress field were then deduced while the inclusion coexists with the crack Lastly, numerical examples are presented to discuss the dependence of dynamic stress concentration factor (DSCF) around the elastic inclusion on different parameters.

Introduction

Elliptical inclusion commonly appears in matrix in engineering. For stress concentration easily occurs around the inclusion in composite material containing arbitrary inclusion, cracks are usually produced in the vicinity of the inclusion. Study on interaction of inclusion and cracks under SH wave can provide references for intensity design of composite materials and non-destructive testing etc., which has been reported by numerous specialists[2-5]. The problem of SH-wave scattering caused by an elliptical inclusion and a crack in elastic half-space, which can be regarded as blast-resistance problem, is investigated in this paper.

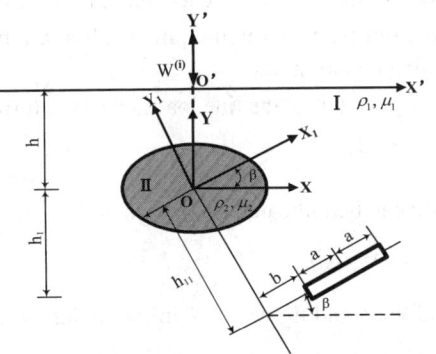

Fig.1.The infinite half space model with an elliptical inclusion and a crack under incident vertically SH-waves

Statement of the Problem

The elastic model with an elliptical inclusion and a crack under incident SH-wave is given as Fig.1. Media I and media II have different material constants (ρ_1,μ_1; ρ_2,μ_2). Three coordinates are indicated in the figure and the relation of them is defined by

$$x' = x, y' = y - h, x_1 = x\cos\beta + y\sin\beta,$$
$$y_1 = y\cos\beta - x\sin\beta, h_{11} = (h_1 + b\sin\beta)/\cos\beta. \quad (1)$$

The Scattering Waves around the Elliptical Inclusion

Governing equation. In an isotropic medium, study on the scattering of SH-wave is the easiest problem of scattering of elastic wave. Inducing comformal mapping function $z = \omega(\eta)$ ($\eta = Re^{i\theta}$), the displacement W impacted by the incident wave should satisfy the governing equation[1]

$$\frac{1}{\omega'(\eta)\overline{\omega'(\eta)}}\frac{\partial^2 W}{\partial\eta\partial\overline{\eta}}+\frac{k^2}{4}W=0 \qquad (2)$$

in which, $k=\omega/c_s$, ω is the circular frequency of the displacement function, $c_s=\sqrt{\mu/\rho}$ stands for the velocity of the shear wave. ρ and μ are the mass density and shear modulus of the medium respectively.

In polar coordinate, the corresponding stresses are given by

$$\tau_{rz}=\frac{\mu}{R|\omega'(\eta)|}(\eta\frac{\partial G}{\partial \eta}+\overline{\eta}\frac{\partial G}{\partial \overline{\eta}}),\quad \tau_{\theta z}=\frac{i\mu}{R|\omega'(\eta)|}(\eta\frac{\partial G}{\partial \eta}-\overline{\eta}\frac{\partial G}{\partial \overline{\eta}}) \qquad (3)$$

The scattering wave in District I. Based on the symmetry of the scattering wave, multi-polar coordinates is used to construct the scattering wave in the medium induced by the elliptical inclusion, which should satisfy the governing equation (2) and Sommerfeld radiation condition for infinite distance beside the zero-stress condition at the horizontal interface.

In the complex cordinate $(z,\overline{z})$, $W_1^{(s)}$ can be expressed by

$$W_1^{(s)}=\sum_{n=-\infty}^{+\infty}A_n\left\{H_n^{(1)}[k_1|w(\eta)|]\left[\frac{w(\eta)}{|w(\eta)|}\right]^n+H_n^{(1)}[k_1|w(\eta)-2ih|]\left[\frac{w(\eta)-2ih}{|w(\eta)-2ih|}\right]^{-n}\right\} \qquad (4)$$

where A_n are the unknown coefficients, determined by the boundary condition of the elliptical inclusion.

The standinging wave in District II. In the complex plane, the standing wave inside of elliptical inclusion is deduced as

$$W_{II}^t=\sum_{n=-\infty}^{\infty}B_n J_n\left(k_2|w(\eta)|\right)\left[\frac{w(\eta)}{|w(\eta)|}\right]^n \qquad (5)$$

where B_n is the unknown coefficient determined by the boundary condition.

Green's Function

The Green's function G_1 denotes an essential solution to the displacement field for an elastic half-space containing an elliptical inclusion while bearing out-plane harmonic linie loads at arbitrary point, which is expressed as $e^{-i\omega t}$ and satisfies the governing equation (2).

In a half elastic space, the incident wave $G^{(i)}$ excited by the out-plane line loads takes the form of

$$G_1^{(i)}=\frac{i}{4\mu_1}H_0^{(1)}(k_1|w(\eta)-w(\eta_0)|) \qquad (6)$$

The reflected wave field caused by horizontal interface can be indicated as $G^{(r)}$, then

$$G_1^{(r)}=\frac{i}{4\mu_1}H_0^{(1)}(k_1|w(\eta)-\overline{w(\eta_0)}-2ih|) \qquad (7)$$

The scattering wave $G_1^{(s)}$ excited by the inclusion and the standing wave G_{II}^t in the inclusion take the forms of Eq.(4) and Eq.(5), respectively.

In the complex plane $(z,\overline{z})$, the boundary condition can be expressed as

$$\begin{cases}G_1=G_{II}^t \\ \tau_{rz,I}=\tau_{rz,II}^t\end{cases} \qquad (8)$$

Substitution of Eqs.(6), (7) and the expressions of $G_1^{(s)}$ and into Eq.(3), the corresponding stresses can be derived, then substituting these expressions into boundary conditions (8), the coefficients A_n and B_n will be solved.

The total wave field is

$$G_1 = G^{(i)} + G^{(r)} + G_1^{(s)}$$

$$= \frac{i}{4\mu} H_0^{(1)}(k_1 |w(\eta) - w(\eta_0)|) + \frac{i}{4\mu} H_0^{(1)}(k_1 |w(\eta) - \overline{w(\eta_0)} - 2ih|) \quad (9)$$

$$+ \sum_{n=-\infty}^{+\infty} A_n \left\{ H_n^{(1)}[k_1|w(\eta)|] \left[\frac{w(\eta)}{|w(\eta)|}\right]^n + H_n^{(1)}[k_1|w(\eta) - 2ih|] \left[\frac{w(\eta) - 2ih}{|w(\eta) - 2ih|}\right]^{-n} \right\}$$

Scattering of SH-wave by Elliptical Inclusion and Crack near Interface

As shown in Fig.1, the incident steady SH-wave defined as $W^{(i)}$ can be given by

$$W^{(i)} = W_0 e^{-\frac{ik_1}{2}[(\omega(\eta)-ih)-(\overline{\omega(\eta)}+ih)]} \cdot e^{-i(\omega t + \pi/2)} \quad (10)$$

where W_0 is amplitude of incident wave.

The scattering wave $W_1^{(s)}$ and the standing wave W_{II}^t excited by the elliptical inclusion can be described as Eqs.(4) and (5), respectively. And the process of solving A_n and B_n is the same as that of the Green's function discussed in preceding paper.

$$\begin{cases} W^{(i)} + W_1^{(s)} = W_{II} \\ \tau_{rz,I} = \tau_{rz,II} \end{cases} \quad (11)$$

The total wave field of domain I can be obtained

$$W_1 = W^{(i)} + W_1^{(s)} \quad (12)$$

The total stresses can be also performed. If additional stresses which have same magnitude but opposite in direction are applied at the same point, the ultimate stresses will be zero. Therefore, when a pair of forces with the same magnitude but opposite direction are loaded at the region where the crack will be created, the resultant forces will be zero, then a crack is created.

Then we can obtain the total wave field in domain I when a crack coexists with the inclusion

$$W_I^{(t)} = W^{(i)} + W_I^{(s)} - \int_{(b,-h_1)}^{(2a+b,-h_1)} \tau_{\theta z,I} G_1 dz_1 \quad (13)$$

Dynamic Stress Concentration Factor(DSCF)

Usually the dynamic stress concentration factor (DSCF) $\tau_{\theta z}^*$ can be written as

$$\tau_{\theta z}^* = \left| \tau_{\theta z}^{(t)} / \tau_0 \right|, \quad (14)$$

where $\tau_{\theta z}^{(t)}$ is the stress around the outer boundary of the elliptical cavity; τ_0 stands for the largest amplitude of the incident stresses, $\tau_0 = W_0 \mu_1 k_1$.

Results and Discussion

Numerical examples are provided here to discuss influence rule of various parameters on dynamic stress concentration factor(DSCF) around the elliptical inclusion. The expression of dynamic stress concentration factor(DSCF) is difined as Eq. (14).

Fig.2 to Fig.5 give distribution of $\tau_{\theta z}^*$ around the elliptical inclusion as different parameters under incident SH-waves. Fig.2 illustrates that the greater the ratio of wave number($k^* = k_1/k_2$)and modulus ($\mu^* = \mu_2/\mu_1$) are, the greater $\tau_{\theta z}^*$ around the inclusion becomes, when $k^* = 4.0, \mu^* = 16.0$, the maximum of $\tau_{\theta z}^*$ can achieve about 2.33, increasing by 109% than $k^* = 0.5, \mu^* = 0.25$. Variation of h affects very slightly $\tau_{\theta z}^*$, as shown as in Fig.3. In Fig.4, it can be intuitively seen that as wave number

k_1 increases, $\tau^*_{\theta z}$ around the elliptical inclusion fast concentrate at two sides of the inclusion, and smaller and smaller in magnitude. Besides, the angle of the crack β has weak influence on dynamic stress concentration factor around the inclusion, as described by Fig.5.

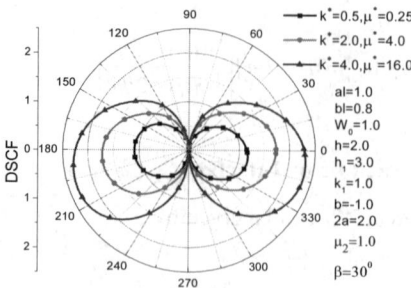

Fig.2 Distribution of DSCF around inclusion with k^* and μ^*

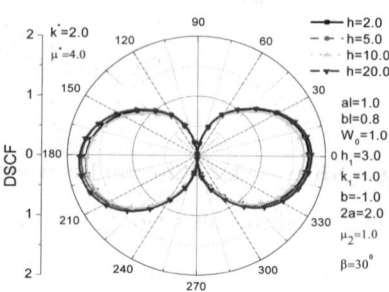

Fig.3 Distribution of DSCF around inclusion with h

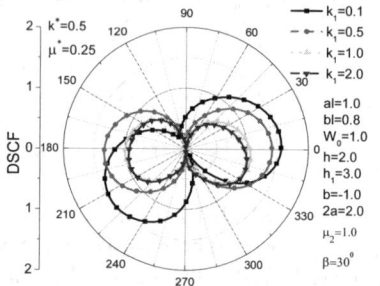

Fig.4 Distribution of DSCF around inclusion with k_1

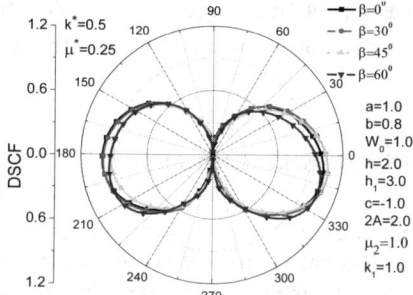

Fig.5 Distribution of DSCF around inclusion with β

Acknowledgement

This Work was Supported by National Natural Science Foundation of China (No. 10972064), the Fundamental Research Funds for the Central Universities Under Grant (No. HEUCFZ1125).

References

[1] Pao Y H, Mow C C: *Diffraction of Elastic Wave and Dynamic Stress Concentration* (Crane and Russak, New York, 1973).

[2] Zailin Yang, Dankui Liu: submitted to Acta Mechanica Solida Sinica (2002).

[3] Zailin Yang, Peilei Yan: submitted to Journal of Key Engineering Materials (2008).

[4] Zailin Yang, Peilei Yan, and Diankui Liu: submitted to Acta Mechanica Sinica (2009).

[5] Zailin Yang, Huanan Xu, Meijuan Xu, Baitao Sun: submitted to Key Engineering Materials (2010).

Influence of welding conditions on residual stresses of multi-pass tube sheet welds

Hong Li[1,a], Yong Zheng[1,b] and Li Li[2,c]

[1]College of Aerospace and Civil Engineering, Harbin Engineering University, Harbin 150001, China

[2]College of Materials Science and Chemical Engineering, Harbin Engineering University, Harbin 150001, China

[a]leeh2005@sohu.com, [b]562600287@qq.com, [c]lili_heu@hrbeu.edu.cn

Keywords: Elastic-plastic-model; Double ellipsoid heat source; Curved multi-pass welding line; Welding condition; Residual stress

Abstract. Residual stresses and residual plastic strains of the welded structures are the products of nonlinear behaviors during welding. The residual stresses will cause errors during the assembly of the structure and injure the beauty of appearance of the structure. Based on an elastic-plastic-model, finite element numerical simulation of a representative tube sheet penetration assembly with loop welding line joined by multi-pass welding is carried out and the influence of welding conditions on residual stresses of the tube sheet welds is studied in this paper. Nonlinear three dimensional transient temperature fields and real-time dynamic stresses field are analyzed by FEM. The heat source is modeled as a moving heat flux following a double ellipsoid distribution and the temperature-dependent properties of materials are considered. The method of birth and death of element in finite element analysis is applied to simulate the gradual growth of weld pass metal. It is shown that welding sequence, size of groove welding and weld toes dressing will obviously change the magnitude of the residual stresses of tube sheet welds.

Introduction

Welding process has always played a major role in industrial production, especially in the automotive, maritime and aerospace industries. Despite many advantages, welding has some process-specific disadvantages, such as thermal expansion and shrinkage, micros structural transformations, residual stresses and component distortions. Residual stresses and distortions of a welded structure will reduce the strength of the structure especially when the process is multi-pass welding [1]. The accurate prediction and control of welding residual stresses have been essential.

There are many researches have been done about welding residual stresses [2-5]. The especial contribution of this paper is that loop welding lines based on multi-pass and different welding conditions, such as welding sequence, size of groove welding and weld toes dressing are studied while the heat source is modeled as a moving heat flux with a double ellipsoid distribution.

Based on the elastic plastic analysis, a 3D numerical simulation of representative tube sheet penetration assembly with loop welding line during multi-pass welding is carried out in this paper. The simulation process consists of calculations of nonlinear heat transfer and real-time dynamic stress. Two different welding sequences, two different angles of groove welding and melting mend to the welding toe or not are considered during multi-pass welding.

Penetration Assembly Model

The tube sheet welds with penetration assembly comes from industrial production. The half model shown as Fig.1 is a reasonable simplification instead of the whole model. A finite element model of transient process is established to simulate the moving along loop welding line of the heat source for multi-pass welding. The number of the multi-pass welding shown as Fig.2 is six. There are two different kinds of welding sequences. One sequence is up and below alternative (up-below-up-below) and the other is one-side first, then the other (up-up- below--below). The two different angles of the groove welding in this model are 30 degree and 50 degree respectively. In order to get a better distribution of the residual stress near the welding line, weld toes dressing is necessary by reheating the weld toes after the first thermo-mechanical coupling analysis in finite element simulation. The model of weld toes dressing is shown as Fig.3.

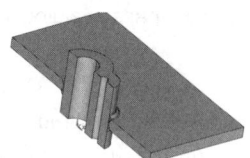 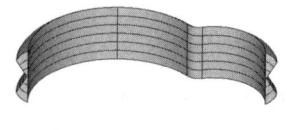 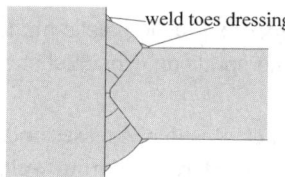

Fig.1 Welding mode   Fig.2 Multi-pass welding lines   Fig.3 Weld toes dressing model

Welding Residual Stresses Analysis

There are very large varieties in temperature and stress in the structure in the process of welding. The distribution of the temperature and the temperature transformation with time are calculated first. The most significant factors affecting the analysis are the heat input rate, the moving speed of the heat source, the thickness of the plate, the geometry of the heating line and the properties of the material. In the process of analysis, the temperature dependent properties of materials are considered and the heat source moved along loop welds following a double ellipsoid distribution. The method of birth and death of element in finite element analysis software is applied to simulate the gradual growth of weld pass metal. The distribution of real-time multi-pass welding stresses analyzed by thermal-elastic–plastic FEM is changed following the moving of heat source.

When cooling to the environment temperature, the distributions of residual stress fields in tube sheet welds for two different welding sequences are consistent, but the magnitudes of residual stresses of them aren't equal. The Mises stresses for alternative sequence are less than those for one-side sequence. So alternative sequence should be selected to do the thick muti-pass welding.

When the size of groove welding is different, the highest temperature and residual stresses fields of the tube sheet welds are changed. The Mises stresses for angles of 30 degree and 50 degree of groove welding are shown as Fig.4 and Fig.5 respectively. It is shown that the maximum of residual stress of 30 degree groove welding is less than that of 50 degree groove welding for this model.

Next, the effect of weld toes dressing is studied. After modifying welding line of the finite element model, heat the weld toes again and do the thermo-mechanical coupling analysis once more so as to get a better distribution of the residual stress. The contrast of the residual stresses is clearly exhibited by stress plot. The residual stresses before weld toes dressing are shown as Fig.6 and that after weld toes dressing are shown as Fig.7.

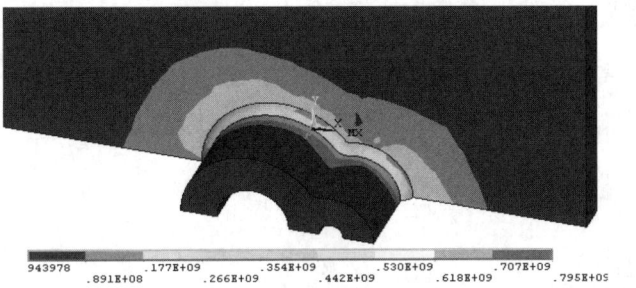 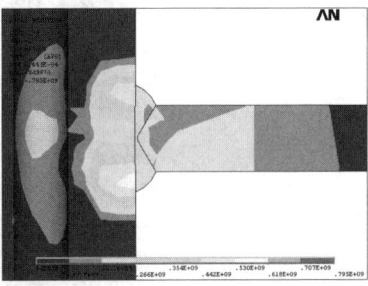

Fig.4 Mises stresses for angle of 30 degree

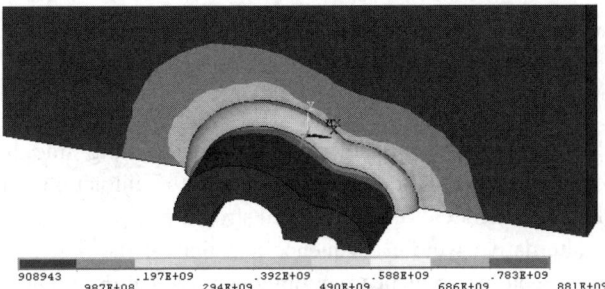 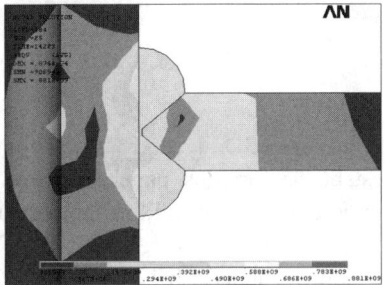

Fig.5 Mises stresses for angle of 50 degree

The maximum of residual stresses before weld toes dressing happens at the junction of tube and welding line. The magnitude of stress is also larger near welding line on the sheet and it reduces rapidly far away from the welding line. After weld toes dressing, the change of the maxiumu of residual stress field is not very obvious, but the distribution of residual stress inside the model is different and the surface of weld metal near weld toes turns to more uniform and slick.

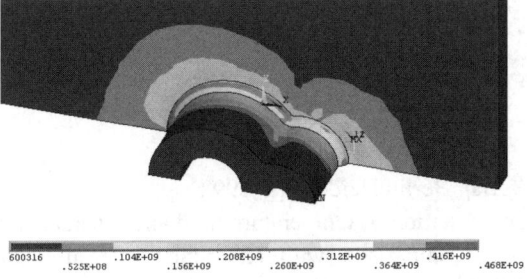 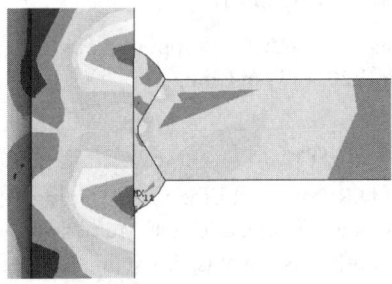

Fig.6 Residual stresses before weld toes dressing

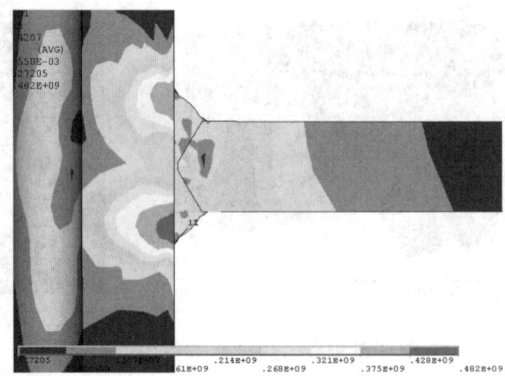

Fig.7 Residual stresses after weld toes dressing

Summary

A feasible 3D dynamic simulation method of multi-pass welding with loop welding line is established in this paper, which provides theory foundation and instruction to simulation on three-dimensional welding temperature field and stress field of complicated structures.

By analyzing the residual stress of the alternative welding sequence and that of the one-side welding sequence, it is shown that alternative welding sequence can obviously reduce the highest temperature and the residual stresses, so it is superior to one-side welding sequence. The change of angle of the groove welding will lead to an obvious variety of the highest temperature and residual stresses fields of the tube sheet welds. Furthermore, weld toes dressing is helpful to get a better distribution of the residual stress near the welding line.

The influence of welding conditions on residual stresses of multi-pass tube sheet welds discussed in this paper and the way of predicting the residual stresses expressed in this paper can have guides at design stage and can be used in accuracy management.

Acknowledgement

This research is supported by the natural science foundation of Heilongjiang Province (No.E201113) in China.

References

[1] D.Radaj. Heat Effects of Welding. Springer Verlag Berlin Heidelberg ,1992

[2] Dean Deng, Hidekazu Murakawa. Numerical simulation of temperature field and residual stress in multi-pass welds in stainless steel pipe and comparison with experiment measurements. Computational Materials Science, 2006 (37):269-277P

[3] Chang Doo Jang, Seung II Seo and Dae Eun Ko. A Study on Prediction of Deformations of Plates Due to Line Heating Using a Simplified Thermal Elasto-Plastic Analysis. Journal of Ship Production. 1997,13(1):22-27P

[4] Martin Becker. Nonlinear Transient Heat Conduction with Application to Welding. ASME 32nd National Heat Transfer Conf. Vol.9 Manufacturing and Materials Processing, 1997

[5] S.A.A. Akbari Mousavi,R. Miresmaeili: Experimental and Numerical Analyses of Residual Stress Distributions in TIG Welding Process for 304L Stainless Steel. Journal of Materials Processing Technology. 208, (2008), p383–394

Study on Mode I Quasi-static Growing Crack in a Rigid-viscoelastic Material Interfacial Crack

Yong Yang[1,a], Zailin Yang[1,b], Liqiang Tang [1,c]

[1]College of Aerospace and Civil Engineering, Harbin Engineering University, Harbin 150001, China

[a]yangharbin@163.com, [b]lining0304@yeah.net, [c]tlq8854@126.com

Keywords: pressure-sensitive dilatant material, stress and strain field, quasi-static growing crack

Abstract. A mechanical model of the pressure-sensitive dilatant material is established in order to investigate the viscous effect in quasi-static growing crack-tip field. The constitutive equations on the pressure-sensitive dilatant material are deducted. Through asymptotic analysis, it is shown that in the stable creep growing stage, the elastic-deformation and the visco-deformation are equally dominant in the near-tip field, as $r^{-1/(n-1)}$. And for the mechanical model of a rigid-viscoelastic material interfacial crack, through numerical calculate, the asymptotic solution of the continuous separate variable formal of stress, strain and displacement. The significance of the material parameter in the crack-tip field is discussed. The study of the paper will provide theoretical references for preventing the glide of the mud rock and reducing the arises of the failure of casings of well.

Introduction

Under high temperature, many metals show viscoelastic properties, which are closely related to environmental factors and strain rate. Engineering practice also shows that time-dependent deformation of materials have great effects on the structural strength, rigidity and life duration. So visco-elasticity has drawn much attention in the mechanical and engineering fields.

In 1968, Hutchinson[1], Rice and Rosengren[2] studied the asymptotic crack tip field for plane strain stationary crack independently. When elastic deformation is neglected, the so-called HRR field is obtained for power law hardening materials. Hui and Riedel[3] addressed the quasi-static mode III and mode I crack problem where elastic-power-creeping constitutive equations are introduced to model the rate-dependent materials. HR field was obtained, where stress and strain possess the same singularity, as $r^{-1/(n-1)}$, with n, the power law exponent. HR field is self-government which means that crack tip stress and strain fields are not dependent on the applied load, and no geometrical parameters related to the crack are included. Once the crack propagation velocity $\dot{a}$ is given, crack tip fields can be fully determined. Gao[4,5] gave a simplified elasto-visco-plastic model in which it was assumed that during the elastic deformation stage, viscous effects can be ignored. When plastic deformation occurs, viscous effects are taken into account. For mode I crack, solutions with logarithm singularity and power singularity were given separately. Tang[6] investigated the mode I crack tip field for incompressible elasto-viscous materials.

At present, more studies on the crack tip focus on the incompressible material. However, incompressibility is an ideal assumption. In most cases, considering the material compressibility is more reasonable. Crack tip fields on mode I quasi-static growing crack in a rigid-viscoelastic material interfacial crack are given in present paper where the distribution law of the stress fields and the strain fields are discussed.

1. Basic Equations

Fig. 1 shows a mechanical model of a planar crack steadily propagating along the positive direction of X, in which the common origin is at the crack tip of planar crack, XOY is the fixed coordinate system, xoy is the moving coordinate system, and the moving rectangular and polar coordinate system denoted respectively by (x,y) and (r,θ). Supposing that the crack grows at the steady velocity V, the material derivative of any physical quantity Φ in steady field can be expressed as

$$D = V\left[\frac{\sin\theta}{r}\frac{\partial}{\partial\theta} - \cos\theta\frac{\partial}{\partial r}\right]. \tag{1}$$

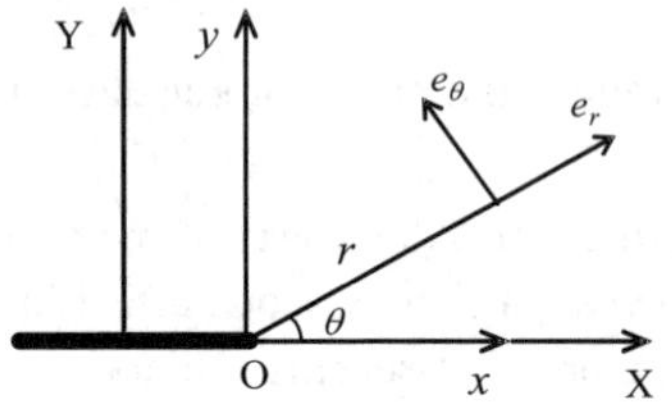

The strain rate of the visco-elastic material is composed of elastic and viscous strain rate, named as

$$\dot{\varepsilon}_{ij} = \dot{\varepsilon}^e_{ij} + \dot{\varepsilon}^c_{ij}. \tag{2}$$

where, $\dot{\varepsilon}^e_{kk}$ and $\dot{\varepsilon}^c_{kk}$ is respectively expressed as elastic volume strain ratio and viscous volume strain ratio. The case where elastic deformation is compressible and viscous deformation is incompressible is studied in this paper, as $\dot{\varepsilon}^e_{kk} \neq 0, \dot{\varepsilon}^c_{kk} = 0$. The constitutive equations are given by

Fig.1 Coordinate with crack

$$\dot{\varepsilon}_{ij} = \frac{1+v}{E}\dot{s}_{ij} + \frac{1-2v}{3E}\dot{\sigma}_{kk}\delta_{ij} + \frac{3}{2}B\sigma_e^{n-1}s_{ij}, \tag{3}$$

Where, the modulus of elasticity E, the exponent of creep strain n, the material parameters of power-law creep material B, and the equivalent stress is $\sigma_e = \left[(3/4)(\sigma_r - \sigma_\theta)^2 + 3\sigma_{r\theta}^2\right]^{1/2}$. The case that elastic deformation is compressible is considered, so the Poisson's ratio $v < 0.5$. The equations of motion and geometry are

$$\nabla \cdot \boldsymbol{\sigma} + \mathbf{F} = \rho \ddot{\boldsymbol{u}}, \quad \boldsymbol{\varepsilon} = \frac{1}{2}(\boldsymbol{u}\nabla + \nabla\boldsymbol{u}), \quad \nabla \times \dot{\boldsymbol{\varepsilon}} \times \nabla = 0. \tag{4}$$

2. Governing Equations

When $r \to 0$, supposed that the form of separate variable of potential function of displacement and separate variable of potential function of stress are

$$u_r = r^{s_1-1}f_1(\theta,n), \quad u_\theta = r^{s_1-1}f_2(\theta,n), \tag{5}$$

$$\sigma_r = r^{s-2}[P(\theta) + F(\theta)], \sigma_\theta = r^{s-2}[P(\theta) - F(\theta)], \sigma_{r\theta} = r^{s-2}T(\theta). \tag{6}$$

In stable creep growing stage, elastic-deformation and visco-deformation are equally dominant in the near-tip field, as a result, the stress and the strain have the same singularity, as $s = s_1 = s_2 = (2n-3)/(n-1)$.

Introducing the relevant quantities to equations of motion and constitutive equations, by noticing that the derivation laws (1) of material derivative and by dimensionless arrangement, as:

$(s-2)P(\theta) + sF(\theta) + T'(\theta) = M^2\{\sin^2\theta \tilde{f}_1'' + (2-s)\sin(2\theta)\tilde{f}_1' + (s-2)[\sin^2\theta + (s-1)\cos^2\theta]\tilde{f}_1 - 2\sin^2\theta \tilde{f}_2'' + (s-2)\sin(2\theta)\tilde{f}_2'\}$;

$P'(\theta) - F'(\theta) + sT(\theta) = M^2\{\sin^2\theta \tilde{f}_2'' + 2\sin^2\theta \tilde{f}_1'' + (s-2)[\sin^2\theta + (s-1)\cos^2\theta]\tilde{f}_2 - (s-2)\sin(2\theta)\tilde{f}_1 + (2-s)\sin(2\theta)\tilde{f}_2'\}$;

$\sin\theta(s-2)(\tilde{f}_1' - \tilde{f}_2) - \cos\theta(s-1)(s-2)\tilde{f}_1 = \sin\theta(1-2v)P'(\theta) + \sin\theta(1+v)F'(\theta) - 2\sin\theta(1+v)T(\theta) + (2v-1)\cos\theta(s-2)P(\theta)$

$-(1+v)\cos\theta(s-2)F(\theta) + \frac{3}{2}A \cdot 3^{\frac{(n-1)}{2}}\left[F^2(\theta) + T^2(\theta)\right]^{\frac{(n-1)}{2}}F(\theta)$;

$$\sin\theta\left[2\tilde{f}_1'+\tilde{f}_2''+(s-2)\tilde{f}_2\right]-\cos\theta(s-2)(\tilde{f}_1+\tilde{f}_2')=\sin\theta\{[(1-2\nu)P'(\theta)-(1+\nu)[F'(\theta)-2T]\}+\cos\theta(s-2)[(2\nu-1)P(\theta)$$

$$+(1+\nu)F(\theta)]-\frac{3}{2}A\cdot 3^{\frac{(n-1)}{2}}\left[F^2+T^2\right]^{\frac{(n-1)}{2}}F(\theta)\,;$$

$$\sin\theta\left[\tilde{f}_1''+2(s-2)\tilde{f}_1+(s-4)\tilde{f}_2'\right]-\cos\theta(s-2)\left[\tilde{f}_1'+(s-2)\tilde{f}_2\right]=(1+\nu)\left[2\sin\theta T'(\theta)+4\sin\theta F(\theta)-2\cos\theta(s-2)T(\theta)\right]$$

$$+A\cdot 3^{\frac{(n-1)}{2}}\left[F^2(\theta)+T^2(\theta)\right]^{\frac{(n-1)}{2}}T(\theta)\,. \tag{7}$$

3. Numerical calculation and Discussions

In the petroleum engineering, the rate of water in the mud rock when water injecting will increases, the coefficient of friction will decrease, the rigidity of the mud rock will decrease, the quasi static shear gliding will possibly happened between the layers, and it will lead to the failure of casings of well. Therefore, studying the solution of the sigularity field on the crack tip in rigidity/elasticity bimaterial layer has theoretical significance.

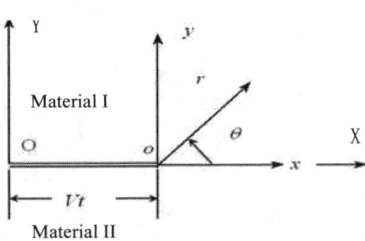

Fig.2 Mechanical model

Solving the upper governing differential equations for Fig.2 needs seven boundary conditions.

$$(s-2)(s-1)\tilde{u}_r(\pi)-f''(s-2)-f'-f(s-2)s+\nu[(s-2)(s-1)sf]-\tilde{\sigma}_e^{n-1}[f''+sf-\nu(s-1)sf]=0;$$

$$-(s-2)\tilde{u}_r'(\pi)-(s-2)[1+\nu(s-1)]\tilde{u}_r(\pi)+(s-2)(s-1)(1-\nu^2)sf+\tilde{\sigma}_e^{n-1}(1-\nu^2)(s-1)f=0;$$

$$-(s-2)\tilde{u}_r'(\pi)-(s-2)^2\tilde{u}_\theta(\pi)-2(s-2)(1-s)f'+2\tilde{\sigma}_e^{n-1}(1+\nu)(1-s)f=0;$$

$$\tilde{u}_r=0\,(\theta=0);\ \sigma_\theta(r,\pi)\leq 0;\ \sigma_{r\theta}(r,\pi)=-\eta\sigma_{\theta\theta}(r,\pm\pi);\ u_\theta(r,\pi)=0;\ \tilde{u}_r(\pi)=1. \tag{8}$$

Here η is the coefficient of sliding friction. There are not enough conditions of determining solution to solve the controlling equations, then the problem should be transformed into initial value, adopting two parameters shooting method to solve it by Runge–Kutta method.

(1) By the coefficient of sliding friction decreasing, the amplitude of spherical compression stress will decrease (Fig.3).

(2) By E of the upper viscoelasticity decreasing, the amplitude of the crack tip will increase.

(3) By the coefficient of sliding friction decreasing, the amplitude of strain field at the crack tip will increase obviously.

Tab.1: Values of parameters (n,η) and $f(\pi)$

No.	n	η	$f(\pi)$
1	5	0.45	-1.2952381
2	5	0.4	-1.0527394
3	5	0.35	-0.8380952
4	5	0.3	-0.7138531
5	5	0.25	-0.6095238
6	5	0.2	-0.5475346

4. Conclusions

It is summarized that, in oil field, by the rate of water in the mud rock increasing, the coefficient of friction between the layers will decrease, the rigidity of the mud rock will also decrease, the amplitude of the crack tip between the layers will increase, the rate of the shear stress will ascend, the destructive effect to the oil well will be enhanced. In this paper, Preliminary attempt is done for studying the mechanism of the failure of casings of well by applying the fracture mechanics. More work should be done for further study.

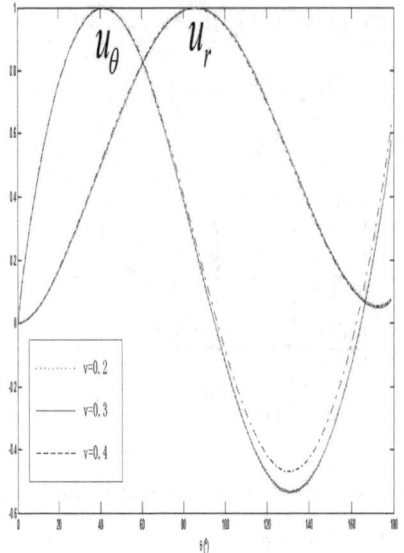 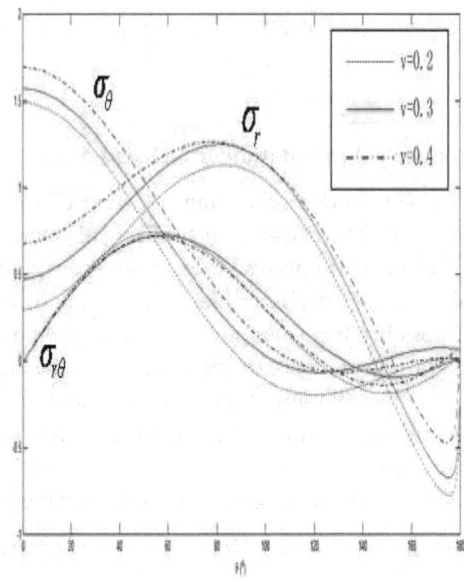

Fig.3: Curves of the displacments with $n = 5, \eta = 0.35$ Fig.4 Curves of the stresses with $n = 5, \eta = 0.35$

Acknowledgments

The authors acknowledges the support by the fundamental research funds for the central universities (HEUCF100209).

Literature References

[1] Chen Huifa, in: *Constitutive Relations of Civil Engineering Material : Plasticity and Modeling*[M].Wuhan: Central China Science and Technology University Publising, 2001.

[2] Wang Ren, K.C. Hwang and Zhu Zhaoxiang. *Progress in Plasticity Mechanics*.[M]. Beijing:Chinese Railway Publising, 1988.

[3] Lade P. V. Elasto-plastic Stress-Strain Theory for Cohesionless Soil with Curved Yield Surface[J]. Int. J. of Solids and Structure.1977,3:1019-1035.

[4] CHEN H.J., TANG L.Q. *Crack Tip Asymptotic Field in Porous Materials* [M]. Beijing: Tsinghua University Press,2000.

[5] CHENETER M.E. THOMTSON T.W. Perforation Stability in Low-permeability Gas Reservioirs [J]. SPE/DOE, 1990,5(1):63-69.

Analysis of interface deformation of steel-concrete-steel sandwich beam

Xia Peixiu [1,a], Zou Guangping [2,b], Chang Zhongliang [3,c]

[123]Harbin Engineering University, Harbin 150001,China

[a]xiapeixiu@hrbeu.edu.cn,[b]zouguangping@hrbeu.deu.cn,[c]czl19820228@163.com

Key words: steel-concrete-steel sandwich beam; interface slip; deformation; theoretical expression

Abstract: The effect of the interface slip is neglected in most studies on calculating deflection of sandwich beams. By taking a simply supported sandwich beams under uniformly distributed loads as an example, simplified analytical models of the interface slip are established, and corresponding clculation formulas of interface slip between steel panels and concrete and section curvatures are derived. The formula for deflection of sandwich beams are then presented. This formula reflects the relationship of influence each other between the interface slip and deflection.

1 Introduction

The interface slipping of steel-concrete-steel(SCS) sandwich beam will reduce the SCS sandwich beam stiffness, and also increase the beam curvature and deformation. In recent decades, the detail experimental and numerical simulation analysis about SCS sandwich beam[1~9] have been studied carefully, but its theory analysis is very limited. In 1994, BCSA established the SCS sandwich beam design specification, considering the influence of interface slip between the steel plate and concrete, the stiffness converted method[8] was adopted to calculate the deformation of SCS sandwich beam. Wright & Oduyemi[9] analyzed the deformation of sandwich beam which neglected the interface slipping between cracked concrete and between upper steel plate and concrete .Sandwich beam was simplified as the sandwich struction with only one interface. SCS sandwich beam with two interface, its concrete in tension zone will failure under lower tensile stress and generate crack, so the SCS sandwich beam will bear load with the upper plate, bottom plate and concrete in compression zone. In this paper, base on the stress state of sandwich beam and consider the relative slipping to establish computation model to solve the interface and sectional curvature of sandwich beam, and deduce the sandwich beam deflection formula under the influence of two interface slipping.

2 Computational model

According to structure characteristics of theSCS sandwich beam，we made the following assumptions:1) The steel panels and concrete are isotropic elastomer; 2) Horizontal shearing stress is proportional to the relative slip between the interface of steel plates and concrete; 3) Steel plates and concrete have the same sectional curvature; 4) Before and after deformation, steel and concrete strain respectively accord with normal sections assumptions and ignore shear deformation of components; 5) In the tension zone the concrete cracked and quit worked. Pressurized concrete section wastransfer to steel section. Micro-segment analysis model is shown in figure 1.

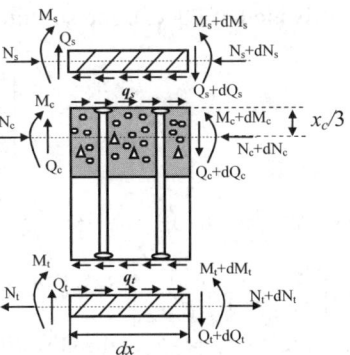

Fig.1 Micro analysis model

3 Calculation formula of the interface slip and deformation

Balance equation of upper and under steel plate and concrete unit.

$$\begin{cases} dM_t - Q_t dx - q_t dx \frac{1}{2} t_t = 0 \\ dM_s - Q_s dx - q_s dx \frac{1}{2} t_s = 0 \\ dM_c - Q_c dx - q_t dx \left(h_c - \frac{1}{3} x_c \right) - q_s dx \frac{1}{3} x_c = 0 \\ dN_t + q_t dx = 0 \\ dN_s + q_s dx = 0 \\ dN_c + q_t dx - q_s dx = 0 \end{cases} \quad (1)$$

Add the formulas in Eq. (A.1), Considering: $Q=Q_t+Q_s+Q_c$, $\phi = \dfrac{M_t}{E_s I_t} = \dfrac{M_s}{E_s I_s} = \dfrac{M_c}{E_s I_{oc}}$, We get:

$$\frac{d\phi}{dx} = \frac{1}{EI}(Q + q_t h_{tc} + q_s h_{sc}) \quad (2)$$

Where: E_s, E_c is elastic modulus of steel and concrete, $\alpha_0 = E_s/E_c$, I_c, I_t, I_s is moment of inertia of concrete steel and bottom plate and upper plate. $EI = E_s I_t + E_s I_s + E_s I_{oc}$, $I_{oc} = \dfrac{b^3}{12\alpha_0}$, $h_{tc} = \dfrac{t_t}{2} + h_c - \dfrac{x_c}{3}$, $h_{sc} = \dfrac{t_s}{2} + \dfrac{x_c}{3}$, x_c is the distance of the centroidal axis and the lower edge of uper steel plate, can be calculated as $x_c = -\alpha_0(t_s + t_t) + \left[\alpha_0^2(t_s + t_t)^2 - \alpha_0(t_s^2 - 2t_t h_c - t_t^2)\right]^{1/2}$.

Eq. (3) gives the relative slip strain ε_s, ε_t of the interface between steel plate and the concrete:

$$\begin{cases} \varepsilon_s = \dfrac{ds_s}{dx} = \varepsilon_{sc} - \varepsilon_{cs} = h_{sc}\phi - \dfrac{N_s}{E_s A_s} + \dfrac{N_c}{E_s A_{oc}} \\ \varepsilon_t = \dfrac{ds_t}{dx} = \varepsilon_{ct} - \varepsilon_{tc} = h_{tc}\phi - \dfrac{N_t}{E_s A_t} - \dfrac{N_c}{E_s A_{oc}} \end{cases} \quad (3)$$

Where: A_c, A_t, A_s are sectional area of the concrete steel and the bottom plate and the upper plate, $A_{oc}=bx_c/\alpha_0$ is steel plates sectional area which concrete is converred to.

Derivation of Eq. (3), and substitut $q_s = k_{Ls} s_s$, $q_t = k_{Lt} s_t$ into Eq. (2):

$$\begin{cases} \dfrac{d^2 s_s}{dx^2} = a_1 Q + b_1 s_s + c_1 s_t \\ \dfrac{d^2 s_t}{dx^2} = a_2 Q + b_2 s_s + c_2 s_t \end{cases} \quad (4)$$

Where: $a_1 = \dfrac{h_{sc}}{EI}$, $a_2 = \dfrac{h_{tc}}{EI}$, $b_1 = k_{Ls}\left[\dfrac{h_{sc}^2}{EI} + \dfrac{1}{E_s A_{oc}} + \dfrac{1}{E_s A_s}\right]$, $b_2 = k_{Ls}\left[\dfrac{h_{sc} h_{tc}}{EI} - \dfrac{1}{E_s A_{oc}}\right]$, $c_1 = k_{Lt}\left[\dfrac{h_{sc} h_{tc}}{EI} - \dfrac{1}{E_s A_{oc}}\right]$, $c_2 = k_{Lt}\left[\dfrac{h_{tc}^2}{EI} + \dfrac{1}{E_s A_t} + \dfrac{1}{E_s A_{oc}}\right]$, $k_{Ls}=k_L/l_p$ is the slip stiffness of unit beam long between the concrete and the upper plate; $k_{Lt}=k_L/l_p$ is the slip stiffness of unit beam long between the concrete and the bottom plate; l_p is the lengthways distance of shearforce connection.

When concentrat load on simply supported sandwich beams (as shown in fig.2), boundary conditions: when $0<x\leq l/2$, and when $x=0$, $ds_s/dx=0$, $ds_t/dx=0$; when $x=l/2$, $s_s=0$, $s_t=0$; when $l/2<x\leq l$, and wen $x=l$, $ds_s/dx=0$, $ds_t/dx=0$; when $x=l/2$, $s_s=0$, $s_t=0$; substitut into Eq. (4):

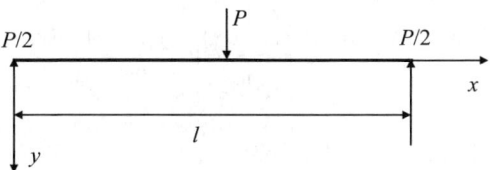

Fig.2. Mid-span concentrated load

$$s_s = \begin{cases} \dfrac{\theta_\beta P}{2(\gamma_\alpha-\gamma_\beta)}\dfrac{\cosh(\alpha x)}{\cosh\frac{\alpha l}{2}} - \dfrac{\theta_\alpha P}{2(\gamma_\alpha-\gamma_\beta)}\dfrac{\cosh(\beta x)}{\cosh\frac{\beta l}{2}} + \dfrac{tP}{2n} & 0<x\leq l/2 \\ -\dfrac{\theta_\beta P}{2(\gamma_\alpha-\gamma_\beta)}\dfrac{\cosh(\alpha x-\alpha l)}{\cosh\frac{\alpha l}{2}} + \dfrac{\theta_\alpha P}{2(\gamma_\alpha-\gamma_\beta)}\dfrac{\cosh(\beta x-\beta l)}{\cosh\frac{\beta l}{2}} - \dfrac{tP}{2n} & l/2<x\leq l \end{cases} \quad (5)$$

$$s_t = \begin{cases} \dfrac{\gamma_\alpha\theta_\beta P}{2(\gamma_\alpha-\gamma_\beta)}\dfrac{\cosh(\alpha x)}{\cosh\frac{\alpha l}{2}} - \dfrac{\gamma_\beta\theta_\alpha P}{2(\gamma_\alpha-\gamma_\beta)}\dfrac{\cosh(\beta x)}{\cosh\frac{\beta l}{2}} - \dfrac{(na_1+tb_1)P}{2nc_1} & 0<x\leq l/2 \\ -\dfrac{\gamma_\alpha\theta_\beta P}{2(\gamma_\alpha-\gamma_\beta)}\dfrac{\cosh(\alpha x-\alpha l)}{\cosh\frac{\alpha l}{2}} + \dfrac{\gamma_\beta\theta_\alpha P}{2(\gamma_\alpha-\gamma_\beta)}\dfrac{\cosh(\beta x-\beta l)}{\cosh\frac{\beta l}{2}} + \dfrac{(na_1+tb_1)P}{2nc_1} & l/2<x\leq l \end{cases} \quad (6)$$

where: $m=b_1+c_2$, $n=b_1c_2-b_2c_1$, $t=a_2c_1-a_1c_2$, $\alpha=\sqrt{\dfrac{m+\sqrt{m^2-4n}}{2}}$, $\beta=\sqrt{\dfrac{m-\sqrt{m^2-4n}}{2}}$, $\gamma_\alpha=\dfrac{\alpha^2-b_1}{c_1}$, $\gamma_\beta=\dfrac{\beta^2-b_1}{c_1}$, $\theta_\alpha=\dfrac{na_1+\alpha^2 t}{nc_1}$, $\theta_\beta=\dfrac{na_1+\beta^2 t}{nc_1}$。

Eq. (5) and Eq. (6) was substituted into Eq. (2). and get axial force of the section of bottom plate and upper plate :

$$\begin{cases} N_s = \mu_1\alpha\dfrac{\sinh(\alpha x)}{\cosh\frac{\alpha l}{2}}\dfrac{P}{2} - \mu_2\beta\dfrac{\sinh(\beta x)}{\cosh\frac{\beta l}{2}}\dfrac{P}{2} + \omega_1 x\dfrac{P}{2} \\ N_t = -\mu_3\alpha\dfrac{\sinh(\alpha x)}{\cosh\frac{\alpha l}{2}}\dfrac{P}{2} + \mu_4\beta\dfrac{\sinh(\beta x)}{\cosh\frac{\beta l}{2}}\dfrac{P}{2} - \omega_2 x\dfrac{P}{2} \end{cases} \quad (7)$$

where : $\omega_1=k_{Ls}\dfrac{a_1c_2-a_2c_1}{b_1c_2-b_2c_1}$, $\omega_2=k_{Lt}\dfrac{a_1b_2-a_2b_1}{b_1c_2-b_2c_1}$, $\mu_1=k_{Ls}\dfrac{(\gamma_\alpha c_1-c_2)\theta_\beta}{(b_1c_2-b_2c_1)(\gamma_\alpha-\gamma_\beta)}$, $\mu_2=k_{Ls}\dfrac{(\gamma_\beta c_1-c_2)\theta_\alpha}{(b_1c_2-b_2c_1)(\gamma_\alpha-\gamma_\beta)}$, $\mu_3=k_{Lt}\dfrac{(\gamma_\alpha b_1-b_2)\theta_\beta}{(b_1c_2-b_2c_1)(\gamma_\alpha-\gamma_\beta)}$, $\mu_4=k_{Lt}\dfrac{(\gamma_\beta b_1-b_2)\theta_\alpha}{(b_1c_2-b_2c_1)(\gamma_\alpha-\gamma_\beta)}$。

According to the balance equilibrium condition and $\phi=\dfrac{M_t}{E_sI_t}=\dfrac{M_s}{E_sI_s}=\dfrac{M_c}{E_sI_{oc}}$:

$$\phi(x)=\dfrac{1}{EI}(M(x)-h_{sc}N_s-h_{tc}N_t) \quad (8)$$

Eq. (7) was substituted into Eq. (8). and sectional curvature of sandwich beams was given:

$$\phi(x)=\dfrac{(2-\zeta)Px}{4EI}-\dfrac{\chi_1\alpha P}{4EI}\dfrac{\sinh(\alpha x)}{\cosh\frac{\alpha l}{2}}+\dfrac{\chi_2\beta P}{4EI}\dfrac{\sinh(\beta x)}{\cosh\frac{\beta l}{2}} \quad (9)$$

Then the deflection of the sandwich beams was given:

$$y(x)=\int\left[\int\phi(x)dx\right]dx+H_1 x+H_2 \qquad (10)$$

Eq. (9) was substituted into Eq. (10)., boundary conditions: $x=0$, $y(x)=0$; $x=l/2$, $dy(x)/dx=0$; $x=l$, $y(x)=0$, The deflection of the sandwich beams was given under condition of concentrating load:

$$y(x)=\frac{(2-\zeta)P(4x^3-3l^2 x)}{96EI}-\frac{\chi_1 P}{4\alpha EI}\frac{\sinh(\alpha x)}{\cosh\frac{\alpha l}{2}}+\frac{\chi_2 P}{4\beta EI}\frac{\sinh(\beta x)}{\cosh\frac{\beta l}{2}}+\frac{(\chi_1-\chi_2)Px}{4EI} \qquad (11)$$

Where, $\chi_1=\mu_1 h_{sc}-\mu_3 h_{tc}$, $\chi_2=\mu_2 h_{sc}-\mu_4 h_{tc}$, $\zeta=\omega_1 h_{sc}-\omega_2 h_{tc}$。

4 Conclusion

In this paper, the formula which considered the two interface relative slip between steel and concrete is derived, this will has the application and theoretical significance for analysis theSCS sandwich beam deflection.

References:

[1] T. M. Roberts, D. N. Edwards, R. Narayanan. Testing and Analysis of Steel-Concrete-Steel Sandwich Beams[J]. J. Construct. Steel Res, 1996, 38(3): 257-279, 2003, 5(4): 5—13.
[2] B. McKinley, L.F. Boswell. Behaviour of double skin composite construction[J]. Journal of Constructional Steel Research, 2002, 58 : 1347–1359.
[3] M.Xie, N. Foundoukos, J.C. Chapman. Static tests on steel–concrete–steel sandwich beams[J]. Journal of Constructional Steel Research, 2007, 63: 735–750.
[4] N. Foundoukos, J.C. Chapman. Finite element analysis of steel–concrete–steel sandwich beams[J]. Journal of Constructional Steel Research, 2008, 64: 947–961.
[5] M. Xie, J.C. Chapman. Static and fatigue tensile strength of friction-welded bar–plate connections embedded in concrete[J]. Journal of Constructional Steel Research, 2005, 61: 651–673.
[6] Wang Zuen, Wang Yongxi. Test analysis of the composite of the concrete and the steel plates[J]. Journal of shanghai institute of railway technology, 1995, 16(2): 71-77. (in Chinese)
[7] Xia Zhicheng, Xu Fei, Wang Yongxi. Experiment studies on vertical shear strength of steel-concrete-steel sandwich composite slab[J]. Journal of Harbin university of civil engineering and architecture, 1995, 28(5): 178-185. (in Chinese)
[8] Xie M, Foundoukos N, Chapman JC. Experimental and numerical investigation on the shear behaviour of friction-welded bar–plate connections embedded in concrete[J]. Journal of Constructional Steel Research, 2005, 61:625–649.
[9] H.D.Wright, T.O.S. Oduyemi. Partial interaction analysis of double skin composite beams[J]. Journal of Constructional Steel Research, 1991, 19(4): 253–283.

Reliability analysis for stiffened plate on the maximum entropy method

HE Jian[1, a], CHEN Xiaoyan[1, b]

[1]Department of Aerospace and Civil Engineering, Harbin Engineering University, Harbin 150001, China

[a]hejian@hrbeu.edu.cn, [b]chenxiaoyanshui@163.com

Key words: maximum entropy method; near-field explosion; cross stiffened plate; failure probability

Abstract. Stiffened plate is widely used in vessel structure because of its high bearing capacity and low weight so the research of failure probability for stiffened plate under explosion load has important engineering meaning. Stiffened plate under near-field explosion is taken as research subject, dynamite density and yield stress of plate are selected as random variables, the original values of one hundred groups of random variables are gotten through the random number generation program, and the moments of random variables are obtained. Based on failure criterion of displacement ductility, the performance function of structure is established, probability density function of performance function is fitted using maximum entropy method then the failure probability of stiffened plate structure is obtained. So as to solve the problem of calculate failure probability when the sample size is small and the probability density function is unknown.

Introduction

Plentiful experimental researches and theoretical analyze of the reliability problem of shell structure under the explosion shock loading were carried out in domestic and overseas [1]. Cross stiffened plate is a common form in shell structure and it is widely used in vessel structure. However, the vessel, as the operational equipment platform in a surface or subsurface, is easier to be attacked by missile weapon. So the ability of the stiffened plate structure on the explosion and shock resistance is directly related to the vitality of vessel. Therefore, it is important to research the reliability of stiffened plate under explosion load [2, 3].

The simple values of random variable

Generate random number on distributing symmetrical with method of multiply and residual in a range [0, 1] [4]:

$$y_{i+1} = \alpha y_i (\mod M), \quad r_i = y_i / M, \quad (i = 1, 2, 3 \cdots). \tag{1}$$

where α, M are constants of predefine; r_i is random number. FORTRAN language is used to calculate one hundred random numbers, take $y_1 = 773311$, $\alpha = 655393$, $M = 33554432$.

According to fifty groups of x_i order, translate x_i into normal distribution:

$$y_i = \mu + \sigma x_i \quad (i = 1, 2, \cdots 50). \tag{2}$$

Dynamite density ρ_e and yield stress of plate σ_s are selected as random variable; the random numbers of dynamite density and yield are produced by the method. The first four moment parameters of random variable are obtained with the simple values of random variable table 1 to show.

Table 1 The first four moment parameters of random variable

Random variable	Mean value μ	Standard deviation σ	Coefficient of skew C_{sX}	Coefficient of peak C_{kX}
Dynamite density ρ_e	1531.81 kg/m^3	44.35 kg/m^3	-0.2137	2.4631
Yield stress σ_s	235.77 MN/m^2	5.46 MN/m^2	-0.2839	2.7448

Maximum entropy method

Shannon introduced the thermodynamic entropy to information theory in 1948. If the random event is continuously distributive and its probability density function is $f_X(x)$, the Shannon entropy is:

$$H = -c\int_{-\infty}^{+\infty} f_X(x)\ln f_X(x)\,dx. \tag{3}$$

Shannon entropy is the measure of uncertainty before the event happen; after it happen, it is the information measure that be got from the event. So Shannon entropy is also called information entropy, it is the measure of uncertainty or information rate [5].

There is an analysis that can make information entropy get maximum in all probability distribution under given conditions; this is called Jayne's maximum entropy theory. Make origin moments of the first m moments of random variable x as restrictions that make Eq.3 have the largest value under the following conditions:

$$v_{Xi} = E(X^i) = \int_{-\infty}^{+\infty} x^i f_X(x)\,dx, \quad i = 0,1,\cdots,m. \tag{4}$$

With the Lagrange multiplier method, Eq.3 and Eq.4 can be used to introduce a correction function:

$$L = -c\int_{-\infty}^{+\infty} f_X(x)\ln f_X(x)\,dx + \sum_{i=0}^{m}\lambda_i\left[\int_{-\infty}^{+\infty} x^i f_X(x)\,dx - v_{Xi}\right]. \tag{5}$$

where $\lambda_0, \lambda_1, \cdots, \lambda_m$ are pending constant. At the stable points, $\partial L / \partial f_X(x) = 0$, let $a_0 = 1 - \lambda_0/c$, $a_i = -\lambda_i/c (i=1,2,\cdots,m)$, the maximum entropy probability density function is:

$$f_X(x) = \exp(-\sum_{i=0}^{m} a_i x^i). \tag{6}$$

where $a_0, a_1, \cdots, a_m$ are pending constant.

Eq.4 is equivalent to give the central moment of X:

$$\mu_{Xi} = E[(X-\mu_X)^i] = \int_{-\infty}^{+\infty}(x-\mu_X)^i f_X(x)\,dx \quad i=0,1,\cdots,m. \tag{7}$$

where $\mu_X = v_{X1}$ is the mean of X.

The first four moment of X can be obtained that:

$$\mu_{X0}=1,\ \mu_{X1}=0,\ \mu_{X2}=\sigma_X^2,\ \mu_{X3}=C_{sX}\sigma_X^3,\ \mu_{X3}=C_{kX}\sigma_X^3. \tag{8}$$

where σ_X is standard deviation, C_{sX} is coefficient of skew, C_{kX} is coefficient of peak.

In order to avoid each center moments have great difference, transform random variable X into standard random variable $Y = (X - \mu_X)/\sigma_X$, then each center moments of X and Y have relation that:

$$\mu_{X^i} = E[(X-\mu_X)^i] = E[(\sigma_X Y)^i] = \sigma_X^i E(Y^i) = \sigma_X^i \mu_{Y^i} = \sigma_X^i v_{Y^i}, \quad i = 0,1\cdots m. \tag{9}$$

With Eq.8 and Eq.9 show that $\mu_Y = 0, \sigma_Y = 1, v_{Y^i} = \mu_{Y^i}$, and the first four moment of Y can be obtained that:

$$v_{Y0} = 1, \ v_{Y1} = 0, \ v_{Y2} = 1, \ v_{Y3} = C_{sY} = C_{sX}, \ v_{Y4} = C_{kY} = C_{kX}. \tag{10}$$

Determine the system parameters of standard random variable Y are:

$$c_0 = -\frac{4C_{kY} - 3C_{sY}^2}{10C_{kY} - 12C_{sY}^2 - 18}, \ c_1 = -\frac{C_{sY}(3+C_{kY})}{10C_{kY} - 12C_{sY}^2 - 18}, \ c_2 = -\frac{2C_{kY} - 3C_{sY}^2 - 6}{10C_{kY} - 12C_{sY}^2 - 18}, \ d = c_1. \tag{11}$$

Assume the basic function of structure is $Z = g_X(X)$, where $X = (X_1, X_2, \cdots, X_n)^T$ and the statistical parameters of X_i are $\mu_{X_i}, V_{X_i}, C_{sX_i}$ and C_{kX_i}, the first four moments are $\mu_{X_i 1}, \mu_{X_i 2}, \mu_{X_i 3}$ and $\mu_{X_i 4}$. The Taylor series expansion for Z on the checking point x^* and takes to the quadratic term and:

$$Z_Q = g_X(x^*) + (X-x^*)^T \nabla g_X(x^*) + \frac{1}{2}(X-x^*)^T \nabla^2 g_X(x^*)(X-x^*). \tag{12}$$

The first four moments of Z can be approximate calculated by Eq.12 and the mean of Z_Q is:

$$\mu_{Z_Q} = E(Z_Q) = g_X(\mu_X) + \frac{1}{2}\sum_{i=1}^n \frac{\partial^2 g_X(\mu_X)}{\partial X_i^2} \mu_{X_i 2}. \tag{13}$$

The calculate formulas of center moments from second to forth are:

$$\mu_{Z_Q i} = E[(Z_Q - \mu_{Z_Q})^i], \ i = 2,3,4 \tag{14}$$

Transform Z into standard random variable $Y = (Z - \mu_Z)/\sigma_Z$, random variable Y satisfies the constraints of Eq.10 and its maximum entropy density function $f(y)$ is also the form of Eq.6. Take Eq.6 and Eq.10 into Eq.4, integral equations can be obtained:

$$\int_{-\infty}^{+\infty} y^i \exp\left(-\sum_{j=0}^m a_j y^j\right) dy = v_{Y^i}, \ i = 0,1\cdots m. \tag{15}$$

where $a_0, a_1 \cdots a_m$ of $f(y)$ can be solved.

The failure probability of structure is:

$$p_f = \Pr(Z \le 0) = \Pr\left(Y \le -\frac{\mu_Z}{\sigma_Z}\right) = \int_{-\infty}^{-\frac{\mu_Z}{\sigma_Z}} \exp\left(-\sum_{i=0}^m a_i y^i\right) dy. \tag{16}$$

Finite interval include zero is used to replaced the infinite integral interval of Eq.15 and the finite interval which is below $-\mu_Z/\sigma_Z$ is used to replaced the integral lower limit of Eq.16.

Example

Stiffened plate of experiment is selected as the model and with fixed supported at its four sides. The dimension of bearing area is 250mm×250mm, the thickness of plate is 3mm, the width of plate is 3mm and the height of plate is 10mm, explosive distance is 1m.

The basic function of structure can be expressed like this:

$$g = \mu - \frac{B + \sqrt{B^2 + 4AC}}{2Ax_y}. \tag{17}$$

where $A = \dfrac{\pi^2 \sigma_s t}{8}\left(\dfrac{a}{b}+\dfrac{b}{a}\right)$, $C = \dfrac{8ab}{\rho t H^2} A_i^2 M_e^{\frac{4}{3}}$, $B = \dfrac{8\sigma_s t^2}{4\sqrt{3}}\left[\dfrac{2a}{b}+\dfrac{b}{a}+\dfrac{1}{2\left(-b/2a+\sqrt{b^2/4a^2+3/4}\right)}\right]$,

$x_y = 0.00127 \times q_s l^4 / D$, μ is the ultimate ductility ratio of displacement and 6~8 is always used, ρ is the density of steel, H is explosive distance, uniform load $q_s = 12\sigma_s t^2 / l^2$, D is the bending stiffness of plate and $D = Et^3 / 12(1-v^2)$, E is elastic module, v is Poisson's ratio, t is the thickness of plate.

Maximum entropy is used to fit the probability density function of basic function:

$$f(x) = e^{(a_0 - a_1 x - a_2 x^2 - a_3 x^3 - a_4 x^4)}. \tag{18}$$

where $a_0 = 0.03587$, $a_1 = 0.09916$, $a_2 = 0.35142$, $a_3 = -0.25163$, $a_4 = 0.98505$.

The minimum probability of the structure can be calculated by Eq.16: $P_f = 0.017\%$.

Summary

The reliability of stiffened plate under explosion load is analyzed and dynamite density and yield stress of plate are selected as random variables, generate random number on distributing symmetrical with method of multiply, the maximum entropy method is used to determine the probability density function of random variable and the failure probability of stiffened plate is calculated.

Example shows that the maximum entropy method can full use the existing sample dates and use each moment of samples to deduce the density function, decrease the dependence of existing dates and classical probability distribution. The probability calculation process of multivariable complexity system is simplified greatly.

This project is financially supported by Graduate Cultivation Fund of Harbin Engineering University. And this paper is funded by the International Exchange program of Harbin Engineering University For Innovation-oriented Talents cultivation.

Reference

[1] Bedair O.K: *Analysis of Stiffened Plates under Lateral Loading Using Sequential Quadratic Programming* (SQP) .Computers & Structure, (1997), p.63-80
[2] ZHANG Xin, WANG Shan and CHEN Zhenyong: *Research on Dynamic Responses of Stiffened-plate under Underwater Oscillatory Explosion.* Journal of System Simulation,(2007), p. 257-260(In Chinese).
[3] Choi B H,Hwang M,Yoon T, et al. *Experimental study of inelastic buckling strength and stiffness requirements for longitudinally stiffened panels.* Engineering Structure, (2009), p.1141-1153.
[4] Malvern L J. *Review of static and dynamic properties of steel reinforcing bars.* ACI Materials Journal,(1998),p.609-616.
[5] Zhang Ming. *Structural reliability analysis: method and procedures.* Science publishing House, (2009),p.88-96.(In Chinese)

An Enhanced Time-reversal Method for Impact Damage Monitoring on Plate-like Structures

Chunlin Chen[1, a], Yulong Li[1, b] and Fuh-Gwo Yuan[2, c]

[1]School of Aeronautics, Northwestern Polytechnical University, Xi'an, Shanxi, 710072, PR China;
[2]Department of Mechanical and Aerospace Engineering, North Carolina State University, Raleigh, NC 27695, USA

[a]email: chenchunlin@mail.nwpu.edu.cn, [b]email: liyulong@nwpu.edu.cn, [c]email: yuan@ncsu.edu

Keywords: impact location detection, time-reversal method, impact damage, calibration spacing, plate-like structure

Abstract. Based on the self-focusing property of time-reversal (T-R) concept, a time focusing parameter was suggested to improve the impact source identification method developed in authors' previous work. This paper presents a further study on monitoring relatively high energy impact events which caused induced damage on structures. Numerical verifications for a finite isotropic plate and a composite plate under low velocity impacts are performed to demonstrate the versatility of T-R method for impact location detection with induced plastic deformation and delamination damage on metallic and composite structures respectively. The focusing property of T-R concept was adequately utilized to detect impact/damage location. The results show that impact events with various features can be localized using T-R method by introducing the time focusing parameter. It is suited to monitor serious impact events on plate like structures in practice in future.

Introduction

Impact monitoring techniques have being investigated widely to improve the safety and reliability of aerospace structures, especially for composite structures threatened by low velocity impacts. The most popular triangulation method for impact location detection is based on the time of arrival of a given wave mode among the distributed sensors and its velocity information [1,2]. However, it is difficult to obtain the time of arrival and wave velocity information accurately in practice. Daniel et al. [3] developed an impact detection method based on the analysis of the energy that has flown over a monitored area. Park and Chang [4] investigated an impact location detection method by mapping the power distribution over the entire structure with uniformly distributed sensors, requiring many sensors. Neural network [5] has been used to estimate impact location for its versatility, yet a large amount of training data may be required. Time-reversal method [6,7] has been recently used in structural health monitoring for its self-focusing property. An impact source identification method based on T-R concept has been developed by the authors [8,9,10]. In summary, most studies focused on elastic response of the structure, implying no damage induced by impact events on structures. This paper presents a further study on monitoring impact events which generate damage on structures using an enhanced T-R method. Numerical verifications for a finite isotropic plate and a composite plate under low velocity impact are performed to demonstrate the capability of T-R method for locating impact events with induced damages.

Enhanced T-R method for impact location detection

An impact source identification method based on T-R concept has been developed in several papers by authors [8,9,10]. The plate is first calibrated (or characterized) by transfer functions at discrete locations on the plate surface prior to impact monitoring. Upon impact, the strain response from sensor locations and the transfer functions are computed in a T-R process to determine the impact location. Mathematically the T-R procedure in frequency domain can be expressed by

$$f_i(\xi,\omega) = G_i^*(\xi,\omega)s_i(\omega) = G_i(\xi,\omega)s_i^*(\omega) \tag{1}$$

where $G_i(\xi,\omega)$ is the transfer function which represents the relationship between the calibrated impact force at ξ and the strain response by the ith sensor, $f_i(\xi,\omega)$ is the reconstructed signal contributed by the *measured* strain $s_i(\omega)$ from impact, the superscript (*) denotes the complex conjugate of the variable.

The plate is meshed into a finite number of element areas whose vertices are uniform distributed calibration locations, $\mathbf{x}_{jk} \equiv (x_j, y_k)$ on the plate. The T-R method is a temporal and spatial focusing technique, the maximum peak amplitude of reconstructed signals through transfer functions will refocus back to the impact location at the time of the impact. Based on its temporal focusing property, a time focusing parameter γ is used here to improve the T-R method for impact location detection. The time focusing parameter at calibration locations from all the sensors can be expressed by

$$\gamma(\mathbf{x}_{jk}) = \underset{0 \le t \le T}{\text{Max}}[f(\mathbf{x}_{jk},T-t)], \quad f(\mathbf{x}_{jk}) = \sum_{i=1}^{n}\left\{f_i(\mathbf{x}_{jk},T-t)\Big/\underset{t_1 \le t \le T}{\text{Max}}[f_i(\mathbf{x}_{jk},T-t)]\right\} \tag{2}$$

where f is revised expression of the reconstructed signal from all sensors, n is the number of sensors, T is the time duration of T-R process. Proper choice of T and n has been discussed in [10]. The selection of t_1 is based on T and the time duration of impact force, in this paper $t_1=0.5T$.

The initial impact location is determined by selecting the maximum γ among calibration locations. Based on its focusing property, a finer search for impact location is then conducted in the most probable impact area. Among the four areas sharing the approximate impact location, the impact area is chosen by the maximum γ at the calibration location and three other calibrations with relatively large γ. The finer estimated impact location $\mathbf{x}=(x_e, y_e)$ is determined by

$$x_e = \sum_{i=1}^{4} x_i \gamma^2(x_i, y_j) \Big/ \sum_{i=1}^{4} \gamma^2(x_i, y_j), \quad y_e = \sum_{j=1}^{4} y_j \gamma^2(x_i, y_j) \Big/ \sum_{j=1}^{4} \gamma^2(x_i, y_j) \tag{3}$$

Numerical results

To verify the enhanced T-R method for detecting the impact source with induced damages, a finite isotropic plate and a composite plate under low velocity impacts are performed by commercial FEM software ABAQUS to generate synthetic data. The FEM is merely used to generate data to validate the T-R method, which can be readily replaced by sensor data in practice.

Impact monitoring on a metal plate. An aluminum plate is of dimension 420mm×420mm×1mm and the four sensors are mounted near the corner of the plate by 40mm from each side. The plate is simply supported on all edges. An ideal bilinear elastic-plastic constitutive relation of metal materials is used. The mechanical properties of the aluminum are $E = 72\text{GPa}$, $v = 0.3$, $\rho = 2700\text{Kg/m}^3$, $\sigma_0 = 150\text{MPa}$, $E_T = 20\text{GPa}$. The aluminum plate is modeled using shell elements (S4R).

Due to the symmetry of the plate, only impacts encountered in a quarter of the plate are modeled. It is noted that the forces used to calibrate aluminum plate are a half-sinusoidal wave with time duration of 100 μs, all impact forces are a half-sinusoidal wave with the 1ms time duration. Fig. 1 shows the locations marked open circles indicating the calibration points with spacing 40mm. To demonstrate the enhanced T-R method for localizing impacts, impact events with features are tested as follows: (1) nine impacts with peak force 40N at A_1-A_9 were detected; (2) two impacts with 40N peak force and 60N peak force at B_1 and B_2 respectively are detected simultaneously as multiple site impacts; (3) impact forces with various peak forces (80N, 120N, and 160N) at C_1 are monitored; (4) impact force with 40N peak force at D_1 on the damaged plate is localized. To simulate the damage occurred before impact monitoring, the Young's modulus of the material in damaged area retains one percent of the pristine material, and the size of the damage area $d_1d_2d_3d_4$ is 6mm×12mm.

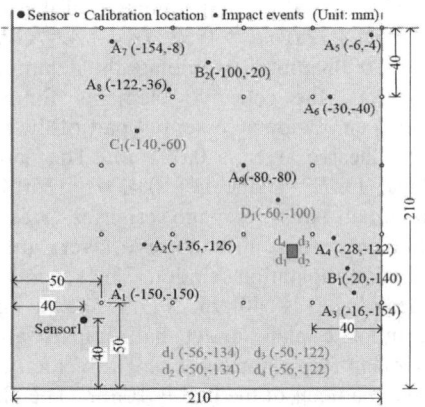

Fig. 1 Calibration and impact locations

Impact events at various locations, A_1-A_9, were detected successfully. The average error of the impact location is 6.7mm and the standard deviation of them is 1.4mm. Compared with the results in [10], the standard deviation reduces significantly. Fig. 2 shows the T-R focusing image of two impacts at B_1 and B_2 simultaneously. The maximum γ occurred at (-40, -120)mm, one of four right calibration locations which was nearest to B_1. Two impact areas are detected at the neighboring region of B_1 and B_2. The error of impact location detection is 40.5mm and 36.0mm for B_1 and B_2

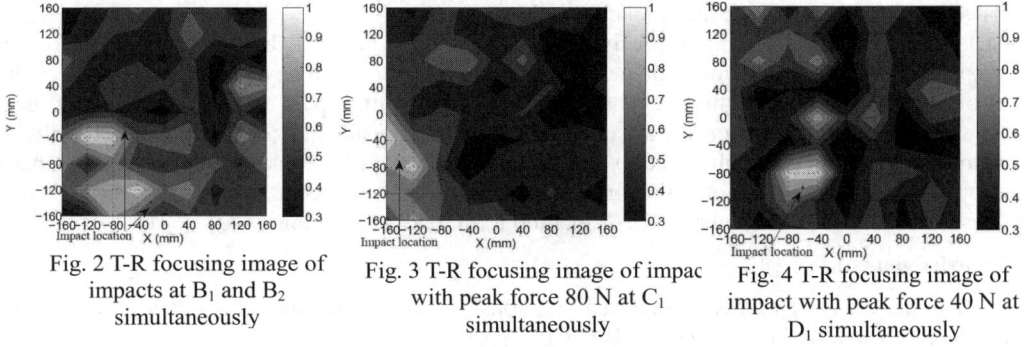

Fig. 2 T-R focusing image of impacts at B_1 and B_2 simultaneously

Fig. 3 T-R focusing image of impact with peak force 80 N at C_1 simultaneously

Fig. 4 T-R focusing image of impact with peak force 40 N at D_1 simultaneously

respectively. Fig. 3 shows the detection result of the impact with 80N peak force at C_1. The location detection error is 1.5mm, 49.2mm, and 79.7mm when the peak force of the impact is 80N, 120N, and 160N respectively, and the size of plastic deformation area caused by impact is roughly 4mm×4mm, 8mm×8mm, and 12mm×12mm respectively. The impact induced damages can be detected. Both the location detection error and the impact damage area increase with the increase of impact energy. Fig. 4 shows the impact detection for the impact at D_1. The impact detection error is 8.9mm. The effect of damage $d_1d_2d_3d_4$ on transfer functions is negligible for impact detection.

Impact monitoring on a composite plate. A composite plate is of dimension 260mm×260mm×0.6mm and the four sensors are mounted near the corner of the plate by 40mm from each side. The plate is simply supported on all edges. To simplify the simulation, only delamination damage is considered in impact. It is made by four ply with stacking sequence of

[45/-45/45/-45]. The parameters of the ply are $E_1 = 135$GPa, $E_2 = 8.8$GPa, $v_{12} = 0.33$, $G_{12} = 4.47$GPa, $G_{12} = G_{23} = 4$GPa, $\rho = 2000$Kg/m^3. To simulate the delamination damage under impacts, the cohesive elements for the adhesive layers between ply are an essential part of the model. The thickness of adhesive layer is 0.001mm. The parameters of adhesive layer are $E_h = 3$GPa, $v_h = 0.4$, $\rho = 1250$Kg/m^3. The quadratic nominal stress damage criterion was used for cohesive elements where the cohesive layers are defined in terms of traction-separation relation. The spacing between calibration location s is 40mm. Fig.5 shows the FEM model of composite plate under ball impact and the delamination damage (top right corner). The velocity of the ball is 2m/s. The diameter of the ball is 10mm. The element type of ball is 3D discrete rigid element. The impact location is (-20, 20)mm. Fig.6 shows the T-R focusing image of impact detection. The maximum γ occurred at (-40, 0)mm and the error in finer detection is 7.1mm. The impact location can be detected successfully though delamination damage occurred in impact area.

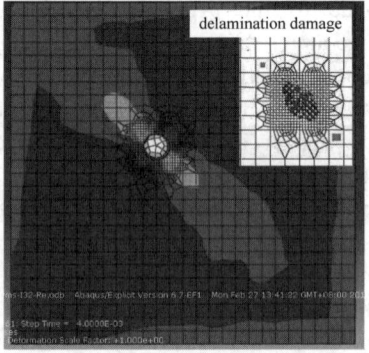

Fig. 5 Impact model of composite plate and delamination damage

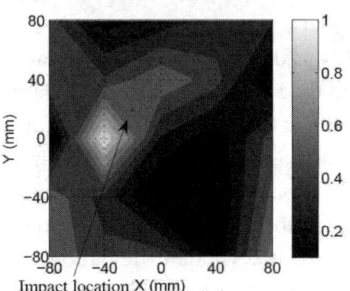

Fig. 6 Impact detection on composite plate with delamination damage

Conclusion

An enhanced T-R method has been developed to monitor the high energy impact on plate-like structures which causes the plastic deformation of a metal and delamination of a laminated composite. The results show that the stability of T-R method for impact detection can be enhanced by the time focusing parameter. Multiple site impacts can be localized simultaneously. The impacts have induced the plastic deformation on metal plate and the delamination damage on composite structures can be detected successfully. The error of impact location detection increases with the increase of impact energy. The effect of local damage on the transfer function is not significant, thus the transfer functions of the structures with no damage could be sequentially used to monitor impact events when unknown local damages have developed in structures. It is noted that only transfer functions are needed to be calibrated and a few sensors are requested in this enhanced T-R method for impact monitoring.

Acknowledgement

This paper is partially supported National Natural Science Foundation of China (No.10932008) and the 111 project (No.B07050)

Reference

[1] M. Meo, G. Zumpano, M. Piggott and G. Marengo: Compos. Struct. Vol. 71 (2005), p. 302-6
[2] T. Kundu, S. Das, S.A. Martin and K.V. Jata: Ultrasonics Vol.48 (2008), p. 193-201
[3] D. Guyomar, M. Lallart, L. petit and X. Wang: J. Sound Vib. Vol. 330 (2011), p. 3270-83
[4] J. Park, S. Hu and F.K. Chang: AIAA Journal Vol. 47 (2009), p. 2011-21
[5] J.R. LeClerc, K. Worden, W.J. Staszewski and et al.: J. Sound Vib. Vol. 299 (2007), p. 672-82
[6] C.H. Wang, J.T. Rose and F.K. Chang: Smart Mater. Struct. Vol. 13 (2004), p. 413-23
[7] L. Qiu, S.F. Yuan, X. Zhang and Y. Wang: Smart Mater. Struct. Vol. 20 (2011): 105014
[8] C. Chen and F.G. Yuan: Smart Mater. Struct. Vol. 19 (2010): 105028.
[9] C. Chen, Y. Li and F.G. Yuan, in: *Proceedings of the 8th International Workshop on Structural Health Monitoring*, edtied by F. K. Chang, DEStech Publications, Inc. (2011).
[10] C. Chen, Y. Li and F.G. Yuan: submitted to Shock and Vibration (2012).

Experimental study on residual strength of panels with multiple site damage

Yan Xiaozhong, Wang Shengnan, WangWen

School of Aeronautics, Northwestern Polytechnical University, Xi'an 710072, China

yxzyx-163@163.com, wangshna@nwpu.edu.cn, wendywang.cool@163.com

Keywords: Wing Skin; MSD; Residual Strength; Initiation Life

Abstract. The fatigue test of typical bolt-hole panels with five open holes in a line and the residual strength test of the panels with multiple site crack (MSD) were carried out. The crack initiation life of the test panels and the residual strength of the panels with different MSD crack sizes were obtained by the tests. The application of the net section yield criterion in the determination of residual strength of the panel containing multiple site damage was studied by comparing the calculated results with the test results. The conclusion is that the criterion meets the engineering accuracy requirement in the case of smaller MSD cracked area or smaller MSD crack sizes.

Introduction

Multiple site damage (MSD) is a source of widespread fatigue damage (WFD) that is characterized by the simultaneous presence of fatigue cracks in the same structural element[1]. The structure containing MSD no longer meets the required damage tolerance when the size or density of MSD cracks in the structure increases to a certain degree. One major concern is that MSD cracking could reduce the residual strength of the structure, and if the residual strength would fall below the design load, the consequences could be amazing. The reduction in structural residual strength due to MSD may be typically 20-40% and even up to 50%[2,3]. The study on the residual strength of structure with MSD is one of the focus in aircraft structural integrity[4,5,6].

An extensive experimental program was conducted by the Boeing Company under the funding of the Federal Aviation Administration(FAA), National Aeronautics and Space Administration (NASA), and the United States Air Force Research Laboratory (USAF/RL) to investigate the effects of multiple site damage (MSD) on the residual strength of typical fuselage splice joints. The experimental results were usually used to validate the analytical prediction using various methodologies, including the crack-tip-opening angle(CTOA), crack link-up, net section yield and T criteria[7], etc. The European project part of the 'Structural Maintenance of Ageing Aircraft' (SMAAC) carried out the experiments and calculations on the residual strength of the lap board containing multiple cracks[8]. It is available that fatigue test verifies that the net section yield criterion of multiple crack residual strength predict the residual strength of the actual aircraft structure[9].

The MSD initiation, MSD crack growth and MSD residual strength are now three main parts of study on MSD. One of the most important characteristics of MSD is that the three parts are close correlated. The reduction of the structural residual strength depends on some factors such as the geometry of the joint and the size, distribution and propagation type of the MSD cracks[10].

The work presented here has been performed as part of the MSD evaluation for large civil aircraft structure funded by china large civil aircraft project, where the MSD initiation of panels with one row of five open holes were tested under typical stress ratio and constant amplitude loading and the tensile residual strength tests were conducted for three given cracking distribution patterns. Based upon the test measurement, the residual strength criterion of MSD was then studied.

Test Specimens and Facility

Test Specimens. Two types of panel are tesed, total 30 specimens are grouped into TPYE A, shown in Fig. 1, of twelve specimens with countersink and there is no initial crack, TYPE B, shown in Fig. 2, of eighteen specimens without countersink and there are three given MSD cracking distribution patterns, 6 specimens for each pattern. The specimen is 184mm width × 8mm thickness and the specimen material is 2324-T39.

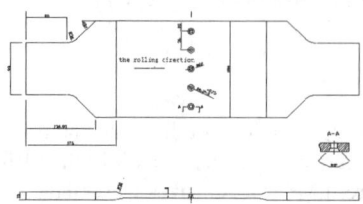
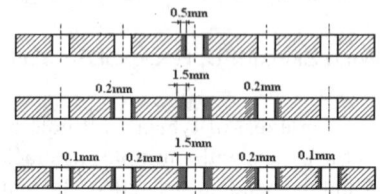

Fig.1 Specimen of TYPE A   Fig.2 Specimen of TYPE B

Test Facility. The fatigue test is conducted by MTS Teststar±600kN fatigue test machine as shown in Fig. 3.

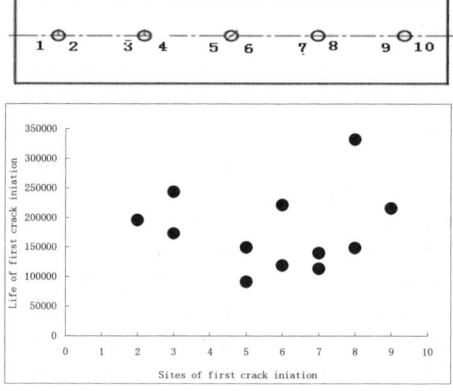

Fig.3 The MTS fatigue test machine   Fig.4 Crack initiation site and life of test TYPE A

Residual strength tests

Test results. The MSD crack initiation of TYPE A is tested under the fatigue loading of the peak stress S_{max}=120MPa and stress ratio R=0.06, followed are the residual strength test for TYPE A and TYPE B. The Test measurements of residual strength of MSD panel are listed in Table 1.

Table 1 Test measurements of residual strength of MSD panel

test specimen	Residual strength/kN	test specimen	Residual strength/kN	test specimen	Residual strength/kN	test specimen	Residual strength/kN
A-1	558	A-9	460	B-1-6	546	B-3-2	452
A-2	522	A-10	539	B-2-1	414	B-3-3	306
A-3	514	A-11	463	B-2-2	467	B-3-4	419
A-4	432	A-12	389	B-2-3	415	B-3-5	457
A-5	468	B-1-2	361	B-2-4	397	B-3-6	453
A-6	470	B-1-3	546	B-2-5	504		
A-7	480	B-1-4	542	B-2-6	397		
A-8	468	B-1-5	535	B-3-1	449		

Analysis of test results. Against high incidence of MSD in structure, what we are concerned is no longer that MSD has occurred, but when it happened. The crack initiation test has been shown that first crack may initiate at different site of the structure under fatigue loading, as shown in Fig. 4. We assume that MSD cracks initiate when cracks initiate at two or more sites in the same structure element. When MSD cracks develop to some point at which the residual strength of the structure does not meet the provisions in the literature[11], the WFD occurs. It can be seen from Fig. 4 that the scatter of crack initiation lives at different sites are relatively larger, so the determination of MSD initiation is important to the evaluation of WFD of structure.

Also, the test results of TYPE A in Tab. 1 show that, when the leading crack is the same, the smaller are the MSD cracks, the higher is the residual strength of structure containing MSD.

Comparison between test and predicted values of residual strength

Residual strength criterion of MSD structure. There are more studies about the residual strength criteria of multiple site damage. Compared with the structure with single crack, the residual strength of structures with multiple site damage is low significantly, and the critical sizes of cracks which result to catastrophic failure occurs are reduced greatly. As the net section yield criterion is easy to be used and suitable for engineering analysis, it is most likely to be the practical criterion for residual strength analysis of multiple site damage[9].

Net section yield criterion. The stress on the structure of the high toughness materials makes the whole net section yield before the fracture occurs, and leads to final structural failure. In engineering applications, when the net section stress on a cross-section of the structure is equal to or greater than the material yield stress, the structural failure occurs[12]. For the flat panel, the net section criterion can be expressed as

$$P_{net} = \left[W - \left(\sum D + \sum a_i\right)\right] t \sigma_{ys}$$

Where P_{net} is the predicted failure load, σ_{ys} is the yield strength of material, W is the width of the flat panel, D is the hole daimeter, a_i is the size of the crack.

Comparison between the test and the predicted values of residual strength. The values of residual strength predicted by the net section yield criterion are listed in Tab. 2. It can be seen from Tab. 2 that the predicted values are smaller than the test values. When the cracked cross-sectional area is small or the crack sizes are relatively small, the interaction between MSD cracks is small, the predicted values are then smaller than the test values, otherwise when cracked cross-sectional area is large, the values predicted by net section yiled criterion are largeer than the test values. This is because when the cracked cross-sectional area is larger or the crack sizes are larger and the interaction between multiple cracks is larger, so the critical damage size is greatly reduced. A wide range of MSD reducs greatly the residual strength of the structure. No matter the predicted values are higher or lower than the test values, the error between them is less than 8%.

Table 2 Comparison between the test and the predicted values of residual strength of MSD

test specimen	A_1 /mm2	A_{net} /mm2	Test value/kN	Predicted value/kN	test specimen	A_1 /mm2	A_{net} /mm2	Test value/kN	Predicted value/kN
4-1	56	1131	538	507	5-1-5	125	1107	535	496
4-2	44	1143	522	512	5-1-6	127	1105	536	495
4-3	52	1135	514	509	5-2-1	302	930	414	416
4-4	232	955	398	428	5-2-2	186	1046	467	468
4-5	155	1032	459	462	5-2-3	300	932	415	417
4-6	159	1028	463	461	5-2-4	242	990	398	444
4-7	132	1055	473	473	5-2-5	132	1100	504	493
4-8	163	1024	454	459	5-2-6	308	924	397	414

4-9	154	1033	460	463	5-3-1	217	1005	449	450
4-10	28	1159	539	519	5-3-2	226	1006	452	451
4-11	180	1007	458	451	5-3-3	425	807	346	362
4-12	405	882	378	395	5-3-4	290	942	419	422
5-1-2	124	1108	504	496	5-3-5	222	1010	457	453
5-1-3	116	1116	526	500	5-3-6	234	998	453	447
5-1-4	138	1094	522	490					

Where A_l is area of structural crack surface, A_{net} is area of the cross-section of the structure.

Conclusion

In structure containing MSD subjected to fatigue loading, the appearance of the first crack initiation is random, and the scatter of the initiation life is relatively larger. The determination of MSD initiation is important to the evaluation of WFD of structure. In case that the leading crack is the same, the smaller are the MSD cracks, the higher is the residual strength of structure containing MSD.

The net section yield criterion is a relatively simple fracture parameter and can be easily applied to the evaluation of residual strength of structure with multiple site damage. This evaluation is also simple and maybe save a lot of computing time and money.

References

[1] McGuire, J., and Foucault, J., "Recommendations for Regulator Action to Prevent Widespread Fatigue Damage in the Commercial Airplane Fleet," Final Rept. of the Airworthiness Assurance Working Group, 1999.

[2] De Wit R, Fields RJ, Low III SR, Harne DE, Foecke T. Fracture Testing of Large-Scale Thin Sheet Aluminium Alloy. NISTIR 5661. US Department of Commerce, 1995.987).

[3] Gruber ML, Wilkins KE, Worden RE. Investigations of fuselage structure subject to widespread fatigue damage. In: Bigelow Ceditor. Proceedings from the FAA-NASA Symposium on the Continued Airworthiness of Aircraft Structures. DOT/FAA/AR-97/2. GA, USA, 1996, pp. 439-60.

[4] R. Citarella. Non-linear MSD crack growth by DBEM for a riveted aeronautic reinforcement[J]. Advances in Engineering Software. 40 (2009) 253–259

[5] R. Citarella, G. Cricrì. A two-parameter model for crack growth simulation by combined FEM–DBEM approach[J]. Advances in Engineering Software. 40 (2009) 363–377

[6] Zhang Jianyu, Bao Rui, Zhang Xiang, Binjun FEI. A Probabilistic Estimation Method of Multiple Site Damage Occurrence for Aircraft Structures[J]. Procedia Engineering, 2 (2010) 1115–1124(in Chinese)

[7] Ching-long H sua, James Lo a, Jin Yu a, Xiao-gong Lee b, Paul Tan. Residual strength analysis using CTOA criteria for fuselage structures containing multiple site damage. Engineering Fracture Mechanics. 70 (2003) 525–545

[8] Roberto Galatolo, Karl-Fredrik Nilsson. An experimental and numerical analysis of residual strength of butt-joins panels with multiple site damage. Engineering Fracture Mechanics 68(2001) 1437-1467

[9] Wang Zhizhi, Chen Li, Nie Xuezhou. Application for Residual Strength Criterion of Multiple Cracks to XX Aircraft. 543-545(in Chinese)

[10] Schijve J.Multiple-site damage in aircraft fuselage structres. Fatigue Fract Engng Mater Struct 1995:329-44

[11] Civil aircraft structure durability and damage tolerance design manual[M]. 2002(in Chinese)

[12] Cherry M C , Mall S , Heinemann M B , et al . Residual strength of unstiffened aluminum panels with multiple site damage[J]. Engineering Fracture Mechanics, 1997, 57(6): 7012713.

Damage Analysis of 2D Woven Composite Laminates Containing an Open-Hole under Tensile Loadings

Xiaoqiong Zhang[a], Weiguo Guo[b], Deshuan Kong[c]

School of Aeronautics, Northwestern Polytechnical University, Xi'an 710072, China

[a]jone1108@live.cn, [b]weiguo@nwpu.edu.cn, [c]646612027@qq.com

Keywords: woven composite; open-hole tension; failure model; damage analysis

Abstract. In order to understand damage mechanism, the influences of lay-up construction of laminates and environgment on tension behavior of 2D woven composite laminates with an open-hole, which was manufactured by a new technology, uniaxial tension tests are performed in 3 different environments on 4 kinds of lay-up specimens, using a WE-50 electromechanical universal material testing machines. The fracture of specimens are analysed through micrographic observations. The result show that there is a large difference both in tensile strength and damage mechanism due to different kinds of lay-up specimens: 1) the tensile strength of specimens that only with ±45 degree laminated is much lower than other samples with different kinds of layup and its tensile stress-strain curves presents nonlinear; 2)The failure modes and damage mechanism determines the strength of specimens; 3)The change of environment had a certain effect on the mechanical behaviors of materials, in this paper, it will cause the tensile strength of speicmens decreasing.

Introduction

Owing to their excellent behaviour, fiber-reinforced woven composite materials have been commonly used for high-performance structures and structural components in automotive, marine aerospace and energy industries. Due to composites often need to be drill a hole in use process, it will cause a large stress concentration and induces a new free edge on the hole. Hence the mechanical property of composite with an open-hole has strongly attracted researcher interest.

Early work in the field, as summarized by Chang, etc. [1-4], the tensile strength of braided composites with a circular hole were studied and failure phenomena were explained, the numerical model was also established. In recent years, more emphases mainly focus on the numerical investigation of composite with an open-hole. Flatscher, Schuecker and Pettermann [5-8] did the study of reinforced laminates' damage model and constitutive equation by finite element analysis. Lapczyk [9] established a damage model for predicting failure in fiber-reinforced materials. Now the mechanical properties of composite under different environment have strongly attracted concerns of researches.

In this study, tensile testing on 4 different kinds of specimens with an open-hole in different environments is carried out. Experimental results are analysed and discussed, several valuable conclusions will be summarized.

Materials and experimental procedures

Materials. In this paper, all specimens are made of carbon fiber by twill-weave method, the surface of specimen and its main size are represented in Fig.1. There are 4 different kinds of lay-up specimens, specific lay-up parameters of each kind of specimen were showed in table 1.

Experimental procedure. The test device as shown in Fig. 2(a). Three strain gauges numbers as 1, 2, 3 were attached on the front and back surface of specimen. On the back surface the strain gauge numbered as 2 was attached corresponding to strain gauge 1. The distance between strain gauge and the centre of specimen is 40mm as shown Fig. 2(b).

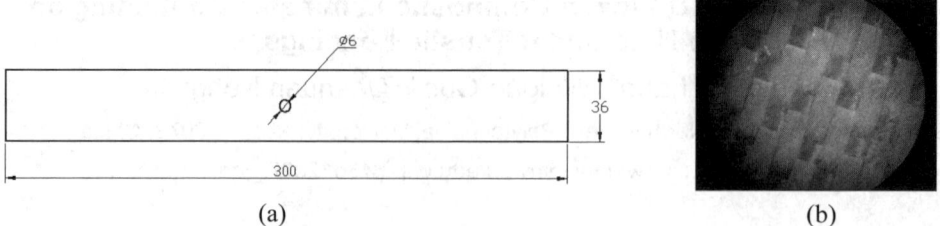

(a)   (b)

Fig.1(a) specimen size; (b) the surface of specimen

Table 1 Lay-up parameters of each kind of specimen

Serial number	Layup parameters	Number of layers	Thickness/mm
A1	[(±45)/(0/90)(±45)$_2$/(0/90)/(±45)]s	12	2.76
B1	[(±45)/(0/90)(±45)/(0/90)/(±45)]s	10	2.3
C1	[(±45)/(0/90)$_2$(±45)]s	8	1.84
D1	[(±45)]$_8$	8	1.84

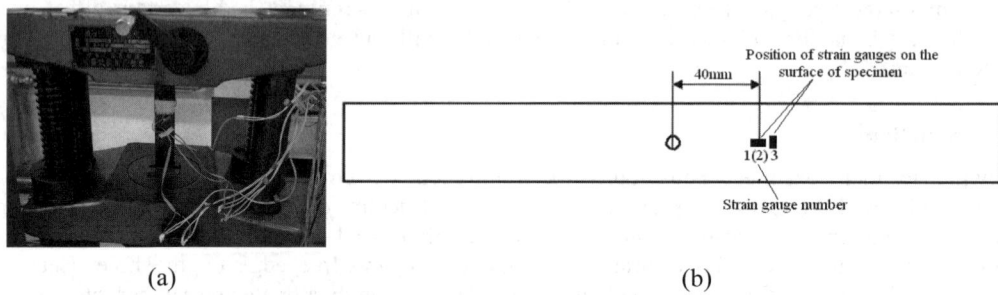

(a)   (b)

Fig. 2(a) Testing device; (b) Position of strain gauge

Testing at room temperature. At room temperature, stress and strain of A1-C1 these 3 kinds of specimens are linear correlation during the tensile testing, see Fig. 3(a). The average failure strength of A1-D1 specimens are 336.82MPa, 318.64MPa, 326.84MPa and 197.39MPa, respectively. The tensile strength of D1 specimens that only with ±45 degree laminated is much lower than other specimens with different kinds of lay-up and its tensile stress-strain curves presents nonlinear (Fig. 3b). For high tensile strength of fiber and low interfacial strength between matrix and interfacial, shear fracture will be formed in specimens. It will lead to the matrix and interface easily to be separated and lower the tensile strength of D1 specimens.

Testing in dry and cold environment. Before tensile testing, specimens need to be placed in drying cabinet more than 24h. The temperature is -55℃±3℃. The stress-strain curves of specimens in dry and cold environment are very similar to stress-strain curves at room temperature, but decreased the tensile strength, see from table 2. Although many researches and experiments show that modulus and specific strength of carbon fiber changes imperceptibly whatever at room temperature or cold environment, is an ideal materials for low temperature. However, traditional resin matrix may become brittle at low temperatures and will easily fracture, especially at ultra-low temperature environment. Tensile strength in low temperature environment is subjected to the combine influences of fiber and matrix properties, so it caused the tensile strength decreasing.

Testing in wet-thermal environment. Tensile testing of wet-thermal environment, temperature and humidity are 70°C, 85%RH, respectively. Before testing, specimens need to be placed in constant temperature humidity chamber about 3 months. Table 3 is the comparison between tensile strength in room temperature and wet-thermal. As shown in table 3,after specimens mositure absorption saturated, its tensile strength decreased. This is because the properties of matrix in wet-thermal environment will deteriorate on contact with humidity and high temperature, but fiber performance degradation is small. The loading direction of tensile testing decide that the tensile strength is affected by fiber performance, hence, after specimens moisture their tensile strength will decrease. As a general rule, tensile strength of composite is mainly depend on tensile properties of fiber, while carbon fiber is affected by wet-thermal aging slightly, so the tensile strength decline is not obvious.

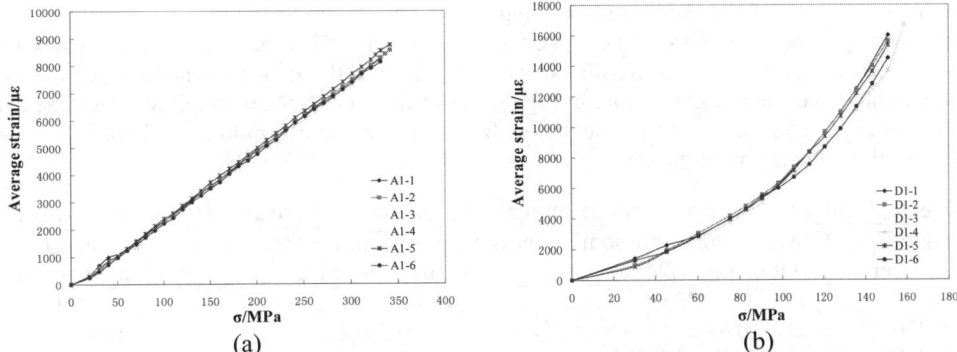

Fig. 3 (a) Stress-stain curves of A1 specimens at room temperature
(b) Stress-stain curves of D1 specimens at room temperature

Table 2 Comparsion between tensile strength in room temperature and dry and cold environment

Serial number	Tensile strength in room temperature/MPa	Tensile strength in wet-thermal environment/MPa	Dry and cold environment factor
A1	336.8223	309.1452	1.0895
B1	318.6393	314.9155	1.0118
C1	326.8418	313.3806	1.0430
D1	197.3883	219.2784	0.9002

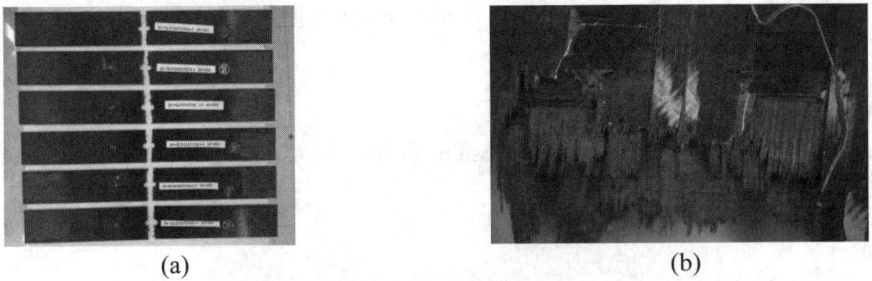

Fig. 4 (a) Failed specimens; (b) Micrograph of a section of the fracture

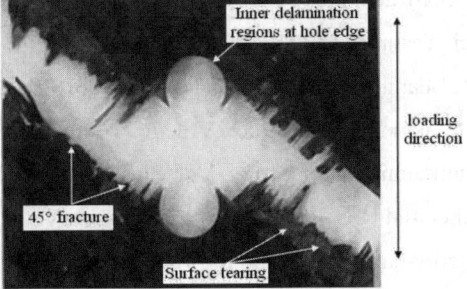

Fig. 5 Micrograph of a section of the fracture in D1 specimen

Damage or failure analyses. After tensile testing in room temperature, the damage region and fracture section of failed specimens are examined with the aid of an optical microscopy. The damage characteristics of A1~C1 specimens are similar with the fractures throughout the specimens. Fractured radial fibers are basically in the same plane, the location of fiber breakage is mainly in the fiber bundle intersection for the stress concentration. It can be seen from Fig. 4 that only a few radial fiber bundles were pulled out and the length of these fiber pull-out were short. While radial fiber pulled out, the weft fibers were carried away, and this leads to the damage of woven structure. However, around the fiber bundles there was no obvious interface failure, this indicated that the combination between fiber and matrix is strong.

The damage characteristics of D1 specimens mainly are fiber breakage, matrix cracking and interface failure. Under uniaxial tensile, specimens are not only axial elongation along the loading direction, but have shear deformation. In Fig. 5, due to the tensile-shear coupling effect, the damage of D1 specimens is started from the hole-edge of stress concentration. And then the damage extended along the layers direction.

Table 3 Comparsion between tensile strength in room temperature and wet-thermal environment

Serial number	Tensile strength in room temperature/MPa	Tensile strength in wet-thermal environment/MPa	Wet-thermal environment factor
A1	336.8223	305.9581	1.1009
B1	318.6393	310.5072	1.0262
C1	326.8418	339.7997	0.9619
D1	197.3883	150.5007	1.3115

Summary

In this paper, tensile results of 4 kinds of specimens at room temperature, dry and cold, and wet-thermal environment are analyzed. The following conclusions were obtained: (1) Although A1, B1 and C1 3 different kinds of specimens their layer ways and layer number are different, the tensile strength of these 3 kinds of specimens are basically similar; The tensile strength of D1 specimens that only with ±45 degree laminated is much lower than other specimens; (2) Tensile strength of these 4 kinds of specimens is affected by temperature and humidity seriously, materials properties will be weakened; (3) The failure modes and damage mechanism determines the strength of samples.

Acknowledgments

This work is supported by First Aircraft Design Institute of China Aviation Industry, authors also thank the help from Prof. Weiguo Guo.

References

[1] L.W. Chang, S.S. Yau and T.W. Chou: Comp. Vol. 18 (1987), p. 233.

[2] H.J. Lin, Y.J. Lee: Comp. Struct. Vol. 21 (1992), p. 155.

[3] P.A. Lagace: Comp. Sci. Technol. Vol. 26 (1986), p. 95.

[4] M.T. Kortschot, P.W.R. Beaumont: Comp. Sci. Technol. Vol. 39 (1990), p. 303

[5] C. Schuecker, H.E. Pettermann: Comp. Struct. Vol. 76 (2006), p. 162

[6] C. Schuecker, H.E. Pettermann: Comp. Struct. Vol. 86 (2008), p. 908

[7] T. Flatscher, C. Schuecker and H.E. Pettermann: Int. J. Fract. Vol. 158 (2009), p. 145

[8] C. Schuecker, H.E. Pettermann: Arch. Comput. Methods. Eng. Vol. 15 (2008), p. 163

[9] I. Lapczyk, J.A. Hurtado: Comp. Part A. Vol. 38 (2007), p. 2333

A New Approach to Determine Dynamic Strength Model Parameters under Taylor Impact Test

F. Xu[1,a], W.G. Guo[1,b,*], Q.J. Wang[1,c] Z.Y. Zeng[2,d]

[1]School of Aeronautics, Northwestern Polytechnical University, 710072 Xi'an, China

[2]No.202 Research Institute of China Ordnance Industry, Xianyang 710099, China

[a]ertertxufeng@126.com, [b]weiguo@nwpu.edu.cn, [c]wqjnwpu@sina.com, [d]zzy202@126.com

Keywords: Taylor impact test, Hopkinson bar test, Johnson-Cook model, numerical optimization.

Abstract: In this paper, to determine the dynamic strength model for steels, a new approach which does not rely on the Hopkinson bar test has been proposed. As the DH36 steel for example, using the results of Taylor impact test and the quasi-static compression test, the initial parameters of Johnson-Cook plastic strength model have been fitted out, then the initial strength parameters have been optimized using the optimization techniques of the sparse Taylor impact cylinder. It has been shown that the optimized results in numerical simulation are consistent with results of Taylor impact test, and the optimized Johnson-Cook model can also well describe flow stress curve fitted from the Hopkinson bar test.

Introduction

Accurate strength parameters for materials are the basis of the research in the numerical simulation study. At present, the dynamic strength parameters are mostly obtained by results of Hopkinson bar equipment [1]. However, Hopkinson bar equipment is expensive and it often has the problems of non-uniform stress and geometric dispersion, which make the determination of strength parameters inaccurate.

Taylor impact test [2] is simpler on operation than the Hopkinson bar test, with the development of theoretical model [3, 4] and experimental techniques [5], Taylor impact test has become an important means of testing and obtaining the dynamic mechanical properties for materials. In recent years, in order to obtain the dynamic strength model parameters efficiently and accurately, Rohr [6], Nussbaum [7] and Martin [8] optimized parameters of dynamic strength model for different steels under Taylor impact test, but all these work need a group of initial strength parameters, which make the direct determination of strength parameters difficult under Taylor impact test.

In this paper, DH36 steel with high strain rate sensitivity and temperature sensitivity [9] is selected, a new approach for determining Johnson-Cook strength parameters, which do not rely on Hopkinson bar equipment, has been presented under Taylor impact test and its numerical simulation.

The determination for initial parameters of Johnson-Cook strength model

The initial fitting of Johnson-Cook model parameters. Johnson-Cook model [10] (J-C model) represents a kind of relationship with the product description of strain, strain rate and temperature, the specific form of function is:

$$\sigma = (A + B\varepsilon^n)(1 + C\ln(\frac{\varepsilon}{\varepsilon_0}))(1 - (\frac{T-T_r}{T_{melt}-T_r})^m) \qquad (1)$$

On the calculation of parameters in equation (1), we set the reference strain rate $\varepsilon_0 = 0.001$ and the reference temperature $T_r = 296K$ in this paper, so the yield strength parameter A, the strain hardening parameters B and n can be fitted by the strain-stress data from quasi-static test, and the temperature softening coefficient m can be calculated by quasi-static test under different temperatures.

In order to get strain rate sensitivity coefficient C, different yield stress at different strain rate should be obtained. In this paper, we are inclined to obtain yield stress at high strain rate under Taylor impact test, so we launched three DH36 Taylor cylinders ($\phi 10 \cdot 60mm$) to the Gr15 target with high hardness using the air gun equipment, then we calculated the yield stress using Taylor theoretical model [2] through the remaining length, undeformed length and impact velocity of three cylinders after test (as the Table 1 shown).

Table 1. Experimental results after the Taylor impact test

No	Velocity (m/s)	Remaining Length(mm)	Undeformed length(mm)	Strain rate (m/s)	Yield stress (MPa)
1	130	56.05	25.72	1896	596
2	141	55.19	23.66	1940	612
3	167	53.15	21.55	2171	635

Using the static yield stress at the strain rate 0.001/s and 0.1/s and dynamic yield stress at the strain rate 2000/s, coefficient C in the equation (1) is fitted out. The initial J-C model parameters are shown in Table 2.

Table 2. Initial J-C model parameters for DH36 steel

A(MPa)	B(MPa)	n	C	m
350	879	0.58	0.01	0.56

The validation of initial parameters. Values of parameters A, B, n and m are fitted in the quasi-static conditions, whether these parameters are the proper coefficients of dynamic strength equation is uncertain. So it is necessary to validate them though Taylor impact simulation. Fig.1 shows plastic deformation comparison between Taylor impact test and its numerical simulation under velocities of 141m/s and 167m/s.

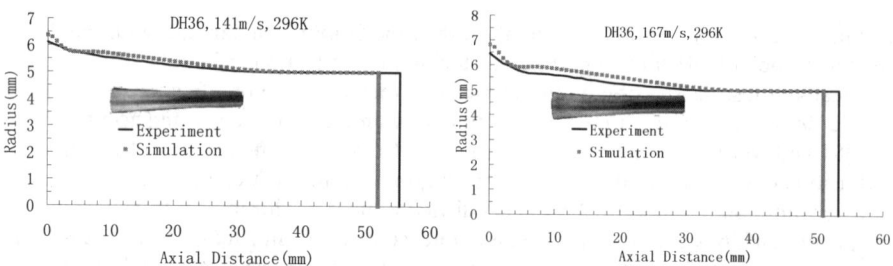

Fig 1. Comparison between Taylor impact test and its numerical simulation

Furthermore, the equation (1), which choose the Table 2 data as its parameters, are compared with the flow stress [9] from Hopkinson bar at the strain rate 2000/s and 7000/s.

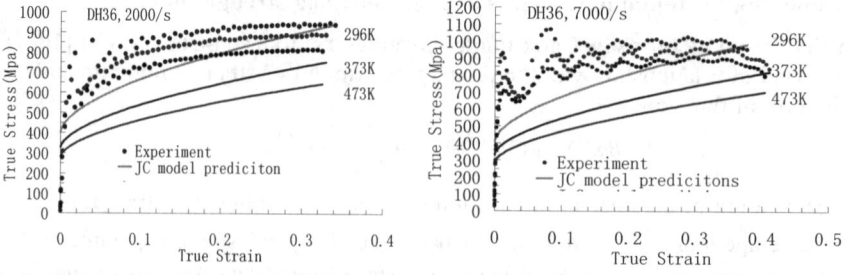

Fig 2. Comparison between initial J-C strength model and Hopkinson compress data

From Fig.1 and Fig.2, We find that the parameters shown in table 2 do not accurately reflect dynamic mechanical behavior during Taylor impact test and Hopkinson compression bar test.

The optimization for initial parameters of Johnson-Cook strength model

The initial parameters optimization. To find the accurate strength parameters for DH36, these parameters will be optimized using the LS-OPT processor in the LS-DYNA. The optimization process is iteratively executed by the comparison between Taylor impact test and its numerical simulation. we choose the sparse Taylor impact as the optimization mode, so the corresponding objective function is as follows:

$$T = \frac{1}{N}\sum_{j=1}^{N}\sqrt{(\frac{L_f^c - L_f^m}{L_f^c})^2 + \sum_{i=1}^{M}(\frac{D_i^c - D_i^m}{D_i^c})^2} \qquad (2)$$

Where L_f and D_i represent the remaining length and radius at different axial distance, c and m represent the actual measured value and corresponding numerical simulation value. M denotes the numbers of measured radius of deformed cylinders, and N denotes numbers of cylinders.

In the LS-OPT processor, five parameters of J-C strength model are treated as design variables, we select the gradient optimization method which could reduce the value of objection function T to realize the optimization. The final optimized results are shown in Table 3

Table 3. the optimized J-C model parameter values for the DH36 steel

Parameter State	A(MPa)	B(MPa)	n	C	m	T
Initial	350	879	0.58	0.01	0.56	0.37
Optimization	578↑	915↑	0.56↓	0.01	0.67↑	0.03↓

From Table 3, the rise of A shows that yield strength is increased at high strain rate, the rise of B and and the descend of n indicate that the original parameters underestimate the strain hardening effects, C values which remain the same indicate that the initial estimate of strain rate effect is accurate, and the increase of m indicate that adiabatic heating in the Taylor impact test strengthen the temperature softening effect.

The evaluation for optimized results. To evaluate optimized J-C model for DH36 steel, these parameters of strength model are used for predict the dynamic mechanical behavior for DH36 steel during Taylor impact test and Hopkinson bar test. The comparative results are as follows:

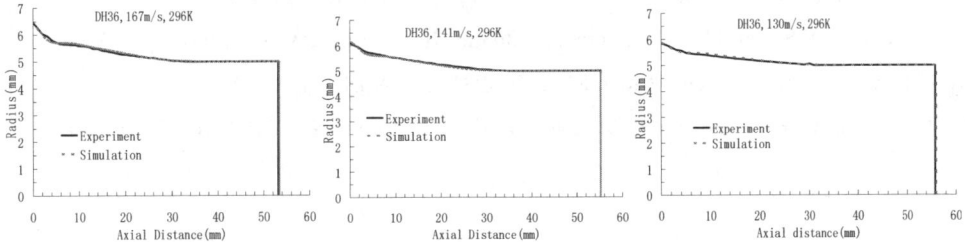

Fig 3. Comparison between Taylor impact test and its numerical simulation after optimization

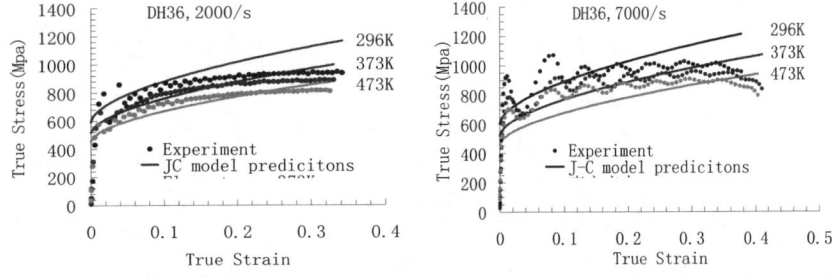

Fig 4. Comparison between the optimized J-C model and the Hopkinson bar results

As the Fig.3 and Fig. 4 shown, the optimized results in numerical simulation are consistent with real experimental results, the adjusttive strength model can also well descript flow stress data[9] obtained from the Hopkinson bar test.

Summary

This paper gives a new method which does not rely on the Hopkinson compression bar to determine the Johnson-Cook strength model parameters for DH36 steel. Although it is shown that obtained Johnson-Cook strength model for DH36 steel can be well evaluated by results of Taylor impact test and Hopkinson bar test, the validation and reliability of this method should be evaluated by other more materials in the future.

Acknowledgement

This work was supported by a grant from the Major State Basic Research Development of China Program (973 program), (No. 613116). Authors also thank the support from the Research Fund of State Key Laboratory of Explosion Science and Technology, Beijing institute of Technology (KFJJ11-11Y).

References:

[1] S.Nemat-Nasser,W.G. Guo:Mech Mater.Vol 37(2005),p.379.

[2] G.I.Taylor: Proc. Roy. Soc. London A.Vol 194(1948),p.289.

[3] J.B. Hawyard, D.Easton and W. Johnson: Int.J.Mech.Sci.Vol 10(1968),p.929.

[4] S.E. Jones, P.P. Gillis and J.C. Foster:J.Appl.Phys.Vol 61(1987),p.499.

[5] D. Eakins, N.N. Thadhani:Int.J.impact.Engng.Vol 34(2007),p.1821.

[6] I.Rohr, H.Nahme and K.Thoma:J.Phys.IV France.Vol 110(2003),p.513.

[7] J.Nussbaum, N.Faerl:Procedia.Engng. Vol 10(2011),p.3453.

[8] M.Martin, T.Shen, N N.Thadhani:Mater.Sci.Engng.Vol 494(2008),p,416.

[9] W.G. Guo, F.F.Shi and F.L.Liu:Acta. J.Ordnance II in Chinese.Vol 30(2009),p.203.

[10] S.Nemat-Nasser,W.G. Guo:Mech Mater.Vol 35(2003),p.1024.

[11] P.J. Maudlin, G.T. Gray and C.M. Cady:Math.Phy&Engng Sci.Vol 357(2011),p.1707.

A Novel Testing Method for Measuring Through-Thickness Properties of Thick Composite Laminates

Li Yutu[a], Zheng Xitao[b], Luo Gui[c]

School of Aeronautics, Northwestern Polytechnical University, Xi'an 710072, China
afamilyviplee@mail.nwpu.edu.cn, bzhengxt@nwpu.edu.cn, clg0225@mail.nwpu.edu.cn

Keywords: through-thickness, thick composites, test methods, mechanical properties

Abstract. The investigation was concentrated on the through-thickness properties of thick-composite laminates, and the development of a novel test method and several specimens with two different shapes and three to four different sizes. The experiments were performed on thick composite specimens; and all through-thickness mechanical properties under compressive and tensile loading with their corresponding failure mechanisms were determined. Overall, the configuration of specimens and test method in this paper are efficient in obtaining the through-thickness properties of thick-composite laminates.

Introduction

Application of composites in aeronautics structures has been developed into primary structure from sub-structure. Meanwhile the thickness of Composite Laminate has been increased from thin-thickness. As a consequence, the requirement of the data through-thickness mechanical properties will be paramount, particularly where the through-thickness strength has a controlling influence on the overall structural strength of the component. Whereas, the majority of composite material characterization to data has concentrated on coupon test specimens between 2mm and 2.5mm thick, which were designed to produce in-plane materials property data[1]. In addition, the effect of three-dimensions brought by the augment of thickness results in the inapplicability of former test method and design. At present, there is another way of obtaining the through-thickness elastic properties which is by ultrasonic methods, either by immersion or by contact. These measurements can be taken for small specimens but local effects can substantially affect the measurements and ultimate properties cannot be achieved by these procedures. Therefore the present requirement has exposed a shortfall in materials characterization skills, and there are no agreed methods for measuring the through-thickness properties.

In response to this shortage, this research will be concentrated on giving the tensile and compressive through-thickness materials properties information required for effective FEA modeling (i.e. $E_{33}, \sigma_{33}, \nu_{31}, \nu_{32}$), and developing a novel test method and several specimens.

Specimen design

To be considered satisfactory for the determination of tensile and compressive material properties, a test coupon should meet four requirements that they are achievement of an acceptably unidirectional and even stress and strain state, both across and along the gauge length, irrespective of the degree of material anisotropy; avoidance of significant stress concentrations adjacent to the gauge length; prevention of bending and Euler buckling in the gauge length when loaded in compression; and being capable of giving consistent results, independent of the operator or laboratory carrying out the test[1].The test coupon for through-thickness testing is constrained by material availability which is limited in thickness. Empirically, it is rare to utilize materials in thickness greater than 40mm for measuring the through-thickness properties. Thereby a figure, 40 mm, was assumed to be the maximum practical height of any candidate through-thickness coupon.

There are some specimens geometries developed by several organizations who have attempted to devise through-thickness testing techniques. Excluding those height greater than 40mm, there are only four tensile and three compressive test coupons remained. These are summarized in Table 1.

Table 1 Through-thickness test methods for thick composite laminates

Test Coupon	Young's modulus	Poisson's ratios	Strength	Loading regime
Short block	Yes	Yes	No	compressive
"C" waist block	No	No	Yes	Tensile or compressive
Parallel waist block	Yes	Yes	Yes	Tensile or compressive
"I" waist block	Yes	Yes	Yes	Tensile or compressive

The geometry, short block, is suitable for through-thickness Young's modulus and Poisson's ratios measurements, but it is unsuitable for strength measurements. In addition, the main advantage of the short block over other configurations is the cost of machining the specimens. "C" waist block specimen cannot provide through-thickness Poisson's ratio information such as v_{32}.

Parallel waist block consists of a relatively short foursquare specimen, with large load surfaces, which are in turn necked down to a short, but parallel, gauge length that its cross-section is foursquare. The advantage is to transfer applied loading into the gauge length with limited stress concentrations. The large load surfaces provide sufficient area to premature adhesive failure when carrying out tensile testing using bonded steel fixtures. The parallel gauge length with square cross-section allows foil strain gauges to be positioned to measure both direct and transverse surface strains. Overall four different gauge length short block test coupons (Figure. 1), because of its affordability, are applied to determining compression through-thickness properties, while three parallel waist block test coupons (Figure. 2) are to generate the tensile testing data.

All tensile testing specimens were bonded to special steel fixtures with an epoxy-based adhesive (TS811) which cured at 80°C for four hours then totally solidified under room temperature for 48 hours. The material considered for specimens was a carbon–epoxy composite. The specimens for all the tests were prepared from unidirectional laminate, with a laminate thickness of 0.125mm.

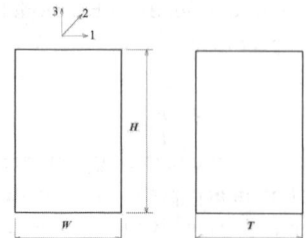

Fig. 1 Geometry of the Short block

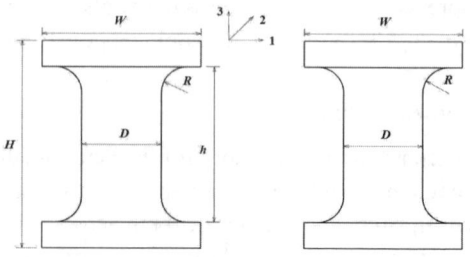

Fig. 2 Geometry of the Parallel waist block

Testing

All tensile testing coupons are parallel waist block while short block is applied to through-thickness compression testing. Each group consisting of five specimens from each of the same thickness materials were prepared for loading in through-thickness tension. The tensile test program was carried out in three stages while the compression testing program was four stages.

Strain gauges were placed on the four sides to generate the data of properties and monitor misalignment. Gauges with an active length of 1mm and 2.8mm were used for specimen with a thickness of 10mm and others respectively. Compression testing and tensile testing were carried out at a loading rate of 0.5mm/min and 0.1mm/min respectively. Figure 3 shows diagrams of the typical testing method.

Result and discussion

Tensile testing. The through-thickness properties of thick composite laminates from the tensile testing are provided in Table 2. As can be seen from testing result, there are three typical failure patterns which are shown in Figure. 4. Majority of specimens from completed tensile tests fractured within or at one end of the gauge length, on a single plane perpendicular to the loading direction. Whilst there were few specimens fracturing at the load surfaces which was bonded to the steel fixture. It was likely to be effected by the misalignment.

Table 2 Result of tensile testing

Specimen and size	Z_t [MPa]	CoV%	E_{33} [GPa]	CoV%	ν_{31}	CoV%	ν_{32}	CoV%
Set A/ 20mm	43.63	25.91	11.18	24.89	0.00634	93.69	0.39	23.47
Set B/ 30mm	45.47	36.04	10.08	10.30	0.00834	72.14	0.44	8.45
Set C/ 40mm	44.15	16.81	10.41	23.23	0.01618	21.69	0.51	5.91

(a) (b) (b) (d)

Fig. 3 (a) Typical tensile testing method
(b) Typical compressive testing method

Fig. 4 Typical failure patterns of tensile specimen

Compressive testing. The through-thickness properties of thick composite laminates from the tensile testing are provided in Table 3. According to the testing result, there are two typical failure patterns which are shown in Figure. 5. All of specimens from completed tensile tests fractured at one end of the specimen, on a single plane perpendicular to the loading direction. The fracture of specimens with thickness of 20mm or 10mm was across the whole specimen while the fracture of the specimens with thickness of 30mm or 40mm just appeared on the loading half part of the specimen. The data of less thick specimen show a great dispersibility.

Table 3 Result of compressive testing

Specimen and size	Zc [MPa]	CoV%	E33 [GPa]	CoV%	v31	CoV%	v32	CoV%
Set A/ 10mm	230.83	4.22	10.82	9.76	0.033	20.40	0.61	33.67
Set B/ 20mm	246.11	4.89	9.47	8.23	0.02	12.40	0.58	7.58
Set C/ 30mm	229.34	8.67	10.11	6.53	0.02	12.57	0.52	1.93
Set D/ 40mm	230.27	7.45	9.68	4.00	0.02	5.87	0.52	3.68

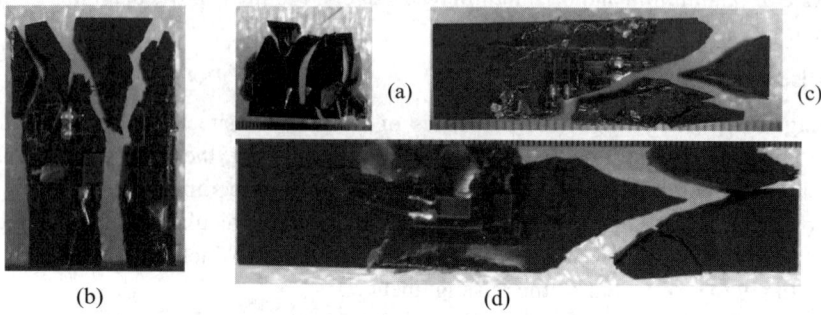

Fig. 5 Typical failure patterns of compression specimen

Conclusion

Specimens and testing methods were developed to measure through-thickness properties of thick composites successfully. The method is applicable to laminates from 10mm thickness upwards

With the thickness increasing, short block specimens' gauge length show an upward trend. Specimens with a thickness of 10mm or less are not suitable to measuring the elastic properties. Local effect show influence on short block apparently. Parallel waist block specimens are sensitive to misalignment and initial damage from machining. It is supposed to prevent premature adhesive failure during tensile testing.

Acknowledge

This investigation was sponsored by key supporting project of graduation design in the year of 2011 at Northwestern Polytechnical University

Reference

[1] R.F. Ferguson, M.J. Hinton & M. J. Hiley, *Determine The Through-thickness of FRP Materials*, Composites Science and Technology 58 (1998) 1411-1420
[2] W R Broughton and G D Sims, *An Overview of Through-thickness Test Methods for Polymer Matrix Composites*, NPL Report DMM(A)148, October 1994
[3] J. L. Abot and I. M. Daniel *Through-Thickness Mechanical Characterization of Woven Fabric Composites* Journal of Composite Materials 2004 38: 543
[4] J. Lee, C. Soutis. *A Study on the Compressive Strength of Thick Carbon Fibre–Epoxy Laminates*. Composites Science and Technology 67 (2007) 2015–2026
[5] Lodeiro, M.J.,Broughton,.*Understanding the Limitations of Through-Thickness Test Methods*, 4th European Conference on Composites: Testing and Standardisation, (1998)pp. 80–90

Numerical Simulation of Compressive Residual Strength for Damaged Composite Laminates

Qu Tianjiao[a], Zheng Xitao[b], Zhang Di[c]

School of Aeronautics, Northwestern Polytechnical University, Xi'an 710072, China

[a]jiao1231561@126.com, [b]zhengxt@nwpu.edu.cn, [c]beyond1907@163.com

Keywords: impact, compressive test, finite element analysis, ultimate compressive load

Abstract. After the low-velocity impact test of composite laminates of T800/BA9916, CAI test and compression test of laminates with a hole have been carried out. Two types of models were set up by the finite element software ABAQUS respectively. The FEA results were good agreement with the testing results. The investigation of models with a hole indicates that the appearance time of ultimate compressive load is earlier than that of fiber breakage expanding to boundary. Moreover, the diameter and the depth of blind hole significantly influence the ultimate compressive load.

Introduction

In aeronautic and aerospace structural components, advanced composites are widely used mainly because of their outstanding qualities, including high strength to weight ratio, low thermal expansion, excellent qualities, outstanding fatigue behavior and the wonderful possibility of tailored design[1]. They may suffer different types of impact damage during their manufacture, assembly maintenance or service life, of which low-energy impact is considered the most dangerous. The damage caused by low-energy impact may escape detection in a routine visual inspection of the impacted surface [2]and this damage has a significant effect on the mechanical properties, of which the most serious invisible damage is delamination which may cause a reduction of the compression strength up to 40-60%. So damage tolerance is an important factor in the design of aeronautic and aerospace components and it is usually studied by determining the compression residual strength after impact (CAI).

In this paper, the low-velocity impact test and compression test after impact of carbon fiber/epoxy composite laminates of T800/BA9916 have been carried out, and the relationships between impact energy and three different parameters (damage area, damage width and dent depth respectively) were obtained. In addition, compression test of laminates with a hole have been carried out similarity. Two models of damaged laminates, one with a hole and the other with a blind hole, were set up by the finite element software ABAQUS respectively. The finite element numerical results were good agreement with the testing results.

Low-velocity impact test

The low-velocity impact tests have been carried out based on the standard ASTM D 7136(Standard test method for measuring the damage resistance of a fiber-reinforced polymer matrix composite to a drop-weight impact event). To investigate the impact damage of 11 pieces of T800/BA9916 composite laminates, there are two types of lay-up sequence as $[45/-45/90/45/0/-45/0/45/-45/0]_{2s}$ regard as A and $[45/-45/90/45/0/45/0/0/-45/0]_{2s}$ regard as B, the geometrical size of specimens are 150×100×5mm. Impact tests are performed using INSTRON DYNATUP impact test machine, which can regulate drop height automatically, avoid second drop, and calculate the impact velocity and energy. The impactor with a hemispherical nose of 12.7mm in diameter is used. In addition, the boundary condition is 125×75mm rectangular simple support. The total damage areas and damage widths have been obtained by ultrasonic C-scan technique, the dent depths have been measured by the depth gauge. The results show that damage area, damage width and depth of dent increase along with the increase of impact energy, but only impact energy- depth of dent shows well orderliness.

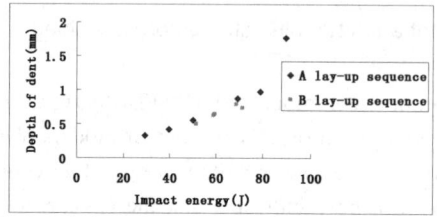

Fig.1 Relationship between Impact energy-depth of dent   Fig.2 The fixture of CAI test

CAI test and compression test of laminates with a hole

Compression tests of impact damaged laminates and laminates with holes[3] have been carried out based on the standard D 7137/D7137M-05(Standard test method for compressive residual strength properties of damaged polymer matrix composite plates). Compression test are performed using electrical almighty machine controlled by computer, and the displacement load velocity is 1.25mm/min in the process. The support fixture is as Fig.2.

CAI test. The test result indicates that during the whole compression test[4], specimens suffer the matrix crack and delamination at first, then the break of some fibers, and the breakage extend to the whole laminates abruptly at last. Fig.3 shows the fracture of a specimen in directions of both width and thickness. Fig.4 shows that the CAI strength decreases along with the increase of dent depths. There is a obvious decrease at first and the CAI strength remains stable till the end.

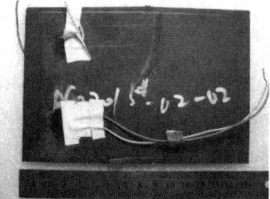
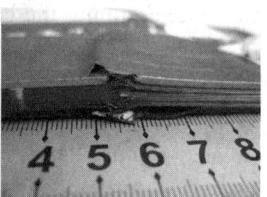
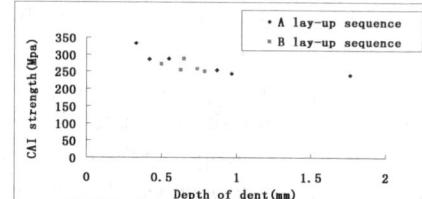

Fig.3 The fracture in width and thickness directions   Fig.4 The relationship between dent depth and CAI

Table 1 Test results of ultimate load and compressive strength

Results Holes' diameter[mm]	Ultimate load[kN]	Compressive strength[Mpa]
50	70.10	140.2
40	76.26	155.0

Compressive test of laminates with a hole. There are two pieces of T800/BA9916 composite laminates with lay-up A in this test. They have the same geometrical size with those impact specimens mentioned above, except for a hole of 40mm or 50mm in diameter in the middle respectively. The experimental results are in Table 1.

FEA models of laminates with a hole and a blind hole

Models with a hole of different aperture and models with a blind hole of different aperture were set up by using the FEA software ABAQUS[5]. As one of the most advanced nonlinear FEA software in the world, ABAQUS can analysis many kinds of problem from simple linear question to challenging nonlinear simulation. In the ABAQUS/CAE module, laminates with a hole were modeled and the compression process was calculated in the ABAQUS/Standard analysis module.

The geometrical size of FEA model is also 150×100×5mm and the lay-up sequence is A lay-up. Considering the nonlinear effect during the compressive process, the Nlgeom(This setting controls the inclusion of nonlinear effects or large displacements and affects subsequent steps) is on. The boundary conditions of the models are U3=0(U3 means the displacement in thickness direction) along the length direction and one side is fix while another side is put on compressive load in the width direction. The matrix cracking, fiber breakage, and delamination are taken into consideration by using SC8R(An 8-node quadrilateral in-plane general-purpose continuum shell, reduced integration with hourglass control, finite membrane strains) to simulate each ply.

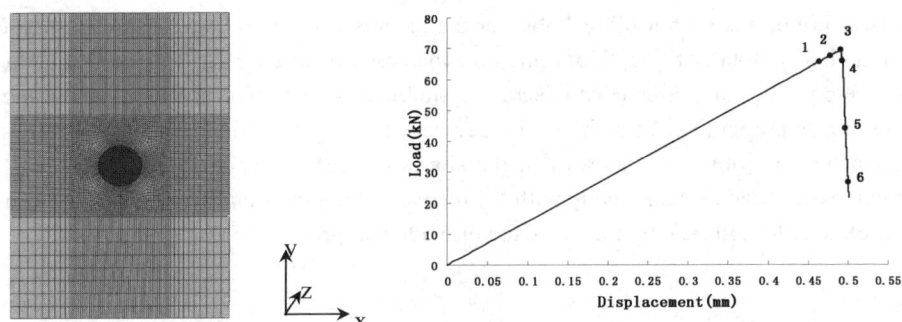

Fig.5 Model of laminates with a hole   Fig.6 Displacement and load curve of 50mm diameter' hole

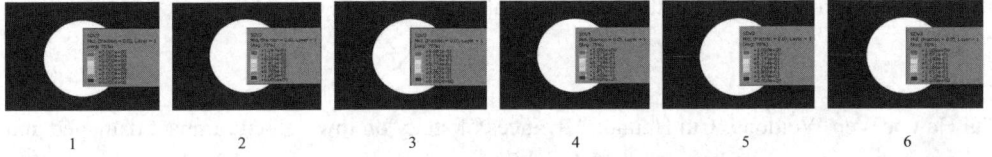

Fig.7 The damage evaluative process of 0° plies

FEA results of laminates with a hole. The numerical results correspond well with the testing results, which indicates that the finite element model is reasonable and can be used to simulate the compression response of composite laminates. Fig.6 shows that the load increases linearly along with the increases of displacement until it reaches to point 1. After point 1, because of the damage of some elements resulted by the stress concentration, the increase trend goes to gentle gradually. The load has a sudden reduction after point 3 which is the ultimate compressive load, and then the whole laminates break absolutely. For the 6 pictures in Fig.7, the displacements and loads are corresponding to the 6 points in the curve of the Fig.6. Fig.7 shows the damage evaluative process of 0° plies, it indicates that the appearance time of ultimate compressive load is earlier than that of fiber breakage expanding to the boundary, so as other typical angle plies. In the Fig.8, it can be contained that the decrease trends of two curves are similar and the compressive strength of two lay-up sequences decrease along with the increase of diameters of hole. At first, the strength decrease acutely and the decrease trends go to stable after the diameter of hole is bigger than 20mm. In a way, the compressive strength incline to a fixed value even though the diameter of hole is increasing gradually.

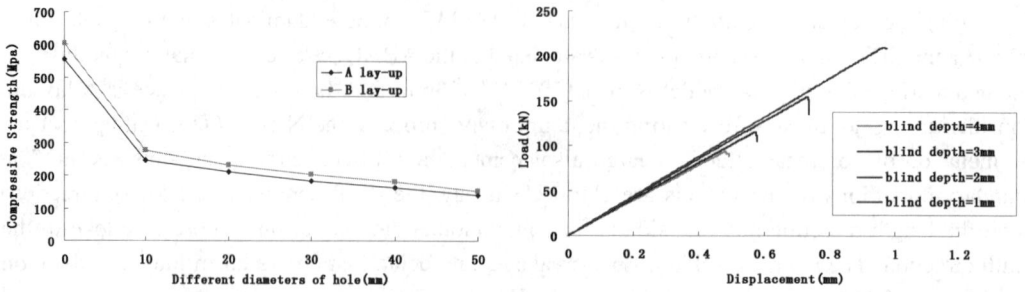

Fig.8 The diameters of hole –compressive strength curves   Fig.9 The displacement-load curves

FEA results of laminates with a blind hole. For the models with blind holes, there are 4 types of depths of blind hole in total. The depth of blind hole increases symmetrically from central ply to top ply and bottom ply. The compressive load decreases nonlinearly along with the depth increasing and the decrease trends are similar. In addition the decrease trend goes gently when the depth of the blind hole reaches to 3mm. Fig.9 shows that the slopes of four curves are almost the same. The ultimate compressive load decrease along with the increase of depth of blind hole. The diameter and the depth of blind hole significantly influence the ultimate compressive load.

Summary

Only impact energy - dent depth curves show well orderliness. The appearance time of ultimate compressive load is earlier than that of fiber breakage expanding to the boundary. The ultimate compressive load decreases along with the increase of depth of blind hole

Reference

[1] Cui Haipo, Wen Weidong, Cui Haitao: Advances of study on low velocity impact damaged and residual strength of composite laminates. Material Science& Engineering, Vol.23(2005),p. 466-472.
[2] S. Sanchez-Saez, E. Barbero, R. Zaera, C. Navarro: Compression after impact of thin composite laminates. Composites Science and Technology, Vol.65 (2005), p. 1911–1919.
[3] V.J.Hawyes, P.T.Curtis, C.Soutis: Effect of impact damage on the compressive response of composite laminates. Composites: Part A 32(2001),p.1263-1270
[4] Cheng Xiaoquan, Zhang Zilong, Wu Xueren: Post-impact compressive strength of small composite laminate specimens [J]. Acta materiae compositae sinica, Vol 19 (2002),p. 8-12.
[5] Zhuang Zhuo.etc.: Finite element analysis and application based on ABAQUS [M]. Beijing, TsingHua University Press(2009).

A Research on Characterization for Damage Tolerance of Composite Laminates

Zhang Di[a], Zheng Xitao[b], Cheng Linan[c]

School of Aeronautics, Northwestern Polytechnical University, Xi'an 710072, China

[a]beyond1907@163.com, [b]zhengxt@nwpu.edu.cn, [c]chenglinan4324@163.com

Keywords: composite laminates, Knee-point, damage tolerance, influence coefficient

Abstract. This paper mainly presents a method that can characterize the damage tolerance capability of composite laminates based on the knee-point feature. Based on the experimental study and numerical result, the Knee-point mechanism has been investigated. A new influence coefficient, δ, was introduced, which is applied to calculating the influence of every pile's damage to the residual strength of the whole laminates. Finally, a conclusion can be drawn that the damage of 0° piles can represent the damage of the whole laminates, whilst the damage area of 0° piles can be applied to characterizing the damage tolerance capability of composite laminates.

Introduction

A well-known problem with composite structure is the poor resistance and tolerance to accidental impact by foreign objects, and the resulting damage, often in the form of delaminations, matrix cracking and fiber failures, may severely reduce the structural strength and stability. In fact, researchers usually care the residual strength of composite laminates after impact, so the damage tolerance capability becomes an integrated part of the capability of composite laminates. Impact damage tolerance is the ability of material/structure system to perform satisfactorily with impact damage present [1]. In order to assessing the impact damage tolerance of composite laminates, considerable amount of research has been done in those areas which are impact and compression after impact of composite laminates. [2] Among them a very significant discovery is the Knee-point phenomenon.[3]

The work presented in this paper focus on finding a way to use the Knee-point feature to characterize the damage tolerance capability of composite laminates, which is convenient in engineering application.

1. Experimental Investigation

The composite laminates used in this study was T800BA/9916 carbon/epoxy. The specimens for all the tests were prepared from an 40-ply laminate with a symmetric ply $[45/-45/90/0/0/45/0/0/-45/0]_{2s}$ and size of 5mm × 100mm × 150mm. The low-velocity impact test and compression after impact test of composite laminates have been carried out. And the relationships between CAI (compression after impact) and damage area, damage width and dent depth were obtained, respectively.

Impact testing. According to the previous experimental data, the laminate's Knee point occurred when the impact energy met 60J [4]. Based on the ASTM D 7136, which is a standard test method for measuring the damage resistance of fiber-reinforced polymer matrix composite to a drop-weight impact event, thereby, a spherical surface impactor with a diameter of 16.00mm was used, and the impact energy was 60.0J. The dent depth was obtained by depth gauge, the damage area and damage width were obtained by C-scan technique.

CAI testing. After the impact testing, based on the ASTM D7137, which is a standard test method for compressive residual strength properties of damaged polymer matrix composite plates, CAI test was carried out, and the relationships between CAI and damage area, damage width and dent depth were obtained, as Fig.1.2 and 3.

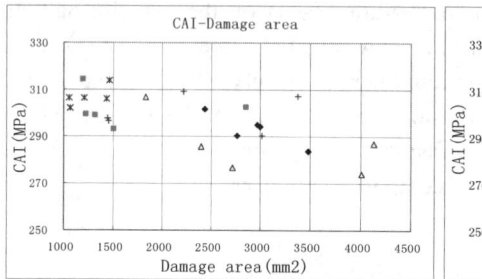
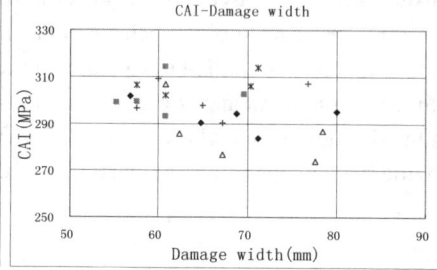

Fig.1 CAI and damage area   Fig.2 CAI and damage width

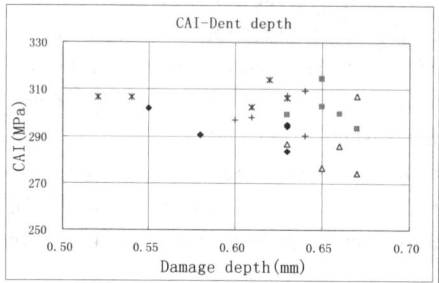
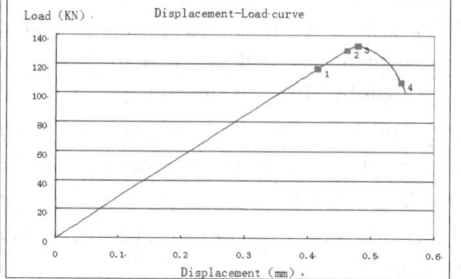

Fig.3 CAI and dent depth   Fig.4 The displacement-load curve

As the Fig.1-3 show, among the three damage parameters, taking the relaxation of dent depth and the dispersion of damage width into consideration, it strongly recommend to utilize damage area as the damage parameter for this kind of laminates. Here the damage area is averagely 2500mm^2.

2. Numerical Simulation

In this part, combined the experimental study with numerical result, the Knee-point mechanism has been investigated. Models with a hole of different aperture and models with a blind hole of different aperture were analyzed, and the compression responses of composite laminates are numerically simulated.

Models with a hole of different aperture. Fig.4 demonstrates the displacement-load curve when the aperture is 25mm. In order to analyze the damage evolution, four representative points were picked out on this curve.

At point 1 and point 2, there is only a small amount of matrix damage in the vicinity of 45°plies and 90°plies. At point 3, when the load meets the maximum, as Fig.5-6 show, fiber damage begins to appear in the 0°plies, and delamination begins to occur in the 45°and 90°plies. At point 4, as Fig.7-8 show damages increased rapidly.

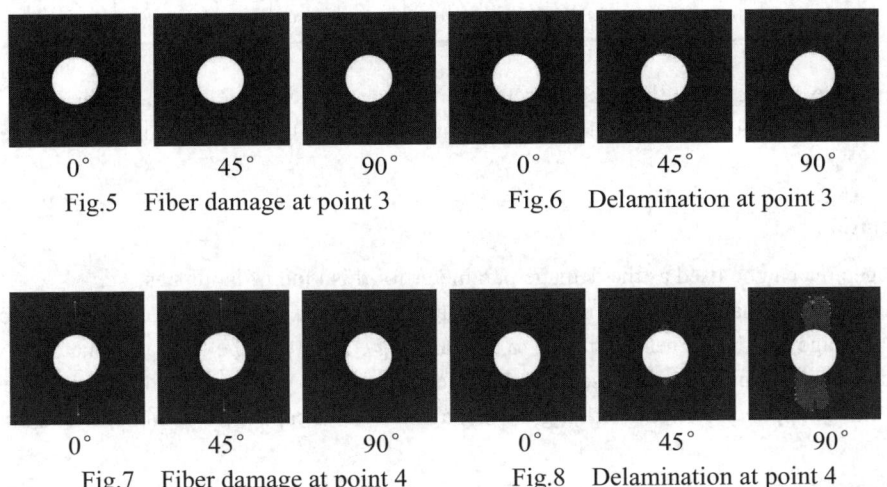

0°　　　45°　　　90°　　　　0°　　　45°　　　90°
Fig.5　Fiber damage at point 3　　Fig.6　Delamination at point 3

0°　　　45°　　　90°　　　　0°　　　45°　　　90°
Fig.7　Fiber damage at point 4　　Fig.8　Delamination at point 4

So, although delamination is the main reason that causes the carrying capacity of laminates to reduce, 0°plies' fiber damage is a sign of when the laminate starts to losing its carrying capacity.

Models with a blind hole of different aperture. By setting blind holes' height, we can use Eq. 1. to get each ply's influence coefficient δ to the laminate's compression residual strength. Table.1 shows the influence coefficient of each ply when the aperture is 20mm and 30mm respectively.

$$\delta_i = \frac{C_{blind(i)} - C_{blind(i+1)}}{C_{blind(0)} - C_{blind(n)}} \times 100\% \qquad (i = 0, 1, \Lambda, n) \tag{1}$$

$C_{blind(0)}$, $C_{blind(i)}$ and $C_{blind(n)}$ represent the laminate's compressive strength with a non-destructive condition, a blind hole of and a through-hole respectively.

Table 1 Influence coefficient of each ply

	45(1)	-45(2)	90(3)	0(4)	0(5)	45(6)	0(7)	0(8)	-45(9)	0(10)
20[mm]	0.30	0.03	0.77	9.89	7.69	4.40	7.14	1.65	7.69	10.44
30[mm]	0.50	0.60	0.10	9.30	5.00	2.00	10.50	4.35	7.15	10.5
	0(11)	-45(12)	0(13)	0(14)	45(15)	0(16)	0(17)	90(18)	-45(19)	45(20)
20[mm]	10.44	7.69	1.65	7.14	4.40	7.69	9.89	0.77	0.03	0.30
30[mm]	10.5	7.15	4.35	10.50	2.00	5.00	9.30	0.10	0.60	0.50

From Table 1 we can get the total influence coefficient of each kind of ply, among them 0°plies' damage has the biggest influence to the carrying capacity of the whole laminate, it is nearly 80 percent.

3. Conclusion

1, Damage area can be used as the damage parameter for this kind of laminates.
2, 0°plies' fiber damage is a sign of when the laminate starts to losing its carrying capacity and 0°plies' damage has the biggest influence to the carrying capacity of the whole laminate.
3, The damage of 0° piles can represent the damage of the whole laminates whilst the damage area of 0° piles can be applied for characterizing the damage tolerance capability of composite laminates.

References

[1] Andereas P.Christoforu: Impact dynamics and damage in composite structures. Composite structures 52(2011) 181-188.

[2] CUI Hai-po, WEN Wei-dong, CUI Hai-tao: Advances of Study on Low Velocity Impact Damaged and Residual Strength of Composite Laminates. Journal of Materials Science & Engineering 2005,23(3):466-472.

[3] SHEN ZHEN, YANG shengchun and CHEN Puhui: Experimental study on the behavior and characterization methods of composite laminates to withstand impact. Acta Materiae Composite Sinica 1000-3851(2008)05-0125-09.

[4] Li Zejiang: Analysis of the Influence Factors on Damaged Composite Laminate Residual Strength.

The simulation of low-velocity impact on composite laminates with the damage model based on strain

Chunhao MA [1,a], Fei Xu [2,b*]

[1]UAV Research and Development Center, Northwestern Polytechnical University, Xi'an, China

[2]School of Aeronautics, Northwestern Polytechnical University, Xi'an, China

[a]machunhao@126.com, [b]xufei@nwpu.edu.cn

*Corresponding author, email: xufei@nwpu.edu.cn

Keywords: Composites laminates; Low-velocity impact; Damage model; Delamination; Cohesive

Abstract: This paper proposed a composite damage model including the damage initiation and evolution based on strain to predict the composite intralaminar damage under impact loading. In the numerical simulation, the user material subroutine VUMAT and the cohesive-zone model are chosen to describe the composite damage model and the delamination of interfaces between different plies. ABAQUS software is used to simulate the low-velocity impact of different thickness composite laminates. It is found that the delamination shape and area, the contact force and the deflection of the impactor obtained by the numerically simulation agree well with the experimental results.

1. Introduction

As a result of the wide application of laminated composites, the study of low-velocity impact on composite laminates becomes more and more important. The damage forms of composite panels are mainly intralaminar fiber failure, matrix failure, and the delamination of interfaces between different plies. The well known Chang-Chang failure criteria [1] based on stress are widely used for the prediction of impact damage in composites. However, the stresses will change fiercely once the damage occurs in the composite panels during impacting. The strains change more gently, so the criteria based on strain are proposed to predict the composite intralaminar damage. In this paper, the composite damage model based on strain was proposed to simulate the low-velocity impact of composite structure.

2. The composite damage model based on strain

2.1 Damage initiation criteria

Damage initiation refers to the onset of degradation at the material properties. The composite strain-based failure criteria (as defined in the expressions below) are derived based on Chang-Chang failure criteria [1] and one-dimensional stress-strain relationship. These criteria consider four different damage initiation mechanisms: fiber tension, fiber compression, matrix tension and matrix compression.

$$\text{Fiber tension}: e_{ft} = \left(\frac{\varepsilon_{11}}{X_T^\varepsilon}\right)^2 + \left(\frac{\varepsilon_{12}}{S_{12}^\varepsilon}\right)^2 + \left(\frac{\varepsilon_{13}}{S_{13}^\varepsilon}\right)^2 \geq 1 \tag{1}$$

$$\text{Fiber compression}: e_{fc} = \left(\frac{\varepsilon_{11}}{X_C^\varepsilon}\right)^2 + \left(\frac{\varepsilon_{12}}{S_{12}^\varepsilon}\right)^2 + \left(\frac{\varepsilon_{13}}{S_{13}^\varepsilon}\right)^2 \geq 1 \tag{2}$$

$$\text{Matrix tension}: e_{mt} = \left(\frac{\varepsilon_{22}}{Y_T^\varepsilon}\right)^2 + \left(\frac{\varepsilon_{12}}{S_{12}^\varepsilon}\right)^2 + \left(\frac{\varepsilon_{23}}{S_{23}^\varepsilon}\right)^2 \geq 1 \tag{3}$$

$$\text{Matrix compression}: e_{mc} = \frac{1}{4}\left(\frac{-\varepsilon_{22}E_{22}}{S_{12}^\varepsilon G_{12}}\right)^2 + \left[\frac{\varepsilon_{22}}{Y_C^\varepsilon}\left(\left(\frac{E_{22}}{4G_{12}}\right)^2 \frac{Y_C^{\varepsilon 2}}{S_{12}^{\varepsilon 2}} - 1\right)\right] + \left(\frac{\varepsilon_{12}}{S_{12}^\varepsilon}\right)^2 \geq 1 \tag{4}$$

In the above equations, σ_{ij} and ε_{ij} are the nominal stress and strain, S_{ij} is the shear strength, X_t and X_c are the longitudinal tensile and compressive strength, Y_t and Y_c are the transverse tensile and compressive strength, $X_T^\varepsilon, X_C^\varepsilon, Y_T^\varepsilon, Y_C^\varepsilon, S_{12}^\varepsilon, S_{13}^\varepsilon$ and S_{23}^ε are the failure strain components based on one-dimensional stress-strain relationship and defined as follows: $X_T^\varepsilon = X_T/E_{11}, X_C^\varepsilon = X_C/E_{11}, Y_T^\varepsilon = Y_T/E_{22}, Y_C^\varepsilon = Y_C/E_{22}, S_{12}^\varepsilon = S_{12}/G_{12}, S_{13}^\varepsilon = S_{13}/G_{13}, S_{23}^\varepsilon = S_{23}/G_{23}$.

2.2 Damage evolution criteria
In Fig.1, the damage variable $d_{ii}(\varepsilon_{ii})$ which changes with the strain is defined to describe the composite damage evolution behavior [2]:

$$d_{ii}(\varepsilon_{ii}) = \frac{\varepsilon_{ii}^F}{\varepsilon_{ii}^F - \varepsilon_{ii}^O}(1 - \frac{\varepsilon_{ii}^O}{\varepsilon_{ii}}) \quad (5)$$

Where, ε_{ii}^O is the fiber or matrix failure initiation strain when the corresponding initiation criterion reaches the value of one. ε_{ii}^F is the completely failure strain and defined as follows[3]:

$$\varepsilon_{11}^{Ft} = \frac{2G_{1t}^c}{X_T l^*}, \varepsilon_{11}^{Fc} = \frac{2G_{1c}^c}{X_C l^*}, \varepsilon_{22}^{Ft} = \frac{2G_{2t}^c}{Y_T l^*}, \varepsilon_{22}^{Fc} = \frac{2G_{2c}^c}{Y_C l^*} \quad (6)$$

Fig.1 damage variable $d_{ii}(\varepsilon_{ii})$

Where, $G_{1t}^c, G_{1c}^c, G_{2t}^c$ and G_{2c}^c intralaminar fracture energy for fiber tension, fiber compression, matrix tension and matrix compression, respectively. l^* is the characteristic length.

2.3 The degradation of material stiffness
Process of material stiffness degradation begins when the strains satisfy certain damage initiation criteria in section 2.2. According to the different forms of damage, material stiffness will be degraded and the corresponding material modulus will be multiply by $(1-d_{ii}(\varepsilon_{ii}))$ as the new damaged material modulus.

Degradation of material modulus:
(1) Fiber tension or compression : E_{11}, G_{12}, G_{13}
(2) Matrix tension or compression : E_{22}, E_{33}, G_{12}, G_{23}

The damage parameter $(1-d_{ii}(\varepsilon_{ii}))$ will change with the strain from one (represents the undamaged material) to zero (represents the completely failure material). Actually, the damage parameter $(1-d_{ii}(\varepsilon_{ii}))$ will not be reduced to zero, but be kept at a residual value which can describe the residual strength even when the material is complete failure in the experiment.

3. Cohesive-zone model
The cohesive-zone model in Abaqus software is used to predict the interface delamination between different plies, and the damage initiation and evolution are described as follows [4]:

(1) Damage initiation
Quadratic nominal stress criterion is used to predict delamination damage initiation in this paper. The damage will initiate when a quadratic interaction function involving the nominal stress ratios (as defined in the expression below) reaches a value of one.

$$e_t = \left(\frac{\langle \sigma_n \rangle}{\sigma_{nc}}\right)^2 + \left(\frac{\sigma_t}{\sigma_{tc}}\right)^2 + \left(\frac{\sigma_s}{\sigma_{sc}}\right)^2 \geq 1 \quad (7)$$

Where, σ_n, σ_t and σ_s represent the normal stress and two shear stress, σ_{nc}, σ_{tc} and σ_{sc} represent the peak values of the nominal stress when the deformation is either purely normal to the interface or purely in the first or the second shear direction, respectively. The symbol $\langle \rangle$ represents the Macaulay bracket with the usual interpretation.

(2) Damage evolution
A power law fracture criterion is used to simulate the mixed mode interlaminar damage evolution:

$$e_t = \left(\frac{G_I}{G_I^c}\right)^\alpha + \left(\frac{G_{II}}{G_{II}^c}\right)^\beta + \left(\frac{G_{III}}{G_{III}^c}\right)^\gamma \geq 1 \quad (8)$$

Where G_I, G_{II} and G_{III} refer to mode I, II and III energy release rate, respectively. G_I^c, G_{II}^c and G_{III}^c refer to the mode I, II and III critical energy release rate, respectively. α, β and γ are the parameters determined by experiments.

4. Application to composite plates under impact
4.1 Experimental test
The impact test of two thickness CYCOM-977-2-35-12K-HTS composite panels is performed on the Instron Impulse 9250 machine. The plies of 3.93mm and 2.096mm thickness composite panels are [+45/90/-45/+45/0/-45/0/$\overline{0}$]s and [+45/-45/0/0/90/0/-45/+45], respectively. The size of the composite panels is 150mm×100mm and the impact energy is 12J.
4.2 Finite Element Simulation
4.2.1 Finite element model and material properties
Fig.2 shows the low-velocity impact FE model of composite panels in the experimental test. Composite laminates are modeled using 8-node linear brick solid elements with self-compiled VUMAT composite damage material model. The impactor was simplified as a rigid body with a mass of 5.713 kg, and the steel frame is used to model the boundary support in the experiment.

Fig.2 FE model

The CYCOM-977-2-35-12K-HTS has the material property of E_{11}=137.5Mpa, E_{22}=E_{33}=7.91GPa, v_{12}=0.37, G_{12}=4.09GPa, X_T=2293MPa, X_C=1516MPa, Y_T=85.9MPa, Y_C=267.3MPa, S_{12}=S_{13}=S_{23}=106 Mpa, G^c_{1t}=91.6N/mm, G^c_{1c}=79.9N/mm, G^c_{2t}=0.22 N/mm, G^c_{2c} = 1.1N/mm, G^c_I= 0.133N/mm, G^c_{II} =G^c_{III} = 0.459 N/mm.

4.2.2 Numerical simulation results
The numerical results obtained by FE model were compared to those obtained experimentally. The deflection of impactor, contact force and the interface delamination were investigated, and the intralaminar damage initiation and propagation was also observed in detail.

(1) Numerically predicted damage
The intralaminar damage model allowed the various forms of intralaminar failure to be investigated as the impact event progressing. Fig. 3 shows a cross-section matrix tension failure of the 3.93mm thickness plates under the impact energy 12J. Fig. 3a is a snapshot at 0.6ms when the strains satisfy the matrix tension initiation criterion. It can be found that the matrix tension failure firstly occurs at the position (black in Fig. 3a) directly under the impactor. In Fig. 3b, at the time 0.7 ms, matrix tension failure had spread directly on the back face of composite plates. Then the failure spread from the back face to the impact face at time 1.0ms (Fig. 3c). Fig. 3d is a snapshot at 2.9 ms, corresponding to maximum impactor deflection in the numerical impact simulation, extensive matrix tension failure was observed and varying in the thickness direction.

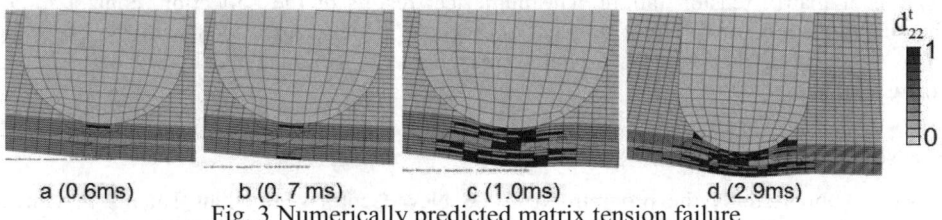

a (0.6ms)　　　b (0.7 ms)　　　c (1.0ms)　　　d (2.9ms)
Fig. 3 Numerically predicted matrix tension failure

(2) Time histories of deflection and contact force
Fig.4 and Fig.5 show the deflection and contact force of impactor for the 3.93mm and 2.096mm thickness composite panels under the impact energy 12J, respectively. Three specimens (test specimen-1, specimen-2 and specimen-3) were tested to confirm the experimental deflection and contact force for each thickness composite panels. It is found that numerical time histories of deflection and contact force agree well with the experimental results.

(3) Damage of delamination
As shown in Fig.6, the simulated damage areas and shapes of delamination agree well with the experimental results: the projection areas of delamination in numerical simulation are 25mm× 22.6 mm and 23mm×21.4mm for the 3.93mm and 2.096mm thickness composite panels, respectively; while the corresponding experimental C-Scan damage areas are 25mm ×27mm and 21mm×20mm.

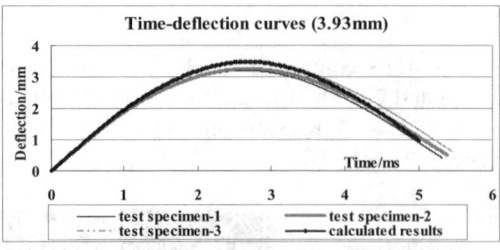
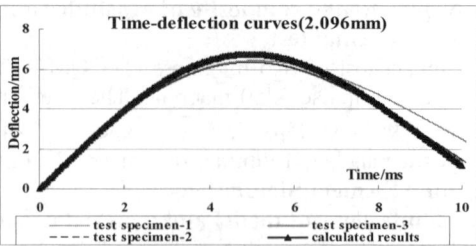

Fig.4 The time-deflection curves for two thickness composite panels

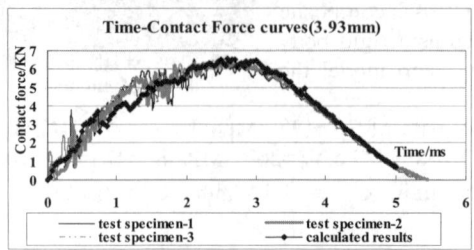
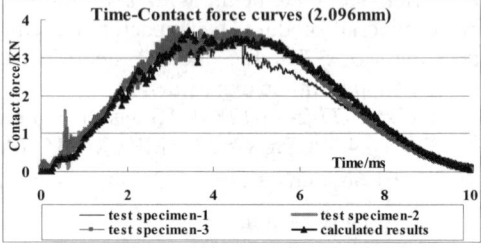

Fig.5 The time-contact force curves for two thickness composite panels

 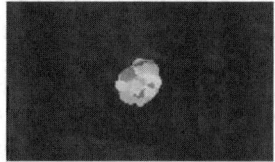    

a) 3.93mm (right: C-scan damage 25mm×27mm)   b) 2.096mm (right: C-scan damage 21mm×20mm)
Fig.6 Simulated and experimental projected damage areas of delamination

5. Conclusions

The work presented the composite damage model based on strain and cohesive-zone model to predict the intralaminar and interlaminar damage of composite laminates under the low-velocity impact. In the numerical simulation, the composite damage model well predicted the initiation and propagation of matrix tension failure. The numerical results of the deflection, contact force and interface delamination for different thickness composite panels, agree well with the experimental results, which proved that this FE model is reasonable to simulate the low-velocity impact of composite laminates.

Acknowledgments

The work is supported by the program for 2008 New Century Excellent Talents in University (NCET080454) and the Aeronautical Science Foundation of China (20102F53071).

References

[1] Chang F, Chang K: Journal of Composite Materials, 21:834-55(1987)

[2] A. Faggiani, B.G. Falzon: Composites, 41:737-749 (2010)

[3] M.V. Dondon, L. Iannucci: Composites and Structures, 86:1232-1252(2008)

[4] Li Z, Science Technology and Engineering, Vol.10,No.5:1170-1174(2010)

Investigation into Damage of Stainless Steel Mesh/AL Plate Multi-Shock Shield under Hypervelocity AL-Spheres Impact

Guan Gongshun[1, a], Pu Dongdong[1, a] and Ha Yue[1, a]

[1]P. O. Box 3020, Science Park, Harbin Institute of Technology, Harbin, 150080, P. R. China

[a]ggsh@hit.edu.cn

Keywords: Hypervelocity impact; Stainless steel mesh; Damage; Crack; Experiment

Abstract. A series of hypervelocity impact tests on stainless steel mesh/aluminum plate multi-shock shield were practiced with a two-stage light gas gun facility. Impact velocity was approximately 4km/s. The diameter of projectiles was 6.4mm. The impact angle was 0°. The fragmentation and dispersal of hypervelocity particle against stainless steel mesh bumper varying with mesh opening size and the wire diameter were investigated. It was found that the mesh wall position, diameter of wire, separation distance arrangement and mesh opening had high influence on the hypervelocity impact characteristic of stainless steel mesh/aluminum plate multi-shock shields. When the stainless steel mesh wall was located in the first wall site of the bumper it did not help comminuting and decelerating projectile. When the stainless steel mesh wall was located in the last wall site of the bumper, it could help dispersing debris clouds, reducing the damage of the rear wall. Optimized design idea of stainless steel mesh/aluminum plate multi-shock shields was suggested.

Introduction

Hypervelocity impacts on spacecraft in low earth orbit by meteoroids and space debris poses a threat to space missions [1]. Therefore, the design of spacecraft for an earth-orbiting mission must take into account such impact effects on the spacecraft structure. Whipple shield [2, 3] can decrease the momentum flux density of after-impact residual fragments acting on the rear wall. Comparing Whipple shield structure, metallic mesh shield structure is better resistance shielding systems because of its low mass and high flexibility [4, 5]. A number of experiments elucidating the integral dispersing and fragmentation properties of meshes due to their interaction with a projectile were also carried out as well [6]. In this paper, by hypervelocity impact experiment, performance of stainless steel mesh/aluminum plate multi-shock shield was investigated. For different shield configurations with the same stainless steel mesh, the results indicated that the first stainless steel mesh wall did not help enhancing the performance of shields, and the last stainless steel mesh wall did not help reducing the damage area on the rear wall. For same shield configurations with different stainless steel mesh, the results indicated that mesh weave size was the important parameters for enhancing the performance of shields. The filmed stainless steel mesh made for weakening the kinetic energy of debris clouds.

Experimental Set-Up

Test bumpers. The impact tests were performed onto metallic meshes multi-shock shield, each of them consisting of one 304 stainless steel mesh bumper and three 0.5mm thickness 2A12 aluminum alloy bumpers with a 3mm thickness 5A06 aluminum alloy rear wall. The overall space between bumper and rear wall was 100mm. The spherical projectile was made of 2017 Aluminum alloy. The meshes of 304 stainless steel mesh were 20, 40, 80 and 160 respectively. The details of the mesh bumpers are shown in Table 1. The geometry parameters D and S in Table 1 are diameter of the wires and spacing between the wires respectively. For an equivalent areal density of 0.5mm thickness 2A12 aluminum alloy plate 0.14g/cm^2, 304 stainless steel mesh areal densities of three-layer 20 meshes, three-layer 40 meshes, five-layer 80 meshes and six-layer 160 meshes were approximately 0.14g/cm^2. Stainless steel mesh multi-shock shields in the test are shown in Fig.1 to Fig.3.

Test facility. For this study, 2017-T4 aluminum alloy spheres were launched, with the protection of sabots, at velocities approximately 4km/s. The diameter of projectile was 6.4mm. The impact angle was 0° for all the tests. The projectile velocities were measured by magnetic induction. The uncertainty in these measurements is approximately ± 2%. The pressure of nitrogen gas in the first stage reservoir was 12MPa. The pressure of hydrogen gas in pump tube was 1.1MPa. The pressure in test chamber was approximately 300Pa. After each test, a digital vernier caliper was used to measure the perforation diameter of bumper and the diameter of crater distribution areas on the rear wall of aluminum. The images of the rear wall were processed to identify the damage zones and the distribution and size of craters on the rear walls.

Table 1 Details of mesh bumpers

No.	Mesh	D (mm)	S (mm)	Areal density (g/cm2)
1	20	0.20	1.07	0.049
2	40	0.14	0.50	0.046
3	80	0.07	0.25	0.030
4	160	0.04	0.12	0.025

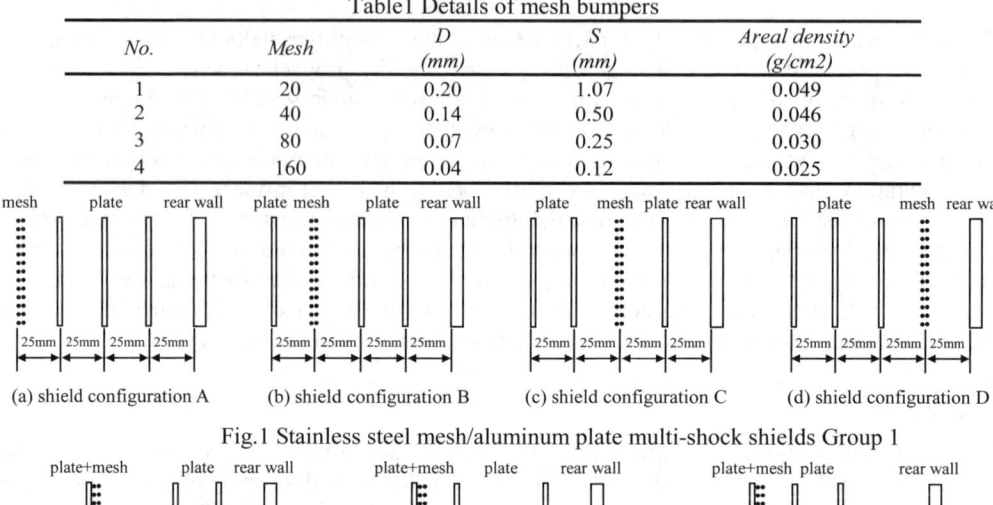

Fig.1 Stainless steel mesh/aluminum plate multi-shock shields Group 1

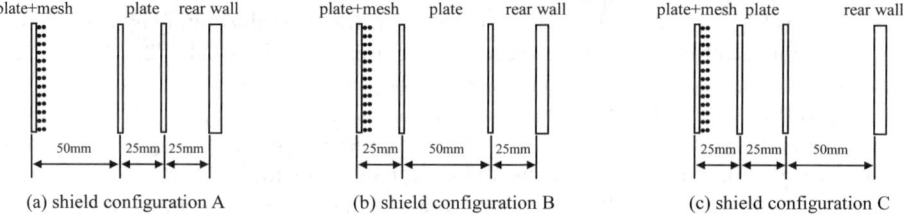

Fig.2 Stainless steel mesh/aluminum plate multi-shock shields Group 2

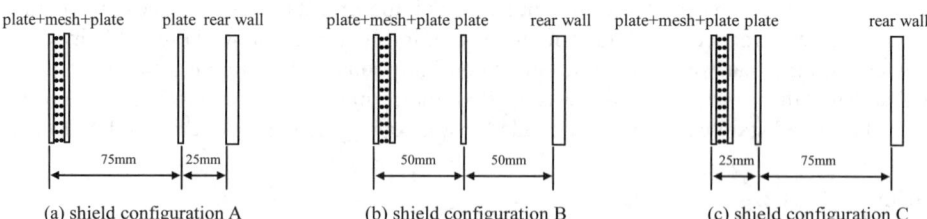

Fig.3 Stainless steel mesh/aluminum plate multi-shock shields Group 3

Results and Discussion

Table 2 presents the data from the impact tests. In Table 2, The geometry parameters d_p in Table 2 is projectile diameter; the parameters V, d_p, D_h, D_{99}, P_s and H_b are the projectile velocity, the projectile diameter, the diameter of maximal hole on rear wall, the diameter of covering 99% of spray area, the sink depth of rear wall front surface, and the bulge height of rear wall back surface respectively.

After projectile penetrated through the first wall of bumper, debris clouds comprising particles of projectile and bumper was brought behind the bumper. Debris clouds impact can generate craters, spallations and perforation on the rear wall. The rear wall damage is characterized by front surface sink and back surface bulge of the rear wall surrounding the impact center and a large splash of finely fragmented and liquid material. The representative test results are presented in Fig.4.

Table2 Damage results of stainless steel mesh/aluminum plate multi-shock shields

No.	V (km/s)	d_p (mm)	D_h (mm)	D_{99} (mm)	p_s (mm)	H_b (mm)	Rear wall damage
1A-40	4.08	6.4	4.30	61	5.33	3.82	Perforation
1B-40	4.25	6.4	-	68	6.32	4.06	No perforation
1C-40	4.08	6.4	-	58	6.87	5.40	No perforation
1D-40	4.01	6.4	-	75	7.02	4.78	No perforation
2A-40	4.17	6.4	-	65	5.89	4.20	No perforation
2B-40	4.03	6.4	-	73	4.20	2.38	No perforation
2C-40	4.01	6.4	-	90	3.23	2.16	No perforation
3A-80	4.01	6.4	1.00	108	7.28	6.04	Perforation
3B-80	4.03	6.4	-	130	6.21	5.44	No perforation
3C-80	4.14	6.4	1.00	157	6.19	4.18	Perforation
3A-160	4.06	6.4	-	115	6.15	5.22	No perforation
3B-160	4.10	6.4	4.59	130	7.65	4.98	Perforation
3C-160	4.14	6.4	1.00	163	6.99	4.78	Perforation
3A-20	4.06	6.4	4.45	122	5.61	4.94	Perforation
3B-20	4.10	6.4	-	129	4.27	2.84	No perforation
3C-20	3.97	6.4	5.39	150	5.02	4.80	Perforation
3Af-20	4.10	6.4	-	153	5.53	3.06	No perforation
3Bf-20	4.08	6.4	-	136	4.27	2.90	No perforation

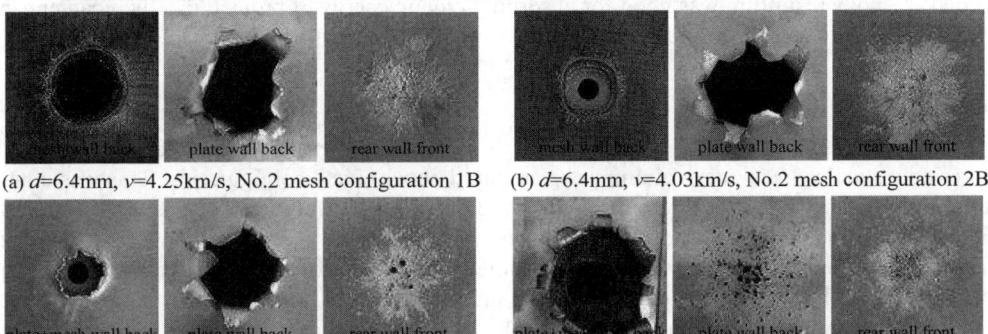

(a) d=6.4mm, v=4.25km/s, No.2 mesh configuration 1B (b) d=6.4mm, v=4.03km/s, No.2 mesh configuration 2B

(c) d=6.4mm, v=3.97km/s, No.1 mesh configuration 3C (d) d=6.4mm, v=4.10km/s, No.1 mesh configuration 3Af

Fig.4 Representative test results

Very thick bumpers will terminate the threat from hypervelocity projectiles in cratering events, while relatively thin shields will be penetrated, causing various degrees of projectile vaporization, melting and collisional fragmentation, the degree and extent of which is highly dependent on the areal density.

For group 1(see Fig.1), when impact condition was constant, the results showed that the rear wall was perforation failure only when the 304 stainless steel meshes wall located the site of the first bumper, and the rear wall of shields with other stainless steel mesh wall location were all no failure, and only sink and bugle of the rear wall were occurred. The results indicated that stainless steel mesh bumper can enhancing the performance of shields from hypervelocity impact, but the resist character was different for the same shield configurations with different stainless steel mesh bumper location. The results in table 2 showed that the damage area on the rear wall increased as spacing between mesh wall and rear wall increased if there is not any other prevention wall between mesh wall and rear wall. Because of time after time comminution, deceleration and dispersion of the projectile from anterior aluminum plates, the size of debris was very small. The mesh wall makes for small debris passing. At the same time, it can comminute remnant biggish debris.

For group 2(see Fig.2), when impact condition was constant, the rear walls of group 2 shields were all no failure. Therefore, comparing the same areal density 2A12 aluminum plate, the primary comminuting projectile capability of stainless steel mesh wall was less than that of 0.5mm thickness 2A12 aluminum alloy plate. The results in table 2 showed that the damage size and damage area on the rear wall of shields with different separation distance arrangement are different. When the separation distance arrangement was 50mm-25mm-25mm, the sink depth and the bulge height on the rear wall of shields were largest, and the damage area on the rear wall was smallest. When the separation distance arrangement was 25mm-25mm-50mm, the sink depth and the bulge height on the rear wall of shields were smallest, and the damage area on the rear wall was largest. Because of impact continuity, the primary fragmentation and comminution of projectile would be sufficient.

For group 3(see Fig.3), when impact condition was constant, perforation size and damage on the rear wall of shields with different weave size were different. These results suggest the finer meshes are to be preferred, as long as they possess sufficient mass to induce substantial projectile fragmentation. There is an optimal ratio of mesh opening size to projectile diameter for the same mesh material. An epoxy resin film was attached on the 20 meshes 304 stainless steel mesh of group 4. The overall areal density of bumper was $0.5769g/cm^2$. The rear walls of shields were all no failure. Comparing the identical shield without epoxy resin film, the damage of the first bumper with epoxy resin film was more graveness. The large petalled avulsion occurred on the front surface and the back surface of the first bumper (see Fig.4d). There were a lot of tiny perforations round the petalled hole. These showed that the first bumper dissipated more kinetic energy of projectile, and the primary fragmentation and comminution of projectile was more sufficient. Therefore, the stainless steel mesh attached an epoxy resin film was good for absorbing kinetic energy of projectile, which can enhance the prevention performance of mesh multi-shock shield from hypervelocity projectile impact.

Summary

This paper discussed the results obtained from all impact tests performed by launching aluminum spheres onto stainless steel mesh/aluminum plate multi-shock shield configuration. This was done to search for the damage and prevention characteristic of stainless steel mesh. In this study, effects of stainless steel mesh wall position, separation distance arrangement and mesh weave size on damage of the rear wall were analyzed. At the same time, the function of the epoxy resin film attached stainless steel mesh was investigated. The results indicated that the first stainless steel mesh wall did not help enhancing the performance of shields, and the last stainless steel mesh wall did not help reducing the damage area on the rear wall. The filmed stainless steel mesh made for weakening the kinetic energy of debris cloud. The optimized mesh/plate multi-shock shield for preventing debris impact should be that: the first bumper possesses the ability to dissipate sufficient kinetic energy of projectile, thus comminuting, decelerating and dispersing the debris more readily; spacing between the last bumper and the rear wall should be propitious to diffuse debris cloud; the mesh wall is not fit for the first bumper of shield.

References

[1] P. P. Bernhard, E. L. Christiansen and J. L. Crews: Int. J. Impact Eng. Vol. 17(1995), p. 57.

[2] W. Schonberg and J. E. Williamsen: Int. J. Impact Eng. Vol. 20(1997), p. 711.

[3] G. S. Guan, B. J. Pang, W. Zhang and Y. Ha: Int. J. Impact Eng. Vol. 35(2008), p. 1541.

[4] N. N. Myagkov, V. A. Goloveshkin, and A. V. Sulimov: Int. J. Impact Eng. Vol. 36(2009), p. 468.

[5] M. Higashide, M. Tanakab and Y. Akahoshi: Int. J. Impact Eng. Vol. 33(2006), p. 335.

[6] F. Horz, M. J. Cintala, R. P. Bernhard and T. H. See: Int J Impact Eng. Vol. 17(1995), p. 431.

Numerical Simulation of Hypervelocity Impact on Mesh Bumper Causing Fragmentation and Ejection

Guan Gongshun[1, a] and Niu Ruitao[1, a]

[1]P. O. Box 3020, Science Park, Harbin Institute of Technology, Harbin, 150080, P. R. China

[a]ggsh@hit.edu.cn

Keywords: Mesh bumper; Hypervelocity impact; Fragmentation; Ejection; Simulation

Abstract. In order to study the fragmentation of projectile and ejection of debris clouds caused by hypervelocity impacting mesh bumper, simulation of aluminum sphere projectile hypervelocity normal impacting aluminum mesh bumper was practiced with SPH arithmetic of LS-DYNA soft. The diameter of projectile was 4mm. Impact velocities of aluminum spheres were varied between 2.2km/s and 6.2km/s. The impact angle was 0°. The relationship between the debris clouds characteristic of projectile and the impact position on aluminum mesh bumper was studied. The effect on fragmentation of projectile from different combination mode of aluminum mesh bumper was analyzed. The results showed that the morphologies of the debris cloud varied with the impact position when a projectile impacted the mesh bumper. The debris clouds as palpus was found, and some local kinetic energy concentrated appeared in the debris clouds. Debris clouds distribution was more uniform when projectile impacted wire across point on the mesh bumper. Debris clouds had more diffuse area and less residual kinetic energy when mesh bumper was combined with interleaving mode. Mesh bumper combined with interleaving mode was helpful in enhancing the protection performance of shields.

Introduction

The increasing of space debris has become a serious hazard to the spacecraft in low earth orbit. Hypervelocity impacts on spacecraft by space debris can damage flight-critical systems, which can in turn lead to catastrophic failure of the spacecraft [1]. An efficient method for protecting the spacecraft from being destroyed by hypervelocity impact of space debris is designing the shield structure for the spacecraft [2]. Comparing plate bumper shield, mesh bumper shield is the potential lightweight shield to protect spacecraft in low-Earth orbit against collisional damage [3]. Earlier studies revealed that single mesh comminute hypervelocity projectile with efficiencies comparable to contiguous target [4]. Multiple interaction of projectile fragments with any number of meshes should lead to increased comminution, deceleration and dispersion of the projectile, such that all debris exiting the mesh stack possesses low specific energies that would readily be tolerated by many flight systems[5]. In this paper, numerical simulation of hypervelocity impacting mesh bumper was practiced with LS-DYNA soft. The study involved the differences in the morphology, the distribution of velocity and kinetic energy of the debris clouds generated by hypervelocity impacts. This study was aimed on mastering the hypervelocity impact characteristics of the mesh bumper and applying effectively it in development of lightweight bumper.

Numerical Simulation Set-Up

Geometric model and equation of state. The catia software was used to construct geometric model of mesh bumper used in simulation (see Fig.1). The Shock equation of state was chosen for this study. The Johnson–Cook strength model was selected because it is empirically based and because it attempts to model the behavior of materials subjected to large strains, high-strain rates, and high temperatures. It is of the form

$$\sigma = (A + B\varepsilon^n)(1 + C\ln\varepsilon^*)(1 - T^{*m}) \qquad (1)$$

where σ is the effective stress, ε is the effective plastic strain, ε^* is the dimensionless plastic strain rate, and T^* is the homologous temperature, defined as $(T_i - T_{room})/(T_{melt} - T_{room})$. A, B, C, n and m are the material constants determined from an empirical fit of flow stress data (as a faction of strain, strain rate and temperature) to Eq. (1) [6]. The first bracketed expression gives the stress as a function of strain when $\varepsilon^*=1.0\text{s}^{-1}$ and $T^*=0$ (for room temperature laboratory experiments). The second and third bracketed expressions represent the effects of strain rate and temperature respectively. The third bracketed term is a thermal softening term where the yield stress drops to zero at the melting temperature T_{melt}. Material parameters of Johnson-cook constitutive model are shown in Table 1.

Table 1 Material parameters of Johnson-cook constitutive model

Material	A/(GPa)	B/(GPa)	n	C	m	T_r/(K)	T_m/(K)
2017-T4	0.270	0.426	0.34	0.015	1.0	300	775
5052	0.265	0.426	0.34	0.015	1.0	300	650
5A06	0.265	0.426	0.34	0.015	1.0	300	864

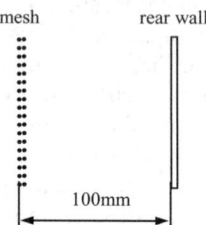

Fig.1 The geometry model of aluminum mesh   Fig.2 The diagram of aluminum mesh shield structure

Validation of numerical simulation. SPH arithmetic of LS-DYNA soft was used to simulate aluminum sphere projectile hypervelocity normal impacting aluminum mesh shield. The Shock Equation of state and the Johnson-Cook strength model along with standard library data for the model parameters were used to simulate the material behavior. The failure criterion of maximum principal stress was used in tension only. The strength limit was 2.5GPa. The shield configuration in the simulation is shown in Fig.2. One impact experiment was used to validate the numerical model. The results of the simulation and experiment are shown in Fig. 3.

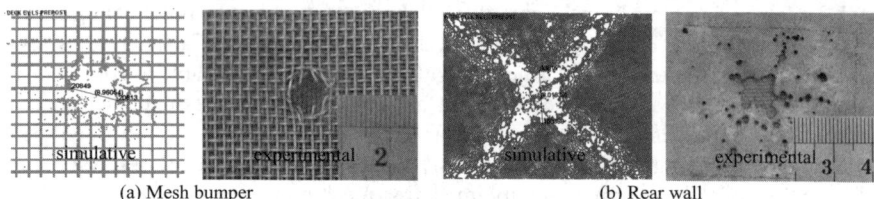

(a) Mesh bumper   (b) Rear wall

Fig.3 Simulative and experimental results

For the penetration of mesh bumper, maximal hole diameters are 8.96mm (simulation) and 8.64mm (experiment) respectively. The error is 3.7%. For the penetration of rear wall, maximal hole diameters are 8mm (simulation) and 8.02mm (experiment) respectively. The error is 0.25%. The comparison of damage state with simulation results and experimental results, the damage pattern on the rear wall is approximately same (see Fig.3). Therefore, numerical simulation method can be used to forecast hypervelocity impact damage on the rear wall of mesh bumper shield.

Results and Discussion

Three impact positions on the mesh bumper in this study were shown in Fig.4, including blank center point, cross point and line middle point. Fig.5 shows morphologies of debris clouds from hypervelocity aluminum spheres impact on the aluminum mesh bumper for three impact positions. In this simulation, the diameter of projectile was 4mm. the impact angle was 0°, the impact velocity was 4.2km/s. All images were taken with 1.0μs stop time.

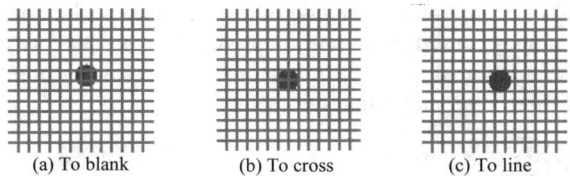

Fig.4 Impact position of projectile center

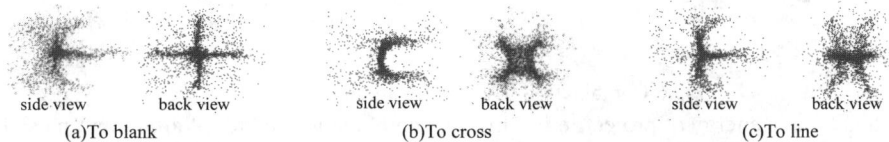

Fig.5 Debris cloud configurations for different impact point

Debris clouds distribution was different when hypervelocity aluminum spheres impact different position of mesh bumper. For the blank center point, five debris clouds palpus were observed, including one center palpus and four periphery palpus. The center palpus was more protrudent than others. The center debris clouds palpus possessed a majority of kinetic energy from hypervelocity impact projectile, which would make the rear wall fail easily. Thereby, bumper showed infirm protection performance. For the cross point, only four periphery debris clouds palpus appeared. The kinetic energy from hypervelocity impact projectile was dispersed to periphery debris clouds, which made bumper show better protection performance. For the line middle point, six periphery debris clouds palpus were observed and kinetic energy of projectile was more dispersed.

Table 2 Velocities of debris clouds as palpus for different impact velocities

V /(km/s)	V_{front} /(km/s)	V_{core} /(km/s)	$(V_{front}-V)/V$ /(%)
2.2	2.60	2.20	18
3.2	3.75	3.25	17
4.2	5.20	4.45	23
5.2	6.40	5.30	23
6.2	7.70	6.10	24

Table 3 Residual kinetic energy of projectile for different impact velocities

V /(km/s)	E_{k0} /(J)	E_k /(J)	E_k/E_{k0}
2.2	221	175	0.79
3.2	470	372	0.79
4.2	810	640	0.79
5.2	1240	990	0.80
6.2	1760	1420	0.81

Velocities of debris clouds as palpus for different impact velocity of projectile are shown in table 2. In Table 2, V is the impact velocity of projectile. V_{front} and V_{core} are velocities along impact direction of projectile in the front and center of the debris clouds as palpus respectively. It is found that V_{front} is larger approximately 20% than V, and V_{core} almost equal V.

Residual kinetic energy of projectile for different impact velocities are shown in table 3. In Table 3, V is the impact velocity of projectile. E_{k0} and E_k are initial kinetic energy and residual kinetic energy of projectile respectively. For different impact velocities, it is found that E_k/E_{k0} is almost constant.

In this study, two kinds of combination mode of aluminum mesh bumper were adopted to study morphologies of debris clouds caused by aluminum projectile hypervelocity impacting multi-meshes bumper, including superposition mode (see Fig.6a) and interleaving mode (see Fig.6b). Mesh bumper is composed of three layer aluminum meshes.

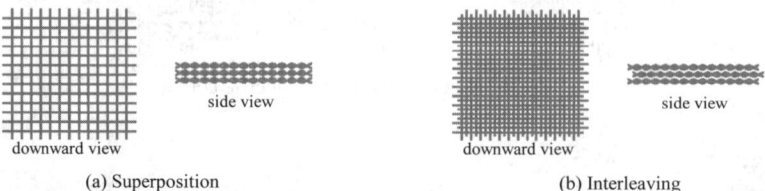

(a) Superposition   (b) Interleaving

Fig.6 Combination modes of Al mesh bumper

Residual kinetic energy of projectile for different combination modes of aluminum mesh bumper is shown in table 4. In Table 4, E_{k1} and E_{k2} are residual kinetic energy of projectile for Superposition mode mesh bumper and interleaving mode mesh bumper respectively. For different impact velocities, it is found that E_{k1} are all larger than E_{k2}. The results indicated that interleaving mode mesh bumper can enhance the performance of shields against hypervelocity impact.

Table 4 Residual kinetic energy of projectile for different combination modes of aluminum mesh bumper

V /(km/s)	E_{k0} /(J)	E_{k1} /(J)	E_{k2} /(J)
2.2	221	175	171
3.2	470	372	361
4.2	810	640	620
5.2	1240	990	955
6.2	1760	1420	1360

Summary

This paper discussed the results obtained from all calculation performed by SPH arithmetic of LS-DYNA soft simulating aluminum sphere onto aluminum mesh shield. This was done to search for fragmentation of projectile and ejection of debris clouds when aluminum spheres hypervelocity impact aluminum mesh. The effect on fragmentation of projectile from different combination modes of aluminum mesh bumper was analyzed. The results indicated that the debris clouds configuration from aluminum sphere hypervelocity impacting aluminum mesh was different with the different impact position on mesh bumper. The debris clouds as palpus was found in the front of debris clouds, and some local kinetic energy concentrated appeared in the debris clouds. Debris clouds distribution was more uniform when projectile impacted across point on mesh bumper. Debris clouds had more diffuse area and less residual kinetic energy when mesh bumper was combined with interleaving mode, and that was helpful in enhancing the protection performance of shields.

References

[1] A. D. Kiureghian, P.V. Geysekns and M. R. Khalessi: Int J Impact Eng. Vol. 19(1997), p. 571.

[2] G. Gongshun, P. Baojun, Z. Wei and H. Yue: Int. J. Impact Eng. Vol. 35(2008), p. 1541.

[3] E. L. Christiansen and J. H. Kerr: Int J Impact Eng. Vol. 14 (1993), p. 169.

[4] F. Horz and M. J. Cintala: NASA TM 104749, (1992).

[5] F. Horz, M. J. Cintala, R. P. Bernhard and T. H. See: Int. J. Impact Eng. Vol. 17(1995), p. 431.

[6] M. B. Corbett: Int J Impact Eng. Vol. 31 (2006), p. 431.

Analysis of Crack Arrest by Electromagnetic Heating in Metal with Oblique-Elliptical Embedding Crack

FU Yuming[a], ZHOU Hongmei[b], Wang Junli[a], ZHENG Lijuan[a]

School of Mechanical Engineering, Yan Shan University, Qinhuangdao 066004

[a]mec9@ysu.edu.cn, [b]zhou.hong163@163.com

Keywords: spatial oblique-elliptical embedding crack; crack arrest by electromagnetic heating; current density; temperature; pulse current

Abstract. Ellipse crack tip can be passivated when pulse current is perpendicular to crack surface of metal structure which can achieve the purpose of crack. In practice, it is hard that the loading current direction perpendicular to the crack surface because of the crack existence of different position and orientation. Theory and numerical simulation analyze the effect of crack arrest by electromagnetic heating based on metal structure with oblique-elliptical embedding crack. And current density and temperature distribution of crack tip are derived. The result shows that due to heat concentration around the oblique-elliptical embedding crack tip, the crack tip temperature exceeds the melting point of materials and small welded joints are formed by metal melting. Thermal compressive stress which is generated near the crack tip can prevent the crack propagation effectively. Thus, electromagnetic heating is proved to be an effective method of preventing general crack propagation.

Introduction

When strong pulse current put into metal structure with crack, the conductor near the crack tip generate the Joule heat. The current flow around and focus make the material temperature increasing more than the melting point of the material and causes the crack tip passivation, which prevent the crack to expand and achieve the purpose of crack arresting [1-4]. When the pulse discharge is perpendicular to the crack surface, the electromagnetic heat is effective. In practice, location and orientation of crack are uncertain, so current direction is difficult to perpendicular to the crack plane. This paper is about theory and numerical simulation of electromagnetic thermal effects on buried oblique crack and the study provides a reference for the buried crack arresting with electromagnetic heating.

1 Modeling

The modeling is a cylinder with an oval oblique crack. The coordinate system is shown in Figure 1, where the crack surface is vertical to yoz surface, an angle of xoz is θ, and an angle of xoy is $\frac{\pi}{2} - \theta$. The long axis of the ellipse is a, the short axis is b.

The method for temperature field of oval oblique crack is transformed to each coordinate plane of the crack tip temperature field.

Oval in the plane of projection can be described as $z=0, \frac{x^2}{b^2} + \frac{y^2}{a^2 \cos^2 \theta} = 1$. In the plane of xoz is $y=0, \frac{x^2}{b^2} + \frac{z^2}{a^2 \sin^2 \theta} = 1$. In the plane of yoz is $x=0, z = y \tan \theta, y \in (0, a \cos \theta)$.

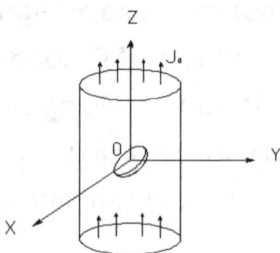

Fig.1 The metal component with a buried inclined crack

2 Temperature field distribution

2.1 Current density distribution in xoy plane

Current density distribution is resulted from literature [5]

$$J_z(x,y,0) = \frac{J_0}{E(k)} \left\{ \frac{ab^2}{\sqrt{Q(\xi)}} - \left[E(u) - \frac{snucnu}{dnu} \right] + 1 \right\} \tag{1}$$

$E(k)$ is the second class of complete elliptic integrals. k is modulus of elliptic functions.

$$k^2 = \frac{b^2 - a^2 \cos^2 \theta}{b^2} \tag{2}$$

$$Q(\xi) = \xi \left(a^2 \cos^2 \theta + \xi \right) \left(b^2 \cos^2 \theta + \xi \right) \tag{3}$$

$$\xi = \frac{2ab}{\sqrt{b^2 \sin^2 \varphi + a^2 \cos^2 \theta \cos^2 \varphi}} r \cos^2 \frac{\phi}{2} \tag{4}$$

Where r is the distance of a point to the origin. ϕ is the angle between r and xoy plane. φ is the angle of a point on the oval boundary.

2.2 Current density distribution in xoz plane

Water flow around the oval plate can be applied to solving the distribution of current density. Velocity potential of the metal structure with oval-shaped crack at the moment of pulse discharge is

$$F(z) = \frac{J_0}{\sigma^*} \left[ze^{-i\alpha} + \frac{(b + a\sin\theta)^2}{(b - a\sin\theta)^2 a^2} e^{i\alpha} - e^{-i\alpha} \right) \left(\frac{z}{2} - \sqrt{\left(\frac{z}{2}\right)^2 - \left(\frac{b - a\sin\theta}{2}\right)^2 a^2} \right) \right] \tag{5}$$

Where J_0 is current density; σ^* is conductivity; α is the angle between current density and x-axis.

$$z = b\cos\alpha + i(a\sin\theta)\sin\alpha \tag{6}$$

In this article, take $\alpha = 90°$.

$$F(z) = \frac{J_0}{\sigma^*} \left[\left(a\sin\theta + 1 \right) + \frac{(b + a\sin\theta)^2}{(b - a\sin\theta)^2 a^2} i \right) \left(\frac{z}{2} - \sqrt{\left(\frac{z}{2}\right)^2 - \left(\frac{b - a\sin\theta}{2}\right)^2 a^2} \right) \right] \tag{6}$$

Function of current density is expressed as

$$J_{x2} = \frac{J_0}{2}\left(\sqrt{4b^2 - 5a^2 \sin^2\theta}\right) \qquad (7)$$

$$J_{z2} = \frac{J_0}{2} a \sin\theta \qquad (8)$$

2.3 Current density distribution in yoz plane

$$J_{y3}^2 = 16 J_0^2 \sin^2\theta \left(\sqrt{\frac{(y^2-z^2)(y^2-z^2+r_0^2)+4y^2z^2}{2(y^2-z^2+r_0^2)^2+8y^2z^2}} + \frac{\sqrt{[(y^2-z^2)(y^2-z^2+r_0^2)+4y^2z^2]^2+(2yzr_0^2)^2}}{2(y^2-z^2+r_0^2)^2+8y^2z^2} + \cos\theta \right)^2 \qquad (9)$$

$$J_{z3}^2 = 16 J_0^2 \sin^2\theta \left(\begin{array}{c} \dfrac{-(y^2-z^2)(y^2-z^2+r_0^2)-4y^2z^2}{2(y^2-z^2+r_0^2)^2+8y^2z^2} \\ + \dfrac{\sqrt{[(y^2-z^2)(y^2-z^2+r_0^2)+4y^2z^2]^2+(2yzr_0^2)^2}}{2(y^2-z^2+r_0^2)^2+8y^2z^2} \end{array} \right) \qquad (10)$$

2.4 Heat source power expression

$$Q_T = \frac{j_x^2 + j_y^2 + j_z^2}{\sigma_t}$$

$$\approx \frac{16 J_0^2}{\sigma_t} \times \left\{ \sin^2\theta \left(\frac{r\sqrt{2(\cos^2\phi\cos^2\varphi+\sin^2\phi)}\cos\theta}{r_0} + \cos^2\theta + \frac{r^2(\cos^2\phi\cos^2\varphi+\sin^2\phi)}{r_0^2} \right) + \frac{1}{E^2(k)} \left\{ \frac{ab^2}{\sqrt{Q(\xi)}} - \left[E(u) - \frac{snucnu}{dnu} \right] + 1 \right\}^2 \right\} \qquad (11)$$

2.5 Temperature field

In this paper, temperature distribution caused by circular line heat source can be seen as the superimposition of an infinite number of point heat sources. Temperature distribution caused by circular line heat source in a period of time is

$$T(x,y,z,t) = \frac{1}{2(\sqrt{\pi a_T t})^3} \int_0^{\pi/2} \frac{Q_T}{\rho c_p} \times a\sqrt{1-k^2\cos^2\phi} \times \exp\left[-\frac{(x-a\cos\phi)^2+(y-b\sin\phi)^2+z^2}{4a_T t}\right] d\phi \qquad (12)$$

3 Numerical Simulation

The material is 45 steel. The material's physical parameters at normal temperature are: the linear expansion coefficient is $\alpha = 11.6 \times 10^{-6} \text{H/m}$, specific heat capacity $c_p = 480 \text{J}/(\text{kg}\cdot{}^\circ\text{C})$, conductivity $\sigma_t = 1.3 \times 10^7 /(\Omega\cdot\text{m})^{-1}$, conductivity temperature coefficient $a_T = 1.1 \times 10^{-5} \text{m}^2/\text{s}$, thermal conductivity $\lambda = 39.4\text{W}/(\text{m}^2 \cdot \text{K})$, surface diffusion coefficient $\alpha_0 = 40\text{W}/(\text{m}^2 \cdot \text{K})$, elastic modulus $E = 2.09 \times 10^{11}\text{Pa}$, Poisson's ratio $\mu = 0.26$, density $\rho = 7.85 \times 10^3 \text{kg}/\text{m}^3$, and melting point $1350\,°\text{C}$. The length of model is 100mm, and radius is $R = 8\text{mm}$. Crack center and the center of the model coincide. $a = 5\text{mm}$, $b = 3\text{mm}$, $\theta = 45°$.

Simulate result of the temperature is shown in Figure.2. Figure 2 shows strong flow around occurs near the crack tip after the pulse discharge, and the instantaneous temperature is increased rapidly up to $1833\,°\text{C}$, more than the melting point so as to passivate the material.

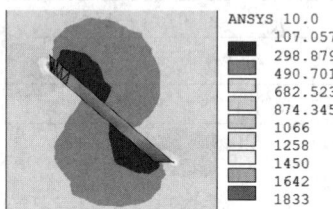

Fig.2 Elliptical cross section of temperature distribution

4 Conclusion

Temperature field of the metal structure with a buried oval oblique crack during pulse discharge is analyzed using decomposition method. Temperature field of oval oblique crack is transformed to each coordinate plane of the crack tip temperature field. Theory and numerical simulation show that during pulse discharge, flow of current still form in the oblique crack tip. Temperature exceeds the melting point of the material and the tip is passivated. Pulse discharge can effectively prevent crack propagation.

Acknowledgement

This paper is supported by National Natural Science Foundation of China No.51075351 and No.51105325.

References

[1] Sumi.Naobumi. Dynamic Thermal Stresses in a Finite Circular Plate with a Penny Shaped Crack Generated by Impulsive Heating. Nippon Kikai Gakkai Ronbunshu. A Hen Transactions of the Japan Society of Mechanical Engineers, Jan, 61(1995).

[2] Bai Xiangzhong, Fu Yuming. Advances in Research on Elect romagnetic Heating Effects to Stop Crack Propa2 gation in Metal Components. International Journalof Nolinear Science and Numerical Simulation. ISRAELand U K, 9(2003).

[3] Bai Xiangzhong, Fu Yuing, Zheng Lijuan. Experimental research on electromagnetic heating effects to stop crack propagation in metal components. Third International Conference on Experimental Mechanics, (2001).

[4] Bai Xiangzhong, Fu Yuming, Hu YD, Zheng LJ. Temperatue field near crack tip in a current-carrying plate under the repeated action of Pulse current. International Journal Nonlinear Science and Numerical Simulation ISRAEL and UK, 9(2003).

[5] Fu Yuming, Zheng Lijuan, Tian Zhenguo, LiWei. Temperature field at time of pulse current discharge in metal structure with elliptical embedding crack. Applied mathematics and mechanics, 29(2008).

Effect of Characteristic Parameters of Exponential Cohesive Zone Model on Mode I Fracture of Laminated Composites

Guowei Zhu[a], Yuxi Jia[*,b], Peng Qu[c], Jiaqi Nie[d], Yunli Guo[e]

Key Laboratory for Liquid-Solid Structural Evolution & Processing of Materials (Ministry of Education), Shandong University, 17923 Jingshi Road, Jinan 250061, China

[a]yhzuiai2006@yahoo.cn, [b]jia_yuxi@sdu.edu.cn, [c]qproc@163.com, [d]jiaqi0204@126.com, [e]yunlinihao@126.com

Keywords: Exponential cohesive zone model; Composites delamination; Interfacial performance

Abstract: Delamination is a particularly dangerous damage mode of high performance laminated composites. In order to describe the composites ductile cracking and its progressive evolution accurately, the adjusted exponential cohesive zone model (CZM) is adopted, which correlates the tensile traction with the corresponding interfacial separation along the fracturing interfacial zone. At first the adjusted exponential CZM is used to simulate the mode I delamination of the standard double cantilever beam (DCB). The simulated results are in good agreement with the corrected beam theory and the corresponding experimental results. Then in order to research how the interfacial properties influence the mode I fracture, the interfacial strength and the critical energy release rate are studied. The main results are obtained as follows. The interfacial strength plays a crucial role in the laminated composites delamination onset, and it affects the peak load significantly if there is not a pre-crack. Once the delamination propagation begins to occur in the laminated composites, the responses of the load-displacement plots are relatively insensitive to the interfacial strength, and only the critical energy release rate is of critical importance. Furthermore, the peak load increases with the increase of the critical energy release rate and interfacial strength.

1. Introduction

Delamination is one of the most important types of damages in laminated fiber-reinforced composites due to its relatively weak interlaminar strength. A number of researchers have developed some relevant techniques to predict the interlaminar fracture behaviors from the experimental and numerical standpoints. The methods such as the virtual crack closure technique and J-integral are widely used. However, these techniques cannot be used to predict the initiation of the delamination [1], therefore they are restricted to the problems that the initial crack position is known in advance and the crack propagation path is also specified.

The use of a cohesive zone model (CZM) which is based on the damage mechanics can overcome some of the aforementioned drawbacks. Therefore, it has attracted a growing interest to describe laminated composites delamination particularly. Cotterell B. *et al* investigated the delamination of composite laminates using the trapezoidal model [2]. Hallett S.R. *et al* analyzed the composites delamination using the bilinear form [3]. Goyal V.K. *et al* simulated the delamination process using the exponential constitutive law [4]. Among these CZMs, the exponential CZM is widely used because of some advantages over other types of CZMs [5]. Namely, the tractions and their derivatives are continuous, which is attractive from a computational viewpoint. However, it is barely used to predict the delamination onset without the initial crack.

In the present work, the adjusted exponential CZM is applied to describe the laminated composites interface, and a three-dimensional (3D) finite element model is constructed to analyze the mode I delamination of laminated composites. At first, a typical double cantilever beam (DCB) test is simulated, and the simulated results are compared with the theoretical and experimental results to verify the feasibility of the adjusted exponential CZM. Then in order to research how the interfacial properties affect the delamination initiation and propagation, the simulations of laminated composites with or without initial cracks are performed. The main fracture parameters, namely the interfacial strength (σ_{max}) and the critical energy release rate (G_c), are studied.

2. Mechanical Model

2.1 Geometrical Model and Boundary Conditions

The commercial software ANSYS is used to generate a 3D mechanical model (Fig. 1) for the mode I DCB. The total length (X-direction) L=100mm, width (Z-direction) b=20mm, and thickness (Y-direction) $2h$=3mm, and the initial crack length a=30mm. Since the interface is thin enough with respect to the overall geometrical dimensions, its thickness is taken as exactly zero [1].

The finite element model is constructed, using a 3D 20-node structural solid element to model the two laminated plates, and using a single layer of 3D 16-node cohesive elements to model the interface. To meet the symmetry [3], the nodes at Z=0 are constrained in the Z-direction. The end of DCB (X=L) is fully constrained. The load is applied by the displacement method: the Y-direction displacement v=-10mm is applied to the lower edge (X=0, Y=-1.5) and v=10mm is applied to the upper edge (X=0, Y=1.5), respectively.

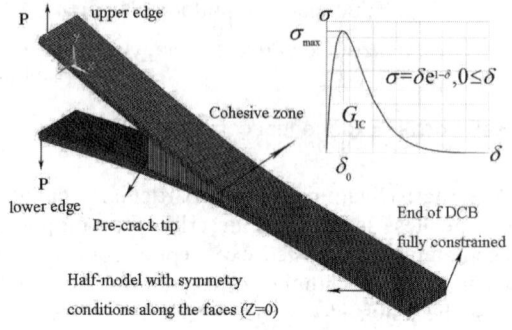

Fig.1 Finite element mesh and boundary conditions for mode I DCB

2.2 Exponential CZM for Interface Failure

The cohesive zone model used in this paper is the exponential relation (see Fig.1) proposed by Xu and Needleman [6, 7]. It is based on the interfacial potential ϕ:

$$\phi(\Delta_n, \Delta_t) = \phi_n + \phi_n \exp\left(-\frac{\Delta_n}{\delta_n}\right)\left\{\left[1 - r + \frac{\Delta_n}{\delta_n}\right]\frac{1-q}{r-1} - \left[q + \left(\frac{r-q}{r-1}\right)\frac{\Delta_n}{\delta_n}\right]\exp\left(-\frac{\Delta_t^2}{\delta_t^2}\right)\right\}, \quad (1)$$

with $q = \frac{\phi_t}{\phi_n}$, $r = \frac{\Delta_n^*}{\delta_n}$, where Δ_n and Δ_t denote the normal and tangential displacement jump, respectively. Δ_n^* is the value of Δ_n after the complete shear separation under the condition of zero normal tension. δ_n and δ_t are the characteristic lengths which relate to the separations for the maximum traction and the maximum shearing, respectively. ϕ_n and ϕ_t are the areas under the normal and shearing traction-separation curve, respectively.

$$\phi_n = e\sigma_{\max}\delta_n, \quad \phi_t = \sqrt{\frac{e}{2}}\tau_{\max}\delta_t. \quad (2)$$

In the case q=1, namely $\phi_n = \phi_t$, the physically realistic coupling behavior can be obtained [7]. At this time, the adjusted form of Eq. (1) is given by:

$$\phi(\delta) = e\sigma_{\max}\delta_n\left[1 - \left(1 + \frac{\Delta_n}{\delta_n}\right)\exp\left(-\frac{\Delta_n}{\delta_n}\right)\exp\left(-\frac{\Delta_t^2}{\delta_t^2}\right)\right]. \quad (3)$$

Only the normal traction and separation are active for the mode I stress field, and the expressions of the normal tractions at the interface can be obtained by differentiating Eq. (3):

$$T_n = e\sigma_{\max}\Delta_n \exp\left(-\frac{\Delta_n}{\delta_n}\right)\exp\left(-\frac{\Delta_t^2}{\delta_t^2}\right), \quad (4)$$

Δ_n and Δ_t are the variables in this relation, and the only parameters which determine the form of the traction-separation curve are δ_n, δ_t and $\sigma_{\max}$.

3. Results and Discussion

3.1 Verification of Exponential CZM

The typical mode I DCB experimental test was reported in Ref. [4]. They measured the mode I delamination fracture toughness of unidirectional T300/977-2 carbon-fiber reinforced epoxy laminated composites. In Fig. 2, the typical load-displacement curve from their experimental results is reproduced. With the same geometry and material mechanical properties, the simulated results using the adjusted exponential CZM are plotted in the same figure, so is the analytical results based on the corrected beam theory.

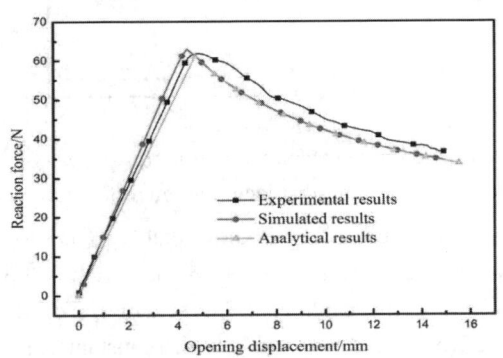

As shown in Fig. 2, in the loading process before the delamination initiation, the three load-displacement relations are all linear, and their slopes are close to each other. When they reach the peak loads, the laminated composites delamination begins to occur. In the delamination propagation, the simulated results are in good agreement with the analytical predictions, but they are both lower than the experimental results. It is necessary to point out that the maximum error is less than 7.96%, thus the adjusted exponential CZM can describe the mode I delamination of laminated composites with reasonable accuracy.

Fig.2 Comparison among the simulated, analytical and the corresponding experimental results for DCB specimen.

3.2 Effect of Exponential CZM Parameters

The responses to the cohesive parameters are assessed using the data listed in Table 1.

Table 1 Material properties for the simulation of laminated composites mode I delamination [1]

Ply	E_{11}=135.3[GPa]	$E_{22}=E_{33}=9$[GPa]	$v_{12}=v_{13}=0.24$	$v_{23}=0.46$	$G_{12}=G_{13}=5.2$[GPa]
Interfacial property	$G_{IC}=0.28$[N·mm^{-1}]		$\sigma_{max}=15$[MPa]		

Note: E is the Young's modulus, v is the Poisson's ratio, G is the shear modulus, G_{IC} is the critical energy release rate particularly.

Fig. 3(a) shows the sensitivity of the load-displacement plot to different interfacial strengths σ_{max}. When there is not an initial crack and σ_{max} increases from 15MPa to 35MPa, the loading forces which cause the delamination initiation increase significantly from 199.45N to 349.40N. As long as the steady state appears, namely the laminated composites delamination begins to propagate gradually, the fracture energy is equal to G_{IC}. At this time, there is no difference among the four load-displacement curves. By comparison, when there is a 30mm-long initial crack, the response of the load to the displacement is nearly the same though σ_{max} is changing. With the increase of the opening displacement, the loading forces increase linearly at first. When the displacement is about 2.2mm, they all reach the peak load and the delamination propagation begins. Because the capacity of material burdening reduces, the loading forces degrade along with the further opening displacements.

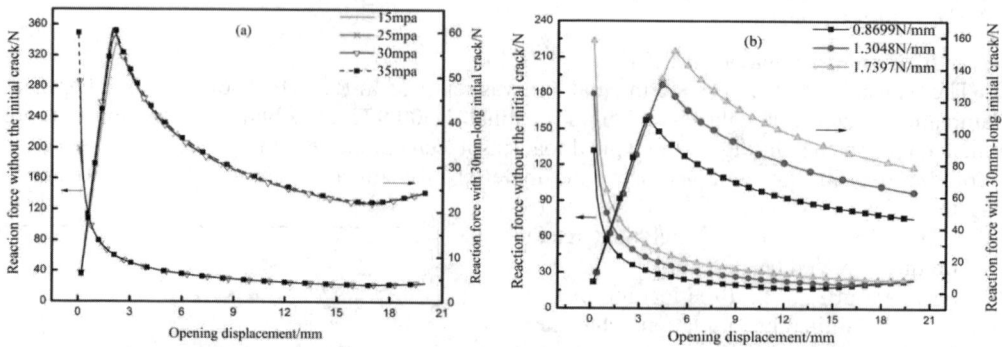

(a) different interfacial strengths, $G_{IC}=0.28N/mm$ (b) different critical energy release rates, $\sigma_{max}=15Mpa$

Fig.3 Load-displacement curves obtained from the simulation of DCB mode I fracture

From Eq. (2), we know that G_{IC} is closely connected with δ_n for laminated composites mode I delamination. When δ_n is 0.004, 0.006 and 0.008mm, G_{IC} is 0.8699, 1.3048 and 1.7397 N/mm, correspondingly. The sensitivity of the load-displacement plot to G_{IC} is illustrated in Fig.3(b). When σ_{max} is kept the constant 15MPa, the peak loads increase as G_{IC} increases. This trend is similar to the trend in the effect of σ_{max} on the peak load. Especially when there is the 30mm-long pre-crack, the initial slope of the corresponding load-displacement curve is almost constant in the elastic range, but they are different in the post-peak regions (delamination propagation) of the load-displacement curves.

4. Conclusions

The mode I delamination of laminated composites has been studied numerically. Based on the damage mechanics, the adjusted exponential CZM is used to describe mathematically the mechanical characteristics of the delamination process, overcoming the physical discontinuousness when the failure of crack tip occurs. The simulated results are in excellent agreement with the analytical solutions derived from the corrected beam theory, so the simulation method can describe the composites mode I delamination with reasonable accuracy. The results show that σ_{max} has a dominant effect on the delamination initiation, and it influences the peak load significantly if there is not a pre-crack. Once the delamination propagation begins to occur in the laminated composites, the responses of the load-displacement plots are relatively insensitive to σ_{max}, whereas G_{IC} is of critical importance. Furthermore, the peak load rises as G_{IC} and σ_{max} increase.

Acknowledgements: This work was supported by the National Key Basic Research Program of China (2010CB631102), the National Natural Science Foundation of China (50973056, 51173100), and the Natural Science Foundation of Shandong Province (JQ201016).

References:

[1] G. Alfano, M. A. Crisfield. Int. J. Numer. Meth. Eng. Vol. 50(2001), p. 1702
[2] G. S. Amrutharaj, K. Y. Lam, B. Cotterell. Theor. Appl. Fract. Mec. Vol. 24(1995),p. 58
[3] P. W. Harper, S. R. Hallett. Eng. Fract. Mech. Vol. 75(2008),p. 4779
[4] V. K. Goyal, E. R. Johnson, C. G. Davila. Compos. Struct. Vol. 65(2004),p. 295
[5] M. J. VandenBosch, P. J. G. Schreurs, M. G. D. Geers. Eng. Fract. Mech. Vol. 73(2006),p. 1223
[6] M. Kulkarni, D. Carnahan, K. Kulkarni, D. Qian. Compos. Part B-Eng. Vol. 41(2010),p. 416
[7] X-P Xu, A. Needleman. Model. Simul. Mater. Sc. Vol. 1(1993),p. 120

Extended Finite Element Method for Fracture Mechanics and Mesh Refinement Controlled by Density Function

Yi Cen

Mechatronics Department, Guangdong Industry Technical College, Guangzhou, P.R.China

cydavid@163.com

Keywords: Extended finite element method; Mesh Refinement; Density Function.

Abstract. This paper discusses the combination of element enrichment by mesh refinement controlled by density function with the extended finite element method and its application in fracture mechanics. Extended finite element method (XFEM) is an effective numerical method for solving discontinuity problems in the finite element work frame. A numerical example of fracture mechanics is analyzed at the end of this paper to show the application of the above method.

1. Introduction

A great number of structural accidents are caused by various internal cracks. The existence and propagation of these cracks deteriorate the strength and the security of the structures. Thus in order to facilitate structural design and simulation, it is significant to study the generation and propagation of the cracks. This paper introduced a method which allows us to achieve an accurate modeling result of the mode I crack during the loading process by applying the enrichment of the shape function and mesh refinement controlled by density function in the vicinity of the crack tip. A numerical example is presented at the end of the paper to show the application of this method

2. Extended finite element method (XFEM)

XFEM is based on partition of unity method where special functions being adopted to describe the field behaviors are incorporated in finite element spaces to reflect the features of interest. The XFEM begins with standard finite element procedures. In standard finite element, the approximation reads:

$$u(x) = \sum_{i=1}^{n} N_i u_i . \qquad (1)$$

Where N_i is the shape function of node i, u is the vector of degree of freedom of node i. The above approximation is only suitable for problems with continuous medium, but not for problems with cracks. In order to obtain solutions accurately, an interpolation technique which is more suitable for discontinuous problems base on partition of unity method is introduced.

$$u(x) = \sum_{i \in m} N_i(x) u_i + \sum_{j \in m_d} N_j H(x) a_j + \sum_{k \in m_t} N_k \sum_{a=1}^{4} F_a(x) b_k^a . \qquad (2)$$

Where m is the set of regular nodes, m_d and m_t are the sets of enriched nodes u_i represents the degree of freedom of all nodes, and a_j and b_k are added degree freedom of enriched nodes. $H(x)$ is the jump function used to reflect the discontinuity:

$$H(x) = \begin{cases} 1 & \text{Nodes above the crack} \\ -1 & \text{Nodes under the crack} \end{cases} \quad (3)$$

$F_a(x)$ is near tip enrichment function. It reflects the stress singularity of crack tip. The near tip enrichment function can be defined in terms of local crack tip coordinate system (r,θ).

$$F_a(r,\theta) = \left\{ \sqrt{r}\sin\frac{\theta}{2}, \sqrt{r}\cos\frac{\theta}{2}, \sqrt{r}\sin\theta\sin\frac{\theta}{2}, \sqrt{r}\sin\theta\cos\frac{\theta}{2} \right\} \quad (4)$$

So the variational form of boundary value problem is:

$$\int_\Omega \sigma \cdot \delta\varepsilon d\Omega = \int_\Omega f^b \cdot \delta u d\Omega + \int_\Gamma f^t \cdot \delta u d\Gamma \quad (5)$$

3. Mesh refinement technique for XFEM

In order to improve the solution accuracy by using XFEM around important discontinuities and smoothness irregularities in fracture mechanics analysis, element refinement is used. A general uniform refinement over all the elements may cause inaccurate results around the crack tip. Instead of using uniform refinement, refinement controlled by density function can improve the solution accuracy around the crack tips.

A Density Function $\overline{f}$ over an open, connected set $\Omega \subseteq R^n$, for some positive integer n, is an almost everywhere positive possibly unbounded function in $C_{fin}^\infty(\Omega)$, satisfying

$$\int_\Omega \overline{f}(x)dx = 1 \quad (6)$$

Where x is the distance between certain point in the domain and the crack tip(singularity). The family of such functions will be denoted by $\overline{f}(\Omega)$. Any nonnegative function $f:[a,b] \to R_+$ that has only countable many zeros can be normalized as:

$$\overline{f}(x) = \frac{f(x)}{\int_a^b f(\tau)d\tau} \quad (7)$$

The corresponding cumulative distribution function $F:[a,b] \to [0,1]$ is defined by

$$F(x) = \int_a^x \overline{f}(\tau)d\tau \quad (8)$$

The mesh refinement controlled by density function reduces the displacement error around the in the vicinity of the crack tip. The elements in other region can achieve accurate result even with initial coarse mesh.

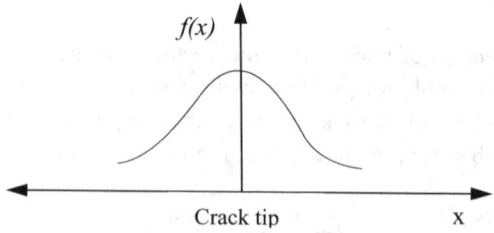

Figure.1 Density function

And the maximum element size $h(x)$ is the inverse of the density function.

4. Numerical Example

A thin plate is subject to an evenly distributed load on its edge. The dimension of the plate is shown in fig.2 (a). The upper edge of the plate is subject to a uniform shell edge load of σ =1000N/m while the bottom edge is fixed.

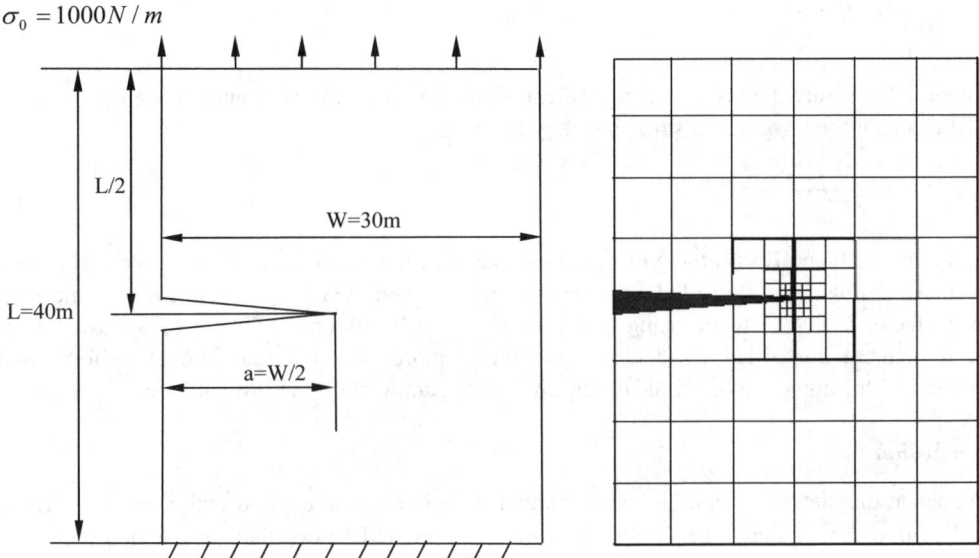

Figure.2 Plate geometry and mesh refinement. a) Dimension of the plate specimen. b) Mesh refinement controlled by density function

We then compute the stress intensity factor of the model problem with mesh refinement controlled by density function and mesh with global refinement. The two results are compared with the exact value. The density function specified in this problem is $\overline{f} = 0.1x^{-0.9}$, where x is the distance from certain point on the plate to the crack tip.

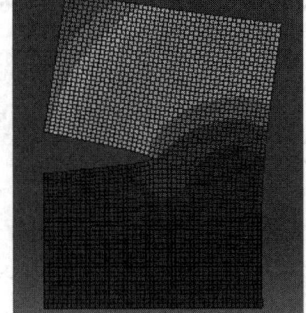

Figure.3 Comparison of Refinement schemes and solution results a) Refinement controlled by density function. b) Global uniform mesh refinement

For finite tensile plate, the exact stress intensity factor KI is calculated using:

$$KI = \left[1.12 - 0.23\left(\frac{a}{2w}\right) + 10.56\left(\frac{a}{2w}\right)^2 - 21.74\left(\frac{a}{2w}\right)^3 + 30.42\left(\frac{a}{2w}\right)^4\right]\frac{\sigma\sqrt{\pi a}}{2} \quad (9)$$

The Mode I Stress intensity is computed by using the displacement extrapolation method, which can obtain ideal accuracy in this problem.

$$KI = \mu\sqrt{\frac{2\pi}{r}}\frac{u_y^a - u_y^b}{2(1-v)} \quad (10)$$

Where u_y^a, u_y^b are the vertical displacement of two points on the plate and $\mu = E/2(1+v)$
The XFEM solution error of the stress intensity factor is:

$$error = \frac{KI_{num} - KI_{exact}}{KI_{exact}} \quad (11)$$

Combing Eq.10 and Eq.11, the XFEM solution error of the stress intensity factor is: 2.11% while using mesh refinement controlled by density function, and XFEM solution error of the stress intensity factor is 8.97% while using global uniform mesh refinement. The results show that the mesh refinement controlled by density function is more affective than global uniform mesh refinement in this numerical example to improve the accuracy of the XFEM solution.

5. Conclusion

In this paper, the method of combining the extended finite element method with a mesh refinement technique is demonstrated. The numerical example presented shows that the method achieves a more accurate result comparing to the global uniform mesh refinement. On the other hand, the computational cost of this method is acceptable. Further research will focus on the calculation of the density function in order to achieve better simulation performances.

Reference

[1] N. Moës, J. Dolbow and T. Belytschko: A finite element method for crack growth without remeshing, Int. J. Numer. Meth. Engng., Vol.46 (1999), pp. 131-150.

[2] Soheil Mohammadi: Extended Finite Element Method for Fracture Analysis of Structures (Blackwell, 2008).

[3] R. M. Dreizler, E. K. U. Gross: Density Functional Theory (Springer, Berlin 1990).

[4] S. C. Brenner, L. R. Scott: The Mathematical Theory of Finite Element Methods (Springer, New York 2002).

[5] G. Dhatt, G. Touzot: The Finite Element Method Displayed, (J. Wiley & Sons, New York 1984).

[6] T. Beck: Real-space Mesh Techniques in Density-Functional Theory, Rev. Mod. Phys. 72 (4) (2000) pp. 1041–1080.

[7] J, Hugger: The Theory of Density Representation of Finite Element Meshes. Examples of Density Operators with Quadrilateral Elements in the Mapped Domain, Computer Methods in Applied Mechanics and Engineering [JIF=1.252] 109 (1993), pp.17-39.

Effects of Surface Roughness on the Fatigue Life of Alloy Steel

W.L. Xiao[1, a], H.B. Chen[1, b], Y. Yin[1, c]

[1]CAS Key Laboratory Mechanical Behavior and Design of Materials, Department of Modern Mechanics, University of Science and Technology of China, Hefei, Anhui 230026, China

[a]wlxiao@mail.ustc.edu.cn, [b]hbchen@ustc.edu.cn, [c]yinyan@mail.ustc.edu.cn

Keywords: surface roughness, fatigue life, power function curve

Abstract: Surface roughness characterizes the micro-geometric appearance variation and significantly affects the fatigue properties of machined specimen. Low cycle fatigue (LCF) tests of alloy steel specimens with different surface roughness were carried out in this paper to investigate the effect of roughness on fatigue life. The dumbbell plate specimens were tested in uniaxial stress-controlled mode on the hydraulic servo machine at room temperature. Obviously discrepant lifetime results corresponding to different surface roughness implied that the greater the roughness was, the lower the fatigue life was. An approximate power function relationship between the roughness and the fatigue life was established through the fitting of the experimental data.

Introduction

Surface roughness, which is an unavoidable phenomenon at machining, is usually strictly required when the processed materials are applied in structural components subjected to cyclic loads. The reason is that the fatigue life of structures is highly dependent on the surface quality. Several research works with respect to the influence of machined surface roughness on the fatigue life have been published in the past decade. Early studies [1] used empirical reduction factors which modified the endurance limit of the material to evaluate the effects. Sigmund and Bjørn [2] and Suraratchai, et al. [3] simulated the stress state of micro surface profile and established a numerical method for fatigue life prediction of components with rough surfaces using the finite element method. Smaga and Eifler [4] investigated the effects of the residual stress induced from surface roughness. However, those researches do not work for low cycle fatigue and the present work is to seek a quantitative relationship between the roughness and the low cycle fatigue life through experiment test.

Material and experimental setup

The material investigated in this paper is alloy steel. Its yield strength is about 420MPa obtained from monotonic tensile test. In the fatigue test, dumbbell plate specimens were used, with a middle parallel length of 14mm and a thickness of 4mm, as show in Fig. 1. The aim using this configuration was to avoid unstability in the fatigue process and conveniently measure the surface roughness. Several group specimens with different surface roughness were created firstly by means of different machining method. Then uniaxial stress-controlled fatigue tests were carried out at room temperature with a stress amplitude of 500MPa and a load ratio of R=-1 using a frequency of 1Hz. The load and displacement change with time and life cycles were recorded by the sensors embedded in the hydraulic servo material test machine, MTS809.

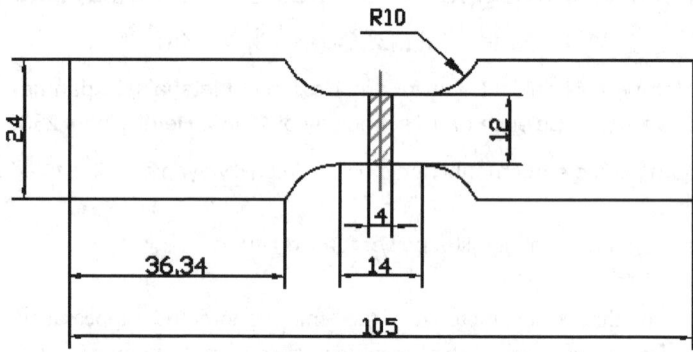

Fig. 1 Configuration of the dumbbell plate specimen

Results

Surface roughness is usually characterized through average geometric parameter named R_a. Thus in this investigation, the values of R_a of the specimens were measured firstly. Based on the measurement results, 34 specimens were divided into 10 groups with each group including 3 or 4 specimens, and each group corresponded to a roughness grade, from minimum 0.1μm to maximum 1.7μm, as listed in the first row of Table 1.

The second row in Table 1 presents the experimental fatigue lives which are the average values of the fatigue lives of 3 or 4 specimens in each group. Obviously, the fatigue lives are less than 10,000, showing that the failure mode of these tests belongs to low cycle fatigue. On the whole, there is a trend that the greater the roughness is, the lower the fatigue life is, although some individual data is biased because of intrinsic discreteness for the fatigue test. In other words, the life has an inverse proportionality with the surface roughness. This phenomenon can be theoretically explained by a concept called the stress concentration factor [5]. It assumes the uneven geometric profile as a lot of side-by-side micro notches lying on the surface of machined specimen. Stress concentration initiates from the notches. Then the stress concentration factor can be defined as

$$K_t = 1 + 2\sqrt{\gamma \frac{R_z}{\rho}} \tag{1}$$

where K_t denotes the stress concentration factor, γ refers to the ratio between spacing and depth of the asperities, which approximately equals to 1, R_z is the 10-point roughness, ρ is the effective profile valley radius of the surface texture. Generally speaking, the machined surface becomes rougher, which will lead to R_z larger and ρ smaller. So it is easy to conclude that K_t has an increscent trend. Then the stress amplitude $\Delta\sigma$ in the notch is the nominal stress amplitude Δe multiplied by K_t, as expressed in Equation (2).

$$\Delta\sigma = K_t \cdot \Delta e \tag{2}$$

Attention that K_t is always greater than 1. Based on the theory of local stress-strain method, fatigue life depends on the local stress-strain field. And a greater local stress-strain leads to a lower fatigue life. In a word, fatigue life is reducing as the surface roughness is increasing under the premise of the same load condition.

Table 1 Experimental fatigue life results corresponding to each roughness grade

Roughness, R_a [μm]	0.1	0.3	0.5	0.7	0.9	1.0	1.2	1.4	1.6	1.7
Fatigue life, N_f [cycle]	3375	2856	2751	2748	2525	2555	2617	2648	2415	2185

The theory of stress concentration factor presents qualitative explanation to the phenomenon that the fatigue life is inversely proportional to the roughness. However, little attention has been paid to the quantitative relationship between them. This research tried to find the character of this kind of relationship through the fatigue experiment and the preliminary results were shown in Fig. 2. The horizontal and vertical axes denote the roughness R_a and the fatigue life N_f, respectively. Ten group data, displayed by the solid circle dots, scatter in the figure. A power function curve was used to fit these data. As a result, the curve is expressed by Equation (3).

$$N_f = 2559 \cdot R_a^{-0.1166} \tag{3}$$

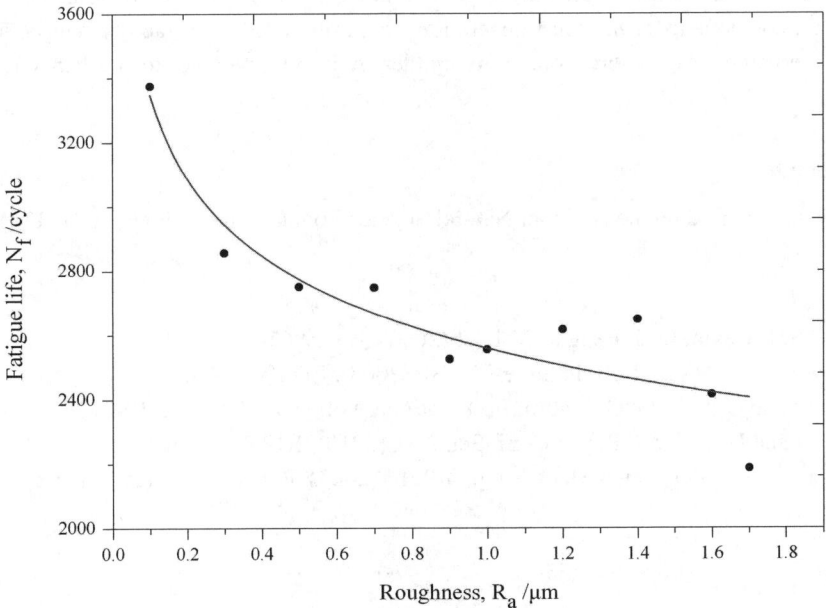

Fig. 2 Relationship between the roughness and the fatigue life

It can be seen that the curve well reflects the relationship between two variables. The result provides some reference significance for engineering application, because the roughness of most structures or components belongs to the range from 0.2μm to 1.6μm.

However, we should stress that this work is our preliminary research on the influence of surface roughness on specimen's low cycle fatigue life and some further works are needed to be done to confirm or improve our preliminary judgments. The first one is the material. Equation (3) is derived from the fatigue process of the material of alloy steel with yield strength 420MPa. Whether other class steels or non-steels follow this rule? The second one is the roughness grade, just from 0.1μm to 1.7μm, not covering wider range of roughness grades. The third one is the most important. The material is subjected to a high level cyclic load, whose peak value exceeds the yield strength. The material accumulates certain amount plastic deformation in the fatigue process and finally fractures after thousands of load cycles. The power function curve implies a specific change trend of fatigue life along with surface roughness in the low cycle fatigue region at room temperature. In our experiment, only a stress amplitude was used. Whether the trend varies if the stress amplitude is replaced by other level stresses but still keeping not beyond the low cycle fatigue region? We imagine that the curve may be more abrupt for a higher level load amplitude and more gently for a lower level load amplitude.

Conclusions

This research tried to investigate the effects of surface roughness on the low cycle fatigue life. Experiments with different surface roughness specimens were carried out at room temperature. Based on the concept of stress concentration factor, qualitative explanation has been presented according to the experimental results. An approximate power function curve was fitted using the experimental data. The results show that there is a certain quantitative relationship between the surface roughness and the low cycle fatigue life. Another potential material that has being used in high heat load components at the third generation synchrotron radiation facility will be studied in our further research. And more roughness grades will be involved to further confirm our conclusions.

Acknowledgment

This work was supported by the National Natural Science Foundation of China (10975130).

References

[1] A. Scott, S. Huseyin, In. J. Fatigue, Vol. 22 (2000), P. 619-630.
[2] K.A. Sigmund, S. Bjørn, In. J. Fatigue, Vol. 30 (2008), P. 2200-2209.
[3] M. Suraratchai, J. Limido, C. Mabru, In. J. Fatigue, Vol. 30 (2008), P. 2119-2126.
[4] M. Smaga and D. Eifler, J. Phys.: Conf. Ser. 240, (2010), IOP Publishing.
[5] D.C. Zhang, X.M. Pei, China Mech. Eng., Vol. 14 (2003), P. 1374-1377 (in Chinese).

Correlation between Alternating Temperature Accelerated Aging and Real World Storage of Composite Propellant

DING Biao[1,a], SHI Pei[2,b], QIU Xin[1,c]

[1]Dept. of Airborne Vehicle Engineering, Naval Aeronautical and Astronautical University, Yantai 264001, China;

[2]Dept. of training, Naval Aeronautical and Astronautical University, Yantai 264001, China;

[a]ding_biao@yahoo.com.cn, [b]radartwo@126.com, [c]854103537@qq.com

Keywords: alternating temperature; real world storage; correlation; elongation

Abstract: Alternating temperature accelerated aging test was designed to ensure that accelerated aging could preferably simulate real world storage of HTPB composite propellant. Mechanical properties of HTPB propellant aged for three different periods were measured and analyzed. The results indicate that, the tensile strength increased and the elongation decreased after accelerated aging, showing the same trend of real world storage; the mechanical properties of HTPB propellant is strongly influenced by the rate of temperature change; alternating temperature accelerated aging test is consistent well with real world storage. The results can be very helpful for solid rocket motor life prediction.

1. Introduction

The aging of composite propellant grains is very complex. During storage, the propellant grains are affected by environmental temperature changes that produce both variable mechanical properties and cyclic thermal stresses. Because of the cyclic thermal stresses, the cumulative damage may occur in micromechanical structure, leading to the macroscopical cracks and failure of grain and motor in the end[1-4].

In this paper, the correlation between alternating temperature accelerated aging and real world storage of composite propellant is mainly researched, which can provide reference for SRM life prediction[5,6].

2. Aging Test

2.1 Sample

The propellant used in the test is the HTPB propellant, stored for 5 years in real world. In order to simulate the motor case, the test fixture was designed (Fig. 1). During testing, the propellant sample was bonded to the faces of the fixture, fixed by four bolts and two locating rods with no stress.

Fig. 1 Test fixture

2.2 Test procedure

(1) Put the fixture with bonded propellant into the seal box. The size of propellant is 14cm×3cm×15cm. Then the seal box is subjected to cyclic temperature environment.

(2) Temperature range: -10℃~60℃.

(3) Three groups were designed to different rate-temperature. The rate-temperature respecti- vely is 10℃/12h, 20℃/12h and 30℃/12h。

(4) The aging time are 13weeks(91d). The sampling time respectively are 5th week, 8th week, and 13th week. In each sampling time, mechanical properties of propellant were measured.

2.3 Test results

The test data processing is according to the composite propellant uniaxial tensile test data processing methods. The data of tensile strength and elongation are shown in table 1. The test temperature is 20℃. The rate of extension is 100mm/min.

Table 1　Uniaxial tensile test data(20℃, 100mm/min)

time	group1		group2		group3	
(d)	σ_m [MPa]	ε_m [%]	σ_m [MPa]	ε_m [%]	σ_m [MPa]	ε_m [%]
0	0.416	75.98	0.524	68.50	0.873	67.21
35	0.587	72.89	0.711	65.29	0.879	63.73
56	0.729	72.16	0.752	62.89	0.909	58.54
91	0.762	69.98	0.667	60.14	1.044	55.14

2.4 Data processing

(1) Aging mathematical model

The data processing of composite propellant aging test usually use 3 mathematical models as follows.

Model 1: $P = P_0 + K \lg t$ （1）

Model 2: $P = P_0 + Kt$ （2）

Model 3: $P = P_0 e^{-Kt}$ （3）

Where, P, property in specific time; P_0, constant; K, rate constant, only a function of temperature; t, aging time.

(2) Calculation method

Under each aging temperature condition, an group data can be obtained between aging time t and property P: t_i, P_i, $i = 1,2,\cdots,n$.

For model 1, assuming that $X = \lg t$, $Y = P$, $a = P_0$ and $b = K$. Then Eq.1 can be expressed by $Y = a + bX$.

For model 2, assuming that $X = t$, $Y = P$, $a = P_0$ and $b = K$. Then Eq.2 can be expressed by $Y = a + bX$.

For model 3, Eq. 3 can change to $\ln P = \ln P_0 - Kt$ by logarithmic transformation, assuming that $X = t, Y = \ln P, a = \ln P_0, b = -K$. Then Eq.3 can be expressed by $Y = a + bX$.

The coefficient a, b and correlation coefficient r can be obtained by least-aquares procedure.

2.5 Test result analysis

(1) Table 1 shows that the tensile strength increased and the elongation decreased after accelerated aging, behaving the same trend of real world storage.

(2) From the view of storage condition and aging mechanics, the reason why the tensile strength increased and the elongation decreased is that chemical reaction creates new hydroxyl group in molecule, raising the amount of functional group[7].Cross link reactions make tensile strength increased and the elongation decreased.; Because of propellant aging in seal box, the moisture in propellant can create hydrolytic chain cleavage of binder matrix, which makes the propellant soft especially at elevated temperature.

(3) The mechanical properties of composite propellant is strongly influenced by the rate-temperature. Table 1 shows that the mechanical properties change slowly in low rate-temperature (3-8 days one cycle). When rate-temperature raised (1-3 days one cycle), the mechanical properties change obviously.

(4) Due to the change trend of elongation is more obvious than that of tensile strength, The elongation is taken as research target. Fitting the data of 3 groups by twos. The coefficient of Eqs $Y = a + bX$ can be obtained by east-aquares procedure, shown in table 2. The coefficient b indicates the relationship of the rate of elongation between different groups.

Table 2 Fitting coefficient

coefficient	a	b	ρ_{XY}
group1 and 2	-38.43	1.411	0.9852
group2 and 3	-34.94	1.498	0.9915
group1 and 3	-88.98	2.064	0.9547

3. Correlation between Accelerated Aging and Real World Storage

3.1 Correlation of temperature load

Since the load environment is different between aging and real world storage after all, the determination of range and rate of temperature plays an important role in accuracy of test result. If the range of temperature is too large, it's prone to change the chemical aging mechanism, on the contrary, it will waste too much time. If the rate-temperature is too great, it's hard to achieve thermal balance, on the contrary, it will waste too much time also.

In order to research on correlation between accelerated aging and real world storage, a typically group temperature data of a depot in Nanhai and a group test data are used for a comparative study. Table 3 shows the Monthly mean temperature data of Nanhai depot from 1981 to 2000.

Table 3 Monthly mean temperature of Nanhai depot

month	1	2	3	4	5	6
mean temp (°C)	15.8	16.4	18.5	22.7	24.2	25.9
month	7	8	9	10	11	12
mean temp (°C)	28.3	29.1	27.9	25.3	19.8	17.8

In order to make sure the practical date and test data are comparable, the time unit is dealt with nondimensionalization. And the two groups data were selected by one-to-one on purpose no to change the trend of temperature. Fig.2 and Fig.3 show the temperature curve respectively.

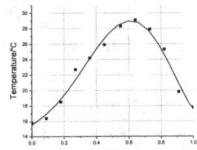 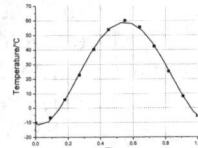

Fig.2 Temperature curve   Fig.3 Test temperature curve of Nanhai depot

3.2 Correlation of mechanical property

The mechanical property of solid rocket motor will degenerate even end in failure during long term storage. The plan of long term monitor of a specific solid rocket motor has been made for many years. A series of data has been obtained through dissecting the motor stored in different period. Table 4 shows the data of tensile strength and elongation obtained through dissecting. The test temperature is at 20°C. The tension rate is 100mm/min.

Table 4 Mechanical properties of propellant

age	6.5	9	10	12
ε_m [%]	32.22	27.54	23.3	18.37
σ_m [MPa]	0.83	0.76	0.75	0.79

It can be seen that the elongation decreased obviously, but the tensile strength changes erratically. The rate of elongation is obtained through data fitting to analyse the correlation between accelerating aging and natural aging. In order to make sure the practical date and test data are comparable, the time unit is dealt with nondimensionalization.

The coefficient of Eqs. $Y = a + bX$ can be obtained by east-aquares procedure, shown in table 5. The coefficient b indicates the relationship of the rate of elongation between two aging conditions.

Table 5 Fitting coefficient

coefficient	a	b	ρ_{XY}
natural aging	49.37	-2.56	-0.9903
aging group 1	75.68	-23.49	-0.9888
aging group 2	68.44	-33.94	-0.9979
aging group 3	67.45	-50.53	-0.9840

It can be seen that the rate of elongation is different at different rate-temperature. The higher is rate-temperature, the larger is rate elongation. According to the relationship of rate elongation, the rate elongation under three aging conditions are higher 9.2 times, 13.3 times, and 19.7 times than that of natural aging respectively. It means that cyclic temperature accelerated aging one day equals to natural aging 9.2days, 13.3days, 19.7days, respectively.

4. Conclusions

(1) The mechanical properties of composite propellant in alternating temperature accelerated aging process represent the same trend of real world storage.
(2) The mechanical properties of composite propellant is strongly influenced by the rate-temperature in accelerated aging process.
(3) Alternating Temperature accelerated aging test is consistent well with real world storage. It can be helpful for solid rocket motor life prediction as a new method.

References

[1] XU Xin-qi,YUAN Shu-sheng,SUI Yu-tang. Stress-strain analysis of propellant grains in storage[J].*Journal of Naval Aeronautical and Astronautical University*,2002,17(3): 313-317.
[2] D Leveque,A Schieffer,A Mavel,et al *Composites Science and Technology* ,2005,65:395-401.
[3] DONG Ke-hai. The evaluation method of remaining-life for charge of solid rocket motor based on cumulative damage[D].Yantai: Naval Aeronautical and Astronautical University,2007.
[4] ZHANG Xing-gao. Study on the Aging Properties and Storage Life Prediction of HTPB Propellant[D]. Changsha: National University of Defense Technology,2009.
[5] Janajreh I, Heller R, Thangitham S Journal of Spacecraft and Rockets. 1994,31(6):1072-1078
[6] Derbalian G, Thomas J M. Probabilistic environmental model for solid rocket motor life prediction[R], 1982, ADA117651
[7] WANG Chun-hua,PENG Wang-da,WENG Wu-jun,et al. Chemical aging mechanisms of HTPB solid propellants and the ways to improve aging-resistance[J].*Chinese Journal of Energetic Materials(Hanneng Cailiao)*,1996,4（3）: 109-116.

Experimental research on fracture toughness of high grade line pipe

Xiong Qingren[1,2a], Zhang Jianxun[1,b], Feng Yaorong[2,c], Wang Haitao[2,d], Ma Qiurong[1,2,e], Xu Zhenzhen[1,f]

1) Xi'an Jiao Tong University, Xi'an, China, 710049
2) CNPC Tubular goods research institute, Xi'an, China, 710065

[a]xiongqr@cnpc.com.cn, [b]jxzhang@mail.xjtu.edu.cn, [c]fengyr@cnpc.com.cn,
[d]wanghaitao008@cnpc.com.cn, [e]qiurongma@cnpc.com.cn, [f]badfish0440@stu.xjtu.edu.cn

Keywords: high grade line pipe; fracture toughness; CTOD test

Abstract: Crack tip opening displacement(CTOD) tests were carried out on different positions of X70, X80 and X100 pipe by three-point bend testing method. The results show that the crack propagation resistance and the tear modulus of base metal, heat affected zone(HAZ), and weld of X70 and X80 pipe are larger than that of X100 pipe. At the same time, The initial CTOD values and the CTOD values withmaximum load of X70 and X80 pipe are larger than that of X100 pipe. The test results suggest that the fracture toughness, crack initiation and propagation resistance of theX70 and X80 pipe are greater than that of X100 pipe, and the mechanismis analyzed finally.

Nowadays X70 pipe has been used widely, and X80 pipe has been put into service already. As to the higher grade linepipe, X100 pipe has been used for trial segment successfully, while X120 pipe has been conducted small scale trial production[1-6]. This article is based on CTOD tests on the base metal, the weld and HAZ of X70, X80 and X100 pipe by three-point bend testing method. Then the fracture toughness is compared among different positions of different grade line pipe to facilitate the engineering application and the research on the high grade linepipe.

1 Experimental procedure

Table 1 shows the size of specimens and the temperature during test. Table 2 and 3 shows the chemical composition and mechanical property of steels separately. The crack of the weld is on the middle of the weld; The crack of the base metal is on the middle of the specimen; The crack of the HAZ is at the intersection point of fusion line and the edge of specimen which is perpendicular to the fusion line, as shown in Fig.1.

The equipment for fatigue precracking is PLS-100 Electro-Hydraulic Servo Testing Machine. And the CTOD of three-piont bend testing was carried out on the CSS-88100 Testing Machine. The span of specimens was 96mm. The test was based on the ASTM E1290-2007 standard[7]. The test began with the loading until the displacement reached the preset value. The loading rate is 1mm/s. After the test, choose the valid data without instability, then fit resistance curvse, and judg the effectiveness of the characteristic values of CTOD finally, such as $\delta_{0.05}$, δ_i, δ_c and δ_u.

2 Results and discussion
2.1 Resistance curves of crack propogation

Fig.2 shows the resistance curves of base metal, HAZ and weld of X80 pipe at different temperatures. Fig.3 and Fig.4 show the crack propagation resistance $tg\alpha$ and tear modulus $T_{\delta R}$ on different positions of X70, X80 and X100 pipe. There is no available data of the crack propagation resistance and tear modulus at the HAZ of X100 pipe when the experimental temperature is -10°C and -30°C.

It can be seen from the results that in -5°C, the crack propagation resistance and tear modulus of X70 pipe are larger than that of X100 pipe, while the fracture toughness is almost the same between X70 and X80 pipe, which indicates that X70 and X80 pipe are of better roughness and higher ability of crack propagation resistance than X100 pipe. The distributing rule of crack propagation resistance and tear modulus is the same in X70, X80 and X100 pipe, which is base metal> HAZ> weld.

Table 1 Specimens and experimental temperature

Grade	Size /mm	Temperature/°C	B /mm	W /mm
X70	1016×21	-5°C	12	24
X80	1219 ×18.4	Room temperature, -10°C, -30°C	12	24
X100	1016×16	Room temperature, -10°C, -30°C	11	22

Table 2 Chemical composition

Grade	C	Si	Mn	Cr	Mo	Ni	Nb+V+Ti
X70	0.048	0.10	1.55	0.083	0.20	0.19	0.072
X80	0.051	0.26	1.83	0.043	0.30	0.30	0.09
X100	0.059	0.23	1.96	0.53	0.0036	0.39	0.043

Table 3 Mechanical property

Grade	Tensile property (at room temperature)			Charp impact energy (−10°C)	
	R_m /MPa	$R_{t0.5}$ /MPa	A / %	Akv / J	SA / %
X70	625	547	26	449	100
	622	558	26	441	100
	625	553	27	435	100
X80	713	565	25	345	100
	715	561	26	340	100
	709	562	26	338	100
X100	873	834	20	250	98
	878	843	17	252	98
	880	840	21	255	100

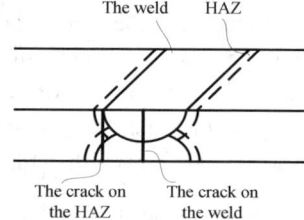

Fig.1 Position of cracks

The data for the HAZ of X70 pipe is decentralized because of the instability. Some of the specimens for the HAZ of X100 pipe are instable at room temperature, while all specimens for the HAZ of X100 pipe are instable at -10°C and -30 °C. It indicates that the increase of plastic deformation is not enough to offset the stress concentration at the crack tip after the crack initiation on the HAZ of X100 pipe. The crack opening displacement and saturated critical value of passivated zone are so low that the crack propagation is instable with small ranges.

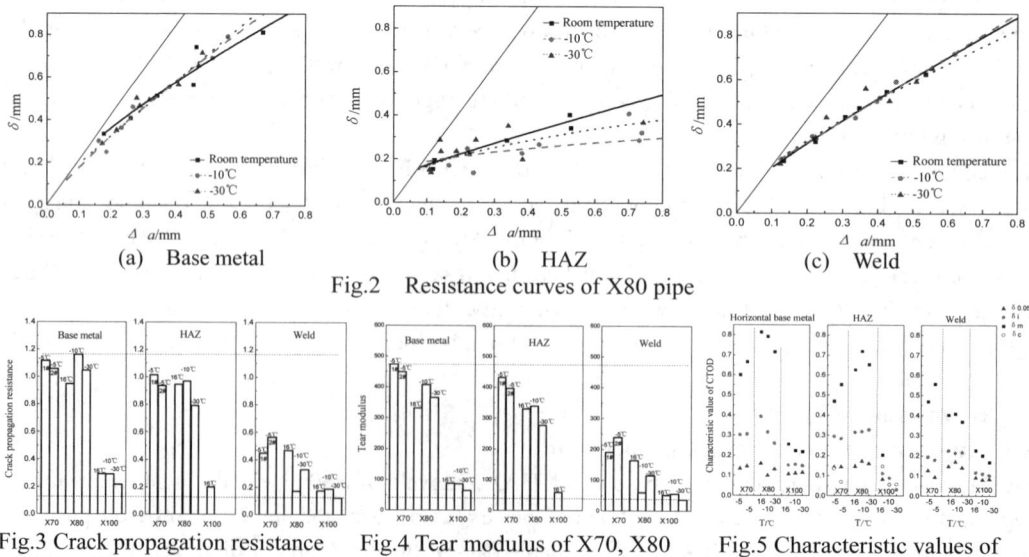

Fig.2 Resistance curves of X80 pipe
(a) Base metal   (b) HAZ   (c) Weld

Fig.3 Crack propagation resistance of X70, X80 and X100 pipe

Fig.4 Tear modulus of X70, X80 and X100 pipe

Fig.5 Characteristic values of CTOD of X70, X80 and X100 pipe

2.2 Characteristic values of CTOD

The characteristic values of CTOD on different positions of X70, X80 and X100 pipe are shown in Fig.5. It was indicated that δ_m of X70 and X80 pipe is larger than that of X100 pipe, so it is of better roughness and higher ability of crack propagation resistance. And δ_i of X70 and X80 pipe are also larger than that of X100 pipe. So X70 and X80 are of higher ability of crack initial resistance as δ_i represents the tolerance limit of crack initial. $\delta_{0.05}$ of X70 and X80 pipe are slightly larger than that of X100 pipe. In short, the ability of crack initial and propagation resistance of X70 and X80 pipe are higher than that of X100 pipe. And the fracture toughness of X100 pipe is lower.

As to the X100 pipe at a temperature, $\delta_{0.05}$, δ_i and δ_m of the base metal is the largest, and that of the HAZ is the smallest. As to the specimens of HAZ of X100 pipe, some of them are instable at room temperature; all of them are instable at -10°C and -30°C; and the data of $\delta_{0.05}$ and δ_m is missing at -10°C and -30°C. The value of stable crack propagation is less than 0.2mm when the specimen is instable. In this situation, the value of CTOD is δ_c. δ_c of HAZ at low temperature is less than that at room temperature, which means it is easy to be instable at low temperature. In short, the toughness of HAZ of X100 pipe is low, especially at low temperature.

2.3 Relevant questions

(1) Fracture toughness of high grade linepipe

The high value of CTOD and good toughness of X80 pipe are closely relevant to the composition, structure, welding parameter and welding material. The specimens of HAZ is stable and its toughness is high at low temperature. Although the value of CTOD and toughness of HAZ of X80 pipe are lower than that of the base metal and the weld, it is still higher than that of the other two kinds of steels. The toughness is not changed with temperature, which indicates that X80 pipe is of greater toughness and higher low temperature toughness.The CTOD value of X70 are close to X80 pipe other than sevaral HAZ specimens of X70 occur instability phynomenon.

Compared with X70 and X80 pipe, the fracture toughness of X100 is lower. It indicates that the toughness of the pipe suffer loss with the increase of strength of the steels.

(2) Fracture toughness of the HAZ

The HAZ specimen of high grade linepipe is prone to occurring instability phynomenon, and the results of HAZ specimens is scattered.

The microstructure of X70 pipe consists of bainite, polygonal ferrite, M-A constituent and pearlite, etal, and its grains are fine. The crack initiate on base metal. The base metal has high toughness so that the passivated zone forming from the crack slide occurs plastic deformation, and then steps are formed. The crack stable propagation zone follows the passivated zone, which is on the fine grained HAZ. When the crack extends to coarse grained HAZ, it occurs instability phynomenon because of the low toughness of the coarse grained HAZ.

Deng et al. analyzed the relationship between decentralize of the CTOD value of HAZ and the position of crack tip[8]. The crack tip on the fine grained region brings larger value of CTOD, while the crack on the coarse grained region brings smaller value of CTOD, which makes the shape of P-V curves different. There are only three test pieces are instable in total 12 test pieces of HAZ, which means the position of crack tip on HAZ has a great effection on fracture toughness.

The microstructures of crack initial zone, stable crack propagation zone and instable crack propagation zone in the instable HAZ specimens of X100 pipe are shown in Fig.6.

Fig.6 shows that there are lots of polygonal ferrite near the fracture of instable HAZ specimen. Liu illustrates that the granular bainite is reinforcements in X100 pipe[9]. It could increase the strength of steel, but suitable amount of ferrite with small grain size and dispersive distribution could increase toughness. However, the amount of ferrite in HAZ specimens of X100 pipe is too much and the ferrite is polygonal. Such ferrite is helpless in increasing HAZ toughnss of X100 pipe, and its restriction to the deformation is too low. As a result, the crack formation energy and the fracture toughness decreases, especially at low temperature.

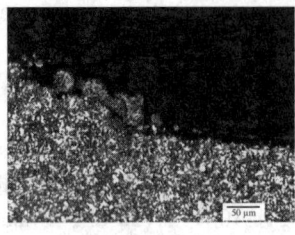 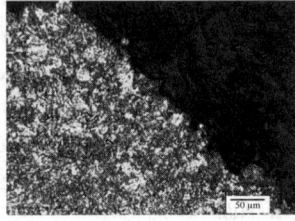 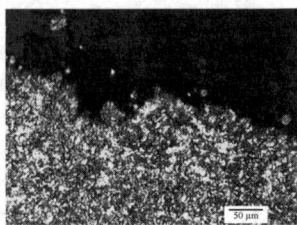

(a) Initial zone　　　　(b) Stable propagation zone　　　(c) Instable propagation zone
Fig.6　Microstructure of HAZ of X100 pipe

3　Conclusion

(1) The ability of crack initial and propagation resistance of X70 and X80 pipe are higher than that of X100 pipe, and the fracture toughness of X100 pipe is lower.

(2) The HAZ specimen of high grade linepipe is prone to occurring instability phynomenon. It is relevant to the positions of crack tip and the microstructure of HAZ. The reason of instability and low fracture toughness of X100 pipe is that there are lots of polygonal ferrite and little reinforcements like granular bainite.

References
[1] HILLENBRAND H G. Development of linepipe in grade up to X100 [C]//Proceedings of the 11th EPRG/PRCI Biennial Joint Technical Meeting on Pipeline Research. Pipeline research council international, Arlington, Massachusetts ,USA, 1997: p.6-10.
[2] HILLENBRAND H G, KALWA C. High strength line pipe for project cost reduction [J]. World Pipelines, 2002, 2(1): p.1-10.
[3] Glover A. Application of grade 555 (X80) and 690 (X100) in Arctic climates [C]//Proceedings of Application & Evaluation of High Grade Linepipe in Hostile Environments Conf. Scientific Surveys Ltd., Yokohama, Japan, Nov, 2002: p.33~52.
[4] Barsanti L. From X80 to X100: know-how reached by ENI Group on high strength steel [C]//Proceedings of Application & Evaluation of High Grade Linepipe in Hostile Environments Conf. Scientific Surveys Ltd., Yokohama, Japan, Nov.2002: p.231 ~ 244.
[5] FAIRCHILD D P. High strength steels – beyond X80 [C]//Proceedings of Application & Evaluation of High Grade Linepipe in Hostile Environments Conf. Scientific Surveys Ltd., Yokohama, Japan, Nov.2002: p.307~321.
[6] BARSANTI L, HILLENBRAND H G, MANNUCCI G, et al. Possible use of new materials for high pressure linepipe construction: an opening on X100 grade steel [C]//Proceedings of 4th International Pipeline Conference. American Society of Mechanical Engineers, Calgary, Alberta, Canada, 2002:　p.287~298
[7] ASTM-E1290-2007. Standard Test Method for Crack-Tip Opening Displacement (CTOD) Fracture Toughness Measurement[S]. Philadelphia, USA, American Society for Testing and Materials, 2007.
[8] Deng Caiyan, Zhang Yufeng, Huo Lixing,　et.al. The fracture toughness of welding jiont of X65 line pipes by CTOD tests [J]. Transactions of the China Welding Institution, 2003,53(3):13-16.
[9] Liu Shouxian. Reseaches about X100 line pipes [D]. Kunming: Kunming University of Science and Technology, Materials and Metallurgy Engineering, 2007.

Stress-Strain Analysis of Composite Propellant under Cyclic Temperature Loads

ZHANG Chunlong, DING Biao, AI Changsheng, HUANG Feng

Dept. of Airborne Vehicle Engineering, Naval Aeronautical and Astronautical University, Yantai 264001，Shandong, China

Keywords: cyclic temperature; stress-strain; simulation; failure

Abstract: Based on integral nonlinear viscoelastic constitutive relationship, stress-strain field was analyzed of composite propellant under cyclic temperature loads to ensure the stress-strain distribution and damageable locality. The simulation conditions were designed in the temperature range from -10℃ to 60℃ at three different rate-temperature. The results indicate that the stress and strain have the same trend of changing. Adhesive interphase is prone to failure because of the alternating stress.

1. Introduction

During storage, the grain of motor was subjected to environmental temperature variations which in turn produced thermal stress and strain in the propellant[1,2]. Large thermal stress gradients induced by rapid temperature changes and extreme thermal conditions may produce stresses and/or strains in excess of the strength and/or strain capacity of the material[3,4]. Because of the cyclic thermal stresses, the cumulative damage may occur in micromechanical structure, leading to the macroscopical cracks and failure of grain and motor in the end.

2. Constitutive eauation

Different from elastic material, viscoelastic material has the characteristics of stress relaxation and creep. The modulus of viscoelastic material is dependent on time and temperature.

It's assumed that the solid propellant is isotropic material, the Possion ratio is constant, the relationship of stress and strain is linear visc elastic, the propellant is the thermorheological simple material[5,6].

The constitutive equations as follows:

$$S_{ij}(t) = 2\int_{-\infty}^{t} G(t-\tau)\frac{\partial e_{ij}(\tau)}{\partial \tau}d\tau \qquad (1)$$

$$\sigma_{kk}(t) = 3\int_{-\infty}^{t} K(t-\tau)\frac{\partial \varepsilon_{kk}(\tau)}{\partial t}d\tau \qquad (2)$$

The equation are modified as follows. a. the time variable t in characteristic function of constitutive equation $E(t)$ and $\mu(t)$, or $G(t)$ and $K(t)$ should be replaced by ξ; b. The strain ε includes the $\alpha\theta$ caused by thermal expansion. α is the coefficient of thermal expansion，$\theta = T - T_0$, T_0 is the initial temperature。Then, the constitutive equations of thermorheological simple material as follows:

$$S_{ij}(t) = 2\int_{-\infty}^{t} G(\xi-\xi')\frac{\partial e_{ij}(\tau)}{\partial \tau}d\tau \qquad (3)$$

$$\sigma_{kk}(t) = 3\int_{-\infty}^{t} K(\xi-\xi')\frac{\partial}{\partial t}[\varepsilon_{kk}(\tau) - 3\alpha\theta(\tau)]d\tau \qquad (4)$$

Where：$G(t)$ is the shear relaxation modulus function；$K(t)$ is the volume modulus function. The shear modulus and volume modulus are expressed by series.

$$G(t) = G_0 \left[1 - \sum_{i=1}^{N} \overline{g}_i^p (1 - \exp(-t/\tau_i)) \right] \quad (5)$$

$$K(t) = K_0 \left[1 - \sum_{i=1}^{N} \overline{k}_i^p (1 - \exp(-t/\tau_i)) \right] \quad (6)$$

3. Numerical Simulation

3.1 Material Attribute and Geometric Model

Finite element analysis had been carried out using 8-noded 3D coupled field body element. Relaxation modulus generated from the results of test was used as material input. Poisson's ratio was assumed as 0.495.

The model of research target was rectangular solid simulating motor structure. As the condition of test, two sides of left and right were fixed. Because of the restraints, loads and geometry were symmetrical, an eighth of model (Fig.1 b)) was analyzed to save calculation time.

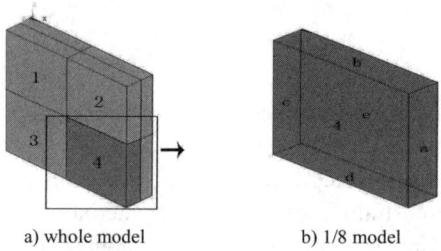

a) whole model   b) 1/8 model
Fig.1   Geometry model

3.2 Boundary Conditions and Loads

The face shown in Fig.1 b) was the fixed boundary. Faces 4 and d were free boundaries. Because of the symmetry of geometry and loads, faces b, c and e were symmetrical boundaries.

The temperature range from -10℃ to 60℃ at three different rate-temperature. The rate-temperature respectively was 10℃/12h, 20℃/12h and 30℃/12h which named group 1, group 2 and group 3.

The initial condition and boundary condition of temperature field were as follows:

Initial condition: $T_{t=0} = T_0$；

Boundary condition: $T = T(x_i, t)$.

3.3 Numerical Simulation Results

(1) Stress-strain analysis under different rate-temperature

The distributions of stress-strain are shown in Fig2~4 when the temperature is at the peak of 60℃ in cycle.

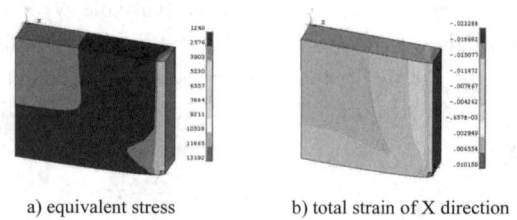

a) equivalent stress   b) total strain of X direction
Fig.2   Stress-strain distribution of group 1 at 60℃

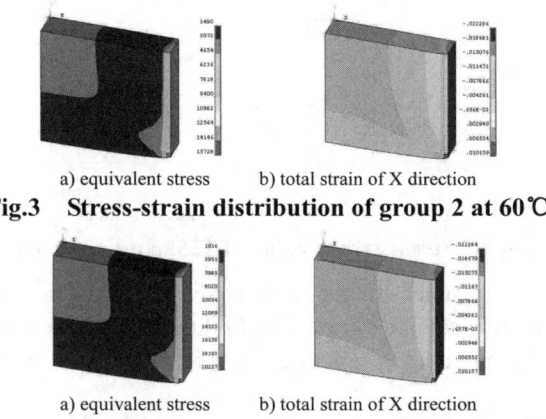

a) equivalent stress   b) total strain of X direction
Fig.3   Stress-strain distribution of group 2 at 60 ℃

a) equivalent stress   b) total strain of X direction
Fig.4   Stress-strain distribution of group 3 at 60 ℃

Fig.2~4 show that the total strain of x direction of the same nodes in different temperature conditions are almost equivalent, but the stress are different. The faster is the rate-temperature, the larger is stress. From this point, we can see that the viscoelastic material is obviously different from elastic material. The mechanical response of viscoelastic material depends on the loading rate. Usually, the faster is loading rate, the more obvious is elastic property. The lower is loading rate, the more obvious is viscous property.

It can also be found that the distribution of stress isoline map approximately identical although the rate-temperature and stress-strain value are different between groups. The maximum value of stress-strain is found at 1255 node of 5039 element (Fig.1 b) through calculation. Fig.5 shows the stress-strain distribution of 1255 node at 60℃.

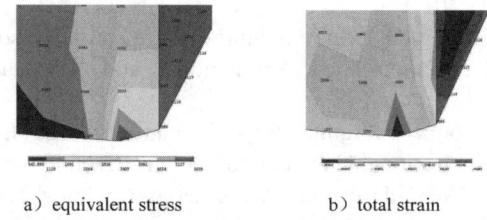

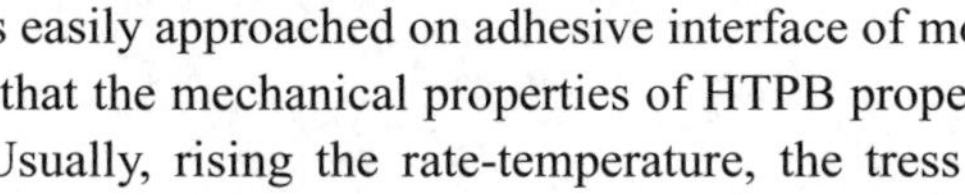

a）equivalent stress   b）total strain
Fig.5   Stress-strain nephogram around 1255 node at 60 ℃ of group 1

The phenomena indicates that although temperature environment of motor storage is different, the damage position of motor approximately identical. Considering the motor structure, it can be presumed that the failure is easily approached on adhesive interface of motor.

It can be seen from above that the mechanical properties of HTPB propellant is strongly influenced by the rate-temperature. Usually, rising the rate-temperature, the tress generated by temperature load will increase and the degeneracy rate of propellant performance will increase as well. It can be concluded that the determination of rate-temperature is very important in propellant accelerated aging process.

(2) Stress-strain analysis of node time history

The failure of motor is usually approached on the adhesive interface or end rather than overall structure. Debonding or cracking are the main failure modes. Thus，stress value at 1255 node of 5039 element is analyzed in a period of group 1 As a comparative study, face 4 with lowest stress is analyzed as well.

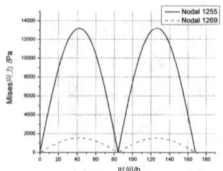
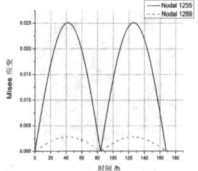

a) equivalent stress   b) equivalent strain

Fig.6   Stress-strain value of 1255 and 1269 node

It can be seen that there are distinct differences in equivalent stress and equivalent strain between two nodes. It shows that the manual debonding and release of stress must be considered in design stage of motor. Meanwhile, avoiding severely changing of temperature is a useful way for storage.

4 Conclusions

(1) The stress and strain have the same trend of changing.
(2) Adhesive interphase is prone to failure because of the alternating stress.
(3) The stress and strain is strongly influenced by the rate-temperature.

References

[1] Xu Xinqi,Yuan Shusheng,Sui Yutang. Stress-strain analysis of propellant grains in storage[J].Journal of Naval Aeronautical and Astronautical University,2002,17(3): 313-317(in chinese)

[2] D Leveque,A Schieffer,A Mavel,et al. Analysis of how thermal aging affects the long-term mechanical behavior and strength of polymer-matrix composites[J].Composites Science and Technology ,2005,65:395-401

[3] Dong Kehai. The evaluation method of remaining-life for charge of solid rocket motor based on cumulative damage[D].Yantai：Naval Aeronautical and Astronautical University,2007(in chinese)

[4] Zhang Xinggao. Study on the Aging Properties and Storage Life Prediction of HTPB Propellant[D]. Changsha：National University of Defense Technology,2009(in chinese)

[5] WANG Chun-hua,PENG Wang-da,WENG Wu-jun,et al. Chemical aging mechanisms of HTPB solid propellants and the ways to improve aging-resistance[J].Chinese Journal of Energetic Materials(Hanneng Cailiao),1996,4（3）: 109-116.

[6] FENG Zhigang, ZHOU Jianping. The Thermal Stress of Solid Rocket Motor Grain in Storage.Journal of Propulsion technology[J].1994(6):42-49(in chinese)

Damage Detection for Structural Health Monitoring Using Ultrasonic Guided Waves

Hongyuan Li[1,a] and Hong Xu[1,b]

[1] School of Energy, Power and Mechanical Engineering, North China Electric Power University, Beijing, China

[a] lihongyuan@ncepu.edu.cn, [b] xuhong@ncepu.edu.cn

Keywords: Guided waves, Dispersion, Matching pursuit, Damage detection

Abstract. The use of ultrasonic guided waves for damage detection suffers from the multi-modes and dispersion. Much attention has been paid to transducer design and excitation frequency chosen to suppress the multiple modes and dispersion. However, little attention has been paid to complex signal processing. In this paper, the dispersive propagation of the guided waves are firstly reviewed. And then the matching pursuit method is introduced as a feature extraction algorithm. In order to present well the characteristic of the guided waves signal, a dispersive dictionary is designed based on the guided waves propagation. A two-stage pursuit method consisted of coarse and fine matching is used. At last, the proposed method is verified by finite element simulation and successfully extracted damage related dispersive pulses from measured noisy signal.

Introduction

Conventional inspection methods using bulk waves for non-destructive testing and monitoring the pipeline interrogate discrete points and are therefore very time consuming, especially for the long length pipeline. An alternative to local inspection is to use guided waves which could propagate a long distance along the wave-guide. Guided waves are interesting for large area inspections since they offer the potential for rapid screening from single transducer position. However, the application of the guided waves suffers from coherent noises caused by the multi-modes and dispersion and the measured noises. The weak damage related feature pulse are usually submerged in the strong noises.

In this paper the matching pursuit method was used to extracted damage related features from noisy measured signal. Firstly, the dispersive propagation of the guided waves are reviewed. And then the matching pursuit method is introduced. The numerical implementation for guided waves of the method is proposed as following. At last the method is verified by numerical simulation.

Dispersive Propagation of Guided Waves

Guided waves propagation properties in pipes are extremely complicated. Fig.1 shows the phase and group velocity dispersion curves for the pipe with the outside diameter of 86mm and the thickness of 5.5mm. The phase velocity dispersion curves provide useful information about wave speeds of single tones and of wavelengths of the modes. The group velocity dispersion curves show the velocity at which finite-time wave packets travel. The guided waves modes travel at different speeds and are dispersive so that the original wave packet is distorted as it travels through the given structure.

Knowing the propagation characteristics of the guided wave mode, the distorted wave pulse after propagating any distance can be modeled.

Assume a pulse $P(t)$ is generated by the transducer at $x=0$ and propagates along a waveguide. The input pulse can be expressed in the frequency domain using Fourier transform.

$$P(x=0, f) = \text{FT}\,[P(x=0,t)] = \int_{-\infty}^{\infty} P(x=0,t)e^{-j2\pi ft}dt. \tag{1}$$

When the pulse arrives at $x=x_0$, each frequency component of the $P(f)$ will be shifted in the phase as[1]:

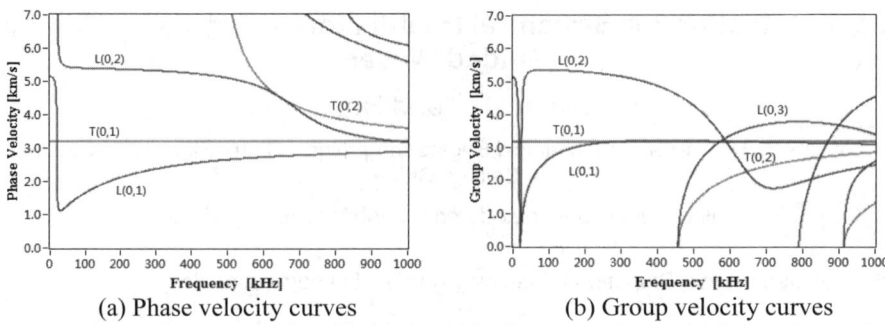

(a) Phase velocity curves (b) Group velocity curves
Fig.1 Dispersion curves of steel pipe (outside diameter: 87mm, thickness: 5.5mm)

$$P(x = x_0, f) = P(x = 0, f)e^{-jk(f)x_0}. \qquad (2)$$

where $k(f)$ is the wave number of the guided wave mode as a function of frequency f.

Consequently, the time-domain signal of the distorted pulse propagated the distance of x_0 can be obtained by inverse Fourier transform as:

$$P(x = x_0, t) = FT^{-1}[P(x = x_0, f)] = \int_{-\infty}^{\infty}[P(x = x_0, f)]e^{j2\pi ft}df. \qquad (3)$$

Eq. (1-3) imply that if the dispersion relationship of wave number $k(f)$ or the group velocity $C_g(f)$ is known, the distorted pulse shape of any given incident pulse propagated arbitrary distance can be predicted.

In order to suppress the dispersion, narrow bandwidth pulses such as Hanning-windowed or Gaussian-windowed sine waveforms are usually used as the incident signals. In this paper, modulated Gaussian pulse (also called Gabor function) is used. The Gabor function $f_{Gp}(t)$ is defined as:

$$f_{Gp}(t) = e^{-t^2/2\sigma^2}\cos(2\pi\xi t + \phi). \qquad (4)$$

where ξ is the center frequency of the pulse, ϕ is phase angle, σ controls the spread of the pulse in time which selected as $5.0/2\pi\xi$ in this paper. The Gabor pulse and the dispersive distorted pulses after propagation predicted by Eq. (1-3) are shown in Fig. 2. The dispersive distorted pulses have the longer time duration and the lower amplitude with the increment of propagated distance.

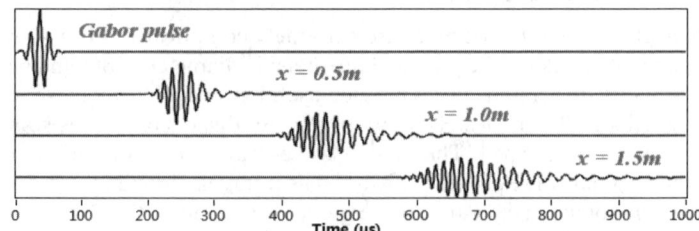

Fig. 2 Gabor pulse and the dispersive distorted pulses of L(0,1) mode

Matching pursuit method

The matching pursuit approach is introduced by Mallat[2] and Qian[3] independently around the same time. The algorithm iteratively projects a signal onto a given redundant dictionary of waveforms and chooses the dictionary atom that is best matched to the signal at each iteration.

To understand the method, considering a signal $f(t)$. The matching pursuit algorithm decomposes the signal as the following equation after M iterations:

$$f = \sum_{i=1}^{M} A_i g_i + R^{M+1} f . \tag{5}$$

where g_i is the atom from the dictionary, A_i is the amplitude of the atom, $R^{M+1}f$ is the residual signal after M iterations. The best matched atom at each iteration is chosen by:

$$g_i = \arg\max_{g_i \in D} \left[\left| \langle R^m f, g_i \rangle \right| / \|g_i\|^2 \right] . \tag{6}$$

where D is the dictionary, $\langle \cdot, \cdot \rangle$ denotes the inner product and $\|\cdot\|$ denotes the 2-norm. The amplitude of the atom can be calculated by:

$$A_i = \langle R^i f, g_i \rangle / \|g_i\|^2 . \tag{7}$$

Numerical Implementation for Guided waves

To implement the matching pursuit method for guided waves, the dictionary must be selected to present well the characteristic of the guided waves signal. Without dispersion, the Gabor pulse propagated after a given time u can be expressed as:

$$g_{[u,\xi,\sigma,\phi]}(t) = e^{-(t-u)^2/2\sigma^2} \cos[2\pi\xi(t-u)+\phi] . \tag{8}$$

Considering the dispersion, the pulse duration get longer with propagating. It is caused by the different group velocity of different frequency component. To describe the dispersion phenomena, a second-order polynomials are used to express the arrive time difference of various frequency components between the center frequency, as

$$\Delta\tau(f) = d_2(f-\xi)^2 + d_1(f-\xi) \tag{9}$$

where d_1 and d_2 are the parameters described the dispersion. According the propagation characteristic of the guided waves described in Eq. (2). The dispersive distorted pulse thus can be written as[4]:

$$g_{[u,\xi,\sigma,\phi,d_1,d_2]}(f) = g_{[u,\xi,\sigma,\phi]}(f) e^{-j2\pi[\frac{d_2}{3}(f-\xi)^3 + \frac{d_1}{2}(f-\xi)^2]} . \tag{10}$$

Where $g_{[u,\xi,\sigma,\phi]}(f)$ denotes the Fourier transform of $g_{[u,\xi,\sigma,\phi]}(t)$. The inverse Fourier transform of $g_{[u,\xi,\sigma,\phi,d_1,d_2]}(f)$ can be used as the atoms in the dictionary to present the guided waves signal.

In order to implement the matching pursuit efficiently, the number of atoms in the dictionary should be optimized for the given input signal. The measured signal can be approximated with six parameters $[u,\xi,\sigma,\phi,d_1,d_2]$. The ξ and σ can be chosen as the same with the input pulse. For the given guided waves mode and propagated distance, the d_1 and d_2 can be fitted as:

$$d_2(f-\xi)^2 + d_1(f-\xi) = x_0 / [C_g(f) - C_g(\xi)] . \tag{11}$$

where C_g is the group velocity of the given mode. The d_1 and d_2 used to make up the dictionary can thus be selected at a small range around the fitted parameters. At each iteration, the parameter u can be first identified coarsely with $\phi, d_1, d_2 = 0$. And then the range of u for fine pursuit is selected nearby the identified u.

Crack detection

In order to verify the proposed method, a pipe with a crack is simulated by finite element model. The outside diameter of the pipe is 87mm, and the thickness is 5.5mm. The half-through full circle crack located at 400mm away from the transducer. The L(0,1) mode of 120kHz are excited by prescribing radial displacements at all the nodes through the thickness at one end of the pipe. The signal is monitored at 200mm away from the transducer. The monitoring signal has been normalized to its maximum amplitude. In order to verify the capability of matching signal in noisy environment, Gaussian noise are added to the received signals as shown in Fig.3(a).

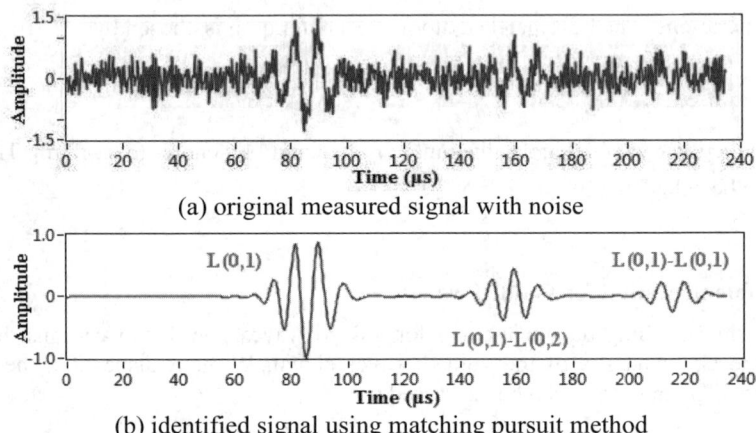

Fig. 3 Crack detection signals at 200mm away from the transducer

The wave packets identified using matching pursuit after three iteration are added as shown in Fig.3 (b). According the group velocities of the L(0,1) and L(0,2) modes, the first wave packet is L(0,1) mode passing from the transducer. The second wave packet is the reflected L(0,2) mode converted form L(0,1) mode at the crack. The third packet is the reflected L(0,1) mode from the crack. From Fig.3 we can see that the features of the crack are identified from the noisy signal accurately.

Summary
A two-stage matching pursuit method consisted of coarse and fine pursuit was used to extracted feature pulses. The dictionary was designed to present the characteristic of the dispersive guided waves signal. The proposed method successfully extracted meaningful pulses from measured noisy signal. The successful application of the present method in the guided-wave inspection was partly due to the good choice of the matching pursuit dictionary.

Acknowledgements
This paper was supported by the NSFC (11074073), NSFC and Shenhua Group Corporation Limited (51134016), and "the Fundamental Research Funds for the Central Universities".

References

[1] P. D. Wilcox: IEEE Trans. on Ultra, Ferroelectrics and Frequency Control Vol. 50 (2003), p.419

[2] Mallat S, and Zhang Z: IEEE Transactions on Signal Processing Vol. 41 (1993), p. 3397

[3] S. Qian and D. Chen: Signal Processing Vol. 36 (1994), p.1

[4] J. C. Hong, K. H. Sun and Y. Y. Kim: IEEE Trans on Ultra, Ferroelectrics and Frequency Control Vol. 53 (2006), p. 592

The Research on Wood Fiber/ Stainless Steel Net Electromagnetic Shielding Composite Board

Su Chu-wang[a], Yuan Quan-ping[b], Gan Wei-xing[c], Huang Jing-da[d], Huang Yuan-yi[e]

Forestry College of Guangxi University, Nanning, 530003, China

[a]glscw58@163.com, [b]asdf67jkl.163@163.com, [c]gweixing@163.com, [d]amaspj@yahoo.cn, [e]a362346875@163.com

Keywords: Wood fiber; stainless steel net; electromagnetic shielding effectiveness; mechanical strength

Abstract. In this paper, the electromagnetic shielding function composite fiberboards were made by filling with stainless steel nets dipped with urea-formaldehyde resin adhesive (UF) and the influence of different mesh and layers of nets on its electromagnetic shielding performance, static bending strength (MOR), modulus of elasticity (MOE) and internal bonding strength (IB) were studied. The results showed that: when the mechanical strength was enough and the frequency was in range of 50MHz to 1GHz, of all the composite fiberboards filled with one-layer stainless steel net, the one filled with 60 mesh was best and the minimum shielding effectiveness (SE) was 36.22 dB; when filled with two-layers nets, the one filled with 80 mesh was best and the minimum SE was 42.54dB; when filled with three-layers nets, the one filled with 60 mesh was best and the minimum SE was 50.77dB.

Introduction

Both wood and pure wooden wood-based panels belong to poor conductor of electricity and scarcely have electromagnetic shielding effectiveness. It would have a certain electromagnetic shielding effectiveness when wood and metal were combined through a suitable method. In recent years, many scholars have done a lot of research on wood-metal electromagnetic shielding materials, and the wood fiber/metal fiber (net) composite materials made of metal fiber of good electrical conductivity was one of development direction of electromagnetic shielding effectiveness material [1-5]. Therefore, it has a certain potential that metal fiber (net) is applied to electromagnetic shielding composite wood-based panels, but the previous process were relatively complicated and its cost were high. In this study, the electromagnetic shielding function composite fiberboards were made of composite of wood fiber and corrosion-resistance stainless steel nets dipped with UF in advance.

Test materials and Methods

Test materials. Wood fiber: it was mainly Pine wood fiber and offered by Guangxi Hualin Forest Industry Co., Ltd; glue consumption was 12%, and it was in good condition.

Stainless steel nets: 20, 40, 60, 80, 100 mesh, purchased from Hebei Xinyuan metal mesh fabric Co., Ltd.

UF: it was offered by Guangxi Hualin Forest Industry Co., Ltd.

Test equipment. Hot-pressing machine: XLB100-D flat vulcanizing machine; multifunctional mechanical testing instrument: microcomputer control electronic universal testing machine.

Test method. First of all, the stainless steel nets were cut into the corresponding specifications and dipped in UF after pretreatment, then dried in the air. Plate thickness was set as 9mm, material density as 760kg/m^3. Stainless steel nets were placed in composite fiberboards according to the set structure. The stainless steel net was placed in the middle when the composite fiberboard was filled with one-layer stainless steel net; the stainless steel nets had better be placed in surface layer when the composite fiberboard was filled with two (or three)-layers stainless steel nets.

Hot-pressing parameters: pressing time was set as 10min, hot-pressing temperature as 170°C, top hot-pressing pressure as 2.5Mpa; and it took a three-paragraph vacuum hot-pressing process.

The physical and mechanical properties of composite fiberboards were tested according to Chinese national medium density fiberboards standard GB/T 11718-1999. And according to the criterion SJ20524-1995" Measurement of Shielding Effectiveness of Materials", the composite fiberboard was kept on cooling 24h, then should be made out the standard disc specimens of diameter $115^0_{-0.5}$ mm; SE of the composite fiberboard would be tested by using DN15115 type of vertical flange coaxial test device, and took the average value of the three specimens. The test specimens were entrusted to Research Institute of Wood Industry, Chinese Academy of Forestry to complete.

Results and Conclusion

Test results and analysis of IB. According to the Chinese national standard, physical and mechanical properties of the composite fiberboard of thickness 9mm were that, static music intensity is more than 23Mpa, and elastic modulus is more than 2700Mpa and inside combinative strength is more than 0.55Mpa. The influence of different mesh and layers on inside combinative strength of composite fiberboards was shown in Table 1.

Table 1 The IB of composite fiberboards

Mesh of stainless steel net [mesh]	IB [Mpa]			Damage location
	One-layer	Two-layers	Three-layers	
20	0.40	0.44	0.47	The one-layer and two-layers were at wood fiber, three-layers was at stainless steel net
40	0.50	0.45	0.44	The one-layer and three-layers were at stainless steel net, two-layers was at wood fiber
60	0.66	0.58	0.56	The one-layer and two-layers were at wood fiber, three-layers was at stainless steel net
80	0.87	0.79	0.62	The one-layer and three-layers were at stainless steel net, two-layers was at wood fiber
100	0.40	0.34	0.23	The one-layer and two-layers were at wood fiber, three-layers was at stainless steel net

It could be seen From Table 1 that the IB of composite fiberboards was rising along with the increase of the mesh of stainless steel nets from 20 to 80mesh, but to 100mesh, the IB was plunged. The reason was that when stainless steel nets were dipped with UF, the more the mesh was, the more resins attached to net was, and the greater surface binding force of net-fiber and fiber-fiber between the net was; but when the mesh was 100mesh and very small, fibers couldn't easily pass through the meshes to form agglutination point with each other and surface binding force of fiber-fiber was abate; along with the increase of the layers of stainless steel nets, the IB of composite fiberboards was reducing.

Test results and analysis of MOR and MOE. It could be seen From Table 2 that, the MOR and MOE of all composite fiberboards could reach the Chinese national standard.

Also, the MOR of composite fiberboards was reducing along with the increase of the mesh of stainless steel nets, and was increasing along with the increase of the layers of stainless steel nets.

Table 2 The MOR and MOE of composite fiberboards

Mesh of stainless steel nets [mesh]	MOR [Mpa]			MOE [Mpa]		
	One-layer	Two-layers	Three-layers	One-layer	Two-layers	Three-layers
20	29.5	50.3	60.5	3590	6172	6926
40	28.2	48	64.3	2899	5571	7786
60	27.0	34.2	56.3	3596	4089	5985
80	28.5	29.9	49.8	4781	4539	5968
100	23.1	28.5	43.8	2999	4362	4205

Test results and analysis of electromagnetic shielding effectiveness
The SE of composite fiberboards with one-layer stainless steel net. SE was divided into five grades in range of 50MHz-1GHz by Chohachiro N [6,7]: ①under 10dB: no effectiveness; ②10-30dB: bad; ③30-60dB: medium; ④60-90dB: good; ⑤above 90dB: excellent.

In different frequency, SE of composite fiberboards filled with one-layer of stainless steel net of different mesh was shown in Fig.1.

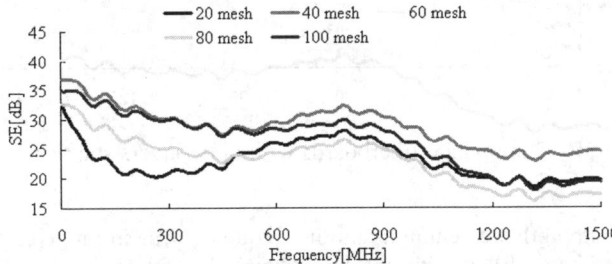

Fig.1 The SE of composite fiberboards with one-layer stainless steel net

It could be seen from Fig.1 that, overall, SE was good in low and medium frequency; and the minimum SE of the composite fiberboards filled with net of 60 mesh was relatively better which was 36.22dB and reached medium level. The one filled with 20 or 80 mesh were worst, and also the one filled with 100 mesh wasn't good; the possible reason is that the more the mesh of stainless steel net was, the more the bonding points of wire-wire that makes contact resistance increasing, so the SE was worse.

The SE of composite fiberboards with two-layers stainless steel nets. In different frequency, SE of composite fiberboards filled with two-layers of stainless steel nets of different mesh was shown in Fig.2.

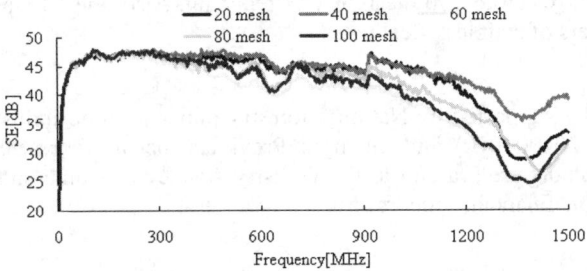

Fig.2 The SE of composite fiberboards with two-layers stainless steel nets

From Fig.2, it could be seen that SE of all composite fiberboards filled with two-layers stainless steel nets of different mesh could reach 40dB above and were relatively stable in range of 50MHz to 1GHz; particularly, the one filled with two-layers nets of 40 mesh was best, and minimum SE was 45.55dB, reaching medium level; among them, the one filled with two-layers nets of 60 mesh was worst; so it was known that SE of composite fiberboards filled with two-layers stainless steel nets wasn't increasing along with the increase of mesh of stainless steel nets. Compared with that from Fig.1, the one filled with two-layer nets was better than the one filled with the one-layer.

The SE of composite fiberboards with three-layers stainless steel nets. In different frequency, SE of composite fiberboards filled with three-layers stainless steel nets of different mesh was shown in Fig.3

From Fig.3, it could be seen that the minimum SE of composite fiberboards filled with three-layers stainless steel nets of different mesh was 41.19dB in range of 50MHz to 1GHz and were relatively stable; SE was becoming more stable along with increase of layers of stainless steel nets; among them, the one filled with stainless steel nets of 60 mesh was best, whose minimum SE was 50.77dB and nearly reached good shielding level in the range of 50MHz to 1GHz; and in the range of 218MHz to 1.5GHz, the minimum was above 60dB, reaching good revel.

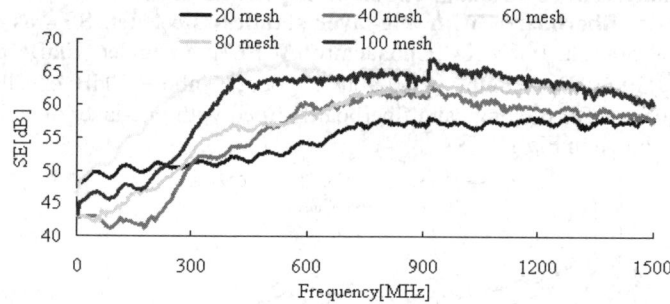

Fig.3 The SE of composite fiberboards with three-layers stainless steel nets

Summary

When the mechanical strength was enough and the frequency was in range of 50MHz to 1GHz, of all the composite fiberboards filled with one-layer stainless steel net, the one filled with 60 mesh was best and the minimum SE was 36.22 dB; when filled with two-layers nets, the one filled with 80 mesh was best and the minimum SE was 42.54dB; when filled with three-layers nets, the one filled with 60 mesh was best and the minimum SE was 50.77dB.

In the entire frequency range, particularly in the medium and high frequency, SE of the composite fiberboards was becoming more stable along with the increase of layers of stainless steel nets.

MOR and MOE of all composite fiberboards could reach the Chinese national standard. MOR was reducing along with increase of the mesh of stainless steel nets and increasing along with increase of the layers of stainless steel nets.

IB of composite fiberboards was increasing with the increase of the mesh of stainless steel nets from 20 to 80mesh; but when to 100 mesh, it was reducing. Also, the IB was reducing along with the increase of the layers of stainless steel nets.

Acknowledgment

This work was partially supported by National forestry public welfare industry scientific research special project (NO: 201004070) and Guangxi Provincial scientific research Foundation [NO: 10100022-26]. The authors are grateful to the Forestry Agency of China and scientific Agency of Guangxi Province for its financial support of this work.

Reference

[1] Liu Xiansen, Fu Feng: Wood Processing Machinery (China). No. 5(2008), p.22-26
[2] Hua Yukun, Fu Feng: Scientia Silvae Sinicae. Vol. 31(1995), p.254-259
[3] Zhang Xianquan, Liu Yixing: China Wood Industry. Vol. 19(2005), p.12-16
[4] Zhang Xianquan, Liu Yixing: China Forest Products Industry. Vol. 31(2004), p.15-19
[5] Zhang Xianquan, Liu Yixing and Li Jian: Journal of Northeast Forestry University (China). Vol. 32(2004), p.26-28
[6] Ma Yuhua, Luo Zhaohui: Anhui Chemical Industry (China). No. 3(2006), p.37-38
[7] Chohachiro N, etc.: The timber Society (Japan). Vol. 35(1989), p.1092-1099

Low Cycle Fatigue Behaviors of TI-6AL-4V Alloy Controlled by Strain and Stress

Ruifeng Wang[1,2,a], Youtang Li[1,b], Huping An[3,c]

[1]Key Laboratory of Digital Manufacturing Technology and Application, Ministry of Education, Lanzhou University of Technology, Lanzhou 730050, China

[2]Lanzhou Jiao Tong University, Lanzhou, China

[3]Lanzhou City University, Lanzhou, China

[a]Wangruifeng0808@163.com, [b]liyt@lut.cn, [c]ahp2004@126.com

Keywords: Ti-6Al-4V alloy; Low cycle fatigue; Strain control; Friction stress; Back stress; Cyclic creep

Abstract. The low cycle fatigue behaviors of TI-6AL-4V alloy controlled by strain were investigated by experiment. The fatigue tests were performed at room temperature, and cyclic strain and stress ratio are 0.1 with triangle load wave. The results show that TI-6AL-4V alloy is soften rapidly under the cyclic tensile stresses and it is harden rapidly under the cyclic compressive stresses during the initial-stage of strain controlled fatigue, and the rates of cyclic soften and cyclic harden are decreased with the fatigue progress. The soften rate is related to the cyclic strain but little to the cyclic stress during the overall fatigue progress. The change of cyclic stress is related to the macro friction stresses. The results of experiment show that obvious cyclic creep occurs under the stress controlled low cycle fatigue conditions, and the magnitude of cyclic creep strain is related to the maximum cyclic stress. The softening of tensile friction stresses is the main factor of cyclic creep.

Introduction

The stress of macroscopic deformation can be divided into the friction stress and back stress. There are several ways to determine the friction stress and back stress, one of them is to use the cycle hysteretic loop of low cycle fatigue. The friction stress and back stress of the fatigue process are obtained by the cycle hysteretic loop of low cycle fatigue that controlled by strain of single crystal copper by Kuhlmann W D [1], and they are obtained cycle hysteretic loop of Ti-6246 alloy by Feaugas X [2], but these are obtained under the condition of the cyclic strain ratio of -1.

Ti-6Al-4V alloy is a moderate intensity (α+β) type titanium alloy and it has the excellent comprehensive performance [3-4]. The typical two-phase (α+β) titanium alloy is used most widely [5]. Ti-6Al-4V alloy is studied and applied at first in the aviation and aerospace industry, it is mainly used to made bearing and fasteners components (i.e, the aircraft fuselage, wing skins, structural beams, joints, bulkheads) and the rotating components with large stresses (i.e. engine fan, compressor disks, blades) [6-7]. It is very important to expand the application of Ti-6Al-4V alloy [8].

The low cycle fatigue behaviors of TI-6AL-4V alloy controlled by strain were investigated by experiment. The fatigue tests were performed at room temperature, and cyclic strain and stress ratio are 0.1 with triangle load wave.

Experimental materials and the experimental methods

Ti-6Al-4V titanium alloy with (α+β) phase was selected in the experiment, which chemical composition and mechanical properties of materials are shown in Table 1 and Table 2.

The experiment was carried out on a Shimadzu fatigue test machine. The four strain fatigues, 0.8%, 1.6%, 2.4%, 2.8%, are used. The maximum cyclic stresses are 600, 650, 700 Mpa, respectively. The stress ratio and stress ratio are 0.1, the waveform is triangular wave, the frequency of strain and stress is 0.17 Hz, 0.50 Hz, respectively. The structure and dimensions of specimen are shown in Fig. 1.

Table 1: The chemical composition of the Ti-6Al-4V alloy

	Al	V	Fe	C	N	H	O
Ti-6Al-4V	5.5~6.8	3.4~4.5	≤0.3	≤0.1	≤0.05	≤0.015	≤0.2

Table 2: The mechanical properties of the Ti-6Al-4V alloy at room temperature

δb(MPa)	δ0.2(Mpa)	δ(%)	ψ(%)
895	825	10	39

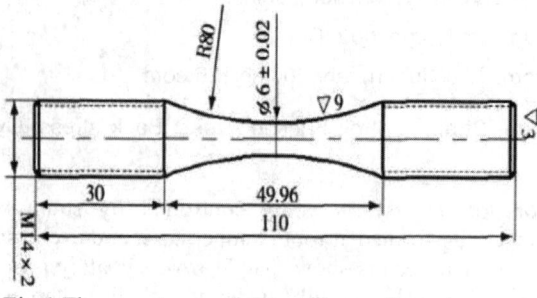

Fig.1 The structure and dimensions of specimen (mm)

Fig. 2 The hysteretic loop of Ti-6Al-4V first cycle of the cyclic stress-strain

Effect of Strain on the Sow-cycle Fatigue Properties

The Hysteretic Loop of Cyclic Stress-strain. The hysteretic loop of first cycle fatigue stress-strain of Ti-6Al-4V alloy under 4 strains conditions are shown in Fig. 2. It can be seen from Fig.2 that the stress increased with the increasing of the cyclic strain. The reverse yield strength of the alloy reduced due to the Bauschinger effect. The certain plastic strain component is existed in all cycle even if the maximum cyclic strain is in 0.8%.

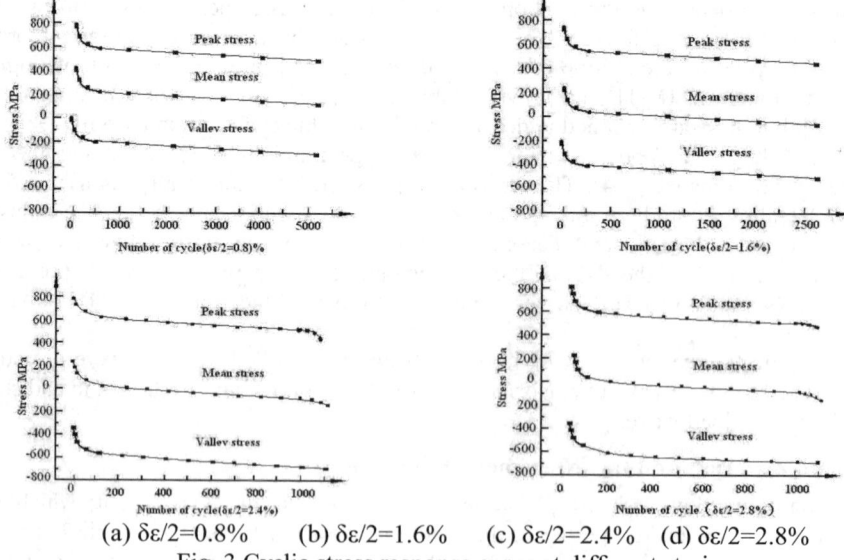

(a) δε/2=0.8% (b) δε/2=1.6% (c) δε/2=2.4% (d) δε/2=2.8%

Fig. 3 Cyclic stress response curve at different strain

The Stress Response Curve. The cyclic stress response curves of Ti-6Al-4V alloy are shown in Fig. 3. It can be known from Fig.3 that the softening of cycle tensile stress is existed significantly in all the cycle process. The higher the cyclic stress is, the faster the softening is. On the other hand, the hardening cycle compressive stress is given in the all cycle process. In the initial cycle, the cyclic tensile stress is soften rapidly, while the cyclic compressive stress is harden rapidly, and the cyclic tensile stress is soften slowly, while the cyclic compressive stress is harden slowly at the late cycle. The hardening rate is lower than the softening rate at the late cycle. There are quite different from that of the symmetric cycle compressive stress (cyclic strain ratio R = -1) [1], the cyclic compressive stress is soften also for the latter.

The changes of the cyclic stress under different strain of Ti-6Al-4V alloy are shown in Fig.4. It can be known from Fig. 4 that the average stress was reduced continuously and the relaxation phenomenon existed, which is due to cyclic tensile stress softening and cyclic compressive stress hardening. Although during the all fatigue process, the softening and hardening phenomenon existed in the cyclic tensile stress and the cyclic compressive stress, respectively, however, the total cyclic stress was reduced continuously; overall the material was softening in the experiment.

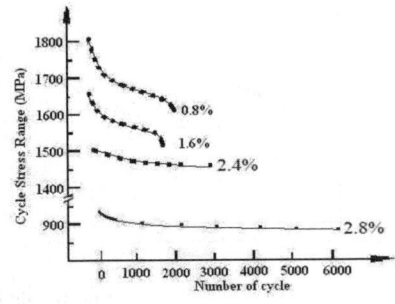

Fig. 4 The cyclic stress under different strain

The Friction Stress and the Back Stress. The macro stress can be divided into the friction stress and back stress. The friction stresses and the back stresses under different fatigue strain of Ti-6Al-4V alloy are shown in Fig.5. The experimental data of half cycle of the tensile and the compression were used to the decomposition of the macroscopic stress. It can be seen from Fig.4 that at room temperature, the friction stress changes continuously while the back stress changes little, which is because the dislocation of Ti-6Al-4V alloy was mainly planar slip, the thermal effect is not evident and the reversibility of slip is good, so the dislocation density changes little. It is considered that titanium alloy softening is related to the shear of the α-2 phase of the alloy [7].

The back stresses are related to the strain, the higher the strain is, the greater the back stress is. When the maximum strain is 1.6%, the back stress of half tensile cycle is greater than that of half compress cycle, while the back stresses are changed little when the maximum strain is 2.4% and 2.8%. The friction stresses under different strain are not equal, because the back stress affects the reverse yield of the alloy. The friction stresses of the tensile half cycle are decreased, while the friction stresses of the compression half cycle are increased. The softening and hardening of the cyclic stress are related to the change of the internal friction stress.

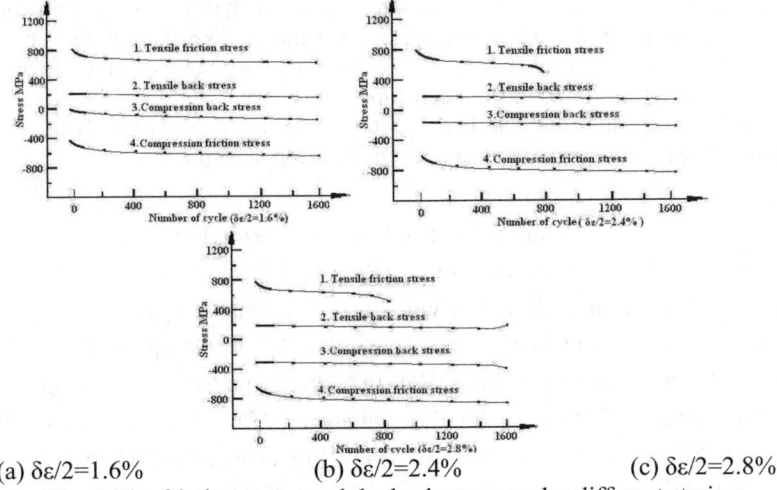

(a) δε/2=1.6% (b) δε/2=2.4% (c) δε/2=2.8%

Fig. 5 The friction stress and the back stress under different strain

Effect of Stress on the Low Cycle Fatigue Properties

The peak and valley value of strain of Ti-6Al-4V alloy under different stress are shown in Fig.6. It can be seen form Fig.6 the cyclic creep of the materials occurs significantly, and the higher the stress is, the more obvious the cyclic creep is. The early speed of the cyclic creep deformation was quickly, but along with the cycles the cyclic creep rate was reduced. The specimen is soften rapidly at the first few cycles, so the cyclic creep strain is increased rapidly, while the cyclic creep rate is decreased at the strain-controlled fatigue of slow cyclic stress softening stage.

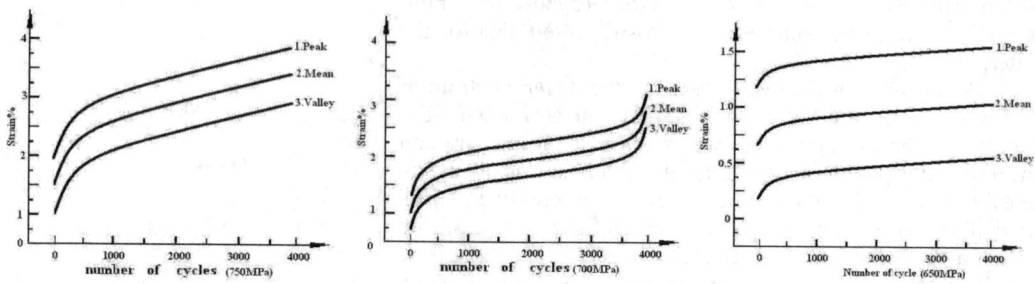

Fig. 6 The cyclic stress-strain and the cyclic creep curve

At the room temperature, the cycle friction stress of the titanium alloy is reduced continuously, and yield strength of the material is decreased, which indicate that under the condition of a constant stress the material is bound to further yield while a certain small amount of plastic deformation will occur in per cycle. The cyclic creep is connected with the cyclic hardening and the cyclic softening effect.

The cyclic creep is related to the average stress and the maximum cyclic stress of fatigue, the maximum cyclic stresses lead to the cyclic micro-plastic strain, while the average stresses lead to the constant stress creep. The cyclic creep is easily taken place in titanium at room temperature, even if the maximum stress is at 90% of the yield stress, However, even if the maximum stress is 700 Mpa, the average stress of it is only 85% of the yield strength, Cyclic creep is the role of the maximum stress.

Conclusion

(1) Ti-6Al-4V alloy is softening in the cyclic of tensile stress, while it is hardening in the cyclic of compress stress. The speed of softening and hardening is rapidly in the initial stage, and then reduced gradually. All the softening and hardening are accelerated with the strain increases.

(2) Under the conditions of different strain, the friction stresses and back stresses are changed with the increasing of the strain. The friction stress of tensile half cycle make the material softening while the friction stress of compress half cycle make the material hardening. The softening effect and the hardening effect of the cyclic stress were caused by the changes of the friction stress.

(3) Under the condition of stress-controlled low cycle fatigue, the Ti-6Al-4V alloy has significant cyclic creep deformation. The cyclic creep was connected with the cyclic stress. The friction stress is the mainly factor to affect the cyclic creep deformation.

References

[1] W.D.Kuhlmann, C.Laird: Mater Sci Engng Vol. 37(1979), p.111
[2] X.Feaugas, M. Clavel: Acta Mater, Vol. 45 (1997), p.2685
[3] Z.L.Yu, S.X.Li, Y.Y.Liu: J Mater Sci, Vol. 40 (2005), p.6049
[4] P.Potozky, H.J.Maier, H.J.Christ: Metall Trans, Vol. A29 (1998), p.2995
[5] Y.Mahajan, H.Margolin. Metall Trans, Vol. 13A (1982), p.269
[6] G.S.Nanjundaswamy, C.Ramchandra, P.K.Sungupta: J Mater Sci Lett, Vol. 17 (1998), p.993
[7] A.L.Helbert, X.Feaugas, M.Clavel: Metall Trans, Vol. 30A (1999), p.2853
[8] L.X Cai, Y.J.Liu, Y.M.Ye. Acta Metallurgiacl Sinica, Vol. 40 (11) (2004), p.1155

Analysis of Stress Singularity near the Tip of Artificial Crack

Youtang Li[a], Huaiqing Li

Key Laboratory of Digital Manufacturing Technology and Application, Ministry of Education, Lanzhou University of Technology, Lanzhou 730050, China

[a]liyt@lut.cn

Key words: artificial crack; bi-material; stress singularity; eigen value; stress extrapolation method

Abstract. A generalized expression of the stress-singularity function at the tip of artificial crack is proposed, and a formula to calculate the stress intensity factor of artificial crack is obtained in the paper. The solutions of stress singularity of a cracked bi-materials beam under uniform tension and bending were computed. The results show that the degree of stress-singularity is determined by the exponent λ at the tip of artificial crack, and the exponent λ is, not only determined by materials parameter of artificial crack but also by angle.

Introduction

A new surface forming method of solid material using crack and its propagation has been put forward and this new method was named crack technique [1]. In the applying research of crack theory, how to make crack quickly is a key problem that meets demand of high efficient production in engineering. Using artificial crack that formed with bi-material may be a feasible way to achieve efficient production.

The problem of a crack perpendicular to and terminating at bi-material interface has been the fully researched theoretically, but the research was restricted only in semi-infinite crack in infinite bi-material medium [2-3]. The crack or notch formed by bi-materials is the main point of fracture theory and the researches of matrix crack that can be used in active fracture have made great progress [4-7]. A method to calculate the stress intensity factor for V-notch of bi-material has been proposed [8]. The stress extrapolation method is used to calculate the intensity factor of artificial crack. As an example, the solutions of stress singularity of a cracked bi-materials beam under uniform tension and bending were computed.

Stress Singularity Eigen-Equation of Bi-Material

The interfacial artificial crack of bi-material is shown in Fig.1. Two plates with elastic module E_1 and E_2, and Poisson's ratio υ_1 and υ_2 are perfectly joined along their common interface. In material 1 there is a notch to the interface, the notch tip is exactly at the interface of the two plates. The Ariy's stress function, $\Phi(r,\theta)$, which satisfies the bi-harmonic equation, can be chosen as following.

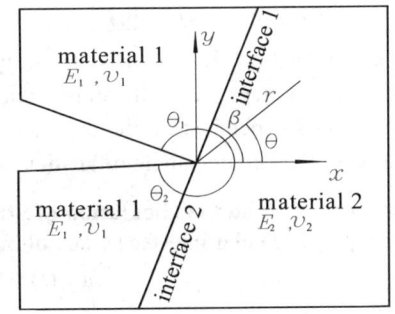

Fig.1 Artificial crack of bi-material

$$\Phi(r,\theta) = r^{\lambda+1} F(\theta) \qquad (1)$$
$$F(\theta) = a\sin(\lambda+1)\theta + b\cos(\lambda+1)\theta + c\sin(\lambda-1)\theta + d\cos(\lambda-1)\theta \qquad (2)$$

In polar coordinates system, the stresses σ_r, σ_θ, $\tau_{r\theta}$ can be expressed as following.

$$\sigma_r = \frac{1}{r^2}\frac{\partial^2 \Phi}{\partial \theta^2} + \frac{1}{r}\frac{\partial \Phi}{\partial r} = r^{\lambda-1}[F''(\theta) + (\lambda+1)F(\theta)] \qquad (3)$$

$$\sigma_\theta = \frac{\partial^2 \Phi}{\partial \theta^2} = r^{\lambda-1}[\lambda(\lambda+1)F(\theta)] \qquad (4)$$

$$\tau_{r\theta} = -\frac{1}{r}\frac{\partial^2 \Phi}{\partial r \partial \theta} + \frac{1}{r^2}\frac{\partial \Phi}{\partial \theta} = -\lambda r^{\lambda-1} F'(\theta) \qquad (5)$$

The material 1 and 2 should have the same stress function given by eq.(1) and different coefficients a, b, c, d. In order to assure the continuity of displacement along the interface, the eigen-values, λ_1 and λ_2 of material 1 and material 2 must be equal in order to assure continuity of displacement along the interface. The stress function in material 1 and 2 can be written as follows

$$\Phi_1(r,\theta) = r^{\lambda+1}[a_1 \sin(\lambda+1)\theta + b_1 \cos(\lambda+1)\theta + c_1 \sin(\lambda-1)\theta + d_1 \cos(\lambda-1)\theta] \qquad (6)$$
$$\Phi_2(r,\theta) = r^{\lambda+1}[b_2 \cos(\lambda+1)\theta + d_2 \cos(\lambda-1)\theta] \qquad (7)$$

Because the boundary of the artificial crack should be free and the stress and displacement in the interface should be continuous. Taking into consideration of the generic condition ($\pi/2 < \gamma < \pi$), a series of equations for a_1, b_1, c_1, d_1, b_2, and d_2 can be obtained. In order to make the homogeneous equation have a solution for the unknown a_1, b_1, c_1, d_1, b_2, and d_2, the determinant of the coefficients of the unknowns should be zero. The eigen-equation for λ can be obtained as follows

$$[1-\beta+2\alpha(1+\alpha-\alpha\lambda^2+\beta(\lambda^2-1))]\sin\pi\lambda + \alpha(\beta-\alpha)\lambda^2\sin(\pi\lambda-2\gamma) - (\alpha+\alpha^2-\alpha\beta)$$
$$\sin[2\lambda(\pi-\gamma)] + \lambda\sin 2\gamma - (1+\alpha)(\alpha-\beta)\sin 2\lambda\gamma + (\alpha\beta\lambda^2 - \alpha^2\lambda^2)\sin(\pi\lambda+2\gamma) = 0 \qquad (8)$$

where

$$\alpha = \frac{\upsilon_1/\upsilon_2 - 1}{1+k_1} \qquad \beta = \frac{\upsilon_1}{\upsilon_2}\frac{1+k_2}{1+k_1} \qquad k_{1,2} = \begin{cases} (3-\upsilon_{1,2})/(1+\upsilon_{1,2}) & \text{for plane stress} \\ 3-4\upsilon_{1,2} & \text{for plane strain} \end{cases} \qquad (9)$$

If the linear symmetrical crack would be considered alone, $\gamma = \pi$ should be substituted into eq.(9), so that the following equation could be obtained.

$$\sin\lambda\pi[\lambda^2(-4\alpha^2+4\alpha\beta)+2\alpha^2-2\alpha\beta+2\alpha-\beta+1-(2\alpha^2-2\alpha\beta+2\alpha-2\beta)\cos\lambda\pi] = 0 \qquad (10)$$

$\lambda=0$ and $\lambda=1$ are the solutions of above equation, if $\lambda=0$, all the stress should be zero, and if $\lambda=1$, all the stress should be constant. Besides $\lambda=0$ and $\lambda=1$, the other solutions (for $0<\lambda<1$) can be obtained as follows

$$\lambda^2(-4\alpha^2+4\alpha\beta)+2\alpha^2-2\alpha\beta+2\alpha-\beta+1-(2\alpha^2-2\alpha\beta+2\alpha-2\beta)\cos\lambda\pi = 0 \qquad (11)$$

The eigen-solution, λ, should be 1/2 only when $0<\lambda<1$, showing the singularity ($1/\sqrt{r}$) of stress near the tip of crack. Thus it can be seen that the stress singularity near the tip of artificial crack of bi-material is dependent on the exponent, λ, and it depends not only on the parameter of material, but also on the opening angle of artificial crack.

Numerical Method to Calculation of Stress Intensity Factor

For the perpendicular interface crack of bi-material, the stress intensity factor can be defined as

$$K_{ij}(t) = \lim_{r\to 0} \sqrt{2\pi} r^{1-\lambda} \sigma_{ij}(r,\theta) \qquad (12)$$

where the parameter, λ, expressing the singularity of stress and it can be solved with eq.(10). According to the definition of stress intensity factor of artificial crack [8], the stress intensity factor of perpendicular artificial crack of bi-material can be written as follows

$$K_1(t) = \lim_{r \to 0}(2\pi r)^{1-\lambda} \sigma_\theta(r,0) \tag{13}$$

where λ can be obtained from eq.(8), and it depends not only on the two materials for artificial crack of bi-material but also on the opening angle of artificial crack. For perpendicular artificial crack of bi-material, the stress intensity factor, K_1, can also be expressed in the form of eq.(13) and it can be regarded as the universal definition of stress intensity factor for crack, perpendicular interface crack with bi-material and perpendicular artificial crack with bi-material. The stress intensity factor is calculated by using the results of stress $\sigma_\theta(r,0)$ extrapolated along the direction of notch. The stress near the tip of crack orientation perpendicular to the crack can be expressed as following

$$\sigma_\theta(r,0) = \frac{K_1(t)}{(2\pi r)^{1-\lambda}}(1+cr) \tag{14}$$

where $K_1(t)$ and c are unknown. The stresses at the point r_1 and r_2 from V-notch tip on the prolongation of V-notch can be obtained as follows

$$\sigma_\theta(r_1,0) = \frac{K_1(t)}{(2\pi r_1)^{1-\lambda}}(1+cr_1) \qquad \sigma_\theta(r_2,0) = \frac{K_1(t)}{(2\pi r_2)^{1-\lambda}}(1+cr_2) \tag{15}$$

The stress intensity factor for mode I problem in bi-material system can be calculated as follows

$$K_1(t) = \frac{(2\pi r_1 r_2)^{1-\lambda}[\sigma_\theta(r_1,0)r_2^\lambda - \sigma_\theta(r_2,0)r_1^\lambda]}{r_2 - r_1} \tag{16}$$

When $\lambda = 1/2$, the above equation can be simplified for the case of single material as follows

$$K_1(t) = \frac{\sqrt{2\pi r_1 r_2}\left(\sigma_\theta(r_1,0)\sqrt{r_2} - \sigma_\theta(r_2,0)\sqrt{r_1}\right)}{r_2 - r_1} \tag{17}$$

Numerical Examples

The artificial crack specimen is shown in Fig.2, the uniform tension and bending is loaded on it, respectively. Material 1 and Material 2 are alloy steel and medium carbon steel, respectively. The parameters are: $a=40$mm, $b=10$mm, $c=5$mm, elastic modulus $E_1 = E_2 = 200$ GPa, Poisson's ratio $v_1 = 0.3$, $v_2 = 0.26$, $P = 10$ MPa. The eigen-value λ can be obtained from eq.(11), and $\lambda = 0.4798$.

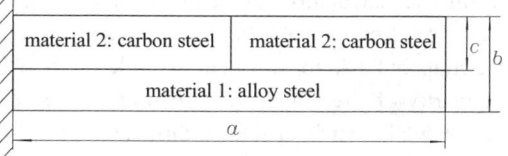

Fig.2 Specimen and its dimension

The stress intensity factors obtained by stress extrapolation method with different r/a under uniform tension are shown in Fig.3. It can be known from Fig.3 that the results of artificial crack are the similar with the crack of single material [9]. For the same width b, the relations of stress

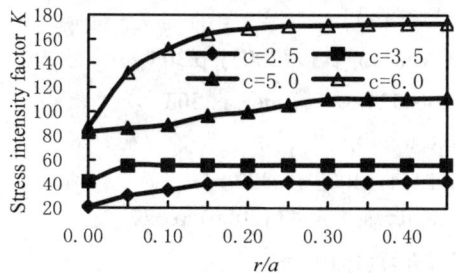

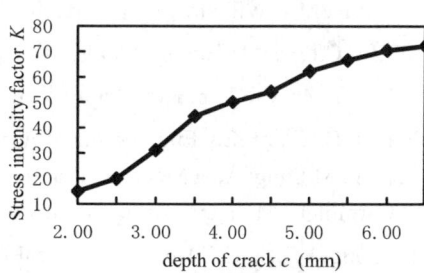

Fig.3 K_I with different r/a under tension   Fig.4 Relation of K_I to depth of crack under tension

intensity factor to the depth of artificial crack under uniform tension are shown in Fig.4. It can be known that the stress intensity factor will be increased with the depth of crack increase under the same boundary conditions.

The stress intensity factors (SIF) obtained by stress extrapolation method with different r/a under bending are shown in Fig.5. It can be known from Fig.5 that the results of artificial crack are the similar with the crack of single material [9]. For the same width b, the relations of SIF to the depth of artificial crack under bending are shown in Fig.6. It can be known that the stress intensity factor will be increased with the depth of crack increase under the same boundary conditions.

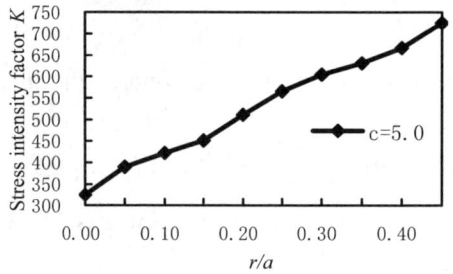
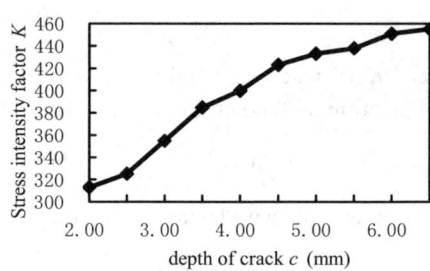

Fig.5 K_I with different r/a under bending   Fig.6 Relation of K_I to depth of crack under bending

Conclusions

1) The stress intensity factor of artificial crack with two materials depends on Poisson's ratio of the two materials and the parameters of specimen. The stress intensity factors of artificial crack are the similar with the crack of single material, which is shown that the crack formed by two materials can be taken as artificial crack.

2) In the calculation of stress intensity factor of artificial crack with two materials, the singularity of stress depends on the exponent, $\lambda-1$, and λ is related not only to the property of the two materials but also to the opening angle of artificial crack. Therefore, it is complicated to model the singularity of stress at the artificial crack tip of with the singularity element and displacement method as well, which would additionally concern the expression displacement field near the artificial crack tip.

References

[1] B.J.Zhao, F.Y.Lang, Q.T.Wei: Engng Fract Mech Vol. 31 (1988), p.923

[2] F.O.Riemelmoser, R.Pippan: Int J Fracture Vol. 103 (2000), p.397

[3] A.R.Zak, M.L.Williams. J Appl Mech, Vol. 30 (1963), p.142

[4] Y.T.Li, C.F.Yan, H.Huang: Materials Science Forum Vols.437-438 (2003), p.309

[5] Y.T.Li, P.Ma, C.F.Yan: Key Engineering Materials Vols.324-325 (2006), p. 503

[6] Y.T.Li, C.F.Yan: Key Engineering Materials Vols.306-308 (2006), p. 7

[7] Y,T,Li , M Song: Acta Mechanica Solida Sinica Vol. 21(4) (2008), p.337

[8] A.Carpinteri, M.Paggi, N.Pugno. Int J Solids and Structures. Vol. 43 (2006), p.627

[9] M. Song, Y,T,Li, H.Y.Dua: Mechanical Design, Vol. 27(9) (2010), p.

An Evaluation on the Restrained Shrinkage of Ultra-High Performance Concrete

Jung-Jun Park[1, a], Doo-Yeol Yoo[2,b], Sung-Wook Kim[1,c] and Young-Soo Yoon[2,d]

[1]Struct. Engng. Research Division, Korea Institute of Construction Technology, Goyang, Korea

[2]School of Civil, Environmental and Architectural Engineering, Korea University, Seoul, Korea

[a]jjpark@kict.re.kr, [b]dooyoul@korea.ac.kr, [c]swkim@ kict.re.kr, [d]ysyoon@korea.ac.kr

Keywords: ultra high performance concrete, restrained shrinkage, tensile strength, stress relaxation

Abstract. Since ultra-high performance concrete (UHPC) is subject to large occurrence of shrinkage at early age due to its low water-to-cement ratio, the mixing of large quantities of powdered admixtures and the absence of coarse aggregates, UHPC presents large risks of shrinkage cracking caused by the restraints provided by the form and reinforcing bars. Accordingly, this study intends to evaluate the shrinkage behavior of UHPC under restrained state by performing restrained shrinkage test using ring-test. The test results reveal that increasing thickness of the inner ring increases the tensile creep at early age leading to the reduction of the average strain and residual stress of the inner ring.

Introduction

UHPC is a cementitious composite improving both strength and toughness by reducing the water-to-cement ratio (W/B) down to 20% and admixing high fineness admixtures and high-elastic steel fiber. However, its low W/B compared to conventional concretes, the admixing of large quantities of admixtures and the absence of coarse aggregates lead to large autogenous shrinkage and large risk of occurrence of shrinkage cracking at early age due to the significant reduction of the cross-section enabled by the increase of strength when applied as structural member [1].

Since the shrinkage cracking behavior of concrete is influenced by the shrinkage development speed and size according to age as well as by the development of strength, stress relaxation and degree of restraint, this behavior is extremely complex. Therefore, the evaluation of the risk of shrinkage cracking should be accompanied by the execution of restrained shrinkage test and the evaluation of the mechanical characteristics [2,3]. Accordingly, this study performs ring-test considering varying thickness of the inner ring as variable to evaluate the shrinkage cracking and restrained shrinkage behavior of UHPC. At the same time, free shrinkage test is also performed on the specimens exhibiting identical cross-sectional dimensions and exposure conditions.

Materials and Mix Design

The test adopts type 1 Portland cement, fine aggregates with grain size smaller than 0.5 mm, filler with granulometry of 2 μm and 98% content of SiO_2, and silica fume (SF) with specific surface area of 200,000 cm^2/g. W/B ratio of 20% is used for the mix of UHPC. Polycarbonate superplasticizer is admixed to secure workability and straight high-elastic steel fiber (density 7.8 g/cm^3, tensile strength 2,500 MPa, length 13 mm, diameter 0.2 mm, shape factor 65) is introduced at a ratio of 2% of the total volume of concrete to improve the tensile strength and ductility. Table 1 lists the mix proportions.

Test Method

The tensile strength was measured from the initial setting using the early age tensile test equipment of Fig. 1(a) and measurement was performed at intervals of approximately 1 to 2 hours. After about 1 day, the form was completely stripped and measurement was conducted until 28 days using the dog-bone test equipment of Fig. 1(b) [4]. Tests were conducted using a UTM (Universal Testing Machine) with maximum capacity of 2,500 kN. Loading was applied at speed of 0.8 mm/min through

Table 1. Mix proportions (ratio in weight)

Nomenclature	W/B	Cement	SF	Filler	Sand	SP	Steel fiber
UHPC	0.2	1	0.25	0.30	1.10	0.012	V_f= 2%

where SF = Silica Fume, SP = Superplasticizer

displacement control. Two LVDTs (Linear Voltage Differential Transformers) were installed at the center of the specimen at which tensile cracks occur to measure the displacement. The mean values of the data measured by the two LVDTs were used. After placing, the specimens were cured in a constant temperature and humidity chamber at temperature of 23±1°C and humidity of 60±5% until the start of the tests.

(a) Early age tensile test (b) Dog-bone test
Fig. 1 Test setup of tensile strength

According to previous research results [3], shrinkage cracking of UHPC cannot be induced by conducting the ring-test as proposed in the specifications of AASHTO PP34-98 because of the outstanding tensile strength development of UHPC and the effect of the tensile creep at early age. Therefore, in this study, the thickness of the concrete ring (t_c) is reduced from the previous 35 mm to 20 mm to improve the degree of restraint and tests were performed by applying various thickness and radius of the inner steel ring. The designation system according to the radius of the inner ring is illustrated at the bottom of Fig. 2. Moreover, the height of the ring-test was reduced from the previous 152 mm to 75 mm to induce uniform drying shrinkage in the section of the concrete ring. The circumference of the concrete ring was sealed using aluminum bond tape and test was carried out by exposing the upper and bottom sides [1]. The test procedure followed the regulations of AASHTO PP34-98. The details of the mold and specimen are depicted in Fig. 2. The test was conducted at constant temperature of 23±1°C and constant humidity of 60±5% after placing.

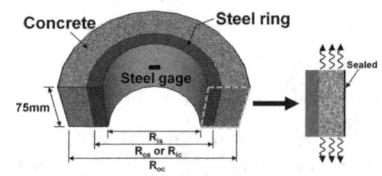

R-NS: R_{is}=127 mm, R_{os}=145.5 mm (t_s=18.5 mm),
R-MS: R_{is}=120 mm, R_{os}=145.5 mm (t_s=25.5 mm),
R-TS: R_{is}=115.5 mm, R_{os}=145.5 mm (t_s=30 mm)
Fig. 2 Restrained ring-test specimens

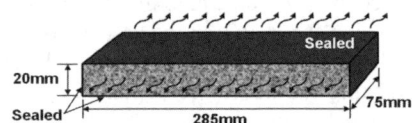

Fig. 3 Prismatic free shrinkage specimen with sealing

For comparison with the ring-test results, the free shrinkage strain and thermal variation of UHPC were measured using dumbbell-shaped gages and a thermocouple embedded in the prismatic mold with dimensions of 20×75×285 mm. The mold was stripped 24 hours after concrete placing and the top, bottom and front sides of the specimen were sealed using aluminum bond tape as shown in Fig. 3 to achieve a volume-to-exposed surface area ratio (V/S) identical to that of the concrete ring. The test was conducted identically to the ring-test under constant temperature and humidity.

Experimental Results and Discussion

Fig. 4 plots the strain of the inner steel ring according to the age. For R-NS featured by the smallest thickness of the inner ring, the largest strain was observed at 4 days with value of about –31 με. The maximum strains of R-MS and R-TS ran around –19 με and –11 με, respectively, and indicate that the strain tends to decrease with larger thickness. Even if identical thickness of concrete ring and exposure conditions were applied, the decrease of the average strain measured in the inner ring with larger thickness of the inner ring can be explained by the fact that the stress developed in the inner

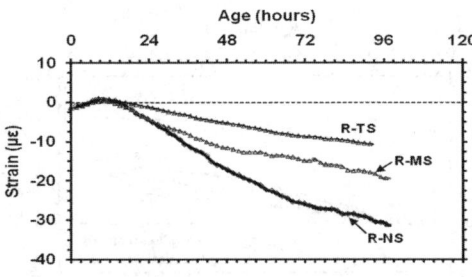
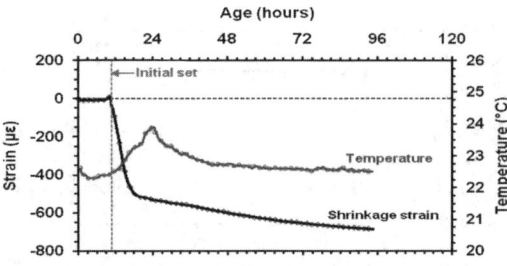

Fig. 4 Measured average strains of the inner steel ring

Fig. 5 Total free shrinkage of the partially exposed specimen

ring decreases with larger thickness of the inner ring while the interface pressure due to the shrinkage of concrete remains unchanged. Test was carried out on a free shrinkage specimen fabricated with identical sectional size and V/S for comparison purpose with the ring-test results. As shown in Fig. 5, the strain and temperature change were measured immediately after placing. The behaviors of the temperature and strain started to differ approximately 1 hour before early setting and, the shrinkage strain experienced sudden increase. The maximum temperature was 23.9°C at about 1 day and the maximum strain reached –686 με. The interface pressure (P_i) provoked by the restrained shrinkage of concrete due to the inner ring acts as a compressive force on the inner steel and causes tensile force of the same size on the concrete ring. Accordingly, assuming that concrete exhibits uniform and linear behavior and that uniform drying shrinkage occurs, the interface pressure can be calculated by the following Eq. (1) using the measured strain of the inner ring [1].

$$P_i = \frac{(r_{os}^2 - r_{is}^2)}{2 r_{os}^2} E_{st} \varepsilon_{st} \quad (1)$$

where r_{is} and r_{os} are respectively the inner and outer radii of the inner steel ring, and ε_{st} and E_{st} are respectively the strain and elastic modulus of the inner steel ring.

The maximum value of the stress in the circumferential direction occurs at the surface of the inner ring and concrete ring. If the interface pressure of Eq. (1) and r_{is} are substituted as the circumferential stress (σ_θ) and arbitrary value r, the equation can be reformulated as the product of the shape factor and elastic modulus and measured strain of the inner ring as expressed in Eq. (2).

$$\sigma_{t\max} = \frac{(r_{os}^2 - r_{is}^2)}{2 r_{os}^2} \frac{(r_{ic}^2 + r_{oc}^2)}{(r_{oc}^2 - r_{ic}^2)} E_{st} \varepsilon_{st} \quad (2)$$

where r_{ic} and r_{oc} are respectively the inner and outer radii of the concrete ring.

The average tensile strength of UHPC at 28 days appears to be approximately 11.8 MPa. The tensile strength tends to increase suddenly as an exponential function at early placing to show reduced increase rate after 1 day and exhibit S-shaped variation to converge to a definite value (Fig. 6). Therefore, the prediction model of the tensile strength expressed in Eq. (3) is applied as proposed in previous studies [4].

$$f_t(t) = f_{t28} \exp\left\{-a \left[\ln(1+(t-t_i))\right]^{-b}\right\} \quad (3)$$

Where f_{t28} is the tensile strength at 28 days (MPa), t_i is the time (days) when the penetration resistance reaches 1.5 MPa, and a and b are regression coefficients. From the regression analysis, $a = 0.204$ and $b = 1.292$.

Fig. 7 compares the residual stress calculated by Eq. (2), the tensile strength measured during the test and the tensile strength obtained by the prediction model (Eq. (3)) [4]. Moreover, the residual stress occurring due to the shrinkage restraint of the inner ring appears to be significantly smaller than the tensile strength of UHPC. Accordingly, shrinkage cracking did not occur in all the specimens. In

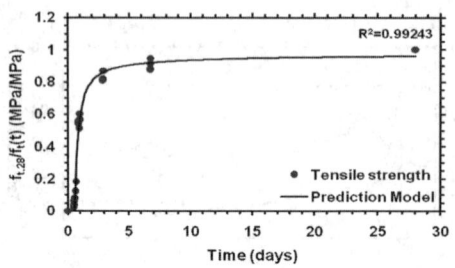
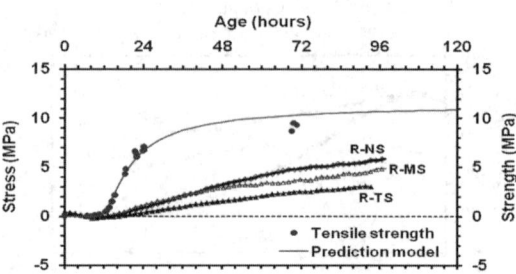

Fig. 6 Tensile strength of UHPC   Fig. 7 Comparison of residual tensile stress and tensile strength of UHPC

addition, the residual stress decreased with larger thickness of the inner ring. This can be explained by the fact that the tensile creep increases at early age for larger thickness of the inner ring despite of the action of identical interface pressure. In other words, insufficiently hardened concrete at early age cannot deform the inner ring when the thickness of the inner ring becomes thicker and induces the increase of the relaxed shrinkage, which causes increase of the tensile creep and stress relaxation.

Conclusions

This study compared the residual stress and tensile strength of UHPC under restrained state by conducting ring-test considering the thickness and radius of the inner ring as variables. The following conclusions can be drawn from the results.

1) Even if identical thickness of concrete ring and exposure conditions are applied, the average strain measured in the inner ring decreases with larger thickness of the inner ring.

2) The results of the tests performed on the free shrinkage specimen fabricated to have sectional size and V/S identical to the concrete ring showed that the shrinkage strain experienced sudden increase about 1 hour earlier than the early setting. The maximum temperature was 23.9°C at about 1 day and the maximum strain reached –686 με.

3) The residual stress of UHPC caused by the restraint of the inner ring appeared to be significantly smaller than the tensile strength. Shrinkage cracking did not occur in all the specimens. Moreover, the residual stress tended to decrease because tensile creep increased at early age with larger thickness of the inner ring.

Acknowledgements

This study was carried out as a partial research of the "Development of Ultra High Performance Concrete for Hybrid Cable Stayed Bridges" in the Korea Institute of Construction Technology. The authors express their gratitude for the support.

References

[1] Korea Institute of Construction Technology: *Development of Ultra High Performance Concrete for Hybrid Cable Stayed Bridges*, Research Report, KICT 2011-067 (2011).

[2] H.T. See, E.K. Attiogbe and M.A. Miltenberger: Shrinkage Cracking Characteristics of Concrete Using Ring Specimens, ACI Materials Journal, Vol. 100, No. 3 (2003), pp. 239-245.

[3] D.Y. Yoo, J.J. Park, S.W. Kim and Y.S. Yoon: Characteristics of Early-Age Restrained Shrinkage and Tensile Creep of Ultra-High Performance Cementitious Composites (UHPCC), Journal of the Korea Concrete Institute, Vol. 23, No. 5 (2011), pp. 581-590.

[4] D.Y. Yoo, J.J. Park, S.W. Kim and Y.S. Yoon: Early Age Setting, Shrinkage and Tensile Characteristics of Ultra-High Performance Fiber Reinforced Concrete, Cement and Concrete Research (2012), Under Review.

Structural Strength Evaluation of a Stainless Railroad Car

Sung-Cheol Yoon [1,a], Jeongguk Kim [1,b], Don-Bum Choi [1,c],
Dong-Hoe Koo [1,d] and Geun-Soo Park [2,e]

[1]Korea Railroad Research Institute, 360-1, Woulam, Uiwang, Kyunggi, South Korea 437-757

[2]Hyundai Rotem Company, 462-18, Sam, Uiwang, Kyunggi, South Korea 437-718

[a]scyoon1@krri.re.kr, [b]jkim@krri.re.kr, [c]eye@krri.re.kr, [d]dhkoo@krri.re.kr, [e]gspark@rotem.co.kr

Keywords: Car body, Stress analysis, Load test , Stress, Natural frequency

Abstract. This study is an experimental research on using stainless steel car body structure for urban railway vehicles. The car body used in the experiment was made of stainless steel. To evaluate the structural characteristics and safety of the stainless car body, static load test was carried out by means of "Performance Test Standard for Electrical Multiple Unit" with the reference code JIS E 7105. The structural safety of the car body was evaluated by implementing vertical load, horizontal compressive load, twist load, and 3-point support tests. Test results verified that the structural safety of a car body was very stable and safe under design load conditions.

Introduction

As a method to save energy and materials, studies on lightweight car structures have been conducted. Stainless steel was first used in the United States in manufacturing railway vehicles 70 years ago. In Japan, lightweight stainless car body is most commonly adopted in the railway vehicle industry, and in Korea, it is currently being manufactured in the metropolitan areas. At first, the main purpose of manufacturing stainless steel car body was to solve corrosion problems and achieve convenience in maintenance of railway vehicles. Eventually, researches on structural improvement for weight reduction were carried out. To evaluate safety, tests on the strength of car body was conducted, targeting vehicles modified in structures and shapes, when standards on the performance test of urban railway vehicles were applied as load conditions of the load test. The urban railway vehicle model for this study is a Commuter DC Electric Traction Vehicle, which corresponds to a medium-sized electric motor unit (hereafter, referred to as EMU) with a length of 17,500 mm and width of 2,800 mm.

Load test of the car body

Stress measurements were conducted by bonding strain gauge to the part in which stress concentration was expected. Areas with high concentration of stress were determined based on the results of structural analysis on the car body. A comparative analysis on the characteristics was carried out by implementing vertical load, horizontal compressive load, 3-point support, and twist load tests. For the frequency measurement, the car body was supported in the same way as in the vertical load test, and an accelerometer was connected to the measuring equipment after being attached to the side seal, and load was applied on the center of the car body. This way, bending and natural frequency was measured [1-3]. As a test vehicle for examining car body load, a motor car with the most severe loading conditions was selected, and its weight and specifications were shown in Table 1 and 2.

Load conditions

Load test on the stainless car body was conducted under the load conditions of vertical load (44.3 ton), horizontal compressive load (11.7 ton), compressive load (60 ton), 3-point support load (11.7 ton), and twist load (4 ton·m). The test loads for the experiment and the calculated weight of stainless steel EMU were shown in Table 2 and 3.

Table 1. Dimension of the body structure

Items	Specifications (mm)
Body length	17,500
Body width	2,750
Roof height	3,600
Distance between the centers of bolsters	12,400
coupler height	880±10(based on the upper side of the rail)

Table 2. Weight of the body structure

Items	Weight
Tare weight	32.8 ton
Bare frame weight	7.5 ton
Bogie weight	13.6 ton
Maximum passenger weight	24.0 ton
Dynamic load coefficient	0.2 g

Table 3. Load conditions

Load conditions	Load
Vertical load	44.3 ton
Horizontal compressive load	Vertical load (11.7 ton) + compressive load(60 ton)
3-point support load	11.7 ton
Twist load	4 ton·m

Fig. 1. Loading test equipment

Mechanical properties of materials of the car body

The bolster and center sill of the car body was manufactured using high tension steel, and the rest of the parts was made of stainless steel.

Table 4. Mechanical properties of materials

Material	Allowable stress	Remarks
STS301L-LT	22 kgf/mm^2	Cross beam, End sill
STS301L-DLT	35 kgf/mm^2	Side window panel
STS301L-ST	42 kgf/mm^2	Entrance frame, Car line Roof panel, Skip plate, Keystone plate
STS301L-HT	70 kgf/mm^2	Cant rail, Side sill, Side post
SMA490B	37 kgf/mm^2	Bolster, Cener sill
STS304	21 kgf/mm^2	Bracket

Evaluation criteria

The strength of the car body was based on the allowable stress shown in Table 4. Stress was set to be less than the allowable stress, and bending frequency was to be more than 10.0 Hz.

Load test results of the car body [4, 5]

Vertical load test

For the vertical load test, the measured points with stress of more than 13.0 kgf/mm² were shown in Table 5. Results showed that all the measured stresses were within the allowable limits, and maximum stress occurred in the connections (strain gauge No. 66) of bolster of under frame and centersill

Table 5. Vertical load

S/G No.	Vertical load [tons]					Material	Allowable stress [kgf/mm²]
	0	5.85	11.7	44.3	Unloading		
66	-0.06	-2.47	-4.57	-16.13	0.02	SMA490B	37
37	0.00	-2.89	-5.02	-15.67	0.05	SMA490B	37
8	-0.05	-2.65	-4.15	-13.10	0.02	STS-DLT	35

Horizontal compressive load test

For the horizontal compressive load test, the measured points with stress of more than 8.0 kgf/mm² were shown in Table 6. All the result values of stress measurement turned out to be within the allowable limits, and maximum stress occurred in the joints (strain gauge No. 38) of bolster and centersill.

Table 6. Horizontal compressive load

S/G No.	0	Vertical load : 11.7 [tons]				Unloading	Material	Allowable Stress [kgf/mm²]
		0	25	50	60			
38	0.00	-1.56	-4.51	-7.37	-8.78	-0.16	SMA490B	37
37	0.02	-4.90	-6.43	-7.83	-8.62	-0.19	SMA490B	37
39	0.02	-4.14	-5.83	-7.33	-8.13	-0.18	SMA490B	37

3-point support test

For the 3-point support test, the measured points with stress of more than 10.0 kgf/mm² were shown in Table 7. Results showed that all the measured stresses were within the allowable limits, and maximum stress occurred in the lower corner of the window (strain gauge No. 31).

Table 7. 3-point support

S/G No.	0	Vertical load: 11.7 [tons]		0	Material	Allowable stress [kgf/mm²]
		4-point support	3-point support			
31	-0.06	-3.18	-13.69	-0.05	STS301L-DLT	35
18	-0.11	0.42	-10.14	-0.16	STS301L-DLT	35
62	-0.06	-3.79	-10.09	-0.13	STS301L-HT	70

Twist load test

For the twist load test, the measured points with stress of more than 2.0 kgf/mm² were shown in Table 8. Based on this result, it was confirmed that all the measured stresses were within the allowable limits, and maximum stress occurred in the lower corner of the door (strain gauge No. 31).

Table 8. Twist load

S/G No.	0	4 [ton · m]	Unloading	Material	Allowable stress [kgf/mm^2]
31	0.06	-2.63	0.03	STS301L-DLT	35
18	0.05	-2.45	0.01	STS301L-DLT	35
37	0.05	-2.34	0.03	SMA490B	37

Installation of the strain gage
The strain gauges were mounted in the part in which high stress was expected under each load conditions. On the assumption that structure was symmetrical in the horizontal direction, the strain gauges were mounted in 1/2 areas collectively, and the strain gauge was also mounted in 1/2 zone of the underframe. As such, a total of more than 70 strain gauges were mounted.

Natural frequencymeasurement test
For the bending frequency test, the car body was supported by vertical supports, and to generate vibrations in the car body, load was applied momentarily on the center of the upper surface of the underframe using hydraulic cylinder. The vibration waveform was measured by a frequency recorder. The bending and natural frequency was recorded at 15.5 Hz.

Conclusions

To evaluate the structural safety of the stainless steel car body in this study, vertical load, horizontal compressive load, twist load, 3-point support, and bending and natural frequency tests were carried out. The test results were as follows:
1) In the vertical load test, a maximum stress of -16.13 kgf/mm^2 occurred in connections of the bolster of the underframe and centersill. In the horizontal compressive load test, a maximum stress of -8.78 kgf/mm^2 occurred in the joints of the bolster and centersill.
2) In the 3-point support test, a maximum stress of -13.69 kgf/mm^2 occurred in the lower corner of the window. In the twist load test, a maximum stress of -2.63 kgf/mm^2 occurred in the lower corner of the door.
3) In the vertical load test, the maximum stress occurred in the connections of the bolster and centersill, and in case of the 3-point support test, the maximum stress occurred in the lower corner of the window. Based on this result, all the measured stresses were within the allowable stress range of the materials used, thereby verifying the safety of the material's strength.
4) Bending and natural frequency of the car body was approximately 15.5 Hz, which met the criteria of more than 10.0 Hz.

References

[1] MOCT, "Performance Test Standard for EMU," Load Test for Body Structure, 2000.

[2] MOCT, "Regulation for Safety Standard for EMU," Safety Standard for Structure, 2000.

[3] MOCT, "Standard Specification for EMU," 1998.

[4] S.C. Yoon, J. Kim, C.S. Jeon, W.K. Kim, and K.Y. Choe: Key Engineering Materials, Vols. 417-418 (2010), p. 101-104.

[5] Japanese Industrial Standards, "Test methods for static load of body structure of railway rolling stock," E 7105, 1989.

Study of the Characteristic of Composite Sandwich Panel with Cut-out under Compression Load

Jong Woong Lee [1,a], Cheol Won Kong [1,b], Young Soon Jang [1,c] and Young Shin Lee [2,d]

[1]Korea Aerospace Research Institute, 45 Eoeun - Dong, Yuseong-Gu, Daejeon, 305-333, Korea

[2]Chungnam National University, 99 Daehak-ro, Yuseong-Gu, Daejeon, 305-764, Korea

[a]jwlee@kari.re.kr, [b]kcw@kari.re.kr, [c]ysjang@kari.re.kr, [d]leeys@cnu.ac.kr

Keywords: Composite sandwich panel, Cut-out, Compression test

Composite materials are used in aerospace structures due to their considerable bending stiffness and strength-to-weight ratio. A composite sandwich is composed of face-sheets and an aluminum core. Launch vehicle structures have cut-out areas through which to check the status of the payload and electric equipment. Cut-out areas reduce the structural stiffness of the launch vehicle structure. The shapes of the cut-outs are circular, square, oval_V(in which the major axis is vertical) and oval_H(in which the major axis is horizontal). In this paper, compression tests and FEM analysis were performed for composite sandwich panels with various types of cut-outs. To verify the failure strength, the results of compression tests and FEM analysis were compared. Results of compression tests and FEM analysis show that the oval_V shape has the maximum load and strain compared to the other shaped cut-outs.

Introduction

A sandwich construction is applied to wing skins and fuselages, along with many other structures in aeronautical applications, due to its increased stiffness and reduced weight. In this paper, a face-sheet composed of plain-woven composite laminate that uses prepregs (HPW193/RS3232) and a core material of aluminum honeycomb (Al5052/F40-0.0025) is used. Launch vehicle structures have cut-out areas to check the status of the payload and electric equipment. Cut-out areas reduce the structural stiffness of the launch vehicle structure. Changes in the shape of cut-outs can cause a difference in load support ability of the structure. Compression tests and FEM analysis were performed on composite sandwich panels with various types of cut-out, but with the same overall area.

Table 1 Dimension of specimens and material property

	Height	Width	Thickness	Shape of cut-out	Material	Material property[1]
Specimen 1	550 [mm]	500 [mm]	28.6 [mm]	No cut-out	Face-sheet	E_1=62 [Gpa], E_2=62 [Gpa], E_3=10 [Gpa], G_{12}=4.25 [Gpa], G_{23}=3.0 [Gpa], G_{13}=3.0 [Gpa], v_{12}=0.045, v_{23}=0.045, v_{13}=0.045, X_t=680 [MPa], X_c=697 [MPa], S=69 [MP]
Specimen 2	550 [mm]	500 [mm]	28.6 [mm]	Square		
Specimen 3	550 [mm]	500 [mm]	28.6 [mm]	Circle		
Specimen 4	550 [mm]	500 [mm]	28.6 [mm]	Oval_H	Core	E_1= 8.27 [Mpa], E_2=1.31 [Mpa], E_3=1276 [Mpa], G_{12}=0.0001 [Mpa], G_{23}=117 [MPa], G_{13}=296 [MPa], v_{12}=0.75, v_{23}=0.0001, v_{13}=0.0001
Specimen 5	550[mm]	500 [mm]	28.6 [mm]	Oval_V		

Compression test

Fig. 1 shows the configuration of the compression test. The upper loading fixture is composed of a hemisphere and a plate. A compression load was applied to the specimen using the upper loading fixture and the hemisphere. First, the specimen was fixed between the upper loading fixture and the base plate. For uniform contact with the specimen, the upper loading fixture was not tightened by a fixing bracket. The upper loading fixture was able to rotate freely in the hemisphere.
In order to ensure close contact between the loading fixture and specimen, the preload was applied and the fixing bracket was then tightened. The upper loading fixture was controlled by displacement. The compression load was 0.01 mm/sec with displacement control. Twenty strain gauges were used to acquire the data of each specimen. Six displacement gauges were used to acquire the axial displacement and radial displacement.

Fig. 1 Test configuration   Fig. 2 Test results of specmen1   Fig. 3 Test results of specmen2

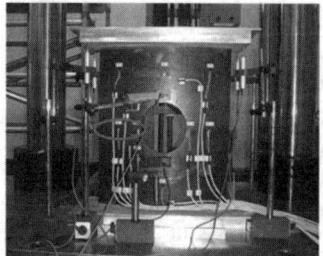 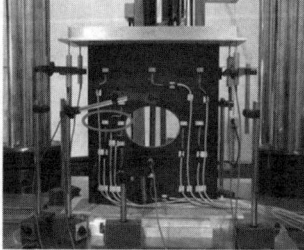 

Fig. 4 Test results of specimen3   Fig. 5Test results of specimen4   Fig. 6 Test results of specimen5

Figs. 2, 3, 4, 5 and 6 show the test results of the specimens and the red circle represents where failure occurred. Figs. 7, 8, 9, 10 and 11 show the load-strain curve of the specimens. The failure mode of the specimens was a face-sheet failure under compressive load [2].

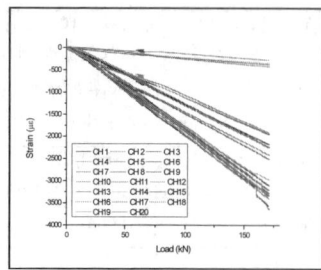 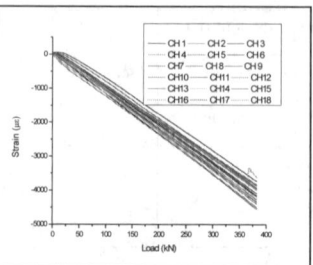 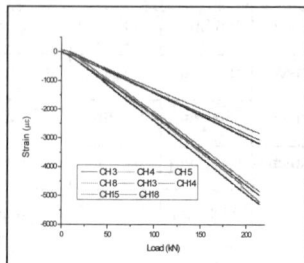

Fig. 7 Strain results of specimen1   Fig. 8 Strain results of specimen2   Fig. 9 Strain results of specimen3

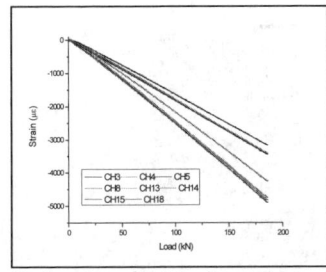
Fig. 10 Strain results of specimen4

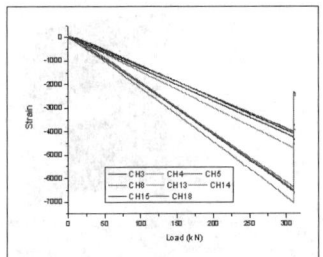
Fig. 11 Strain results of specimen5

The failure load and strain of specimen 1 were 381.6 kN and 4545 μs, respectively. Structural failure occurred in the face-sheet which is located below the upper fixture. Specimen 2 had a square cut-out. The failure load and strain of Specimen 2 were 171.4 kN and 3657 μs. Specimen 3 has circular cut-out. The failure load and strain of Specimen 3 were 251.1 kN and 5290 μs. Specimen 4 had an oval cut-out. The major axis of the oval was in the horizontal direction. The failure load and strain of Specimen 4 were 181.6 kN and 4900 μs. Specimen 5 had an oval cut-out. The major axis of the oval was in the vertical direction The failure load and strain of Specimen 5 were 312.3 kN and 7000 μs. All specimen failures occurred in the face-sheet which is located on the edge of the cutout.

FEM analysis

In FEM analysis, ANSYS was used. The composite sandwich structure has a thick core between the face-sheets and various modeling methods are used for FEM analysis. In this paper, the composite sandwich panel is modeled using SHELL91 element. The element of SHELL91 have eight nodes and the material properties are entered at each layer. First-order shear deformation theory is applied. Analysis results are shown in Fig. 12 - Fig. 16. The face-sheets of the composite sandwich panels are carbon/epoxy fabric and the thickness of each face-sheet is 3.2 mm. The core of the composite sandwich panels is an aluminum honeycomb and its thickness is 25.4 mm. The bottom of the FEM model has a fixed boundary condition and axial force is applied to the top of the FEM model.

The strain of specimen 1 was 4000 μs at the place where the failure occurred. The compression load and displacement were 390 kN and 2.16 mm respectively. The strain of specimen 2 was 3200 μs at the place where the failure occurred. The compression load and displacement were 180 kN and 1.18 mm. The strain of specimen 3 was 5600 μs the edge of the cut-out. The compression load and displacement were 220 kN and 1.45 mm. The strain of specimen 4 was 4600 μs at the place where the failure occurred. The compression load and displacement were 200 kN and 1.44 mm. The strain of specimen 5 was 6100 μs at the place where the failure occurred. The compression load and displacement were 320 kN and 1.98 mm.

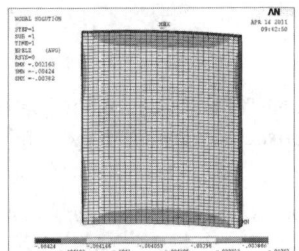
Fig. 12 FE results of specimen1

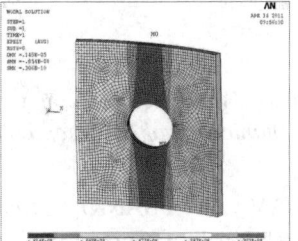
Fig. 13 FE results of specimen2

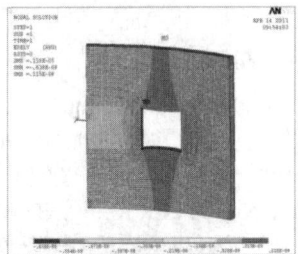
Fig. 14 FE results of specimen3

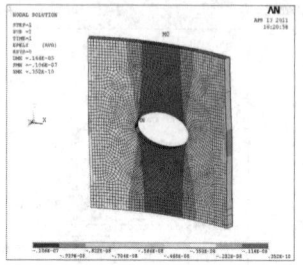 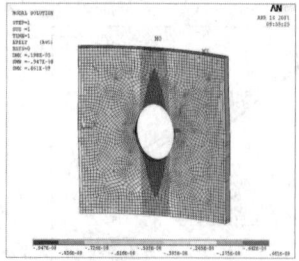

Fig. 15 FE results of specimen4   Fig. 16 FE results of specimen5

Conclusion

The results of analysis are compared to the results of the test in Table 2. In the test results, failure load and strain measured in the failure area were verified. Compared to the test results, the failure loads and strains from the analyses are similar to the test results. The specimens with various cut-out shapes confirmed the failure load compared to the specimen which had no cut-out. The cut-out area of all specimens is the same. The failure load of specimen 2 is 44.9 %(171.4 kN) of specimen 1(100%, 381.6 kN). The failure loads of specimens 3, 4 and 5 are 56.3 %(215.1 kN), 48.7 %(186.1 kN) and 81.8 %(312.3 kN) of specimen 1 respectively. Results of compression tests and FEM analysis show that the oval_V shape can bear the maximum load and strain compared to the other specimens with cut-outs.

Table 2 Test and analysis results

	Load [kN]	Strain [µs]	Disp. [mm]
Specimen 1 Test	381.6	4545	2.62
Analysis	390	4000	2.16
Specimen 2 Test	171.4	3657	1.62
Analysis	180	3200	1.18
Specimen 3 Test	215.1	5290	3.19
Analysis	220	5600	1.45
Specimen 4 Test	186.1	4900	2.96
Analysis	200	4600	1.44
Specimen 5 Test	312.3	7000	3.08
Analysis	320	6100	1.98

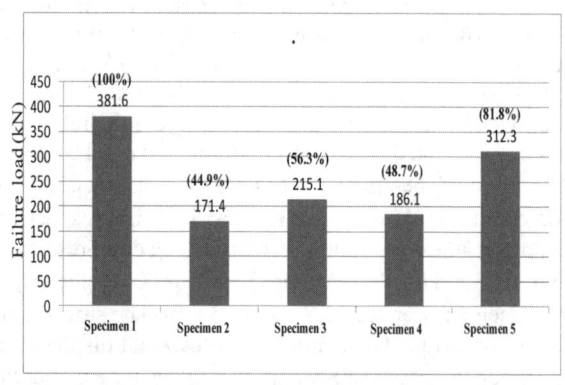

Fig. 17 Comparison of failure load

References

[1] C. W. Kong, S. W. Eun, J. S. Park, H. S. Lee, Y. S. Jang, Y. M. Yi and G. R. Cho, *The effect of honeycomb core on the mechanical properties of composite sandwich*, Key Engineering Materials, Vol. 297-300, (2005)

[2] JOHN L. AVERY III AND BHAVANI V. SANKAR, *Compressive Failure of Sandwich Beams with Debonded Face-sheets*, Journal of Composite Materials, Vol. 34, No. 14/2000, pp. 1176-1199 (1999)

Size Effect of PHE prototype on high-temperature structural integrity

Kee-nam Song[1,a]

[1]P.O.Box 105, Yuseong, Daejeon, 305-600, Korea

[a]knsong@kaeri.re.kr

Key Words: Process Heat Exchanger (PHE) Prototype, High-temperature Structure Analysis, Small-scale Gas Loop, Very High Temperature Reactor (VHTR)

Abstract. PHE (Process Heat Exchanger) is a key component for transferring the high-temperature heat generated from a VHTR (Very High Temperature Reactor) to a chemical reaction for massive production of hydrogen. Recently, Korea Atomic Energy Research Institute (KAERI) has manufactured a small-scale and a medium-scale PHE prototype made of Hastelloy-X of high-temperature alloy and a performance test on the PHE prototype is scheduled in a small-scale nitrogen gas loop established at KAERI. In this study, in order to compare the high-temperature structural integrity of the PHE prototypes under the normal test condition of the gas loop, high-temperature structural analyses on the PHE prototypes were carried out and the analyses results were compared to each other. As a result of comparisons, the high-temperature structural integrity of the medium-scale PHE prototype gets worse due to higher thermal expansion by a size effect.

Introduction

Hydrogen is considered a promising future energy solution because it is clean, abundant, and storable, and has high-energy density. One of the major challenges in establishing a hydrogen economy is how to produce massive quantities of hydrogen in a clean, safe, and economical way. Among the various hydrogen production methods, nuclear hydrogen production is garnering attention worldwide since it can produce hydrogen, a promising energy carrier, without an environmental burden. Research demonstrating the massive production of hydrogen using a VHTR designed for operation at up to 950℃ has been actively carried out worldwide [1,2,3].

The nuclear hydrogen program in Korea is strongly considering producing hydrogen by employing a Sulfur-Iodine (SI) water-splitting hydrogen production process [1]. Recently, KAERI (Korea Atomic Energy Research Institute) has manufactured a medium-scale PHE prototype made of Hastelloy-X of high-temperature alloy and a performance test on the PHE prototype is scheduled in a small-scale nitrogen gas loop at KAERI. In this study, in order to compare the high-temperature structural integrity of the PHE prototypes under the normal test condition of the gas loop, high-temperature structural analyses on the PHE prototypes were carried out and the analyses results were compared to each other.

Finite Element (FE) Modeling on a Medium-Scale PHE Prototype

Overall structure. The PHE prototype which is composed by primary and secondary flow plates as shown in Fig. 1 is designed as a hybrid concept [4]. A schematic view of the PHE prototype is illustrated in Fig. 2. From Figs. 1 and 2, the hot nitrogen gas channel has a compact semicircular shape, similar to a printed circuit heat exchanger, and is designed to withstand the high pressure difference between loops, while the cold sulfuric acid gas channel has a plate fin shape with sufficient space to install and replace the catalysts for sulfur trioxide decomposition. All parts of the PHE prototype are made from Hastelloy-X. Grooves of 1.0 mm diameter are machined into the flow plate for the primary coolant (nitrogen gas). Waved channels are bent into the flow plate for the secondary coolant (SO_3 gas). Twenty flow plates for the primary and secondary coolants are stacked in turn for the small-scale PHE prototype, while forty flow plates are stacked for the medium-scale PHE prototype, and are bonded along the edge of the flow plate using a solid-state diffusion bonding method. After stacking and bonding the flow plates, the outside of the PHE is covered with the Hastelloy-X plate of 3.0 mm thickness.

FE Modeling. Figure 3 shows the overall dimensions and each part of the PHE prototypes via the 3-D CAD modeling. Based on Fig. 3, FE modeling using I-DEAS/TMG Ver. 6.1 [5] was carried out and analyses such as a thermal analysis and structural analysis are carried out using ABAQUS Ver. 6.8 [6]. For the sake of simplicity and understanding the overall behavior of the PHE prototypes, the FE model of the small-scale PHE prototype is composed of 546,764 2-D linear quadrilateral shell elements and 911,012 3-D linear solid elements made of 830,304 brick elements, 80,348 wedge elements, and 360 tetrahedron elements, while 680,772 2-D linear quadrilateral shell elements and 870,696 3-D linear solid elements including 66,456 tetrahedron elements are used in the FE model of the medium-scale PHE prototype. Figure 4 shows the boundary conditions of the primary/secondary flow plates for a thermal analysis and a structural boundary condition for structural analyses to gauge the stiffness of pipelines connected to the PHE prototype in the gas loop. The material properties of Hastelloy-X [7] are used in the FE model.

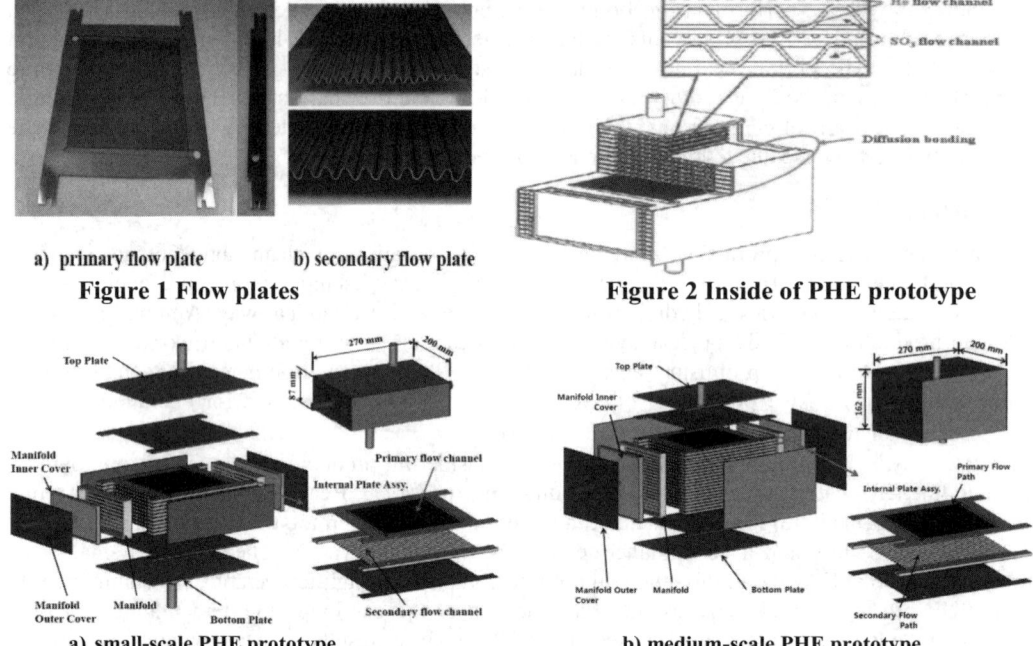

a) primary flow plate   b) secondary flow plate
Figure 1 Flow plates

Figure 2 Inside of PHE prototype

a) small-scale PHE prototype   b) medium-scale PHE prototype
Figure 3 Parts of PHE prototypes

a) thermal boundary condition   b) structural boundary condition
Figure 4 Boundary conditions

Analysis

Thermal analysis. Figure 5 shows the thermal analysis results of the outside surface of the PHE prototype under the normal test condition of the gas loop. From Fig. 5, the maximum temperature of the outside surface is about 836.26 ℃.

Coolant pressure. Coolant pressures of 3.0MPa and 0.1MPa are acting on the primary and secondary flow plate, respectively.

Stress-Strain curves for elastic-plastic structural analysis. Figure 6 represents the bilinear stress-strain curves for high-temperature elastic-plastic structural analysis [7].

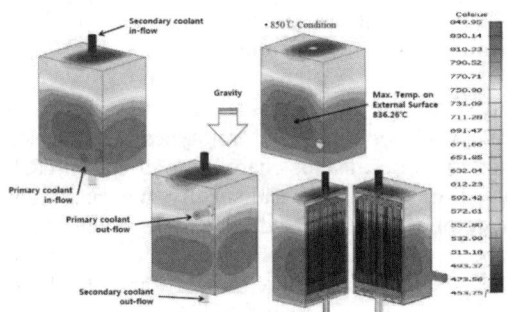

Figure 5 Temperature contour　　　　　　　**Figure 6 Bilinear stress-strain curves**

High-Temperature structural analysis. Based on the thermal analysis result, high-temperature elastic and elastic-plastic structural analyses were performed for the small-scale and medium-scale PHE prototypes.

Figure 7 represents the elastic stress contours at the pressure boundary of the small-scale and the medium-scale PHE prototypes. From the analysis results shown in Fig. 7, the maximum local stress of 331.23 MPa occurs near the edge between the top plate and side plate for the medium-scale PHE prototype, while 272.33 MPa does for the small-scale PHE prototype, even in that the maximum stresses exceed the yield stress. From the elastic analyses, the reason of higher stress for the medium-scale PHE prototype is that higher thermal expansion occurs and consequently results in higher stress and wider regions of high stress.

Figure 8 represents the elastic-plastic stress contours at the pressure boundary of the small-scale and the medium-scale PHE prototype. From the analysis results shown in Fig. 8, the stress level is higher and high stressed regions are wider than those for the small-scale PHE prototype, even in that the maximum stresses exceed the yield stress. From the elastic-plastic analyses, the reason of higher stress for the medium-scale PHE prototype is same for the elastic analysis.

From the results of Figs. 7 and 8, higher stress and wider region of high stress occur for the medium-scale PHE prototype compared to the small-scale PHE prototype.

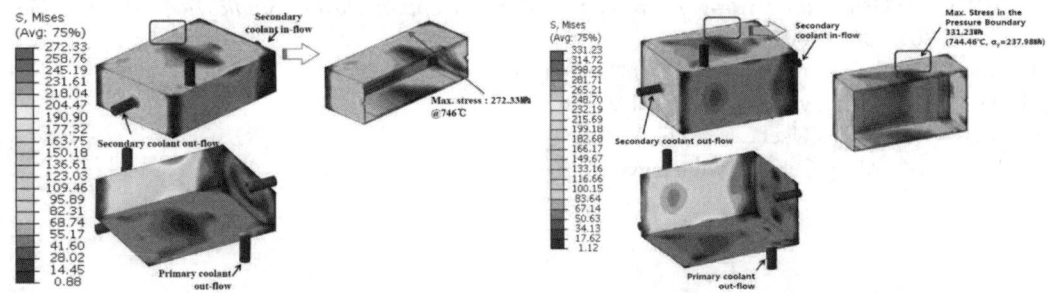

　　　　a) small-scale PHE prototype　　　　　　　　　　b) medium-scale PHE prototype
Figure 7 Elastic stress contours of the PHE prototypes

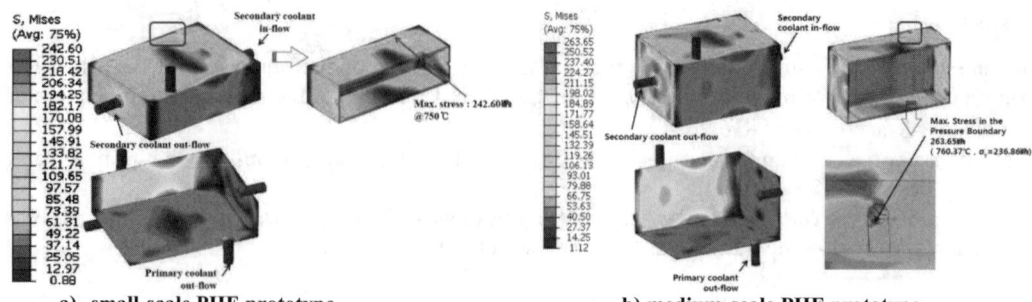

a) small-scale PHE prototype   b) medium-scale PHE prototype
Figure 8 Elastic-plastic stress contours of the PHE prototypes

Even in that the maximum stresses exceed the yield stress for the small-scale and medium-scale PHE prototypes, the high-temperature structural integrities of the PHE prototypes seem to be maintained under the normal test condition of the gas loop, because of chamfering effect on each edge. That is to say, since edges of the PHE prototype are chamfered realistically, the maximum stress will be decreased to some extent when considering the chamfered edges. However, some measure to modulate the thermal expansion of the medium-scale PHE prototype should be recommended for both re-designing the pipelines of the gas loop more flexibly and the PHE prototype for the sake of safety.

Summary

To understand the macroscopic behavior of a small-scale and a medium-scale PHE prototype, FE modeling, thermal analysis, and high-temperature elastic and elastic-plastic structural analyses were carried out under the normal test condition of the gas loop. As a result of these analyses, we draw the following conclusions.
1. Due to a size effect, higher stress and wider region of high stress occur for the medium-scale PHE prototype compared to the small-scale PHE prototype.
2. Due to a chamfering effect on each edge, the high-temperature structural integrities of the small-scale and medium-scale PHE prototypes seem to be maintained under the normal test condition of the gas loop. However, some measure should be recommended for both re-designing the pipeline of the gas loop and the medium-scale PHE prototype for the sake of safety.

References

[1] Chang, J. H. *et al.*, Nuclear Engineering and Technology, Vol. 39, No. 2 (2007), p. 111.
[2] US DOE, *Financial Assistance Funding Opportunity Announcement*, NGNP Program (2009).
[3] AREVA, *NGNP with Hydrogen Production Pre-conceptual Design Studies Report*, Doc. No. 1209052076-000 (2007).
[4] Kim, Y. W., R.O.Korea patent # 10-0877574 (2008).
[5] I-DEAS/TMG Analysis User Manual Version 6.1 (2009).
[6] ABAQUS Analysis User's Manual, Version 6.8 (2009).
[7] Hastelloy-X Alloy website, www.haynesintl.com.

High-temperature structural analysis on the small-scale PHE prototype

Kee-nam Song[1,a]

[1]P.O.Box 105, Yuseong, Daejeon, 305-600, Korea

[a]knsong@kaeri.re.kr

Key Words: Process Heat Exchanger (PHE), High-temperature Structural Analysis, Very High Temperature Reactor (VHTR), Small-Scale Gas Loop

Abstract. PHE (Process Heat Exchanger) is a key component for transferring the high-temperature heat generated from a VHTR (Very High Temperature Reactor) to a chemical reaction for massive production of hydrogen. Korea Atomic Energy Research Institute has established a small-scale nitrogen gas loop for the performance test on VHTR components and has manufactured a small-scale PHE prototype made of Hastelloy-X of high-temperature alloy. A performance test on the PHE prototype is underway in the gas loop. In this study, in order to evaluate the high-temperature structural integrity of the PHE prototype under the test condition of the gas loop, structural analysis on the PHE prototype was carried out to gauge the stiffness of pipelines connected to the PHE prototype in the gas loop.

Introduction

Hydrogen is considered a promising future energy solution because it is clean, abundant, and storable, and has high-energy density. One of the major challenges in establishing a hydrogen economy is how to produce massive quantities of hydrogen in a clean, safe, and economical way. Among the various hydrogen production methods, nuclear hydrogen production is garnering attention worldwide since it can produce hydrogen, a promising energy carrier, without an environmental burden. Research demonstrating the massive production of hydrogen using a VHTR designed for operation at up to 950℃ has been actively carried out worldwide [1,2,3].

The nuclear hydrogen program in Korea is strongly considering producing hydrogen by employing a Sulfur-Iodine (SI) water-splitting hydrogen production process. An intermediate loop that transports the nuclear heat to the hydrogen production process is necessitated for the nuclear hydrogen program as shown in Fig. 1. The PHE which generates process gas such as H_2O, O_2, SO_2, and SO_3 is a key component that utilizes the heat from a nuclear reactor to produce hydrogen. Recently, KAERI (Korea Atomic Energy Research Institute) manufactured a small-scale PHE prototype made of Hastelloy-X of high-temperature alloy and a performance test on the PHE prototype is scheduled in a small-scale nitrogen gas loop at KAERI. In this study, in order to evaluate the high-temperature structural integrity of the PHE prototype under the test condition of the gas loop, a structural analysis on the PHE prototype was carried out to gauge the stiffness of pipelines connected to the PHE prototype in the gas loop.

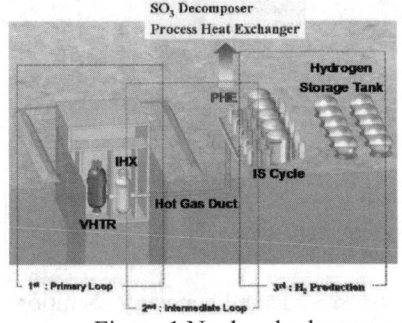

Figure 1 Nuclear hydrogen system

a) primary flow plate   b) secondary flow plate

Figure 2 Flow plates

Finite Element (FE) Modeling on a Small-Scale PHE Prototype

Overall structure. The small-scale PHE prototype which is composed by primary and secondary flow plates shown in Fig. 2 is designed as a hybrid concept [4]. That is to say, the hot nitrogen gas channel has a compact semicircular shape, similar to a printed circuit heat exchanger, and is designed to withstand the high pressure difference between loops, while the cold sulfuric acid gas channel has a plate fin shape with sufficient space to install and replace the catalysts for sulfur trioxide decomposition. All parts of the PHE prototype are made from Hastelloy-X. Grooves of 1.0 mm diameter are machined into the flow plate for the primary coolant (nitrogen gas). Waved channels are bent into the flow plate for the secondary coolant (SO_3 gas). Twenty flow plates for the primary and secondary coolants are stacked in turn, and are bonded along the edge of the flow plate using a solid-state diffusion bonding method. After stacking and bonding the flow plates, the outside of the PHE is covered with the Hastelloy-X plate of 3.0 mm thickness.

FE Modeling. Figure 3 shows each part of the PHE prototype via 3-D CAD modeling. Based on Fig. 3, FE modeling using I-DEAS/TMG Ver. 6.1 [5] was carried out and analyses such as a thermal analysis and structural analysis are carried out using ABAQUS Ver. 6.8 [6]. For the sake of simplicity and understanding the overall behavior of the PHE prototype, the FE model is composed of 546,764 2-D linear quadrilateral shell elements and 911,012 3-D linear solid elements made of 830,304 brick elements, 80,348 wedge elements, and 360 tetrahedron elements. Figure 4 shows the boundary conditions of the primary/secondary flow plates for a thermal analysis under a test condition of 850℃. The material properties of Hastelloy-X [7] are used in the FE model.

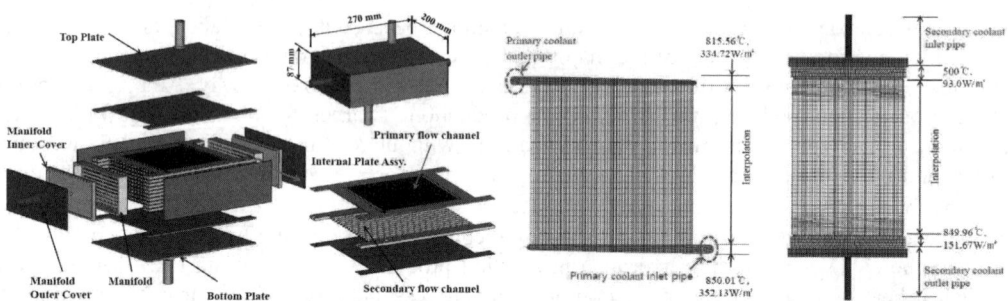

Figure 3 Parts of a small-scale PHE prototype   Figure 4 Thermal boundary condition

Analysis

Thermal analysis. Figure 5 shows the thermal analysis results of the outside surface of the PHE prototype under fixed test conditions of the gas loop. From Fig. 5, the temperature distribution is nearly symmetrical along the vertical axis, and the maximum temperature of the outside surface is about 837.15℃.

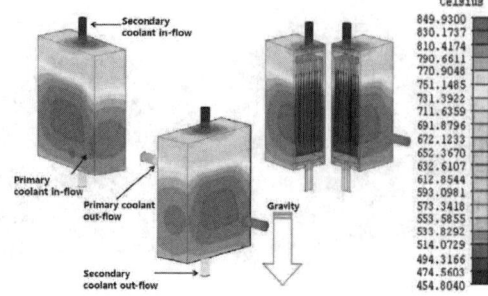
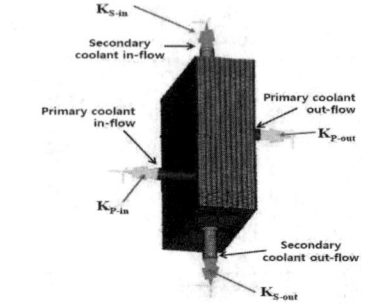

Figure 5 Temperature contour   Figure 6 Structural boundary condition

Coolant pressure. Coolant pressures of 3.0MPa and 0.1MPa are acting on the primary and secondary flow plate, respectively.

Preliminary structural analysis. A preliminary high-temperature structural analysis was performed based on the thermal analysis results and imposing a spring stiffness equivalent to 1.0m pipeline length on each end of the primary/secondary flow pipelines shown in Fig. 6 as displacement constraints for the sake of convenience. Figures 7 and 8 show stress contours by thermal load and pressure load, respectively. According to Figs. 7 and 8, an interesting finding is that thermal load is far more influential on the high-temperature structural integrity of the PHE prototype than pressure load. Therefore, thermal expansion and thermal stress are very important for the structural integrity evaluation of the PHE prototype in the gas loop.

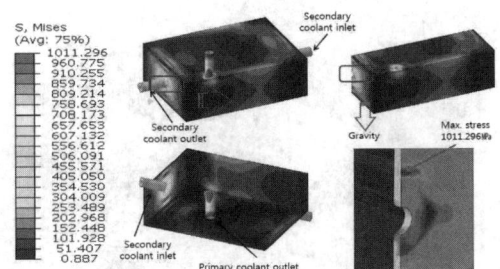   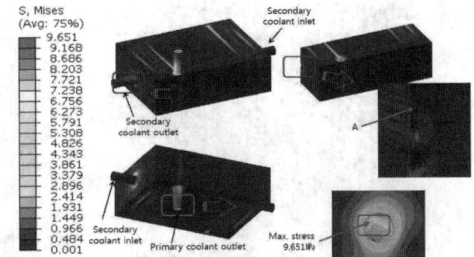

Figure 7 Stress contour by thermal load   Figure 8 Stress contour by pressure load

Boundary condition considering the gas loop pipeline stiffness. The most important aspect of properly predicting the stress and deformation of the PHE prototype under the gas loop test condition is to determine how to impose the displacement constraint condition on each end of the primary/secondary flow pipelines. In order to evaluate the structural integrity of the PHE prototype installed in the gas loop, it is necessary to impose realistic displacement constraints on each end of in-flow/out-flow pipelines connected to the PHE prototype.

Figure 9 represents a setup of the PHE prototype in the gas loop. The PHE prototype is connected with various tubes, such as a U-tube, elbow tube, and straight tubes for the sake of modulating thermal stresses under the gas loop test condition. Pipelines such as U-tube and elbow tube could modulate the thermal expansion due to a bending deformation, while the straight tube could do it due to an extension. In this study, equivalent spring stiffness which comes from the bending deformation and the extension of the connecting tubes is imposed as a boundary condition on each end of in-flow/out-flow pipelines connected to the PHE prototype. The procedure to get the equivalent spring stiffness is as follows.

The spring stiffness by the extension of the tubes, k_{ex}, can be obtained from the length (L_{ex}), cross-section area (A), and material elastic modulus (E) of the straight pipeline as follows,

$$k_{ex} = \frac{AE}{L_{ex}} \qquad (1)$$

The spring stiffness by the bending, k_b, can be obtained from the length of the arm for bending deformation (L_b), and the second moment of the pipe cross-section area (I_b) and the material elastic modulus of the U-tube or elbow tube as follows,

$$k_b = \frac{3EI_b}{L_b^3} \qquad (2)$$

Then, the equivalent spring stiffness, k_{eq}, can be obtained from the spring stiffness by the extension and by the bending deformation as follows,

$$\frac{1}{k_{eq}} = \frac{1}{k_{ex}} + \frac{1}{k_b} \qquad (3)$$

Structural analysis. Based on the thermal analysis result, a high-temperature elastic structural analysis was performed imposing the spring stiffness from Eq. (3) on each end of the primary/secondary flow pipelines. Figure 10 represents the stress contour and scaled deformed shape at the pressure boundary of the PHE prototype. The maximum local stress of 272.33 MPa occurs near the edge between the top plate and side, even in that the maximum stress exceeds the yield stress (239.7 MPa at 746℃) by 13.6%. Since edges of the PHE prototype are chamfered realistically, the maximum stress will be decreased to some extent when considering the chamfered edges. Therefore, the high-temperature structural integrity of the PHE prototype seems to be maintained in the gas loop test condition.

Looking at the deformed shape in Fig. 10, it is speculated that the deformed shape comes from the temperature distribution in Fig. 5, in the lower part at higher temperatures expands while the upper part at lower temperatures contracts.

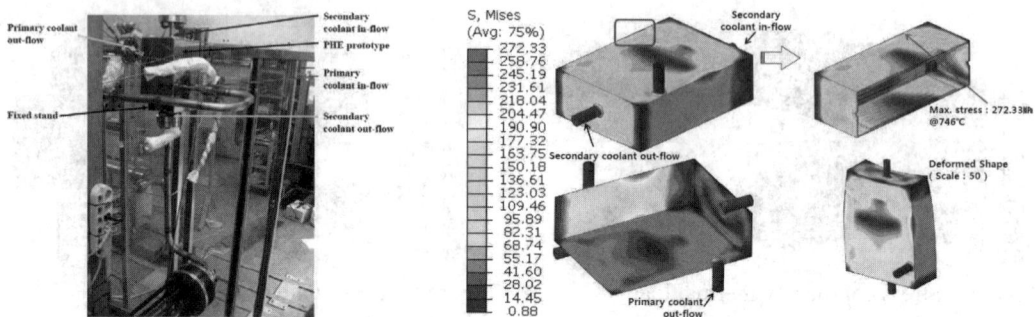

Figure 9 Set-up of a PHE in the gas loop   Figure 10 Stress contour by thermal/pressure load

Summary

To understand the macroscopic behavior of the small-scale PHE prototype, FE modeling, thermal analysis, and high-temperature elastic structural analyses were carried out, considering the stiffness of pipelines connected to the PHE prototype in the small-scale nitrogen gas loop. As a result of these analyses, we draw the following conclusions.

1. Thermal load is far more influential on the high-temperature structural integrity of the PHE prototype than pressure load. Therefore, thermal expansion and thermal stress are very important for the structural integrity evaluation of the PHE prototype in the gas loop.

2. In considering the stiffness of the gas loop pipeline, the high-temperature structural integrity of the PHE prototype seems to be maintained in the gas loop test condition, even though the maximum local stress at the pressure boundary of the PHE prototype exceeds the yield stress of Hastelloy-X.

References

[1] Chang, J. H. *et al.*, Nuclear Engineering and Technology, Vol. 39, No. 2 (2007), p. 111.
[2] US DOE, *Financial Assistance Funding Opportunity Announcement*, NGNP Program (2009).
[3] AREVA, *NGNP with Hydrogen Production Pre-conceptual Design Studies Report*, Doc. No. 1209052076-000 (2007).
[4] Kim, Y. W., R.O.Korea patent # 10-0877574 (2008).
[5] I-DEAS/TMG Analysis User Manual Version 6.1 (2009).
[6] ABAQUS Analysis User's Manual, Version 6.8 (2009).
[7] Hastelloy-X Alloy website, www.haynesintl.com.

Characterization of ductile failure behavior of the ferritic steel using damage mechanics modeling approach

Pawel Kucharczyk[1,a], Sebastian Münstermann[1,b]

[1]Department of Ferrous Metallurgy, RWTH Aachen University, Intzestr. 1, 52056 Aachen, Germany

[a]pawel.kucharczyk@iehk.rwth-aachen.de, [b]sebastian.muenstermann@iehk.rwth-aachen.de

Keywords: ductile damage and failure behavior, crack initiation, damage mechanics model, FEM simulation.

For the characterization of ductile failure of steels a coupled damage mechanics model has been derived from the combination of different types of damage approaches. The main feature of the applied model is that it considers the influence of the following two factors on the hardening behavior: stress triaxiality (defined as the pressure-to-equivalent stress ratio) and the Lode angle (related to the third invariant of the deviatoric stress tensor). Additionally, the effect of temperature and strain rate was taken in consideration. Furthermore, a new definition of the onset and the progress of damage in the material were introduced. In order to indicate, when the damage starts to play a role in the failure process, the micro-crack initiation was estimated using the direct current potential drop (DCPD) method. The numerical simulation allowed for the determination of the symmetrical ductile crack initiation locus. The consecutive microstructure degradation was described using a dissipation-energy-based damage evolution law. This approach was implemented into ABAQUS/Explicit by means of a user's material subroutine (VUMAT). The ductile failure behavior was experimentally and numerically investigated for the ferritic steel 22NiMoCr3-7. The procedure used for the calibration of the material parameters and for the estimation of the crack initiation locus curve is presented.

Introduction

A ductile failure is characterized by pronounced plastic deformations which involve significant plastic strains. The modeling of this failure behavior requires a precise description of the material plasticity starting from the crack initiation, its propagation through the material to the final fracture. The classical theory of metal plasticity based on the von Mises or Tresca formulations assumes that the effect of hydrostatic pressure on the flow potential is insignificant. Furthermore, it postulates that the flow stress is independent of the third stress invariant of the deviatoric stress tensor. The scientific findings from last few years show, however, that these both quantities should be considered for the precise description of plasticity, especially, of the real materials [1-4].

Damage mechanics model

The model for the simulation of ductile fracture applied in this paper has been derived from the combination of different types of damage approaches. It is based on the formulations given by Bai and Wierzbicki [5], which include the effect of the following two factors: stress triaxiality η (pressure dependence) and the Lode angle Θ (related to the third invariant of the stress deviator) on the constitutive description of the material. The yield potential function proposed by Bai and Wierzbicki is given by Eq. 1.

$$\sigma_{yld} = \overline{\sigma}(\overline{\varepsilon}_p) \cdot [1 - c_\eta(\eta - \eta_0)] \cdot \left[c_\theta^s + (c_\theta^{ax} - c_\theta^s)\left(\gamma - \frac{\gamma^{m+1}}{m+1}\right) \right] \qquad (1)$$

For the estimation of the basic flow curve $\sigma(\varepsilon_p)$ a reference test (for example tensile, upsetting or torsion test) can be chosen arbitrarily, however, the selected test type determines the further procedure for the parameter calibration. In this paper the basic flow curve was estimated in tensile test with smooth bar specimens and extrapolated using Hollomon equation. This curve is corrected by two terms which consider the effect of stress triaxiality and the Lode angle on the yield potential. The stress triaxiality η is defined as a ratio of hydrostatic pressure to the equivalent stress and its impact can be estimated in tensile tests with different notched bar specimens (in this work 3 different notch geometries were applied). The Lode angle Θ describes the location of the yielding point in the deviatoric stress plane; for its definition see [5]. For the estimation of the effect of the Lode angle on the yield potential flat grooved specimens (3 different groove radii) were investigated.

Ductile damage of steel consists of different stages, which have to be included in the numerical description. In order to find the moment when the damage contributes to the failure process different empirical approaches can be applied. In the models known from the literature [6-10] the strains at the fracture point are usually included in the damage description. The model presented here utilizes the micro-crack initiation as the onset of the damage process. Furthermore, in order to consider the damage-induced softening of the material a damage evolution low based on the energy dissipation is implemented. A damage indicator D characterizes the accumulated damage and reaches a critical value D_{cr} at the final fracture. The implementation of crack initiation and the new description of damage evolution law allows for the precise simulation of the crack propagation through the material.

Another innovative component of the proposed model is a new definition of the influence of strain rate and temperature on the yield potential. The description of the strain rate impact covers the mechanisms of dislocation slip as well as the dislocation damping, so that this model can be applied to very high strain rates typical for impact loading conditions. The implementation of temperature effect allows for the simulation of Charpy tests at different temperatures. The yield criterion used in the developed model and the definition of D variable are given respectively by Eq. 2 and 3. The meaning of the symbols can be found in [11].

$$\sigma_{yld} = \left[\bar{\sigma}(\bar{\varepsilon}_p)\cdot\left(c_1^{\dot{\varepsilon}}\cdot\ln\dot{\varepsilon}+c_2^{\dot{\varepsilon}}\right)+c_3^{\dot{\varepsilon}}\cdot\dot{\varepsilon}\right]\cdot\left[c_1^T\cdot\exp\left(-c_2^T\cdot T\right)+c_3^T\right]\cdot\left[1-c_\eta(\eta-\eta_0)\right]$$
$$\cdot\left[c_\theta^s+\left(c_\theta^{ax}-c_\theta^s\right)\left(\gamma-\frac{\gamma^{m+1}}{m+1}\right)\right]\cdot(1-D) \qquad (2)$$

$$\dot{D} = \begin{cases} 0 & \text{for} \quad \varepsilon \leq \varepsilon_i \\ \dfrac{\sigma_{yld}\cdot L\cdot\dot{\varepsilon}}{2\cdot G_f} & \text{for} \quad \varepsilon_i < \varepsilon < \varepsilon_f \\ D_{cr} & \text{for} \quad \varepsilon = \varepsilon_f \end{cases} \qquad (3)$$

In order to estimate the initiation of a micro crack a DCPD method accompanied the performed tests. This coupled approach is widely used in fracture mechanics and is based on a change of electrical potential due to decreasing sample cross section caused by the micro crack initiation. The following numerical simulations of the performed tests allow for the determination of critical strains, stress triaxiality and Lode angle at the moment of crack initiation. Based on the obtained simulation results a symmetrical crack initiation locus curve can be calculated using the least squares approach.

Experimental and simulation results

The ductile failure behavior was investigated using a reactor pressure vessel steel 22NiMoCr3-7. Its chemical composition is given in Table 1.

Table 1: Chemical composition of steel 22NiMoCr3-7 (mean mass contents in %)

C	Si	Mn	P	S	Cr	Mo
0.20	0.20	0.91	0.005	0.002	0.40	0.49
Ni	Al	Nb	Cu	Ti	V	Co
0.802	0.017	0.002	0.034	0.001	0.005	0.010

For the calibration of parameters given in Eq. 2 and 3 as well as for the estimation of ductile crack initiation locus experimental tests with different specimens accompanied by DCPD-measurements were carried out. The parameter calibration was performed in numerical simulations in order to achieve the best correlation between simulation results and results of the tensile tests with un-notched and notched round bar specimens (parameters c_η, η_0, c_θ^t, D_{crit}, G_f), plane strain specimens (parameters c_θ^s, D_{crit}, G_f), shear tests (parameter c_θ^s) and upsetting tests (parameters c_θ^c, m).

The ductile crack initiation locus defines the critical strains ε, stress triaxiality η and Lode angle θ at the moment of the crack initiation. The variation of samples geometries leads to the different stress states in the stressed body so that different sample types are necessary for the proper estimation of this curve. In the present work 3 different notched round bar and 3 notched plane strain specimens were investigated. The obtained crack initiation locus curve is presented in Fig. 1.

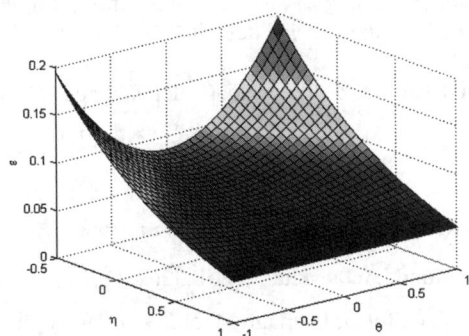

Fig. 1: Crack initiation locus curve for steel 22NiMoCr3-7

Fig. 2 shows exemplary force-displacement responses obtained at different steps of parameter calibration for the notched round bar specimen (R=1.5 mm) and notched plain-strain specimen (R=3 mm). For both specimen types the effect of the 3rd invariant of stress deviatior (Lode angle) on the flow potential as well as advanced damage after crack initiation can be easily seen. Based on this procedure the parameters set for the investigated material was determined and is given in Table 2. It should be noted that the flow potential of this particular material shows extremely small dependence on the hydrostatic pressure so that the parameter c_η was assumed to be zero.

Table 2: Calibrated parameters set

c_η	η_0	c_θ^s	c_θ^t	c_θ^c	m	D_{crit}	G_f
0.00	0.33	0.95	1.00	1.00	5.00	0.10	300

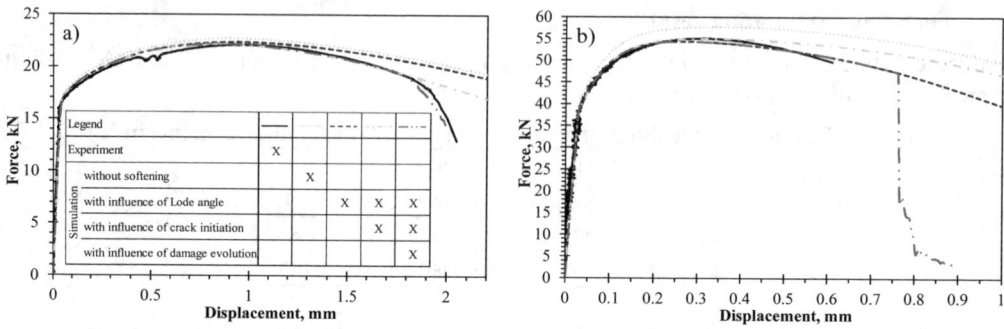

Fig. 2: Parameter calibration for: a) notched round bar specimen (R=1.5 mm), b) notched plain-strain specimen (R=3 mm)

At the present stage of the model development the parameters, which describe the influence of temperature and strain rate on the yield potential, are being still estimated. The validation of the model by the Charpy test is planned to be performed soon.

Conclusions

Based on the performed scientific work the main conclusions are as follows:
- The presented procedure of the parameter calibration clearly shows the influence of the 3rd invariant of stress deviatior (Lode angle) on the flow potential. Implementation of the onset and progress of damage allowed for the precise description of the ductile failure in the ferritic steel 22NiMoCr3-7.
- The investigated material shows a very small dependence of the flow potential on the hydrostatic pressure.

References

[1] M. Brunig: Int. J. Plasticity, Vol. 15(11) (1999), p. 1237-1264

[2] W.A. Spitzig and O. Richmond: Acta Metall., Vol. 32(3) (1984), p. 457-463

[3] X. Gao, G. Zhang, and C. Roe: Int. J. Damage Mech., Vol. 19(1) (2009), p. 75-94

[4] F. Yang, Q. Sun, and W. Hu: Acta Metall. Sinica, Vol. 22(2) (2009), p. 123-130

[5] Y.L. Bai and T. Wierzbicki: Int. J. Plasticity, Vol. 24(6) (2008), p. 1071-1096

[6] Y. Bao and T. Wierzbicki: J. Eng. Mater. Technol., Vol. 126(3) (2004), p. 314-324

[7] Y. Bao and T. Wierzbicki: Int. J. Mech. Sci., Vol. 46(1) (2004), p. 81-98

[8] M. Dunand and D. Mohr: Int. J. Solids Struct., Vol. 47(9) (2010), p. 1130-1143

[9] J.R. Rice and D.M. Tracey: J. Mech. Phys. Solids, Vol. 17(3) (1969), p. 201-217

[10] G.R. Johnson and W.H. Cook: Eng. Fract. Mech., Vol. 21(1) (1985), p. 31-48

[11] J. Lian, M. Sharaf, F. Archie and S. Münstermann: submitted to Int. J. Damage Mech. (2011)

Thermal and Static Analysis of an Insulation Block from Recycled Polymer HDPE for Solution of Thermal Bridges in Wall-footing Detail

Jan Pencik[1, a], Libor Matejka[2, b] and Lukas Matejka[3, c]

[1,2,3]Brno University of Technology, Faculty of Civil Engineering, Institute of Civil Engineering, Veveri 331/95, 602 00 Brno, Czech Republic

[a]pencik.j@fce.vutbr.cz, [b]matejka.l@fce.vutbr.cz, [c]matejka.lukas@gmail.com

Keywords: material recycling, FEA, thermal bridge, HDPE, viscoelastic

Abstract. With the sustainable construction the emphasis is placed on saving energy, reducing of consumption of natural resources, extending the life cycle of recycling, etc. One of the important groups of materials that can be reused are polymers. Polymers and waste polymers can be used as a base material for products used in civil engineering. One of these products, which were developed, is an insulation block from modified recycled polymer HDPE for direct solution of thermal bridges in wall footing detail. Design of the insulation block has been done using the MAP method together with long-time experimental testing of specimens and in a testing wall in scale 1:1. In the mathematic modeling the installation block was assessed in terms of statics and thermal technology. Static assessment was performed using Standard Solid rheological model, which represents the most accurate approximation of long-time behaviour.

Introduction

In the research project MSM 0021630511 "Progressive building materials with utilization of secondary raw materials and their impact on structures durability" addressing the possibility of use of a wide range of high-volume industrial and building wastes in the building industry was solved a partial topic regarding the use of waste polymers or combination of waste polymers with other materials for various building materials, structural elements and products. Waste polymers in various types of PP, PE, HDPE etc. generally form a significant group of wastes. These materials belong to the group of thermoplastic materials and they can be easily recycled [1] thanks to thermal plasticity [2]. Another advantage of using recycled polymers consists in the possibility to improve their resistance against atmospheric ageing and thermal and mechanical properties [3].

Insulation Block from Recycled Polymer HDPE

One of possible areas of use of waste polymers in the construction industry is their use for the solution of thermal bridges. Within the solution of the topic of interruption of thermal bridges a new product has been developed – *insulation block* to interrupt thermal bridges in wall footing, i.e. in place between the foundation and the masonry and inner floor (Fig. 1a). The product developed has been subject to intellectual property protection by patent – patent EU 1918471 [4].

The insulation block, actively interrupts the thermal flow in the place of wall footing by its interference (Fig. 1a). It is formed by plates of constant thickness with locks (Fig. 1b) that fit into each other and prevent relative mutual displacement of plates in the transverse direction. Insulation block is made of modified recycled polymer HDPE, denote as mHDPE. By modification lower thermal conductivity $\lambda = 0.33$ [W/mK] has been achieved [5]. The example of the wall arrangement with an insulation block is shown at (Fig. 1c).

The design of the insulation block is carried out using Modelling – Analysis – Prediction method. Mathematical modelling (FEA thermal and static analysis) is performed by the help of ANSYS system depending on designed material. The FEA analysis is supplemented with help of long-time experimental testing in scale 1:1 i.e. testing in a real cut of structure part.

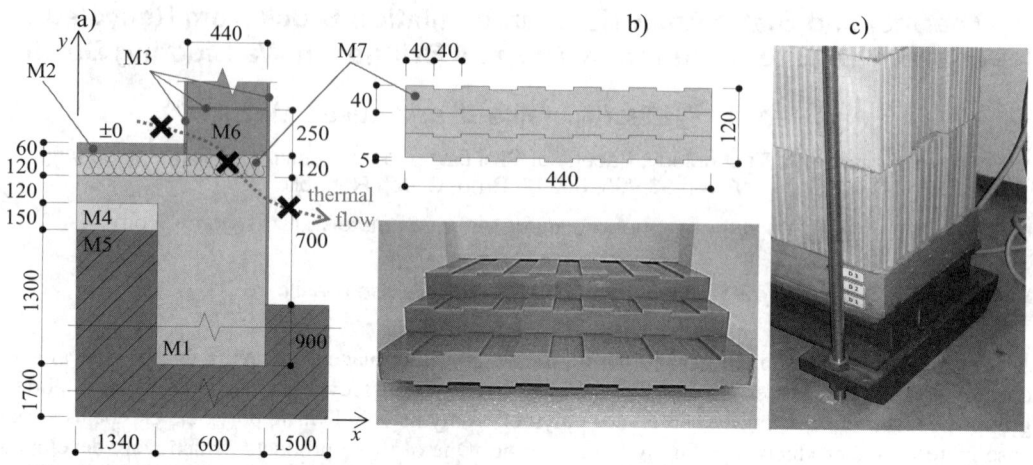

Fig. 1: Wall footing detail: active interruption of thermal flow in wall footing using an insulation block (a), plates of an insulation block (b) and insulation block in a real structure (c)

FEA: Thermal Analysis

Thermal analysis of the detail of the wall footing was performed by solution of heat diffusion equation which is a differential equation governing 2D heat conduction and is given by the expression [4]

$$\frac{\partial}{\partial x}\left(k_x \frac{\partial T}{\partial x}\right) + \frac{\partial}{\partial y}\left(k_y \frac{\partial T}{\partial y}\right) + Q = 0 \qquad (1)$$

where k_x and k_y is the thermal conductivity in x and y direction, Q the internal heat energy generated within a body and the temperature gradient. The solution was found using the ANSYS system as a steady state ($\partial/\partial t = 0$) and without a heat generation ($Q = 0$).

No.	Material	λ [W/mK]
M1	Concrete	1.23
M2	Floor layer	1.01
M3	Plaster	0.99
M4	Gravel	2.30
M5	Soil	2.30
M6	Masonry blocks	x: 0.11/y: 0.68
M7	Insulation block	0.33

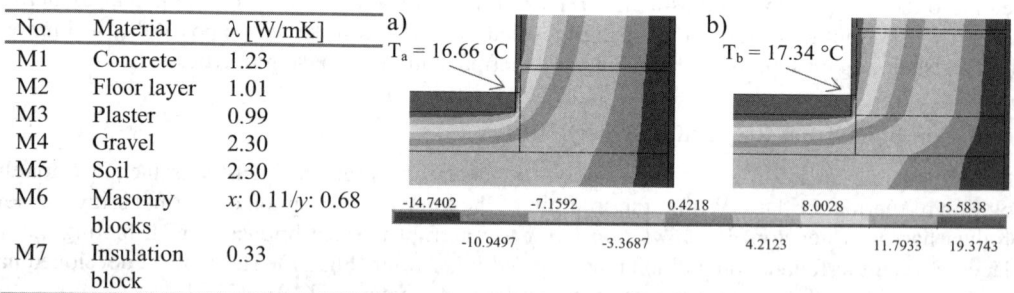

Fig. 2: Thermal conductivities of materials (material number in Fig. 1a) and thermal field [°C] in corner detail: without (a) and with (b) an insulation block

Detail analyses were performed for different variants of solution of the composition of perimeter wall. Positive insulation effect of the insulation block on the indoor temperature (Fig. 2; difference T_a:T_b is 4.1%), and thus the reduction in the energy consumption for heating the building is documented on the example of the classical foundation of the cladding of masonry blocks at the wall footing without and with the insulation block (Fig. 2). Boundary conditions are considered according to design code CSN 73 0540. The temperature considered in the interior was +20 °C, with the heat transfer coefficient $R_{si} = 0.13$ m²K/W. The temperature considered in the exterior was -

15 °C, with the heat transfer coefficient of exterior $R_{se} = 0.04$ m²K/W. The soil temperature at a depth of 3 m was +5 °C. Thermal conductivities of materials λ [W/mK] are shown in (Fig. 2). Orthotropic thermal conductivity is considered in masonry blocks as opposed to other materials.

Long-time Experimental Testing

Material behaviour of mHDPE, which is one of the viscoelastic materials, is time-dependent and also depends on the type of loading. Stress-strain diagram (Fig. 3a) and the initial material characteristics, Young modulus E and Poisson ration μ are determined using experimental tests. The determined initial values for level of assumed load were $E = 242.23$ MPa and $\mu = 0.35$.

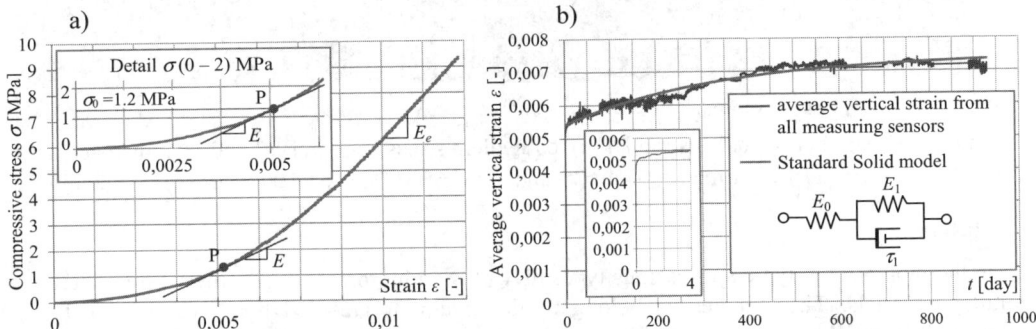

Fig. 3: Stress-strain σ-ε diagram of mHDPE (a), dependence of average vertical strain on time with viscoelastic fit using Standard Solid viscoelastic model (b)

Long-term behaviour of the insulation block made from mHDPE and determination of creep curvature and for compress loading is performed using a stress frame at a scale of 1:1 i.e. testing in a real cut of masonry. The determined creep curvatures are used to predict the behaviour of the material using statistic FEA conducted with updated real data about the behaviour of mHDPE.

The tested insulation block is formed, as in thermal and statics FEA, by 3 plates of width 440 mm and height 40 mm (Fig. 1b), the height of an insulation block is 120 mm and the length is 2000 mm. The insulation block and masonry are loaded by a pressure with an intensity of 1.2 MPa. Vertical movements in the quarters and half of the length of insulation block are continuously measured. Long-time measurement ($t = 930$ days) is shown in graph (Fig. 3b) where average vertical strain ε from all measuring sensors is drawn together with compensation of temperature influence and results obtained using Standard Solid viscoelastic material model for material (Fig. 3b) [6], which represents the most accurate approximation of long-time behaviour. Using measured data the ratio of strain to applied stress, which is the compliance $D(t) = \varepsilon(t)/\sigma_0$ is given by

$$D(t) = \frac{1}{E_0} + \frac{1}{E_1}\left(1 - e^{(-t/\tau_1)}\right) \qquad (2)$$

where $E_0 = 219.7643$ MPa, $E_1 = 590.0475$ MPa, $\tau_1 = 347.6301$ day and $\sigma_0 = 1.2$ MPa. Progress of relaxation modulus $E(t)$ using an equilibrium modulus E_∞ goes to infinity is given by the expression

$$E(t) = E_\infty + (E_0 - E_\infty)e^{(-t(E_0 + E_1)/\tau_1 E_1)} \qquad E_\infty = \left(E_0^{-1} - E_1^{-1}\right)^{-1} \qquad (3)$$

FEA: Static Analysis

Description of the initial behaviour of an insulation block and its prediction was performed using mathematical modelling in the ANSYS system. The analysis was conducted using 2D parametric analysis model created by means of quadrilateral plane finite elements with plane strain behavior (Z strain = 0) (Fig. 4a). Connection of the plates of an insulation block is modeled by means of contact finite elements with standard behavior, the contact occurs only in case of compression stress.

The analyses considered compress loading of the system by forces derived from the ultimate strength of masonry blocks 1.2 MPa with initial eccentricity 0.025 m. The (Fig. 4) shows a total displacement U_{SUM} and an equivalent stress σ_{EQV} in time $t = 0$ day using determined initial values and in time $t_\infty = 365 \cdot 10^3$ days (100 years) using measurement data and (3); $E(t_\infty) = 160.13$ MPa.

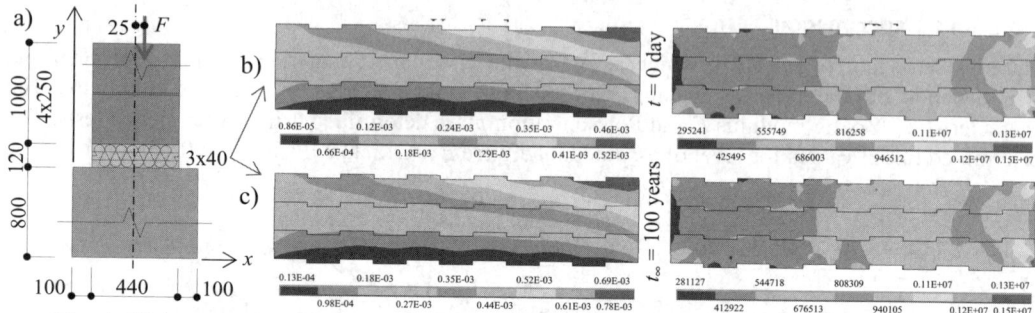

Fig. 4: 2D parametric analysis model (a), equivalent stress σ_{EQV} in [Pa] and total displacement U_{SUM} [m] in time $t = 0$ day (b) and $t_\infty = 365 \cdot 10^3$ days = 100 years (c)

Conclusion

Waste thermoplastic polymers in various types may play an important role in the future selection of building materials. These materials belong to the group of recycled materials due to thermal plasticity. It is possible to improve their thermal-technical and mechanical properties.

One of possible areas of use of waste polymers, presented in form of mHDPE, is use for the solution of thermal bridges in wall footing using developed an insulation block. The insulation block, actively interrupts the thermal flow and meets requirements for proper functionality and stability for the entire service life resulting from long time experimental testing of specimens and of a section of a real wall. Functionality in a life-cycle of an insulation block also results from FEA thermal and static analysis. It is possible to conclude that for mHDPE material the Standard Solid viscoelastic material model represents the most accurate approximation of behaviour of the long-time loaded material by a constant pressure.

Acknowledgement

This work was financially supported by grant FAST-S-12-20/1650 and FAST-S-12-39/1730.

References

[1] L.A. Utracki: *Polymer blends handbook*. Kluwer academic publishers, Netherlands, 2002, ISBN 1-4020-1114-8.
[2] L.E. Nielsen and R.F. Landel: *Mechanical properties of polymers and composites*. Marcel Dekker, USA, 1996, ISBN 0-8247-8964-4.
[3] T.A. Osswald, G. Menges: *Material Science of Polymers for Engineers*. Hansen Publishers, USA, 2003, ISBN 3-446-22464-5.
[4] VUT v Brne, L. Matejka, J. Pencik, European Patent, EP 1918471 (2009).
[5] J. Pencik and L. Matejka: *Numerical modeling of behavior of an insulation block from recycled polymers*. (Advances and Trends in Structural Engineering, Mechanics and Computation, Netherlands : CRC Press. ISBN 978-0-415-58472-2).
[6] E.J. Barbero: *Finite element analysis of composite materials*. CRC Press, UK, 2008, ISBN 1-4200-5433-3.
[7] M.T. Shaw, W.J. MacKnight: *Introduction to polymer viscoelasticity*. Wiley-Interscience, USA, 2005, ISBN 0-471-74045-4.

Diagnosis of Damage in a Steel Tank Model by Shaking Table Harmonic Tests

Daniel Burkacki[1,a] and Robert Jankowski[1,b]

[1]Faculty of Civil and Environmental Engineering, Gdansk University of Technology,
ul. Narutowicza 11/12, 80-233 Gdansk, Poland

[a]daniel.burkacki@op.pl, [b]jankowr@pg.gda.pl

Keywords: Steel tank, damage diagnosis, experimental study, shaking table, harmonic tests

Abstract. Diagnosis of damage in civil engineering structures has recently become an important issue in the safety assessment procedure. Among a number of different approaches, a method of measuring the changes in natural frequencies is one of the most effective indicators of global damage. It has been successfully applied to relatively small structures, however, the tests on large structures are very difficult and the practical application of the method still requires further investigations. The aim of the present paper is to show the results of the shaking table experimental study concerning the diagnosis of damage in a model of cylindrical steel tank with self-supported roof which is filled with liquid. During the tests, the base of the structure was excited under the harmonic loading with variable frequency. The tests were repeated for different stages of damage, which was introduced in the model by easing the bolts of structural supports as well as by cutting the welds between the shell and roof as well as between roof elements. The results of the study show a characteristic decrease in the natural frequencies for the case of structural supports with reduced stiffness (global type of damage). On the other hand, cutting the welds (local type of damage) has lead to the considerable increase in the power spectral density values for higher vibration modes.

Introduction

The safety of civil engineering structures is one of the most import issues of building industry and the assessment of the damage-involved structural response has become of major concern to engineers (see, for example, [1,2]). Different methods of diagnosis of damage have recently been intensively studied [3]. Among them, a method of measuring the changes in natural frequencies is considered to be one of the most effective indicators of global damage. It has been successfully applied to relatively small structures, such as compact buildings. However, the tests on large structures, i.e. long bridges, large tanks, are very difficult and the practical application of the method still requires further studies [4]. This concerns also the experimental investigations on scaled models conducted on shaking tables, which are useful in earthquake engineering [5], since their results are very limited [6].

The aim of the present paper is to show the results of the shaking table tests focused on the diagnosis of damage in a model of cylindrical steel tank with self-supported roof which is filled with liquid. In the study, damage of the model structure was introduced by easing the bolts of structural supports as well as by cutting the weld between the shell and the roof and also between roof elements. During the tests the base of the structure was excited by the harmonic loading with variable frequency ranging from 1.0 Hz up to 60 Hz and the structural response was measured.

Experimental setup

A shaking table located at the Gdansk University of Technology was used in the experimental study. It is a unidirectional device with the platform dimensions of 2.0×2.0 m which allows for testing the specimens with a maximum mass of 1000 kg.

The subject of the current study concerns the cylindrical steel tank with self-supported roof (compare [7]). The real tank has a total capacity of 32,000 m³ and has been constructed in the northern Poland. The experimental model was made of stainless steel and scaled based on a scale equal to 1:33 (see Fig. 1a). Its diameter and the total height is equal to 1.5 m and 0.7 m, respectively. The thickness of the bottom plate, shell and roof is equal to 3 mm, 1.2 mm and 1.2 mm, respectively. The structure was fixed by eighteen M10 bolts to the platform of the shaking table (see Fig. 1b). The investigation was carried out for the tank model fully filled with water (water height of 486 mm). The total weight of such a model was equal to 945 kg. Five accelerometers were mounted at different places of the tank so as to measure the structural response during the tests (see Fig. 1a for their locations).

The following different structural damage cases, described by different types of damage often observed in the reality, were considered in the study:

A – undamaged tank;

B – tank with reduced stiffness of structural supports after easing the support bolts (bolts were unscrewed by 5 mm) – see Fig. 1b;

C – tank with circumferential cut of weld between shell and roof (length of 50 cm) – see Fig. 1c;

D – tank with circumferential cut of weld between shell and roof (length of 50 cm) as well as with radial cut of weld in roof (length of 20 cm) – see Fig. 1c.

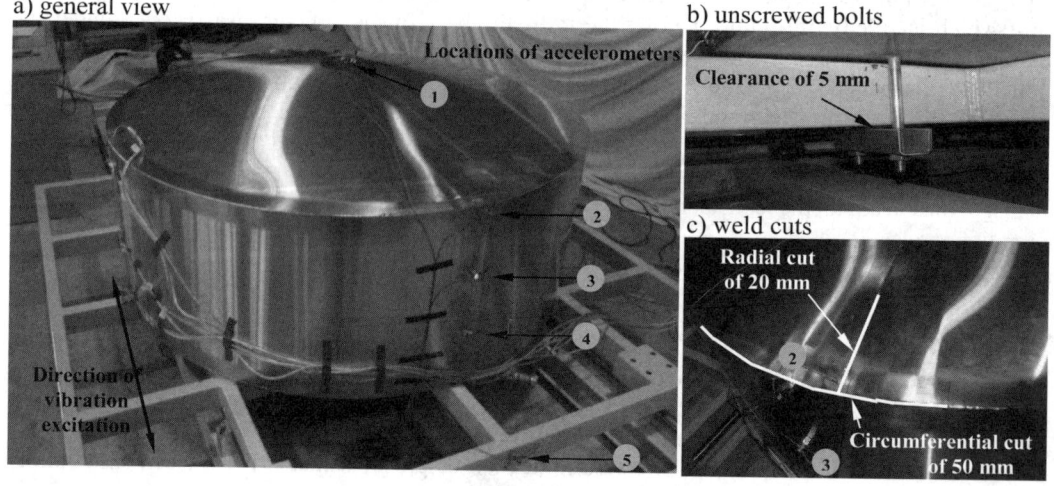

Fig. 1. Setup of the experiment

Results of the experiment

The shaking table experimental study was carried out under the harmonic loading. During the tests, the frequency of the excitation began at 1.0 Hz and it was uniformly increased up to 60 Hz (sweep-sine tests).

The response time histories were subjected to the fast Fourier transform in order to obtain the exact values of the natural frequencies of the tank model at different stages of damage. The examples of the results of the analysis for the accelerometer no. 2 (see Fig. 1a for its location) are shown in Fig. 2. It can be seen from the figure that the forms of the Fourier spectra can be directly related to the level and type of damage introduced in the structure. The results of the study show a characteristic decrease in the natural frequencies when the structural supports with reduced stiffness

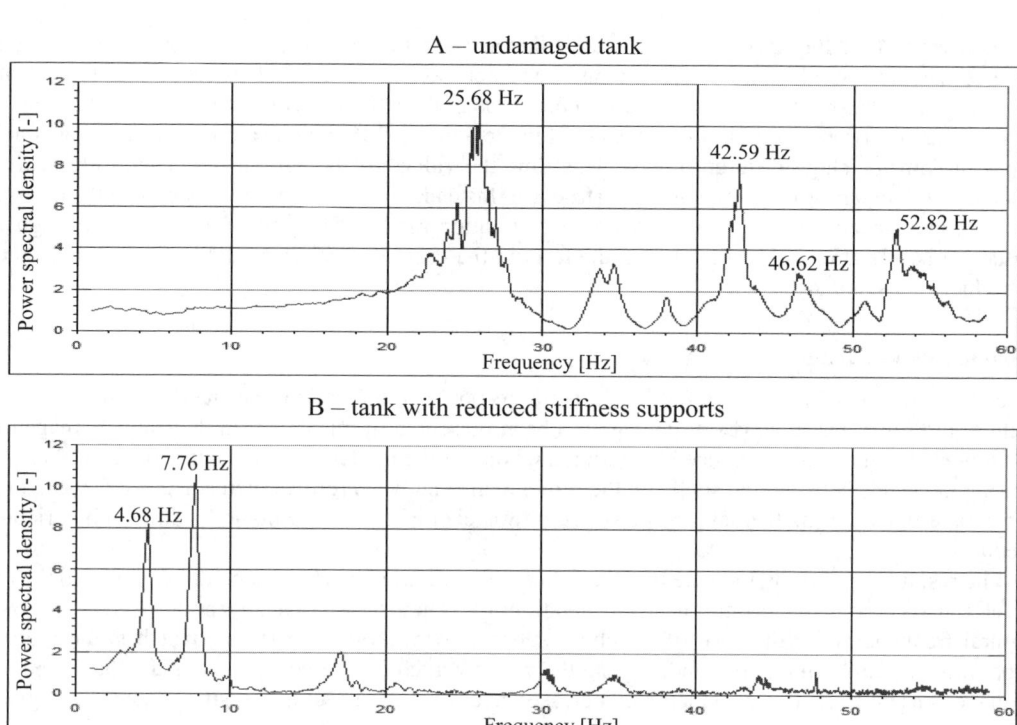
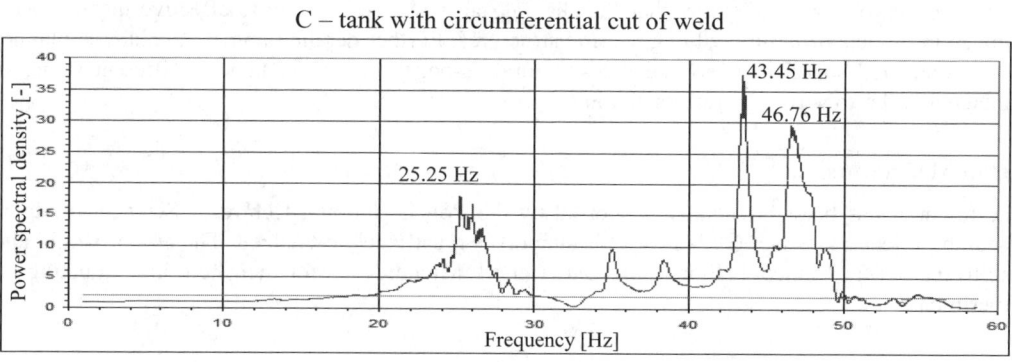
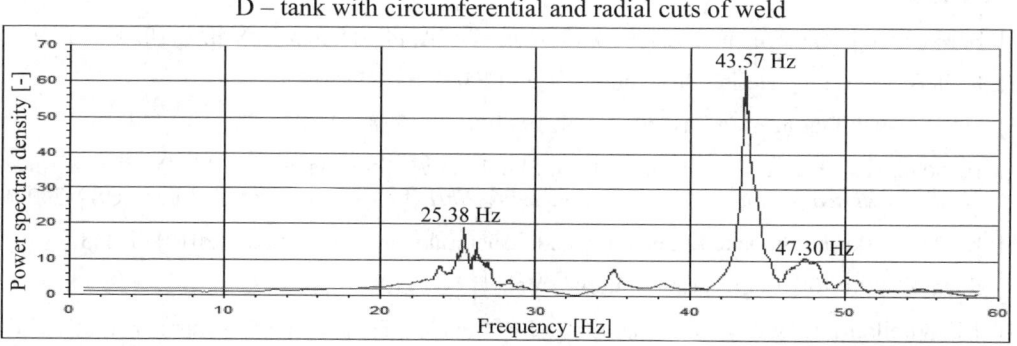

Fig. 2. Fourier spectra for the model at different stages of damage (accelerometer no. 2)

are concerned (see Fig. 2a) since this type of damage can be considered as the global one. In the case of the first vibration mode, for example, the decrease in the natural frequency for type B of damage with relation to the undamaged structure is as high as 81,8%. On the other hand, the results of the study for type C and D of damage (see Fig. 2c and Fig 2d), which can be considered as local types of damage, show different trends. This time the values of the natural frequencies are nearly unchanged comparing to the undamaged structure. Instead, we can observe the significant increase in the power spectral density values for higher vibration modes. This increase results from higher values of accelerations observed during the tests in the case of accelerometers located close to the introduced local damage.

Concluding remarks

The experimental study concerning the structural response of the cylindrical steel tank model filled with liquid for different stages of damage has been presented in this paper. In the study, damage of the model structure was introduced by easing the bolts of the model structural supports as well as by cutting the welds between the shell and the roof and also between roof elements. During the shaking table experimental tests the base of the structure was excited by the harmonic loading with variable frequency.

The results of the study indicate that the forms of the Fourier spectra can be directly related to the level and type of damage introduced in the structure. They show a characteristic decrease in the natural frequencies for the case of structural supports with reduced stiffness, which is the global type of damage. On the other hand, cutting the welds, which can be considered as the local type of damage, has lead to the considerable increase in the power spectral density values for higher vibration modes.

The results of the study show that the shaking table tests can be quite effective in diagnosing damage in models of large civil engineering structures. Further detailed numerical study is planned to be conducted so as to determine the structural response of a steel tank at different stages of damage in order to verify the results obtained.

Acknowledgments

The research has been financially supported by the Polish National Centre of Science through a research project no. N N506 121240. This support is greatly acknowledged. The author would also like to thank Mr. Tomasz Falborski and Mr. Henryk Michniewicz for their help in conducting the experiment.

References

[1] S. Mahmoud, X. Chen and R. Jankowski: Journal of Applied Sciences Vol. 8 (2008), p.1850.

[2] R. Jankowski: Engineering Structures Vol. 31 (2009), p. 1851.

[3] O.S. Salawu: Engineering Structures Vol. 19 (1997), p. 718.

[4] H. Sohn, C.R. Farrar, F.M. Hemez, D.D. Shunk, D.W. Stinemates and B.R. Nadler: *A review of structural health monitoring literature: 1996–2001* (Los Alamos National Lab., USA 2003).

[5] R. Jankowski: Earthquake Engineering and Structural Dynamics Vol. 39 (2010), p. 343.

[6] R. Jankowski: Key Engineering Materials Vol. 417-418 (2010), p. 157.

[7] J.C. Virella, L.A. Godoy, L.E. Suárez and J.B. Mander: Engineering Structures Vol. 25 (2003), p. 877.

Numerical Analysis of a Steel Frame Building with Soft-storey Failure under Ground Motion Excitation

Wojciech Migda[1,a] and Robert Jankowski[1,b]

[1]Faculty of Civil and Environmental Engineering, Gdansk University of Technology, ul. Narutowicza 11/12, 80-233 Gdansk, Poland

[a]wojtek@migda.de, [b]jankowr@pg.gda.pl

Keywords: Soft-storey failure, steel frame building, impact load, ground motion.

Abstract. It has been observed during earthquakes that the soft-storey failure of an upper floor of a building results in large impact load acting on structural members of the lower storeys. It may further lead to progressive collapse of the whole structure substantially intensifying human losses and material damages. The aim of this paper is to show the results of a numerical analysis focused on the behaviour of multi-storey steel frame building that suffers from a soft-storey failure under ground motion excitation. A numerical model of the structure was created in FEM computer software and was exposed to an impact that would have been generated after a soft-storey failure due to falling of the upper floors. During the analysis, the whole structure was exposed to ground motion excitation and different moments have been chosen for the impact so as to estimate the most critical moment for the structure. The results of the study show that not only the value of the impact force is crucial but also the moment when the impact occurs. This is due to the fact that horizontal deflection of the supporting members (steel columns) varies during the time of the excitation. It has been observed that the most critical moment for the building for being subjected to a vertical impact is when the horizontal deflection is close to its peak.

Introduction

The so called soft-storey failure an intermediate storey of a building is a very common failure mechanism observed during major earthquakes [1]. From the structural point of view, a soft-storey failure generates a high impact load acting on structural members of the lower storeys. It may therefore result not only in damage at only one level of the structure but it may also lead to the progressive collapse of the whole building [2]. If the progressive collapse is initiated, human as well as material losses are significantly intensified.

The effects of impact loads acting on buildings during ground motions have been studied by structural engineers for over twenty years now. However, almost all of the analyses so far conduced have concerned the horizontal interactions between insufficiently separated buildings, often referred as the earthquake-induced structural pounding (see, for example, [3-9]). Only a few studies have been focused on the effects of vertical impact acting on the columns of the lower stories after a soft-storey failure (see [10]).

The aim of the present paper is to show the results of a numerical analysis concerning the behaviour of multi-storey steel frame building that suffers from a soft-storey failure under ground motion excitation. A model of the structure has been exposed to an impact that would have been generated after a soft-storey failure due to falling of the upper floors. During the analysis, different moments have been chosen for the impact so as to estimate the most critical moment for the building.

Numerical analysis

A four-storey steel frame building, with plan dimensions of 31.68 m by 19.44 m and a storey height of 3.6 m, has been considered in the analysis. The detailed numerical model of the structure has been created using a commercial Finite Element Method (FEM) software MSC Marc (see Fig. 1).

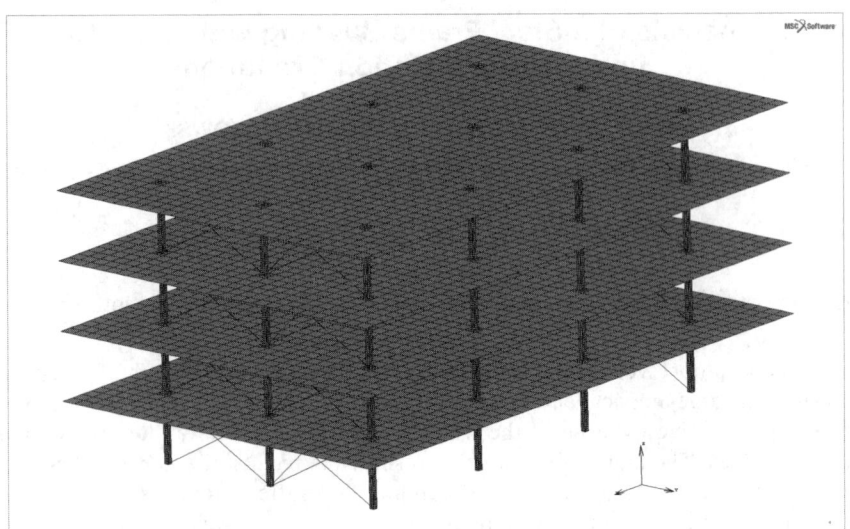

Fig. 1. Numerical model of the steel frame building

The slabs with thickness of 0.35 m have been modelled as rigid shell elements with the use of 1320 elements for each slab. Each steel column (all having the cross section of HEB 240) has been modelled using 180 flat shell elements. Diagonal steel bracings have been modelled by beam elements. The sine wave acting at the foundation of the building in its transverse direction has been used to simulate the ground motion excitation. In order to investigate the behaviour of the building subjected to the soft-storey failure, the collapse of the arbitrary chosen third storey has been initiated at different moments during the response of the structure (see Fig. 2). The impact load due to soft-storey failure has been simulated as a vertical load acting on the slab of the second storey with a time history obtained from the experimental study (see [10]). The examples of the results, for impact at $t=1.1$ s, $t=1.2$ s (peak horizontal response - see Fig. 2) and $t=1.4$ s, are presented in Fig. 3-5. They show the difference in the equivalent von Mises stress values at the node number 52170 (joint between the second storey column and slab) with relation to the results obtained for $t=1.0$ s, which has been considered as the reference time point in the analysis.

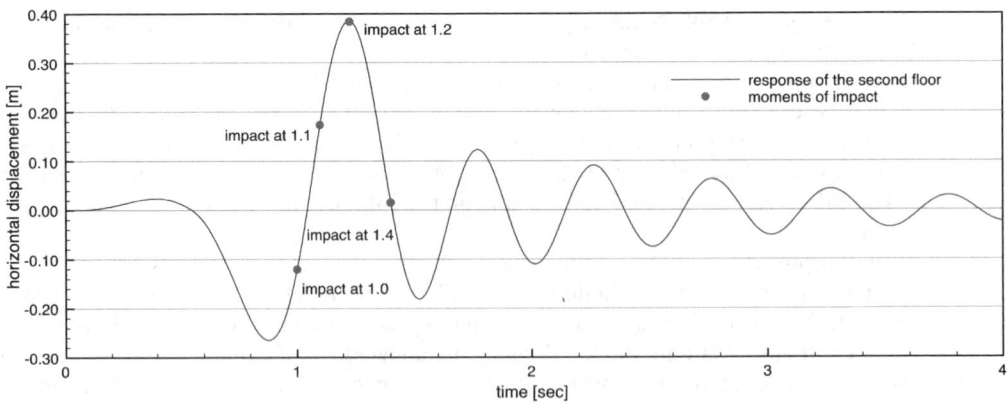

Fig. 2. Horizontal response of the second floor and moments of the soft-storey failure initiation

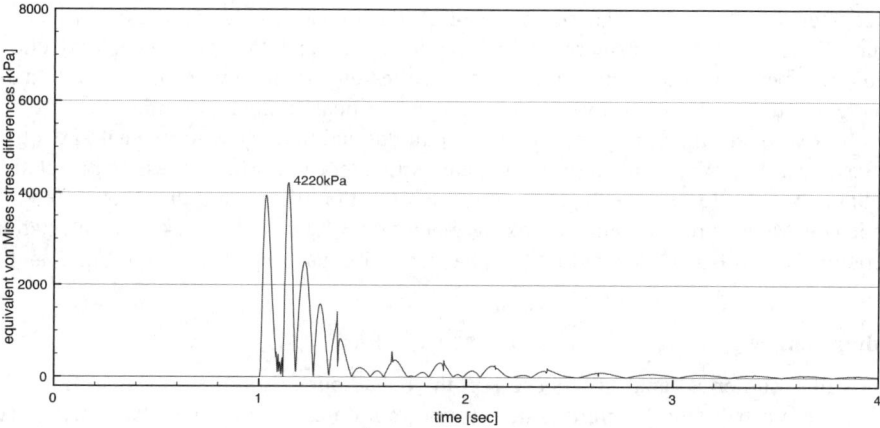

Fig. 3. Von Mises stress differences for impact at $t = 1.1$ s with relation to impact at $t = 1.0$ s

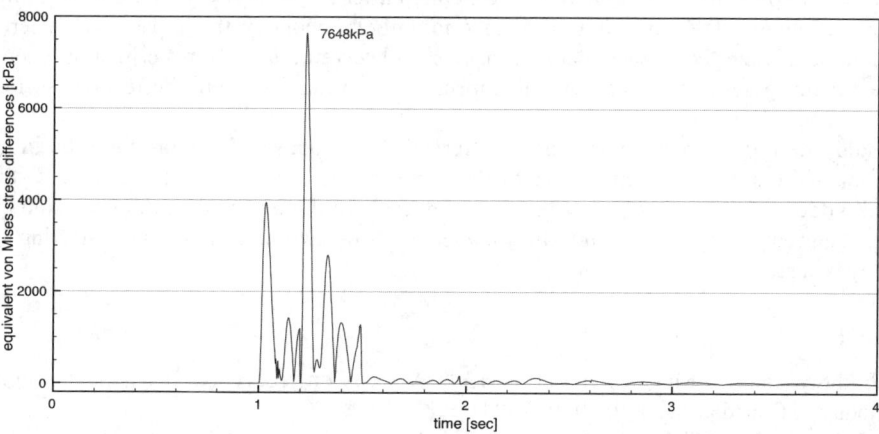

Fig. 4. Von Mises stress differences for impact at $t = 1.2$ s with relation to impact at $t = 1.0$ s

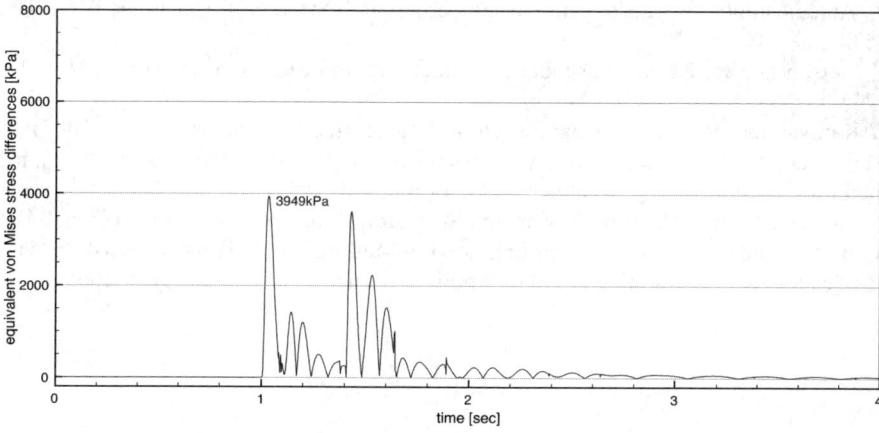

Fig. 5. Von Mises stress differences for impact at $t = 1.4$ s with relation to impact at $t = 1.0$ s

It can bee seen from Fig. 3-5 that the moment of the impact has a significant influence on the behaviour of the structure considered. This is due to the fact that horizontal deflection of the supporting members (steel columns) substantially varies during the time of the excitation as can be seen in Fig. 2. It has been observed that the most critical moment for the structure for being subjected to a vertical impact is when the horizontal deflection is close to its peak, i.e. at $t = 1.2$ s. For this case (see Fig. 4) the increase in the peak equivalent von Mises stress, with relation to the results obtained at $t = 1.0$ s, is as high as 7648 kPa. On the other hand, the peak difference in the equivalent von Mises stress for impact taking place at $t = 1.1$ s and $t = 1.4$ s, with relation to the results obtained at $t = 1.0$ s, is equal to 4220 kPa and 3949 kPa, respectively (see Fig. 3 and Fig. 5).

Concluding remarks

The results of a numerical analysis focused on the behaviour of multi-storey steel frame building, that suffers from a soft-storey failure under ground motion excitation, have been presented in this paper. A numerical model of the structure has been created in FEM computer software and has been exposed to an impact that would have been generated after a soft-storey failure due to falling of the upper floors. The results of the study show that not only the value of the impact force is crucial but also the moment when the impact occurs. It has been observed that the most critical moment for the structure for being subjected to a vertical impact is when the horizontal deflection is close to its peak.

The study described in this paper has concerned the response of a four-storey building under unidirectional ground motion excitation in the form of sine wave. Further numerical studies on structural models of different types of buildings under three-dimensional earthquake excitations are therefore required in order to extend our knowledge on behaviour of buildings suffering from the soft-storey failure.

References

[1] S. Elkholy and K. Meguro, in: Proc. 13th World Conference on Earthquake Engineering, Vancouver, Canada, Paper No. 930 (2004).
[2] M. Talaat and K.M. Mosalam: Earthquake Engineering and Structural Dynamics Vol. 38 (2009), p. 609.
[3] S.A. Anagnostopoulos: Earthquake Engineering and Structural Dynamics Vol. 16 (1988), p. 443.
[4] B.F. Maison and K. Kasai: Earthquake Engineering and Structural Dynamics Vol. 21 (1992), p. 771.
[5] C.G. Karayannis and M.J. Favvata: Structural Engineering Mechanics Vol. 20 (2005), p. 505.
[6] S. Mahmoud, X. Chen and R. Jankowski: Journal of Applied Sciences Vol. 8 (2008), p.1850.
[7] R. Jankowski: Engineering Structures Vol. 31 (2009), p. 1851.
[8] R. Jankowski: Earthquake Engineering and Structural Dynamics Vol. 39 (2010), p. 343.
[9] S. Mahmoud and R. Jankowski: Key Engineering Materials Vol. 417-418 (2010), p. 513.
[10] W. Migda and R. Jankowski: Journal of Applied Sciences Vol. 12 (2012), p. 466.

Relation between structural size and the discretization density of brittle homogeneous lattice models

Miroslav Vořechovský[1,a], Jan Eliáš[1,b]

[1]Institute of Structural Mechanics, Faculty of Civil Engineering, Brno University of Technology, Brno, Czech Republic

[a]vorechovsky.m@fce.vutbr.cz, [b]elias.j@fce.vutbr.cz

Keywords: lattice model, brittle elements, mesh size, bent beams, size effect

Abstract. This paper contains the results of an investigation into the effect of the discretization of lattice models. The study is performed with homogeneous models where all elements share the same strength. Elemental constitutive law is linearly-brittle, meaning that elements behave linearly but are completely removed from the structure as soon as they reach the limit of their strength. The relation between structural size and discretization density is studied with unnotched beams loaded in three point bending (modulus of rupture test). We report the results for regular discretization and irregular networks obtained via Voronoi tessellation. This is carried out for two types of models: these being with and without rotational springs (normal and shear springs are always present). The numerically obtained dependence of strength on discretization density is compared to the analytical size effect formula.

Introduction and brief model description

The lattice representation of material is a well-established approach to modeling the failure of brittle and quasibrittle materials. Material is treated as a set of discrete rigid-like elements inter-connected by springs. In the classical version of these models, the connections behave in an elasto-brittle manner until some failure criteria are reached [1-3]. Then, the elements are completely removed from the structure. The response of these brittle models appears to be highly dependent on discretization density [4]; this dependence can be avoided by the application of material mesostructure [5].

In this contribution we focus only on homogeneous brittle lattice models (i.e. without mesostructure). Thus, all lattice connections share the same elastic parameters and strength criterion. We assume that varying the network density in homogeneous models corresponds to changes in the structural size of the structure modeled. These mesh-density effects are studied for specimens that fail due to a crack initiated from a smooth surface. In particular, we have performed numerous simulations with unnotched three-point-bent specimens. The span $S = 3D$, where D is the specimen depth.

The model used is a rigid-body-spring network in accordance with a paper by Bolander & Saito [6]. The fracture criteria are taken from the same article, i.e. a Mohr-Coulomb surface with tension cut-off is adopted. The solution proceeds in linear elastic steps that are scaled so that one connection breaks at each step [7]. The irregular network is generated by Voronoi tessellation on a set of pseudo-randomly placed nuclei with limited maximal mutual distance l_{min}. By changing l_{min} one can control the mesh density. The IRN model (with an irregular network) is compared to a network with locally regular geometry (REN). In the REN model, the crack can only propagate along the axis of symmetry through regularly placed squared elements of exact l^{min} size. The rest of the specimen is meshed by a lattice of irregular geometry. Consequently, the obtained nominal forces are scattered.

Two different versions of the mechanical model are studied. They differ in how internal forces (between rigid bodies) are transmitted through the connections of adjacent facets. In the first model type (denoted NS), only normal and shear springs act. In the NSR model type, rotational springs transferring local bending moments are also added. However, only stresses in normal and shear springs contribute to the fracture criteria in both model types.

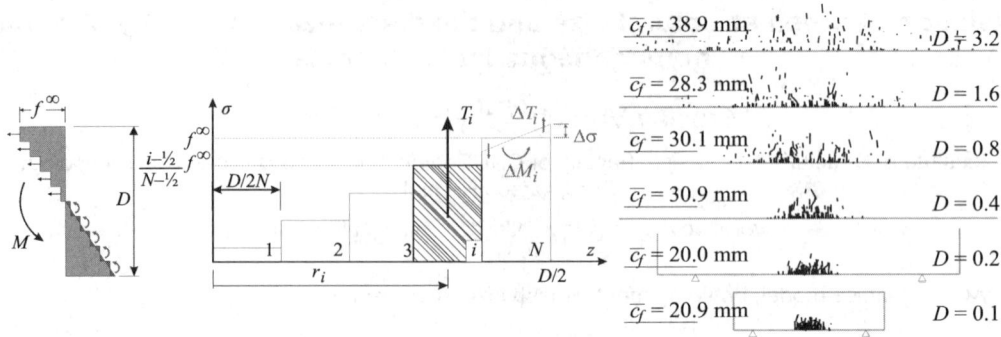

Figure 1: Left: On derivation of the peak moment in a bent specimen. Right: Crack patterns at the peak load for various sizes of the IRN beam. The left horizontal lines indicate the average height c_f.

Size effect simulations and formulas

The size of a concrete specimen typically affects the observed nominal strength. Several sources of this phenomenon have been documented [8]; the statistical and deterministic size effects are the two most significant. Since there is no internal length in our constitutive law/model we can represent varying size by varying network density. The characteristic size (depth) D is kept constant at a reference size $D_0 = 0.1$ m, whereas the network density l^{min} is varied; we can mimic the varying of the intrinsic size D by writing $D = D_0 \cdot l_0^{min} / l^{min}$, where l_0^{min} is the selected reference mesh density.

Since we are dealing, in fact, with models of the same size, it is not necessary to report the size dependence on nominal strength (nominal stress at peak load). It suffices to report the loading forces $F(D)$. On the other hand, however, the lengths (e.g. the crack length) must be recalculated.

Let us now deliver a closed-form expression for the observed size effect. Consider the midspan rectangular cross-section BD. The depth is discretized into $2N$ rigid bodies' contacts of the same size, and therefore the stress profile is a piecewise constant function along the depth D and approximates the actual linear profile. When the outermost spring reaches the extreme tensile stress f^∞, the cross-section reaches its maximum bending moment M. Due to the symmetry along the neutral axis we can only consider the lower half of the depth (N elements) and calculate the bending moment as a doubled sum of force contributions multiplied by the corresponding arms. Each i-th force contribution can be written as (Figure 1 left): $T_i = (i-1/2) BD f^\infty / 2N(N-1/2)$. Each such force has the following arm from the neutral axis: $r_i = D(i-1/2)/2N$. The resisting moment is double the sum for moment contributions:

$$M(N) = 2\sum_{i=1}^{N} T_i r_i = \frac{BD^2}{4N^2} \frac{f^\infty}{N-1/2} \sum_{i=1}^{N}\left(i-\frac{1}{2}\right)^2 = \frac{BD^2}{6} f^\infty \left(\frac{2N+1}{2N}\right) \quad (1)$$

As N tends to infinity, the bending moment converges to the well-known value $M^\infty = f^\infty BD^2 / 6$. The external bending moment equals $M = F/2 \cdot 3D/3$. Equalizing these two expressions and considering that $l^{min} = D/(2N)$, one ascertains that:

$$F = \frac{2BD}{9} f^\infty \left(\frac{2N+1}{2N}\right) = \underbrace{\left(\frac{2BD}{9}\right) f^\infty}_{F^\infty} \left(1 + \frac{l_0^{min}}{D}\right) = F^\infty \left(1 + \frac{l_0^{min}}{D}\right) = F^\infty \left(1 + \frac{D_b}{D}\right) \quad (2)$$

We can introduce a new length constant, $D_b = l_0^{min}$. Then, Eq. (2) becomes identical with Bažant's size effect formula for type 1 deterministic size effect (see pages 41–43 of [8]). The increment to the asymptotic force F^∞ is inversely proportional to D and therefore diminishes for large D (or small l^{min}). What remains to be clarified is the choice of the extreme stress f^∞. An obvious choice would be the direct tensile strength ($f_t^\infty = 5$ MPa) of the model. This is because very large specimens fail at the initiation of a crack right at the midspan bottom face, which must therefore equal the tensile strength, yielding the asymptotic force $F_l^\infty = 11.11$ kN. Unfortunately, the stress profile in not perfectly linear in reality. The real stress profile is affected by wall-like stress distribution (the span of the beam is only $3D$) and by the local compressive stress concentration around the point load. The nonzero Poisson's ratio causes additional deviation from the linear stress profile; see [9] or [10]. As an approximation, we used a nonlinear least-square fitting procedure to determine the two free parameters D_b and F^∞ in Eq. 2. The fits are plotted in Figure 2 and compared to the computed data.

Now, what if the rotational springs are employed? In each element (or contact area), its spring adds a new additional moment ΔM_i; see the last strip in Figure 1 left. These contributions are equal. In each bin there is a pair of forces ΔT_i that represent two triangles (below and above the constant stress σ_i. Each of these triangles are as long as half of the strip ($=D/(4N)$) and the maximum stress difference is $\Delta \sigma$. The stress $\Delta \sigma$ is one half of the difference between the current strip and the adjacent strip: $\Delta \sigma = (\sigma_i - \sigma_{i-1})/2 = f^\infty/(2N-1)$. The pair of forces ΔT_i representing the two triangles are:

$$\Delta T_i = \frac{1}{2} \Delta \sigma \cdot B \cdot \frac{D}{4N} = \Delta \sigma \frac{BD}{8N} = \frac{f^\infty}{2N-1} \cdot \frac{BD}{8N} \qquad (3)$$

Each of these two forces act over the distance of $D/(6N)$ from the "neutral" state and form an additional increase in the total bending moment. The magnitude of each such moment contribution (in each strip-bin) is twice the arm multiplied by the force ΔT_i: $\Delta M_i = 2 \cdot \Delta T_i \cdot D/6N$. In total, there are $2N$ such partial moments over the whole cross-section and therefore the total moment increment is $2N \times$ the contribution ΔM_i and $\Delta M = 2N \cdot \Delta M_i = f_\infty BD^2/12N(2N-1)$. The moment increment is not reflected in the failure condition. Transforming it into the increment of maximal force gives:

$$\Delta F = \frac{4}{3D} \Delta M = F^\infty \frac{1}{2N(2N-1)} = F^\infty \frac{l_0^{min}}{D} \cdot \frac{l_0^{min}}{D - l_0^{min}} \qquad (4)$$

Adding this increment to the total force from Eq. 2 yields the upgraded size dependence. The increment ΔF increases the maximal load especially for small sizes because it is proportional to the inverse of D^2 while the increment to the asymptotic stress in Equations 2 was only inversely proportional to D. The irregularity of the network geometry (IRN) allows the model to choose the "weakest" area to initiate and propagate the crack. Contrary to REN, where the rupture of the first connection leads to the failure of the whole beam, the load applied to break the first spring in the IRN model (elastic limit in Fig. 2) is, on average, much lower than the peak forces. The peak forces in IRN models are greater than those of REN models, whereas the elastic limits are lower in IRN models.

Qualitatively, however, both force dependencies of IRN are similar to REN and follow the tendency proposed by Eq. 2 and Eq. 4, respectively. The deviations for larger specimens (finer mesh densities) are caused by the local stress deviation described earlier, meaning namely the stress fluctuations in the lowermost layer caused by Poisson's ratio.

Instead of one crack, many small cracks are created inside the bottom area of the IRN specimen (Figure 1b) and the model allows for the redistribution of forces after many such local ruptures. These cracks do not form a continuous line at maximal load. The fact that the zone has, on average, approximately the same height for all sizes (over size ratio 1:32) supports our claim that the data can be approximated reasonably well by Bažant's size effect formula, which is similar to Eq. 2.

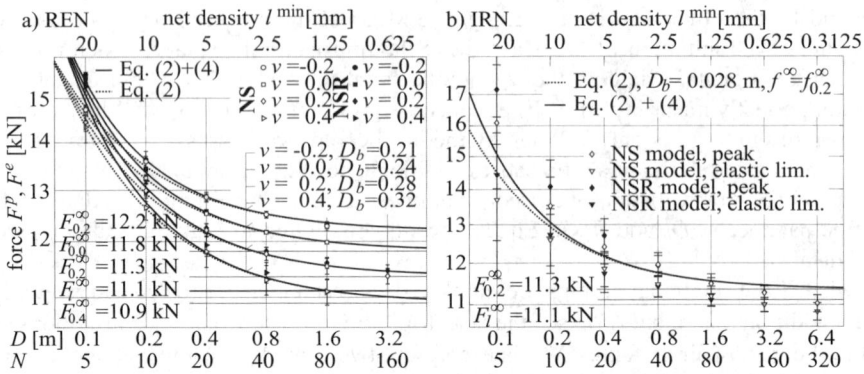

Figure 4: Dependency of the peak load on the network density (or structural size D) and comparison to the size effect formulas (Equations 2 and 4). a) REN meshes with a different Poisson's ratio. b) Plot of elastic limits and peak loads of beams with an irregular network. Average values and standard deviations are computed from 50 realizations for every size.

Summary

The effect of the discretization of homogeneous lattice models was studied on three-point bending simulations. We report the results for regular and irregular geometry. The dependence of strength is compared to derived size effect formulas and we show that the fineness of the discretization of specimens of the same size can mimic variations in the size of lattice models with the same discretization.

Acknowledgement. This research was conducted with the financial support of the Czech Science Foundation, projects GACR P105/10/J028 and GD103/09/H08.

References

[1] G. Lilliu, J.G.M. van Mier: Eng. Fract. Mech. Vol. 70 (2003), p. 927–941.

[2] J.E. Bolander, H. Hikosaka, W.-J. He: Eng. Computation. Vol. 15 (1998), p. 1094–1116.

[3] J. Eliáš, H. Stang: Int. J. Fracture, available online, (2012), DOI 10.1007/s10704-012-9677-3

[4] A. Jagota, S. J. Bennison: Model. Simul. Mater. Sc., Vol. 3 (1995), p. 485–501.

[5] J. Eliáš, M. Vořechovský: Key Eng. Mat., Vol. 488-489 (2012), p. 29-32

[6] J.E. Bolander, S. Saito: Eng. Fract. Mech., Vol. 61 (1998), p. 569–591.

[7] J. Eliáš, M. Vořechovský, P. Frantík: Eng. Fract. Mech., Vol. 77 (2010), p. 2263–2276.

[8] Z.P. Bažant: *Scaling of Structural Strength* (Elsevier Butterworth-Heinemann, 2005).

[9] M. Vořechovský, J. Eliáš, in: Computational Modelling of Concrete Structures (EURO-C 2010), edited by Bicanic et al, CRC Press (2010), ISBN 978-0-415-58479-1, p. 419–428.

[10] M. Vořechovský, J. Eliáš, in: Proceedings of the Tenth International Conference on Computational Structures Technology, Civil-Comp Press (2010).

MATCHING ASYMPTOTIC METHOD IN PROPAGATION OF CRACKS WITH DUGDALE MODEL

Dang Thi Bach Tuyet[1,a], Jean Jacques Marigo[1,b], and Laurence Halpern[2,c]

[1]Laboratoire de Mécanique des Solides (UMR 7649), École Polytechnique, 91128 Palaiseau cedex, France

[2]LAGA, Institut Galile, University Paris XIII, 93430 Villetaneuse, France

[a]tuyet@lms.polytechnique.fr, [b]marigo@lms.polytechnique.fr, [c]halpern@math.univ-paris13.fr

Keywords: Brittle fracture, matching asymptotic, anti-plane case, Dugdale model.

Abstract. The goal of this work is to apply the matching asymptotic method combined with a variational approach to study the initiation and the propagation of a cohesive crack from the tip of a preexisting notch following the Dugdale cohesive force model when the characteristic length of the material (included in the Dugdale model) is small by comparison with the characteristic length of the body.

1. Introduction

Dugale model is based on the assumption that the surface energy density ϕ depends on the displacement jump, such as in the mode III, the displacement field at equilibrium $\mathbf{u}$ is antiplane, *i.e.*

$$\mathbf{u}(\mathbf{x}) = u(x_1, x_2)\mathbf{e}_3$$

the surface energy density is formulated :

$$\phi(\llbracket u \rrbracket) = \begin{cases} G_c \llbracket u \rrbracket / \ell_c & \text{if} \llbracket u \rrbracket \leq \delta_c \\ G_c & \text{if} \llbracket u \rrbracket \geq \delta_c \end{cases} \quad (1)$$

In Eq.1, G_c denotes for the critical energy release rate of the Griffith theory, whereas δ_c is an internal length characteristic of the cohesive forces model. $\llbracket u \rrbracket$ denotes for the jump of the displacement. The so-called cohesive foces given by the ratio G_c/δ_c is denoted σ_c with $\sigma_c = \frac{G_c}{\delta_c}$. From Eq.1, the normal stress, such as σ_{32}, on the crack lips is equal σ_c if $\llbracket u \rrbracket < \delta_c$ and eliminates if $\llbracket u \rrbracket > \delta_c$. Thus, the crack lips are divided into two zone : a cohesive zone where the cohesive forces are equal to σ_c and a non cohesive zone where $\llbracket u \rrbracket > \delta_c$ where having non cohesive forces. Here, we are interested in studying the progress of initiation as well as propagation of a crack in the geometry of the notch characterized by the parameter $\epsilon = \tan(\pi - \frac{\omega}{2})$, ω being the angle of the notch, see Fig. 1. We will consider two scale of coordinates: the "macroscopic" coordinates $\mathbf{x} = (x_1, x_2)$ and the "microscopic" coordinates $\mathbf{y} = \mathbf{x}/\ell = (y_1, y_2)$, where ℓ is a small characteristic length of the crack inside the body. The tip of the notch is taken as the origin of the space. With supposing that crack appears inside the body follows a straight-path which derives from the notch tip and the axis x_1 is chosen in the same direction of crack propagation. The unit vector orthogonal to the (x_1, x_2) plane is denoted $\mathbf{e}_3$.[0.1cm]

2. The real problem

The natural reference configuration of the *sound* two-dimensional body is denoted by Ω_0. Assuming that a crack denoted by Γ appears inside Ω_0. The cohesive zone is denoted by Γ_c and the non-cohesive zone is denoted by Γ_0. They are governed by their lengths of ℓ and h : $\Gamma = \Gamma_0 \cup \Gamma_c$ with $\Gamma_0 = (0, l) \times \{0\}$ and $\Gamma_c = [l, \ell) \times \{0\}$. The associated body containing the crack Γ is denoted by $\Omega_\Gamma = \Omega_0 \setminus \Gamma$. The two edges of the notch are denoted by Γ^+ and Γ^- in Fig. 1. When one uses polar coordinates (r, θ), the pole is the tip of the notch and the origin of the polar angle is the edge Γ^-. Accordingly, we have : $r = |\mathbf{x}|$, $\Gamma^- = \{(r, \theta), 0 < r < r^*, \theta = 0\}$, $\Gamma^+ = \{(r, \theta), 0 < r < r^*, \theta = \omega\}$. With assuming without body forces, u must be an harmonic function : $\Delta u = 0$ in Ω_Γ

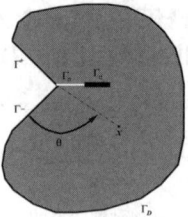

Fig. 1: The domain Ω_Γ for the real problem

The edges of the notch are free while the lips of the crack are submitted to a cohesive forces :

$$\frac{\partial u}{\partial \nu} = 0 \quad \text{on} \quad \Gamma^\pm \cup \Gamma^0, \qquad \frac{\partial u}{\partial \nu} = \sigma_c \nu \quad \text{on} \quad \Gamma_c \tag{2}$$

In (2), ν denotes the unit outer normal vector to the domain. The remaining part of the boundary of Ω_Γ is denoted by Γ_D where the anti-plane displacement is prescribed such that the load is beyond the fracture threshold requires us to consider the initiation and the propagation of a non-cohesive part of the lips of the crack. Specifically, we have :

$$u = U_\infty \quad \text{on} \quad \Gamma_D \tag{3}$$

The linearity allows us to decompose the original problem into two problems corresponding to the value of U_∞ on the boundary and σ_c on the crack lips independently. The two problems will be denoted by the non-cohesive problem and the cohesive problem. We denote u^∞ be the solution of the non-cohesive problem and u^c be the solution of the cohesive problem.

The non-cohesive problem Finding the solution u^∞ such that it satisfies the set of the following equations:

$$\begin{cases} \Delta u^\infty = 0 & \text{in} \quad \Omega_\Gamma \\ \dfrac{\partial u^\infty}{\partial \nu} = 0 & \text{on} \quad \Gamma_N^+ \cup \Gamma_N^- \cup \Gamma_0 \cup \Gamma_c \\ u^\infty = 1 & \text{on} \quad \Gamma_D \end{cases} \tag{4}$$

The cohesive problem. Finding the solution u^c such that it satisfies the set of the following equations:

$$\begin{cases} \Delta u^c = 0 & \text{in} \quad \Omega_\Gamma \\ \dfrac{\partial u^c}{\partial \nu} = 0 & \text{on} \quad \Gamma_N^+ \cup \Gamma_N^- \cup \Gamma_0 \\ \mu \nabla u^c = -\mathbf{e}_2 & \text{on} \quad \Gamma_c \\ u^c = 0 & \text{on} \quad \Gamma_D \end{cases} \tag{5}$$

The displacement of the real problem is found through the solution u^∞ of Eq. 4 and the solution u^c of Eq. 5. It can be read as :

$$u(\mathbf{x}) = U_\infty u^\infty(\mathbf{x}) - \sigma_c u^c(\mathbf{x}) \tag{6}$$

When the length ℓ is small comparision with the characteristic length of the body, MA method is applied because of the existing the two overlapped singularities near the notch tip and the crack tip.

3. Matching asymptotic approach

The solution is found by two asymptotic expansions in terms of ℓ. The first one, called the inner expansion, is valid in the neighborhood of the tip of the notch, ie. $r \ll 1$, while the other, called the outer expansion, is valid far from this tip, ie. $r \gg \ell$. These two expansions are matched in an intermediate zone. We approaching both of the problems Eq. 4 and Eq. 5 by using the MA expansion with respect to the asymptotic variable ℓ in order to obtain the more correct solution in vicinity of the crack. With Eq. 4, we have a homogeneous Neumann conditions on the crack lips, while in the Eq. 5 the Neuman conditions on the crack lips are constant.

MA approach for the non-cohesive problem. Supposing the outer expansion and the inner expansion of the solution u^∞ is expanded in the series of ℓ. We denote u^∞_{out} for the outer expansion of u^∞ and the u^∞_{in} for the inner expansion of u^∞.

$$u^\infty_{out}(\mathbf{x}) = \sum_{i \in \mathbb{N}} \ell^{i\lambda} U^i(\mathbf{x}) \quad \text{and} \quad u^\infty_{in}(\mathbf{x}) = \sum_{i \in \mathbb{N}} \ell^{i\lambda} V^i(\mathbf{y}) \tag{7}$$

Each of the expanded terms such as U^i and V^i are determined by inserting Eq. 7 into Eq. 5. Even though the expansion of u^∞_{out} is valid far from the crack, U^i must be defined in the whole outer domain Ω_0 which corresponds to the sound body. However, the behavior of U^i in the neighborhood of $r = 0$ is singularity. Even though inner expansion valid only in the neighborhood of $r = 0$, V^i must be defined in the infinite inner domain Ω^∞. Moreover, the conditions at infinity is missing in the set of equations governed for V^i. The singularity at $r = 0$ of U^i and the conditions at infinity of V^i will be given by the matching conditions.

MA approach for the cohesive problem. Being similar to the non-cohesive problem. We find the expanded forms of the outer expansion and inner expansion of the solution u^c. We denote u^c_{out} for the outer expansion of u^c and the u^c_{in} for the inner expansion of u^c. They can be read as:

$$u^c_{out}(\mathbf{x}) = \ell \sum_{i \in \mathbb{N}} \ell^{i\lambda} \bar{U}^i(\mathbf{x}) \quad \text{and} \quad u^c_{in}(\mathbf{x}) = \ell \sum_{i \in \mathbb{N}} \ell^{i\lambda} \bar{V}^i(\mathbf{y}) \tag{8}$$

The process of finding the expanded terms, ie. $\bar{U}^i$ and $\bar{V}^i$ in Eq. 8 is similar to process for the non-cohesive problem.

This work will be mentioned precisely in our article "Matching asymptotic method and nucleation of a defect of a notch" submitted in ACOME 2012. In the following section, we will introduce the criteria of minimum energy for determining the two tips of the cohesive zone and the non-cohesive zone.

4. The minimum energy approaching

At equilibrium sate, the total energy of the body can be read as:

$$\mathcal{E}(U_\infty, l, \ell) = \frac{1}{2} \int_{\Omega_\Gamma} \nabla u . \nabla u dx - \int_{\partial \Omega} \nabla u \nu U^\infty ds + \int_{\Gamma_c} \sigma_c [\![u]\!] dx_1 + 2G_c(\ell - l) \tag{9}$$

where the last two terms which are considered as the surface energy of the body including the surface energy of the cohesive zone and non-cohesive zone. The evolution of the tip of non-cohesive zone and the tip of cohesive zone corresponding to the external load U_∞ must be such that the total energy $\mathcal{E}(U_\infty, ., .)$ of the body obtain local minimum at (l, ℓ) for a given U_∞. The minimum energy criteria can be read as:

$$\frac{\partial \mathcal{E}}{\partial \ell}(U_\infty, l, \ell) = 0 \quad \text{and} \quad \frac{\partial \mathcal{E}}{\partial l}(U_\infty, l, \ell) = 0 \tag{10}$$

In other words, it proposes the tips of cohesive zone and non-cohesive zone are determined such that the total energy release rates due to the propagation of them in turn vanish. We denotes the energy release rates due to the propagation of the tip of the cohesive zone by $\mathcal{G}_\ell$ and the energy release rates due to the propagation of the tip of the non-cohesive zone by and $\mathcal{G}_l$. They can be formulated as following:

$$\mathcal{G}_\ell = -\frac{\partial \mathcal{E}}{\partial \ell}(U_\infty, l, \ell), \quad \mathcal{G}_l(u) = -\frac{\partial \mathcal{E}}{\partial l}(U_\infty, l, \ell) \tag{11}$$

Calculation of the energy release rate $\mathcal{G}_\ell$. Taking the first derivative in Eq. 11 meets some difficulties because of the existing the singularity at the tip of the cohesive zone. Following the idea of mapping the variable crack domain onto a fixed crack domain, we can compute $\mathcal{G}_\ell$. Supposing the crack in small, we use a smooth map to transform the neighborhood of the crack at the state of $\Gamma_{\bar{\ell}}$ ($\bar{\ell} = \ell + \delta \ell$) into the neighborhood of the crack at the state of Γ_ℓ. Let $\mathbf{v}(\mathbf{y}) = v(\mathbf{y})\mathbf{e}_1$ be a smooth vector where $v(\mathbf{y})$ is defined :

$$v(\mathbf{y}) = \begin{cases} -\sqrt{(y_1-1)^2 + y_2^2} + 1 & \text{if } \sqrt{(y_1-1)^2 + y_2^2} \leq 1 \\ 0 & \text{if } \sqrt{(y_1-1)^2 + y_2^2} > 1 \end{cases} \quad (12)$$

applying the MAE into the formulating of $\mathcal{G}_\ell$, using the inner expansion, finally, we obtain :

$$\mathcal{G}_\ell = \frac{\mu}{\ell} \int_{B(1,1)\backslash\Gamma_1} \left(\frac{\partial u^{in}}{\partial y_j} \frac{\partial u^{in}}{\partial y_1} \frac{\partial v}{\partial y_j} - \frac{1}{2} \frac{\partial u^{in}}{\partial y_j} \frac{\partial u^{in}}{\partial y_j} \frac{\partial v}{\partial y_1} \right) ds - \int_{\Gamma_c} \sigma_c [\![u^{in}]\!] \frac{\partial v}{\partial y_1} dy_1 \quad (13)$$

where $B(1,1)$ is notation for a zone defined by $\{(y_1,y_2) \in \Omega^\infty \text{ such that } (y_1-1)^2 + y_2^2 \leq 1\}$. Because of Eq. 6, accordingly, the stress field σ and the SIF K_{III} can be expressed as :

$$\sigma(\mathbf{x}) = U_\infty \sigma^\infty(\mathbf{x}) - \sigma_c u^c(\mathbf{x}) \quad \text{and} \quad K_{III} = U_\infty K^\infty_{III} - \sigma_c K^c_{III} \quad (14)$$

where $(\sigma^\infty(\mathbf{x}), K^\infty_{III})$ and $(\sigma^c(\mathbf{x}), K^c_{III})$ are the stress and the SIF at the tip of the cohesive zone coressponding to u^∞ and u^c. Besides, because the cohesive forces are constant, they do not change the form of the singularity at the tip of the crack of the cohesive zone. The criteria governing the propagation of the cohesive crack tip $\mathcal{G}_\ell$ is equivalent to SIF $K_{III} = 0$. Irwin formula give us that $\mathcal{G}_\ell \sim K^2_{III}$, moreover, Eq. 14 leads to :

$$U_\infty = \sigma_c \sqrt{\mathcal{G}^c_\ell / \mathcal{G}^\infty_\ell} \quad (15)$$

with $\mathcal{G}^c_\ell$ and $\mathcal{G}^\infty_\ell$ are defined as in Eq. 13 but u^{in} is replaced by corresponding u^∞ and u^c.

Calculation of the energy release rate $\mathcal{G}_l$. There is no singularity at the tip of non-cohesive zone, so $\mathcal{G}_l$ can obtained by taking the derivative of $\mathcal{E}(U_\infty, l, \ell)$ as in Eq. 9 with respect to l under the integration. It can be read as :

$$\mathcal{G}_l = 2\sigma_c [\![u]\!](l) - 2G_c \quad (16)$$

besides, Eq. 10 and Eq. 11 lead to :

$$[\![u]\!](l) = G_c/\sigma_c \quad \text{or} \quad U_\infty [\![u]\!]^\infty(l) - \sigma_c [\![u]\!]^c(l) = \frac{G_c}{\sigma_c} \quad (17)$$

Accordingly, we obtain the system of two coupled non linear equations of Eq. 15 and Eq. 17 which determine the tips of the cohesive zone and the non-cohesive zone.

References

[1] B. Bourdin, G. A. Francfort, and J.-J. Marigo. *The variational approach to fracture*. Springer, 2008.

[2] M. Dauge, S. Tordeux, and G. Vial. Selfsimilar perturbation near a corner: matching versus multiscale expansions for a model problem. In *Around the research of Vladimir Maz'ya. II*, volume 12 of *Int. Math. Ser. (N. Y.)*, pages 95--134. Springer, 2010.

[3] P. Grisvard. *Elliptic problems in non smooth domains*. Number 24 in Monographs and Studies in Mathematics. Pitman, 1985.

[4] J.-B. Leblond. *Mécanique de la rupture fragile et ductile*. Collection Études en mécanique des matériaux et des structures. Editions Lavoisier, Paris, 2000.

[5] J.-J. Marigo. Initiation of cracks in Griffith's theory: an argument of continuity in favor of global minimization. *Journal of Nonlinear Science*, 20(6):831--868, 2010.

Formability evaluation of Non-Crimp Carbon Fabric by non-contact 3D deformation measurement system

Kazuto TANAKA[1,a], Kazuya KANAZAWA[2],
Shinichi ENOKI[3] and Tsutao KATAYAMA[1,b]

[1] Professor, Dept. of Biomedical Eng., Doshisha Univ. (Japan)

[2] Graduate Student, Dept. of Mechanical Eng., Doshisha Univ. (Japan)

[3] Associate Professor, Dept. of Mechanical Eng., Nara National College of Technology (Japan)

[a] ktanaka@mail.doshisha.ac.jp, [b] tkatayam@mail.doshisha.ac.jp.

Keywords: Non-Crimp Fabric (NCF), Non-contact 3D deformation measurement, Stitch, Stitching Parameter.

Abstract. Non-Crimp Carbon Fabric (NCF) consists of unidirectional plies which are kept together by stitching yarns arranged in a number of different orientations relative to the fabric production direction. It is reported that NCF possesses excellent drape performance compared to woven fabrics. However there is not a clear criterion of a drape evaluation on the drape characteristic of the NCF. In addition, it is not clarify that stitch pattern and stitch tension influence on the drape characteristic of the NCF. Moreover, in existing bias extension test, measurement of shear angle is based on the pin-jointed net (PJN) approximation. The PJN approximation doesn't takes into consideration the fiber sliding and the effect of the stitched parameters of the NCF. In this study, the bias extension test based on the measurement of shear angle by non-contact 3D deformation measurement system was conducted to evaluate the drape performance of the NCF. We made a proposal of the formability evaluation index based on the measurement results. Moreover, the 3D draping tests were conducted onto hemisphere geometry and regular tetrahedron, in order to verify availability of the formability evaluation index. The availability of the formability evaluation index was verified.

Introduction

Non-Crimp Fabric (NCF) have attracted much attention of the automotive industry and their usage is growing rapidly [1]. This structure leads to an advantageous combination of high material properties, low cost processing, and excellent drape performance, thus it is also suitable for moulding composite materials of complex shapes [2]. Common problems in the fabricating process of three-dimensional shells of complex curvature include unwanted features such as wrinkle or tears. The problems depend on the geometry of the mould surface and the type of fabric. Furthermore, during draping and forming of a fabric, there are local variations of fiber orientation, fiber volume fraction and possibly fabric thickness which are going to affect processability and the mechanical properties of the final product. It is reported that NCF possesses excellent drape performance compared to woven fabrics. But, there is not a clear criterion of a drape evaluation. The influence of stitch type and stitch tension to the drape characteristic of NCF is not clarified. Until now, the picture frame test [3-5] and the bias extension test [6-9] have been proposed as the drape evaluation to the textile preform like woven fabrics. Especially, the bias extension test has been considered to be an alternative to the picture frame test. The reasons are as follows. The bias extension test can be carried out easily, the influence of the boundary condition to the result is small and the bias extention test obtains high reproducibility results. However, measurement of shear angle is based on the pin-jointed net (PJN) approximation [10]. The fibers sliding and the effect of the stitching parameters of NCF were could not consider in PJN. Therefore, the development of a new measuring method of the shear angle which takes into consideration the fiber sliding and the proposal of formability evaluation index are needed. In this study, the bias extension test based on measurement of shear angle by non-contact 3D deformation measurement system was conducted to evaluate the drapability of NCF. Measured shear angle proposed the formability evaluation index. Moreover, the draping test onto hemisphere geometry and regular tetrahedron was conducted and the availability of formability evaluation index calculated by the bias extension test was verified.

Material and Experimental Procedure

Material. Bidirectional carbon fiber Non-Crimp Fabrics (Benny-Toyama Co. Ltd., Japan) used in this study. The specimen size is 50mm×150mm in bias extension test. The specimen size is 300mm×300mm in 3D draping test. The directions of carbon fiber bundles are 0°/90°. The areal density of the fiber bundles is 300g/m^2. The stitching yarn is polyester sewing thread. The stitching pattern is tricot stitch. The stitch tensions are Normal, Tight and Loose.

Bias extension test. The set-up consists of specimens installed in the jaws of a tensile testing machine with the fibers in the warp and weft directions initially oriented at ±45°from the loading direction. During testing, the central portion of the specimen undergoes a pure interplay shear deformation, where the angle between the fibers decreases gradually until it reaches a locking angle at which the sheet theoretically undergoes an out of plane deformation to from a wrinkle. In this study, the bias extension test was conducted by universal testing machine (Autograph AG-100kNX, Shimadzu Co. Ltd.) and measurement of shear angles of carbon fiber bundles was conducted by non-contact 3D deformation measurement system (ARAMIS®, GOM mbH).

Draping test for 3D shapes. Fig.1 shows Schematic view of draping tests by hemisphere punch. Fig.2 shows Schematic view of draping tests by tetrahedral punch. The draping tests were conducted by universal testing machine (Autograph AG-100kNX, Shimadzu Co. Ltd.) and measurement of shear angles of carbon fiber bundles was conducted by non-contact 3D deformation measurement system (ARAMIS®, GOM mbH). The displacement rate was set for 10mm/min.

We conducted the draping tests by two kinds of punch shapes. One of the shapes is hemisphere and another shape is tetrahedral. Diameters of the hemisphere punch shape are 50mm. Each blank holder for the hemisphere punches has 52mm diameter hole. Another shape of the tetrahedral punch is regular tetrahedron of 100mm on a side. A blank holder for the tetrahedral punch is regular triangle hole of 120mm on a side. In order to measure by non-contact deformation analysis, random pattern was made on the square specimen(300mm×300mm in size) . The random pattern was produced by an air spray paint mixed with calcium carbonate and ethanol. During test, the state of deformation was measured by non-contact 3D deformation measurement system (ARAMIS®, GOM mbH).

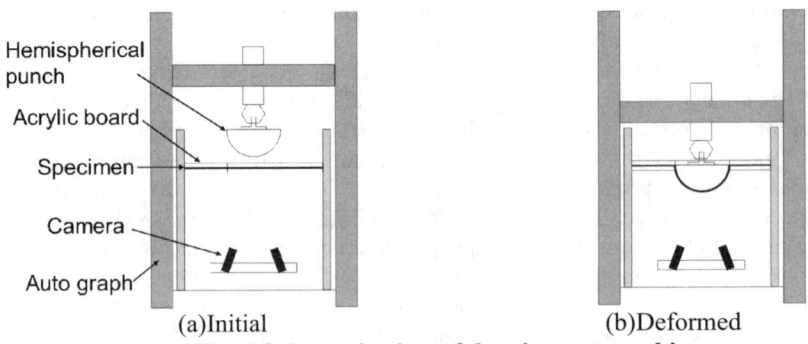

Fig. 1 Schematic view of draping test machine

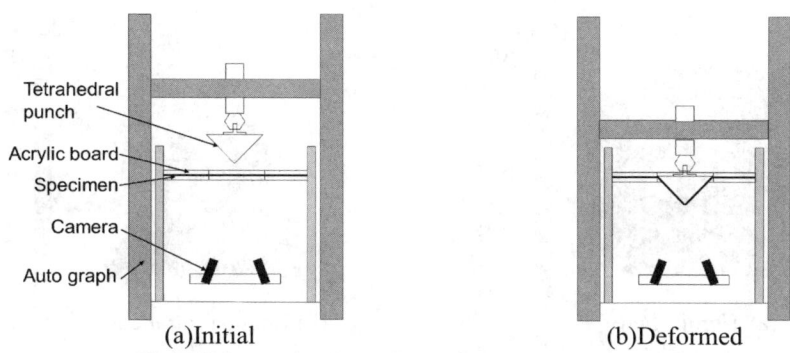

(a)Initial (b)Deformed
Fig.2 Schematic view of draping test machine

Maximum shear angle measuring procedure. Maximum shear angle is defined by angle that did not result in defect. Therefore, maximum shear angle was assumed to be formability evaluation index. Maximum shear deformation angle was measured from the last image, which become impossible to analyze by defect.

Results and Discussions
The availability of formability evaluation index.
Fig. 3 shows the availability of formability evaluation index measured by the bias extension test. Fig. 4 shows the maximum shear angle after 3D draping test. Fig. 5 shows the contour of shear angle by hemisphere punch and tetrahedral punch after draping test in normal stitch tension. In the draping test of hemisphere punch, shearing angle in directions of 0° and 90 is the smallest of all area from Fig.5. However, the defect was confirmed in directions of 0° and 90° after draping test. When the maximum shearing angle of bias extension test was exceeded in the direction 45°, these defects were caused. Therefore, formability of NCF can be evaluated by maximum shear angle. Significant difference was not seen between maximum shear angle of the hemisphere punch and maximum shear angle of bias extension test. Therefore, the draping performance of the hemisphere can make a valuation by the maximum shear angle. However, there was a significant difference was seen between maximum shearing angle of the regular tetrahedron draping test and maximum shear angle of bias extension test. In draping test of regular tetrahedron punch, the defect was caused in the direction of 0° and 90°. However the defect caused in draping test of tetrahedral punch before maximum shear angle of bias extension test was exceeded. The defect was caused in the top part of regular tetrahedron. The defect was not due to shear deformation. The cause of the defect is the local change. Therefore, the index is effective for a uniform deformation and not effective for the local change.

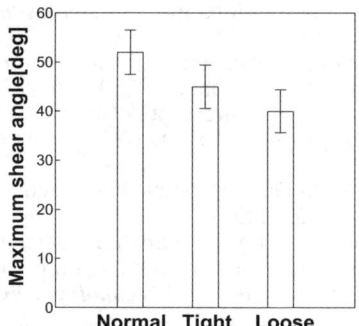

Fig.3 Maximum shear angle of each Specimen (bias extension test)

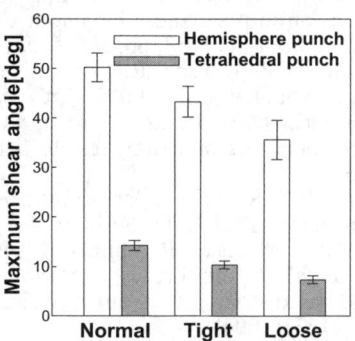

Fig.4 Maximum shear angle of each punch

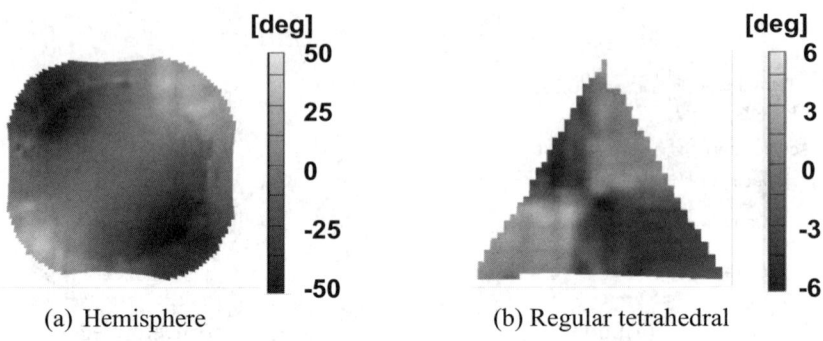

(a) Hemisphere (b) Regular tetrahedral

Fig. 5 Contour of shear angle

Conclusion

In this study, the bias extension test based on the measurement of shear angle by non-contact 3D deformation measurement system was conducted to evaluate the drape performance of the NCF. The measurement results proposed the formability evaluation index(maximum shear angle). Moreover, the draping test onto hemisphere geometry and regular tetrahedron was conducted. And the availability of formability evaluation index calculated by the bias extension test was verified. The investigation yielded the following conclusions:
1. The defect had been caused in the examination to tetrahedral shape before the limit shearing transformation angle was exceeded. The defect caused in the top part of regular tetrahedral. The defect was not due to shear deformation.
2. The index is effective for a uniform deformation and not effective for the local change.

References

[1] M. Ozawa, H. Satake, "Development of CFRP Body for LFA", *Jidoushagijyutu*, Vol. 64, pp. 52-57,(2009).
[2] T.C. Truong, M. Vettori, S. Lomov, I. Verpoest, "Carbon composites based on multi-axial multi-ply stitched performs. Part 4: Mechanical properties of composites and damage observation", *Composites: Part A*, Vol. 36, pp. 1207-1221, (2005).
[3] P. Boisse, B. Zouari, J.L. Daniel, "Importance of in-plane shear rigidity in finite element analysis of woven fabric composite preforming", *Composite: Part A*, Vol. 37, pp. 2201-2212, (2006).
[4] S. V. Lomov, M. Barburski, Tz. Stoilova, I.Verpoest, R.Akkerman, R.Loendersloot, R. H. W. tem Thije, "Carbon composites based on multi-axial multi-ply stitched performs. Part 3: Biaxial tension, picture frame and compression tests of the preform", *Composites: Part A*, Vol. 36, pp. 1188-1206, (2005).
[5] J. Launay, G. Hivet, A.V. Duong, P. Boisse, "Experimental analysis of the influence of tensions on in plain shear befaviour of woven composite reinforcement", *Composites Science and Tecnology*, Vol. 68, pp. 506-515, (2008).
[6] H. Kong, A.P. Mouritz, R. Paton, "Tensile extension properties and deformation mechanisms of multiaxial non-crimp fabrics", *Composite Structure*, Vol. 66, pp. 249-259, (2004).
[7] P. Potluri, D.A. Perez Ciurezu, R.B. Ramgulam, "Measurement of meso-scale shear deformations for modeling textile composites", *Composites: Part A*, Vol. 37, pp. 303-314, (2006).
[8] G. Creech, A.K. Pickett, "Meso-modelling of Non-Crimp Fabrics composites for coupled drape failure analysis", *J Mater Science*, Vol. 41, pp. 6725-6736, (2006).
[9] N. Takano, M. Zako, R. Fujitsu, K. Nishiyabu, "Study on large deformation characteristics of knitted fabric reinforced thermoplastic composites at forming temperature by digital image-based strain measurement technique", *Composites Science and Technology*, Vol. 64, pp. 2153-2163, (2004).
[10] G. Lebrun, M.N. Bureau, J. Denault, "Evaluation of bias-extension and picture-frame test methods for the measurement of intraply shear properties of PP/glass commingle fabrics" *Composites Structures*, Vol. 61, pp. 341-352, (2003).

Modelling micro-damage in granular solids

M. Buonsanti[1,a], G. Leonardi[1,b] and F. Scopelliti[1,c]

[1]MECMAT University of Reggio Calabria, Feo di Vito, 89100 Reggio Calabria, Italy

[2]DIMET University of Reggio Calabria, Feo di Vito, 89100 Reggio Calabria, Italy

[a]michele.buonsanti@unirc.it, [b]giovanni.leonardi@unirc.it, [c]francesco.scopelliti@unirc.it, [d]francis.cirianni@unirc.it

Keywords: Impact; FEM; Airport pavement; Aircraft landing, RVE analysis.

Abstract. The prediction of the spacing and opening of cracks in asphalt or concrete pavements, and particularly in airports (runway, taxiway and apron) is important for the durability assessment. A basic problem is the spacing of parallel planar cracks from a half space surface, approached and solved by numerous authors by means of macro-scale computational models. The calculated values of crack spacing are in relatively good agreement with the values reported in observations on asphalt concrete pavements. The constituents of granular solids are, fundamentally, made of grain in contact and, these materials are highly discontinuous and non-homogeneous with two or three phases (solid, voids with air or water), and finally binding among solid parts. The aim of this paper is to suggest a micromechanical approach in granular material solids, focusing the attention on a simple RVE (representative volume element) based on two rigid particles linked through an adhesive material (bitumen). Our final aim is to propose a micro-damageability parameter (interface loss) supposing the adhesion decreasing under the action of prescribed tangential and normal relative displacement. The reduction is attributed by progressive damage and comes with energy dissipation and moreover we assume unilateral contact conditions for normal displacement and Coulomb friction for the tangential displacement.

Introduction

Fragmentation, i.e. the breaking of particulate materials into smaller pieces is a ubiquitous process that underlies many natural phenomena and industrial processes. The length scales involved in it range from the impact evolution of deformable solids at macro-scale until the fragmentation at micro-scale [1]. In most of the realizations of fragmentation processes the energy is imparted to the system by impact, i.e. typical situations are in the specific engineering applications, as the ground contact of the airplane wheel during the landing [2, 3].

In this case the problem appears in more length of scale, since having the macro effects, large stress distribution and stress waves-train. Moreover, the effects at micro-scale regarding damaging and fracture in the body constituents as granular materials and their adhesive mixture.

The most striking observation about fragmentation is that the size distribution of fragments shows power law behaviour independent on the microscopic interactions and on the relevant length scales [4]. Beside the general interest in fragmentation processes one can also mention other fields where fracture and fragmentation of solid particles due to impact play an important role.

It is well known that in the flow of granular materials a large part of the kinetic energy of the grains is dissipated in the vicinity of their contact zone during the collisions. In the present paper we want to elaborate the impact fracture and fragmentation of solids at low imparted energy suggesting a simple approach way through a RVE as modified lattice model.

Micromechanics approach

According to Fremond [5] the basic unilateral contact theory does not allow resistance to tension but, the resistance to traction is due to microscopic bonds among the solids in contact.

The relations describing the static contact problem are summarized below where, T_{ij}, f_i, E_{ij}, u_i and $W(E)$ are respectively, the stress tensor, the body forces, the strain tensor, the displacement field and the stored deformation energy:

$$T_{ij,j} + f_i = 0 \qquad \text{in } \Omega$$
$$E_{ij}(u) = 1/2\left(u_{i,j} + u_{j,i}\right) \qquad \text{in } \Omega \qquad (1)$$
$$T_{ij} = \partial W(E)/\partial E \qquad \text{in } \Omega$$

Valid in the volume Ω, while in the boundary we have:

$$s_n = 0 \Rightarrow t = 0 \qquad \text{on } \partial\Omega$$
$$-s_n > 0 \Rightarrow -t = \partial w(d_t)/\partial d_t \qquad \text{on } \partial\Omega \qquad (2)$$

Where s_n, t, $w(.)$ and d_t are respectively: the surface forces normal component, the surface forces tangential components, the contact potential and the contact displacement field. Significant difficulties arise in the mathematical and numerical treatment of the boundary value problem and for this a micro approach way results more expedient. In the our case we consider two or more solids which glued together and the microscopic bond are constituting the adhesion. When the material which make the adhesion surface fails, then the bond breaking.

An opportune description of the adhesion framework and his evolution should be focused on the bonds between particles. Initially we will to remember the formulation of the contact problem [6], as Eq. 1 and Eq. 2 then focus the applications in the granular materials. For more clarity to the question we referring to Cambou [7].

Granular materials are made up of grains in contact and, again these kinds of materials are highly discontinuous and nonhomogeneous with two or three constituents namely, solids and voids, which may be air, water or ligand.

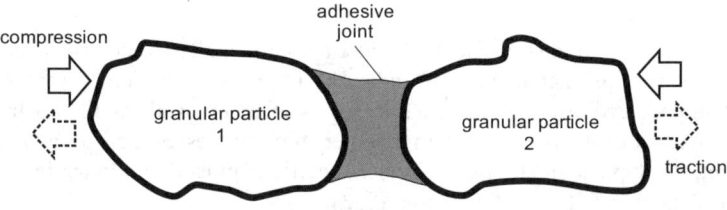

Fig. 1 Modelling the RVE

It's our opinion that an efficient approach to the detailed explanation at question may be regarded in the particle and lattice models as proposed in [8]. Here we consider two elementary particles in contact through an adhesive joint done in high elastic material, while the granular particles can be considered as rigid. Two types of actions must be conceived over the *RVE* that is, pure traction or compression neglecting flexure and shear.

We follow the volume element theory *RVE*, it's possible to represent a non-homogeneous solid with periodic microstructure. Particularly in the transition toward the micro-scale our *RVE* can be represented by more granular elements joint by means of an asphalt mixture, so considerations are applied on the contact area among two granular elements.

Results and conclusions

In this section our target will be to investigate the *RVE* (representative volume element) behaviour when external forces are growing at macro-scale. We consider a *RVE* as formed by two granular particles (Fig. 1), where the first one lies fixed to the bottom and the second one free to move but constrained through adhesion bond joining either of the two particles.

The latter, made of the same material, are glued together by a viscoelastic thin sheet (i.e. asphalt). The fundamental theory governing the matter has been developed in literature as it regards three different behaviours, in the complexity of the phenomenon. From the physical point of view the peeling is just an initial aspect, followed by sheet detachment and finally by bond breaking. In speaking terms, we have a contact dynamic problem with adhesion [5].

Our aim is to investigate the *RVE* (representative volume element) behaviour performing a very simple numerical procedure by means finite element methods. The considered elementary model has deformable particles glued by means hyperelastic adhesive materials.

Starting from the stress distribution on the macro-model [9], considering the opportune scale effects, the stress distribution of the micro-model of Fig. 1 is investigated.

The considered macro-structure of airport flexible pavement has been idealized as closed systems consisting of four layers; so it was decided to model the surface, base, sub-base and sub-grade material using three dimensional finite elements. The pavement section is comprised of asphalt concrete and crushed aggregate, as shown in Fig 2.

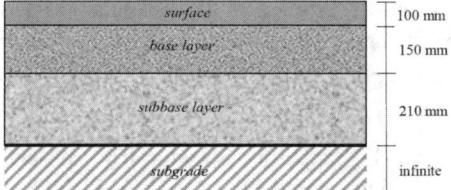

Fig. 2 Flexible Pavement section

Starting from the stress distribution on the macro-model [2, 9], considering the opportune scale effects, the stress distribution of the micro-model of Fig. 1 is investigated.

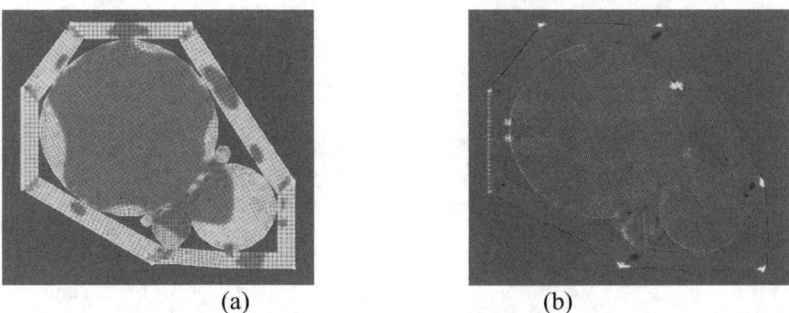

(a)   (b)
Fig. 3 Maximum stress field in compression (a) and traction (b)

Fig. 3 shows the analysis results of the micro-model with the maximum stress values in traction and compression.

The proposed approach shows how the surface effects play a basic role in the constitutive degradation of the layer.

In particular any aspects of surface damage in flexible pavements can be sided as scale effects over non homogeneous materials and the micromechanics of the problem can be utilized for maintenance models calibration.

References

[1] F. Kun, H. J. Herrman: Transition from damage to fragmentation in collision of solids, in http:arXiv: cond-mat/9810315v1, (1998).

[2] M. Buonsanti, F. Cirianni, G. Leonardi, F. Scopelliti: Impact dynamics on granular plate, Proceedings of the 8th International Conference on Structural Dynamics, EURODYN (2011), 4 - 6 July 2011, Leuven (Belgium). ISBN 978-90-760-1931-4.

[3] M. Buonsanti, G. Leonardi: A Finite Element Model to Evaluate Airport Flexible Pavements Response under Impact, Applied Mechanics and Materials, 138-139, pp.257-262, (2011).

[4] I. Afek, E.Bouchbinder, E. Katzav, J. Mathiesen, I. Procaccia: Void formation and roughening in slow fracture, Physical Review E, 71, 6, (2005).

[5] M. Frémond in: *Non-Smooth Thermo-mechanics*, Springer, N.Y., (2001).

[6] J.J. Kalker: On the Contact Problem in Elastostatics, Unilateral Problems in Structural Analysis, CISM n°288, Springer-Verlag, Berlin (1983).

[7] B. Cambou: Micromechanical Approach in Granular Materials, Behaviour of Granular Materials, CISM n°385, Springer-Verlag, N.Y., (1998).

[8] Z.P. Bazant, J. Planas in: *Fracture and Size Effects in brittle and Quasi-brittle Materials*, CRC Press, N.Y., (1998).

[9] M. Buonsanti, G. Leonardi, F. Scopelliti: Structural impact and dynamics response in brittle materials, Pavements Cracking, edited by I.L. Al-Qadi, T. Scarpas, A. Loizos, CRC Press, London, (2008).

Load and environmental effects on the corrosion behavior of a Ti6Al4V alloy

Sergio Baragetti[1,2,a], Alessandro Medolago[1,2,b]

[1]Department of Design and Technology, University of Bergamo, Viale Marconi 5, 24044, Dalmine, Italy

[2]GITT - Centre on Innovation Management and Technology Transfer, University of Bergamo, Via Salvecchio 19, Bergamo 24129, Italy

[a]sergio.baragetti@unibg.it, [b]alessandro.medolago@unibg.it

Keywords: quasistatic incubation tests, aggressive environment, Ti6Al4V.

Abstract. This paper focuses on the static mechanical behavior of Ti-6Al-4V titanium alloy when exposed to several aggressive environments. Flat samples, with very light notches (i.e. $K_t = 1.16$), were tested under static loads, in inert environment and aggressive solutions: the samples gage sections were exposed to air and immerged in a NaCl solution (3.5%) and in a methanol solution (95%). The results of this experimental tests were analyzed and then compared with previous data coming from fatigue tests, carried out in paraffin oil, air, 3.5 wt.% NaCl, and in a methanol solution (95%), with the intent to decouple the effect of the alternating load and of the aggressive environment.

Introduction

The use of Ti-6Al-4V titanium alloy in high-performances applications has been one of the best known and most promising challenges over the last decades. Its spreading diffusion in many of the most demanding fields, like the aerospace, naval, automotive and defense industries, is sustained by its good corrosion resistance, low density and high mechanical properties. Considering the key role played by this material, its fatigue resistance represents a critical issue: a deeper awareness of the mechanical behavior of components made of Ti-6Al-4V, while operating in corrosive media, would be fundamental to prevent dramatic failures due to environment-assisted fatigue or notch-induced fatigue crack growth. While the fatigue strength of Ti-6Al-4V specimens tested in laboratory air has been widely investigated [1-10], the effect of a NaCl solution has not been already completely understood. So, the purpose of this paper is to investigate the crack nucleation and propagation mechanisms of this alloy, in case of quasistatic loads (10^{-3} MPa/s), and to evaluate the influence of an aggressive environment over these phenomena. Experimental data coming from quasistatic tensile tests, carried out on specimens exposed to laboratory air, to a NaCl solution (3.5wt%) and to a methanol solution (95%), will be reported. These data will be then compared with literature ones, referring to axial-fatigue tests [7, 8] performed on similar specimens exposed to the same environments, with the intent to decouple the effect of the alternating load and of the aggressive environment.

Material and methods

The chemical composition (average wt.%) of the Ti-6Al-4V hot rolled plate, which the fatigue samples were machined from, was: 5.97 Al, 4.07 V, 0.20 Fe, 0.19 O, 0.003 C, 0.015 H, 0.05 N and Ti bal. Fig. 1 a) shows the dimensions of the samples tested during the quasistatic incubation tests. The samples were formed with the notch axis parallel to the rolling direction. The notches were machined by milling at very low speed, to keep the residual stresses as low as possible, and a notch root radius of 30 mm was obtained. This value is associated with a stress concentration factor, calculated with FE models, of 1.16. The samples were then solution treated at 925°C (1 h) and vacuum annealed at 700°C (2 h) for stabilization (STOA treatment). The tensile properties along the transversal direction – i.e. that of load application – after the STOA treatment were determined using a universal testing machine and are reported in Table 1.

Table 1. Mechanical characteristics of the Ti-6Al-4V before and after the STOA treatment.

	Mechanical properties		
Plate	UST [MPa]		1000
	YS [MPa]		958
Tensile test after STOA	UST [MPa]		947
	YS [MPa]		900

A testing system was built specifically to test these samples: this system was equipped with special grips in order to bypass the parasite action caused by any misalignment or deformation of the system. The test procedure was the following: the sample was mounted on the system and then a static load was applied and incremented following a "step loading" procedure. Every hour the load value was increased of 5 MPa and monitored using a load cell.

The application of the aggressive environment, to the sample notch area, was done as shown in Fig. 1 b); the aggressive solution (methanol and NaCl) was continuously renewed by means of a pumping circuit. The tests results were then compared with previous experimental data, obtained by fatigue tests carried out on samples having similar geometry and in the same environments [7, 8]. These axial fatigue tests were carried out at a frequency of 10 Hz and with a load ratio $R = 0.1$.

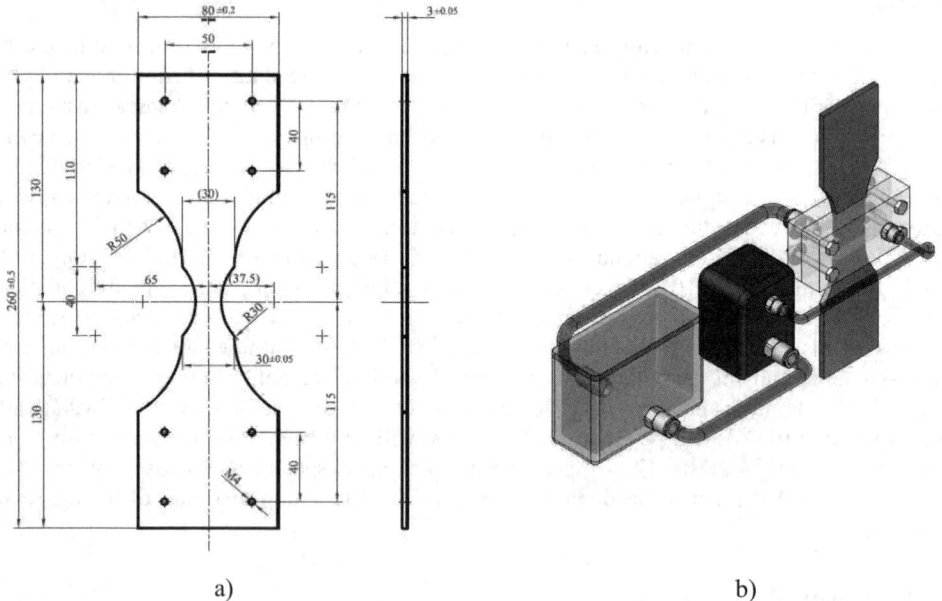

a)   b)

Fig. 1. Geometry of the tested samples a) and a scheme of the pumping circuit for the application of the aggressive environment b).

Results and discussion

Quasistatic very low load increase incubation tests were carried out on samples with $K_t = 1.16$, in order to evaluate the environmental effects on the resistance of this alloy. A very low strain rate (10^{-8} per sec rate) was adopted for this tests and for this reason an increment of 5 MPa/hour was chosen. The initial nominal stress in the minimum cross section was chosen lower than the material yield strength: a reduction of 27% was applied. The main parameters of the quasistatic tests are summarized in Table 2.

Table 2. Quasistatic incubation tests main parameters.

Yield strength	YS [MPa]	900	STOA treatment
Minimum cross section	A [mm^2]	45	
Nominal initial stress	s_n [MPa]	657	27% reduction of YS
Initial load	F_i [N]	29,250	
Stress increment	D_s [MPa/hour]	5	
Stress increment rate	$\dot{\sigma}$ [MPa/s]	1.38·10^{-3}	
Strain rate	$\dot{\varepsilon}$ [1/s]	1.26·10^{-8}	

Similar values of the failure stress, from the quasistatic incubation tests, were found between the samples tested in air and NaCl (Table 3). These two samples anyway showed a higher failure stress than the one tested in the methanol solution (Table 3). In Fig. 2 a) these experimental results are shown and compared. It can be seen that, notwithstanding a shorter exposition time for the sample tested in methanol (Table 3), the methanol aggressive solution has decreased the material's resistance much more than the NaCl aggressive solution. On the contrary, the failure stress coming from the sample tested in air is very near to the Yield Stress. The methanol fatigue data were taken from previous experimental campaigns [11].

Table 3. Summary of the quasistatic (q-s) incubation tests results.

	Time [h]	$\sigma_{Fail\ q-s}$ [MPa]	$\sigma_{max\ fatigue}$ [MPa]	$\Delta\%$ YS/q-s	$\Delta\%$ q-s/Fatigue	$\Delta\%$ YS/Fatigue
Air	50	895	619 [7, 8]	0.6	30.8	31.2
Nacl 3.5%	48	885	389 [7, 8]	1.7	56.0	56.7
Methanol 95%	33	810	239 [11]	10.0	70.5	73.4

The comparison (Fig. 2) between the quasistatic experimental results and the ones coming from the fatigue tests, reported in previous works [7, 8], shows that the Ti-6Al-4V quasistatic strength decreases proportionally to the aggressiveness of the environment (Table 3). This matter can be found also in the results coming from the former fatigue tests. Moreover it can be seen that the reduction due to the combination of the dynamic load with the aggressive environment represents the majority of the strength loss with respect to the YS value.

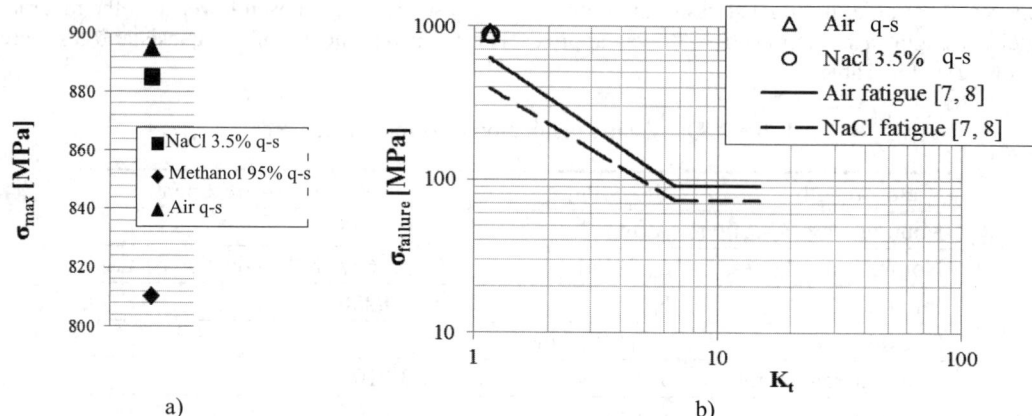

Fig. 2. Comparison between the quasistatic incubation tests (q-s) results, in different environments a), and comparison between these data and the fatigue ones b) [7, 8].

Summary

The resistance of Ti-6Al-4V under quasistatic applied loads (10^{-3} MPa/s) was evaluated in air and in two different aggressive environments (NaCl 3.5% and Methanol 95%). Tests were carried out on smooth specimens ($K_t=1.16$). The material quasistatic strength decreases proportionally to the aggressiveness of the environment and the strength reduction under a quasistatic load is much lower than the one obtained under a fatigue one.

Acknowledgements

The authors wish to thank prof. dr. A.K. Vasudevan of ONR (Office of Naval Research) – US Navy, for the discussion and his precious suggestions. The research was funded by ONR (Office of Naval Research) – U.S. Navy (Contract Number: N00014-08-1-1197).

References

[1] R.J. Morrissey, D.L. McDowell, T. Nicholas: International Journal of Fatigue Vol. 21 (1999), p. 679.
[2] N.E. Frost, K.J. Marsh, L.P. Pook: Metal Fatigue (Clarendon Press, UK 1974).
[3] R.O. Ritchie, B.L. Boyce, J.P. Campbell, O. Roder, A.W. Thompson, W.W. Milligan: International Journal of Fatigue, Vol. 21 (1999), p. 653.
[4] T. Nicholas: Fatigue (David L Davidson symposium. Warrendale. The Minerals, Metals & Materials Society, 2002).
[5] G.K. Haritos, T. Nicholas, D. Lanning: International Journal of Fatigue Vol. 21 (1999), p. 643.
[6] R.J. Morrissey, T. Nicholas: International Journal of Fatigue Vol. 27 (2005), p. 1608.
[7] S. Baragetti, S. Cavalleri, F. Tordini: Advances in Fracture and Damage Mechanics X (2011), p. 502.
[8] S. Baragetti, S. Cavalleri, F. Tordini: Proceedings of 11th International conference on the Mechanical Behavior of Materials Vol. 10 (2011), p. 2442.
[9] D.B. Lanning, T. Nicholas, G. K. Haritos: International Journal of Fatigue, Vol. 27 (2005), p. 45.
[10] R.S. Bellows, S. Muju, T. Nicholas: International Journal of Fatigue Vol. 21 (1999), p. 687.
[11] S. Baragetti: proceedings of the VI ASST 2012 Symposium, in course of publication.

Fatigue crack nucleation and growth mechanisms for Ti6Al4V in different environments

Sergio Baragetti[1, 2, a], Cristian Foglia[1, b] and Riccardo Gerosa[3, c]

[1]Department of Design and Technology, University of Bergamo, Viale Marconi 5, 24044, Dalmine, Italy

[2]GITT - Centre on Innovation Management and Technology Transfer, University of Bergamo, Via Salvecchio 19, Bergamo 24129, Italy

[3]Department of Mechanical Engineering, Politecnico di Milano, Via La Masa 1, 20156, Milano, Italy

[a]sergio.baragetti@unibg.it, [b]cristian.foglia@unibg.it, [c]riccardo.gerosa@polimi.it

Keywords: Ti6Al4V, fatigue crack growth, microstructure, K_t effect, corrosion.

Abstract. In this work fracture surfaces of Ti-6Al-4V flat samples subjected to fatigue tests were examined by means of a scanning electron microscope. SEM analyses allowed to observe in detail the morphology of the fracture surfaces, in order to identify the crack nucleation zones and the crack growth mechanisms accurately. The surface morphologies were examined in order to compare the alloy behavior, considering different test environments and stress concentration factor values (K_t). The analyses resulted in the observation of different corrosion fatigue micro-mechanisms related to the stress concentration factor and to the test environment.

Introduction

Ti6Al4V is one of the most commonly used titanium alloys in the military and civil aviation industry, because of its good corrosion resistance and high strength-to-mass ratio. Thus, considering the high performance applications of this material, in the last years the fatigue behavior has been investigated more and more deeply [1-9]. In this work, fatigue data coming from previous experimental campaigns [11, 12] were rearranged and related to SEM analyses of the fracture surfaces of the samples. These fatigue tests were performed with the aim to study the fatigue behavior of notched Ti6Al4V titanium alloy flat samples in two different environments: laboratory air and 3.5 wt.% NaCl solution. The specimens, obtained from the same plate, were machined with different root radii in order to obtain K_t values varying from 2.55 to 13.34. Axial fatigue tests (R=0.1) were carried out and the step loading procedure [7, 8] was used to estimate the fatigue limits for a constant life of $2 \cdot 10^5$ load cycles. Crack nucleation and growth mechanisms were monitored by measuring the crack length, with the replica method. The aim of this paper is to study the morphology of the fracture surfaces of the tested samples and calculate the crack propagation rates in order to better understand the corrosion fatigue mechanisms in aqueous environments.

Material and methods

This study was carried out on a Ti6Al4V titanium alloy with the following average chemical composition (wt.%): 5.97 Al, 4.07 V, 0.20 Fe, 0.19 O, 0.003 C, 0.015 H, 0.05 N and Ti bal. The geometry of the fatigue dogbone V-notched flat samples is shown in Fig. 1. They were machined from a hot rolled plate, so as to have the notch axis parallel to the rolling direction (L - direction) of the parent Ti6Al4V plate. A solution treatment at 925°C (1 h) and vacuum annealing at 700°C (2 h) for stabilization were carried out after the machining operations. The notches were formed by milling at very low cutting speed to limit the residual stresses, then by electrical discharge machining (EDM) to obtain the precise values of the following notch root radii (ρ): 0.06, 0.26, 1.50 and 2.50 mm. The associated stress concentration factors K_t, determined by finite element modeling, were respectively: 13.34, 6.63, 3.10 and 2.55. All the samples were finally stress relieved at 700°C in vacuum for 1 hour before being tested under fatigue. The tensile strength along the transversal direction (T - direction), after the solution and overaging treatment (STOA), was determined with tensile tests (INSTRON 1273®).

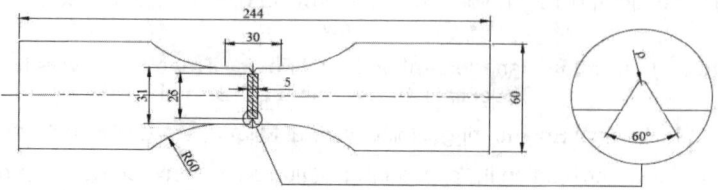

Fig. 1. Sample geometry used for the fatigue tests (dimensions in mm).

Axial fatigue tests (R=0.1) were carried out with a universal testing machine (BRT T1000®) at a frequency of 10 Hz and with an upper limit of $2 \cdot 10^5$ cycles. The samples were exposed to laboratory air and to a recirculating 3.5 wt.% NaCl solution. Small acetate strips were used to detect the presence of cracks nucleated from the notches tip. A watertight cell with a suitable recirculation system was assembled for the tests in NaCl solution, so that the sample notch area was completely immerged in the corrosive medium. The aggressive solution was continuously pumped through the recirculation circuit. The limiting maximum stress was calculated by means of a step - loading technique (Eq. 1) [7, 8]. The maximum stress limit at the complete fracture was calculated doing a linear interpolation between the maximum stress, applied to the sample, during the last step ($\sigma_{max, final}$) and the maximum stress of the previous one ($\sigma_{max, prior}$). Referring to Eq. 1, N_f is the number of cycles performed during the last loading block before failure occurred. The stresses in Eq. 1 are the nominal ones, considering the sample's minimum cross section.

$$\sigma_{FL,max} = \sigma_{prior,max} + \frac{N_f}{2 \cdot 10^5}(\sigma_{final,max} - \sigma_{prior,max})$$ (Eq. 1)

Further details on the fatigue tests and results can be found in [11, 12].
A selection among the tested samples fracture surfaces were observed by means of a SEM microscope (ZEISS – EVO50), with the intent of identifying the main differences in the nucleation and failure mechanisms for different environments and K_t values.

Results and discussion

The tensile and yield strength along the T - direction of Ti6Al4V after STOA were, respectively, 990 and 945 MPa. The Young's modulus was 110,000 MPa and the elongation at break was 16%.
Fig. 2 shows the bimodal α and β microstructure along the T direction. The image analysis showed that the amount of primary α grains was estimated equal to about 50% and the average grain size was about 30 µm.
By means of the replica method, the crack growth rate was calculated (Table 1) for all the tested conditions. Finally the fracture surfaces were observed by SEM. The comparison between the two test environments resulted in no appreciable difference among the fracture surface morphology. This is probably due to the crack propagation rates, generally too fast to make a general corrosion process start. In Fig. 3 the crack propagation rates were related with the NaCl tested samples fracture surface features, considering the minimum and the maximum investigated K_t values (i.e. K_t=2.55 and K_t=13.34).

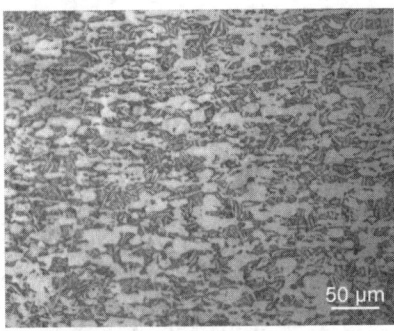

Fig. 2. Microstructure in the T direction (500X).

Table 1. Limiting maximum stress ($R=0.1$) at failure and average crack growth rate for all the samples tested (for a constant life of $2 \cdot 10^5$) [11, 12].

	$\sigma_{max\ failure}$ [MPa]				Average da/dN [m/cycle]			
K_t	2.55	3.1	6.63	13.34	2.55	3.1	6.63	13.34
Laboratory air	265	204	119	84	$1.2 \cdot 10^{-6}$	$4.2 \cdot 10^{-7}$	$2.4 \cdot 10^{-7}$	$9.9 \cdot 10^{-8}$
NaCl solution	191	147	115	84	$7.0 \cdot 10^{-6}$	$4.2 \cdot 10^{-6}$	$4.1 \cdot 10^{-7}$	$1.8 \cdot 10^{-7}$

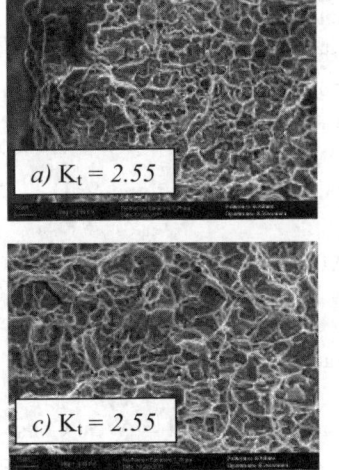
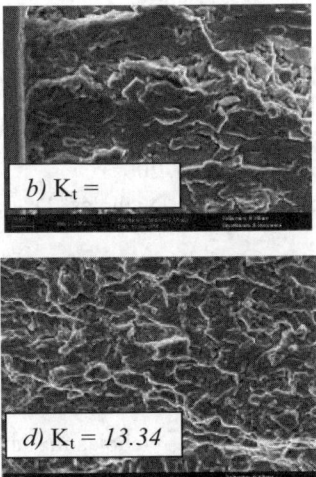

Fig. 3. Fracture surfaces at the notches tip (NaCl 3.5%): a) K_t=2.55, da/dN ≈ 10^{-6} m/cycle; b) K_t=13.34, da/dN ≈ 10^{-8} m/cycle. Fracture surfaces 15 mm far from the notches tip (NaCl 3.5%): c) K_t=2.55, da/dN ≈ 10^{-5} m/cycle; d) K_t=13.34, da/dN ≈ $3 \cdot 10^{-7}$ m/cycle.

In the K_t=2.55 samples, very high fatigue crack growth rates were observed and the surface morphology was similar to the one of a ductile static fracture, characterized by many dimples. On the other hand, the samples with K_t = 13.34 were characterized by different fracture surface morphologies: as the crack growth rate increases, the fracture surface gets rougher and rougher.

Summary

The fracture surfaces of Ti6Al4V flat notched samples subjected to fatigue tests in aggressive environment were examined in this paper by means of a scanning electron microscope. SEM analyses allowed to study the morphology of the fracture surfaces in order to identify the crack nucleation zones and the crack growth mechanisms. Flat Ti6Al4V samples with different K_t were tested under axial fatigue (R=0.1) in two different environmental conditions: laboratory air and a 3.5 wt.% NaCl solution. A step loading technique was implemented to evaluate the maximum fatigue limit, for a constant life of $2 \cdot 10^5$ load cycles. Crack growth was monitored using the surface replica method. The metallographic analysis, after the STOA treatment, showed a bimodal alpha-beta microstructure. The saline environment affected the fatigue behavior only for low K_t values (corresponding to high applied loads) and the same behavior was observed considering the average crack propagation rate. SEM analyses showed no significant difference, in the surface morphology, between air and NaCl solution. The fracture surface resulted rougher as the crack propagation rate increased.

Acknowledgements

The authors wish to thank prof. dr. A.K. Vasudevan of ONR (Office of Naval Research) – US Navy, for the discussion and his precious suggestions. The research was funded by ONR (Office of Naval Research) – U.S. Navy (Contract Number: N00014-08-1-1197).

References

[1] R.J. Morrissey, D.L. McDowell, T. Nicholas: International Journal of Fatigue Vol. 21 (1999), p. 679.
[2] N.E. Frost, K.J. Marsh, L.P. Pook: *Metal Fatigue* (Clarendon Press, UK 1974).
[3] R.O. Ritchie, B.L. Boyce, J.P. Campbell, O. Roder, A.W. Thompson, W.W. Milligan: International Journal of Fatigue, Vol. 21 (1999), p. 653.
[4] T. Nicholas: *Fatigue* (David L Davidson symposium. Warrendale. The Minerals, Metals & Materials Society, 2002).
[5] G.K. Haritos, T . Nicholas, D. Lanning: International Journal of Fatigue Vol. 21 (1999), p. 643.
[6] R.J. Morrissey, T. Nicholas: International Journal of Fatigue Vol. 27 (2005), p. 1608.
[7] D.B. Lanning, T. Nicholas, A. Palazotto: International Journal of Fatigue, Vol. 27 (2005), p. 1623.
[8] D.B. Lanning, T. Nicholas, G. K. Haritos: International Journal of Fatigue, Vol. 27 (2005), p. 45.
[9] D.C. Maxwell, T. Nicholas in: Fatigue and fracture mechanics, edited by West Conshohocken, Vol. 29 of ASTM STP 1321, p. 626–41 (1999).
[10] R.S. Bellows, S. Muju, T. Nicholas: International Journal of Fatigue Vol. 21 (1999), p. 687.
[11] S. Baragetti, S. Cavalleri, F. Tordini: Proceedings of 11th International conference on the Mechanical Behavior of Materials Vol. 10 (2011), p. 2442.
[12] S. Baragetti, S. Cavalleri, F. Tordini: Advances in Fracture and Damage Mechanics X (2011), p. 502.

Modeling Delamination of Interfacial Corner Cracks in Multilayered Structures

Badrinath Veluri[a] and Henrik Myhre Jensen[b]

Department of Mechanical Engineering, Aarhus School of Engineering

Aarhus University

Dalgas Avenue 4, Aarhus C 8000 Denmark

[a]E-mail: vb@agse.dk , [b]Email: hmj@agse.dk , Web page: http://www.science.au.dk

Keywords: Microelectronics, Interface fracture, thin layers, coatings, Corner cracks, Delamination, Mode Mixity.

Abstract. Multilayered electronic components, typically of heterogeneous materials, delaminate under thermal and mechanical loading. A phenomenological model focused on modeling the shape of such interface cracks close to corners in layered interconnect structures for calculating the critical stress for steady-state propagation has been developed. The crack propagation is investigated by estimating the fracture mechanics parameters that include the strain energy release rate, crack front profiles and the three-dimensional mode-mixity along the crack front. The developed numerical approach for the calculation of fracture mechanical properties has been validated with three-dimensional models for varying crack front shapes. A custom quantitative approach was formulated based on the finite element method with iterative adjustment of the crack front to estimate the critical delamination stress as a function of the fracture criterion and corner angles.

Introduction

Performance of electronic components is driven by the system integration i.e. the interconnection between the silicon die and its package. Delamination between the interface initiating from edges and corners of the bimaterial systems or the laminated composites is an important problem in their respective applications. Crack driving forces may be intrinsic stresses caused during the fabrication, or stresses due to temperature variations and thermal expansion mismatch. A comprehensive review of the mechanics of two-dimensional interface delamination under plane stress, plane strain and axisymmetric problems can be found in [1]. Experimental observations state that the fracture toughness for a crack at an interface is heavily dependent on the relative amount of the opening stress field (Mode 1- K_I) to the shear stress fields (Mode II and III - K_{II} & K_{III}) at the crack tip [2, 3]. In the current investigations, a 2-D finite element iterative method has been developed to analyze the critical stress intensity factors locally along the crack front. The interface crack geometry considered in the present work is as illustrated in Fig. (1); a thin film bonded to a substrate contains an interface crack of length L close to a corner with the included opening angle γ.

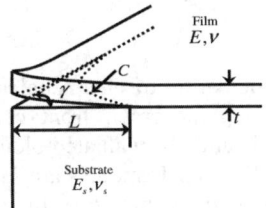
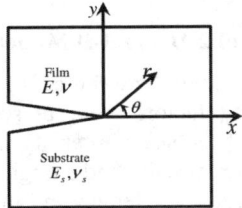

Fig.1. Crack at the interface between a thin film and a semi- infinite substrate located close to a corner.

Fig. 2. Crack on the interface of two dissimilar materials

Mixed-Mode Interface Fracture

The asymptotic stress field around the crack tip for a given crack on the bimaterial interface as shown in Fig. (2) for 2-D in-plane and anti-plane problems [1, 4] and is given by

$$\sigma_{ij} = \frac{1}{\sqrt{2\pi r}}\left\{\text{Re}\left[Kr^{i\varepsilon}\right]\sigma_{ij}^{I}(\theta,\varepsilon) + \text{Im}\left[Kr^{i\varepsilon}\right]\sigma_{ij}^{II}(\theta,\varepsilon) + K_{III}\sigma_{ij}^{III}(\theta)\right\} \quad (1)$$

where r and θ are the in-plane polar coordinates, K is the complex interface stress intensity factor for the in-plane modes, $K = K_I + iK_{II}$, K_{III} is the Mode III stress intensity factor. In Eq. (1) $\sigma_{ij}^{I}, \sigma_{ij}^{II}$ and σ_{ij}^{III} are known functions of the corresponding stress components for each mode, and ε is the bi-material parameter given by

$$\varepsilon = \frac{1}{2\pi}\ln\left(\frac{1-\beta}{1+\beta}\right) \quad (2)$$

Here, β is the second Dundurs' parameter [5], characterize the elastic mismatch of the bimaterial. The mode dependent fracture criterion applied in the present investigation was formulated in [3] as

$$\left(\frac{1-v^2}{E} + \frac{1-v_s^2}{E_s}\right)\frac{K_I^2 + \lambda_2 K_{II}^2}{2} + \left(\frac{1}{\mu} + \frac{1}{\mu_s}\right)\frac{\lambda_3 K_{III}^2}{4} = G_{1c} \quad (3)$$

The stress intensity factors (SIF's) K_I, K_{II} and K_{III} varies along the crack front C, λ_2 and λ_3 are factors that range between 0 and 1 adjusting the relative contribution of Modes II and III to the fracture criterion. The interface fracture criterion G_{1c} in Eq. (3) can be related to micromechanical models of the interface fracture [3]. Under the assumption that the in-plane extent of the crack is large compared to the layer thickness, Eq. (3) reduces to

$$\sigma_{nn}^2 + \frac{2\lambda}{1-v}\sigma_{nt}^2 = \sigma_c^2 = \frac{2EG_c^*}{(1-v^2)t} \quad \text{where } \lambda = \frac{\lambda_3}{1+(\lambda_2-1)\sin^2\psi} \quad \text{and} \quad G_c^* = \frac{G_{1c}}{1+(\lambda_2-1)\sin^2\psi} \quad (4)$$

The fracture criterion is affected through the angle $\psi = \psi(\alpha,\beta)$ due to the elastic mismatch of the system which affects the singular fields at the crack tip [1].

The fracture criterion in Eq. (4) based on normal (σ_{nn}) and Shear (σ_{nt}) stresses are imposed locally along the crack front. The results for these stress components σ_{nn} and σ_{nt} are calculated by the finite element method. Locally along the crack front C, the normal stresses σ_{nn} induces the Mode I and Mode II in a fixed proportion characterized by ψ and the shear stresses σ_{nt} induces the Mode III. The corresponding stress intensity factors of the three modes given by the following equations

$$\left(\frac{1-v^2}{E} + \frac{1-v^2}{E_s}\right)(K_I^2 + K_{II}^2) = \left(\frac{1-v^2}{E}\right)\sigma_{nn}^2 t \quad \text{and} \quad \frac{1}{2}\left(\frac{1}{\mu} + \frac{1}{\mu_s}\right)K_{III}^2 = \frac{1}{\mu}\sigma_{nt}^2 t \quad (5)$$

Comparison of 2-D and 3-D Model

A comparison study between a full 3-D and 2-D numerical calculation of energy release rate and stress intensity factors were presented for straight-sided triangular crack front configuration. Distribution of the stress intensity factors (SIF's) for mode I, II and III evaluated along the crack front for quarter-circular and straight sided crack fronts from 2-D and 3-D models are shown in Fig. (3) for different crack length and thickness ratios. Fig. (4) shows the comparison of strain energy release rate for the straight sided corner crack with $\lambda = 1$ and $v = 1/3$.

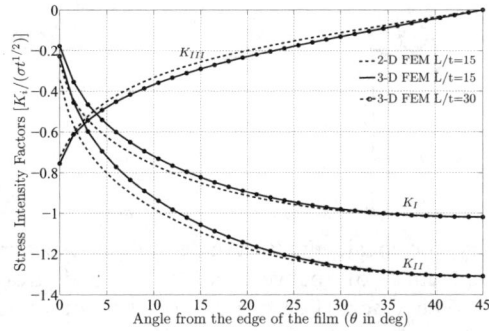
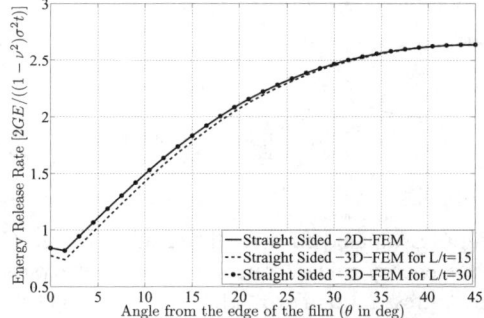

Fig.3. Comparison of stress intensity factors estimated from 2-D and 3-D FEM models for straight-sided crack front for a crack length to thickness of the film ratio $L/t = 15$ and 30.

Fig.4. Normalized Strain energy release rate comparison for the straight sided crack front estimated from 2-D and 3-D FEM models for a crack length to thickness ratio of $L/t = 15$ and 30.

Modeling the Crack Tip Profile for Steady-State Delamination Criteria

The model Problem considered was as illustrated in Fig. (5), which was divided into sub problems 1 and 2. The Problem 1 is trivial where in a linearly varying tensile stress in the film and the substrate is stress-free that does not give rise to singular stress field. The of Problem 2 with a stress field in the cracked body due to normal stress σ_{nn} and tangential stress σ_{nt} at the edge of the film was solved by the finite element method assuming plane stress conditions.

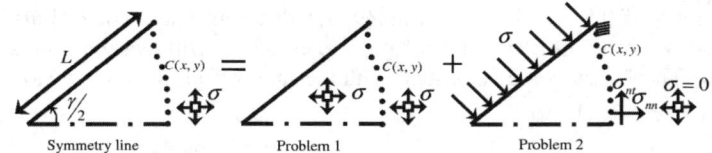

Fig.5. Schematic of the Study case and its reduced sub cases as problem 1 and 2.

For any given initial shape of the crack front, a given stress level and a given interface fracture criterion the stresses σ_{nn} and σ_{nt} are calculated [6]. In general Eq. (7) will not be satisfied so the shape is adjusted iteratively in order to minimize e defined as

$$e = \sum_{i=1}^{N} e_i = \sum_{i=1}^{N} W_i G_i^2 \qquad (6)$$

where N is the number of nodes along the crack front. In Eq. (6), W_i are weighting functions, and the object functions G_i are defined by Eq. (7) as

$$G_i = \sigma_c^2 - \sigma_{nn(i)}^2 - \frac{2\lambda}{1-\nu}\sigma_{nt(i)}^2 \qquad (7)$$

The nodal coordinates at the crack front which are determined by finite element iterative method at the crack front thus minimization the error e by Newton-Raphson iteration. The iterations are terminated when the root mean square of error e_{rms} as defined in Eq. (8) gets below the specified error 1% as shown in Fig. (6) for a value of $\lambda = 0.3$ and with stress ratio of $\sigma/\sigma_c = 1.2$.

$$e_{rms} = \sqrt{\frac{\sum_{i}^{N} W_i \left(\sigma_c^2 - \sigma_{nn(i)}^2 - \frac{2\lambda}{1-\nu}\sigma_{nt(i)}^2 \right)^2}{N}} \qquad (8)$$

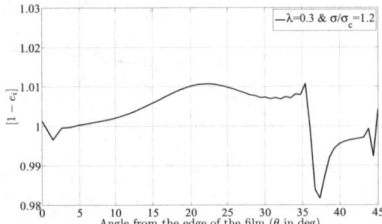 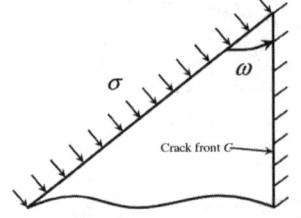 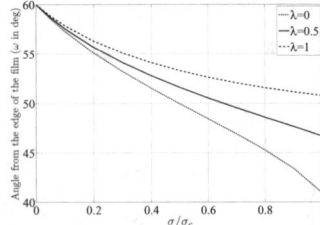

Fig.6. Estimated error for the steady state energy release rate for $\lambda = 0.3$ for a stress ratio of $\sigma/\sigma_c = 1.2$.

Fig.7. Geometry for local analysis at point of intersection between crack front and stress free edge.

Fig.8. Angle of intersection between crack front and stress free edges a function of stress ratios.

The geometry for sub problem 2 in Fig. (5) on this scale is shown in Fig. (7). The angle of intersection is denoted ω. Closed form solution can be obtained for this problem and is given by

$$\sigma_{nn} = -\frac{2\sigma\cos 2\omega}{1-\nu+(1+\nu)\cos 2\omega} \quad \text{and} \quad \sigma_{nt} = -\frac{(1-\nu)\sigma\sin 2\omega}{1-\nu+(1+\nu)\cos 2\omega} \qquad (9)$$

On substitution of these stresses into the fracture criterion in Eq. (7), a relationship between the residual stresses in the film, σ the angle of intersection, ω and the parameter λ entering the fracture criterion is obtained. Fig. (8) shows the angle of intersection of the crack front with the free edge as a function of the stress ratio, σ/σ_c and the parameter λ in the fracture criterion.

Results and Discussion

The numerical results generated the crack front profiles at steady-state delamination are discussed below for stress ratio of 0.8 are shown in Fig. (9) for different values of λ. Fig. (10) depicts the results for the crack front profiles for the 90° corner angle with varying stress ratios for the parameter $\lambda = 0.0$. These results are consistent with the fact that all the energy stored in the bonded film is released at $\sigma/\sigma_c = \sqrt{(1+\nu)/2}$

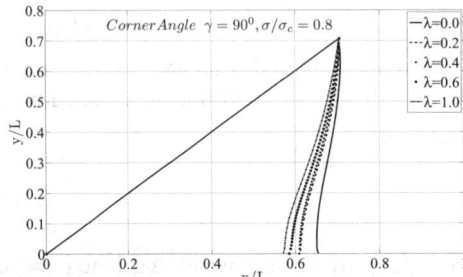 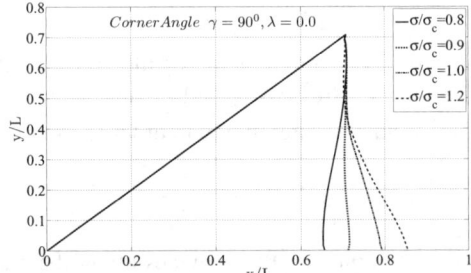

Fig.9. Shape of crack front at steady-state delamination for 90° corner angles and the stress ratio of 0.8 with varying parameter λ.

Fig.10. Shape of crack front at steady-state delamination for 90° corner angle and the varying stress ratios for a value of $\lambda = 0.0$.

References

[1] Hutchinson, J.W., and Suo, Z., 1991, "Advances in Applied Mechanics", Elsevier, pp. 63-191.
[2] Cao, H. C., and Evans, A. G., 1989, "An Experimental Study of the Fracture Resistance of Bimaterial Interfaces", Mechanics of Materials, 7(4) pp. 295-304.
[3] Jensen, H. M., 1990, "Mixed Mode Interface Fracture Criteria", Acta Metallurgica Et Materialia, **38**(12) pp. 2637-2644.
[4] Rice, J. R., 1988, "Elastic Fracture Mechanics Concepts for Interfacial Cracks", Journal of Applied Mechanics, **55**(1) pp. 98-103.
[5] Dundurs, J., 1969, "Edge Bonded Dissimilar Orthogonal Elastic Wedges", Journal of Applied Mechanics, **36**(2) pp. 650-652.
[6] Veluri, B., Jensen, H. M., 1969, "Steady-state Propagation of Interface Corner Crack", Submitted to International Journal of Solids and Structures.

Crack modelling using the Material Point Method and a strong discontinuity approach

Irene Guiamatsia[1,a], Giang Nguyen[1,b]

[1]School of Civil Engineering, The University of Sydney, Sydney, 2006, Australia

[a]irene.guiamatsia@sydney.edu.au, [b]giang.nguyen@sydney.edu.au

Keywords: Material point method, strong discontinuity, cohesive crack, fracture.

Abstract. Modern numerical techniques utilised to model crack propagation tend to be optimized for tracking the evolution of a single crack. Real fracture processes are however complex, involving the initiation and propagation of opening (activated) cracks, while other may close (deactivate) and undergo frictional dissipations. Accounting for the correct loss of energy (through debonding and friction) is essential to achieving a realistic description of the fracture process. One common strategy has been to make small adaptations to traditional techniques to tackle multiple cracking, in effect relying on extensive complicated computational algorithms. A typical example is the use of cohesive models in combination with the eXtended finite Element method where cracks, sometimes intersecting, need to be defined explicitly. In this study the Material Point Method is used for the analysis of fracture propagation. Crack states, as internal variables, are stored within the material points and mapped as strong discontinuities to the elements during the Lagrangian phase of the solution. Consequently, material points carrying cracks of different sizes and orientations are allowed to cohabit within the same element, yielding a natural description of the fracture/fragmentation process. The three-point bending test is used to demonstrate the features of the new approach.

Introduction
Understanding how cracks initiate and propagate is crucially important to assess the load bearing capacity of structures. Cracks are displacement discontinuities, generally numerically idealized, at the mesoscale, either as diffuse or discrete, depending on their spatial distribution. From a finite element perspective, it is straightforward to model the damage due to diffuse cracking as a change in the constitutive response of the material [1]. However, for some applications, for example in composite delamination, rock explosion (mining), surgery simulation, etc., the explicit representation of cracks is necessary as subsequent interaction of the new fracture surfaces may occur. One approach is to embed the discontinuous displacement field in the finite element by enriching the latter with a corresponding assumed enhanced strain, e.g. [4].
The Material Point Method (MPM), Sulsky et al. [2], is a hybrid Lagrangian/Eulerian approach that combines the positive aspects of particle-based and element-based formulations to yield a reliable and accurate approach for solving problems involving large-deformation and fracture, commonly witnessed during the fragmentation of rocks followed by the flow of granulated materials. Consideration of kinematic discontinuities is not entirely straightforward in the framework of the MPM, due to the back-and forth mapping that is performed between particles and background FE mesh, imposing a single-valued velocity field onto the finite element grid. One of the previous attempts to model cracks with the MPM, proposed by Nairn [3], consists precisely of allowing nodes near cracks to assume multiple velocity fields, but this involves complicated searching if more than one crack happens to be located within a single element.
In the work reported here, cracks are associated with material points, like other history-dependent state variables. The displacement discontinuity and accompanying relaxation in the material is captured by adopting a strong discontinuity approach (SDA) whereby the displacement jump of a cracked material point is mapped to the element's nodes as an enhanced strain field.

Integrating the SDA within the MPM framework
The strong discontinuity approach [4] in the finite element (FE) modelling of a crack is illustrated, in Figure 1, in opposition to the smeared crack concept. The correct behavior of the cracked element (i.e. softening of the cracked area and unloading of the bulk) is achieved by augmenting the

standard equilibrium equations with the condition of traction continuity at the interface. The displacement field is expressed as an additive decomposition of a regular field and an enhanced field that represents the displacement discontinuity:

$$\mathbf{u}(\mathbf{x}) = \hat{\mathbf{u}}(\mathbf{x}) + \bar{\mathbf{u}}(\mathbf{x}) = \hat{\mathbf{u}}(\mathbf{x}) + f(\mathbf{x})[\mathbf{u}]_{\Gamma_d} \quad (1)$$

In Equation (1), $\hat{\mathbf{u}}(\mathbf{x})$ is the displacement field resulting from the bulk elasticity, $\bar{\mathbf{u}}(\mathbf{x})$ is the enhanced displacement field and $[\mathbf{u}]_{\Gamma d}$ is the displacement jump along the discontinuity surface Γ_d. The function $f(\mathbf{x})$ distributes the discontinuity between the 'negative' (Ω^-) and positive (Ω^+) sides of the domain. For instance, one could choose:

$$f(\mathbf{x}) = H_{\Gamma_d}(\mathbf{x}) + (r-1) \quad (2)$$

where $0 \leq r \leq 1$ and $H_{\Gamma d}$ is the Heaviside function at the discontinuity surface Γ_d. Here, the general form $f(\mathbf{x})[\mathbf{u}]_{\Gamma d}$ is kept to represent any suitable mapping of the displacement discontinuity from the singularity to the continuum.

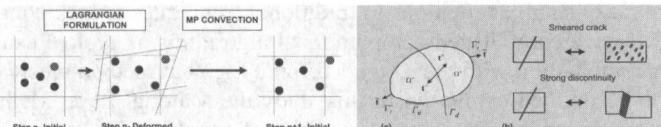

Figure 1: Left: MPM computational flow; Right: Cracked body (a) and alternative FE modelling concepts (b)

Using the virtual displacement field, of the form $\delta\mathbf{u} = \delta\hat{\mathbf{u}} + f\delta[\mathbf{u}]_{\Gamma d}$, in the weak form of equilibrium equations written for two parts of the solid, Ω^+ and Ω^-, the following equations are derived [5, 7] (neglecting the body force and dynamic effects):

$$\int_{\Omega \backslash \Gamma_d} \nabla^s(\delta\hat{\mathbf{u}}) : \sigma d\Omega = \int_{\Gamma_t} \delta\hat{\mathbf{u}} \cdot \bar{\mathbf{t}} d\Gamma \quad (3)$$

$$\int_{\Omega \backslash \Gamma_d} f\nabla^s\left(\delta[\mathbf{u}]_{\Gamma_d}\right) : \sigma d\Omega = \int_{\Gamma_d} \delta[\mathbf{u}]_{\Gamma_d} \cdot \mathbf{t} d\Gamma + \int_{\Gamma_t} f\delta[\mathbf{u}]_{\Gamma_d} \cdot \bar{\mathbf{t}} d\Gamma \quad (4)$$

It can be seen that the solid behavior (Eq. 3) is augmented by the weak form of traction continuity across the discontinuity Γ_d (Eq. 4). Within the context of the material point method, our crack Γ_d effectively comprises several cracked material points (MPs). Their collection of openings, denoted as [**u**], contribute to the jump $[\mathbf{u}]_{\Gamma d}$ across Γ_d. While the first term in the right hand side of (Eq. 4) can then be written as a sum of virtual works at individual cracked MPs, the other terms of that equation require the regularization of the function f over a certain volume for the purpose of the spatial discretization. A linear representation is adopted as follows [7]:

$$\bar{\mathbf{u}}(\mathbf{x}) = f(\mathbf{x})[\mathbf{u}]_{\Gamma_d} \approx \Lambda(\mathbf{x})[\mathbf{u}] \Rightarrow f(\mathbf{x})\delta[\mathbf{u}]_{\Gamma_d} = \Lambda(\mathbf{x})\delta[\mathbf{u}] \quad (5)$$

Since field variables are described by their nodal values in the FE method, the nodal value of the enhanced field is obtained by using the mapping between node (i) and crack (c) using the discrete form of $\Lambda(\mathbf{x})$:

$$\bar{\mathbf{u}}_i = \Lambda_{ic}[\mathbf{u}]_c, \text{ and corresponding variation } \delta\bar{\mathbf{u}}_i = \Lambda_{ic}\delta[\mathbf{u}]_c \quad (6)$$

Using the above approximations (see details in [7]), from (3-4), the linearised system of algebraic equations in terms of the total displacement increment $\Delta\mathbf{u}$ and the displacement jump increment $\Delta[\mathbf{u}]$ can be written as:

$$\begin{cases} \mathbf{K}_{uu}\Delta\mathbf{u} - \mathbf{K}_{u[u]}\Delta[\mathbf{u}] = -\mathbf{F}_u^{int} + \mathbf{F}_u^{ext} \\ \mathbf{K}_{[u]u}\Delta\mathbf{u} - \left(\mathbf{K}_{[u][u]} + \mathbf{K}_{dd}\right)\Delta[\mathbf{u}] = -\mathbf{F}_{[u]}^{int} + \mathbf{F}_{[u]}^{ext} \end{cases} \quad (7)$$

where the $\mathbf{K}_{uu}$, $\mathbf{K}_{u[u]} = (\mathbf{K}_{[u]u})^T$ and $\mathbf{K}_{[u][u]}$ are stiffness matrices representing, respectively, the elasticity of the bulk, the coupling between bulk and interface, and an equivalent 'continuum stiffness' of the interface. $\mathbf{K}_{dd}$ is the tangent stiffness of the decohesive interface that is calculated according to a coupled damage/friction interface constitutive model that is described at length in [7, 8]). $\mathbf{F}_u^{int}$, $\mathbf{F}_u^{ext}$, $\mathbf{F}_{[u]}^{int}$ and $\mathbf{F}_{[u]}^{ext}$ are internal and external force terms corresponding to the total ($\mathbf{u}$) and enhanced ($[\mathbf{u}]$) fields. With further assumptions that the crack is far away from force boundaries, as well as there is no contribution of the crack to internal forces at incipience of the discontinuity, the latter two terms can also be neglected. Henceforth, the displacement discontinuity is extracted from the second equation and substituted in the first to obtain the condensed form:

$$\Delta[\mathbf{u}] = \left(\mathbf{K}_{[u][u]} + \mathbf{K}_{dd}\right)^{-1} \mathbf{K}_{[u]u} \Delta \mathbf{u} \Rightarrow \left[\mathbf{K}_{uu} - \mathbf{K}_{u[u]}\left(\mathbf{K}_{[u][u]} + \mathbf{K}_{dd}\right)^{-1} \mathbf{K}_{[u]u}\right] \Delta \mathbf{u} = -\mathbf{F}_u^{int} + \mathbf{F}_u^{ext} \quad (8)$$

Mapping of the strong discontinuity: The role of the mapping matrix Λ, which discretises the function $f(\mathbf{x})$, is to ensure that the energy dissipated by the (cohesive) crack opening is matched by the bulk unloading resulting from Eq. (2). Obviously there may be different choices depending on the numerical schemes. In this work, the element's shape functions φ_i are used for its construction: organising the nodes into positive and negative sides of the crack, the contribution of each node to the enhanced field is determined as per the following equations (examples are shown in Figure 2 with $\varphi_i(\mathbf{x}_p)$ denoting the value of shape function φ_i at the material point $\mathbf{x}_p$; further details in [7]):

$$\Lambda_{ip} = \begin{cases} \dfrac{\varphi_i(\mathbf{x}_p)}{\Phi^+(\mathbf{x}_p)}, i \in \Omega^+ \\ -\dfrac{\varphi_i(\mathbf{x}_p)}{\Phi^-(\mathbf{x}_p)}, i \in \Omega^- \end{cases}, \text{ with } \begin{aligned} \Phi^+(\mathbf{x}_p) &= \sum_i \left(\varphi_i(\mathbf{x}_p)\right)\Big|_{i=1,N, \mathbf{x}_i \in \Omega^+} \\ \Phi^-(\mathbf{x}_p) &= \sum_i \left(\varphi_i(\mathbf{x}_p)\right)\Big|_{i=1,N, \mathbf{x}_i \in \Omega^-} \end{aligned} \quad (9)$$

Thus, the sum of nodal weights on Ω^+ is equal to 1, and that for Ω^- is -1. This way, nodal displacements r and r-1 imposed, respectively, onto the positive and negative sides of the crack, deliver a displacement jump of unity, as expected. For the simple case of a 4-node element with a single cracked MP at its centroid, Figure 2, the 8 x 2 mapping matrix is:

$$\Lambda = \begin{bmatrix} -\tfrac{1}{2} & 0 & \tfrac{1}{2} & 0 & \tfrac{1}{2} & 0 & -\tfrac{1}{2} & 0 \\ 0 & -\tfrac{1}{2} & 0 & \tfrac{1}{2} & 0 & \tfrac{1}{2} & 0 & -\tfrac{1}{2} \end{bmatrix}^T$$

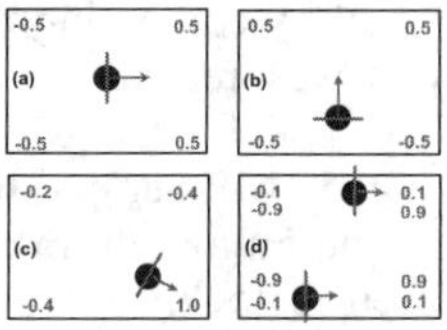

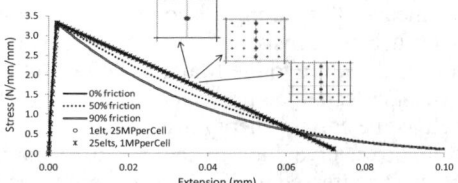

Figure 2: Mapping coefficients

It is easy to verify that the condensed (tangent) stiffness matrix for the case illustated in Figure 2 is identical to that obtained using a smeared crack approach. However, the advantage of the SDA is to be less prone to snap-back instabilities, thanks to the additional degree of freedoms introduced in this work at the cracked MPs. The particular choice of mapping utilised here results in the total energy dissipated by the fictitious crack - composed of several cracked MPs - being equal to the sum of their individual dissipations. For an objective representation of the energy lost in the fracture process, the length associated with each cracked MP is taken to be that of a crack traversing the entire element, and the fracture toughness is scaled (divided) by the number of current cracked MPs in the element.

Figure 3: Predicted response of a single element.

Results

The interface law was first validated with a single element (20mm x 10mm), carrying one material point at its centre and pulled apart in tension. The material and interface properties were: Young modulus E=30000MPa, Poisson's ratio ν=0.2, interface stiffness K_0=10^6N/mm, mode I fracture energy G_I=0.124N/mm, mode I tensile strength N=3.33MPa. The stress is plotted against the element's extension in Figure 3, showing the expected final displacement, δ_f= 0.074mm, for linear softening. The longer-tail response resulting from the coupling with friction is also shown. Next, the mesh-independence of the mapping was demonstrated using the same element, modelled with either a single element with several MPs, or with several elements carrying one MP each, cf. Figure 3. All scenarios, expectedly, resulted in the same prediction of the global response. This formulation was then used to predict the response of a classical benchmark for concrete fracture: the three point bending test of Petersson [6] (to which the reader is referred to for specimen geometry and material properties; Figure 4).

The MPM model made use of the problem's symmetry and only half of the specimen was analysed. Accordingly, half the nominal value of fracture toughness, G_f=124N/m, was specified. With a mesh size of 20mm x 20mm, the response obtained with 4-node linear elements and using 1 and 4 MPs per cell exhibited a marked lack of convergence with the number of MPs, as shown by the strong discrepancy between the predictions shown in Figure 4.

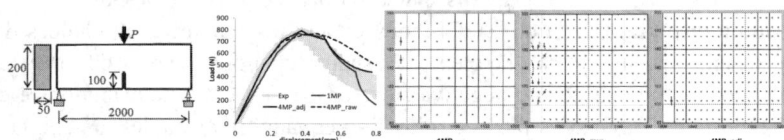

Figure 4: Geometry and predicted response for the three-point bending specimen

This occurred because the maximum principal stress, used as the criterion for initiating cracks, was always found at the element corner and the corresponding computed normal was far from perpendicular to the specimen midsection. As a result, stress locking occurs, due to the fact that the stress σ_{xx} is still able to transfer across the discontinuity even after complete failure of the interface. Henceforth, the global response of the structure during the damage process was stiffer than expected, whenever more than one MP per cell was utilised. This problem was overcome (Figure 4, 4MP_adj) by computing the normal to cracks based on the stress state at the element's centroid. The analysis was repeated with the adjusted normal, using a range of MPs per cell for the 20mm x 20mm mesh, and compared with the converged 1MP per cell (obtained with smaller mesh sizes -10mm x 10mm and 5mm x 5mm). The linear elements near the loading point appeared to lock, for the models with more than one MP per cell, reflected by the plateau in the load/displacement curve. Overall, the predictions were in good agreement with the experiment, as reported in Figure 5.

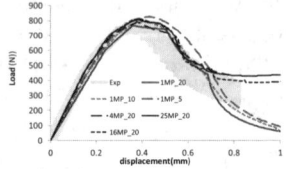

Figure 5: Load vs displacement plot

Summary

By means of a strong discontinuity representation of the crack using a consistent particle-to-element mapping scheme, this work successfully added a new powerful feature to the Material Point Method. Because the explicit crack is treated at the material level, the technique presented here accurately captures the discontinuous kinematics of cracks, as well as the correct energy dissipation, using relatively few lower-order elements. Stress locking due to incorrect prediction of the crack orientation was found to severely affect the numerical results and was addressed in a simple way in this early stage of the research. More advanced strategies, e.g. crack orientation based on a nonlocal stress field [9], combined with the use of different types of finite element will be explored to overcome the issue, before the application of the proposed method in the study of dynamic fragmentation.

References

[1] Kachanov M. (1980) J. Engng. Mech. Div. **106**: 1039-1051.
[2] Sulsky D, Chen Z, Shreyer HL (1994). Comput. Methods Appl. Mech. Engrg. **118**: 179-196.
[3] Nairn JA (2003) Comput. Model. Eng. Sci. **4**: 649-663.
[4] Simo JC, Oliver J. Fracture and damage in quasi-brittle structures, pp. 25-39. Bazant ZP et al., editor, 1994.
[5] Alfaiate J, Simone A, Sluys LJ (2003). Int. J. Solids Struct. **40**: 5799-5817.
[6] Petersson PE. Report TVBM-1006. Lund Institute of Technology, Lund, Sweden, 1981.
[7] Guiamatsia I, Nguyen GD, Sulsky D. MPM implementation of a strong discontinuity approach. In preparation.
[8] Guiamatsia I, Nguyen GD (2012). Compos. Sci. Technol. **72**: 269-277.
[9] Simone A, Sluys LJ. Comput. Methods Appl. Mech. Engrg. **193**(27-29): 3015-3033.

Numerical Simulation of Crack Tip Behavior under Fatigue Loading

Y.G. Xu[1,a], W. Tiu[2,b] and Y.Z. Xu[2,c]

[1]School of Engineering and Technology, University of Hertfordshire, AL10 9AB, UK

[2]High North Technology Centre, Narvik University College, N-8505 Narvik, Norway

[a]y.2.xu@herts.ac.uk, [b]william.tiu@hin.no, [c]yxu@hin.no

Keywords: fatigue, numerical simulation, crack closure, dislocation distribution technique

Abstract. Fatigue damage is a localized phenomenon controlled by the near-tip crack behavior. This paper presents an application of a dislocation distribution technique to the simulation of crack tip behavior under fatigue loading. A centre-cracked tension specimen under uni-axial fatigue loading is used in the study. Crack opening and plastic deformation around the crack tip are simulated by distributions of dislocation dipoles in crack plane and four inclined planes ahead of the crack tip. Climb dislocation dipole is used to model the opening and closing of the crack while glide dislocation dipole is used to simulate the backward and forward slip in the inclined planes during loading and unloading of the fatigue cycle. Stress field around the crack tip is obtained by the superposition of the contributions of the applied external load and the distributed dislocation dipoles. Correct boundary conditions of the model are achieved by employing a quadratic programming technique to minimize a properly constructed non-negative object function. It is found that the simulated crack closure variations under the constant amplitude fatigue load agree well with the result of a previously developed modified strip yield model with an appropriate constraint factor.

Introduction

Many engineering structures such as aircraft or nuclear reactors suffer fatigue damage due to cyclic loading over many years of their service lives. A major challenge for the industry is the reliable prediction of fatigue crack propagation lives which can be influenced by many factors such as loading conditions, material properties, and structure geometries. A major breakthrough in improving the accuracy of the crack growth driving force is the discovery of crack closure phenomenon, a concept first proposed by Elber in 1970 [1]. Unlike the ideal zero-width saw cut static crack which closes when the stress intensity factor $K=0$ and opens when $K>0$, Elber's experimental observation that a fatigue crack can be mechanically closed at a far-field tensile load first established that crack growth rate is not only influenced by the conditions ahead of the crack tip, but also by the nature of crack face contact behind the crack tip. Such premature contact between the surfaces of a fatigue crack with load transferred through the contact area is generally referred to as crack closure, which shields the range of the externally applied stress transmitted to the crack tip and hence reduce the crack growth driving force.

Various micro-structural and micro-mechanical factors can influence this premature contact of the crack surfaces, which explains the great capacity of the crack closure concept in rationalizing various fatigue behaviours observed in real life [2-4]. It is however worth noting that, while the crack closure concept has enjoyed the dominance in fatigue research for the last 40 years, the identification and its real significance on crack growth remain controversial among the leading researchers in the fatigue community. There is neither agreement between the closure levels identified from different techniques [5-7] nor consensus about the relevance of crack closure in crack growth, with some even questioning its very existence when the plane strain near-tip constraint is approached [8-12]. This paper aims to achieve a further understanding of the crack closure through detailed simulation of the crack tip behaviour under fatigue loading. A dislocation distribution technique has been presented to simulate a centre-cracked fatigue speccimen. It is found that the simulated crack closure variations under constant amplitude fatigue load agree well with the result of a previously developed modified strip yield model with an appropriate constraint factor.

Model Construction

Fig. 1 is the schematic diagram of the physical model to simulate crack closure under plane strain condition of a centre-cracked tension (CCT) specimen under uni-axial fatigue loading σ_∞ in y direction. To simulate the in-plane plastic deformation under plane strain condition, two pairs of planes (planes BC & BD and EF & EG) were introduced ahead of the tips of a centre crack EAB of length of 2a. Considering the symmetry of the problem, these planes are of the same length b and have the same angle θ with the crack plane. During loading and unloading phase of the fatigue cycle, forward and backward slips occur in these planes. The extent of the slip depends on the sample geometry, external load, material property, and the related crack closure level.

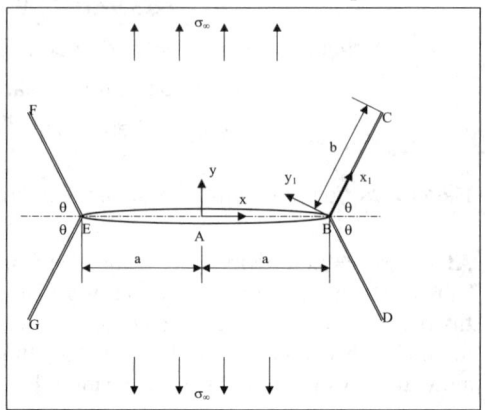

Fig.1 CCT specimen under fatigue loading.

Fundamental Solutions of the Near-Tip Stresses

Crack opening and plastic deformation around the crack tip are simulated by proper distributions of dislocation dipoles in crack plane (EAB) and four inclined planes. Climb dislocation dipole b_{yy} is used to model the opening and closing of crack EAB while glide dislocation dipole $b_{x_1y_1}$ is required to simulate the backward and forward slip in the inclined planes. Stress field in the model is obtained by the superposition of the contributions of the applied external load and the distributed dislocation dipoles. Fig.2 serves as an example of the simulation of stress $\bar{\sigma}_{yy}$ at Point H(x, 0) on Segment AB due to a dislocation dipole b_{yy} at $(\xi, 0)$ on AB and another dislocation dipole b_{yy} of the same strength located at a symmetrical point $(-\xi, 0)$ on AE

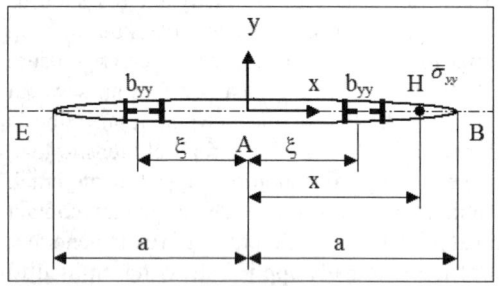 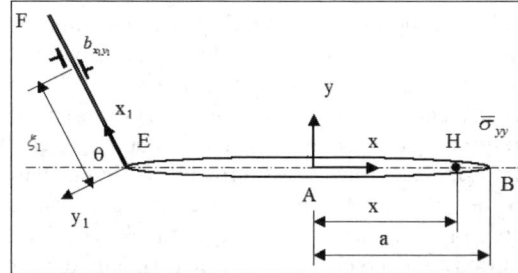

Fig.2 Stress $\bar{\sigma}_{yy}$ at point H due to dislocation dipoles on segments AB and AE.

Fig. 3 Stress $\bar{\sigma}_{yy}$ at point H due to a dislocation dipole on segment EF.

Fundamental equations for the stress at point H due to the dislocation dipoles in Fig.2 can be expressed as follows [13]:

$$\bar{\sigma}_{yy}(x,0) = \frac{2\mu}{\pi(\kappa+1)} \left[b_{yy}(\xi,0) \cdot (L_{yy}^{yy})_1 + b_{yy}(-\xi,0) \cdot (L_{yy}^{yy})_2 \right] \quad (1)$$

where μ is the shear modulus of the material and κ is Kolosov's constant, equal to $(3-4\nu)$ in plane strain. ν is the Poisson's ratio of the material.

$$(L_{yy}^{yy})_1 = \frac{1}{(x-\xi)^2} \quad ; \quad (L_{yy}^{yy})_2 = \frac{1}{(x+\xi)^2} \quad (2)$$

Since these two dipoles have the same strength, $\bar{\sigma}_{yy}(x,0)$ at point H is expressed as:

$$\bar{\sigma}_{yy}(x,0) = \frac{2\mu \cdot b_{yy}}{\pi(\kappa+1)} \left[\frac{1}{(x-\xi)^2} + \frac{1}{(x+\xi)^2} \right] \tag{3}$$

Fig.3 shows the $\bar{\sigma}_{yy}$ at point H(x, 0) on segment AB due to a dislocation dipole $b_{x_1 y_1}$ at $(\xi_1, 0)$ in the local co-ordinate system $x_1 E y_1$ of segment EF. Co-ordinates of point H in the local co-ordinate system $x_1 E y_1$ are: $x_1 = -(a+x) \cdot \cos(\theta), \quad y_1 = -(a+x) \cdot \sin(\theta)$ (4)

Relative to the dipole $b_{x_1 y_1}$ at $(\xi_1, 0)$:

$\bar{x}_1 = -(a+x) \cdot \cos(\theta) - \xi_1, \quad \bar{y}_1 = -(a+x) \cdot \sin(\theta), \quad r_1^2 = \bar{x}_1^2 + \bar{y}_1^2$ (5)

Stress tensor at point H in the local co-ordinate system $x_1 E y_1$ [13]:

$$\begin{Bmatrix} \bar{\sigma}_{x_1 x_1}(x_1, y_1) \\ \bar{\sigma}_{y_1 y_1}(x_1, y_1) \\ \bar{\sigma}_{x_1 y_1}(x_1, y_1) \end{Bmatrix} = \frac{2\mu \cdot b_{x_1 y_1}}{\pi(\kappa+1)} \cdot \begin{bmatrix} L_{x_1 x_1}^{x_1 y_1} \\ L_{y_1 y_1}^{x_1 y_1} \\ L_{x_1 y_1}^{x_1 y_1} \end{bmatrix} \tag{6}$$

Transform this stress tensor to the stress component $\bar{\sigma}_{yy}$ in global co-ordinate system by rotating $x_1 E y_1$ with $(\pi + \theta)$ anti-clockwise, we have:

$\bar{\sigma}_{yy}(x,0) = \sin^2(\pi+\theta) \cdot \bar{\sigma}_{x_1 x_1} + \cos^2(\pi+\theta) \cdot \bar{\sigma}_{y_1 y_1} - 2\sin(\pi+\theta)\cos(\pi+\theta) \cdot \bar{\sigma}_{x_1 y_1}$

$$\bar{\sigma}_{yy}(x,0) = \frac{2\mu \cdot b_{x_1 y_1}}{\pi(\kappa+1)} \cdot \left[\frac{\sin^2(\theta)[2\bar{x}_1 \bar{y}_1(3\bar{x}_1^2 + \bar{y}_1^2)] + \cos^2(\theta)[2\bar{x}_1 \bar{y}_1(\bar{x}_1^2 - 3\bar{y}_1^2)]}{r_1^6} \right.$$
$$\left. - \frac{\sin(2\theta) \cdot (\bar{x}_1^4 - 6\bar{x}_1^2 \bar{y}_1^2 + \bar{y}_1^4)}{r_1^6} \right] \tag{7}$$

Same procedure has been used to determine the stress at point H due to the dislocation dipoles on inclined planes EG, BC, and BD. The shear stresses on the inclined planes have been determined in the same way. Real stress components are from combined contributions of both the external loading and the dislocation dipoles distributed along the crack plane and slip planes.

$$\sigma_{ij} = \tilde{\sigma}_{ij} + \bar{\sigma}_{ij} \tag{8}$$

where $\tilde{\sigma}_{ij}$ is caused by the external loading and $\bar{\sigma}_{ij}$ is caused by dislocation dipoles.

Simulation of Crack Growth and Crack Closure

To simulate a growing fatigue crack, a layer of residual stretch is appended to the crack surface to account for the fact that the crack is growing through plastically deformed material. This stretched material is modelled by climb dislocation dipoles of the strength of $(b_{yy})_o$ behind the crack tip.

Boundary element method is employed to solve the stress and deformation of the model. To increase the efficiency of the model solution, varied element lengths are used with smaller elements being distributed near the crack tip. Boundary conditions will be enforced at collocation points which are positioned at the centre of each boundary element. Correct boundary conditions of the model are achieved by employing a quadratic programming technique to minimise a properly constructed non-negative object function. The object function is minimised, subject to various constraints on the variables in the problem, with a standard quadratic programming subroutine called E04NAF in FORTRAN library. To investigate crack closure variations during fatigue crack growth, An iterative procedure is employed to capture the load level which just fully opens the crack.

Results and Discussions

Fig. 4 shows the plane strain crack closure variations for propagating cracks under constant amplitude (ΔK) fatigue loading of stress ratio of 0.0. It can be seen from the results that crack closure levels stabilise after the crack has propagated by about 4 times the baseline plastic zone size, which agrees with the criterion adopted in ASTM load shedding procedure in obtaining reliable da/dN curves [14]. It is, however, noticed that the stabilised closure values are slip plane angle dependent in the current model. Compared with current model results, predicted crack closure from the modified strip yield model with a constraint factor of three [15] lies in-between the results of the current model with slip plane angles of 60° and 45°. It is obvious that the result from the modified strip yield model with a reduced constraint factor will favour the current model of greater slip plane angles.

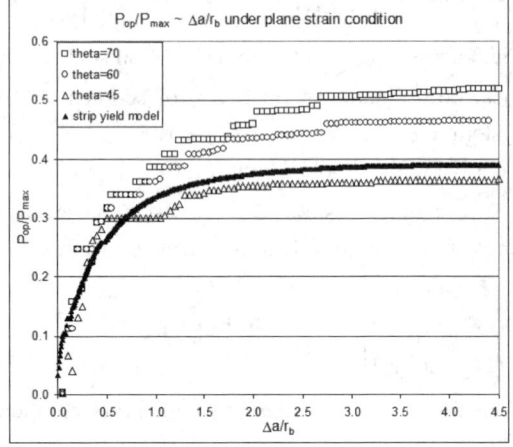

Fig.4　Plane strain crack closure under various slip plane angles.

Also worth noting is the fact that the way of passing the crack tip plastic deformation to the wake of the crack at the peak load also has a large influence on the crack closure variations. At present, a simple geometry relation is used to relate the thickness of the stretched layer and the crack tip plastic deformation. Further work would be required to study the effect of selected slip plane angle and the physical relationship between the stretched layer and the crack tip plastic deformation on model performance, particularly for the study of variable amplitude fatigue behaviour.

References

[1] W. Elber (1970), *Eng. Fract. Mech.,* Vol.2, pp.37-44.
[2] J. C. Newman, W. Elber (Editors) (1988), *ASTM STP 982*.
[3] R. C. McClung, J. C. Newman (Editors) (1997), *ASTM STP 1343*.
[4] Y.G. Xu, L. Wang, Y. Chen, W. Tiu (2012), *Key Engineering Materials*, Vols. 488-489, pp. 545-548.
[5] Y.G. Xu, P. J. Gregson, I. Sinclair (2000), *Materials Science and Engineering: A*, Vol.284, pp.114-125.
[6] E. P. Phillips (1993), NASA Technical Memorandum 109032, Langley Research Center, Hampton, VA.
[7] S. Stoychev, D. Kujawski (2003), *Fatigue Fract Engng Mater Struct*, Vol.26, pp.1053-1067.
[8] N. Louat, K. Sadananda, M. Duesbery, A. K. Vasudevan (1993), *Met. Trans.*, A24, pp.2225-2232.
[9] K. Sadananda, A. K. Vasudevan (2003), *Fatigue Fract Engng Mater Struct*, Vol.26, pp.835-845.
[10]　P. C. Paris, D. Lados, H. Tada (2008), *Eng. Fract. Mech.*, Vol.75, pp.299-305.
[11]　S. Pommier (2002), *Eng. Fract. Mech.*, Vol.69, pp.25-44.
[12]　J. Toribio, V. Kharin (2011), *European Journal of Mechanics A/Solids*, Vol.30, pp.105-112.
[13]　D. A. Hills, P. A. Kelly, D. N. Dai, and A. M. Korsunsky, (1996), *Solution of Crack Problems: The Distributed Dislocation Technique,* Kluwer Academic Publisher.
[14]Annual Book of ASTM Standard, (1996), Vol.03.01, *ASTM E647-95a*, pp.565-601.
[15]Y.G. Xu, (2001), Ph. D Thesis, Southampton University.

Delamination Threshold Load of Composite Laminates under Low-Velocity Impact

Y.G. Xu[1,a], Z. Shen[1,b], W. Tiu[2,c], Y.Z. Xu[2,d], Y. Chen[1,e] and G. Haritos[1,f]

[1]School of Engineering and Technology, University of Hertfordshire, AL10 9AB, UK

[2]High North Technology Centre, Narvik University College, N-8505 Narvik, Norway

[a]y.2.xu@herts.ac.uk, [b]z.shen2@herts.ac.uk, [c]william.tiu@hin.no, [d]yxu@hin.no, [e]y.k.chen@herts.ac.uk, [f]g.haritos@herts.ac.uk

Keywords: low-velocity impact, composite laminate, delamination threshold load

Abstract. A key factor affecting the use of carbon fibre reinforced composite laminates is the low velocity impact damage which may be introduced accidentally during manufacture, operation or maintenance of the component. Among the several barely visible impact damages, interlaminar delamination is the dominant failure mode and may reduce the post-impact compressive strength of the component significantly. This paper focuses on the study of the delamination threshold load (DTL) above which significant increase of delamination and thus large reduction of the residual compressive strength of the component may occur. Instrumented drop weight tests were carried out under various impact energy levels to determine the delamination threshold load. Efforts are directed to the study of the laminate thickness effect on the reliability of the detection of the DTL. The validity of the concept of DTL has been investigated and possible implications on the measurement of the DTL has been discussed. It is demonstrated that DTL exists but its detection requires proper testing conditions.

Introduction

The desire to reduce structural weight has been the driving force behind the research and development of new technologies and materials in the aerospace industry. A popular and effective solution to the ever-demanding weight-saving requirement is to replace the conventional metal alloys with composite materials for aircraft primary structures. Compared with conventional metal alloys, composite materials have higher specific strength and specific stiffness due to their high strength, high stiffness, and low density. They can also be formed into virtually any shape and offer optimized mechanical properties through a combination of fibre, matrix and interface conditions [1]. The overall structural weight of the aircraft can therefore be reduced without compromising the stiffness and strength of its structure by using composite materials.

A major concern associated with the effective use of composite laminates is the substantial reduction in the structural strength due to the low-velocity impact which can be introduced accidentally during the manufacture, operation or maintenance of the component [2-5]. This can be attributed to the inherent brittleness of both the carbon fibre and the epoxy matrix materials which can only absorb impact energy in elastic deformation and through damage mechanisms, and not via plastic deformation as most conventional ductile alloys do. The low-velocity impact creates damage which may involve local indentation, matrix cracking, fibre matrix debonding, delamination, and fibre breakage. Even no damage can be observed on the surface at the point of impact, barely visible impact damage (*BVID*) such as matrix cracking and delamination can occur and contribute up to 60% loss in a laminated composite component's compressive strength [3]. This paper focuses on the study of the delamination threshold load (DTL) through a series of instrumented drop weight tests. Efforts are directed to the study of the laminate thickness effect on the reliability of the detection of the DTL to clarify some important issues associated with the concept of DTL.

Experimental
The autoclave moulding technique was used for the fabrication of the sample. The unidirectional material used in this study is the carbon prepreg Cytec *977-2-35%-12KHTS-268-600* supplied by the Centre of Composites, Airbus UK. The impact test was conducted in accordance to the ASTM standard [6].

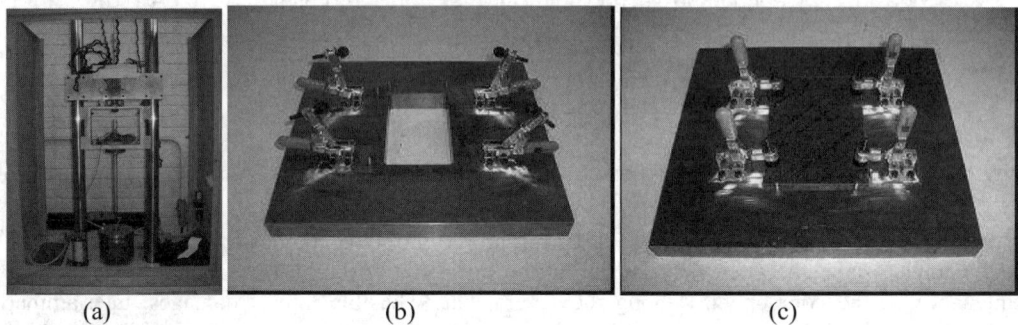

Fig.1 (a) Instrumented drop-weight impact test rig, (b) base plate to support the test sample, (c) base plate with the sample clamped

Fig. 1 shows the impact test rig in the Material Test Lab for the instrumented drop weight impact test. The impactor head has a shape of semi-sphere with a diameter of *20mm*. The total mass of the drop-weight (impact rod + support frame) is *11.8kg*. Impact force history was recorded with Picoscope PC oscilloscope (3000 series). Laminates of layup configurations of $[0/90_2/0]_s$ (2mm), $[0_2/90_3/0]_s$ (3mm), $[0_2/90_3/0_2/90]_s$ (4mm), and $[0_2/90_3/0_2/90_2/0]_s$ (5mm) have been tested under various impact energy levels to catch the possible delamination threshold load.

Results and Discussions
The load history for the low-velocity impact event can provide important information regarding damage initiation and propagation. It has been documented by many investigators that sudden load drops on the impact force histories are associated with the loss of stiffness of the laminate due to the initiation of laminate level damages such as the delamination [2, 7-8]. Detailed impact load histories have been obtained under various test conditions in the current research to examine the features of those curves and investigate whether there exists a *DTL* for the laminate concerned.

Fig.2 shows the impact force results of the 4mm thick specimen under four impact energy levels. It can be seen that there is no noticeable drop of the impact force before reaching the peak impact force for the impact test under the impact energy of 3J. Clear drops of the impact force can be seen for the 4mm thick laminate under the impact energies of 6J and 12J. The DTL so obtained is around 4kN. The same DTL level measured under the two different impact energy levels demonstrates that DTL does exist for the low-velocity impact of composite laminates.

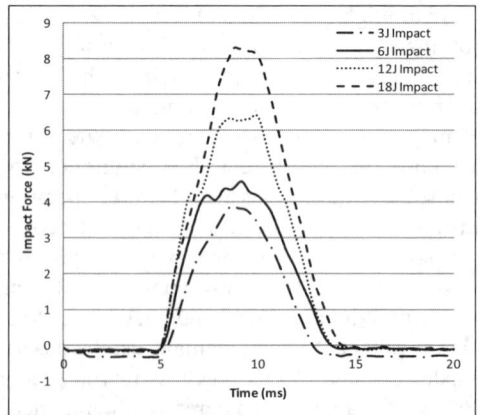

Fig.2   DTL of 4mm laminates.

It is however interesting to note that the impact force history under the impact energy of 18J doesn't show a clear dip in the impact force. This may look like in contradictory to the concept of the DTL. Repeated tests were carried out to check the results and similar results were recorded. The current explanation is that the relative impact of the delamination initiation on the stiffness of the sample becomes less significant under higher impact energy levels. As a result, care should be taken in the detection of the DTL using the impact load histories.

Fig.3 shows the laminate thickness effect on the impact force history during the low-velocity impact. Laminates of four different thicknesses were tested under the impact energy of 6J. It can be seen clearly that the maximum impact force will increase when the laminate thickness is increased. The impact duration decreases with the increase of the laminate thickness. This is expected as the increase of the laminate thickness makes the sample stiffer.

It is however noticed that no obvious impact force fluctuation can be observed for either the thin (2mm) or the thick (5mm) samples. For the 5mm thick sample, we may argue that the current impact energy level might not be high enough to trigger the delamination. The result of the 2mm thick laminate does seem a bit odd at first sight. Again, repeated tests were carried out to check the results and similar results were recorded. The current explanation is that most of the impact energy will be absorbed by the elastic deformation for the thin laminate. Bending stress in the thin laminate is not high eough to trigger the delamination. It is also possible that, for the thin laminate, the relative effect of the delamination initiation on the impact energy absorption becomes less significant compared with the energy absorbed by the elastic deformation.

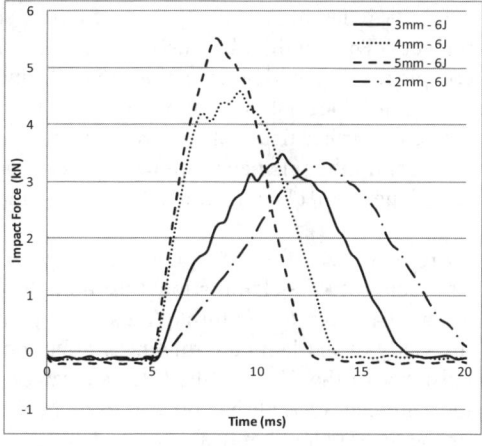

Fig.3 Thickness effect on impact force.

Fig.4 shows some further evidence on the issue of determining the DTL for the thinner laminate under low-velocity impact. The 2mm thick laminate has been tested under impact energy from 1J to 10J. It can be seen clearly that, while the maximum impact force is increasing with the increase of the impact energy level, no clear indication of the initiation of delamination can be reliably determined based on the impact force history. As such, alternative methods should be considered to determinate the DTL for the thin laminates. It is speculated that there might exist an optimum ratio between the impact energy and the laminate stiffness (thickness) to determine the DTL reliably.

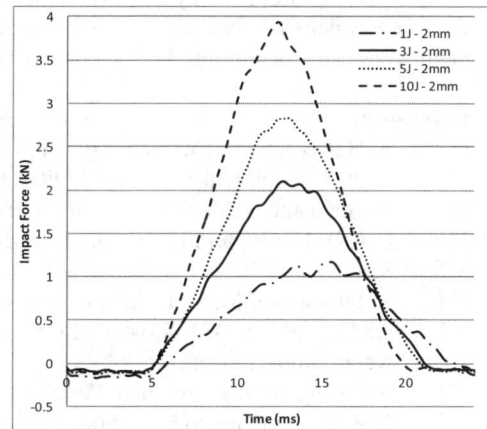

Fig.4 Impact force history of 2mm laminate.

Fig.5 shows the effect of repeated impact on the impact force history of composite laminates under low-velocity impact. 4mm thick laminates were first impacted under the impact energy of 6J and 12J, respectively. The same laminate was then subjected to a second impact under the same impact energy level. It can be seen clearly that the second impact on the same laminate doesn't affect the impact duration. It is however interesting to note that the impact force curve becomes much smoother compared with the results during the first impact, showing no clear growth of the initial delamination under the same repeated impact energy level. Another interesting finding is that the maximum impact force actually has been increased during the second impact, which is no expected. Repeated tests show similar trends for the laminates of different thickness and under difference impact energy levels. No clear explanation is available for this finding at the moment. It is speculated that the residual stress built up during the curing process of the laminate in the autoclave may be associated with this result. The compressive residual stress in the laminate might be released during the first impact through the initial delamination. The re-balance of the internal stress in the laminate makes the laminate stiffer and gives the higher peak force for the second impact. This will obviously bring in further complexity to the determination of DTL and may even introduce some ambiguity of the DTL concept. Further detailed research will be required on this. Finite element simulation is a suitable technique to check and validate above assumption.

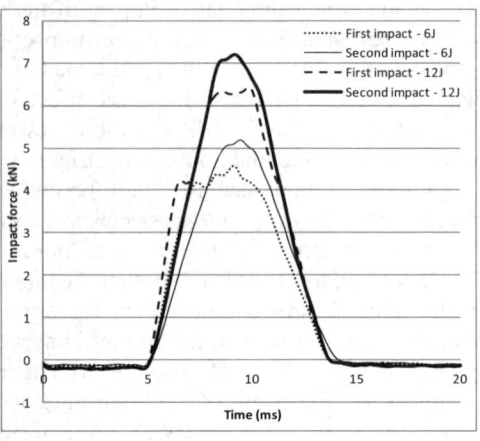

Fig.5 Effect of second impact on DTL

Summary
The concept of delamination threshold load (DTL) of composite laminates under low-velocity impact has been tested through a series of instrumented drop weight tests. The main focus is about the study of the laminate thickness effect on the reliability of the detection of the DTL. While the existence of the DTL for the laminate can be seen clearly from some of the test results, the realiable measurement of DTL has proven to be far from strightforward. Both the laminate thickness and the impact energy level may affect the detection of the DTL. There might be an optimum ratio between the impact energy and the laminate stiffness (thickness) to determine the DTL. The residual stress from the curing process may also affect the behavior of the laminate under low-velocity impact. It is therefore concluded that care must be taken in interpreting the impact force result and further detailed research is required for the concept of DTL.

References
[1] B. Harris (1999). Engineering Composite Materials, 10M Communications Ltd.
[2] W.J. Cantwell and J. Morton (1991), Composites **22**(5): 347-362.
[3] Z. Guan and C. Yang (2002), Journal of Composite Materials **36**(7): 851-871.
[4] G.A.O. Davies and D. Hitchings, et al. (2006), Composite Science and Technology **66**: 846-854.
[5] R. Olsson and M. V. Donadon, et al. (2006), Int. J. Solids and Structures **43**: 3124-3141.
[6] ASTM_D7136/D7136M-05 (2005), Fiber-reinforced polymer matrix composite to a drop-weight impact event.
[7] X. Zhang (1998), Proc Instn Mech Engrs: Part G **212**: 245-259.
[8] G.A. Schoeppner and S. Abrate (2000), Composites: Part A **31**: 903-915.

Inelastic seismic behavior without and with over-resistance effects of 10-story building RC damaged due to the 1985 earthquakes in Mexico City

Jorge A. Avila [1, 2, a]

[1] Institute of Engineering, National University of Mexico (UNAM)

[2] Faculty of Engineering, National University of Mexico (UNAM)

Ciudad Universitaria, Coyoacan 04510, México, D.F.

[a] javr@pumas.iingen.unam.mx

Abstract. The observed damage level in field, due to the 1985 earthquakes, is compared in a 10-story RC building in Mexico City with the analytical predicted behavior without and with the available over-resistance effects. The results were compared to those obtained from the conventional seismic analysis and to the observed damage behavior after the earthquake. Facing the observed behavior in many buildings, the necessity of studying in detail the available over-resistance effects merged, in order to widely explain the seismic-resistance behavior participation of such structures. The over-resistance effects are included: slab steel, deformation hardening period of reinforcement steel, average real stresses of concrete and steel, compression slab participation and concrete confinement. Lineal and non-lineal step-by-step dynamic analysis are made. The soil-structure interaction and the P-Δ effects are included in the analysis. The analytical periods are compared to those experimentally obtained. A very good congruency between the analytically predicted behavior and the observed damage level after the earthquake is obtained. The direction and the stories with maximum damages match with the direction and stories with maximum deformations obtained from the analysis. It is noticed that the structure has a superior lateral resistance capacity regarding to the given in the conventional design.

Introduction

Even the structures behavior located in the soft zone in Mexico City, due to the 1985 earthquakes, are qualified as satisfactory, there were some problems in some of them, specially the buildings between 7 to 17 levels. The structural behavior shows that these structures count with a certain over-resistance range that has been given and that was possibly the cause that a great number of buildings have not collapsed, even though they presented severe damages. Facing the observed behavior in many buildings, the necessity of studying in detail the available over-resistance effects merged, in order to widely explain the seismic-resistance behavior participation of such structures. The inelastic response of a structure that suffered damages in the 1985 earthquake was analyzed in this work in front of the SCT-EW record, 19th September 1985 earthquake. The results were compared to the obtained from the conventional seismic analysis and to the observed damage behavior after the earthquake.

Structure description

The earthquake resistant system was based on frames in the longitudinal direction. The short direction head lines had four shear walls, and the internal lines had only frames (see fig. 1). The foundation was semi-compensated with a 6.425 meters deep rigid box, a foundation beams grid and 22 meters length friction piles. Its construction year was between 1970 and 1971. During the project, the structure was considered type A (important).

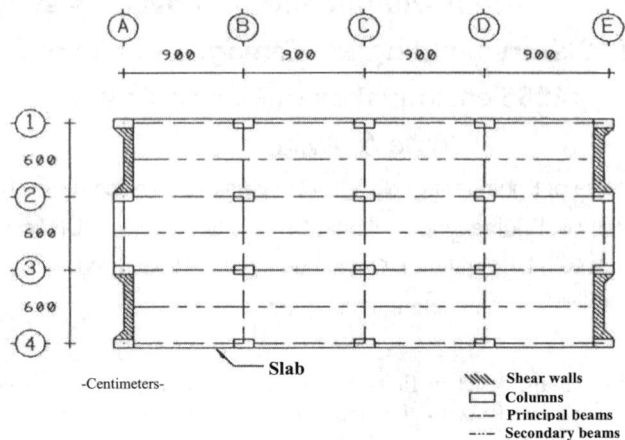

Figure 1: Structural plant-type (dimensions in centimeters)

Damages description. There were only longitudinal direction damages, between ground level and level 6. The evidence of plastic hinges in the frame beam extremes in this direction was evident. Plastic hinges were observed in the base of the columns located in the ground level as well as diagonal fissures in some 3-4 and 5-6 stories columns.

Over-resistance effects. The studied over-resistance sources were: 1) Slab steel (additional to the beam); 2) Hardening effect because of the reinforcement steel strain (EPD); 3) Average real stress in steel and concrete; 4) Slab participation in the beam positive flexural moment resistance; 5) Concrete core confinement. The hardening zone consideration by reinforcement steel strain, was one of the most important; the results in a resistance level show significant increase, according to case 1. This type of comparison is also made in columns..

Vibration periods. The longitudinal direction vibration periods for the fixed base condition in ground level (PB) and in slab foundation level result practically the same, 1.67 and 1.69 seconds, respectively; when including the influence of the soil-structure interaction effects an increase of little more than the 10% was obtained, getting to 1.84 seconds (see table 1). The measured period (2.1 seconds) shows a great flexibility. The difference between can be attributed to the suffered damage in this direction for the lateral stiffness lost, regarding the maximum damage direction. The transversal direction vibration periods variation for the two fixed base types results practically null, with 1.00 second for both conditions. Nevertheless, the period difference between the fixed base condition and the condition in which the soil-structure interaction effects were taken, results significant because of the increase of 30%. Comparing this last and the measured one result nearly the same which is congruent so that this direction does not present damages.

Inelastic responses

Studied cases. The selected cases characteristics, from a total of 21 inelastic step-by-step analyzed cases, were: A (without confinement, EPB model, rectangular beam, V3%, C1.5%); B (without confinement, Takeda model, rectangular beam, V3%, C1.5%); C (with confinement, Takeda model, rectangular beam, V3%, C1.5%); D (with confinement, EPB model, rectangular beam, V3%, C1.5%); E (with confinement, Takeda model, "T" beam, V3%, C1.5%); F (with confinement, EPB model, "T" beam, V3%, C1.5%); G (with confinement, EPB model, rectangular beam, V3%, C1.5%, EP); H (with confinement, Takeda model, rectangular beam, V3%, C1.5%, EP). EPB: elastic-plastic bilinear hysteretic model, V3% and C1.5%: 3% and 1.5% slopes given to the program to take notice the deformation hardening effect in beams and columns, respectively, and EP: reinforcement steel and concrete average real stresses.

Table 1: Vibration periods of longitudinal and transversal directions

Mode	Base condition		
	Fixed in ground level	Fixed in foundation level	Soil-structure interaction
1	1.67	1.69	1.84
2	0.55	0.56	0.63
3	0.33	0.33	0.42

Note: T1 (measured period) = 2.1 seconds

Mode	Base condition		
	Fixed in ground level	Fixed in foundation level	Soil-structure interaction
1	1.00	1.01	1.31
2	0.27	0.27	0.51
3	0.21	0.21	0.26

Note: T1 (measured period) = 1.3 seconds

b) Transversal direction

Curves of base shear force–roof lateral displacement. In this study non-linear static analysis were made until taking the structure in both directions, to its collapse condition, for a determined failure mechanism (see fig. 2). The employed resistances in the analysis were those obtained in cases A, F and G. For the longitudinal direction seismic coefficients of 0.14 and 0.23 were obtained, for cases A and G, respectively. In the step-by-step inelastic analysis, the results were 0.15 and 0.25. Regarding the transverse direction, the over-resistance effects employed were clearly shown, from 0.22 for the nominal case, and 0.30 for the case G. In fig. 3 the curves of base shear force-roof lateral displacement for the case G are compared, obtained from the inelastic dynamic (step-by-step) and static (Pushover) analysis, longitudinal direction.

Local ductility maximum demands in beams and columns. The results for case E present bigger similitude to the physically observed case; case A is presented for being the case in which the conventional criteria for the resistance calculation is supported. Case H resulted very similar to case E. For the three cases, A, E and H, the local ductility maximum demands "μ_L" were also calculated by level, for girders and columns. The developed maximum demands in girders were concentrated in the first level, and in the inferior extremes in columns of ground level; in columns the values that result are small, not very important. The structural element resistances determined in a nominal way resulted to be quite lower of their average real values.

Conclusions

In general, a good congruency between the calculated behavior and the observed damage level after the earthquake. The vibration periods showed that the structure presented a great flexibility in one of its directions. The direction and the stories with maximum damages match with the direction and stories with maximum deformations obtained from the analysis. It is noticed that the structures have a superior lateral resistance capacity regarding to the given in the conventional design; calculating the inelastic seismic responses with the nominal resistances could get us to a greatly overestimated non-linear behavior value, global and locally. The mechanism that tends to be formed in each case, independently of the resistance type, matches with the design philosophy "weak beam-strong column", the most part of plastic hinges are formed in the beam extremes.

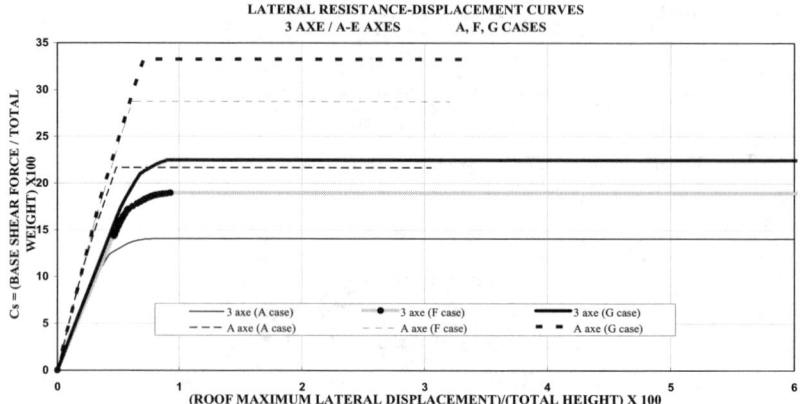

Figure 2:   Curves of base shear force–roof lateral displacement, non-linear static analysis, lines A-E (transversal direction), line 3 (longitudinal direction), cases A, F and G

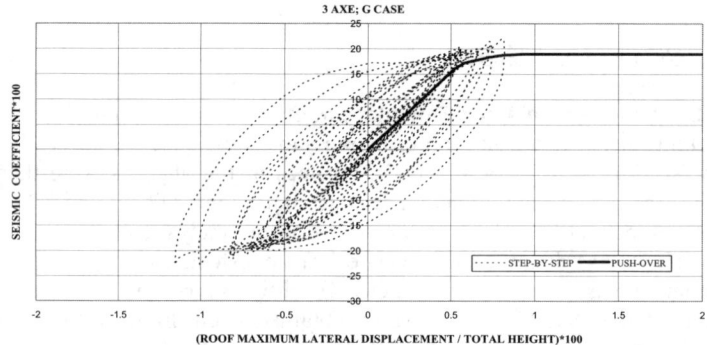

Figure 3:   Curves of shear force-roof lateral displacement, case G, inelastic step-by-step and static (Pushover) analysis, longitudinal direction

References

[1]   Diario Oficial de la Federación, "Reglamento de Construcciones de la ciudad de México y Normas Técnicas Complementarias para Diseño por Sismo, y para Diseño y Construcción de Estructuras de Concreto" (2004).

Accurate description of near-crack-tip fields for the estimation of inelastic zone extent in quasi-brittle materials

Václav Veselý[1,a], Jakub Sobek[1,b], Lucie Šestáková[1,c] and Stanislav Seitl[2,d]

[1]Institute of Structural Mechanics, Faculty of Civil Engineering, Brno University of Technology, Brno, Czech Republic

[2]Institute of Physics of Materials, Academy of Sciences of the Czech Republic, v. v. i., Brno, Czech Republic

[a]vesely.v1@fce.vutbr.cz, [b]sobek.j@fce.vutbr.cz, [c]sestakova.l@fce.vutbr.cz, [d]seitl@ipm.cz

Keywords: crack, near-crack-tip fields, Williams series, higher-order terms, inelastic zone, quasi-brittle fracture.

Abstract. A description of stress and displacement fields by means of the Williams power series using also higher-order terms is the focus of this paper. Coefficients of this series are determined via the over-deterministic method from the results of conventional finite element (FE) analysis. A study is conducted into the selection of the FE node set whose results are processed in this regression technique. Coefficients up to the twelfth term were determined with high precision. The effect of the position of the FE node set on the accuracy of the values of the higher-order term coefficients is reported.

Introduction, motivation

Accurate approximation of the stress field in a cracked body is essential when estimating the extent of the zone of the current onset of failure (part of the entire zone of inelastic material behaviour, the fracture process zone − FPZ) ahead of the tip of a propagating macro-crack in quasi-brittle materials. The method of determining the zone is similar to the technique utilized to evaluate the crack-tip plastic zone (PZ) in the case of e.g. metals [1]. However, in this case this is provided that the proper failure condition is met for quasi-brittle materials, which are in this research represented by cementitious composites commonly used in the building industry. As the zone of inelastic material behaviour is considerably larger (by several orders of magnitude) in the case of quasi-brittle materials [2],[3],[4] than that of metals (due to their much lower tensile strength), the stress field must be considered even at a larger (than usual) distance from the crack tip. So, to obtain an accurate approximation of the stress field in a cracked body at locations beyond the immediate vicinity of the crack tip what is referred to as multi-parameter fracture mechanics is employed [5],[6],[7]. This theory is based on the analytical description of the stress and deformation fields in a body with a crack by means of the Williams power series [8]. It is therefore necessary to take into account a higher number of terms of the power series in order to describe the fields in the more distant surroundings of the crack tip. Coefficients of the higher-order terms of the power expansion are calculated in this work using the over-deterministic method [9] (a regression technique based on least squares formulation) with the results of FE analysis.

The ongoing research by the authors, part of which is presented in this paper, is mainly focused on finding the appropriate level of accuracy for the stress field description (i.e. the relevant number of terms of the Williams series). The aim is that the (current) inelastic zone extent is constructed sufficiently accurately by means of a semi-analytical technique based on the application of a proper failure condition to the stress state over a body reconstructed analytically by the Williams formula (for the calculated values of its coefficients). Note that the technique for estimation of the extent of the zone of the current onset of failure is a subroutine of a procedure currently under development [6] that shall eventually provide the whole extent (shape, size) of the FPZ, i.e. not only the region where the material is currently starting to fail at a certain stage of loading but also the region where the material is already softening. The procedure combines several approaches to material failure modelling; besides the above-mentioned theories of multi-parameter fracture mechanics and plasticity it also employs the classical nonlinear fracture models for concrete.

Accurate approximation of the stress and displacement fields in a cracked body

Multi-parameter fracture mechanics; the Williams series. Multi-parameter fracture mechanics is based on utilization of several initial terms of the Williams power series approximating the stress and displacement fields in a cracked body [8]. Williams' solution for an elastic homogeneous 2D body with a crack provides the following formulas for the stress tensor and displacement vector:

$$\sigma_{ij} = \sum_{n=1}^{\infty} \frac{n}{2} r^{\frac{n}{2}-1} A_n \, {}^{\sigma}\!f_{ij}(n,\theta), \, i,j = \{x,y\}, \qquad u_i = \sum_{n=1}^{\infty} r^{n/2} A_n \, {}^{d}\!f_i(n,\theta,E,\nu), \, i = \{x,y\}. \qquad (1)$$

Here, r and θ are polar coordinates centred at the crack tip (considering the direction of crack propagation along the positive x-axis), ${}^{\sigma}\!f_{ij}(n,\theta)$ and ${}^{d}\!f_i(n,\theta,E,\nu)$ are known functions which can be found in classical textbooks on fracture mechanics (e.g. [1]); E and ν are Young's modulus and Poisson's ratio. Coefficients A_n are functions of relative crack length $\alpha = a/W$, where a is the crack length and W is the specimen's width. They can be expressed as dimensionless functions (with regard to loading) as follows:

$$g_n(\alpha) = \frac{A_n(\alpha)}{\sigma} W^{\frac{n-2}{2}} \quad \text{for } n = 1,3,4\ldots,N \quad g_2 = t(\alpha) = \frac{4A_2(\alpha)}{\sigma}, \qquad (2)$$

where $\sigma = P_{sp}/(BW)$, BW is the cross-sectional area of the specimen and P_{sp} is the splitting component of the loading force (see Fig. 1).

Determination of coefficients of higher-order terms. Coefficients of higher-order terms of the series are determined from the results of FE analysis using a least-squares based regression technique referred to as the over-deterministic method (ODM). Typically, displacements in selected nodes of the FE model (together with their coordinates) serve as inputs to the over-deterministic procedure utilizing Eq. (1). The principle of the method dictates that the number N of terms of the power series must not exceed twice the number k of the FE nodes from which the calculated values of displacements u and v are taken for the regression computation. The ODM is described in detail in [9]; it has also been utilized recently by the authors of this paper [7],[10].

Example – wedge splitting testing of a quasi-brittle specimen

A wedge splitting test (WST) geometry [11] was chosen as an illustrative example since numerous specimen shapes and boundary conditions (both in the areas of load imposition and the specimen's supports) are commonly used in laboratories, see e.g. [6], [12]). Note that this test geometry is mainly used in the field of testing quasi-brittle building materials, especially cementitious composites. In this case, a WST specimen prepared from a standard testing cube (edge length equal to 100 mm) is considered (for detailed dimensions see Fig. 1a). Loading is transferred to the specimen via loading platens and two alternative forms of support are investigated: i) one central support (marked as 1S) and ii) two supports (2S) placed (approximately) under the centres of gravity of the specimen halves. Due to loading through the wedge two components of the loading force (marked as 2F) are introduced (applied on the outer surface of the loading platens).

Determination of dimensionless functions of higher-order terms. For the estimation of the extent of the crack-tip inelastic zone during the fracture process through the ligament of the notched body by the above-mentioned method, the stress field needs to be analytically expressed (with sufficient accuracy over the necessary portion of the specimen's volume). Therefore, coefficients of the terms of the Williams series up to a proper number must be known; moreover, they must be

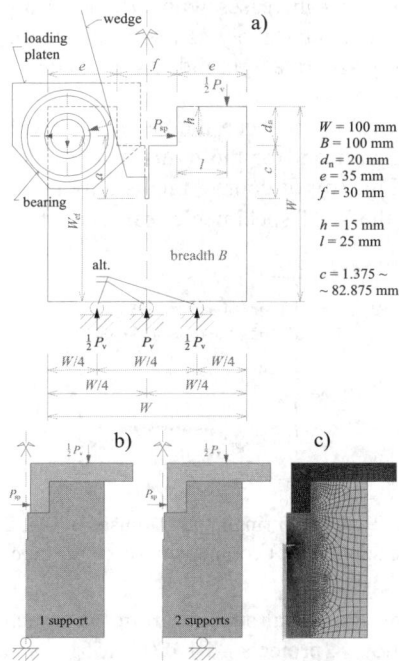

Fig. 1. a) WST geometry, specimen dimensions, decomposition of imposed loading; b) loading scheme of the problem with two types of boundary conditions (1S vs. 2S); c) FE model used in calculations

expressed as the function of the (relative) crack length. It is convenient to use their dimensionless/load-normalized values (without dependence on the load on the cracked specimen).

The absolute and corresponding dimensionless values of the Williams-series-term's coefficients A_n and g_n, respectively, were determined using the ODM for a set of selected relative crack length values, $\alpha = a/W_{ef}$, ranging from very short to very long cracks. A study was conducted which investigated the influence of the number and position of nodes of the FE mesh whose results were used as inputs to the ODM. The considered nodes were taken from a circular ring centred on the crack tip. The study includes three alternative options for the radius R of this ring as well as three alternative options for the angle Θ corresponding to the number of nodes along the ring. The FE calculations were conducted using the ANSYS computational system [13]; the ODM was programmed in MathCad package and its results were processed in MS Excel.

Investigation of the FE nodal set selection – the influence of angular distance Θ. The half-space around the crack tip (when modelling only one of the symmetrical halves of the specimen) was tangentially divided into 10, 16 and 24 equal portions (with angles Θ of 9.0, 5.625, and 3.75 degrees, respectively). This results in an FE mesh which at a specified radius R creates a ring with 21, 33, and 49 FE nodes, respectively. The FE mesh in the crack tip vicinity was created from special (singular) triangular elements (with shifted side nodes); for the rest of the mesh 8-node isoparametric quadrilateral elements were used (see Figs. 1c and 2).

Investigation of nodal set selection – the influence of radial distance R. The radiuses of the ring from which FE nodes were selected for the purpose of saving and then processing the FE analysis results (using ODM), were considered with the values 0.5, 2 and 8 mm, respectively. This resulted in the covering of the interval of the relative crack length values $\alpha \in \langle 0.075, 0.975 \rangle$, $\langle 0.1, 0.95 \rangle$, and $\langle 0.175, 0.9 \rangle$, respectively, with a step of 0.025 considered in the conducted calculations.

Discussion of results and conclusions

Convergence analysis was performed for the results of the ODM. The accuracy of the values of the higher-order term coefficients was evaluated. The results for each required number of terms N of the series (terms up to $N = k$ were considered in this study; here k is the number of FE nodes) were specified by their value for the largest N (i.e. 21, 33 and 49, respectively), as the dependences of the coefficient's values on the number of terms exhibited a stabilizing tendency (reported also in [9],[10]). A filter of 2% was specified for these normalized values.

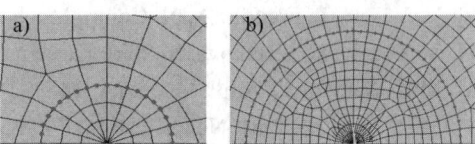

Fig. 2. FE mesh around the crack tip for a variant with: a) ring radius $R = 0.5$ mm with 21 nodes, b) ring radius $R = 2$ mm with 49 nodes

The evaluation of the accuracy of the higher-order term coefficients can also be performed very simply just based on the visual observation of the course of the coefficient's value as the function of the relative crack length. Fig 3a and 3b show gradual deviations from the smoothness of the curves corresponding to the smaller ring radiuses.

It is observed that the accuracy of the values of the coefficients increases with the increasing number of FE nodes in the ring; however, this occurs much more intensively as the radius of the ring increases. This is shown in Fig. 4, where the last sufficiently accurate term index is plotted as a function of both the number of FE nodes in the ring and the ring radius.

It should be also noted that the deviation in the coefficient values between the geometries with one and two supports is weaker than expected. It becomes significant for the terms of the index higher than approx. $n = 10$ in the investigated geometry (see Fig. 3c); of course, this is only in the case of large cracks ($\alpha > 0.8$), i.e. when the crack approaches the WST specimen's rear face where the specimen is supported.

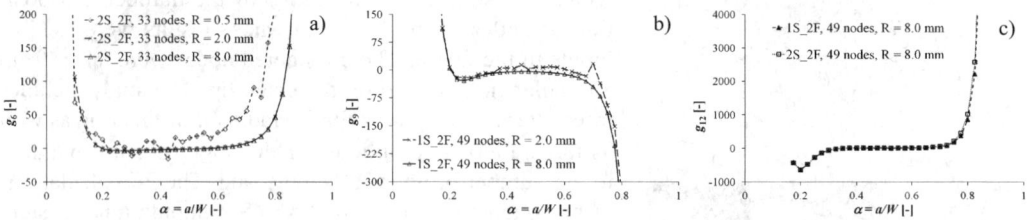

Fig. 3. a) and b) Dimensionless $g_6-\alpha$ and $g_9-\alpha$ functions for 33 and 49 nodes in rings with radiuses of 0.5, 2 and 8 mm and 2 and 8 mm, respectively (a – two supports, b – one support); c) comparison of $g_{12}-\alpha$ functions for WST geometries with one and two supports

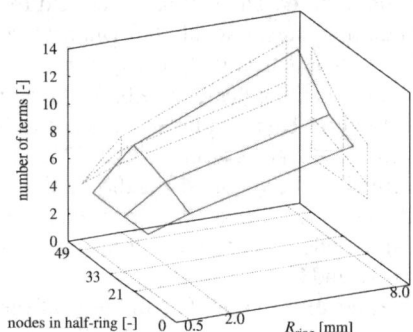

Fig. 4. Index of the last accurate term as a function of the number of nodes and the ring radius

Acknowledgement: Financial support from the Czech Science Foundation (projects P105/11/1551 and P105/12/P417) is gratefully acknowledged.

References

[1] T.L. Anderson: *Fracture mechanics: Fundamentals and applications* (CRC, Boca Raton 2005).

[2] S.P. Shah: Engng. Fract. Mech., 35 (1990), p. 3.

[3] K. Otsuka, H. Date: Engng. Fract. Mech., 65 (2000), p. 111.

[4] J.G.M. van Mier: *Fracture processes of concrete: assessment of material parameters for fracture models* (CRC, Boca Raton 1997).

[5] B.L. Karihaloo, H.M. Abdalla, Q.Z. Xiao: Engng. Fract. Mech., 70 (2003), p. 2407.

[6] V. Veselý, P. Frantík: Appl. Comp. Mech., 4 (2010), p. 237.

[7] V. Veselý, L. Šestáková, S. Seitl: Key Eng. Mat., 488-489 (2012) p. 399

[8] M.L. Williams: J Appl. Mech. (ASME), 24 (1957), p. 109.

[9] M.R. Ayatollahi, M. Nejati: Fatigue Fract. Engng. Mater. Struct., 34 (2011), p. 159.

[10] L. Šestáková: In proc. of *Applied Mechanics 2011*, IPM ASCR Brno, Czech Rep., p. 211

[11] H.N. Linsbauer, E.K. Tscheg: Zement und Beton, 31 (1986), p. 38.

[12] S. Seitl, V., Veselý, L. Řoutil: Computers and Structures 89 (2011), pp. 1852

[13] ANSYS User's manual version 10.0, Swanson Analysis System, Inc., Houston, 2005.

Application of the Mesh Superposition Technique to the Study of Delaminations in Composites Thin Plates.

A. Sellitto[1,a], R. Borrelli[2,b], F. Caputo[1,c], A. Riccio[1,d], F. Scaramuzzino[1,e]

[1]Second University of Naples, Aerospace and Mechanics Engineering Department
Via Roma 29, Aversa (CE) 81031, Italy

[2] Italian Aerospace Research Center (CIRA), Aerostructural Design and Analysis Lab.
Via Maiorise, Capua (CE) 81043, Italy

[a]andrea.sellitto@unina2.it, [b]r.borrelli@cira.it, [c]francesco.caputo@unina2.it,
[d]aniello.riccio@unina2.it, [e]francesco.scaramuzzino@unina2.it

Keywords: Global-Local, Superposition Technique, Davies-Zhang, Delamination, Composite Panel.

Abstract. Laminated composite structures are increasingly finding more applications in various fields thanks to their lower weight if compared with other materials of the same strength. Nevertheless, composites thin plates show a critical behavior in terms of damage propagation mechanisms when subjected to (low velocity) impact. Indeed they tend to produce delaminations which can be hardly detected by optical inspections and can affect the global load carrying capability, leading to a premature structural collapse. The aim of this paper is to assess the capabilities of the Davies- Zhang approach (introduced in 1994 and aimed to the estimation of both the delamination initiation impact load and the size of the impact induced delaminations) by using a multiscale FE model based on the mesh superposition technique. Indeed the impact area has been modeled layer-wise with an element per layer while the rest of the structure has been modeled at laminate level by layered elements by means of a homogenization approach for the determination of the equivalent laminate material properties. The impact induced delamination area has been determined by adopting stress-based criteria. The results (in terms of delamination initiation impact force and delamination size) have been compared to the ones obtained by adopting the Davies-Zhang approach.

Introduction

A lot of research activities on composites have been focused on their impact loading behavior. Traditionally, the nature of impact damages has been investigated through experimental drop tests; nevertheless tests may be misleading if compared with a full scale real structure. Furthermore no manufacturer is willing to perform so many impact tests over a complete aircraft or any of its full scale parts like wing or fuselage; for this reason it would be advisable to employ a predictive strategy for the impact damages.

In 1994 Davies and Zhang [1] developed an approximated formulation for the estimation of both the delamination initiation impact load and the magnitude of the impact induced delaminations, represented, respectively, by equations 1 and 2:

$$P_c^2 = \frac{8\pi^2 E t^3 G_c}{9(1-v^2)} \qquad (1)$$

$$A = \pi r^2 = \frac{9}{16\pi t^2}\left(\frac{P}{\tau}\right)^2 \qquad (2)$$

where P_c is the critical threshold force for instant delamination, E is the flexural modulus of the laminate, t is the thickness, G_c the mode II critical energy release rate, v the poisson's ratio, r the radius of the damaged area A (supposed to be circular) and τ the allowable interlaminar shear stress.

A valid alternative to the Davies-Zhang formulation is the detail numerical analysis of the domain through the Finite Element Method [2]; however, a very fine mesh refinement is required in the impacted area, where delamination will onset. In order to reduce the computational cost without compromises on the accuracy of results, a global/local approach could be adopted: a very refined mesh can be adopted in the critical region, whereas a coarser mesh can be used in the rest of the domain. Several global-local coupling methods can be found in literature [3-7], some of which have been implemented in finite element commercial codes [4-7]. Indeed in [5-7] a global-local approach based on Multi-Point-Constraints (MPS) has been used in conjunction with fracture mechanics approaches to predict the delamination growth in composite stiffened panels. In the present paper, the Superposition Technique has been adopted as global/local method [8, 9].

By the superposition method, it is possible to model the local domain at ply-level. In such a way by means of stress-based criteria it is possible to determine the lamina failure associated to delaminations onset. The adopted stress-based criteria are shown hereafter [10]:

$$\left(\frac{\sigma_z}{Z_T}\right)^2 + \left(\frac{\sigma_{xz}}{S_{xz}}\right)^2 + \left(\frac{\sigma_{yz}}{S_{yz}}\right)^2 \geq 1 \qquad (3)$$

$$\left(\frac{\sigma_z}{Z_C}\right)^2 + \left(\frac{\sigma_{xz}}{S_{xz}}\right)^2 + \left(\frac{\sigma_{yz}}{S_{yz}}\right)^2 \geq 1 \qquad (4)$$

where σ_{ij} are the stress components in each layer and the denominators are the strengths in the relevant directions; equation 3 and equation 4 are applicable, respectively, for delamination in tension ($\sigma_z > 0$) and delamination in compression ($\sigma_z < 0$).

The Superposition Method

The superposition method (also referred to as s-refinement method) is a refinement technique which consists in the creation of locally refined domains over an existing global coarse domain by superimposing independent local models on one or more portions of the global model, without modifying the global model discretization itself.

The total displacement on a domain where the superposition method has been applied is obtained by the sum of two different contributions: the *global* displacements, i.e. the displacements of the global underlying elements, without the influence of the local superimposed elements and the *local* displacements, i.e. the displacements of the local elements superimposed on the underlying global ones. the local displacements are considered as global incremental displacements.

The definitions of total displacement and total strain relative to a single element e are given by the following expressions:

$$\mathbf{u} = \mathbf{u}_G + \mathbf{u}_L = \mathbf{N}_G \mathbf{Z}_G^{(e)} + \mathbf{N}_L \mathbf{Z}_L^{(e)} \qquad (5)$$

$$\boldsymbol{\varepsilon} = \boldsymbol{\varepsilon}_G + \boldsymbol{\varepsilon}_L = \mathbf{B}_G \mathbf{Z}_G^{(e)} + \mathbf{B}_L \mathbf{Z}_L^{(e)} \qquad (6)$$

where $\mathbf{u}$ and $\boldsymbol{\varepsilon}$ are the displacement and the strain vector respectively, $\mathbf{N}$ and $\mathbf{B}$ are the matrix of the shape functions and the matrix of its derivate, respectively. $\mathbf{Z}$ is the vector of nodal DoFs, subscripts G and L represent the quantities of the global model and the local model respectively and superscript *(e)* represents the quantities referring to the single element domain $\Omega^{(e)}$.

Like other displacement-based approach, the governing finite element equation for the single element $\Omega^{(e)}$ can be obtained minimizing the total potential energy functional:

$$\Pi_{TPE}^{(e)}(\mathbf{u}) = \frac{1}{2}\int_{\Omega^{(e)}} \boldsymbol{\varepsilon}^T \mathbf{C}\boldsymbol{\varepsilon}\, d\Omega - \int_{\Omega^{(e)}} \boldsymbol{\varepsilon}^T \mathbf{C}\boldsymbol{\varepsilon}^0\, d\Omega - \int_{\Omega^{(e)}} \mathbf{u}^T \mathbf{F_B}\, d\Omega - \int_{S_\sigma^{(e)}} \mathbf{u}^T \mathbf{T}\, dS - \sum_i \mathbf{u}_i^T \mathbf{R}_i \quad (7)$$

where $\mathbf{C}$ is the elasticity matrix, $\boldsymbol{\varepsilon}^0$ is the initial strain vector, $\mathbf{F_B}$ is the body force vector, $\mathbf{T}$ is the surface traction force vector, $\mathbf{R}_i$ are the external concentrated load vectors and $S_\sigma^{(e)}$ represents the surface where traction forces are applied. Furthermore, $\mathbf{u}$ is the displacement vector, $\mathbf{u}_i$ are the displacement vectors corresponding to the external load vectors [9]. If equations 5 and 6 are substituted in equation 7 and the first derivatives with respect to the nodal degrees of freedom $\mathbf{Z}_G^{(e)}$ and $\mathbf{Z}_L^{(e)}$ are set to zero, the resulting expression can be written adopting a matrix formulation:

$$\begin{bmatrix} \mathbf{K}_G^{(e)} & \mathbf{K}_C^{(e)} \\ \mathbf{K}_C^{(e)T} & \mathbf{K}_L^{(e)} \end{bmatrix} \begin{Bmatrix} \mathbf{Z}_G^{(e)} \\ \mathbf{Z}_L^{(e)} \end{Bmatrix} = \begin{Bmatrix} \mathbf{Q}_G^{(e)} \\ \mathbf{Q}_L^{(e)} \end{Bmatrix} \quad (8)$$

where the matrices $\mathbf{K}_G$, $\mathbf{K}_L$ are the stiffness matrices corresponding to the global and local DoFs and $\mathbf{K}_C$ is the coupling stiffness matrix between the two domains. The vectors $\mathbf{Q}_G$ and $\mathbf{Q}_L$ are the force vectors applied to the global and local DoFs respectively.

An Application: Composite Panel Subjected to Transversal Load

The superposition has been applied to an 8-layer 0° plies panel manufactured from a current aerospace material (Hexcel T800/924) and subjected to a transversal load, in order to study the behavior of the panel at ply-level. The geometrical and material data are listed in Table 1.

Table 1. Geometrical and Material Panel's Data

Geometry			Material							
Length [mm]	Width [mm]	t [mm]	E_1 [GPa]	E_2 [GPa]	v_{12}	G_{12} [GPa]	Z_T [MPa]	Z_C [MPa]	τ [MPa]	G_{IIC} [Jm^{-2}]
360	360	1	155.21	8.57	0.36	7.40	50	-250	115	575

All the external sides have been clamped, while a concentrate force equal to the delamination initiation impact load of equation 1 has been applied to the centre of the panel. The prediction of the delaminated area is obtained from the equation 2; this area is compared to the one obtained from the finite element model. To correctly compute the damaged area in the finite element model, the criteria in equation 3 and 4 should be applied on each ply: for this reason a discretization at ply-level is necessary. However such a uniform refined discretization is computationally expensive; therefore two different models have been employed over the panel: a global coarser model and a local more refined one that has been superimposed to the global one. The panel has been discretized in its global domain with one element in the thickness direction; a critical region, identified with the center of the plate, has been superimposed with a local mesh discretized with 8 elements (one element per layer) in the thickness direction and 20 elements in the length and width directions. Figure 1 shows the location of the local domain in the global one (left), and the interface between the global and the local domains. Figure 2 shows the magnitude of the delamination and a comparison between the delaminated area, computed numerically by adopting equation 3 and equation 4, and the one predicted by the Davies-Zhang theory. The gray-scale depends on the number of layers that have experienced a failure due to delamination in that location.

Referring to Figure 2, the surface that experiences at least one layer delaminated is wide 7.75 mm^2, while the surface that experiences all its 8 layers delaminated is wide 1.5 mm^2: the area predicted by Davies-Zhang is wide 4.5 mm^2 and is perfectly between these values.

As a conclusion, the capabilities of the Davies-Zhang approach have been well assessed by using a multiscale FEM model based on the mesh superposition technique.

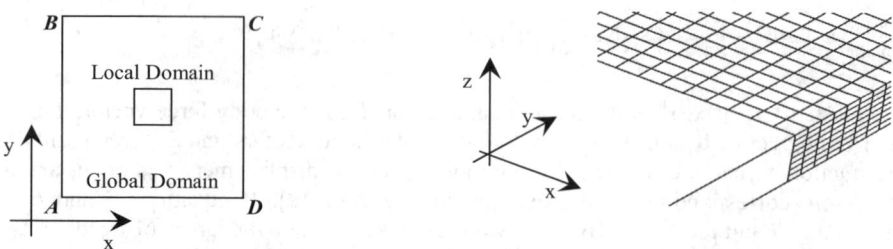

Figure 1. Global and Local Domains.

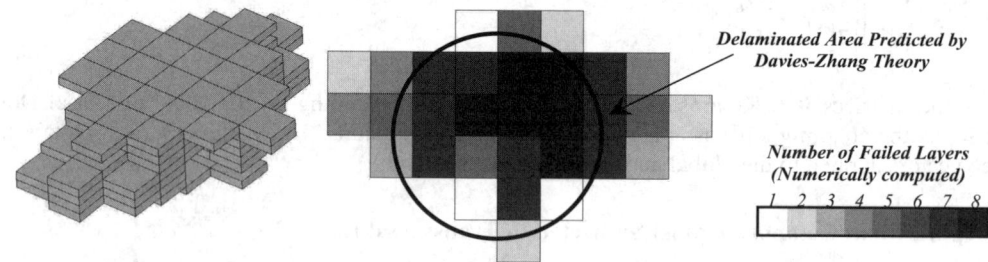

Figure 2. Delamination (left); comparison between computed and predicted delamination (right).

References

[1] G. A. O. Davies, X. Zhang: *Impact damage prediction in carbon composite structures*, International Journal of Impact Engineering Vol. 16, No. 1, pp. 149-170, 1995.

[2] E. Armentani, F. Caputo, R. Esposito, G. Godono, *Evaluation of energy release rate for delamination defects at the skin/stringer interface of a stiffened composite panel*, Engineering Fracture Mechanics, Vol. 71/4-6, pp 885-895, 2004.

[3] H.-G. Kim: *Interface element method: Treatment of non-matching nodes at the ends of interfaces between partitioned domains*, Comput. Methods Appl. Mech. Engrg. 192 (2003) 1841-1858.

[4] H. Alesi, V. M. Nguyen N. Mileshkin: *Global/Local Postbuckling Failure Analysis of Composite Stringer/Skin Panels*, AIAA Journal Vol. 36, No. 9, September 1998.

[5] Pietropaoli, E. , Riccio, A.: *A global/local finite element approach for predicting interlaminar and intralaminar damage evolution in composite stiffened panels under compressive load*, Applied Composite Materials 18 (2011) 113-125

[6] Pietropaoli, E. , Riccio, A.: *Formulation and assessment of an enhanced finite element procedure for the analysis of delamination growth phenomena in composite structures*, Composites Science and Technology, 71 (2011) 836-846.

[7] Riccio, A. , Giordano, M., Zarrelli, M.: A linear numerical approach to simulate the delamination growth initiation in stiffened composite panels, Jou. Comp. Mat. 44 (2010).

[8] J. Fish: *The s-version of the finite element method*, Computers & Structures Vol. 43, No. 3, pp. 539-547, 1992.

[9] J. W. Park, J. W. Hwang, Y. H. Kim: *Efficient finite element analysis using mesh superposition technique*, Finite Elements in Analysis and Design 39 (2003) 619-638.

[10] M. M. Shokrieh, L. B. Lessard and C. Poon: *Tree-Dimensional Progressive Failure Analysis of Pin/Bolt Loaded Composite Laminates*, Paper presented at the 83rd Meeting of the AGARD SMP on "Bolted/Bonded Joints in Polymeric Composites", held in Florence, Italy, 2-3 September 1996, and published in CP-590.

Deterministic and Probabilistic Earthquake Scenarios for the Seismic Risk Analysis of URM Buildings

Avila-Haro J. A.[1,a], González-Drigo J.R.[1,b], Vargas Y. F.[2,c], Pujades L. G.[2,d] and Barbat A. H.[1,e]

[1]RMEE Technical University of Catalonia, Barcelona, Spain

[2]ETCG Technical University of Catalonia, Barcelona, Spain

[a]jorge.avila-haro@upc.edu, [b]jose.ramon.gonzalez@upc.edu, [c]yeudy.felipe.vargas@upc.edu, [d]lluis.pujades@upc.edu, [e]alex.barbat@upc.edu

Keywords: unreinforced masonry, deterministic spectra, probabilistic spectra, risk analysis.

Abstract. Barcelona, as well as a large number of cities in the Mediterranean basin, has a housing stock composed of a large number of unreinforced brick masonry buildings. Motivated by different factors, the enlargement of the city (Eixample in Catalan) was held from the second half of the 19th century and the beginning of the 20th, a period in which a large number of buildings of this type were built, many of which are still used as dwellings.

Although the buildings were built individually, some of them are linked to adjacent buildings by the side walls. This feature leads to the analysis of the buildings as isolated structures and also as an aggregate.

Barcelona is located in a seismic region of low to moderate hazard, with macroseismic intensity between the grades VI and VII of the European macroseismic scale EMS'98. Based on the deterministic and probabilistic response spectra for the different types of soils present in Barcelona obtained in the work of Irizarry (2004), the seismic risk of four individual buildings and an aggregate is evaluated. The buildings are modeled and analyzed using the TREMURI program and MATLAB routines under the guidance of RISK-UE project.

Introduction

The modern cities of the Mediterranean basin accumulate a large stock of infrastructures, resulting in a concentration of important socioeconomic value. The construction period, typology, materials and code levels in an important number of buildings of this stock, lead us to evaluate the seismic hazard and vulnerability of some of them.

In the case of Barcelona city, originally founded by Romans between the 15 to 10 years B.C., the enlargement of the city (example in Catalan) took place during the second half of the 19th century and the beginning of the 20th century. During this period of expansion, most of the buildings were constructed with unreinforced brick masonry and without any seismic consideration. Nowadays, many of those ancient buildings are still used as dwellings, largely surpassing their initially supposed service life.

Sometimes, the constructive habits of the time allowed the construction of the buildings not only as isolated structures but also as buildings assembled to each other as aggregates, constituting the so called blocks.

Following the guidance of the RISK-UE project [1], the aim of this work is to assess the seismic risk and vulnerability of the isolated and aggregated buildings. The structural analysis is performed with the TREMURI [2] program and some MATLAB (Matlab, v.2009b, The Mathworks), routines that allow us to obtain the different results concerning fragility and damage.

Based on the work and results obtained by Irizarry (2004), we have been able to analyze the buildings for a deterministic and a probabilistic earthquake scenario for the corresponding type of soil where these buildings are located.

The buildings

The studied buildings correspond to a block section located in Muntaner Street in the city of Barcelona, Spain. The four isolated buildings correspond to the numbers 153, 155, 157 and 159, and the aggregate corresponds to the buildings with the number 153 and 155, whom happen to be identical (Fig. 1).

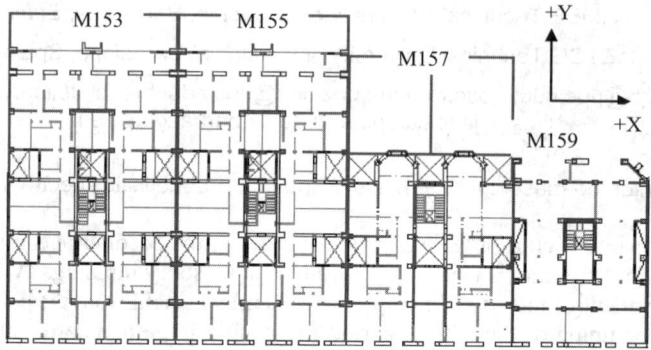

Fig. 1 Floor plan of the analyzed buildings

The buildings are quite representative of the typology of buildings existent in the Eixample district, presenting bearing walls as main resistant elements and, in some cases in the base levels, masonry or cast iron columns. The evaluated structures have 7 levels.

The façade walls present thickness from 45 or 60 cm in the base level, to 30 cm in the upper levels. In the case of the side walls the thickness of the first level is 30 cm, being reduced to 15 cm for the consequent levels. The internal walls tend to present the same thickness than the side walls in each level including the staircases core walls (Fig. 2). There are some distribution walls which structural function is neglected due to their low thickness of 10 cm or lower.

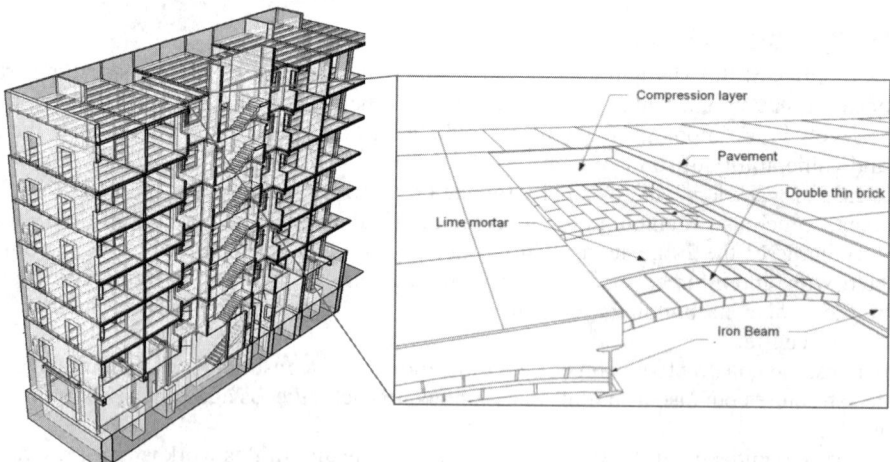

Fig. 2 Section cut of one of M153 building and its floor system

We can observe in the different type of walls the presence of big openings that work as doorways or windows, which loads are discharged via iron or wood lintels or via arches. In the case of the floors, for this particular construction period it is usual to find one-way timber floors, wood beams and brick vaults or, like in our buildings, iron beams and brick vaults (Fig. 2). The size and separation of the beams depends on the total distance that they cover, usually having separations between 60 cm and 120 cm.

The demand

Barcelona is located in a low-to-moderate seismic region. The data related to the demand used in this work includes the 5 percent-damped demand spectra proposed in the RISK-UE project, originally deduced by the University of Genoa [3]. The different soil types present in the city and the main features for the acceleration response spectra for the deterministic and probabilistic earthquake scenarios were obtained from the previous studies performed by Cid (1998) and Irizarry (2004). The analyzed buildings are located in the Eixample neiborghood, wich is placed in the soil zone II.

Soil Zone	Parameters							Scenario
	pga [g]	T_B	T_C	B_C	d	T_D	B_D	
II	0.194	0.10	0.22	2.45	1.43	2.20	0.09	Deterministic
	0.141	0.10	0.23	2.50	1.28	2.21	0.14	Probabilistic

Table 1. Parameters for the deterministic and probabilistic scenarios [4]

The capacity and fragility

The procedure followed in order to assess the capacity and fragility is based on the capacity spectrum method, which provides a graphical representation of the structure's force-displacement capacity curve and compares it to the response spectra representations of the seismic demand [5, 6].

The RISK-UE guidelines [1] define an undamaged state (NO) and four damage states, d_{si}, which are: Slight (SL), Moderate (MO), Severe (SE) and Collapse (CO). For the purpose of this work, the chosen parameter in order to define the seismic action will be the spectral displacement, S_d, and the damage states are defined from the parameters of the bilinear representation of the capacity spectrum, as follows: $\overline{ds}_1 = 0.7D_y$; $\overline{ds}_2 = D_y$; $\overline{ds}_3 = D_y + 0.25(D_u - D_y)$; and $\overline{ds}_4 = D_u$, where, D_y is the displacement at the yielding point and D_u is the displacement at the ultimate point.

It is assumed that the fragility curves will follow a lognormal cumulative probability function and that the seismic damage of the buildings will follow a binomial probability distribution (or an equivalent beta distribution). It is also assumed that for each damage state, the probability of reaching or exceeding that particular damage state will be 50%.

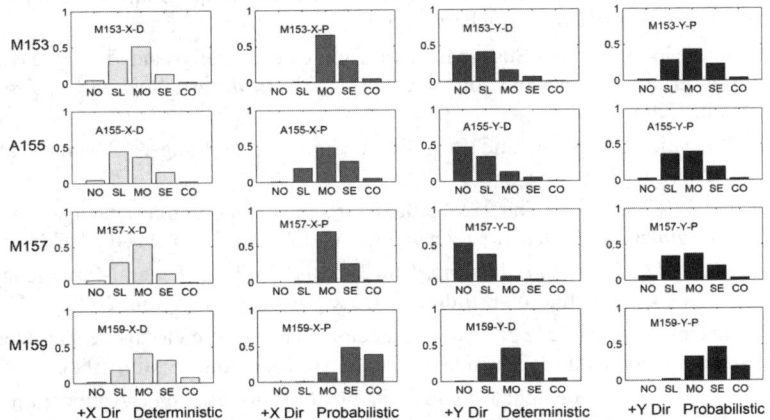

Fig. 3 Damage states distributions for each analyzed building, direction and scenario

Following the mentioned methodology and assumptions we can observe in Fig. 3 the damage states distributions for each building it its two directions, +X and +Y, and for both seismic scenarios, deterministic and probabilistic.

Discussion and Conclusions

For both directions, the probabilistic scenario presents higher damage values and worst performance than the deterministic scenario, in consistency with the corresponding *pga* values. For both scenarios, we can observe a better performance in the +Y direction rather than in the +X direction (short direction).

In the case of the aggregate buildings, and comparing the results to those obtained for the isolated building M153, we can observe a significant improvement especially in the +Y direction, suggesting a positive effect for this construction habit.

For the rectangular plan buildings (M153 and M157) we can notice that the damage state probabilities, in the +X direction, are centered between the slight and moderate states for the deterministic scenario, and between moderate and sever for the probabilistic scenario, meanwhile for +Y direction the damage is centered between the no damage and the slight states, for the deterministic scenario, and between slight and moderate for the probabilistic scenario. Buildings with a more squared floor plan (M159) show closer values of fragility.

The results indicates that the expected damage associated to these unreinforced masonry buildings of Barcelona is predominantly moderate

Acknowledgements

This work was partially funded by the Spanish Science and Innovation Ministry, by the European Commission with FEDER funding through the investigation projects: CGL2008-00869/BTE, SEDURECCONSOLIDER: CSD2006-00060, INTERREG: POCTEFA 2007-2013/73/08 and MOVE-FT7-ENV-2007-1-211590.

References

[1] Z. V. Milutinovic and G. S. Trendafiloski, "WP4: Vulnerability of Current Buildings," in *RISK-UE Project Handbook*, 2003, p. 111.

[2] S. Lagomarsino, A. Galasco, A. Penna and S. Cattari, *TREMURI: Seismic Analysis Program for 3D Masonry Buildings (User Guide)*, Technnical Report ed., U. o. Genoa, Ed., Genoa, Italy, 2008.

[3] S. Lagomarsino, A. Galasco and A. Penna, "Pushover and dynamic analysis of URM buildings by means of a non-linear macro-element model," in *International Conference on Earthquake Loss Estimation and Risk Reduction*, Bucharest, 2002.

[4] J. Irizarry, X. Goula and T. Susagna, "Evaluación de la peligrosidad sísmica de la ciudad de Barcelona en términos de aceleración espectral," in *2o Congreso Nacional de Ingeniería Sísmica*, Málaga, Spain, 2003.

[5] ATC-40, "Seismic Evaluation and Retrofit of Concrete Buildings," Seismic Safety Comission, Redwood City, CA, 1996.

[6] P. Fajfar and P. Gaspersic, "The N2 Method for the seismic damage analysis of RC buildings," *Earthquake Engineering & Structural Dynamics*, vol. 25, no. 1, pp. 31-46, 1998.

[7] FEMA/NIBS, *HAZUS Technical Manual SR2*, Vols. 1, 2, 3, Federal-Emergency-Management-Agency and National-Institute-of-Building-Sciences, Eds., Washington, 2002.

[8] J. Cid, "Zonación Sísmica de la ciudad de Barcelona basada en métodos de simulación numérica de efectos locales," Universitat Politècnica de Catalunya, Barcelona, Spain, 1998.

[9] J. Irizarry, "An Advanced Approach to Seismic Risk Assessment. Application to the Cultural Heritage and the Urban System Barcelona," Universidad Politécnica de Cataluña, Barcelona, Spain, 2004.

[10] A. J. Kappos, "Seismic damage indices for RC buildings: evaluation of concepts and procedures," *Progress in Structural Engineering and Materials*, vol. 1, no. 1, pp. 78-87, 1997.

Micromechanics damage analysis in fiber-reinforced composite material using finite element method

Chayun Kimyong[1,2,a], Sontipee Aimmanee[1,b], Vitoon Uthaisangsuk[1,c] and Wishsanuruk Wechsatol[1,d]

[1]Department of Mechanical Engineering, King Mongkut's University of Technology Thonburi, Bang Mod, Thung Khru, Bangkok 10140, Thailand

[2]Faculty of Science and Engineering, Kasetsart University Chalermphrakiat Sakon Nakhon Province Campus, Sakon Nakhon 47000, Thailand

[a]pom_kku@yahoo.com, [b]sontipee.aim@kmutt.ac.th, [c]vitoon.uth@kmutt.ac.th, [d]wishsanuruk.wec@kmutt.ac.th

Keywords: fiber-reinforced composite; FE analysis; damage; failure.

Abstract. Fiber-reinforced composite materials (FRC) are used in a wide range of applications, since FRC exhibits higher strength-to-density ratio in comparison to traditional materials due to long fibers embedded in a matrix material. Failures occurred in FRC components are complicated because of the interaction of the constituents. The aim of this study is to investigate damage behavior in a unidirectional glass fiber-reinforced epoxy on both macro- and micro-levels by using finite element method. The Hashin's criterion was applied to define the onset of macroscopic damage. The progression of the macroscopic damage was described using the Matzenmiller-Lubliner-Taylor model that was based on fracture energy dissipation of material. To examine the microscopic failure FE representative volume elements consisting of the glass fibers surrounded by epoxy matrix with defined volume fraction was considered. Elastic-brittle isotropic behaviour and the Coulomb-Mohr criterion were applied for both fiber and epoxy. The results of the macroscopic and microscopic analyses were correlated. As a result, damage initiation and damage development for the investigated FRC could be predicted.

Introduction

Failures occurred in FRC components are mostly due to three principle mechanisms, micro-cracking in matrix, debonding of interface between fiber and matrix, and delamination [1,2]. These failure processes are the important key factor for designing and developing advanced FRC. Several failure criterion models have been introduced for the FRC material. Two well-known failure criteria for FRC are the Tsai-Wu and Hashin models. However they are different from each other. The Tsai-Wu criterion [3] presents a single mode of failure in the average sense like Von Mises criterion. On the other hand, the Hashin criterion [1] considers failure in FRC along the fiber and matrix directions separately. Each mode is examined either under tensile or compressive load. Another more involved anisotropic failure criterion was developed by Christensen [4]. This model is similar to the Hashin's criteria, but there is no limitation of a three dimensional calculation. Additionally, an orthotropic damage model was proposed in [5] for predicting failure initiation in FRC by Hashin's criterion and post-failure response of the brittle fiber-reinforced material was calculated with the Matzenmiller-Lubliner-Taylor (MLT) model. These failure and damage studies are basically used to describe failure development on lamina level, which only refers to the behavior of FRC at the macro-scale.

To be able to understand the strength and failure of FRC better, failure and damage behavior in a unidirectional glass fiber-reinforced epoxy was investigated on both macroscopic and microscopic levels by using finite element method (FEM) analysis. Numerical tensile tests of composite laminated plate with a hole at the middle were carried out. On the macroscopic level, the Hashin's criterion was applied to determine the onset of damage as a function of the effective stress space. The progression of the macroscopic damage was described using the MLT model based on fracture energy dissipation of material during the damage process. The damage propagation was governed by the equivalent displacement with close relation to loading condition. Depending on damage parameters of the MLT model the degradation of the elastic moduli of fiber and matrix could be calculated. To examine the microscopic failure behavior FE representative volume elements (RVEs) containing glass fibers surrounded by epoxy matrix with a defined volume fraction was used. At this level, the fiber and epoxy are assumed to be isotropic and brittle materials. The Coulomb-Mohr criterion [6,7] was then applied to both constituents in order to characterize the local damage initiation. A correlation between the results of the macroscopic and microscopic FE simulations was done. As a result, damage development in the investigated FRC could be predicted. This study provides a better understanding of the failure mechanism in the FRC that is necessary for further designing and optimizing a novel composite material.

Finite Element Modeling

The composite material investigated in this study was a symmetric cross-ply laminate of three layers of unidirectional E-glass (21xK43 Gevetex)-epoxy, [0°/90°/0°]. Each layer had a thickness of 0.4 mm. Material properties data used in the finite element simulation were taken from experimental testing in [8] and tabulated in Table 1. The numerical tensile tests of a composite plate including a circular hole in the middle were performed. The hole was intended to localize strain and damage evolution to be in the vicinity of the hole area. The uniaxial tensile load was applied in the x-direction. An eighth of the composite laminate was modeled with symmetry on the planes x-y and y-z. The dimension of the sample and used boundary conditions were illustrated in Fig. 1. RVE models were generated along a half of the entire thickness of the macroscopic model (a full layer of 0° ply and a half layer of 90°-ply as illustrated in Fig. 2c) at several locations close to the hole area, where damage onset and propagation was expected to occurred initially, as designated in Fig. 1b. The 62% fiber volume fraction that is reported in [8] was also defined in the RVE for the microscopic model. The macroscopic deformation tensors calculated from the chosen locations were applied as the boundary conditions of the RVE simulations.

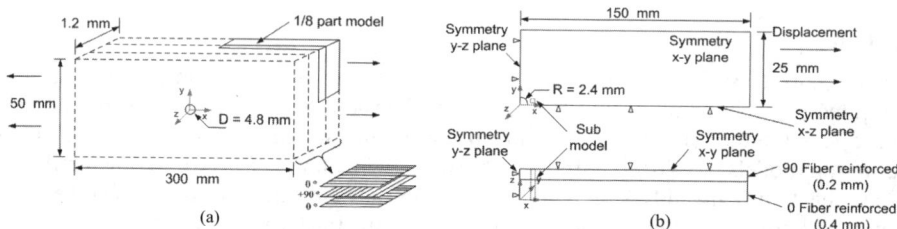

Figure 1 (a) Dimension of composite specimen and (b) 1/8 model of the sample and applied boundary conditions.

Table 1 Effective material properties of E-glass (21xK43 Gevetex) Epoxy.

E_1 (GPa)	E_2 (GPa)	G_{12} (GPa)	v_{12}	X_T (MPa)	X_C (MPa)	Y_T (MPa)	Y_C (MPa)	S_{12} (MPa)	V_f	G_{IC} (J m^{-2})
53.5	17.7	5.8	0.78	1140	570	35	114	72	0.62	165

Damage Criteria

On the macro-level, failure initiation in the FRC was described by the Hashin's criterion. In table 1, the quantities E_1, E_2, G_{12}, and v_{12} are the longitudinal modulus, transverse modulus, in-plane shear modulus, and major Poisson's ratio of the composite laminate, respectively. The quantities X_T, X_C, Y_T, Y_C, and S_{12} are the longitudinal tensile strength, longitudinal compressive strength, transverse tensile strength, and transverse compressive strength, and in-plane shear strength of the FRC, respectively. G_{IC} is the strain energy release rate which was used as the effective fracture energy of the FRC. Here, strength of FRC was considered in four modes: fiber tension (F_{ft}), fiber compression (F_{fc}), matrix tension (F_{mt}), and matrix compression (F_{mc}). When these values reached the value of 1 in Hashin criterion, damage started to initiate in the corresponding element. Subsequently, material degradation took place and the corresponding moduli of the damaged element decreased. By continuous loading damage evolves in accordance with the increasing parameters d_{ft}, d_{fc}, d_{mt}, and d_{mc} in the same sense as the parameters defined for the failure criterion. With increasing d value material stiffness became less. A completed failure occurred when d reached the value of 1.

On the micro-level of the FRC, failure in fiber and matrix could be separately calculated using the RVE model. Material properties were defined individually for the fiber and epoxy. The Coulomb-Mohr (CM) criterion was used to describe failure developed in the fiber and matrix. The material modulus instantly became zero when the CM criterion was reached. By this manner, local stress-strain responses on the fiber and epoxy matrix could be investigated. The individual material properties for the investigated fiber and epoxy taken from [8] are given in Table 2. The indices f and m stand for fiber and matrix, respectively.

Table 2 Individual material properties for glass fiber and epoxy.

E_{f1} (GPa)	G_{f12} (GPa)	X_{fT} (GPa)	X_{fC} (GPa)	E_m (GPa)	G_m (GPa)	Y_{mT} (MPa)	Y_{mC} (MPa)
80	33.33	2.15	1.45	3.35	1.24	80	120

Results

In the macroscopic calculations, the damage principally occurred in matrix rather than in fiber as the strength of the fiber is higher. In the 0° layer stress development in matrix was not as intense as that in the 90°-fiber layer. At the point near the hole the where the first failure took place, the failure in mode F_{mt} propagated rapidly in the epoxy matrix in the 90°-fiber layer. The progression of the damage initiation and damage evolution in matrix tension mode of a critical element from both layers were plotted with the applied displacement in Fig. 2a. RVE simulation was carried out at the same local area of the macro-simulation. The effective damage of the entire RVE could be considered as an increasing of the tensile equivalent plastic strain on the tension cutoff or when the PEEQT value deviated from zero as shown in Fig 2b. The overall load-displacement curve was presented in Fig. 2d. Obviously, slope of the curve was decreased around the displacement of 0.32 mm that could be interpreted as a overall failure occurrence. The damage onset calculated from the RVE simulations well correlated with this failure point. However, this point could be referred to the point where the damage evolution parameters first became a constant. The damage distributions in macro- and micro-model shortly after the failure occurrence were illustrated in Fig. 2c. It can be seen that failure mostly propagated along the matrix area located between each fiber.

Summary

In this study, the failure and damage in macro- and micro-level of E-glass (21xK43 Gevetex) Epoxy laminate with a hole were investigated by FEM. The matrix failure in tensile mode was initially observed at a point near the hole circumference in the 90° layer and the damage propagated rapidly from this point when the displacement was slightly increased. The macroscopic analysis showed the significant overall damage in the force-displacement relations when the specimen was elongated about 0.3 mm. This correlates fairly well with the failure in the microscopic investigation, from which the equivalent plastic strain becomes nonzero around 0.24 mm.

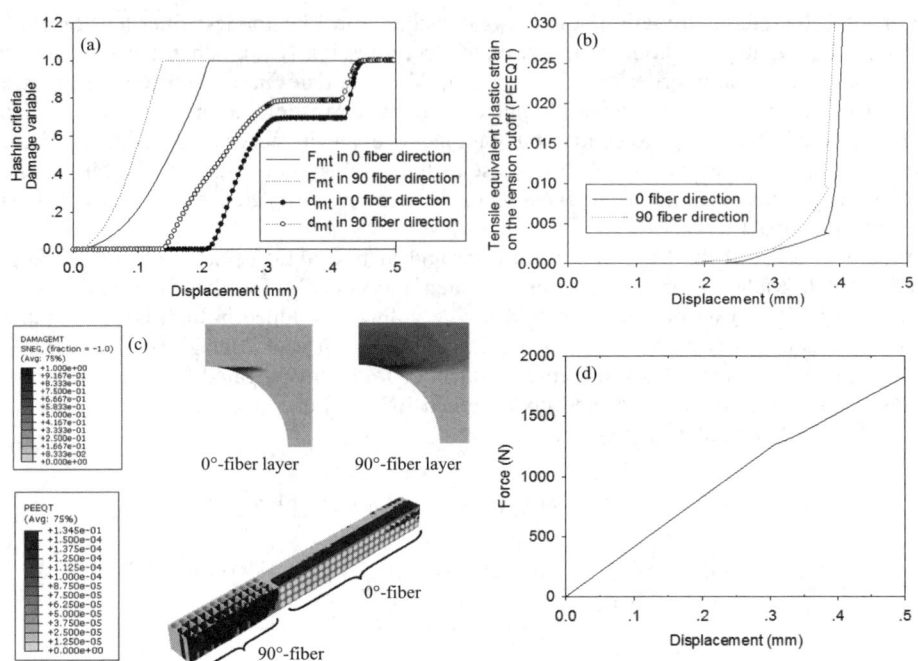

Figure 2 (a) Macroscopic damage initiation and evolution in tensile matrix mode (F_{mt} and d_{mt}), (b) effective damage development in RVE, (c) damage distribution in macro- and micro-model at the displacement of 0.35 mm, and (d) load-displacement curve.

References

[1] Davila, C.G. and Camanho, P.P., 2003, "Failure criteria for FRP laminates in plane stress", Tech. Rep. NASA/TM-2003-212663.

[2] Matzenmiller, A., Lubliner, J. and Taylor, R.L., 1994, "A constitutive model for anisotropic damage in fiber-composites", Mechanics of Materials, vol.208, pp.125-152.

[3] Tsai, S. W. and Wu, E. M., 1971, "A General Theory of Strength for Anisotropic Materials," J. Comp. Mater. 5, 58-80.

[4] Christensen, R. M., 1997, "Stress Based Yield/Failure Criteria for Fiber Composites," Int. J. Solids Structures, 34, 529-543; see also J. Engr. Mats and Technology, 1998, 120, 110-113.

[5] Ireneusz, L. and Hurtado, J.A., 2007, "Progressive damage modeling in fiber-reined materials", Composite Part A, vol.38, pp.2333-2341.

[6] Hoek, E., 1990, "Estimating Mohr-Coulomb Friction and Cohesion Values from the Hoek-Brown Failure Criterion". International Journal of Rock Mechanics and Mining Sciences, Vol.27, pp. 227-229.

[7] Boming Zhang., Zhong Yang., Xinyang Sun and Zhanwen Tang, "A virtual experimental approach to estimate composite mechanical properties: Modeling with an explicit finite element method", Composite Materials Science, 2010, vol.49, pp. 645-651.

[8] Soden, P. D., Hinton, M. J. and Kaddour, A. S., 1998, "Lamina properties, lay-up configurations and loading conditions for a range of fibre-reinforced composite laminates", Composite Science and Technology, vol.58, pp.1011-1022.

On the direction of a crack initiated from an orthotropic bi-material notch composed of materials with non-uniform fracture mechanics properties

Tomáš Profant[1,a], Jan Klusák [2,b], Oldřich Ševeček[1,c] and Michal Kotoul[1,d]

[1]Brno University of Technology, Technická 2896/2, 616 69 Brno, Czech Republic

[2]Institute of Physics of Materials, Academy of Sciences of the Czech Republic, Žižkova 22, 616 62, Brno, Czech Republic

[a]profant@fme.vutbr.cz, [b]klusak@ipm.cz, [c]sevecek@fme.vutbr.cz, [d]kotoul@fme.vutbr.cz

Keywords: Orthotropy, bi-material, notch, generalized stress intensity factor, complex potentials, initiated crack, two-state integral.

Abstract. The assessment of conditions of crack initiation in a tip of a bi-material notch composed of two orthotropic materials is dealt. The assessment of the bi-material orthotropic notch stability criteria based on standard linear elastic fracture mechanics can lead to incorrect results due to a change of fracture mechanics properties. The change of the fracture mechanics properties are taken into account in the discussed stability criterion. It is shown that the criterion of this kind can qualitatively and quantitatively influence the results, and it contributes to more reliable assessment of components with geometrical and/or material discontinuity.

Introduction

Components of modern engineering constructions are often composed of materials with different mechanical properties. Parts of this kind allow the properties to reach the standard that cannot be gained by means of homogeneous materials. On the other hand, the mismatch of elastic properties of particular material parts leads to additional stress concentrations, thus the bi-material interfaces can be responsible for the final failure of the whole construction. These stress concentrations may also be accelerated by the geometry of the interface, such as the notches. Under the assumption of the linear elastic fracture mechanics, the paper deals with the assessment of conditions of crack initiation in a tip of a bi-material notch composed of two orthotropic materials.

In cases of bi-material notches the absolute value of the singular stress distribution is expressed by means of generalized stress intensity factor H [MPa m$^{1-\delta}$]. Due to its unit depending on the stress singularity exponent $1 - \delta$, it is not possible to compare the value of H with the fracture toughness K_{IC}, which is material characteristic. The criteria for crack initiation direction and critical loading, under which a crack will initiate, are based on finding the extreme of a controlling quantity and on its comparison with critical value depending on K_{IC}, see [1]. In case of material orthotropy the fracture mechanics properties change with direction of supposed crack initiation direction, thus the change of fracture toughness is taken into account.

The stress distribution around orthotropic bi-material notch

The necessary step for the crack initiation assessment is detailed knowledge of the stress distribution near the stress singularity. Within determination of the stress near the orthotropic bi-material notch tip, the Lekhnitskii-Eshelby-Stroh (LES) formalism is used, [2]. The complex potentials satisfying the equilibrium and the compatibility conditions as well as the linear stress-strain dependence and given boundary conditions are the basis for the determination of stress and deformation fields. According to the LES theory for an orthotropic material, the relations for deformations and stresses can be written as follows

$$u_i = 2\Re\{A_{ij}f_j(z_j)\}, \sigma_{2i} = 2\Re\{L_{ij}f'_j(z_j)\}, \sigma_{1i} = -2\Re\{L_{ij}\mu_j f'_j(z_j)\}, \tag{1}$$

where $\Re$ denotes the real part, μ_j are the eigenvalues of the material, and matrix elements A_{ij} and L_{ij} depend on material stiffness's and μ_j, [3, 4]. In the case of the studied notch, the potential $f_j(z_j)$ has the form

$$f_j(z_j) = H z_j^\delta v_j, \tag{2}$$

where H is the generalized stress intensity factor (GSIF), v_j is a complex eigenvector corresponding to the real eigenvalue δ, where $1 - \delta$ is the exponent of the stress singularity at the notch tip. Eigenvector v_j and eigenvalue δ are the solution of the eigenvalue problem leading from the prescribed notch boundary and compatibility conditions, [3, 4]. In most practical cases, there are two singular terms corresponding to two stress singularity exponents $1 - \delta$. Note that for the final determination of the stress field in the bi-material notch vicinity the GSIFs can be estimated by means of analytical-numerical approaches based on the path independency of the two state integral, so-called ψ-integral, [3, 4],

$$H = \frac{\Psi(u^h, r^{-\delta}\tilde{u}(\theta))}{\Psi(r^\delta u(\theta), r^{-\delta}\tilde{u}(\theta))}, \tag{3}$$

where u_i and $\tilde{u}_i$, respectively, are the regular and auxiliary basis function of Williams asymptotic expansion depending on the eigenvalues δ and $-\delta$, respectively, and corresponding eigenvectors v_j and $\tilde{v}_j$, respectively, as the solution of the eigenvalue problem mentioned above. The symbol u_j^h represents the numerical solution evaluated along the remote integration path along the notch tip. In order to derive the stability criterion, the mean value of the tangential stress $\sigma_{\theta\theta}$ over a certain distance d is useful and can be expressed as follows

$$\bar{\sigma}_{\theta\theta}(\theta) = \frac{1}{d}\int_0^d \sigma_{\theta\theta}(r,\theta)\mathrm{d}r = H_1 F_{\theta\theta 1}(\theta) + H_2 F_{\theta\theta 2}(\theta), \tag{4}$$

where the functions $F_{\theta\theta i}(\theta)$ are the functions depending on d and given in [3].

Modified maximum tangential stress criterion

Within the failure assessment of singular stress concentrators, the crack initiation and propagation direction is often gained from the criterion of the maximum tangential stress (MTS) [5]. In the case of bi-material notch, the direction of maximum of tangential stress depends on the radial distance r from the notch tip. Therefore the mean value of the tangential stress component given in Eq. 4 is used in order to reduce the dependence. The distance d has to be chosen with respect to the mechanism of a rupture, e.g. as a function of a crack increment length or as a size of material grain or it can be related to a fracture process zone. By analogy with cracks in the homogeneous case it is assumed that the crack at the bi-material notch tip is initiated in the direction θ_0, where $\bar{\sigma}_{\theta\theta}$ has its maximum. Furthermore, it is assumed that the crack is initiated when $\bar{\sigma}_{\theta\theta}(\theta_0)$ reaches its critical value ascertained for a crack in homogeneous media. The potential direction of crack initiation is then determined from the maximum of the mean value of tangential stress, Eg. 4, in both materials.

The notch stability

In order to find the direction and conditions of the crack initiation, the critical value of GSIF has to be ascertained from comparison of critical value of the $\bar{\sigma}_{\theta\theta}$ for the crack in homogenous material under mode I of loading and Eq. 4 under critical conditions. It is further assumed that the ratio $\Gamma_{21} = H_2/H_1$ is constant for a given bi-material configuration and boundary conditions and does not depend on the value of the applied stress σ^{appl}. This assumption is justified, because when

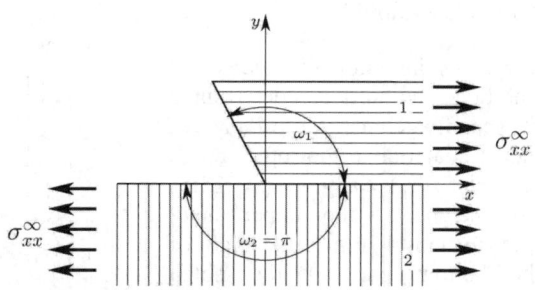

Fig. 1: The geometry and the applied external loadings of the studied bi-material orthotropic notch.

changing the value of the applied stress, only the absolute values of GSIFs change, but their ratio is constant even for the critical values of H_{2C}/H_{1C}. The ratio has no physical meaning, it just represents the contribution of particular singular terms to the stress distribution, and its unit is $[m^{\delta_2-\delta_1}]$. Following the assumption of the same mechanism of a rupture in both cases (crack and notch) we can compare the critical value of Eq. 4 with the $\bar{\sigma}_{\theta\theta}$ for the crack in homogenous material and one obtains the expression

$$H_{1C} = \frac{2K_{IC}}{\sqrt{2\pi d}\left(F_{\theta\theta 1}(\theta) + \Gamma_{21} F_{\theta\theta 2}(\theta)\right)}. \tag{5}$$

The values H_{1C} must be evaluated for both materials. Then the direction of crack initiation is related to the H_{1C} minimum value.

The influence of the directional change of fracture mechanics properties

The angular dependence of the critical value of the SIF with respect to the orthotropic nature of the material is considered. The fracture mechanics orthotropy can be described by the values ascertained in x and y directions and by the relation, [6, 7]

$$K_{IC}(\theta) = \frac{K_{ICx}K_{ICy}}{\left[K_{ICy}^2 - \left(K_{ICy}^2 - K_{ICx}^2\right)\cos^2\theta\right]^{\frac{1}{2}}}. \tag{6}$$

The substitution of Eq. 6 into Eq. 5 leads to the formula

$$H_{1C} = \frac{2K_{ICy}\left[\Gamma_{yx}^2 - (\Gamma_{yx}^2 - 1)\cos^2\theta\right]^{-\frac{1}{2}}}{\sqrt{2\pi d}\left(F_{\theta\theta 1}(\theta) + \Gamma_{21} F_{\theta\theta 2}(\theta)\right)}, \tag{7}$$

where

$$\Gamma_{yx} = \frac{K_{ICy}}{K_{ICx}}. \tag{8}$$

The minimum value of generalized fracture toughness H_{1C} is found in both material regions and in the interface. Depending on the fracture toughness, the global minimum can occur in any part of the notch. Note, that in case of crack initiation into massive material, the direction of minimum H_{1C} can notably differ from the direction of the maximum of tangential stress.

Numerical example

The crack initiation directions and stability conditions are determined for specific geometry, see Fig. 1. of the bi-material orthotropic notch characterized by angle $\omega_1 = 120°$. The height of the upper layer (domain 1) is 4 [mm], the height of the bottom part of the notch (domain 2) is 25 [mm]. The material combinations of orthotropic material parts are $E_x^{mat1} = 50$, $E_y^{mat1} = 100$, $E_x^{mat2} = 400$ and $E_y^{mat2} = 50$ [MPa]. Remaining characteristic are valid for both materials, i.e. $v_{xy} = v_{xz} = v_{yz} = 0.3$, $G_{xy} = G_{xz} = G_{yz} = 30$ [MPa] Following this material properties and geometry, the stress singularity exponents $1 - \delta_1 = 0.495$ and $1 - \delta_2 = 0.194$ can be evaluated. The notch is subjected to the external loading $\sigma_{xx}^\infty = 100$ [MPa], Fig. 1, which implies the following values of the generalized stress intensity factors $H_1 = 0.837$ [MPa m$^{1-\delta_1}$] and $H_2 = 5.333$ [MPa m$^{1-\delta_2}$] corresponding to the

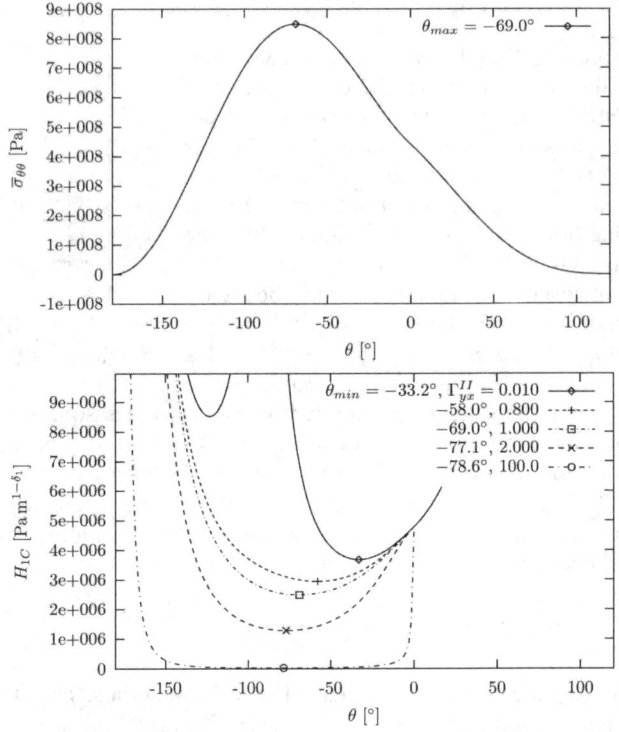

Fig. 2: The maximum value of the tangential stress near the notch tip and minimum values of the generalized fracture toughness of the notch for various values of the ratio Γ_{yx}.

ratio $\Gamma_{21} = 6.371$ [$m^{\delta_1 - \delta_2}$]. There are the values of the tangential stresses $\bar{\sigma}_{\theta\theta}$ and fracture toughness H_{1C} depending on the polar coordinate θ on the Fig. 2. There are also confronted the extremes of both quantities corresponding to the supposed directions of the crack initiation. The mean value of the tangential stress $\bar{\sigma}_{\theta\theta}$ is calculated over the distance $d = 1 \times 10^{-5}$ [mm] and the generalized fracture toughness H_{1C} is given for the $K_{ICy}^{mat1} = K_{ICy}^{mat2} = 10$ [MPa m$^{1/2}$], $\Gamma_{yx}^{mat1} = 1$ and various values of Γ_{yx}^{mat2}.

Acknowledgements: The authors are grateful for financial support through the Research projects of the Czech Science Foundation (101/09/1821, P108/10/2049).

References

[1] J. Klusák and Z. Knésl: Theoretical and Applied Fracture Mechanics 53 (2010), p.89

[2] T.C.T. Ting: *Anisotropic Elasticity* (Oxford University Press, New York 1996)

[3] J. Klusák, T. Profant and M. Kotoul: Key Engineering Materials Vols. 417-418 (2010), p.385

[4] J. Klusák, T. Profant and M. Kotoul: Procedia Engineering 2 (2010), p.1635

[5] F. Erdogan and G.C. Sih: International Journal of Basic Engineering 85(4) (1963), p.519

[6] M.B. Buczek and C.T. Herakovich: Journal of Composite Materials 19 (1985), p.544

[7] A.P. Kfouri: Fatigue & Fracture of Engineering Materials & Structures, 19 (1) (1996), p.27

Fracture Behaviour of Thin Sheet Stainless Steel

Nenad Gubeljak[1,a], Darko Jagarinec[1,b], Jožef Predan[1,c], John Landes[3,d]

[1] University of Maribor, Faculty of Mechanical Engineering, Smetanova u.17, SI-2000 Maribor, Slovenia

[2] metalna IMPRO, Zagrebska c. SI-2000 Maribor, Slovenia

[3] University of Tennessee, MABE Department, Knoxville, Tennessee, USA

[a] nenad.gubeljak@uni-mb.si, [b]djagarinec@gmail.com, [c]jozef.predan@uni-mb.si, [d]landes@utk.edu

Keywords: Stainless Steel, Compact tension specimen, Middle tension specimen, CTOD-R curves.

Abstract. The differences in fracture behavior between the compact tension C(T) and the middle tensile M(T) specimens make structure integrity assessment uncertain. Two different types of specimens C(T) and M(T) specimens made from stainless steel have been used for fracture toughness testing at the room temperature by the principles of the ASTM 1820-05 standard procedure. Stable crack initiation and crack propagation occurred for the C(T) specimens at lower values of crack driving force than for the M(T) specimens. Crack tip opening displacement-CTOD has been directly measured on the surface of specimens by using a stereo-optical grading method. The critical crack tip opening displacement at crack initiation CTODi has been measured as a plastic Stretch Zone Width (SZW) during a post test fractographic inspection. Comparison between the CTOD-R curves of both types of specimens shows some difference between the C(T) and the M(T) specimens, but a more significant difference appeared in the crack driving force, as consequence of different constraint (triaxiality) of the C(T) versus the M(T) specimens. Therefore, the result obtained by test on laboratory C(T) specimens cannot be directly used as fracture toughness material properties in a structure integrity assessment, except as a conservative lower bound estimate.

Introduction

The ductile fracture behavior strongly depends on constraint (stress triaxiality) and plane strain conditions at the crack tip [1]. Ductile fracture behavior is a consequence of the initiation and coalescence of voids in the material [2]. Usually the geometry of the specimen plays a significant role in determining constraint. It is possible to recognize that constraint (stress triaxiality) trends to rapidly decrease when the yield load of structural component is approached. Therefore, in case of where identical material and thickness are used with the same mode of loading, the fracture behavior of different structural components can be different. Usually, fracture behavior of the M(T) specimen is more like that of the structural component, exhibiting less constraint than the C(T) specimen. In this case the fracture behavior of the C(T) specimen is more conservative, e.g. crack starts at lower fracture toughness values than for the M(T) specimen [3]. Reference [3] reports the results of tests on two types of specimens each made from two different materials. These effects demonstrate the transferability problem in fracture mechanics and reliability of structural integrity assessment. However, it leads to safe crack-damage tolerant design of structure.

Material properties and Fracture mechanics testing

The material used for testing is an austenitic stainless steel. Mechanical properties, are yield strength, R_p=280 MPa, tensile strength, R_m=575 MPa and Young´s modulus, E=188 GPa. Mechanical data were obtained from a tensile test. The chemical composition for this material is given in Table 1.

Table 1: Chemical composition in weight of X2CrNiMo17122 stainless steel (W.Nr. 1.4404)

C, %	Si, %	Mn, %	P, %	S, %	Cr, %	Mo, %	Ni,%	others,%	Bal
<0,03	<1	<2	<0,045	<0,015	16,5-18,5	2-2,5	10-13	N<0,11	Fe

Two different types of specimens compact tension C(T) and middle tension M(T) specimens made from stainless steel has been used for fracture toughness testing at the room temperature (Fig. 1).

Fig.1: Testing C(T) and M(T) specimens with optical 3D Deformation system ARAMIS

Dimensions of the C(T) specimens are thickness B=3 mm, width W=50 mm. Dimensions of the M(T) specimens are thickness B=3 mm, width 2W=140 mm and diameter D=6 mm (MT1 and MT2 specimens), D=18 mm (MT4 and MT5 specimens) and D=30 mm (MT7 specimens). In order to prepare specimens for a fracture mechanics test according to ASTM 1820-05 [4], the fatigue pre-cracking of specimens was performed by cyclic loading for the C(T) specimens F_{max}=3,2 kN, R=0,1 and for the M(T) specimens F_{max}=68 kN, R=0,1. Fracture mechanic tests were performed under monotonically increasing load. During the tests the direct technique δ_5 for crack tip opening displacement [5] was measured by using optical grading method with 3D deformation system ARAMIS [6] In order to determine crack tip opening displacement CTOD vs. crack extension Δa curve, the normalization method was applied for the C(T) specimens. R-curves of the M(T) specimens were estimated by fractographic analysis and measurement of plastic stretch zone width-SZW and the final crack extension as is shown in Fig. 2. The R-curves obtained are given in Fig. 3. Fracture mechanics tests show slightly higher of material resistance to crack extension for the M(T) specimens than for the C(T) specimens at crack initiation. The advantage of direct applying of CTOD(δ_5) technique makes possible to create Crack Driving Force curves if is applied load normalized by yield load solutions as L_r non-dimensional parameter. In this case a significant difference appeared in the crack driving force (CDF) as is shown in Fig. 4. It is a consequence of different constraint (triaxiality) of the C(T) versus the M(T) specimens. The experimentally obtained results are confirmed by numerical calculation and 3D finite element method modeling by ABAQUS 6.11.[7]. It is possible to recognize difference in FEM CDF curve that the constraint (stress triaxiality) change trends to decrease rapidly when the yield load of M(T) specimen is approached. It is not so obvious from experiment of the M(T) test. However, the result obtained by test on a laboratory C(T) specimen cannot be directly used as fracture toughness material properties in structure integrity assessment of an M(T) type geometry.

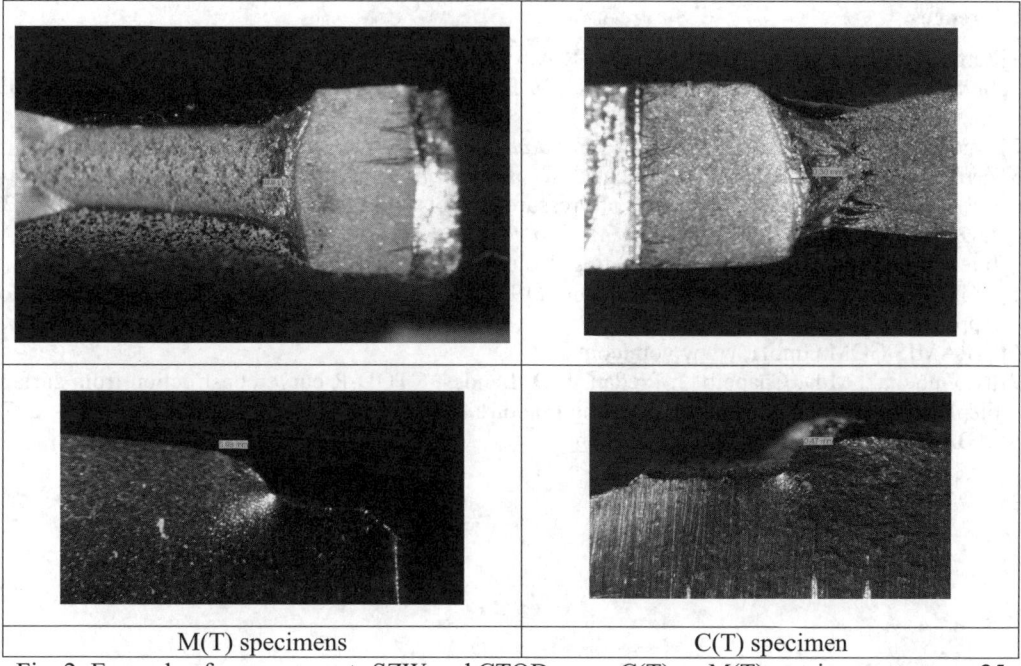

| M(T) specimens | C(T) specimen |

Fig. 2: Example of measurements SZW and $CTOD_{pl}$ on a C(T) an M(T) specimens at mag. x25

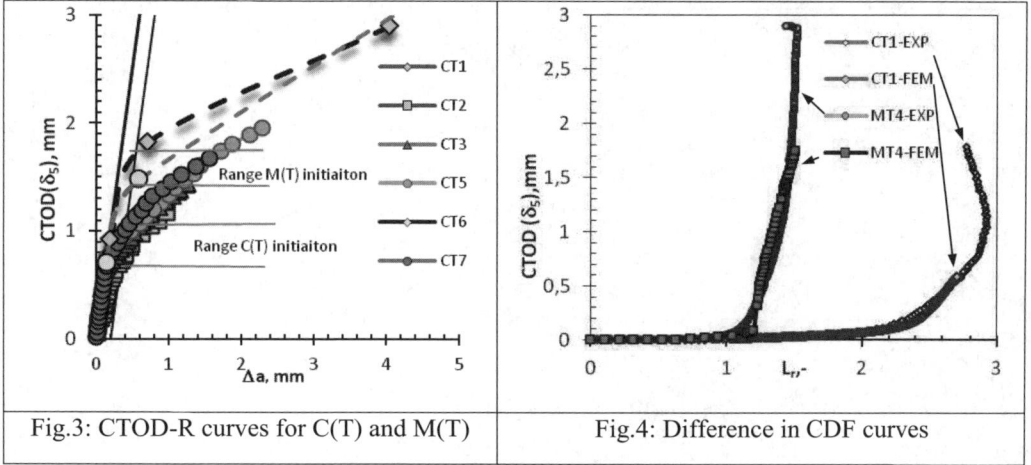

| Fig.3: CTOD-R curves for C(T) and M(T) | Fig.4: Difference in CDF curves |

Conclusions

In the paper the transferability problem in fracture mechanics between the C(T) and the M(T) specimens made from the same material and same thickness has been demonstrate. While differences in CTOD-R curves are in the range from slightly higher behavior of the M(T) specimens and at lower bound of the C(T) specimen, more significant differences appear in crack driving force, as consequence of different constraint (triaxiality) of the C(T) and the M(T) specimens. It confirms that the specimen geometry alone is not sufficient of describe the constraint conditions and hence the resistance to crack extension. Therefore the result obtained by test on a laboratory C(T) specimen cannot be directly used as fracture toughness material properties in structure integrity assessment and can only be used as a conservative lower bound.

References

[1] Zerbst U., Schwalbe K.-H., Ainsworth R.A., »*An Overview of Failure Assessment Methods in Code Band Standards*«, Volume 7, Practical Failure Assessment Methods, Elsevier Ltd., 2003, pp. 1-48
[2] Anderson, T.L.:*" Fracture Mechanics Fundamentals and Applications",* Second edition 1994
[3] Zerbst U., Hamann R., Wohlschlegel A., »*Application of the European Flaw Assessment Procedure SINTAP to pipes*«, Int. J. of Pressure Vessels and Piping, Vol. 77 (2000) pp. 697-702.
[4] ASTM E 1820-01. *Standard Test Method for Measurement of Fracture Toughness.* ASTM International: Metals-Mechanical Testing, Volume 03.01.2004.
[5] GKSS-Forschungszentrum Geesthacht (1991) GKSS-Displacement Gauge Systems for Application in Fracture Mechanics.
[6] ARAMIS GOM GmbH, www.gom.com
[7] N. Gubeljak, M.D. Chapetti, J. Predan, J. D. Landes. CTOD-R curve construction from surface displacement measurements, Engineering Fracture Mechanics, Vol. 78, Issue 11, pp. 2286-2297
[8] ABAQUS solver 6.11. www.abaqus.com

Transient and Steady State Regimes of Fatigue Crack Growth in High Strength Steel

J. Toribio[1,a], B. González[1,b], J.C. Matos[1,c]

[1]E.P.S., Campus Viriato, Avda. Requejo 33, 49022 Zamora, SPAIN

[a]toribio@usal.es, [b]bgonzalez@usal.es, [c]jcmatos@usal.es

Keywords: Cold-drawn steel, Paris curve, R-ratio, Plastic zone size, Overload retardation effect.

Abstract. This paper analyzes the propagation of fatigue cracks in pearlitic steel in two forms, hot rolled bar and cold drawn wire. The experimental procedure consisted of fatigue tests on bars under tensile loading, using steps with decreasing amplitude of stress and constant stress range during each step. The curves plotting cyclic crack growth rate versus stress intensity factor range show a main steady-state regime preceded by transient paths. The steady-state regime is associated with the curves of the Paris regime. The cold drawing process improves the fatigue behaviour of steel by retarding the cyclic crack growth rate, and the propagation rate is not dependent on the R-ratio. The transient branches allow one to calculate the plastic zone size, considering that they are a consequence of the overload retardation effect at each step change, and a unique expression is fitted as a function of $K_{max}\Delta K$ product and of the conventional mechanical properties.

Introduction

Fatigue phenomenon is usually considered as a two load parameter problem [1-3] involving two crack tip driving forces, the stress intensity factor range and the maximum stress intensity factor (ΔK and K_{max}). In fatigue crack growth of fully pearlitic steels, the increase of the load ratio, R-ratio, produces a significant raise of the slope of the Paris-Erdogan curve; this phenomenon coincides with an increase in the amount of cleavage fracture present in the fracture surface [4]. The drawing process of eutectoid steel induces retardation in the propagation of fatigue cracks, due to the microstructural changes producing a rougher microcracking [5].

The main reasons of the retardation in fatigue advance caused by overload are crack closure due to the residual compressive stresses, crack branching and contact between the rough crack surfaces after overload [6]. The retarding effect due to consecutive overloads is found to increase with the number of overloads, until it reaches a maximum [7]. There are several models to calculate the plastic zone size of the crack tip: by means of a simple application of the linear elastic fracture mechanics [8], with the strip yield model [9,10], using the superposition method [11] or according to the condition of positive equivalent plastic strain rate [12].

Experimental procedure

The material was pearlitic steel, chemical composition 0.789% C, 0.681% Mn, 0.210% Si, 0.010% P, 0.218% Cr and 0.061% V, supplied in two forms: firstly, as a hot rolled bar which has not been cold drawn at all and, secondly, as a commercial cold drawn wire which has undergone seven steps of cold drawing and a thermo-mechanical treatment to eliminate residual stresses at its surface.

In fatigue tests the specimens used were cylindrical bars of 300 mm of length, diameters of 11.0 mm and 5.1 mm for the hot rolled bar and for the cold drawn wire, respectively. During the test, the applied load F, the extensometer displacement u (placed on the specimen in a symmetric way to the crack front) and the number of cycles N, were recorded. The test procedure consisted of applying an oscillating tensile loading in the axial direction in successive decreasing steps, with a constant stress range $\Delta\sigma$ during each step and decreasing value from one to another step. The frequency used was 10 Hz, with the shape of a sinusoidal wave and different R-ratios: 0, 0.25 and 0.50. The initial maximal stress was always lower than the yield strength, decreasing in the next steps around 20÷30 % with regard to the maximal stress in the previous step.

Experimental results

In every load change the crack front was modelled as part of an ellipse with centre at the bar surface, of semiaxes a (crack depth) and b, from a set of points taken on that front and using the least squares method. In the central point of the crack front, where a plane strain state is achieved due to the enough constraint there, the curve representing the fatigue crack growth was calculated: cyclic crack growth rate da/dN versus stress intensity factor range ΔK. The value of the ΔK for the studied geometry, the crack size and the type of applied load, is:

$$\Delta K = Y \Delta \sigma \sqrt{\pi a} \qquad (1)$$

The dimensionless stress intensity factor (SIF) Y, calculated by Astiz [13], for the central point of the crack front by using the finite element method, was used depending on the relative crack depth (crack depth divided by the diameter, a/D) and the aspect ratio (ratio between the semiaxes of the ellipse, a/b), with the C_{ij} coefficients on Ref [13].

$$Y = \sum_{\substack{i=0 \\ i \neq 1}}^{4} \sum_{j=0}^{3} C_{ij} \left(\frac{a}{D}\right)^i \left(\frac{a}{b}\right)^j \qquad (2)$$

The procedure followed to obtain the fatigue crack growth curves was that outlined in Fig. 1. First of all, the a/D vs. $1/CED$ plot (where C is flexibility, $C=\Delta u/\Delta F$) and the a/b vs. a/D plot were calculated from the characterization of the crack front and the F-u data, and both curves were fitted providing polynomial equations of the third degree. These fit curves were used to calculate in every instant, from the value of the dimensionless stiffness $1/CED$, the relative crack depth a/D and from this the value of the aspect ratio a/b, so that the geometry characterization of the crack front is available at every single moment of the test. Finally, the cyclic crack growth rate da/dN versus the SIF range ΔK was obtained, where every load step constitutes a different section of the curve, converging all of them asymptotically to a common Paris branch (Fig. 2).

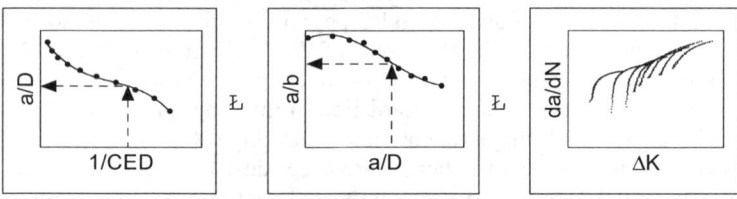

Figure 1. Calculation of the fatigue crack growth rate

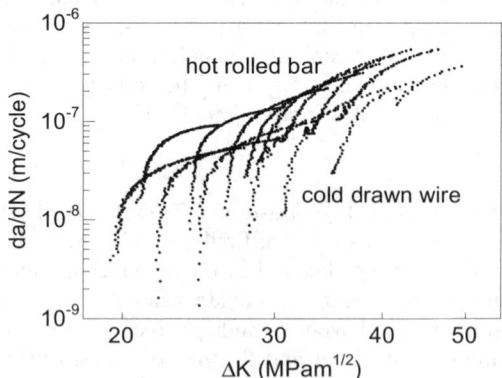

Figure 2. da/dN-ΔK curve: transient paths and steady-state regime

Discussion

Steady-state regime vs. Paris law. Fig. 3 plots the da/dN-ΔK curve in the steady-state region (Paris regime) for both steels, with R-ratios of 0, 0.25 and 0.50. For each steel and for different values of the R-ratio, there is a unique fit of the fatigue crack growth rate curve (da/dN vs. ΔK) in the Paris regime, i.e., for a given material there is a single Paris law plot with independence of the *R*-ratio (it implies a unique *C* and *m* coefficients for each steel under distinct *R*-ratios), although there is shifting of the plot towards lower values of ΔK when the *R*-ratio increases. Furthermore, the bilogarithmic fit is parallel for both steels, hot rolled bar and cold drawn wire, with the same slope and, therefore, the same Paris exponent *m*, but the cold drawn wire curve appears down and to the left (with a lower Paris coefficient *C*) in relation to that of the hot rolled bar. Therefore, the cold drawn prestressing steel wire exhibits fatigue crack growth retardation in relation to the behaviour of the hot rolled bar, i.e., there is an improvement of fatigue performance in the former.

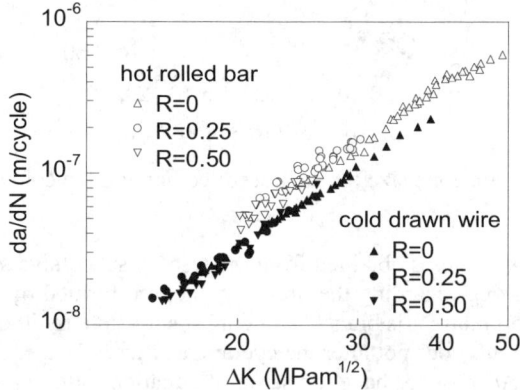

Figure 3. Paris regime for the hot rolled bar and the cold drawn wire

Transient path vs. plastic zone size. An experimental estimation of the plastic zone size was carried out, assuming that this is the responsible for the overload retardation in the tests. The retardation is considered to be produced when the plastic zone emerged at the crack tip due to the last cycles of a step (surrounded by elastic material zone which "sews up" the crack with high compressive residual stresses) is crossed by the crack (which presents a small plastic zone size) as a consequence of the subsequent step's initial cycles. An expression to calculate the plastic zone size, which is similar to that proposed by other authors [8-10], has been studied. Taking into account that the plastic zone size depends on the K_{max} value but also of the *R*-ratio, a dependency on the $K_{max}\Delta K$ product as the best solution is proposed, since it considers both parameters governing the mechanisms in the vicinity of the crack tip. Furthermore, in pearlitic steel, material with strain hardening, the yield strength σ_Y and the tensile strength σ_R have been taken into account in the plastic zone size expression's denominator, both having been obtained on the standard tension test. The proposed expression to calculate the plastic zone size r_p was:

$$r_p = k \frac{K_{max}\Delta K}{\left(\dfrac{\sigma_Y + \sigma_R}{2}\right)^2} \tag{3}$$

In Fig. 4 the experimental results of the plastic zone size, for the hot rolled bar and the cold drawn wire, with different values of the *R*-ratio (0, 0.25 and 0.50), are shown. It is observed that the plastic zone size is greater in the hot rolled bar (with smaller yield strength and tensile strength) than in cold drawn wire, deducting for both materials a unique fit for the plastic zone size, which presents a *k* coefficient of 0.14, similar to that obtained by other authors [8-10].

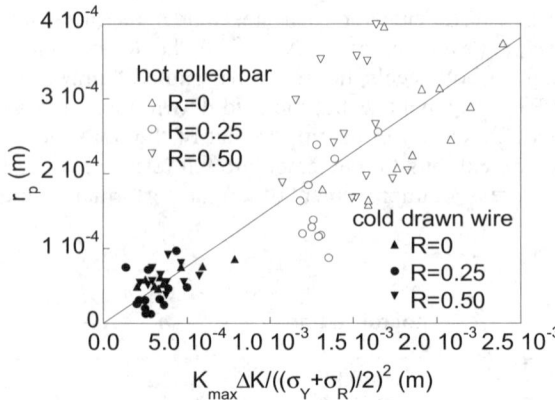

Figure 4. Plastic zone size for the hot rolled bar and the cold drawn wire

Conclusions

(i) The cyclic crack growth curves obtained from the tests in successive steps, with constant $\Delta\sigma$ in every step and decreasing regarding the previous one, are formed by sections which converge asymptotically to a common Paris branch, showing steady-state regime and transient paths.

(ii) The change of the R-ratio does not alter the cyclic crack growth rate, because the Paris curve is the same (the da/dN-ΔK plotting have the same fit equation), although its values tend to lower ΔK when R-ratio increases. The cold drawing process improves the fatigue behaviour of eutectoid steel by retarding the cyclic crack growth rate in the Paris regime.

(iii) In the estimation of the plastic zone size on the basis of overload retardation, apart from the conventional mechanical properties (σ_Y and σ_R), two fracture mechanics parameters (ΔK and K_{max}) have to be considered to completely characterize the fatigue process. The proposed relationship, similar to that obtain by other authors, has a fitting coefficient of about 0.14.

Acknowledgements

The authors wish to acknowledge the financial support provided by the following Spanish Institutions: MCYT (Grant MAT2002-01831), MEC (Grant BIA2005-08965), MICINN (Grant BIA2008-06810 and BIA2011-27870), and JCyL (Grants SA067A05, SA111A07, and SA039A08).

References

[1] K. Sadananda and A.K. Vasudevan: Int. J. Fatigue Vol. 26 (2004), p. 39.
[2] S. Stoychev and D. Kujawski: Int. J. Fatigue Vol. 27 (2005), p. 1425.
[3] J. Zhang, X.D. He and S.Y. Du: Int. J. Fatigue Vol. 27 (2005), p. 1314.
[4] A.B. El-Shabasy and J.J. Lewandowski: Int. J. Fatigue Vol. 26 (2004), p. 305.
[5] J. Toribio, J.C. Matos and B. González: Int. J. Fatigue Vol. 31 (2009), p. 2014.
[6] S. Suresh: Eng. Fract. Mech. Vol. 18 (1983), p. 577.
[7] S. Pommier and M. de Freitas: Fatigue Fract. Eng. Mater. Struct. Vol. 25 (2002), p. 709.
[8] G.R. Irwin, in: Mechanical and Metallurgical Behaviour of Sheet Materials (Proceedings of the 7th Sagamore Ordonance), Section IV, New York, USA (1960), p. 63.
[9] D.S. Dugdale: J Mech. Phys. Solids Vol. 8 (1960), p. 100.
[10] G.I. Barenblatt: Adv. Appl. Mech. Vol. 7 (1962), p. 55.
[11] J.R. Rice, in: Fatigue Crack Propagation (ASTM STP 415), Philadelphia, USA (1967), p. 247.
[12] J. Toribio and V. Kharin: J. Mater. Sci. Vol. 41 (2006), p. 6015.
[13] M.A. Astiz: Int. J. Fract. Vol. 31 (1986), p. 105.

Estimation of Shear Behavior of Ultra High Performance Concrete I Girder without Shear Stirrups

J.W Lee[1, a], C. Joh[2, b], E.S Choi[3, c], I.J Kwak[4, d], B.S Kim[5, e]

[1,2,3,4,5]Korea Institute of Construction Technology 1190, Simindae-Ro, Ilsanseo-Gu, Goyang-Si, Gyeonggi-Do, 411-712 Republic of Korea

[a]duckhawk@kict.re.kr, [b]cjoh@kict.re.kr, [c]eschoi@kict.re.kr, [d]kwakim@kict.re.kr, [e]bskim@kict.re.kr

Keywords: Ultra High Performance Concrete, Shear Behavior

Abstract. Thinner and lighter members can be designed by utilizing the high stiffness and toughness, and high compressive strength of Ultra High Performance Concrete (UHPC), which reaches up to 180MPa. This high strength and ductile tensile behavior of UHPC makes it possible to design the web of the UHPC I Girder without conventional shear stirrups, which makes the UHPC I girder slender, light and economical. However, establishing shear design procedure for UHPC I girders without stirrups requires additional theoretical and experimental studies. This paper investigated shear behavior of UHPC I girder without shear stirrups. The test results show, in spite of no shear stirrups, test specimens have high ductility due to the bridging action of steel fibers against crack opening. UHPC I girders without shear stirrups tested show gradual increase of strength after initial cracking instead of brittle loss of strength as expected from the ordinary reinforced concrete I girders without stirrups. The decrease of the shear span-depth ratio increase the shear strength of the UHPC I girder without stirrups.

Introduction

Compared to conventional concretes, Ultra High Performance Concrete (UHPC) is featured not only by higher compressive and tensile strength but also by its high ductility. The mean compressive strength of UHPC exceeds 180 MPa and its tensile strength reaches 12 MPa. The high ductility of UHPC relies on its high tensile strength and the bridging action of the steel fibers at the cracked section[1][2][3]. Such ductility is advantageous in mitigating the brittle characteristics observed in ordinary concrete and also in reducing the need of shear stirrups. These properties of UHPC open new possibilities to design light, economical and slender UHPC structures. However, the lack of research dedicated to the behavior of UHPC structures constitutes an obstacle for the application of UHPC on field.

Accordingly, this study intends to estimate the shear behavior of UHPC girder without shear stirrups. To that goal, test specimens with various shear span-to-depth ratios and eventual prestressing are fabricated and subject to static loading tests for the analysis of their behavior.

Test specimens and test setup

In order to evaluate the shear behavioral characteristics of UHPC beam without shear stirrups, a total of 6 girder members with I-shaped cross-section were fabricated and subject to static loading tests. The specimens were designed with various shear span-to-depth ratios chosen as variables together with the introduction or not of prestressing. The mix composition of UHPC applied for the fabrication of the girders and the test variables of each specimen are listed in Tables 1 and 2. The compressive strength and elastic modulus of the adopted UHPC are respectively 186.5 MPa and 46.2 GPa, and the mean tensile strength of UHPC obtained through direct tensile test is 13.1 MPa. The cross-section of the specimens is illustrated in Fig. 1. The thickness of the web is 40 mm or 50 mm without reinforcement by shear stirrups. Twelve 15.2 mm SWPC7B steel wires are arranged in the bottom flange to prevent flexural failure and induce shear failure. All the specimens were cured at ambient temperature during 24 hours followed by steam curing at 90°C during 72 hours.

Table 1. Mix composition of UHPC (volume ratio for steel fiber and weight ratio for others)

W/B	Cement	Silica fume	Sand	Filling powder	Superplasticizer	Steel fiber (V_f)
0.2	1	0.25	1.1	0.3	0.016	2 %

Table 2. Details of test specimens

Specimen designation	Span length [mm]	Shear span-to-depth ratio	Prestressing [MPa]
S25-F20-P0	3200	2.5	0
S25-F20-PS	3200	2.5	845
S34-F20-P0	4400	3.4	0
S34-F20-PS	4400	3.4	950
S17-F20-P0	3200	1.7	0
S17-F20-PS	3200	1.7	950
S22-F20-P0	4200	2.2	0
S22-F20-PS	4200	2.2	950

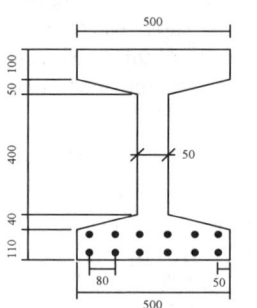
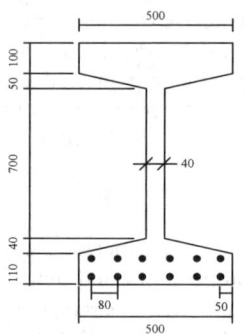

Fig. 1 UHPC girders without stirrups (unit: mm)

An actuator with capacity of 3,500 kN was used to apply load through the 3-point loading test configuration as shown in Fig. 2. The member was simply supported and load was applied at mid-span. Loading was applied by displacement control at speed of 0.017 mm/sec. Three LVDTs were installed to measure the deflection of the member together with 8 wire gages to measure the lateral displacement.

Fig. 2 Fabricated test specimen installed for 3-point loading test

Results of Test

As shown in the load-deflection curves of Fig. 3, the UHPC girder without shear stirrups exhibits a crack behavior in which the load increases continuously until the ultimate state even after the initiation of cracks. Additional cracks developed between the diagonal cracks provoked by the increased load after the initiation of cracks. The inclined cracks in the web propagated to the bottom and upper flanges. Among the diagonal cracks, some cracks developed onto major diagonal cracks due to the increase of the crack width (Figs. 4(a) to 4(h)). Sudden loss of the load did not occur even after the ultimate load and the members showed relatively continuous load distribution capacity. Such load-carrying capacity can be attributed to the bridging action of the steel fibers after cracking. However, specimens S17-F20-P0 and S34-F20-PS experienced sudden loss of the load due to the vertical cracks developed at the center of the member after the ultimate load (Figs. 4(a) and 4(h)).

The test results indicate that the initial crack load and ultimate load decrease according to the increase of the shear-to-span ratio. Moreover, in the case of prestressing, the initial crack load and ultimate load increase, and the angle of the major diagonal cracks appears to be 32.1°, which is smaller than the angle of 39.8° in case of no prestressing.

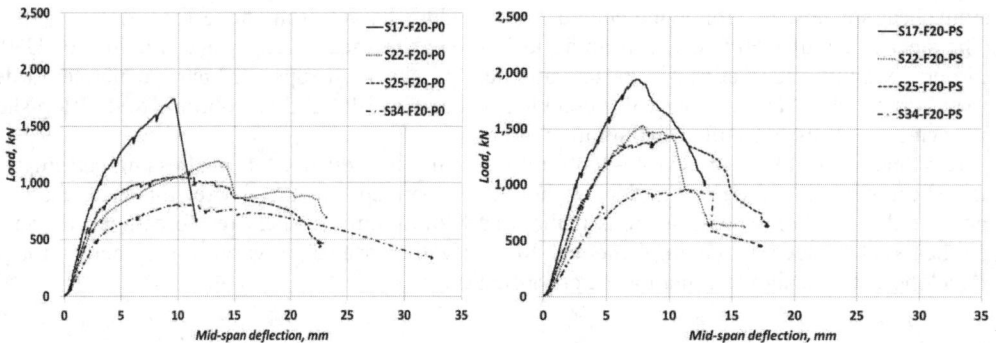

Fig. 3 Load-deflection curves

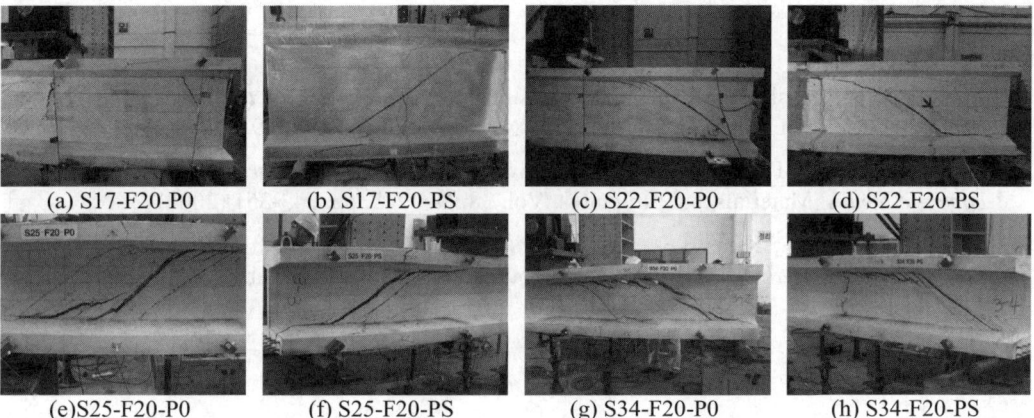

Fig. 4 Failure patterns of the test specimens

Table 3: Shear behavior of the test specimens (at mid span)

Specimens	Initial cracking		Ultimate load		Angle of major diagonal crack	Failure mode
	P_{cr} [kN]	Δ_{cr} [mm]	P_u [kN]	Δ_u [mm]		
S25-F20-P0	600	2.08	1,053	10.18	45.0	
S25-F20-PS	820	3.00	1,433	10.27	31.5	
S34-F20-P0	490	2.84	808	10.58	43.5	
S34-F20-PS	800	4.71	954	11.65	35.0	Shear failure
S17-F20-P0	600	1.75	1,738	9.56	44.2	
S17-F20-PS	1,000	2.56	1,937	7.30	33.7	
S22-F20-P0	580	2.55	1,196	13.61	43.8	
S22-F20-PS	910	3.42	1,529	7.73	35.5	

Conclusions

This paper conducted static loading tests in order to estimate the shear behavior of UHPC girders without shear stirrups. The following conclusions could be derived from the test results.
- The steel fibers of UHPC can contribute to improve the shear strength and ductility of UHPC girders without shear stirrups. In the absence of shear stirrups, ordinary concrete girders experience sudden failure after early cracking whereas UHPC girders continue to develop shear resistance capacity even after the initiation of cracks.
- The reduction of the shear span-to-depth ratio and the introduction of prestressing can improve the shear strength of UHPC girders without shear stirrups. In the future, this will enable to achieve design of slim cross-sections without or with minimized shear reinforcement. However, further studies need to be implemented to verify the effects of various parameters for the development of design formula for the cross-section.

Acknowledgements

This work was supported by "Design Technology for Ultra High Performance Concrete", a research project of the Korea Institute of Construction Technology.

References

[1] Lim, T.Y., Paramasivam P., and Lee. S.L., in: *Analytical model for tensile behavior of steel-fiber concrete*, ACI Materials Journal, Vol. 84, No. 4, pp. 286-298 (1987).

[2] Meda, A., Minelli, F., Plizzari, G.A., and Riva, P., in: *Shear behavior of steel fiber reinforced concrete beams*, Materials and Structures, Vol. 38, No. 277, pp. 343-351 (2005).

[3] Yang, I.H., Joh, C., Kim, B.S., in: *Flexural strength of large scale ultra high performance concrete prestressed T-beams*, Canadian journal of Civil Engineers, Vol. 38, No. 11, pp. 1185-1195 (2011).

A Local Compression Tests of UHPC Anchor Blocks for Post-Tensioning Tendons

Eun Suk Choi[1,a], Jung Woo Lee[1,b], Changbin Joh[1,c], Jong Won Kwark[1,d], Jee Sang Kim[2,e], Yoon Seok Choi[2,f]

[1]Structural Engineering Research Division, Korea Institute of Construction Technology, Goyang, Rep. of Korea

[2]Dept. of Urban and Environmental System Engineering, Seo Kyeong University, Seoul, Rep. of Korea

[a]eschoi@kict.re.kr, [b]duckhawk@kict.re.kr, [c]cjoh@kict.re.kr, [d]origilon@kict.re.kr, [e]zskim@skuniv.ac.kr , [f]cyschoi87@naver.com

Keywords: UHPC, Anchor Blocks, Post-tensioning, Compression Test

Abstract. In the application of Ultra High Performance Concrete (UHPC) to PSC girders by using the post-tensioning system, the high strength and ductility of UHPC in tension can be exploited to substitute the confined reinforcing bars which control the rupture around the anchorage device. The exploitation of such properties is expected to simplify the reinforcing details around the anchorage zone. Taking advantage of UHPC can downsize a cross section with the attributes of high compression and tensile strength. This paper reports the local behavior of UHPC anchor block under compression. Test specimens were made based on mix proportion of K-UHPC (KICT-Ultra High Performance Concrete) developed by the Korea Institute of Construction Technology (KICT). The performance of the anchor block was evaluated according to ETAG-013 (European Technical Agreement Guide No.13) of EOTA (European Organization for Technical Approvals). As the results of the experiment, it is found that the details and reinforcement of UHPC anchorage zone can be simplified with the interconnection effect and the high intensity of the matrix itself.

Introduction

For the design and analysis of post-tensioned PSC members using ordinary concrete, the understanding of the overall structural behavior like bending, shear and torsion, the understanding of long-term behavior like drying shrinkage and creep, and the understanding of the design of the anchoring zone are required. Especially, in the case of tendon anchor blocks, the cross-section needs to be increased compared to ordinary sections of the girder due to the necessity to secure spacing between the anchoring devices and sufficient spacing with the free ends as well as to cope with the occurrence of large bearing forces. Moreover, since the concentrated loads applying on the anchor head are transmitted in the form of bearing force and bursting force at the ends of the anchor block, additional arrangement of reinforcing bars is also required. The application of the K-UHPC (KICT-Ultra High Performance Concrete) developed by the Korea Institute of Construction Technology (KICT) to the anchoring zone using post-tensioning is expected to bring large reduction of the section owing to the high compressive and tensile strengths of UHPC and also to enable simplification or elimination of the reinforcing bars. In this study, PSC anchor block specimens are fabricated using the standard mix of K-UHPC and subject to test in compliance with the ETAG-013 regulations of the EOTA (European Organization for Technical Approvals). Finite element analysis is conducted to evaluate the performance of the anchoring zone of the post-tensioned member applying K-UHPC. Foreign countries have also performed experimental studies on the application of UHPC to the tendon anchor [1], this study intends to estimate the possibility to secure appropriate safety factor for the anchor applying K-UHPC and to assess the possibility to reduce the amount of reinforcement.

Test of Anchor Block

Since tendon anchor blocks show composite behavior similarly to common structures, the test variables chosen for the test are limited to 2 large types and the test results are compared according to these test variables. The first test variable is the level of reinforcement. Accordingly 3 types of specimens are fabricated (Fig. 1) to evaluate the behavior of the anchor block with respect to the level of reinforcement: a non-reinforced specimen (designation: *7st-NTP-NSR-1,2-S1,2,3*) without anchoring device (anchor plate and trumpet guide); a specimen using only anchoring device without confining reinforcing bar (designation: *7st-**TP**-NSR-1,2-S1,2,3*); and, a specimen using anchoring device and reinforcing bars (designation: *7st-**TP**-**SR**-1,2-S1,2,3*). The designation of the specimens is given according to the number of strands – the eventual use of trumpet – the eventual use of reinforcing bars – the volume content of steel fiber – the serial number of the specimen. The second test variable is the volume content of steel fibers. Volume contents of 1% and 2% are selected and their effect is evaluated. The adopted anchoring device is the tensile anchor block VSL Type EC model 6-7 of VSL in which the trumpet and plate are monolithic. The measurement spots were decided according to the results of the finite element analysis. The largest bursting force occurred between positions 0.3b and 0.5b (b: section width) along the axis of the member at the end of the anchor. Beyond 1.0b, the bursting force converged to zero. The measurement spots were determined based on these analytic results. According to the ETAG-013 regulations of EOTA, cyclic loading shall be applied more than 10 times from 0.12Fprg to 0.8Fprg (Fprg: design load) and crack and eventual deformation shall be observed at the load points. At stabilization of the strain, the loading speed shall be 0.5 MPa/sec and, thereafter, loading shall be applied through displacement control at speed of 05 mm/min until failure load [2]. During the fabrication of the anchor block specimens, cylinders for strength measurement test were also fabricated with identical conditions. Strength tests showed that the mean compressive strength runs around 180 MPa, the mean elastic modulus around 46,000 MPa and the direct tensile strength around 10 MPa [3].

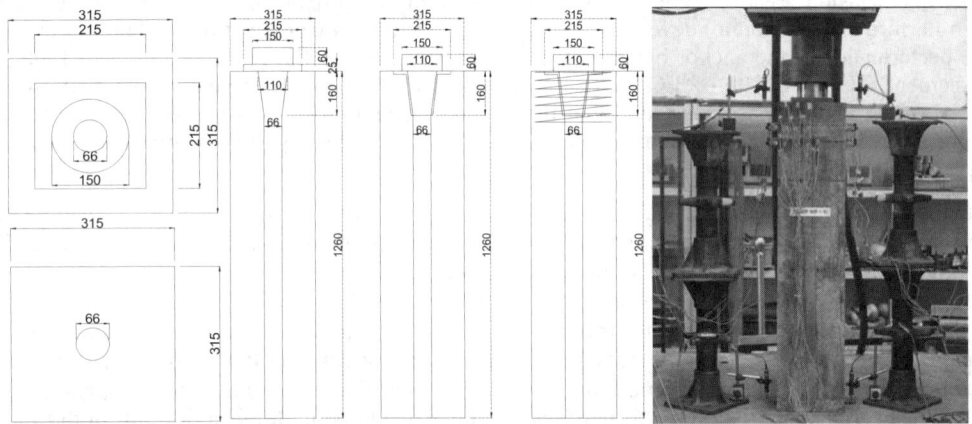

Fig. 1 Dimensions of specimens and test setup

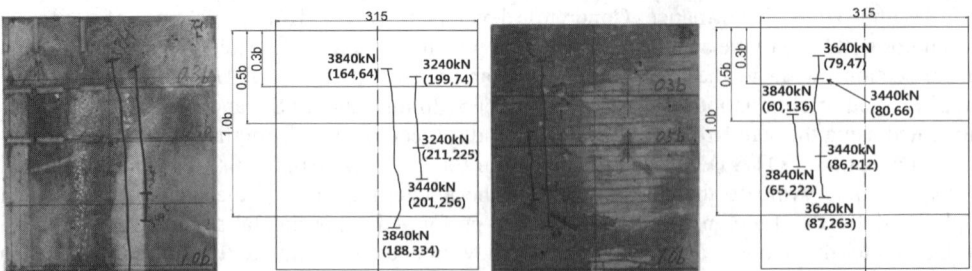

Fig. 2 Crack patterns at both sides of the specimens (non-reinforced specimen, steel fiber 2%)

Test Results
In the tests, loading was applied through a loading plate installed between the anchor head and the specimen. All the specimens in which a loading plate was installed sustained a load of 3,800 kN. The specimens in which the loading plate was removed experienced local failure similar to the punching failure of decks at the opening due to the reduction of load-supporting area. The relation

between the ultimate load (Fu) corresponding to the resistance to vertical crack-induced failure and the design load (Fprg = 1820 kN) is Fu ≥ 1.1Fpk = 2,002 kN and was satisfied by all the specimens, which attested the remarkable performance of the anchor block using K-UHPC. Table 1 lists the position and size of the average maximum tensile stress for each specimen. Fig. 3 plots the stress distribution in the section located at a distance of 110 mm from the left side of the specimen in the direction of the center. The horizontal axis represents the ratio of the section width to the axial distance of the specimen from the anchored end.

Table 1. Position and size of the average maximum tensile stress per specimen

Reinforcement type	Steel fibers 1 %		Steel fibers 2%	
	Max. position	Max. stress (MPa)	Max. position	Max. stress (MPa)
Non-reinforced (-NTP-NSR-)	0.3b	7.2	0.5b	9.5
Trumpet (-TP-NSR-)	0.5b	9.8	0.3b	7.9
Confining reinforcement (-TP-SR-)	0.5b	5.2	0.5b	6.8

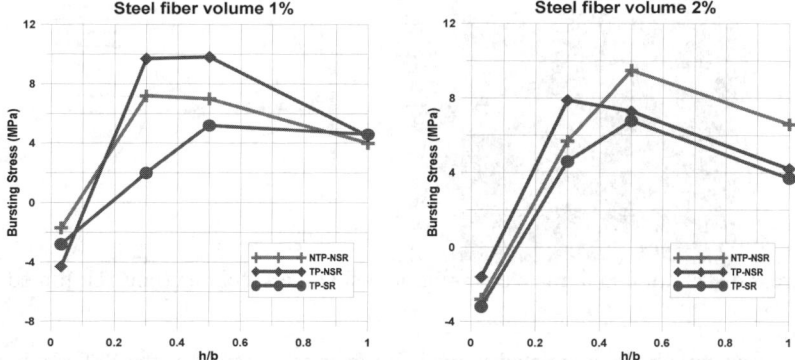

Fig. 3 Stress distribution according to reinforcement device of specimen
(at 110 mm from the left side)

Structural Analysis

The analysis model was classified into no-reinforcement, anchor reinforcement, anchor block and spiral reinforcement. The non-reinforced model was subdivided according to the presence or not of the anchor plate. The anchor block reinforced model adopted the dimensions of the tensile anchor block VSL Type EC model 6-7 of VSL. In addition, the cross-section of all the specimen models was set to 3 sizes (315 mm, 365 mm, 415 mm). The overall shape was modeled using 3D tetrahedron solid elements for K-UHPC and the anchor block and truss elements for the steel reinforcement around the anchor block supported by the finite element software Midas FEA. The material properties adopted for the analysis are compressive strength of 180 MPa and elastic modulus of 45 GPa for concrete, and SM400 for the trumpet, plate and anchor head.

Fig. 4 shows the analysis results for the non-reinforced specimen. The stress distribution at a distance of 10 mm in the axial direction from the anchored end exhibits compressive stress at the edge of the specimen to turn onto a sequence of compressive and tensile stresses toward the center of the specimen with spacing of about 35 mm. The maximum tensile stress is 11.5 MPa at 64 mm and the maximum compressive stress occurs at 37 mm with value of -10.8 MPa. Even if the maximum tensile stress is larger than the 10 MPa resulting from the direct tensile test, tensile cracks did not occur in the anchored end during the tests. The stress distribution of 0.3b increases monotonically toward the center of the specimen to show a maximum tensile stress at 93.55 mm. The stress distribution of 0.5b shows identical pattern to that of 0.3b with a maximum tensile stress of 1.9 MPa at 84 mm. The stress distribution of 1.0b shows invariably value of 0 identically to the stress occurring at the surface. In other words, compressive and tensile stresses occurred in the anchored end due to the disturbance of a local stress field, but the maximum bursting force occurred at the center far away from the anchored end. This confirms the results of conventional structural analysis.

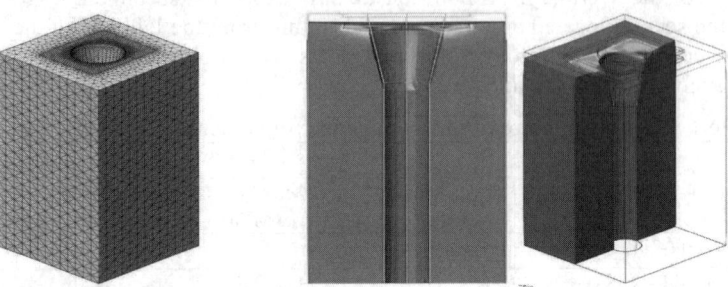

Fig. 3 Finite element analysis model and results

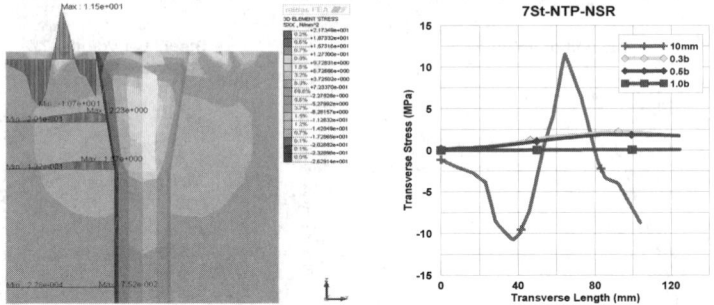

Fig. 4 Stress distribution in a transverse section inside the specimen (non-reinforced specimen)

Conclusions

Test and analysis have been conducted to evaluate the performance of the PSC anchor block using K-UHPC. In view of the test results, the overall characteristics of the stress and deformation did not show particular difference compared to the PS anchor block using ordinary concrete. Therefore, economy and efficiency can be achieved by applying conventional design and analysis techniques so as to secure adequate safety factor and reduce significantly the amount of reinforcement. Considering that this study is limited to the test and analysis of single anchor blocks with straight arrangement of tendons, further studies should extend to the test and analysis of multiple anchor blocks, anchor blocks with various dimensions, curved arrangement of tendons and the interference between anchor devices.

Acknowledgment

This work was supported by "Design Technology for Ultra High Performance Concrete", the research project of Korea Institute of Construction Technology. The authors express their gratitude for the support.

References

[1] Toutlemonde, F., Renaud, J.-C., Lauvin, L., Behloul, M. Simon, A. and Vildare, S., "Testing and analysing innovative design of UHPFRC anchor blocks for post-tensioning tendons", 6th International Conference on Fracture Mechanics of Concrete and Concrete Structures (FRAMCOS-6), Catania, Italy (2007), pp. 17-22

[2] EOTA, ETAG-013: Guideline for European Technical Approval of post-tensioning kits for prestressing of structures. (2002)

[3] KICT: Interim Design Recommendations for Ultra High Performance Concrete and Commentary (2011).

Impact Identification in Composite Stiffened Panels

M. Ghajari[1,a], Z. Sharif Khodaei[1,b] and M. H. Aliabadi[1,c]

[1]Department of Aeronautics, Imperial College London, South Kensington Campus,
Roderic Hill Building, Exhibition Road, SW7 2AZ, London, UK.

[a]m.ghajari@imperial.ac.uk, [b]z.sharif-khodaei@imperial.ac.uk [c]m.h.aliabadi@imperial.ac.uk

Keywords: impact, large mass, small mass, stiffened panel, composites, artificial neural network

Abstract. A number of small mass and large mass impacts on a sensorised aircraft stiffened panel were numerically simulated. Sensor signals and the contact force history were recorded during each impact. A significant difference was noticed between the small mass and large mass impacts with respect to the contact force. To distinguish between these two types of impacts, the Fast Fourier Transform was performed on the sensor signals and a categorisation criterion was defined. Finally, two separate Artificial Neural Networks were trained to approximate the peak contact force for each type of impact. It was found that the performance of these ANNs were better than a single ANN trained for both small and large mass impacts.

Introduction

Composite materials are being increasingly used in modern aircrafts due to their superior properties. However, their weakness in impact damage remains an on-going concern. Two types of damage which can occur in a composite structure can be categorised as Visible Impact Damage (VID) and Barely Visible Impact Damage (BVID), which if not detected on time can result in catastrophic failure. Therefore, an increase of research in Structural Health Monitoring techniques resulting in impact and damage detection can be noticed. In aviation industry, real-time characterisation of impact events is an important problem that has direct influence on design and maintenance costs. By detecting impact location and force magnitude, a condition-based maintenance rather than a schedule-based one can be performed; this results in lower costs. Once the location and magnitude of the impact force is known, a localised damage detection technique can be applied [1],[2] and [3].

In last two decades, various methods have been proposed to determine the impact location and force. For instance, Park et al. [4] identified impacts on composite panels by using numerically obtained transfer functions of some sensor/impact locations. The impact location was estimated through minimising a cost function of forces in three locations. Once the approximate location of the impact was known, the corresponding transfer function was used to reconstrcuct the force. Martin and Doyle [5] used the convolution integral and its properties to reconstuct the impact force history. A response (acceleration) was experimentally recorded and spectral analysis was performed to establish a relation between the Fourier transforms of the impact force and the response. Then, the frequency domain deconvolution and the inverse Fourier transform were employed to obtain the force history.

In a number of studies on impact identification [6-9], Artifitial Neural Networks (ANN) have been used. ANNs are mathematical models that can be trained to model complex nonlinear relationships between the inputs and outputs. Worden and Staszewski [6] used ANNs to quantify the impact force magnitude applied onto a stiffened composite plate and to detect the impact location. To provide training data, 80 impacts were carried out at random locations. The level of the force was kept below 0.1 N. The approach was successful in finding the impact location, but it failed to make an acceptable estimation of the peak impact force.

In a previous study by the authors [10], an ANN was established and trained for detecting impact events on various locations of an aircraft stiffened panel. In contrast to the majority of previous studies, both small mass and large mass impacts were considered in training of the network. The percentage error of the peak contact force was reported around 26%, which is high. This may be related to the fact that one network was established for both small and large mass impacts. In the present paper, a separate ANN is established for each type of impact to detect the peak impact force. Signal processing techniques are employed to reveal any possible difference between sensor signals recorded in both small and large mass impacts; this difference can be used to identify the type of impact.

Impacts on an aircraft stiffened panel

During the service life of an aircraft different impacts can occur. These impacts can fall into two groups: large mass impacts, such as tool drop, or small mass impact such as runway debris. It was pointed out in [11] that two impacts with the same impact energy can have completely different effects on the structure. Therefore, by simply detecting the impact energy we cannot comment on damage initiation or propagation in the structure. However, the contact force at the location of the impactor is related to damage initiation and propagation. Thus instead of impact energy, contact force at the location of impact is detected.

200 impacts of different mass, velocity and location were modelled on a composite stiffened panel. Details of the FE model can be found in [10]. In this work only a third of the panel with 4 sensors were used as depicted in Figure 1. Sensor readings were stored to train an ANN for detecting max contact force. The distributions of contact forces for two different impacts (different mass and velocity) of the same energy (6J) are shown in Figure 2. It can be seen that the contact forces have very different responses and by training the same network with small and large mass impact data, it can result in high error values due to regularisation of the ANN. In this work an attempt has been made to categorise the impact from sensor data and according to the impact type, the appropriate ANN will be used for detecting max contact force.

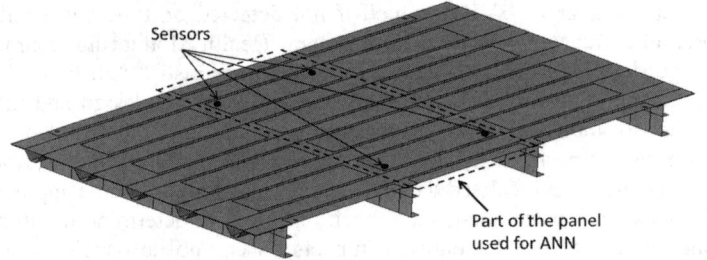

Figure 1 Stiffened panel geometry and lay-up

First step is to find a feature extraction of the sensor signal, which can best distinguish the type of impact. The features used in this work is first to obtain the discrete Fast Fourier Transform (FFT) of the received signals to decompose it into different frequencies using Equation (1).

$$x(n) = \left(\frac{1}{N}\right) \sum_{k=0}^{N-1} X(k) e^{-jk2\pi n/N}, \qquad n = 0 \ldots N-1 \qquad (1)$$

The real part of the FFT of the sensor signals gives the amplitude of the signal. In this example 4 sensors have been used. This signal processing procedure is repeated for all impacts and all sensor readings. For each impact the min amplitude of the FFT of sensor signals are stored and their histogram is plotted in Figure 3. It can be seen that there is clearly a difference between the dominant frequencies of small mass and large mass impacts. By defining a treshold (1 kHz in this case) the impact can be categorised into large or small mass impact from the sensor readings. Once the type of impact is established, by using the appropriate ANN the max contact force can be approximated and checked againts defined damage thresholds for the structure.

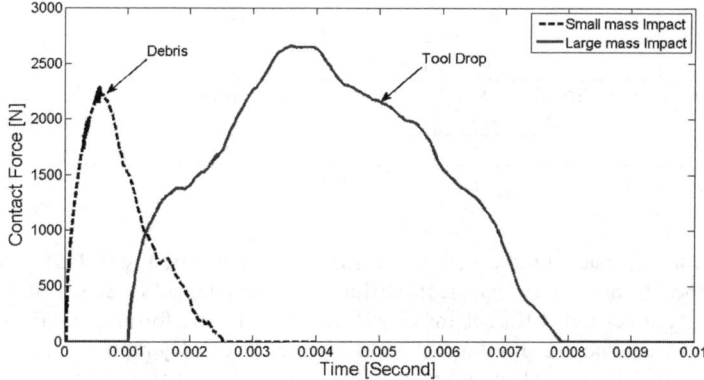

Figure 2 Contact force of 6J Impact with Small and Large mass

In addition what can be concluded from Figure 3 is that the small mass impacts show a very scattered behaviour in terms of the dominant frequency of the sensor signals. This fact influences the training of an ANN in predicting the contact force. However the large mass impacts show a more unified behaviour and are all below the 1 kHz threshold. Thus in this work it is porposed to separate the two networks to better describe the behaviour of different types of impacts, which consequently results in better performance of the trained ANN.

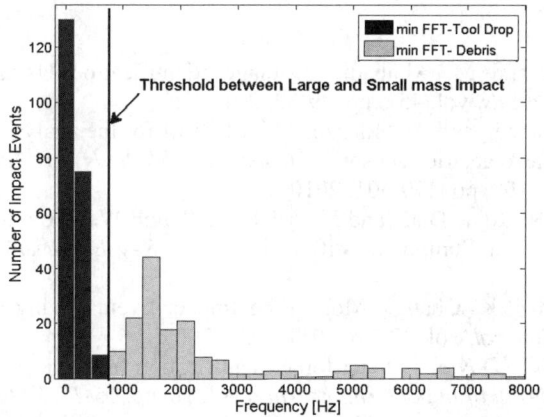

Figure 3 Dominant frequencies of small mass vs larg mass impacts

Artificial neural networks

Two separate ANNs were trained for large and small mass impact events. The established networks are Multi-Layer Perceptron with backpropagtion learning rule and two hidden layers. The inputs to the networks are maxima of the approximation and detail coefficients of the first level Wavelet decomposition of the signal and the peak of the signal.The output of the network is the max contact force. The sensor data was divided into training, validation and test sets. The test set was introduced to the trained network for evaluating the networks performance. The network performance is described as the Mean Square Error (MSE) in percentage and the results for each trained network is presented in

Table 1. As it can be seen the performance of a separate network for the large mass and small impact has significantly improved. There is 12% reduction in the MSE of the approximated max contact force.

Table 1 ANN performance

	ANN for Small Mass Impact	ANN for Large Mass Impact	ANN for Combined Large and Small mass Impacts
MSE	14%	14 %	26 %

Conclusions

Small and large mass impacts onto a composite stiffened panel fitted with PZT sensors have been simulated using the FE method. A significant difference in the impact force was noticed for impacts of the same energy level but different mass and velocity. By performing DFFT on the received signals a dominant frequency is introduced for each impact event. By plotting the min frequency for each impact scenario, it was observed that there is a substantial variance in the dominating frequencies of the small and large mass impacts. By defining a threshold for this frequency, any impact event can be classified into large or small mass impact. Subsequently a separate ANN was trained for each impact type and used for predicting the max impact force. By doing so, a considerable improvement (~12%) was noticed in the performance of separate ANNs for large and small mass impact. Further research needs to be done on defining the threshold for categorising small mass and large mass impact events and to relate the max contact force to initiation and propagation of damage in the structure.

References

[1] Z. Sharif Khodaei and M. Aliabadi, "Damage Identification Using Lamb Waves," *Key Engineering Materials,* vol. 452, pp. 29-32, 2011.
[2] I. Benedetti, M. Aliabadi, and A. Milazzo, "A fast BEM for the analysis of damaged structures with bonded piezoelectric sensors," *Computer Methods in Applied Mechanics and Engineering,* vol. 199, pp. 490-501, 2010.
[3] Z. Sharif Khodaei, R. Rojas-Diaz, and M. Aliabadi, "Lamb-Wave Based Technique for Impact Damage Detection in Composite Stiffened Panels," *Key Engineering Materials,* vol. 488, pp. 5-8, 2012.
[4] J. Park, S. Ha, and F. K. Chang, "Monitoring Impact Events Using a System-Identification Method," *AIAA journal,* vol. 47, pp. 2011-2020, 2009.
[5] M. T. Martin and J. F. Doyle, "Impact force identification from wave propagation responses," *International Journal of Impact Engineering,* vol. 18, pp. 65-77, 1996.
[6] K. Worden and W. J. Staszewski, "Impact location and quantification on a composite panel using neural networks and a genetic algorithm," *Strain,* vol. 36, pp. 61-68, 2000.
[7] W. J. Staszewski, K. Worden, R. Wardle, and G. R. Tomlinson, "Fail-safe sensor distributions for impact detection in composite materials," *Smart Materials and Structures,* vol. 9, p. 298, 2000.
[8] S. Dae-Un, O. Jung-Hoon, K. Chun-Gon, and H. Chang-Sun, "Impact monitoring of smart composite laminates using neural network and wavelet analysis," *Journal of intelligent material systems and structures,* vol. 11, pp. 180-190, 2000.
[9] J. Haywood, P. T. Coverley, W. J. Staszewski, and K. Worden, "An automatic impact monitor for a composite panel employing smart sensor technology," *Smart Materials and Structures,* vol. 14, p. 265, 2005.
[10] M. Ghajari, Z. Sharif Khodaei, and M. Aliabadi, "Impact Detection Using Artificial Neural Networks," *Key Engineering Materials,* vol. 488, pp. 767-770, 2012.
[11] R. Olsson, "Mass criterion for wave controlled impact response of composite plates," *Composites Part A: Applied Science and Manufacturing,* vol. 31, pp. 879-887, 2000.

Electro-mechanical Impedance Technique for Structural Health Monitoring of Composite Panels

Martin Schwankl[1,a], Z. Sharif Khodaei[2,b], M.H. Aliabadi[2,c] and C. Weimer[1,d]

[1]Eurocopter Deutschland GmbH, Willy-Messerschmitt-Straße, 85521 Ottobrunn
[2]Department of Aeronautics, Imperial College London, South Kensington Campus, Roderic Hill Building, Exhibition Road, SW7 2AZ, London, UK.

[a]martin.schwankl@eurocopter.com , [b]z.sharif-khodaei@imperial.ac.uk , [c]m.h.aliabadi@imperial.ac.uk and [d]christian.weimer@eurocopter.com

Keywords: Electro-Mechanical Impedance, Damage detection, Structural Health Monitoring.

Abstract. Numerical modelling of EMI for damage detection has been presented in this paper. The PZT model is validated against the published experimental result for free disk and tied to the structure. The numerical modelling of the PZT patch will result in the admittance measure of the structure. The imaginary part of the admittance measure is used for developing a self-diagnostic sensor system. The real part of the admittance measure was used to develop a damage detection algorithm. Damage detection using EMI method was successfully applied to a simple composite disk and a stiffened panel. The EMI method is suitable for short range damage detection in structural parts with limited or no access.

Introduction

Over the past decades, the aerospace industry has increasingly used composite materials as structural parts due to their outstanding properties over conventional materials, such as high specific strength and stiffness and excellent fatigue resistance. However, the presence of damage can result in significant stiffness losses and drastic failures. At present, the aircraft industry faces this problem with high safety factors as well as short maintenance and inspection intervals resulting in high in-service costs for airlines. Hence, recent research focuses on the development of Structural Health Monitoring (SHM) systems by utilisation of so-called smart sensorised structures capable of detecting damage [1-2] in order to shift the schedule-driven to condition-based maintenance.
One particularly promising SHM technique is the Electro-mechanical Impedance (EMI) method [3] which is cost-effective and categorised under the active near-field techniques. Therefore, a piezo-ceramic (PZT) patch is bonded to or embedded in the monitored structure and excited by a sinusoidal voltage sweep with low amplitude and frequencies varying from 10-500 kHz [4], which excites the surrounding structure to vibrate due to their bonding [5]. The structural response of the interrogated area is measure at the same PZT patch in form of the electromechanical admittance. The presence of damage in the structure will alter its response, thus the change in the electromechanical admittance can be related to the presence of damage.
In this paper a numerical study has been performed to detect damage in composite structures. The study first presents the numerical modelling of the PZT transducers and the validation against published experimental results. It then continues by applying it first to a simple structure and studying the influence of damage size and severity (damage index) on the electromechanical admittance measures. It is then finalised by applying the EMI method to a stiffened panel.

FE model

The electromechanical admittance comprises of the conductance (real part) and susceptance (imaginary part), which can be decomposed to obtain the impedance parameters. The structural impedance $Z_s(\omega)$ is directly related to mass, damping and stiffness as shown in Equation 1:

$$Z(\omega) = \left[j\omega C \cdot \left(1 - \kappa_{31}^2 \cdot \frac{Z_s(\omega)}{Z_s(\omega) + Z_{PZT}(\omega)}\right)\right]^{-1}, Z_s(\omega) = j\omega m(\omega) + c(\omega) - \frac{jk(\omega)}{\omega} \qquad (1)$$

Where $m(\omega)$ the mass, $c(\omega)$ the damping and $k(\omega)$ the stiffness of the structure are. The PZT patch is coupled with the structure and consequently the overall impedance $Z(\omega)$ recorded by the sensor comprises of both the structural impedance $Z_s(\omega)$ and the sensor impedance $Z_{PZT}(\omega)$ as shown in Equation 1 [6]. C denotes the zero-load capacitance and κ_{31}^2 the electromechanical coupling factor. The EMI method proposed in this paper is based on numerical simulations using the commercial FE code ABAQUS together with an in-house code developed in MATLAB. Due to absence of experiments in this study, the FE model has been validated against published results in [7] first for a free PZT patch and second for a patch tied to an Aluminium disk. The PZT disc was excited by a sinusoidal 10V voltage sweep. A mesh convergence investigation was carried out to provide reasonable results within the EMI relevant frequency range. The validations of the free and bonded PZT patches are depicted in Figure 1 and Figure 2. The deviation between the real and simulation impedance signature originates from simplifications such as neglected adhesive layer and electromechanical coupling factor of the patch.

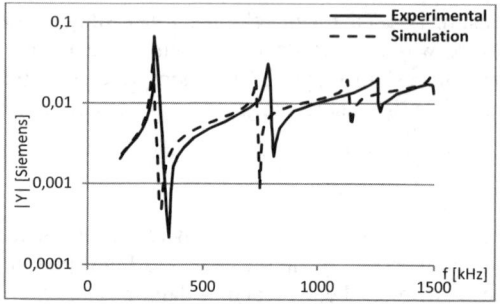
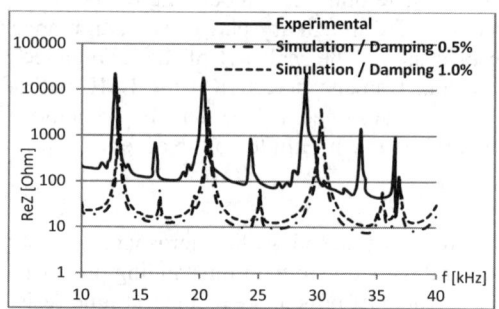

Figure 1: Magnitude of admittance signature of free PZT patch as a function of frequency

Figure 2: Real Part of impedance signature for PZT patch bonded to Aluminium disc

Self-Diagnostic

The gained knowledge was subsequently transferred to a CFRP disc with a $[+45/-45/0/90]_s$ layup consisting of eight HTA/977-2 carbon fibre unidirectional prepregs with 0.252 mm ply thickness. It was reported in [7] that the imaginary impedance signature is indicative for the bonding between PZT patch and surrounding structure. Hence, this property can be used to develop self-diagnostic capability of the sensorised structure.

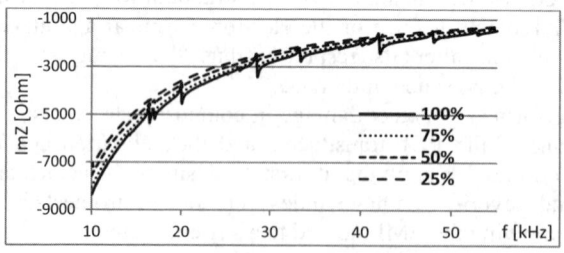
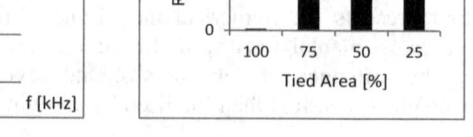

Figure 3: Imaginary Impedance Signature vs. frequency

Figure 4: RMSD PZT patch

Figure **3** shows the deviation of the imaginary parts of the impedance signatures for different percentages of the bottom surface of the PZT patch tied to the structure. Root Mean Square Deviations (RSMD) between the fully tied and debonded PZT patches are presented in Figure **4**. It shows a linear increasing pattern in the RMSD value as the debonded area increases.

Damage detection using EMI

In this study, material degradation (softening) was incorporated into the composite disc while the effect of parameters such as distance to PZT transducer, radius and softening on the Impedance measures were investigated. It can be seen in Figure 5 that as the damage radius increases, the higher the deviation between the pristine and damaged impedance signatures becomes. A similar pattern could be shown for the damage severity and distance variations.

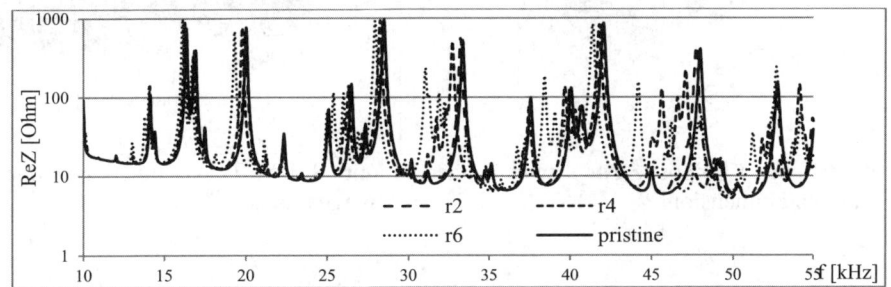

Figure 5: Real part of impedance signature as a function of frequency for different radii and constant distance and softening

When the RMSD values for different material softening were plotted against the damage distance from PZT patch and damage radius, it was observed that the softening and damage size (damage severity) have significant influence on the Impedance signature. Therefore a damage index D has been introduced by Equation 2 and the RMSD plotted as a function of Damage Index in Figure 6.

$$D = \frac{Softening\,[\%] \cdot r^2[mm^2] \cdot \pi}{100} \qquad (2)$$

Additionally, it can be noted that damages closer to the sensor result in a higher RMSD for a similar damage severity.

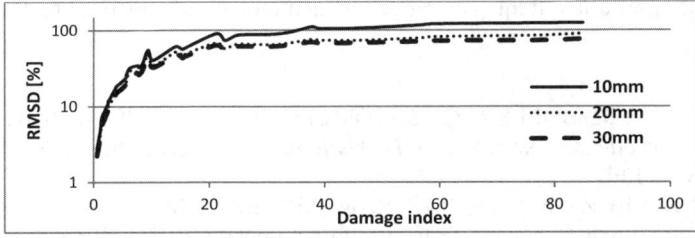

Figure 6: Damage index as a function of frequency range for different damage distances

Application to stiffened panel

The damage detection capability of the EMI method was proved on the simple CFRP disc and in the following applied to a composite stiffened panel with the same layup and material as depicted in Figure 7. At first, one patch was excited in the frequency range of 10-55 kHz while damage distance and radius were varied. It can be seen in Figure 8 that the EMI method was capable of detecting damage in the used frequency range. Moreover, the value of RMSD is increasing with damage severity (radius). However, it is apparent from Figure 8 that the pattern is ambiguous for the distance variation, which is due to the relatively low used frequency range.

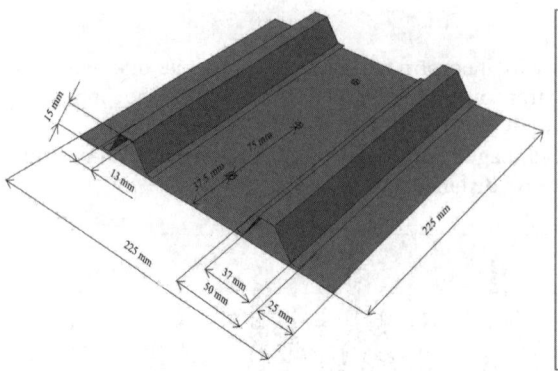 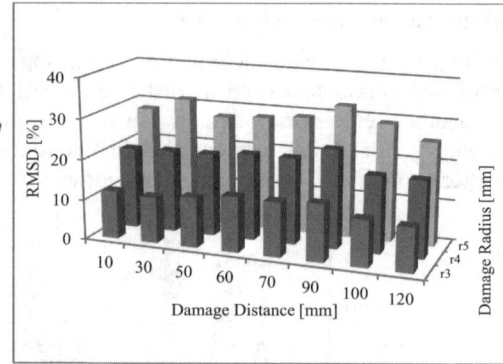

Figure 7: Stiffened Composite Panel with bonded PZT patches and dimensions

Figure 8: RMSD as a function of distance and damage radius

Conclusion

In summary, the numerical simulation of the EMI method was successfully carried out and proved in this work. Beginning with the model validation against experimental data, the gained knowledge was subsequently transferred to the monitoring of composite structures. Therefore, a circular PZT was bonded to a CFRP disc and excited by a sinusoidal voltage sweep. The self-diagnostic capability of the method was shown for detecting debonded transducers and therefore resulting in a reliable SHM system. Furthermore, the damage detection capability with regard to material degradation was successfully modelled. It was established, that the damage area and percentage of softening have a significant influence on the impedance deviation while damage locating capability is mainly related to the sensing area and thus working frequency range. Due to limitations in computational resources, it was not possible within this work to investigate higher frequency ranges. Although the EMI method is particularly promising due to advantages such as high sensitivity, resistance against in-service mass load and vibrations, further research is necessary on the selection of the appropriate frequency range for each monitored structural part.

References

[1] STASZEWSKI, W. J., BOLLER, C., & TOMLINSON, G. R. (2004). *Health Monitoring of Aerospace Structures: Smart Sensor Technologies and Signal Processing.* Chichester: John Wiley & Sons Ltd.

[2] SHARIF KHODAEI Z., ROJAS-DIAZ R. & ALIABADI M.H. (2012). Lamb-wave Based Technique for Impact Damage Detection in Composite Stiffened Panels. *Key Engineering Materials Vols. 488-489 pp 5-8*

[3] LIANG, C., SUN, F., & ROGERS, C. (1994). Coupled Electro-Mechanical Analysis of Adaptive Material Systems - Determination of Actuator Power Consumption and System Energy Transfer. *Journal of Intelligent Material and Systems Structures.*

[4] NAIDU, A., & SOH, C. (2004). Identifying damage location with admittance signatures of smart piezo-transducers. *Journal of Intelligent Material Systems and Structures.*

[5] ANNAMADAS, V., & SOH, C. (2009). Application of Electromechanical Impedance Technique for Engineering Structures: Review and Feature Issues. *Journal of Intelligent Material Systems and Structures.*

[6] GIURGIUTIU, V., & ROGERS, C. (1998). Recent Advancements in the Electro-Mechanical (E/M) Impedance Method for Structural Health Monitoring and NDE. *Proc. SPIE 3329.*

[7] ZAGRAI, A. N. (2002). *Piezoelectric-Wafer active sensor Electro-Mechanical Impedance Structural Health Monitoring.* University of South California.

Dynamic Response Analysis of Lining Structure with Primary Defects for Tunnel under Moving Train Loading

Yuping CUI[1,2,b], Weize SUN[2,a], Jun DONG[2,a], and Fei DONG[2,a]

[1] China Communication Construction Road and Bridge Consultants Co., Ltd, Beijing 100029, China
[2] School of Civil Engineering and Traffic Engineering, Beijing University of Civil Engineering and Architecture, Beijing 100044, China

[a]email: jdongcg@bucea.edu.cn, [b]email: cuiyuping@163.com

Keywords: subway tunnel, FDM, primary defects, dynamic response

Abstract. Considering the background of vibration engineering of lining structure of subway tunnel with primary defects under moving train loading, and according to the measured information of a dynamic loading spectrum of subway in Beijing, dynamic response of the lining structure with differently primary defects caused by neighboring tunnel construction has been investigated for different cases including one-way and two-way trains in this paper. The results show that the curves of dynamic time history analysis are similar with each other for displacement and stress under different cases with differently primary defects. The peak value of displacement and stress under the case of two-way train is much larger than that under the case of one-way train, and the structure primarily depends on tensile strength of concrete in safety. Under the case of one-way train with primary defects would there be no defect by vibration loads.

Introduction

Since 1980s, the vibrations caused by trains across a tunnel have been paid many attentions in the world. Especially in China where the construction of a city subway has entered a new era of rapid development, there are much existed subways and new subways in many big cities such as Beijing, Shanghai and Guangzhou. Although they have improved the traffic conditions of cities efficiently, a series of safety problems about the tunnels attracting much more attentions since some of them have been subjected to the long-term vibration of moving trains. Many scholars from home and abroad studied the problem from different ways. For examples, Gardien [1] and [2] Andersen, investigated dynamic response of the tunnels by 3-dimensional FEM and FE-BE coupled method. It is evident that numerical simulation will be accepted by more and more people. Based on the background of vibration engineering of lining structure of subway tunnel with primary defects under the loads of moving trains, dynamic response of the tunnel structures will be investigated in this paper.

Model and Related Parameters

Engineering Background. ShuangJing station, an island type platform station of three underground layers with two spans, is a station of Beijing Subway Line No.10th, whose length is 181m. The basement storey 1 does mainly belong to equipments layer, and the basement storey 2 and 3 are the station hall and the platform respectively. The parts including south and north are the structures with open cut soil, and the middle span is undermining structure. There is a deformation joint (about 20 mm) between middle span and three underground south/north layers respectively. Due to this factor, a simple model could be established and investigated, as shown in Fig. 1~2.

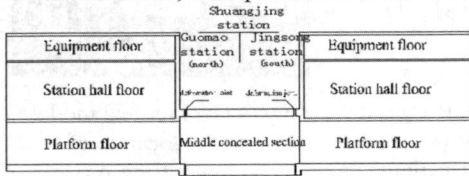

Fig. 1 Structure of the station

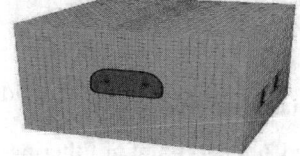

Fig. 2 Structure of the station

Model. As is shown in fig. 2, the selection of model takes account into the boundary effect and effect scope of structure. The length, width and height of the model are 61.3 m, 59.24 m and 41.55 m, respectively. The thickness of soil on the arch structure of tunnels is 13.78m. The geometrical model is divided into 34920 elements with 3-D 10-node structural solid and 6000 elements with 2-D lining plane element.

Boundary Conditions and Parameters. As shown in Table 1, the soil could be regarded as isotropy medium, constitutive model of soil and lining structure are M-C and elastic, respectively. The density, bulk and shear are 2.5g/cm^3, 11.46GPa, and 15.28 GPa, respectively.

Table 1   The physical and mechanical parameter of rock

Soil property	Thickness (m)	density (Kg/m^3)	Deformation (MPa)	Poisson's ratio	Internal friction angle(°)	Cohesion (kPa)
powder soil fill	3.0	2014	14	0.25	24.8	14
miscellaueous fill	6.2	1930	5.6	0.3	12.5	11.0
silry clay	8.0	2010	5.6	0.3	12.5	11
sand	10.0	2000	22	0.35	32	0
gravel	14.55	2220	90	0.3	45	0

The static-boundary scheme proposed by Lysmer and Kuhlemeyer (1969) involves dashpots attached independently to the boundary in the normal and shear directions. The dashpots provide viscous normal and shear tractions given by

$$t_n = -\rho C_p v_n \tag{1}$$

$$t_s = -\rho C_s v_s \tag{2}$$

Where v_n and v_s are the normal and shear components of the velocity at the boundary, ρ is the mass density; and Cp and Cs are the p- and s-wave velocities. Rayleigh damping was originally used in the analysis of structures and elastic continua to damp the natural oscillation modes of the system. The equations, therefore, are expressed in matrix form. A damping matrix C, is used, with components proportional to the mass matrix M and stiffness matrix K:

$$C = \alpha M + \beta K \tag{3}$$

Where α is the mass-proportional damping constant; and β is the stiffness-proportional damping constant. Frequency of oscillation of approximately 11.11 Hertz is observed when the model has been applied by vibration. Damping ratio of experience method selection is 0.005.

Moving Tran Loading Spectrum

Real vibration could be reflected in acceleration spectrum and load spectrum, whose mathematical expressions of the train loads could be given according to D. Alembert principle.

$$P(t) = (m_1 + m_2 + m_3)g + m_1\ddot{Z}_1 + m_2\ddot{Z}_2 + m_3\ddot{Z}_3 \tag{4}$$

Where m_1 is wheel mass, and m_2, m_3 is mass of different spring. $\ddot{Z}_1, \ddot{Z}_2, \ddot{Z}_3$ are the acceleration of different masses, respectively. g is gravity acceleration. As shown in Eq. 4, the load spectrum and acceleration spectrum are interconvertible with each other. The site measurements, which are discrete, need to be corrected by Filter and Baseline Correction.

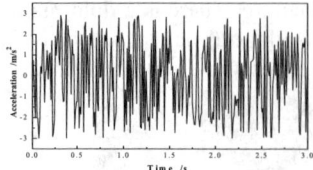
(a) the measured acceleration

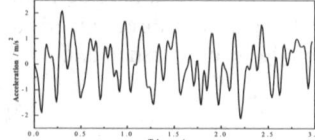
(b) acceleration of Filtering and Baseline Correction

Fig. 3 Contrast figure of Filtering and Baseline Correction

Fig. 4 Geometrical model and the location of monitoring A

Filtering could reduce the number of elements and save computing time, and Baseline Correction could make the final displacement and speed be zero so as to guarantee the accuracy of dynamic calculation.

Vibration Response of Underground Structures

There are two cases considered in this paper. The first one is lining of Line No.10 with or without primary defect which is said the response of line 10 lining affected by one-way vibrating loads before and after line No.7 construct, and the other one is response of line No.10 lining affected by two-way vibrating loads when the lining with or without primary defect. Many literatures [5] [6] show that fierce response take placement at bottom of the rail, inverted arch and the side wall of arch. According to the conclusions referred before, we took the most danger point A for consideration to analysis the vibrating response of the lining. The location of point A is shown in Fig. 4.

Response of One-Way Loads. The displacement spectrum at the point A affected by one-way loads is shown in Fig. 5. As is shown in Fig. 5, the response of displacement and stress are fierce at the beginning of vibrating loads. The vibrating loads make the lining began to vibrate up and down and the whole structure with tendency of downside. The maximum displacement is respectively 1.19mm and 1.10mm in the case of lining with or without primary defect. The curves are shown in Fig. 5 (a). Fig. 5 (b) shows the maximum primary pressure stresses are 0.981MPa and 1.03MPa and the minimum primary strain presses are -1.333MPa and -1.37MPa. The stress of lining structure are much smaller than the limit of concrete stress and the vibrating loads may not cause the lining structure destroy.

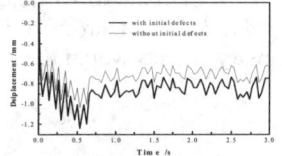 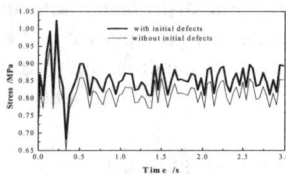

(a) vertical displacement spectrum at the point A
(b) minimum primary stress spectrum of rail bottom
(c) maximum primary stress spectrum of rail bottom

Fig. 5 Dynamic response under one-way vibrating loads at the point A

Response of Two-Way Loads. The following gives the response of two-way loads. The comparisons of response at the point A under one-way and two-way vibrating loads have been done, as shown in Fig. 6 and Table 2 including displacement spectrum and stress spectrum.

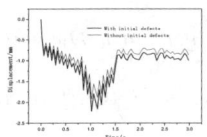 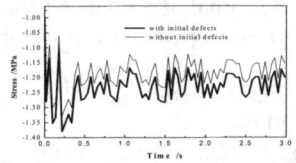 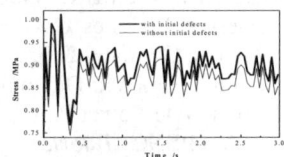

(a) vertical displacement spectrum of point A
(b) minimum primary stress spectrum of rail bottom
(c) maximum primary stress spectrum of rail bottom

Fig. 6 Response of point A under two-way vibrating loads

Table 2 Peak value of point A under two cases

Cases		Response of point A		
		Displacement (mm)	Max primary stress (MPa)	Min primary stress (MPa)
Case one (one-way)	without primary defect	1.10	0.981	-1.333
	with primary defect	1.19	1.03	-1.37
Case two (two-way)	without primary defect	1.97	0.988	-1.338
	with primary defect	2.21	1.038	-1.376

As is shown in Fig. 6, the two-way loads could affect each other. The peak value of displacement is 2.21mm which take place when the two-train run in the same time. The curves in Fig. 6 (b) and (c) show the lining without primary detect. In the graphs the maximum primary stress and the minimum primary stress are all smaller than the limit of concrete, and this will not cause damage.

The displacement in two cases are different, in other words, the two-way loads can affect each other but not so remarkable. In two cases, the maximum strain stresses of point A are respectively 1.03MPa and 1.038MPa reach to 56.0% and 56.5% of the limit of concrete. The maximum pressure stresses are 1.37MPa and 1.376MPa reach to 10.3% and 10.4% of the limit. We can get the same conclusions to literature [7] that the destroy regulation is controlled by strain reinforcement. After down through, the soil is disturbed which make the stress in the layer changed, so the responses of defect are fierce than without defect.

Conclusion

Considering dynamic response of lining structure of subway tunnel under the loads of train, The results are as follows: In different conditions, displacement and stress is changing by the same way at the bottom of tunnel. The maximum of displacement and stress occurred in the early vibration , and this kind of phenomenon recovers soon. Being forced to vibrate up and down in a balanced position. vibration load had little impact each other in two-way tloads. Line No.7 down through line No.10 under the case of two-way train is unfavorable case of line No.10 The largest displacement of vibration is far less than structure deformation limits, strength of lining structure is controlled by the tensile Strength. The peak value of displacement and stress under the case with primary defects is much larger than that under the case without primary defects, Line No.7 down through line No.10 cause soil disturbance and make soil distribute again and lining structure of line No.10 defect. Tensile and compressive stress of lining structure at the bottom of tunnel is far less than the limits strength of concrete, in a word, under the case of one-way train with primary defects would there be no defect by vibration loads.

Acknowledgements

This work was financially supported by 2011Science and Technology Research Project of Beijing Municipal Education Commission (KM201110016012) and by Beijing Higher Institution Engineering Research Center of Structural Engineering and New Materials, BUCEA.

References

[1] Gardien W, Stuit H G: submitted to Journal of Sound and Vibration (2003).
[2] Andersen L, Jones C J C, in: *Proceeding of 5th European Conference Structural Dynamics* (Munich, Germany 2002).
[3] Yumin CHEN and Dingping XU: *Fundamentals and Practical Engineering of FLAC/FLAC 3D* (China WaterPower Press, China 2008).
[4] Wenbin PENG: *Tutorials of FLAC 3D* (China Machine Press, China 2007).
[5] Gui ZHANG, Jian WANG: submitted to Railway Engineering (2008).
[6] Juan HUANG: *Study on the Vibration Response and Fatigue Life of High-speed Railway Tunnels Based on Damage Theory* (Ph.D., Central south University, China 2006).
[7] ChenHua SHI, LiMin PENG: submitted to Journal of Experimental mechanics (2005).

Monitoring Analysis of Tunnel Construction in a New Subway Line Down-Through an Existed Line

Fei DONG[1], Jun DONG[1,a], Yuping CUI[1,2], Weize SUN[1] and Dongyong LI[2]

[1] Beijing University of Civil Engineering and Architecture, Beijing 100044, China
[2] Road and Bridge Consultants Co., Ltd, CCCC, Beijing 100029, China
[a]email: jdongcg@bucea.edu.cn

Keywords: subway tunnel, construction monitoring, sedimentation, numerical simulation

Abstract. For a practical engineering in which the tunnel of Subway Line No.7 between Guangqumen outside Station and Shuangjing station is down through Shuangjing Station of existed Subway Line No.10, numerical simulations of construction process of the tunnel have been investigated by FLAC 3D software and compared with site monitoring data. The sedimentation curves of the earth's surface along the cross-sectional and longitudinal direction are obtained, which show that all results including numerical simulation and site monitoring are accord with the data from the Peck formula. At the same time according to numerical computation, some relations between width and depth of excavating a tunnel are discussed. All these have important significant to theoretical research and engineer practice of subway construction.

Introduction

With municipal subways fast developing in China, there are not only many existed subways in service, but also new subways under construction. Many problems including such security, durability and stability of subway tunnels are being been paid enough attention, for example, influence of new subway tunnels excavation to adjoining existed one or buildings. The new Subway Line No.7 is under construction, which will begin at the origination Beijing Western Railway station and end at the destination Beijing Chemical Factory. All of its interval tunnels and stations are underground, and their length of whole rails is 23.67 km with average interval of 1.14 km. The twin-track tunnel between Guangqumen Outside station and Shuangjing station is under the Guangqumen outside Street which is a busy road. In this interval the tunnel cross the Shuangjing station of subway line No.10. The diameter of the line No.7 tunnel is 6.2m and the distance between it and the exited one is only 1m. The exited station is island-type which is underground structure has three-storey with two spans (part of it has three spans). Both of the north and south part of the station are three-storey structures and the other part between them is concealed tunnel. The vertical section of the station is shown in Fig. 1 as follows. The geology from top to the bottom is powder soil fill, miscellaneous fill, powder soil, silty clay, sand and gravel. The layer property is shown in Table 1.

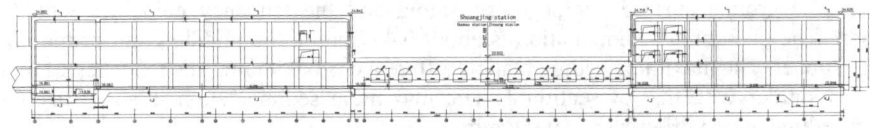

Fig. 1 Vertical section of Shuangjing station

Table 1 Layer Property

Soil property	Thickness (m)	Density (kg/m³)	Deformation modulus (MPa)	Poisson's ratio	Internal friction angle (°)	Cohesion (kPa)
powder soil fill	3.0	2014	14	0.25	24.8	14
miscellaneous fill	6.2	1930	5.6	0.3	12.5	11
powder soil	4.8	2014	14	0.25	24.8	14
silty clay	8.0	2010	5.6	0.3	12.5	11
sand	10.0	2000	22	0.35	32	0
gravel	14.55	2220	90	0.3	45	0

Construction Monitoring

Site Monitoring. According to the Beijing local standard *Technical Code for Monitoring Measurement of Subway Engineering (DB11/490-2007)*, the ground monitoring points are assigned along the tunnel direction. Usually, arrange these points along the midline of the two tunnels and the subway station. The longitudinal spacing of these points can be chose from 5 to 30 meters according to the real condition.

The number of the transverse monitoring section is 2~3 sections at the station and 3~5 at the interval, each section has 7~11 points. The distance between the most outside of these points and the station structure should be more than twice buried deep. The site monitoring points are shown in Fig. 2 and Fig. 3.

Fig. 2 Site monitoring point I

Fig. 3 Site monitoring point II

Simulation for Monitoring. The size of the calculation model is 61.3 m*59.24 m*41.55 m with two monitoring sections. Section 1 is perpendicular to line No.10 tunnel, 30 m apart from the south boundary and section 2 along the line No.7 above the two tunnels at the midline on the ground. The points' distance is 3 m. The computation model and monitoring sections are shown in Fig. 4.

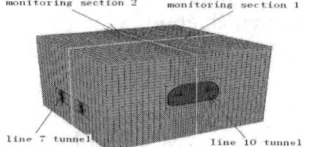

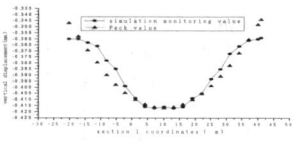

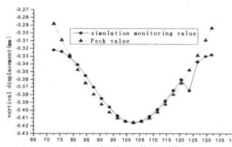

Fig. 4 Computation model and monitoring sections

Fig. 5 Subsiding curve after line 7 tunnel excavated at section 1

Fig. 6 Subsiding curve after line 7 tunnel excavated at section 2

Sedimentation Curves after Excavation. In the simulation, the left tunnel of line No.7 is excavated at first. During this process, keep the distance between the two-excavation faces until the end. After that, the vertical displacements of the points are recorded and the curves are drawn, which is shown in Fig. 5 and Fig. 6. The solid line in Fig. 5 represents the subsidence character of section 1 which is calculated by the software [1]. Because of stress release [4] in the layer after excavation, the ground surface begins to subsiding and the tendency coincides with Gaussian distribution. The maximum sedimentation is about 0.417mm at the middle point of the midline of line No.10 tunnel. The maximum sedimentation difference is 0.058mm. The tendency of monitoring section 2 is similar with that of section 1. The maximum sedimentation is about 0.416mm and maximum sedimentation difference is 0.094mm.

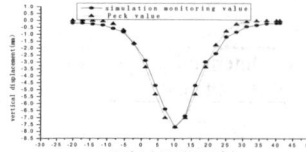

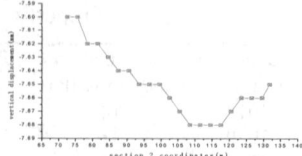

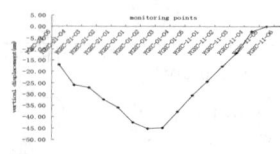

Fig. 7 Subsiding curve only line 10 tunnel excavated at section 1

Fig. 8 Subsiding curve only line 10 tunnel excavated at section 2

Fig. 9 Real subsiding curve

Because of the two sections are respectively perpendicular to line No.10 and No.7 tunnel, the curves both accord with Gaussian distribution at the two sections. Only one Gaussian curve can be seen if only one tunnel excavated. Fig. 7 and Fig. 8 give the subsidence character of the two sections when only line No.10 tunnel excavated. Fig. 9 show the subsidence curve accord to the real data belongs to the Jiulongshan station to Dajiaoting station of line No.7. Monitoring point YGXC-01-03 is located at the middle of the tunnel. From the drawing we can get that the real curve coincides with Peck formula.

According to the result of literature [6], in the isotropy soil, influenced by the excavated the subsiding tendency curve can be considered to be Gaussian curve, the formula is given as follows:

$$y = \omega e^{-x^2/2i^2} \qquad (1)$$

Where y represent the vertical displacement where the point is x meter apart from the tunnel midline, ω represent the maximum vertical displacement, i represent the distance between the turning point of the curve and the tunnel midline. Peck gives the formula to calculate the volume of the subsiding geosynclines:

$$V = \int_{-\infty}^{+\infty} y dx = \sqrt{2\pi} i \omega \qquad (2)$$

The experience formula for the width of the subsiding geosynclines $\Re$ is given as follows:

$$\Re = 2\tan(45^\circ - \varphi/2)Z + D \qquad (3)$$

Where $\Re$ represent the width of the subsiding geosynclines, Z is the burial depth, D is the diameter. The subsiding tendency along with the longitudinal direction can be described as two symmetrical piecewise functions:

$$\begin{cases} S = -A[1 - e^{(x-x_0)B}] + U_0, x < x_0 \\ S' = -A[1 - e^{-(x-x_0)B}] + U_0, x > x_0 \end{cases} \qquad (4)$$

Where A and B are regression coefficient, X_0 and U_0 are the coordinates of the turning point. The dash lines in Fig.5 to Fig.7 represent the curves calculate accord to Eq. 1. Compared any two of them in the same drawing, the curves calculated are coincide with the Peck formula. It is the same to the subsiding geosynclines width.

Influence Range of Excavation Face. Taking the cross point of section 1 and 2 for an example, the curve of relationship between the point's vertical displacement and the distance from the point to excavation faces is drawn, as shown in Fig. 10 and Fig. 11. In the graphs, the positive numbers represent the excavation face haven't passed the point and the negative ones represent passed. The values represent the horizontal distance of the faces and the point.

From the graphs, it is seen that the distance between the faces and point less than 3 times of the tunnel diameter is the main subsiding area and it will be steady when the distance is more than 5~8 times of the tunnel diameter.

The result of reference [7] shows that the sedimentation caused by excavation can be divided into four stages: (1) Tiny deformation stage. This occurred when the distance are 1~1.5 times of the tunnel diameter. This subsidence is about 15%~20% of the total value caused by the layer stress variation and the underground water loses. (2) Rapid deformation stage. From the distance is twice of the tunnel diameter to the face passed the monitoring point. In the coverage of the 3 times of the diameter, the velocity and value of the deformation become larger than the other stages. This subsidence is about 50%~60% of the total value caused by the boundary condition change and the stress redistribution. (3) Slow deformation stage. When the face pass the monitoring point and the distance is 3~5 times of the tunnel diameter. This subsidence is about 15%~20% of the total value caused by the consolidation when the lining becomes a circle one. (4) Steady deformation stage.

When the face passed the monitoring point and the distance is more than 5 times of the tunnel diameter, the main reason is the layer deformation become steadily. This subsidence is about 5%~10% of the total value.

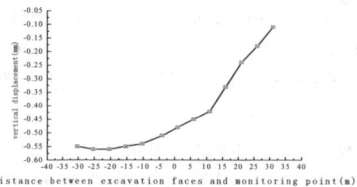 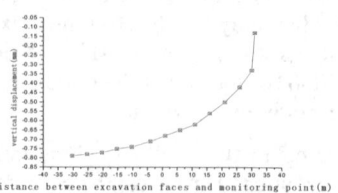

Fig. 10 Relationship of the distance between the excavation faces and the monitoring points (left)

Fig. 11 Relationship of the distance between the excavation faces and the monitoring points (right)

Conclusion

Based on above simulation of line No.7 down through the No.10 at Shuangjing station, it is proved that the regulation of subsiding tendency and the range of excavation are both reliable, described as follows: The subsiding curve is a Gaussian curve and coincide with the Peck formula, which show that the subsiding line caused by excavation can be considered to be a probability statistics law, the equation is $y = \omega e^{-x^2/2i^2}$. The subsiding tendency along the lengthways direction can be described as two symmetrical piecewise functions. The main subsiding stage is rapid deformation stage and the steady subsidence occurred in the distance is 5~8 times that of the tunnel diameter.

Acknowledgements

This work was financially supported by 2011Science and Technology Research Project of Beijing Municipal Education Commission (KM201110016012) and by Beijing Higher Institution Engineering Research Center of Structural Engineering and New Materials, BUCEA.

References

[1] Yumin CHEN and Dingping XU: *Fundamentals and Practical Engineering of FLAC/FLAC 3D* (China WaterPower Press, China 2008).

[2] Bo LIU and Yanhui HAN: *Theory, Practical Engineering and Application Guide of FLAC* (China Communication press, China 2005).

[3] Wenbin PENG: *Tutorials of FLAC 3D* (China Machine Press, China 2007).

[4] Peiling HE and Ting ZHANG etc: *Engineering Geology* (Peking University Press, China 2006).

[5] Minghua ZHAO: *Soil Mechanics and Foundation Engineering* (Wuhan University of Technology Press, China 2004).

[6] Peck,R.B, in: *Petrasovits G, Mecsi J Proceedings of the 7th International Conference on Soil Mechanics and Foundation Engineering*, (Mexico City, Mexico, 1969). State of the Art Volume.

[7] Mengshu WANG: *General Theory of Shallow Burying Subsurface Excavation Technology for Underground Works* (Anhui Educational Press, China 2004).

SMART Platform for Structural Health Monitoring of Sensorised Stiffened Composite Panels

Z. Sharif Khodaei[1,a], M. Ghajari[1,b], M. H. Aliabadi[1,c] and A. Apicella [2,d]

[1]Department of Aeronautics, Imperial College London, South Kensington Campus, Roderic Hill Building, Exhibition Road, SW7 2AZ, London, UK.

[2]Alenia, Viale dell'Aeronautica, Pomigliano d'arco, 80038, Naples, Italy

[a]z.sharif-khodaei@imperial.ac.uk, [b]m.ghajari@imperial.ac.uk, [c]m.h.aliabadi@imperial.ac.uk, [d]caapicella@alenia.it

Keywords: Structural Health Monitoring, Damage detection, Ultrasonic Guided Waves, Electromechanical Impedance, Impact detection, Artificial Neural Network.

Abstract. A SMART Platform is developed based on sensor readings for Structural Health Monitoring of a stiffened composite panel. The platform's main function is divided into three categories: Passive sensing, Active sensing and Optimal sensor positioning. The platform has self-diagnostic capabilities, i.e. prior to its application the health of the sensors and their connection will be checked to avoid any false alarm. Passive sensing results in impact location and force magnitude detection. Active sensing is performed for damage detection. It results in detecting the damage location and severity. Finally the optimal sensor location can be provided given the number of sensors and probability of detection value. This platform is the first step in applying the developed SHM methodologies to real size structures in service load conditions.

Introduction

The concept of Structural Health Monitoring (SHM) techniques for monitoring of composite structures has increasingly being evaluated as a potential method to reduce the maintenance activities and operational costs as well as to improve the aircraft safety and reliability. The main principle of SHM techniques is based on distributed sensor-actuator networks which can continuously monitor the status of the structure with minimal human involvement.

There are a number of commercially available SHM technologies in the market such as SMART Layer by Acellent technologies [1], PAMELA SHM system by Aernnova [2] and Boeing SHM system [3]. However, most of the available research and experiments have been performed on simple structures on ground. Therefore there are still missing steps to allow the SHM systems to be applicable to real commercial aircrafts. During the JTI – Cleansky project SMASH the first step was taken for developing a robust and reliable SHM platform capable of detecting and characterising impact and damage in actual structures under real load conditions. The aim of the SMASH project was to develop methodologies for impact and damage detection for a sensorised composite aircraft panel. A platform was established (see Figure 1) for the project which consists of different modules. The main objective of the platform can be divided into three different modules: 1) Passive sensing: Capable of detecting impact location and magnitude, 2) Active sensing: Capable of damage detection and characterization 3) Optimisation: Finding the best sensor location.

SMART Platform

The developed methodologies are based on sensor technologies composed of PZT, Fibre Optic and hybrid systems. Two main methodologies were adopted for active sensing modules: Guided Ultrasonic Wave (GUW) and Electro-Mechanical Impedance (EMI) methods. GUW methods were applied to sensorised panels employing both Piezoelectric (PZT) transducers and hybrid system of sensors and actuators (PZT actuators and Fibre Optic sensors). However, EMI methods can only be applied to panels with PZT transducers. The damage detection methodologies are based on comparing the current state of the structure to a reference structure without any damage.

Accurate prediction of damage due to impact requires knowledge of impact force and location as well as structural prediction of damage. Two main achievements of the SMART platform are predicting impact location and force magnitude and damage detection and characterisation. To achieve both main goals, the first step was to simulate various impact events on a sensorised composite panel alongside simulating the behaviour of a sensorised panel in terms of sensing and actuating within a computational tool. This resulted in development of a SMART FEM (Finite Element Method) that could replicate experimental impact events on a sensorised panel (see [4]) together with both actuating and sensing abilities for damage detection methodologies (details can be found in [5] and [6]).

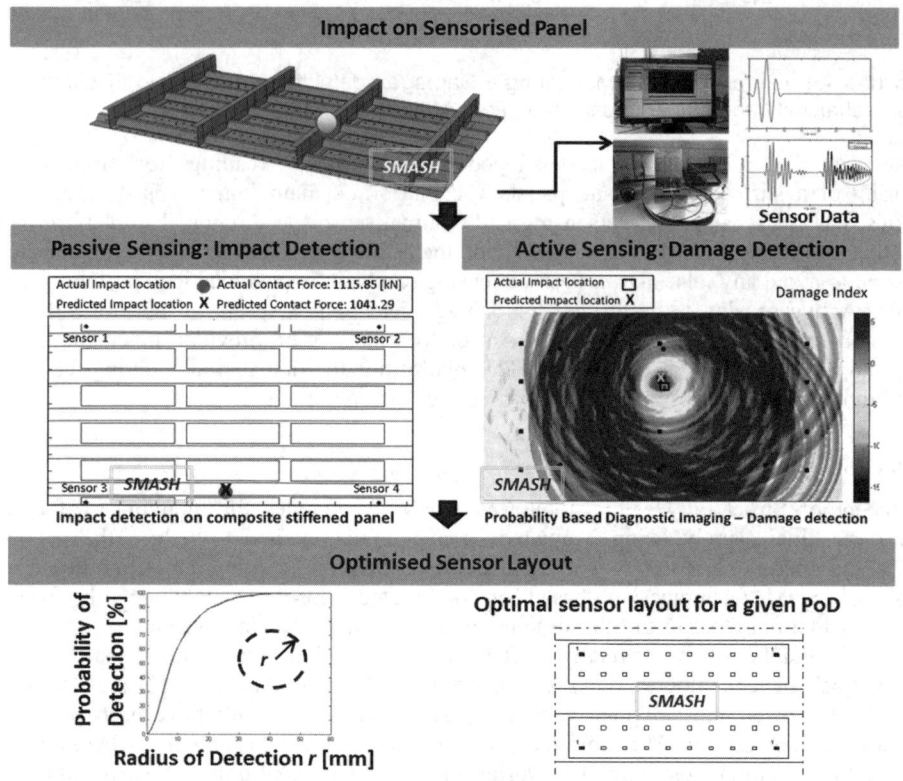

Figure 1 SMART Platform developed during SMASH project

Impact Detection and Identification

Passive sensing refers to the case when the PZT transducers are used as sensors only. Passive sensing module can be used for prediction of an impact event. When an impact event happens on a sensorised panel, it creates surface waves which after propagating through the plate can be sensed by the sensors. These signals can then be used for locating the impactor and approximating the impact force. Once the location of the impact is known, a local damage detection search can be performed. Moreover, if a damage threshold can be defined for the structure in question, approximating the impact force can give information about the probability of existence of damage and whether an impact force of a given magnitude can be hazardous to the structure or not. Thus it can result in reduction in maintenance cost and time. Moreover, having gained the knowledge of impact location and max contact force, a full damage analysis can be performed to achieve more information about the type and extent of damage that such an impact force can produce in a composite structure.

For the SMART platform an Artificial Neural Network has been used to develop a meta-model for approximating the impact location and force magnitude. For details of the ANN development see [4]. An example of Passive sensing resulting from the SMART Platform is shown Figure 1 with the actual and predicted location and magnitude of impact force.

Damage Detection and Characterisation

Active sensing comprises of sensing and actuating which results in damage detection step. Damage detection methodologies which have been developed for the developed SMART Platform can be divided into two main categories as mentioned in the introduction: GUWB and EMIB sensing.

The EMI method for damage detection relies on the electromechanical properties of PZT transducers. In this method, a harmonic voltage is applied to terminals of the transducer and the current is measured. The frequency of the harmonic voltage is varied in a wide range and the impedance (the ratio of the current to voltage) is calculated at each frequency. In case the transducer is attached to a structure, it excites the structure with a frequency that is equal to the frequency of the harmonic voltage. Presence of damage in the structure would alter its natural frequencies. Therefore, damage may be detected by comparing the conductance (real part of impedance) spectrums of the pristine state and the damaged state. EMIB sensing can be used in two important contexts. First context is to use it as a damage detection step, and second is to use it as self-diagnostic step for transducers. PZT transducers are permanently attached to the structure to monitor its health. The attachment quality is crucial for the reliability of the proposed SHM system. Thus, prior to any actuation and sensing by the PZT transducers, their attachment condition should be inspected to avoid false alarms. This inspection is usually referred to as the self-diagnostics. This can be effectively done by EMI method. An example is illustrated here to show the applicability of EMI in self-diagnostics of the structure, see Figure 2. It can be observed that from the magnitude of the imaginary part of the impedance, the degree of the debonding can also be detected.

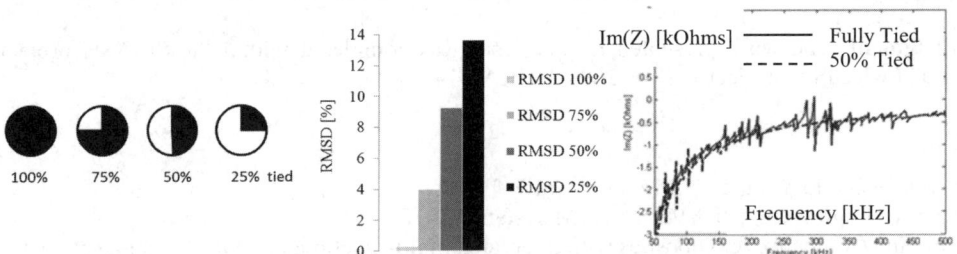

Figure 2 Self-Diagnostic property of the SHM system using EMI method

The GUWB method is based on actuating and sensing of transducers surface mounted on the stiffened panel. The sensing and actuating is performed using a SMART FEM in the absence of experimental data. The propagation properties of Lamb Waves depend on the elastic properties of the surface that guides them. Therefore the presence of damage can be identified from the wave scattering when it travels through the damaged area. The reflections and transmissions depend on characteristic of damage, thus it can lead to damage characterization in addition to detection. The details of the damage detection algorithm based on actuator-sensor signals can be found in [5] and [6]. The severity of the damage (percentage of softening) influences sensor response. Therefore a Damage Index can be introduced which can further characterise its severity. An example of damage detection on a stiffened panel can be found under Active sensing in Figure 1.

Optimisation of Sensor Positioning

Sensor optimisation consists of optimal placement of sensors/actuators in order to detect with high probability and reliability any damage prior to it becoming critical. Thus an optimisation analysis is developed resulting in sensor number and layout which is best capable to detect the impact location on the basis of the sensor signals. It is important to have a minimum effect on the design of composite stiffened panel via the addition of an SHM system; the number of sensors/actuators must remain low while at the same time the probability of detection must remain high.

The optimisation methodology developed for the SMART Platform is based on the Genetic Algorithm (GA). The fitness function, which evaluates each solution, is the performance of the ANN predicting the impact location. The number of sensors is an input to the optimisation module, together with either the max acceptable error or probability of detection. So for any given number of sensors the optimisation analysis will search the best sensor layout corresponding to either a given error limit or probability of detection [7]. To show the performance of the optimization, a part of the stiffened panel has been optimised for finding the optimal sensor layout for four sensors for best impact detection. The Results of the optimised locations for four sensors corresponding to 85% Probability Of Detection (POD) is shown in Figure 1 under optimisation (four corner sensors).

Conclusions

During the European JTI project SMASH, an SHM platform was successfully developed which can lead to robust and reliable health monitoring techniques. The developed SHM methodologies were applied to an actual size aircraft stiffened panel under various impact loads and location (small and large mass impacts). The developed platform is based on numerical models and is able to successfully detect impact location and force magnitude from sensor readings (using ANN). Afterwards, by applying active sensing techniques, damage (softening, debonding) was successfully detected and characterised (based on acting and sensing) in the stiffened panel shown in Figure 1. Finally the optimal sensor locations were provided by the aid of Genetic Algorithm. This is the first step to apply the developed methodologies to actual size aircraft part under real load condition.

Acknowledgment. The work presented in this paper was completed within the SMASH project funded by JTI-CleanSky project JTI-CS-2009-1-GRA-01-005.

References

[1] Acellent Technologies Inc., "SMART LAYER," 2011.
[2] AERnnova Engineering, "PAMELA SHM system," 2010.
[3] The Boeing Company, "Composites with integrated multi-functional circuits," US Patent, 2011.
[4] M. Ghajari, Z. Sharif Khodaei, and M. Aliabadi, "Impact Detection Using Artificial Neural Networks," *Key Engineering Materials,* vol. 488, pp. 767-770, 2012.
[5] Z. Sharif Khodaei and M. H. Aliabadi, "Damage Identification Using Lamb Waves," *Key Engineering Materials,* vol. 452, pp. 29-32, 2011.
[6] Z. Sharif Khodaei, R. Rojas-Diaz, and M. Aliabadi, "Lamb-Wave Based Technique for Impact Damage Detection in Composite Stiffened Panels," *Key Engineering Materials,* vol. 488, pp. 5-8, 2012.
[7] M. H. Aliabadi, "Final Project Report - SMASH," 2012.

Micromechanical characterisation of microstructure in weld heat affected zone of structural steel

Yusuke Shimada [1,a], Yoichi Kayamori [1,b], Shohei Nishida [2,c],
Mitsuhiro Matsuda [2,d] and Kazuki Takashima [2,e]

[1]Steel Research Laboratories, Nippon Steel Corp., 20-1 Shintomi, Futtsu, Chiba, 293-8511, Japan

[2]Kumamoto University, 2-39-1 Kurokami, Kumamoto, Kumamoto, 860-8555, Japan

[a] shimada.yuusuke@nsc.co.jp, [b] kayamori.yoichi@nsc.co.jp, [c] 117d8416@st.kumamoto-u.ac.jp,

[d] matsuda@alpha.msre.kumamoto-u.ac.jp, [e] takashik@gpo.kumamoto-u.ac.jp,

Keywords: Structural steel; Heat affected zone (HAZ); Fracture; Micro-sized mechanical testing

Abstract. Microstructures in the weld heat affected zone (HAZ) can cause a decrease in fracture toughness, and evaluating the effect of microstructures on fracture toughness is helpful in understanding the cause of the fracture toughness decrease. In this study, micro-sized tensile specimens were sampled from base metal and HAZ, and the mechanical properties and fracture behaviours of different steel microstructures were directly investigated by micro-sized mechanical testing.

Introduction

Welding is useful for fabricating large steel structures, but welding heat input can deteriorate fracture toughness in the weld heat affected zone (HAZ). Complicated microstructures are formed by weld thermal history, especially in the case of multi-pass welding, and the mechanical and fracture properties of each microstructure are helpful in understanding the effect of microstructure on fracture toughness. However, usual mechanical test methods such as ASTM E8 [1] and E1820 [2] do not demonstrate the local mechanical properties, but indicate the average mechanical properties of each microstructure. On the other hand, a few micro-sized mechanical test methods have been developed in recent years [3-5], and the authors have applied one of the methods to the evaluation of micromechanical properties in steel.

In this study, tensile tests were conducted using smooth and notched micro-sized tensile specimens sampled from base metal and HAZ, and their deformation and fracture behaviours were investigated by using a white light interferometer and an SEM.

Experimental methods

Steel used in this study was a 500MPa class steel plate for welded structures, SM490A in the Japan Industrial Standard (JIS). Table 1 shows the chemical composition and mechanical properties

Table 1 Steel used in this study.

(a) Chemical composition (mass %).					
C	Si	Mn	P	S	Fe
0.14	0.27	1.39	0.014	0.005	Bal.

(b) Mechanical properties		
Yield stress σ_{ys} (MPa)	Tensile strength σ_{ts} (MPa)	Elongation EL (%)
388	544	36

of the steel. Its microstructure was composed of ferrite and pearlite. A two peak thermal cycle, as shown in Fig. 1, was applied to each smooth rectangular specimen sampled from the plate, and the thermal cycle simulated the inter-critically reheated coarse

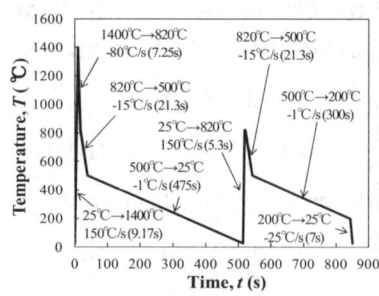

Fig. 1 Heat-treatment pattern.

Fig. 2 Optical Microscope image.

grained HAZ in offshore structures [6]. Fig. 2 exhibits the simulated HAZ microstructure composed of ferrite and bainite. Its prior-austenite grain size was about 200μm, and precipitates were recognised along prior-austenite grain boundaries and bainite lath boundaries.

Micro-sized tensile specimens were cut by focused ion beam machining from the 20μm thickness thin foils manufactured by mechanical and chemical polishing. Fig. 3 shows the SEM images of a smooth specimen and a notched one. The length, width and thickness of the specimens

 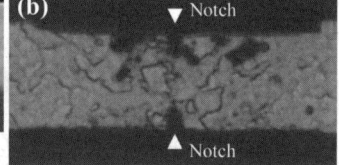

Fig. 3 Micro-sized (a) smooth and (b) notched specimens.

were 50, 20 and equal to or less than 20μm, respectively. As smooth specimens, three specimens were sampled from three different microstructures, 1)base metal, 2)bainite in HAZ and 3)bainite and precipitates in HAZ. In addition, a double edge nocthed micro-sized tensile specimen was also made for investigating fracture behaviour around precipitates, where 4)notches were introduced near precipitates in the bainite HAZ. Fig. 4 demonstrates the micromechanical testing machine used in this study. One side of a specimen was held on the α-β stage, and the other was fixed at a part connected to the load cell and the Z-stage. Load was measured by the load cell, and load line displacement was translated from X-Ystage movement in the tensile direction. Specimen surface roughness was detected by using a white light interferometer, and strain was caliculated from distance between any particular two points defined by the roughness. Micro-sized tensile tests were conducted at room temperature in air at the stroke speed of 0.1μm/sec.

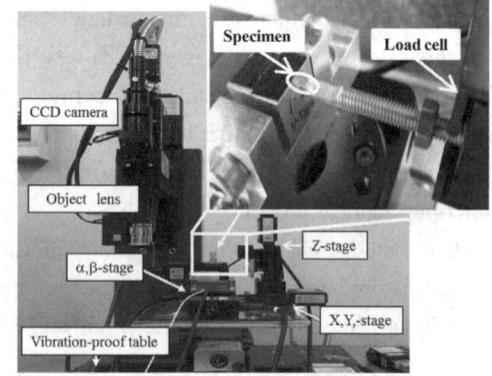

Fig. 4 Micromechanical testing machine.

Experimental results

Smooth specimens. Fig. 5 shows the relathinships between the nominal stress and the nominal strain obtained by macro-sized A2 specimens in JIS and micro-sized specimens. σ_{ys} and σ_{ts} of micro-sized specimens were lower than those of macro-sized ones in the identical microstructure, respectively. In addition, σ_{ys} and σ_{ts} of HAZ microstructures were higher than those of base metal. Furthermore, σ_{ys} and σ_{ts} of a bainite and precipitates micro-sized specimen were higher than those of a bainite micro-sized specimen. Fig. 6 shows white light interferometry images on micro-sized specimen front surfaces just before failure, and SEM images on micro-sized specimen front surfaces and fracture surfaces after failure. Slight necking was observed in the base metal specimen. In the

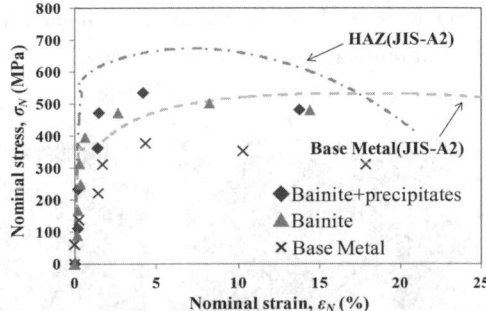

Fig. 5 Relationships between nominal stress and nominal strain in smooth specimens.

bainite specimen, a slip band appeared after the maximum tensile stress, and the specimen was broken along the slip band by shearing. The bainite and precipitates specimen was also broken by shearing, and its slip band appeared around precipitates.

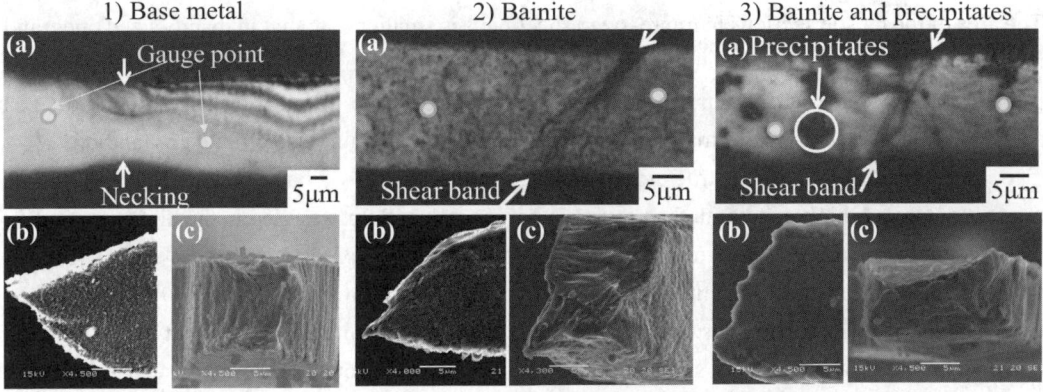

Fig. 6 (a)Specimen front surface white light interferometry images just before failure, (b)Front surface and (c)Fracture surface SEM images after failure.

Micro-sized notched specimen. Fig. 7 shows the change of specimen front surface at three different stages. The initial state was shown in (d-1) and (e-1). Double edge notches opened as shown in (d-2) and (e-2), a crack initiated from the notch root, and the crack propagated into bainaite around precipitates as shown in (d-3) and (e-3).

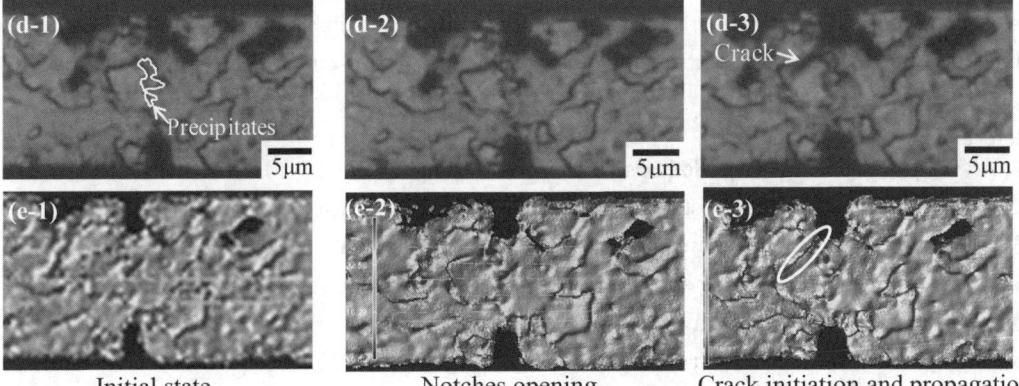

Fig. 7 Change of specimen surface (d)optical microscope images,(e)white light interferometry images.

Fig. 8 shows the SEM image of specimen fracture surface, front surface after failure and the phase distribution image of the front surface of the specimen. A crack did not propagate straight into the ligament, but deviated from precipitates.

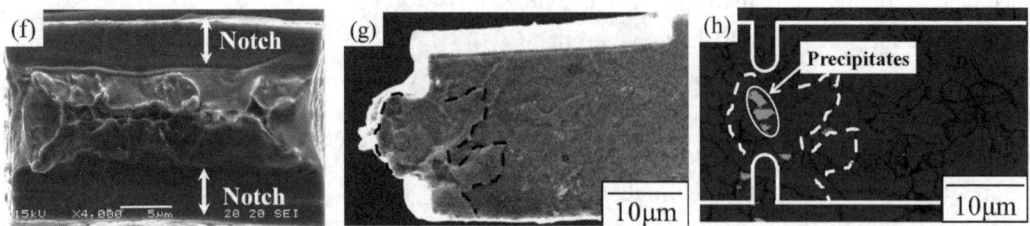

Fig. 8 (f)Fracture surface and (g)Front surface SEM images, (h)Front surface phase distribution image

Discussion

σ_{ys} and σ_{ts} of micro-sized specimens were lower than those of macro-sized ones as shown in Fig.5. The area of the tensile sections in micro-sized specimens is smaller than that in macro-sized ones, and the number of grain boundaries and precipitates, which can be the resistance to dislocation movement, probably affected σ_{ys} and σ_{ts} in the specimens. In addition, the stress triaxiality of micro-sized specimens is lower than that of macro-sized ones, and their constraint also affected σ_{ys} and σ_{ts}.

σ_{ys} and σ_{ts} of HAZ microstructures were higher than those of base metal. Bainite in HAZ microstructure has higher strength than ferrite in base metal, and this resulted in the higher σ_{ys} and σ_{ts} of HAZ microstructures. Furthermore, σ_{ys} and σ_{ts} of a bainite and precipitates micro-sized specimen were higher than those of a bainite micro-sized specimen. As shown in Figs 6 to 8, slip bands and cracks deviated from precipitates, and it is probable that the precipitates are resistant to deformation and cracking. On the other hand, some crack initiation processes around precipitates such as breaking and debonding have been reported [7]. The effects of crack initiation processes around precipitates on strength and fracure properties should be investigated in the future.

Concluding remarks

Micro-sized tensile tests of base metal and HAZ microstructures in steel were carried out, and mechanical properties and fracture behaviors were investigated. Slip bands and cracks deviated from precipitates in HAZ, and it is probable that the precipitates are resistant to deformation and cracking.

References
[1] ASTM E8 : Standard Test Methods for Tension Testing of Metallic Materials
[2] ASTM E1820 : Standard Test Method for Measurement of Fracture Toughness
[3] Y. Mine, K. Hirashita, M. Matsuda, M. Otsu and K. Takashima; Corros. Sci. 53 (2011) 529-533
[4] H. D. Espinosa, B. C. Prorok and M. Fischer; J. Mech. Phys. Solids 51 (2003) 47-67
[5] Julia R. Greer, Warren C. Oliver and William D. Nix; Acta Materialia 53 (2005) 1821-1830
[6] T. Tagawa, R. Miyata, S. Aihara and K. Okamoto; ISIJ 79 (1993) 10 1176-1182
[7] C. L. Davis and J. E. King: Metal. & Mater. Trans. A Vol. 25A (1994)

Two scale-based continuum damage model for brittle materials under thermomechanical loading

Andreas Ricoeur[a], Dimitri Henneberg[b]

University of Kassel, Institute of Mechanics, Mönchebergstr. 7, 34125 Kassel, Germany

[a]ricoeur@uni-kassel.de, [b]d.henneberg@uni-kassel.de

Keywords: refractories, thermal shock resistance, damage, fracture, multiscale modelling, microcracks

Abstract. Ceramic refractory materials initially contain a multitude of defects such as voids, microcracks, grain boundaries etc. Particularly being exposed to high temperatures above 1000 °C the macroscopic properties such as effective compliance, strength and lifetime are essentially determined by microscopic features of the material. The deformation process and failure mechanisms are going along with the creation of new microdefects as well as the growth and coalescence of cracks. A brittle material damage model for dynamic thermomechanical loading conditions is presented in this paper. Representative volume elements (RVE) include microcrack initiation and growth. The material laws are formulated on the continuum level using appropriate homogenisation methods. To demonstrate the potential of the numerical tools, two examples are presented which are taken from applications. Based on experiments, cyclic thermal shock tests at refractory plates are simulated by FEM. To quantify the thermal shock resistance of ceramics, experiments suggested by Hasselman are simulated numerically supplying a critical temperature slope.

Introduction

Refractory products and components find their application especially in iron, steel and foundry industry. Examples are refractory bricks at different shapes for lining of steel casting ladles and blast furnaces. Refractories are generally subjected to combined thermo-mechanical loading. The challenging aim in this research field is to develop structures with materials properties matched for specific applications. Above all, the thermal shock resistance is the one mechanical property, which has to be improved besides other properties such as abrasive or corrosive resistance. This requires an understanding of the influence of effective parameters of the microstructure on the thermal shock resistance. The rapid temperature change at thermal shock first leads to the formation of incipient macroscopic cracks due to the growth of microscopic defects and then to the reduction of the residual strength due to further thermo-mechanical loading cycles. Thus, the damage mechanism of ceramic materials is classified according to the macroscopic phenomena of brittle damage, where the predominant mechanisms are initiation and growth of microcracks.

Various methods have been developed in recent years to predict thermal shock resistance, partly based on experimental data, partly purely theoretical. Among those we find damage models for brittle materials considering microcracks [1] or an indentation quench method [2] with stable and unstable crack growth governed by the combination of residual and thermal stress. There are some analytical und numerical approaches dealing with models using micro-macro coupling [3]. In [4], a micromechanics-based damage evolution law is presented by combining the propagation criterion for a single crack embedded in an infinite matrix (linear-elastic fracture mechanics) and an RVE comprising interacting microcracks in a brittle material (continuum micromechanics). A stable damage evolution law for brittle solids under complex loading conditions is described in [5]. However, fundamental model-based investigations of damage processes in brittle materials or refractories under thermal shock conditions are rare, particularly those going along with numerical multiscale calculations.

The goal of this paper is to present a microcrack based damage model for brittle materials under thermo-mechanical dynamical loading conditions. The material laws are formulated on the macro level using appropriate homogenisation methods and introducing effective material tensors. Within the framework of continuum damage mechanics, defects are introduced into the material law as internal state variables. Local failure occurs if the damage variable reaches a critical value. In order to properly model the thermo-mechanical coupling, the temperature-dependence of material constants is taken into account. Due to the highly dynamic character of the thermal shock, inertia forces are included in the model. Results are presented showing damage patterns under different thermal loading conditions and a Hasselman diagram with the corresponding coefficients calculated for different material properties.

Theoretical fundamentals

A solid continuum with thermomechanical initial and boundary conditions given by stresses $\bar{t}$ and heat flux $\bar{q}$ (Neumann) or displacements $\bar{u}$ and temperature θ (Dirichlet) is considered. To incorporate local microstructural features, mesoscale cell models with linear elastic matrix properties are introduced, generally containing voids, cracks or grain boundaries. The constitutive relation is derived from thermodynamical considerations starting from the complementary elastic energy density potential $\psi(\sigma_{ij}, s)$ and the corresponding Gibbs' fundamental equation

$$d\psi(\sigma_{ij}, s) = \theta\, ds - \varepsilon_{ij}\, d\sigma_{ij} \qquad (1)$$

where stress and entropy are independent and temperature and strain associated variables. Decomposing the potential into two parts – the one representing the energy density for the homogeneous domain and the other that one of the defect phase – and neglecting heat exchange ($ds=0$) we obtain

$$\psi(\sigma_{ij}, f) = \psi^h(\sigma_{ij}) + \psi^d(\sigma_{ij}, f) \qquad (2)$$

introducing a scalar damage variable f. From Eq. (1) the potential for the homogeneous phase is obtained by integration and from Hooke's Law:

$$\psi^h = -\int_0^{\bar{\sigma}_{ij}} \varepsilon_{ij}\, d\sigma_{ij} = \frac{\nu}{2E}\sigma_{ll}\sigma_{jj} - \frac{1+\nu}{2E}\sigma_{ij}\sigma_{ij}. \qquad (3)$$

The potential of the defect phase depends on the damage variable representing void or crack growth, grain boundary delamination or other irreversible processes. If microcrack growth is considered to be the predominant damage mechanism in brittle structures, the potential can be expressed in terms of the energy release rate G:

$$\psi^d = -\frac{1}{\Delta V}\int G(A)\, dA \qquad (4)$$

where ΔV is a representative volume and A denotes the crack surface. The energy release rate depends on the local stress and the crack configuration. A constitutive law is derived from Eq. (2) by partial differentiation:

$$\varepsilon_{ij} = -\frac{\partial \psi(\sigma_{kl}, f)}{\partial \sigma_{ij}} \quad (5)$$

introducing an effective forth order elasticity tensor C^*_{ijkl} and the averaged stress and strain in a representative volume element:

$$\langle \sigma_{ij} \rangle = C^*_{ijkl}(f) \langle \varepsilon_{ij} \rangle \quad . \quad (6)$$

The increase of the internal variable $\dot{f} > 0$ is governed by an evolution equation accounting for a suitable microcrack growth law.

Accounting for thermomechanical loading conditions, the energy balance is solved independently from the dynamic balance equation of momentum leading to the transient differential equation

$$\rho^* c^*_H \frac{\partial \theta}{\partial t} = \lambda^*_{ij} \frac{\partial^2 \theta}{\partial x_i \partial x_j} \quad . \quad (7)$$

The effective thermal properties, i.e. conductivity λ^*_{ij} and specific heat c^*_H, as well as the effective mass density ρ^* in general depend on the internal variable f. For crack configurations in brittle materials, however, it came out to have a minor influence. The coupling of thermal and mechanical fields is achieved including the effective thermal expansion coefficients α^*_{ij}:

$$\langle \sigma_{ij} \rangle = C^*_{ijkl} \left(\langle \varepsilon_{ij} \rangle - \alpha^*_{ij} \Delta \theta \right) \quad . \quad (8)$$

Numerical examples

Cyclic thermal shock tests have been simulated numerically, on the basis of experiments from [6]. Cylindrical AZT- and MgO-C-specimens have been thermally loaded with a local surface heat flux of 42 MW/m². The green area (approx. 4 cm²) in Fig. 1 indicates the heating zone. The rest of the model boundary is assumed to be adiabatic. The loading regime is cyclic, with constant heat flux and subsequent breaks. The blue zones in Fig. 1 represent partly damaged

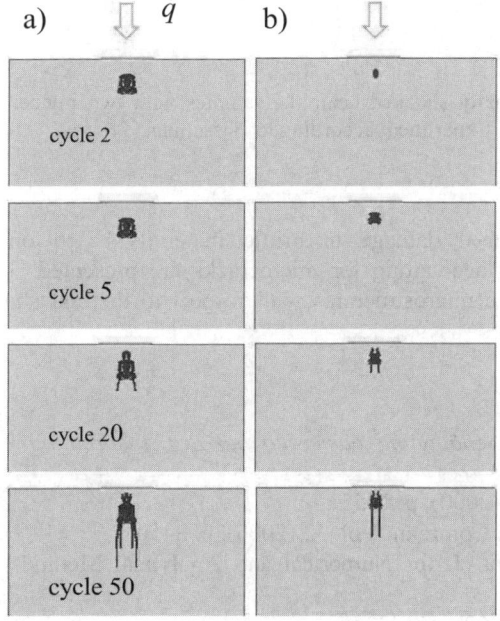

Fig. 1: FE-simulation of cyclic thermal shock tests with a locally absorbed surface heat flux $\dot{q}$, a) 0.5 s thermal loading + 2 s break, b) 0.1 s thermal loading + 2 s break

areas ($f_0 < f < 1$). The red areas indicate fully destroyed zones where $f = 1$ and must therefore be interpreted as macro-cracks. It should be noted that damage initiates within the material, propagating towards the surface with increasing number of load cycles.

In Fig. 2 the results of thermal shock simulations with subsequent three-point-bending are presented. A beamlike specimen is quenched at one surface after homogeneous heating. Afterwards, the bending strength is determined by simulation with the pre-damaged sample. The diagram shows a considerable decrease in residual strength above a critical temperature slope $\Delta\theta_{crit}$. The latter is depending on thermal expansion and elastic modulus. A coefficient R suggested by Hasselman [7] qualitatively describes the thermal shock resistance of the material.

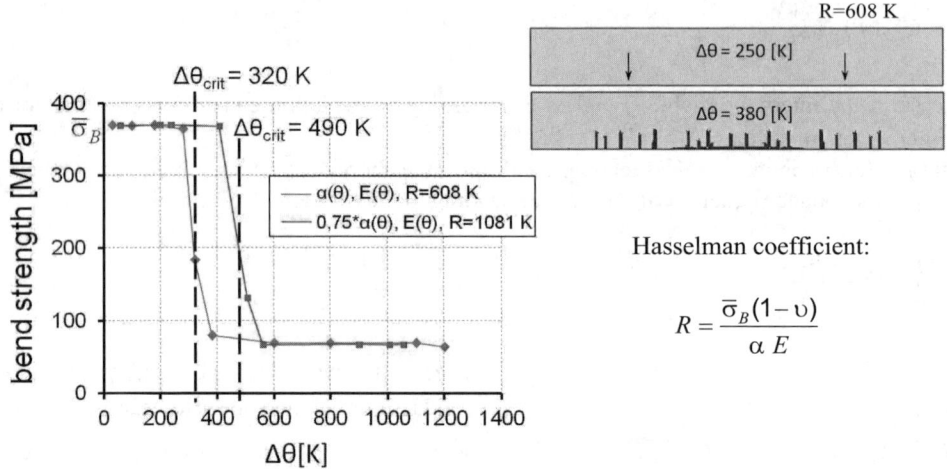

Hasselman coefficient:

$$R = \frac{\bar{\sigma}_B(1-\upsilon)}{\alpha E}$$

Fig. 2: FE-calculations of residual strength of thermally shocked beamlike samples with two images of damage patterns at $\Delta\theta$=250 K and 380 K for R=608 K. Experiment according to Hasselman [7]

Summary

Numerical simulations of thermal shock induced damage in brittle materials based on a micromechanically motivated constitutive model accounting for microcracks are presented. The calculation tools are a helpful means to optimize microstructures with respect to thermal shock resistance and to predicts critical temperature slopes.

References

[1] H. Schütte: *Ein finites Modell für spröde Schädigung basierend auf der Ausbreitung von Mikrorissen* (PhD Thesis 2001).
[2] M. Collin, D. Rowcliffe: Acta mater. Vol. 48 (2000), p. 1655
[3] J.J. Zhou, J.F. Shao and W.Y. Xu: Mech. Res. Commun. Vol. 33 (2006), p. 3450
[4] B. Pichler, C. Hellmich and H.A. Mang: Int. J. for Numerical and Analytical Methods in Geomechanics Vol. 31 (2007), p. 111
[5] X.Q. Feng, D. Gross: Acta Mech. Vol. 139 (2000), p. 143
[6] E. Skiera, C. Thomser, J. Linke, V. Roungos and C. G. Aneziris: Refractries Worldforum Vol. 4 (2012), p. 125
[7] D.P.H. Hasselman: Am. Ceram. Soc. Bull. Vol. 49 (1970), p. 1033

The Applicability Study on the FRP-Concrete Composite Bridge Deck for Cable-Stayed Bridges

Sung Tae Kim[a], Sung Yong Park[b], Keunhee Cho[c], Jeong-Rae Cho[d] and Byung-Suk Kim[e]

[a,b,c,d,e]Structural Engineering Research Division, Korea Institute of Construction Technology, 2311 Daewha-dong Ilsan-Gu Goyang-si, Gyeonggi-Do Republic of Korea

[a]esper009@kict.re.kr, [b]sypark@kict.re.kr, [c]kcho@kict.re.kr, [d]chojr@kict.re.kr, [e]bskim@kict.re.kr

Keywords: FRP; Composite; Deck; Cable-stayed bridges; Bridge

Abstract. This study is related to the FRP-concrete composite bridge deck for cable-stayed bridges developed by the Korea Institute of Construction Technology since 2007. This deck disposes a FRP panel at the bottom and is orthotropic owing to its fabrication through pultrusion process. In the cable-stayed bridge applying precast deck, support conditions occur at the cross beam and edge girder. Therefore, need is to verify the performances in the longitudinal and transverse directions when applying the orthotropic deck to cable-stayed bridges. Accordingly, specimens enabling to verify the performance in each direction are fabricated and subject to structural performance test. Based on the test results, the serviceability and applicability of the FRP-concrete composite deck to cable-stayed bridges are evaluated.

Introduction

The FRP-concrete composite deck for cable-stayed bridges [1] under development at the Korea Institute of Construction Technology (KICT) is shown in Fig. 1. This deck uses concrete exhibiting remarkable compressive performance at the top. At the bottom, the deck applies multi-box panel made of glass fiber composite material, which enables to replace tensile reinforcement and plays the role of permanent form that is free from corrosion. Through such composition, this FRP-concrete composite deck for cable-stayed bridges takes full advantage of the properties of each material and provides durability and lightweight. The structural type of this deck is completely different from conventional decks that were used for cable-stayed bridges.

Here, the FRP panel exhibits orthotropic behavior owing to its fabrication by pultrusion process. If this deck is applied to cable-stayed bridges, the pultrusion direction of the FRP panel coincides with the longitudinal direction of the bridge whereas the direction perpendicular to the pultrusion direction coincides with the transverse direction of the bridge. Accordingly, need is to verify the performance of the deck in the longitudinal direction and the performance in the transverse direction should also be imperatively verified since supporting the moments developed at the connection between the edge girder and the deck.

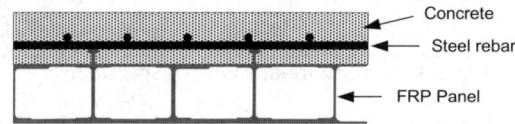

Fig. 1 Conceptual drawing of the FRP-concrete deck section

Fabrication of Specimens

Full-scale specimens were fabricated with various support conditions so as to evaluate the performances in the longitudinal and transverse directions. Table 1 arranges the support conditions of the specimens and the corresponding test objectives.

Table 1. Support conditions of specimens and corresponding test objectives

Designation of specimen	Support conditions	Test objectives
2-side support	Cross beam 2-side fixed	Performance check in the longitudinal direction
3-side support	Cross beam 2-side, edge girder 1-side fixed	Performance check in the transverse direction

Fig. 2 illustrates the sequences of the fabrication process of the specimens. Table 2 lists the mechanical properties of FRP panel used for the specimens. In the case of the direction perpendicular to pultrusion, Because the fabrication of tensile FRP panel specimens were impossible, the analysis results are indicated. Fig. 3 depicts the support conditions applied to each specimen and the cross-sectional shape in each direction.

(a) Pultrusion of FRP panel   (b) Assembling of panel and arrangement of reinforcement   (c) Placing of concrete

Fig. 2 Fabrication process of specimens

Table 2. Mechanical properties of FRP panel

Property		Unit	Bottom flange	Web	Top flange
Pultrusion direction	Tensile strength	MPa	324	268	369
	Tensile elastic modulus	GPa	32.3	19.8	24.5
Direction perpendicular to pultrusion	Tensile elastic modulus (analysis results)	GPa	14.5	13.2	16.1

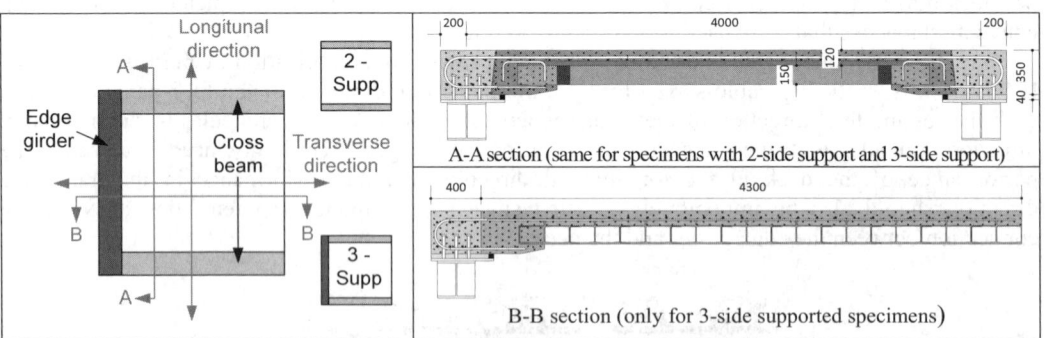

Fig. 3 Cross-sectional shape of specimens

Structural Performance Test and Results

The structural performance test for the FRP-concrete composite deck for cable-stayed bridges proceeded by fixing the cross-beam installed at the bottom of each specimen to the support blocks so as to enable negative moments to develop. Loading was applied at the center of the specimen through a loading plate with size corresponding to the rear wheel load DB-24 and at a speed of 0.5 mm/min. Fig. 4 shows a view of the test. Figs. 5 and 6 illustrate the failure patterns of 2-side supported and 3-side supported specimens, respectively.

For the 2-side supported specimen, quasi-circular crack was observed around the loaded part at load of about 250 kN. Moreover, additional negative moment cracks developed at the ends of the haunch above the cross beams according to the increase of the load to finally result in punching failure of concrete at the loaded part for a load of 1,098 kN.

 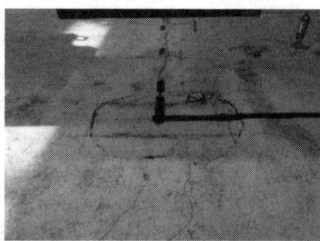 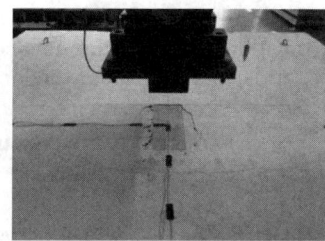

Fig. 4 View of test   Fig. 5 Failure pattern of 2-side supported specimen   Fig. 6 Failure pattern of 3-side supported specimen

For the 2-side supported specimen, cracks occur at the upper surface of concrete at the ends of the transverse haunch under load of about 400 kN. At approximately 450 kN, cracks also developed at the ends of the longitudinal haunch. Thereafter, circular cracks started to develop around the loaded part. Finally, failure through punching of concrete occurred at 1,084 kN similarly to the 2-side supported specimen.

Analysis of Structural Performance Tests Results

Fig. 7 plots the load-deflection curves according to the test results of the 2-side and 3-side supported specimens. Fig. 8 plots the load-crack width curves.

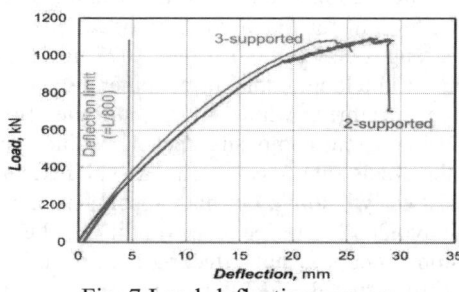 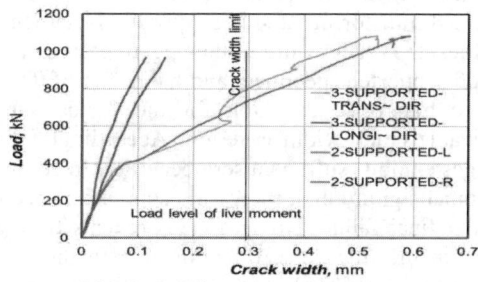

Fig. 7 Load-deflection curves   Fig. 8 Load-crack width curves

The live load moment applying on the 2-side supported FRP-concrete composite deck specimen was obtained by displacing the position of the load through advanced structural analysis to find the location at which the largest moment occurred. The so-obtained value is 47.7 kN·m/m including the impact. The conversion of this value into load considering the boundary conditions of the specimen gives a value of 212 kN. Under this service load state, the deflection of the specimens reaches 2.5 mm for the 2-side supported specimen and 2.58 mm for the 3-side supported specimen, which satisfy the deflection criterion (= $L/800$) of 4.625 mm considering the supported length of 3.7 m of the steel girder [2]. In addition, the crack width shown in Fig. 8 ranges between 0.021 and 0.028 mm for the 2-side supported specimen, and has value of 0.025 mm in the longitudinal direction and 0.030 mm in the transverse direction for the 3-side supported specimen. These values satisfy sufficiently the allowable crack width criterion [3]. Accordingly, the 2-side supported specimen of the FRP-concrete composite deck for cable-stayed bridges can be seen as securing sufficient serviceability.

Furthermore, if strength reduction factors of 0.55 and 0.85 are applied respectively for the FRP-concrete composite section and for the reinforced concrete section in the 2-side supported specimen, the FRP-concrete composite deck for cable-stayed bridges satisfies the flexural design moment as shown below. Here, since the failure load is 1,098 kN and is the failure strength corresponding to punching of the loaded section, the failure strength relative to bending can be assessed to be larger than the failure strength relative to punching. Therefore, using the punching failure strength for the verification of the bending-induced failure appears to be conservative. In this case, the load of 534.4 kN corresponds to the conversion of the design moment considering live and permanent loads.

Reinforced concrete section: $0.85 \times 1{,}098 = 933.3\,kN \geq 534.4\,kN \rightarrow OK$
FRP-concrete composite section: $0.55 \times 1{,}098 = 603.9\,kN \geq 534.4\,kN \rightarrow OK$

Besides, the 3-side supported specimen exhibited larger sectional rigidity than the 2-side supported specimen at early loading as shown in the load-deflection curves of Fig. 7, but both specimens showed similar final failure strength. This reveals that the unavoidable low sectional rigidity of the FRP panel in the transverse direction due to the pultrusion process is not affecting significantly the performance of the deck. Accordingly, the edge girder and details of the connection between the decks adopted for the 3-side supported specimen can be adequately applied in the future.

Summary

KICT has developed a FRP-concrete composite deck for cable-stayed bridges exhibiting long span and enabling large reduction of the weight. Full-scale 2-side and 3-side supported specimens were fabricated and tested to evaluate the performances in the longitudinal and transverse directions.

The test results showed that both specimens satisfy the deflection criterion of 4.625 mm ($=L/800$) with 2.8 mm for the 2-side supported specimen and 2.58 mm for the 3-side supported specimen. Besides, the crack width in the negative moment zone ranges between 0.021 and 0.028 mm for the 2-side supported specimen, and has value of 0.025 mm in the longitudinal direction and 0.030 mm in the transverse direction for the 3-side supported specimen, which satisfy sufficiently the allowable crack width criterion. Accordingly, the FRP-concrete composite deck for cable-stayed bridges exhibits sufficient serviceability. In addition, the 3-side supported specimen exhibited larger sectional rigidity than the 2-side supported specimen at early loading, but both specimens showed similar final failure strength. This reveals that the unavoidable low sectional rigidity of the FRP panel in the transverse direction due to the pultrusion process is not affecting significantly the performance of the deck. Consequently, it can be affirmed that the FRP-concrete composite deck for cable-stayed bridges developed in this study secures sufficient serviceability and applicability for real cable-stayed bridges.

Acknowledgement

This research was supported by a grant from a Strategic Research Project "Development of Deck Systems for Hybrid Cable-Stayed Bridge" funded by the Korea Institute of Construction Technology.

References

[1] Korea Institute of Construction Technology, Super Bridge 200 5th report, "Development of Deck systems for hybrid cable-stayed bridge" (2010)
[2] American Association of State Highway and Transportation Officials, "AASHTO LRFD Bridge Design Specifications", p. 2-11 (2004).
[3] Korea Concrete Institute, Concrete design codes, Republic of Korea (2007).

Improvement of Fatigue Crack Growth Behavior in the Case of the Cracked Specimen with relatively Narrow Width

Md. Shafiul Ferdous[1,a], Chobin Makabe[2,b] and Tatsujiro Miyazaki[2,c]

[1] Graduate School of Science and Engineering, University of the Ryukyus,
1 Senbaru, Nishihara, Okinawa, 903-0213, Japan

[2] Mechanical Systems Engineering Department, University of the Ryukyus,
1 Senbaru, Nishihara, Okinawa, 903-0213, Japan

[a]munazeer_218@yahoo.com, [b]makabe@tec.u-ryukyu.ac.jp, [c]t-miya@tec.u-ryukyu.ac.jp

Keywords: Fatigue, Crack growth, Detecting method, Stop-hole, Inserting pin, Stress concentration, Stress intensity

Abstract. The improvement of acceleration behavior of crack growth was investigated with constant stress amplitude under negative stress ratio $R=-1$. Then a technical method to detect the fatigue crack growth was discussed. For example, when the stress amplitude exceeds a critical value, crack growth rate of overloaded specimen became higher than that of baseline which was obtained by crack growth test without applying overload. In some experimental cases, the acceleration of crack growth was observed and that could be happened on practical cases. Stop-holes were drilled at crack tips or in the vicinity of crack tips to remove the plastic zone and the effect of that on crack growth behavior were investigated. Also, steel pins were inserted into the stop-holes and its effect was discussed. Finite element method (FEM) was used to analyze the stress concentration at the edge of stop-holes. Positions of center of the stop-holes were varied at different distance from the crack tips to investigate the effect of stop-holes on fatigue crack growth. Also, stress intensity of base and stop-holed specimen was calculated. Then, the effect of stop-hole was discussed by both the experimental and analytical results. Specially, it was discussed whether the crack growth behavior was improved or not in the case of relatively smaller width specimen.

Introduction

Fatigue failure is the dominant failure phenomena in various mechanical components that are subjected to cyclic loading even at below yielding stress. Most of the failure accidents are originated from fatigue crack at notched or weakened site due to stress concentration. For maintenance of machine equipment, it is important to detect the crack growth. On some loading conditions, fatigue crack growth could be accelerated with unexpected manner. For example, after applying overload of some loading conditions, crack growth would be accelerated. So, in the present study, the method of detect the crack growth acceleration was investigated. The center of stop-holes were located at crack tips and on the nearby of that. Then the effects of those on crack growth behavior was discussed both experimentally and calculation methods. Specially, the detection of crack growth in narrow width specimen was discussed.

Materials and Testing Procedure

Material used was 0.15% carbon steel. The chemical composition of the material is (wt %): 0.15 C, 0.30 Si, 0.50 Mn, 0.013 P, 0.013 S, 0.05 Ni and balance Fe. The mechanical properties are: 449 MPa tensile strength, 283 MPa lower-yield strength, and 69% reduction of area. The type of specimen employed was a center-cracked plate for which the width and thickness were 20 mm and 4 mm, respectively. A notch of 3 mm in length was cut in the center of the specimen by electrical discharge machine. After being polished by emery paper and metal polisher, 3 mm of initial cracks were introduced from the notch roots by a pull-push hydraulic fatigue testing machine. The crack length, $2a$, was defined as including the notch length. After annealing the specimens at 600°C in a vacuum furnace, the fatigue crack growth tests were carried out using a hydraulic fatigue testing machine with a loading frequency of 10 Hz under laboratory conditions. Crack growth was observed by optical microscopes. The microscopes were connected to image display system devices to make it easy to measure the crack length.

A series of fatigue crack growth tests were done. As shown in Fig.1 (a), the center of drilling stop-holes were located at crack tips (x=0mm) or 0.5mm away from the tips towards the center of specimen (x=-0.5mm) to investigate the arresting technique of the crack growth. Also, pins were inserted into the stop-holes to investigate the effect of plastic hardness and residual stress on the crack growth behavior. To make acceleration condition of crack growth, overload was applied in some cases. The effectiveness of drilling stop-holes on crack growth was discussed with the experimental results and the calculation of the stress concentration factor K_t at the edge of stop-hole using ANSYS. Also, the stress intensity K_I was calculated with different width of specimen by body force method [4]. The calculation models were shown in Fig. 2.

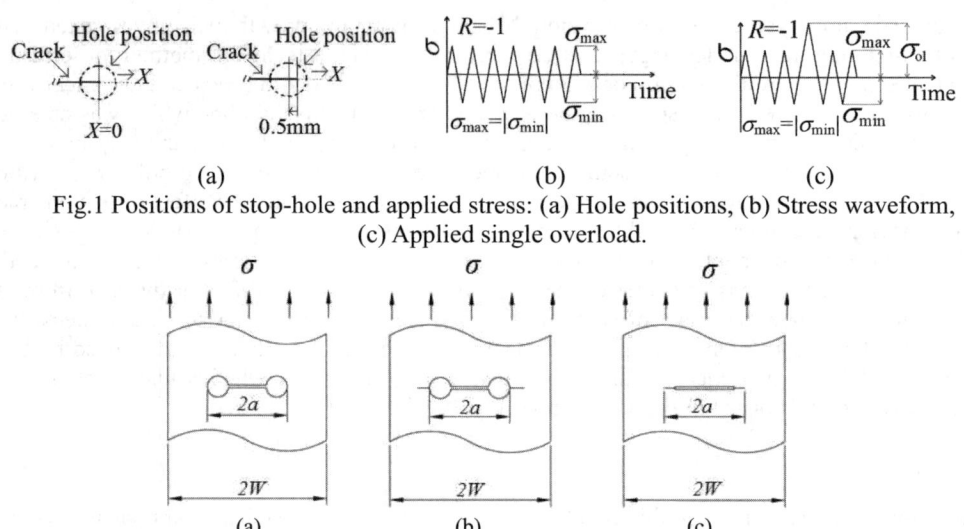

Fig.1 Positions of stop-hole and applied stress: (a) Hole positions, (b) Stress waveform, (c) Applied single overload.

Fig.2 Models to determine K_t and K_I : (a) Model with stop-hole, (b) With stop-holes and crack, (c) With centre crack.

Results and Discussion

Crack growth behavior

Fig. 3 shows the crack growth curves, which is a relationship between half crack length a and number of stress cycles N. Under some loading conditions, the crack growth rate accelerated after overloading. Therefore, the crack growth rate could be faster by an applying overload. The retardation and acceleration of crack growth was observed and discussed in the previous study [5]. Example of the acceleration crack growth behavior after overloading is shown in Fig. 3(a). The purpose of this study is to investigate whether the acceleration fatigue crack growth behavior can be improved or not by applying stop-holes. Now, the arrow in Fig. 3 indicates the overload applying position. Fig. 3(b) shows the crack growth curve and effects of positions of the center of stop-hole on the crack growth behavior. The fatigue crack growth curves of the specimens with stop-holes were compared with the baseline which is crack growth curve of the specimen without stop-holes and crack growth curve of overloaded specimen without holes. The dotted line is data for overloaded specimen and solid line is for baseline. After drilling stop-holes, the value of stress concentration factor at the edges of the holes was reduced in some case and improves the fatigue crack growth rate [1-3]. However, in case of narrow width specimen, stop-hole did not work effectively. The fatigue life became shorter in both the cases of overload and without overloaded specimens. The position of stop-hole affects slightly on the fatigue crack growth behavior. However, the crack growth curves of specimens were not so different when the stop-hole was drilled at $x=0$mm and $x=-0.5$mm. The effect of drilling stop-holes only was not effective in the present case.

When stop-holes were drilled at $x=0$mm and $x=-0.5$mm, fatigue life did not improve for narrow width specimen. But after inserting 4% larger pins into the holes (in Fig. 3(b)), the fatigue life was improved due to the compressive residual stress and hardening in the vicinity of hole edges. Without stop-hole, the total crack length was 6mm. However, when stop-hole was drilled, total length $2a$ of crack and holes were 7mm and 8mm. So, $2a$ of holed specimen were larger than the original cracked length. This is the reason that drilling stop-hole only did not work to improve the fatigue life. However, when pins were inserted into the stop-holes, crack growth behavior had been improved. Around the stop-holes, local hardening and compressive residual stress developed due to inserting pins [2].

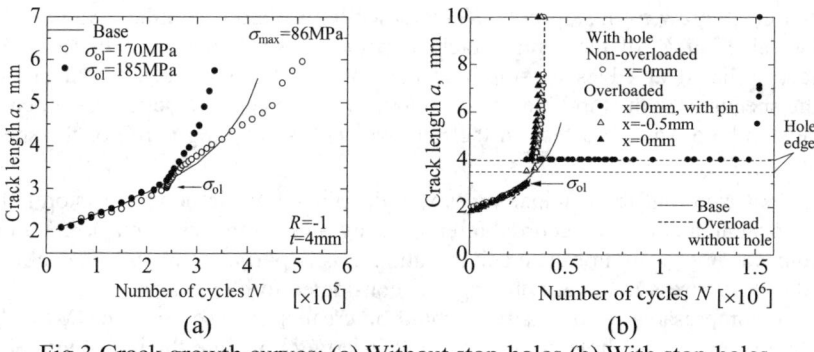

Fig.3 Crack growth curves: (a) Without stop-holes (b) With stop-holes.

Analysis of Stress Concentration and Stress Intensity

Fig.4 shows the calculation result of stress concentration factor K_t where $2W$ is total width and $2a$ is total crack length including stop-holes. ANSYS was used for calculation with some cases of specimen width to investigate the effect of specimen width on K_t values. In the present experimental cases of specimen, total crack length with stop-holes $2a=$ 7mm or 8mm and $2W=$20mm ($W=$10mm). The value of stress concentration is higher when specimen width $2W$ is shorter. From these result, it is concluded that the drilling stop-hole made increasing stress concentration in the case of specimen with narrow width. It is the reason that the crack growth was not improved.

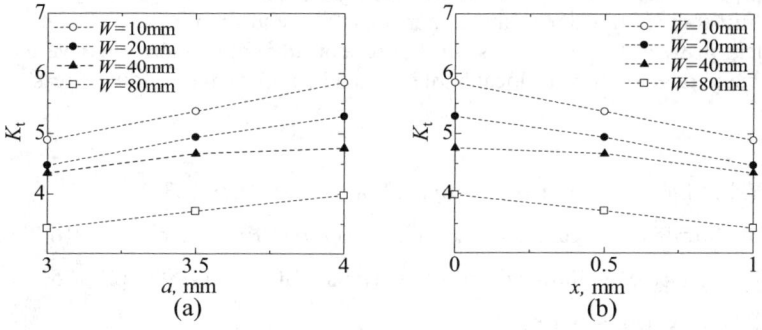

Fig.4 Variation of stress concentration factor, K_t with respect to W: (a) K_t vs. a, (b) K_t vs. x

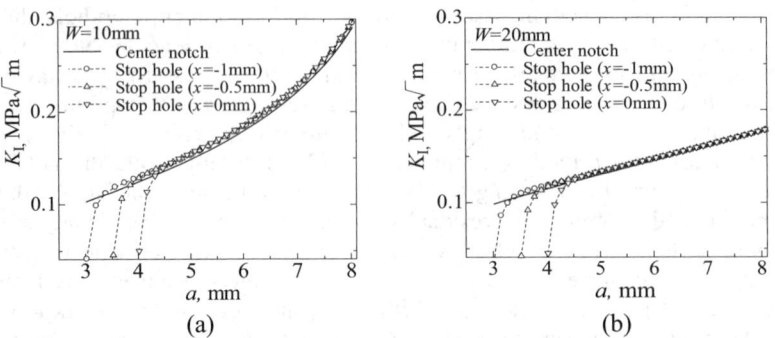

Fig.5 Calculation result of K_I by body force method: (a) $W=10$mm, (b) $W=20$mm

Fig.5 shows the calculation results of stress intensity factor K_I after crack grew from stop-hole edges. The calculation was carried out with $W=10$mm and $W=20$mm by body force method. After crack grew from stop-hole, K_I increased immediately, and these values of K_I became closer to the values of K_I of the center cracked specimen. From these results, it is discussed as follows: crack initiated at edges of stop-hole due to high stress concentration, and after crack initiation, the specimen with stop-hole can be regarded as the center cracked specimen. Therefore, applying the stop-hole was not affect on the improvement of the fatigue life of the specimen with narrow width.

It is expected that the residual stress was distributed at the bottom of stop-holes, and the hardening at local area could be occurred after inserting pins. In this case, crack initiation behavior at hole bottom was improved. For the crack initiation, it is important to repeat the cyclic shear stress. Therefore, the shear stress range at hole edges is considered to be reduced in the case of inserting pins. Also, the compressive residual stress would be created by inserting pins. We calculated the stress distribution around the hole before and after inserting pins into the holes with elastic-plastic conditions by ANSYS. After inserting pin, stress at the bottom of hole became lower when tensile stress was applied. However, this result is not obtained by exact boundary conditions. So, we will calculate that with exact boundary condition in the future.

Conclusions

A series of fatigue tests were conducted to investigate the effect of stop-holes on fatigue crack growth behavior of specimen with narrow width. The stress concentration factor, K_t and stress intensity factor, K_I was calculated to analyze the effect of stop-holes on fatigue crack growth behavior. Main results obtained are as follows:
(1) Where specimen width was narrow, fatigue crack growth behavior was not improved by drilling stop-holes only at crack tips due to the higher stress concentration.
(2) Fatigue crack growth behavior was improved considerably when pins were inserted into stop-holes due to development of local hardening and compressive residual stress.

References

[1] P. S. Song, Y. L. Shieh, Int. Journal of Fatigue: Vol. 26 (2004), p. 1333

[2] C. Makabe, A. Murdani, K. Kuniyoshi, Y. Irei, A. Saimoto, Eng. F Anal: Vol. 16(2009), p. 475

[3] H. Bao, A. J. McEvily, Metallurgical and Mats. Trans: Vol. 26 A (1995), p. 1725

[4] H. Nishitani, D. H. Chen, Key Eng. Materials: Vol. 251-252 (2003), p. 97

[5] C. Makabe, P. Anindito, T. Miyazaki, J. McEvile, J. of Testing and Eva: Vol. 33 (2005), p. 181

Dynamic Crack Problems using Meshless Method

P.H. Wen[1a] and M.H. Aliabadi[2b]

[1]Department of Engineering, Queen Mary, University of London, London, UK, E1 4NS

[2]Department of Aeronautics, Imperial College, London, UK, SW7 2BY

[a]p.h.wen@qmul.ac.uk, [b]m.h.aliabadi@imperial.ac.uk

Keywords: Meshless method, radial basis function, analytical solutions, elastodynamics.

Abstract.. A meshless approximation based on the radial basis function (RBF) is developed for analysis of dynamic crack problems. A weak form for a set of governing equations with a unit test function is transformed into local integral equations. A completed set of closed forms of the local boundary integrals are obtained. As the closed forms of the local boundary integrals are obtained, there are no any domain or boundary integrals to be calculated numerically in this approach.

Introduction

The main purpose of dynamic fracture mechanics is to analyse the crack behaviour under the dynamic load. The boundary element method (BEM) is also a well-established technique for the analysis of certain engineering problem such as fracture mechanics (see Aliabadi [1]). BEM has been particularly successful in solving crack dynamics problems with high level of accuracy [2-8]. In recent years, the computational mechanics community has turned its attention to so-called mesh reduction methods. These mesh reduction methods (commonly referred to as Meshless or Meshfree) have been applied to different types of problems relating to fracture mechanics and composites [9-14]. In this paper, the meshless local integral method is presented for two dimensional dynamic problems. The weak formulations for the governing equations with a unit test function are obtained exactly for the local domain integrals. A numerical inversion technique, Durbin's inversion method, is applied to determine each variable in the time domain.

Meshless local integral equation method

Consider a linear elastic body in three dimensional domain Ω with boundary Γ. The governing equations of motion can be written as

$$\sigma_{ij,j} + f_i = \rho \ddot{u}_i \tag{1}$$

where σ_{ij} are stresses, f_i is the body force, ρ is the density of material and $\ddot{u}_i$ is the acceleration ($i=1,2\ j=1,2$ for 2D) and ($i=1,2,3\ j=1,2,3$ for 3D). By Hook's law for plane stress problem, one has

$$\sigma_{11} = \frac{E}{1-\nu^2}\left(\frac{\partial u_1}{\partial x_1} + \nu\frac{\partial u_2}{\partial x_2}\right), \ \sigma_{22} = \frac{E}{1-\nu^2}\left(\frac{\partial u_2}{\partial x_2} + \nu\frac{\partial u_1}{\partial x_1}\right), \ \sigma_{12} = \frac{E}{2(1+\nu)}\left(\frac{\partial u_1}{\partial x_2} + \frac{\partial u_2}{\partial x_1}\right) \tag{2}$$

For two dimensional problem, where E is Young's modulus, ν is Possion ratio and $\mu = E/2(1+\nu)$ the shear modulus. The boundary conditions are given as

$$\begin{aligned} u_i &= u_i^0 \quad \text{on} \ \Gamma_u \\ t_i &= \sigma_{ij} n_j = t_i^0 \quad \text{on} \ \Gamma_t \end{aligned} \tag{3}$$

in which u_i^0 and t_i^0 are the prescribed displacements and tractions respectively on the displacement boundary Γ_D and on the traction boundary Γ_T, and n_i is the unit normal outward to the boundary Γ. In the local boundary integral equation approach, the weak form of differential equation over a local integral domain Ω_s can be written as

$$\int_{\Omega_s}(\sigma_{ij,j}+f_i-\rho\ddot{u}_i)u_i^* d\Omega = 0 \tag{4}$$

where u_i^* is a test function. By use of divergence theorem, (4) can be rewritten in a symmetric weak form as

$$\int_{\Gamma_s}\sigma_{ij}n_j u_i^* d\Gamma - \int_{\Omega_s}(\sigma_{ij}u_{i,j}^* - f_i u_i^* + \rho\ddot{u}_i)d\Omega = 0 \tag{5}$$

If there is an intersection between the local boundary and the global boundary, a local symmetric weak form in linear elasticity may be written as

$$\int_{\Omega_s}\sigma_{ij}u_{i,j}^* d\Omega - \int_{L_s}t_i u_i^* d\Gamma - \int_{\Gamma_D}t_i u_i^* d\Gamma = \int_{\Gamma_T}t_i^0 u_i^* d\Gamma + \int_{\Omega_s}(f_i - \rho\ddot{u}_i)u_i^* d\Omega \tag{6}$$

in which, L_s is the other part of the local boundary inside the local integral domain Ω_s; Γ_D is the intersection between the local boundary Γ_s and the global displacement boundary; Γ_T is a part of the traction boundary as shown in Figure 1.

The local weak forms in (5) and (6) are a starting point to derive local boundary integral equations if appropriate test functions are selected. A step functions can be used as the test functions u_i^* in each integral domain

$$u_i^*(\mathbf{x}) = \begin{cases} \varphi_i(\mathbf{x}) & \text{at } \mathbf{x} \in (\Omega_s \cup \Gamma_s) \\ 0 & \text{at } \mathbf{x} \notin \Omega_s \end{cases} \tag{7}$$

where $\varphi_i(\mathbf{x})$ is arbitrary function. In the Laplace transform domain, the Laplace transform of function $f(t)$ is defined as

$$\tilde{f}(s) = \int_0^\infty f(t)e^{-pt}dt \tag{8}$$

where p is a Laplace parameter. Equation (6) becomes

$$\int_{\Gamma_s}\tilde{\sigma}_{ij}n_j u_i^* d\Gamma - \int_{\Omega_s}(\tilde{\sigma}_{ij}u_{i,j}^* - \tilde{f}_i u_i^* + p^2\rho\tilde{u}_i)d\Omega = 0 \tag{9}$$

The approximation scheme

Consider a local domain $\partial\Omega_s$ which is the neighbourhood of a point $\mathbf{x}$ $(=\{x_1,x_2\})$ and is considered as the domain of definition of the RBF approximation for the trail function at $\mathbf{x}$ and also called as support domain to an arbitrary point $\mathbf{x}$. Generally the support domain is chosen as a circle centred at $\mathbf{x}$. To interpolate the distribution of function u in the local domain $\partial\Omega_s$ over a number of randomly distributed nodes ξ $[=\{\xi_1,\xi_2,...,\xi_K\}$, $\xi_k=(\xi_{k1},\xi_{k2})$, $k=1,2,...,K$], the approximation of function u at the point $\mathbf{x}$ can be expressed by

$$u(\mathbf{x}) = \sum_{k=1}^{K} R_k(\mathbf{x},\xi_k)a_k = \mathbf{R}(\mathbf{x})\mathbf{a}(\mathbf{x}) \tag{10}$$

where $\mathbf{R}(\mathbf{x}) = \{R_1(\mathbf{x},\xi), R_2(\mathbf{x},\xi),..., R_K(\mathbf{x},\xi)\}$ is the set of radial basis functions centred around the point $\mathbf{x}$ [$=(x_1,x_2)$], $\{a_k\}_{k=1}^K$ are the unknown coefficients to be determined. The radial basis function selected multi-quadrics, for further information see [9]

In the Laplace transform domain, a total number of $L+1$ samples in the transformation space s_k, $k=1,2,...,L$, are selected. Physical values are calculated for these transform parameters and the real value at time t must be obtained by an inverse transform using Durbin[14].

In addition, the dynamic stress intensity factor is evaluated by crack opening displacement (COD) in the Laplace transform domain. For mode I fracture, the stress intensity factor for plan stress problem is related to crack opening displacement, in the transformed domain, as following

$$\tilde{K}_I = \frac{E}{8(1-v^2)}\sqrt{\frac{2\pi}{r_0}}\Delta\tilde{u}_2, \quad \Delta\tilde{u}_2 = \tilde{u}_2^+ - \tilde{u}_2^-. \tag{11}$$

where r_0 indicates the distance between the collocation point and crack tip, $\Delta \tilde{u}_2$ is the crack opening displacement (COD) in the Laplace transform domain.

Numerical example: *A Single central crack in rectangular plate under tension*
Consider a rectangular plate of width $2b$ and length $2h$ with a centrally located crack of length $2a$. It is loaded dynamically in the direction perpendicular to the crack by a uniform tension $\sigma_0 H(t)$ on the top and the bottom. Due to the symmetry, a quarter of plate is considered as shown in Figure 1. Here Poisson's ratio $\nu = 0.3$ and Young's modulus is unit.

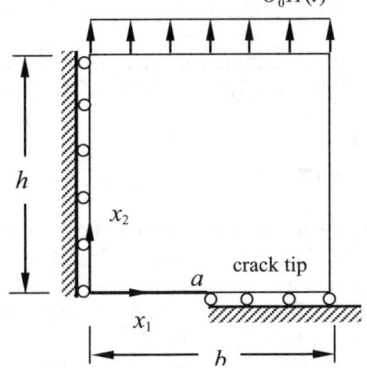

Firstly we observe the accuracy of stress intensity factor with the density of collocation point. The analytical static solution for a square plate $b/h=1$ containing a central crack, if $a/b=0.5$, is $K_I = 1.325\sigma_0\sqrt{\pi a}$ [10] and the result by the second nodal value with 21×21 node distribution is $K_I = 1.331\sigma_0\sqrt{\pi a}$. Therefore, the relative error can be expected to be less than 1% for elastostatic problems. Therefore, we use the second nodal value with 21×21 node distribution to evaluate stress intensity factor in the following examples.

Figure 1. Rectangular plate with a central crack of length $2a$ under tension $\sigma_0 H(t)$.

A rectangular plate is considered in this example, i.e. $b/h = 0.5$, $b/h = 1$ with total numbers of nodes 21×11, 21×21. Figure 2 show the normalized stress intensity factors various against the normalized time $c_1 t/b$. In addition, the results given by Wen *et al* [11] using the mesh free Perov-Galerkin method and the indirect boundary element method (fictitious load method[17]) are presented in the same figures for comparison. Apparently before the arrival time of dilatation wave travelling from the top of plate, the stress intensity factor should remain to be zero. The agreement between these solutions is considered to be good.

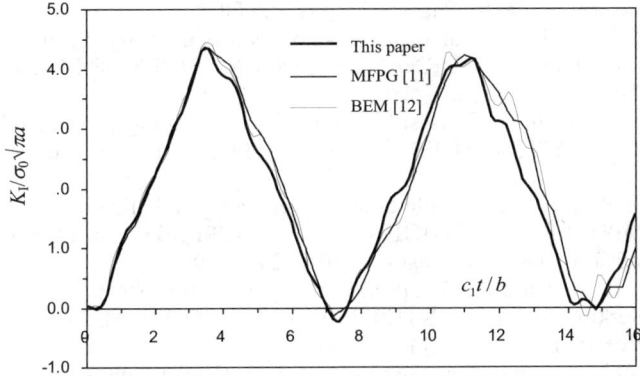

Figure 2. $K_I / \sigma_0 \sqrt{\pi a}$ against the normalized time $c_1 t/b$ for $b/h = 0.5$.

Conclusion

This paper presents the meshless local integral method for two dimensional elastodynamic fracture problems using the Laplace transform technique. Considering a local integral domain with local support domain, the analytical formulations were derived in the Laplace transform domain. Durbin's inversion method was applied to determine all variable in the time domain. The dynamic stress intensity factor of mode I was evaluated by using the COD technique.

References

[1] M.H. Aliabadi The boundary element method, applications in solids and structures, Wiley, Chichester, 2002.
[2] Fedelinski P; Aliabadi MH; Rooke DP Boundary element formulations for the dynamic analysis of cracked structures ENGINEERING ANALYSIS WITH BOUNDARY ELEMENTS Volume: 17 Issue: 1 Pages: 45-56 , 1996
[3] Sollero,P; Aliabadi MH Anisotropic analysis of cracks in composite laminates using the dual boundary element method COMPOSITE STRUCTURES Volume: 31 Issue: 3 Pages: 229-233 1995
[4] Fedelinski P; Aliabadi MH; Rooke DP The laplace transform deem for mixed-mode dynamic crack analysis COMPUTERS & STRUCTURES Volume: 59 Issue: 6 Pages: 1021-1031 , 1996
[5] Portela A; [5] Aliabadi MH; Rooke DP The dual boundary element method – effective implementation for crack problems, INTERNATIONAL JOURNAL FOR NUMERICAL METHODS IN ENGINEERING Volume: 33 Issue: 6 Pages: 1269-1287 1992
[6] Wen PH; Aliabadi MH; Rooke DP Cracks in three dimensions: A dynamic dual boundary element analysis COMPUTER METHODS IN APPLIED MECHANICS AND ENGINEERING Volume: 167 Issue: 1-2 Pages: 139-151, 1998
[7] Albuquerque EL; Sollero P; Aliabadi MH Dual boundary element method for anisotropic dynamic fracture mechanics INTERNATIONAL JOURNAL FOR NUMERICAL METHODS IN ENGINEERING Volume: 59 Issue: 9 Pages: 1187-1205 2004
[8] Saleh, AL; Aliabadi MH Crack growth analysis in concrete using boundary element method ENGINEERING FRACTURE MECHANICS Volume: 51 Issue: 4 Pages: 533-545 1995
[9] Wen P. H.; Aliabadi M. H.; Lin Y. W.Meshless method for crack analysis in functionally graded materials with enriched radial base functions, CMES-COMPUTER MODELING IN ENGINEERING & SCIENCES Volume: 30 Issue: 3 Pages: 133-147 2008
[10] Wen P. H.; Aliabadi M. H. An improved meshless collocation method for elastostatic and elastodynamic problems COMMUNICATIONS IN NUMERICAL METHODS IN ENGINEERING Volume: 24 Issue: 8 Pages: 635-651 2008
[11] Wen P. H.; Aliabadi M. H. Evaluation of mixed-mode stress intensity factors by the mesh-free Galerkin method: static and dynamic JOURNAL OF STRAIN ANALYSIS FOR ENGINEERING DESIGN Volume: 44 Issue: 4 Pages: 273-286 2009
[12] Li L. Y.; Wen P. H.; Aliabadi M. H. Meshfree modeling and homogenization of 3D orthogonal woven composites, COMPOSITES SCIENCE AND TECHNOLOGY Volume: 71 Issue: 15 Pages: 1777-1788 2011
[13] Sfantos G. K.; Aliabadi M. H Multi-scale boundary element modelling of material degradation and fracture COMPUTER METHODS IN APPLIED MECHANICS AND ENGINEERING Volume: 196 Issue: 7 Pages: 1310-1329 2007
[14] Wen PH, Aliabadi MH, A variational approach for evaluation of stress intensity factors using the element free Galerkin method, INTERNATIONAL JOURNAL OF SOLIDS AND STRUCTURES, 2011, Vol:48, Pages:1171-1179, 2011
[15] F. Durbin, Numerical inversion of Laplace transforms: an efficient improvement to Dubner and Abate's method, *The Computer J.*, 17, 371-376, 1975.
[16] D.P. Rooke, D.J. Cartwright, Compendium of Stress Intensity Factors, London, Her Majesty's Stationery Office, 1976.
[17] P.H. Wen, *Dynamic Fracture Mechanics: Displacement Discontinuity Method*, Computational Mechanics Publications, Southampton UK and Boston USA, 1996.

Investigation of Damping Characteristics of Metals by the use of Inverted Torsion Pendulum Method

Toshikatsu ASAHINA [1,a], Tadashi SHIOYA [1,b],
Takahiro TOGURI [1,c] and Masanao SEKINE [2,d]

[1] Department of Mechanical Engineering College of Industrial Technology Nihon University

1-2-1 Izumichou, Narashinoshi Chiba, Japan

[2] Department of Aeronautics and Astronautics of Engineering, the University of Tokyo

7-3-1 Hongo, Bunkyo-ku, Tokyo 113-8656, Japan

email: [a]asahina.toshikatsu@nihon-u.ac.jp, [b] shioya.tadashi@nihon-u.ac.jp,
[c] cith10030@g.nihon-u.ac.jp, [d] sekine@mat.t.u-tokyo.ac.jp

Keywords: Damping characteristics, Inversed torsion pendulum method, Aluminum alloy, Magnesium alloys, Inconel

Abstract. Damping characteristics of metals are measured by the use of inversed torsion pendulum method. As the materials, aluminum alloys, magnesium alloys, Inconel etc are used with the temperature and frequency parameters. Mild steel is employed as a reference material for calibrating the apparatus. The measured damping coefficient of magnesium alloys is larger than that of aluminum alloys indicating that the HCP crystal structure has more damping effect than the FCC structure. The damping of these metals increases with temperature. The manner of increment of damping is investigated to analyses the microscopic mechanism of damping.

Introduction

In recently years, vibration and noise became social issues and there are more attentions paid to vibration damping of mechanical items for prevention of fatigue failure and noise. Therefore, needs are increasing to design qualify vibration-proof metal materials. For instance, it is expected to apply to auto parts, audio and visual equipment and computer parts. However, there are little reports concerning characteristics to absorb vibration of commercialized alloys.

This paper is to study how atmospheric temperature and numbers of torsional vibration would affect logarithmic decrements for commercialized alloys including aluminum, magnesium and Nickel-base superalloy through the Inverted Torsion Pendulum Method.

Materials and experimental procedures

The specimens include aluminum alloy, magnesium alloy, Inconel 718 and mild steel (hereinafter, A7075, AZ31, AZ61, AZ91, Inconel and S45C) machined processed to be a square bar in size of 4×4×130mm. The metal specimens are annealed under a proper condition.

Fig. 1 shows the inverted torsion pendulum test equipment. The specimen are chucked as 105mm. Upon applying the initial displacement to the pendulum to move it on a horizontal plane, the specimen will be distorted. Once a weight are moved to a position as required on the sides of the pendulum, the inertia moment of the pendulum starts to change for obtain more variations of frequencies. In this test, a weight of a mass of 2kg is relocated to the positions of 25mm and 300mm from the specimen placed in the center of the testing equipment for experiments in a range from 0.5 to 5Hz. The maximum (initial) strain was increased by about 0.009% for the metals in order to conduct the test in a macroscopically elastic region of the specimen. A counterweight is placed in the back through a pulley at the superstructure of the testing equipment to set of compressive or tensile loads to be applied to the specimens. Furthermore, under conditions with various testing temperature, tests were conducted in the macroscopically elastic region. The test temperature is a range from 298 to 623K except Inconel. Inconel was tested at 300 ～ 984K. The displacement was detected in the macroscopically elastic region at the rotational parts at the top of pendulum, which the data was recorded on PC with A/D converter.

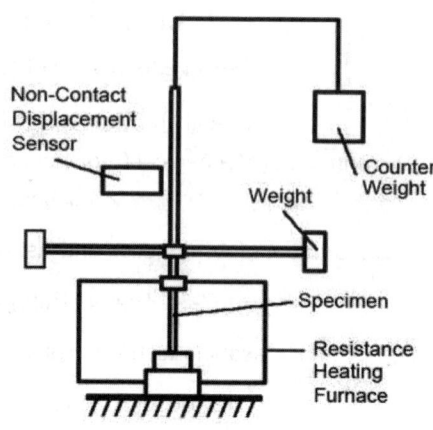

Fig.1 Schematic illustration of inverted torsion pendulum.

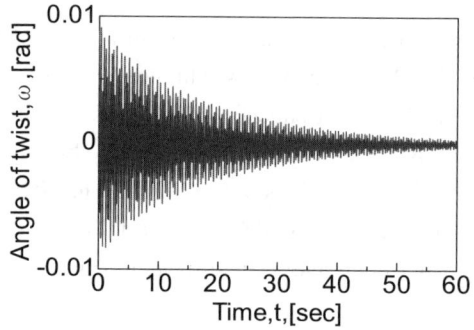

Fig.2 Torsion amplitude of specimen AZ31 at room temperature.

Results and discussion

Fig. 2 shows an example of damped oscillation curve of each material measured at room temperature. The shear elastic coefficient G (hereinafter, G) for each material was obtained based on S45C using the frequency calculated from this curve. G of each material is calculated as figure (1).

$$\frac{\omega_1}{\omega_2} = \sqrt{\frac{G_1}{G_2}} \tag{1}$$

Here, ω_2 and G_2 are the frequency and G of S45C respectively. ω_1 is a frequency obtained from the damped oscillation curve. According to a reference, G of S45C is 78Gpa[1], which shows the difference of the value for S45C are only within 5% and other materials showed also same value. In case of S45C, the frequency was 1.22 Hz at the weight located at 300mm. It was 0.70Hz for A7075 and 0.59Hz for magnesium alloys. For calculation of the damping ratio, its recorded data were used by removing some bands with noise.

Fig. 3 shows the impacts of temperature on G for each material. The difference of 298K and 623K was about 10% for magnesium alloys in terms of G.

The damping oscillation curve is represented by a function of vibration amplitude A and time t. The damping oscillation curve obtained through the test was quantified to an absolute value, and its peak was applied to Figure (2) to calculate the inclination for the logarithmic decrement [2].

$$A = A_0 \exp(-\lambda f t) \cos 2\pi f t \quad (2)$$

A_0, f and λ are the initial vibration amplitude, frequency of vibration, and damping oscillation ratio respectively.

This research is to observe damping free vibration curves to scan the peak per cycle, which the ratio is plotted to calculate these inclination by the least square method. The logarithmic ratio λ and frequency of vibration f is calculated.

Fig. 4 shows the damping ratio of each material. As the melting temperature is high as about 1673K for S45C, the damping ratio was only 1.5% in average as no distinct change of composition was identified in this test.

In terms of A7075, the softening temperature was about 688K for aluminum alloy so that it is considered to be softened at 623K. Therefore, the logarithmic decrement was large since it is easily slip with changes in temperature.

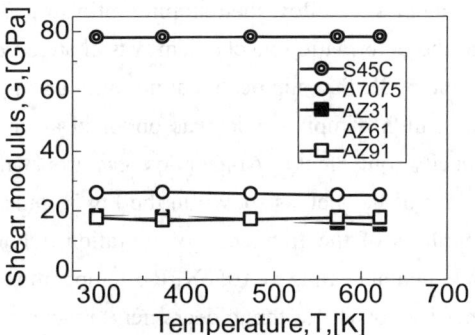

Fig.3 Relation between shear modulus and temperature.

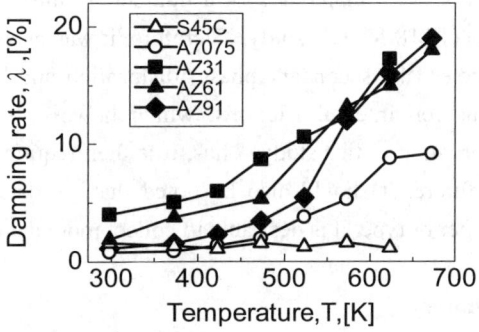

Fig. 4 Relation between damping rate and temperature.

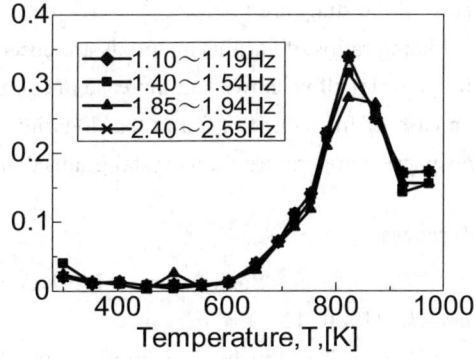

Fig.5 Relation between the logarithmic decrement and the temperature of Inconel annealed at 984K.

According to the research of Kê, it reveals that the peak of damping ratio will be at around 600K for pure aluminum[3]. However, as many crystal grains and chemical elements of alloy are the factors to slip, it is considered that there was no peak since the dislocation movement was limited.

For magnesium alloy, the damping ratio surged at 523K or more with temperature rise. This means that the deformation mechanism was changed due to the temperature rise. Therefore, it is found at around 573K that slip occurs at not only basal plane but also vertical face plane. So it is considered that a high damping ratio was obtained at or above 523K[4]. The magnesium alloy is a damping capacity equivalent to Al-Zn alloy, cast irons and Cr stainless steel 12%.

In case of Inconel, as shown in the Fig.5, under heat treatment at 984K are temperature dependent, regardless of the frequency of vibration, which showed the peak around 800K. In terms of the simulated annealing at 1050K, the temperature dependency was also observed. It is found that the inner friction (logarithmic decrement) reaches the peak at almost same atmospheric temperature for a crystal grains diameter. However, there is a negative interrelationship between the crystal grains diameter and logarithmic decrement, which is not consistent to the result of the preceding research [4]. Although many preceding researches were focusing on relatively pure materials to study, the authors use a material with a number of alloy elements with nickel as its main composition for Inconel 718 for this study. Therefore, it was necessary to consider a function to prevent precipitate particles (the secondary phase) dislocation and boundary slipping which could have a significant impact on internal frictions, which however was difficult to observe with the microstructure observation in this study. Thus, it is then required to observe microstructures of crystals further in the future. It should also be noted that, as data obtained through the test was not sufficient in number or type, it is desirable to collect more data to design a viscoelastic dynamics model.

Summary

The following results are obtained for damping ratio upon measurement applying the inverted torsion pendulum method to A7075 aluminum alloy, AZ31, AZ61, AZ91 magnesium alloys and Inconel 718, as well as its comparison with the basic material S45C.
Results of the study are :
1) Damping ratios of various materials are observed by the inverted torsional pendulum method.
2) Magnesium alloy showed an increase in damping ratio at 523K or more.
3) In case of Inconel, it is found that logarithmic damping ratios reaches the peak at almost same atmospheric temperature for a crystal grains diameter.

References

[1] The Japan Society for Mechanical Engineers, "*Elastic Coefficient of Metal Materials*",(1980),42-43.
[2] T.Shioya, T.Kakiuchi, K.Fujimoto and M.Sekine: "*Estimation of Visco-Elastic Constitutive Equation from Free Oscillation Experimen*", Strength, Fracture and Complexity, 6 (2010)31-50.
[3] Kê,T.S., "*Experimental Evidence of the Viscous Behavior of Grain Boundaries in Metals*", Physical Review, Vol.71, No8(1947), 533-546.
[4] Y. Takahashi, M. Sekine, K. Yamada, K. Fujimoto, "*Effect of Crystal Texture on Internal Friction of Commercially Pure Aluminum*", The Japan Society for Mechanical Engineers (A), 76(2010), 766.

Mechanical Characteristics of Bamboo

Tadashi Shioya[1, a] and Toshikatsu Asahina[1, b]

[1] Department of Mechanical Engineering, College of Industrial Technology, Nihon University

1-2-1 Izumicho, Narashino, Chiba 275-8575, Japan

[a]tshioya@gakushikai.jp, [b]asahina.toshikatsu@nihon-u.ac.jp

Keywords: bamboo, fracture toughness, elastic modulus, attenuation, damping coefficient, inverted torsion pendulum, water concentration.

Abstract. Fracture toughness, stress-strain relation and the damping characteristics of bamboo are investigated. The fracture toughness of bamboo in tearing along the longitudinal direction is measured by the use of newly devised apparatus in which the crack opening displacement is controlled in a constant velocity and a quasi-steady extension of the crack is maintained. The stress-strain relation of bamboo is examined in a reversible elastic range using a conventional tensile test in the longitudinal direction. Repeated tensile loading tests show that the stress-strain relation has a strong non-linear hysteresis and that it converses to a steady loop. The damping of bamboo is measured by the use of inverted torsion pendulum apparatus. The specimen is taken so that damping of twisting longitudinal bar is measured. The damping coefficient of bamboo is much larger than that of metals. The mechanical properties of bamboo are examined in terms of water concentration and fiber density in the bamboo.

Introduction

Bamboo is one of candidate bio-materials for structural use. The strength of the bamboo in the longitudinal direction (fiber direction) is higher than that of most other wood materials. With its lightness and strength, the bamboo has larger specific strength (strength per weight) than most metallic materials. However, there have been few researches on mechanical properties other than the elastic constants and the fracture strength of longitudinal direction or the bending direction.

In this study, mechanical properties of bamboo are extensively examined using three types of experiments. First, conventional tensile test is conducted for a straight specimen of bamboo. Loading and unloading are repeated with changing the velocity to obtain the hysteresis curve. The second type of test is a quasi-steady tearing along the fiber direction to obtain the fracture toughness. The third is a free oscillation test in pure torsion obtaining the damping characteristic of bamboo.

Experiment

Material. The material used in the experiment of this study is mao bamboo or moso bamboo (academic name: phyllostachys edulis), most widely spread in east Asia. Since bamboo is a bio-material, its characteristics differ from specimen to specimen. In this experimental study, the position at the original material from which the specimen is taken is examined. It is found in the following experiments that the difference of characteristic is small among data of the specimens taken from different height positions from the earth and the different facing direction (north, south, east and west) if the specimen is taken from the same bamboo. However, it is found that the specimen radial position and water content have some effect on the mechanical characteristics. The bamboo is a kind of composite material with cellular fibers and matrix. The fiber density of outer side is richer than that of inner side (Fig. 1). The water concentration in the specimen is controlled by aging for several days to weeks.

Stress-Strain Hysteresis Tests.

Uni-axial tensile loading and unloading are repeated to straight bar specimen with rectangular cross-section of 4mm×20mm. An example of measured stress-strain relationship is shown in Fig. 2, showing a strong non-linear hysteresis that converges to a steady reversible loop. It should be noted that the hysteresis loop becomes thinner as the specimen position becomes inner side (less fiber density).

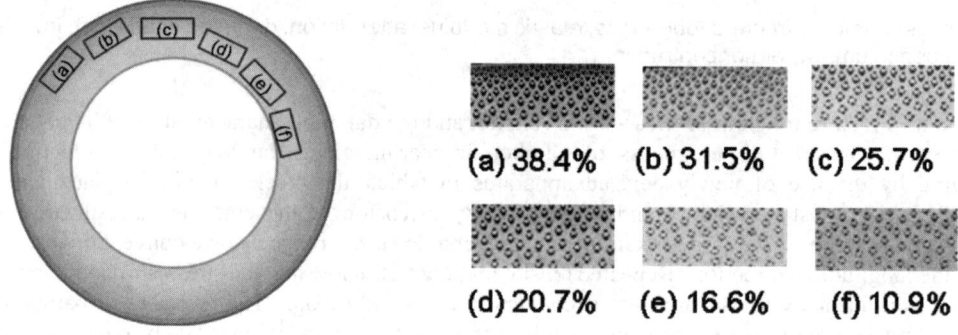

Fig. 1. Specimen bamboo

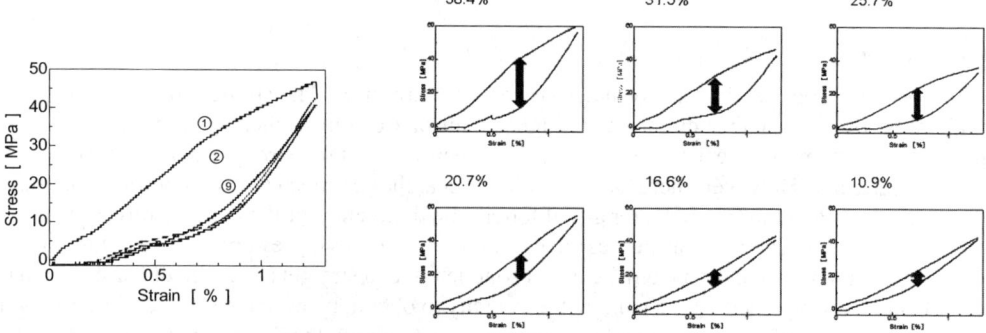

Fig. 2. Repeated stress-strain relationship of bamboo. Effect of fiber density on hysteresis.

Fracture Toughness Tests.

An experiment of quasi-steady crack extension is executed by the use of newly developed device to obtain the fracture toughness in the direction of tearing along the fibers. Figure 3 explains the mechanism of the device. The specimen is a beam shape with a pre-crack in one end at the middle height of beam. A constant increment rate is given to the displacement between the upper and the lower parts of the beam at the pre-cracked end. Under this condition, the released energy rate decreases as the crack extends so that quasi-steady crack extension is maintained. The opening displacement at the end of beam $2y(0)$ is expressed as

$$y(0) = \frac{Pl^3}{3EI} \quad (1)$$

where l is the crack extension length and EI is the bending rigidity of the cracked part of beam. The fracture toughness K_{IC} is calculated by the cantilever beam theory as

$$K_{IC}^2 = \frac{EG_C}{1-\nu} = \frac{P^2 l^2}{bI} \qquad (2)$$

Example of the observed load-displacement is shown in Fig. 4. The crack length l is observed by a video camera. The obtained fracture toughness K_{IC} is also exhibited in Fig. 4.

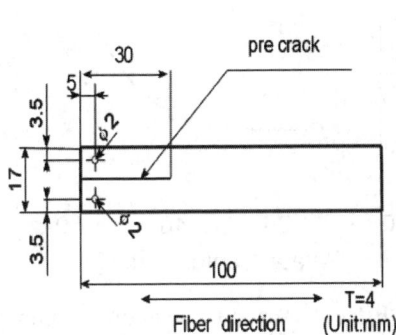
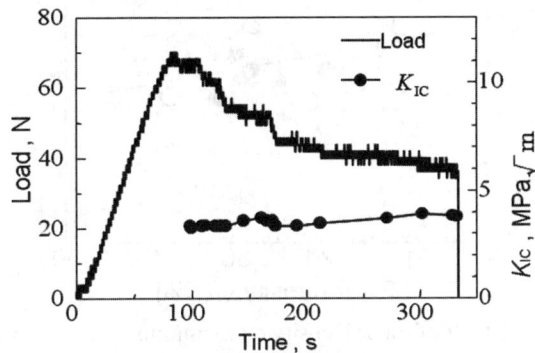

Fig. 3. Fracture Toughness Specimen   Fig. 4. Load-Time Relation with Fracture Toughness

Damping Tests.
The inverted torsion pendulum apparatus used in the experiment is shown in Fig. 5. An example of attenuation wave in free oscillation in the experiment is shown in Fig. 6. Assuming linear small range attenuation, the amplitude A in a free oscillation is expressed as,

$$A = A_0 \exp(-\lambda ft)\cos(2\pi ft), \qquad (3)$$

where A_0 is the initial amplitude at $t=0$, f is the frequency and λ is the damping coefficient per unit wave length time. The peak profile (amplitude) is fitted to logarithmic curve, and the damping coefficient λ is obtained by the logarithmic decay. The experiment was conducted in the frequency range from 0.13 to 0.76 Hz. The obtained damping coefficient per unit wave length is almost unchanged in this range, confirming that the damping is proportional to the strain velocity.

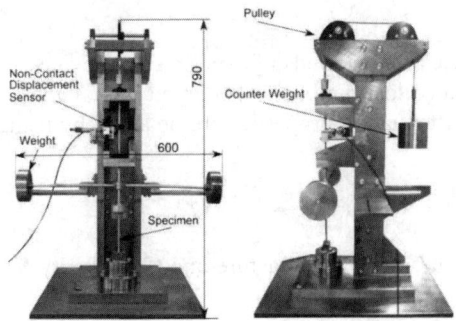
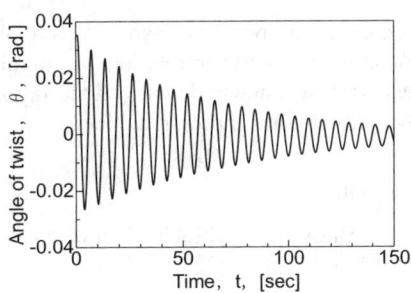

Fig. 5. Inverted Torsion Pendulum Apparatus   Fig. 6. Attenuation Curve in Torsion Oscillation.

The specimens are taken from several radial positions in the same bamboo bar, in order to examine the effect of fiber density. The obtained damping coefficient of bamboo from different radius is shown in terms of fiber density in Fig. 7. It is found that the damping increases as the fiber density decreases. The effect of water concentration is shown in Fig. 8. The effect of water concentration by the use of different aged specimens is shown in Fig. 8. The damping increases as the water concentration increases.

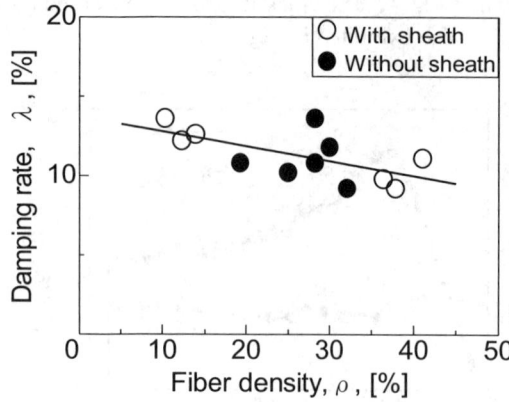
Fig.7. Effect of Fiber Density on Damping

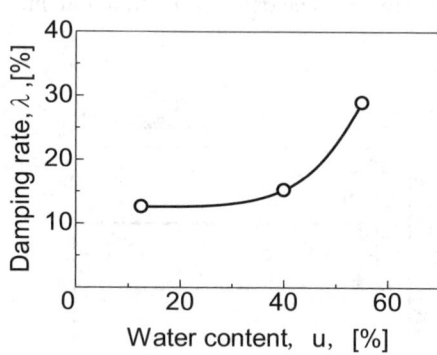
Fig. 8. Effect of Water Content on Damping

Discussions

The strength of bamboo is generally thought to be very high in the longitudinal direction, but relatively weak in the tearing direction. However, it is found in the present experiment that the tearing toughness of bamboo is comparable to that of other wooden materials. It should be noted that the stress strain curve of bamboo in uni-axial tension has strong hysteresis even in small strain range and it converses to the unloading curve. The difference between loading and unloading curves represents a kind of dissipation energy and it depends on the fiber density. The damping of bamboo in the vibration experiment also represents the energy dissipation. However, its dependence on the fiber density is opposite tendency to the uni-axial tension test. It may be interpreted that while the hysteresis in the tension depends on the tension-relaxation manner in the fibers, the damping in the vibration is mainly due to the water content in the material and matrix part contains more water that the fiber part.

Summary

Mechanical properties of bamboo were obtained by use of three kinds of experimental devices. Hysteresis stress-strain relation was obtained by repeated load-unload tests. Fracture toughness was measured by a newly developed tearing apparatus. Attenuation properties are obtained by using an inverted torsion pendulum device.

References

[1] T.Shioya, T. Kakiuchi, K. Fujimoto and M. Sekine: Strength, Fracture and Complexity Vol. 6 (2010), p. 31–50

Relationship between Crack Extension Behavior and Fatigue Life of C/C Composites

Sofyan A. Setyabudi [1,a], Chobin Makabe [2,b], Masaki Fujikawa [3,c]

[1] Graduate School of Engineering and Science, University of the Ryukyus, 1 Senbaru, Nishihara, Okinawa, 903-0213, Japan

[2, 3] Mechanical Systems Engineering Department, University of the Ryukyus, 1 Senbaru, Nishihara, Okinawa, 903-0213, Japan

[a] sasbudi@yahoo.com, [b] makabe@tec.u-ryukyu.ac.jp, [c] fujikawa@tec.u-ryukyu.ac.jp

Keywords: C/C Composites, Fatigue limit, Crack growth behavior, Fracture constraint, Coalescence

Abstract. The fatigue and fracture mechanism of C/C composites material was investigated in notched and smooth specimen. Also, the initiation and growth behaviors of cracks were continuously observed. Specially, the phenomenon of fracture constraint in notched specimen was discussed. This phenomenon was related to the crack growth behavior and shear damage in the matrix. In the present study, fatigue limit was defined by the critical stress level that the specimen was not broken even after applied 1×10^6 stress cycles. When the fatigue life was longer than 1×10^6 stress cycles, the initiated crack was stopped to growth after reaching some critical length. Also, the density of crack initiation was low. When the fatigue life became shorter than that, the cracks extended rapidly and coalescences of cracks were observed. The crack growth behavior was strongly related to the fatigue limit and unstable fracture conditions. Also it is found that the compliance or elastic modulus of the specimen was related to residual fatigue life. Now, an observation of crack in C/C composites was performed effectively in the present study.

Introduction

In the recent years, the utilization of carbon composites material on engineering applications is growing rapidly. The carbon composites material was used in many engineering part. Also, some kinds of composites were replaced from the utilization of metals, especially on the aircraft industry with the basic consideration of weight and fuel saving issues [1]. In this present study, the machineable C/C composites were used in the testing material. There are notched site on many machine components. So, investigation of the effect of notch on the fracture strength is important in such material. In previous study [2, 3], it was discussed that the fracture behavior of C/C composites was different from plastics and metals. The fatigue limit and tensile strength of the notched specimen used were higher than those of the smooth specimens or plane specimens. However, the relationship between the strengths and cracks growth behavior was not discussed in detail, because the crack growth was difficult to observe. In the present study, the crack growth was observed in detail by a special technique from specimen surface. Then the fracture behavior on fatigue process was discussed. The strength of the composites material was depending on the deformation behavior of matrix and fracture behavior of carbon fiber.

Materials and testing procedure

The material used was two-dimensional sheets of C/C composites fabricated using fine-woven carbon fiber laminates supplied by Toyo Tanso, Co. Ltd., with the commercial grade CX-31. The mechanical properties are shown in Table 1. The geometry of the original specimen is a square shape with 130mm length, 30mm width, and 4mm thickness. The stitching fibers were directed with 0°/90° angles to the specimen axis.

Table 1 Mechanical properties

C/C type	Grade	Bulk Density	Flexural strength	Compressive strength	Tensile strength	Ash content
		[g/m³]	⊥*[MPa]	⊥*[MPa]	//* [MPa]	[mass %]
2DC/C	CX-31	1.61	90	249	98	0.02

*⊥: perpendicular to laminate   *//: parallel to laminate

Several configurations of specimens were used to investigate the relationship between crack growth and fatigue life. The slit specimen with 5mm length and 1mm width of slits, the holed specimen with 2mm radius of hole, and the blunt notched specimen with 8mm root radius and 7.5mm length of notch were used. Fig. 1 shows the geometry of these specimens ($2W$=30mm).

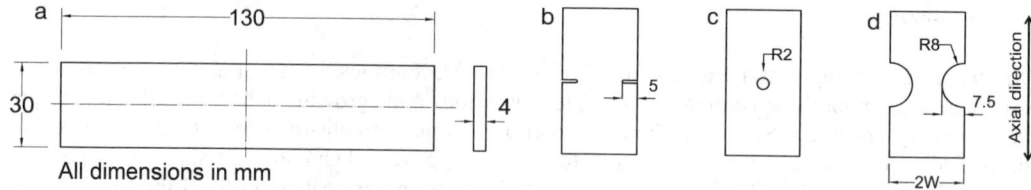

Fig. 1 Geometry of specimens; (a) Original plate, (b) Slit, (c) Holed, (d) Blunt notched specimens

The fatigue test was performed at a frequency of 10 Hz with constant net stress σ. The net stress σ was calculated with the projection-cross section, which was projected in the axial direction at the ligament. To ease the observation of crack initiation and growth, the special film was paste on the specimen surface. The crack length was measured perpendicular to the axial direction by aids of a dial gauge and microscope. Now, the cracks were easier to observe with pasting special film on the surface. Also, a series of static tensile tests were performed before and after the application of cyclic stress by using blunt notched specimen. In these tests, the accumulated damage in the fatigue specimen was investigated by measurement the compliance or the elastic modulus.

Results and discussion

Fig. 2 shows the S-N curves that are the relationship between stress amplitude σ_a and number of cycles to failure N_f. The arrow shows the data where the fatigue tests were stopped at 10^6 cycles. We defined the fatigue limit σ_w as a critical stress where the specimen was not broken at $N=10^6$. The fatigue limit σ_w of the smooth specimen obtained in the previous study [2, 3] was varied from 28 MPa to 30 MPa as shown dashed lines in Fig. 2. In the case of the blunt notched specimen with 8mm of notch radius, the fatigue limit σ_w was within the range of σ_w of the smooth specimens. So, this specimen is regarded the smooth specimen. Because the damage in the matrix was expanded in large area, the fatigue limit σ_w became almost the same between the smooth specimen and blunt notched specimen.

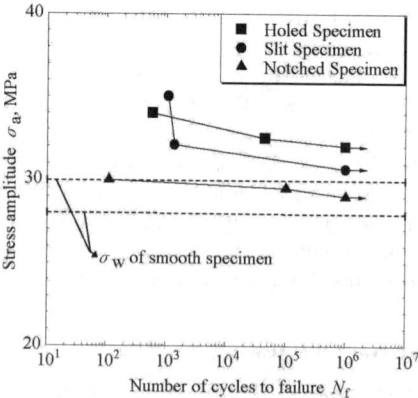

Fig. 2 S-N curves

In the present study, the highest fatigue limit σ_w was obtained in the case of the holed specimen. The order of the specimen shape with high level of fatigue limit σ_w in the present study was the holed specimen, slit specimen and blunt notched specimen (or smooth specimen). This can be explained from the expansion of damaged area and the crack growth behavior. That order is different from the metal case. Also, those behaviors were explained with the fracture constraint in previous study [2]. The shear deformation and succeeding broken behavior of carbon fiber is related to the fracture constraint. However, the relationships between the crack growth behavior and fracture behavior were not discussed exactly in the previous study because the observation of crack was hard to do. So, the observation method is developed in this study by above mentioned method.

The crack growth curves of holed and blunt notched specimens are shown in Fig. 3. The arrow shows the coalescence of the cracks. In the present study, the blunt notched specimen is regarded as the smooth specimen or plane specimen. The crack growth behavior, in which the crack growth was stopped, was different between the blunt notched specimen and holed specimen. It was found that the crack growth was stopped after coalescence of cracks. In the case of holed specimens, after increasing crack length within 2mm or 3mm, crack growth was stopped even if the coalescence of cracks was occurred. The crack growth rate of these cases was lower than that of the broken specimens. Specially, the crack growth rates of broken specimen and unbroken specimen were clearly different in first stage of crack growth. In the case of holed specimen, even when crack growth length was reached about 2mm at $N=1\times10^6$, the specimen was not broken. On the other hands, in the case of blunt notched specimen, there was small difference tending of crack growth rate between the broken specimens and unbroken specimens at the first stage of crack growth. The crack growth length was lower than 0.5mm at $N=1\times10^6$ in the case of unbroken specimen. After the crack growth length was reached 1mm, the crack grew unstably in the case of broken specimen. Those behaviors of crack growth are related to the fatigue limit order of the specimens. Now, the crack length $2a$ in the case of the holed specimen is included diameter hole, and measured length was between the crack tips. The crack length a in the case of the blunt notched specimen was average value of both side of cracks which including the notch length.

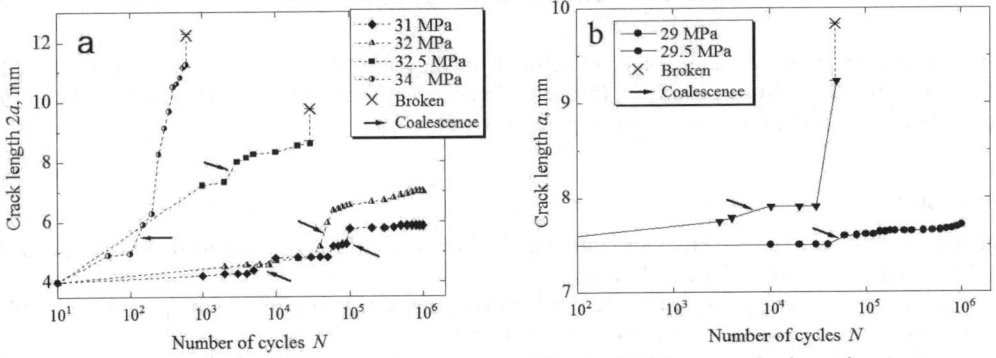

Fig. 3 Crack growth curves; (a) Holed specimens, (b) Blunt notched specimens

To clarify the relationship between the crack extension behavior and the accumulated damage, tensile tests were carried out before and after fatigue tests on the blunt notched specimens under σ_w. In the present study, the deformation within elastic range was investigated. The applied stress during the fatigue test, σ_a, was 25 MPa. Anggit et al. [3] were reported that C/C composites material possesses an inhomogeneous property, which could be responsible for the non-linearity. The inhomogeneity could lead to not only the non-linearity, but also to the alteration of properties such as tensile and fatigue strength. Fig. 4 shows the variation of the elastic modulus after applying cyclic stress. The stress-strain curves in low stress level were approximated by straight lines in those tensile tests. The variation behavior of elastic modulus shows that the damage was accumulated in the matrix by cyclic stress. Even in the case that applied stress was lower than σ_w, the specimen was damaged. However the decreasing of the elastic modulus was saturated after applying some stress cycles. It means that the crack was initiated and some parts of fibers were broken, but the damage did not expand anywhere and breaking of fiber was not expanded other area. Then, the specimen was not broken. This behavior is related to the fatigue limit σ_w.

Fig. 5 shows the examples of the observations of crack growth in the broken and unbroken specimens. In the case of unbroken specimens, cracks were started to grow after applying large numbers cycles and then stopped to grow. In the case of broken specimens, cracks were started with small numbers of stress cycles. After reaching some length of crack, the specimen was broken suddenly. These behaviors of crack growth were related to the fracture constraint [2] which leads to increase the fatigue limit of slit and holed specimens. Also, the deformation behavior of matrix was related to the non-growth behavior of cracks and the final fracture behavior of specimens.

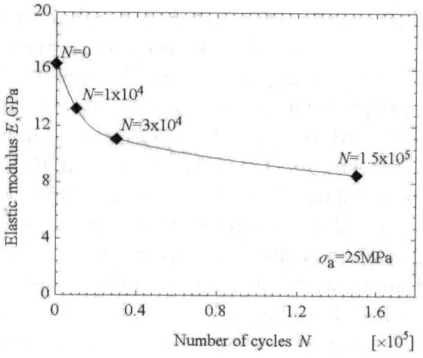

Fig. 4 Variation of elastic modulus

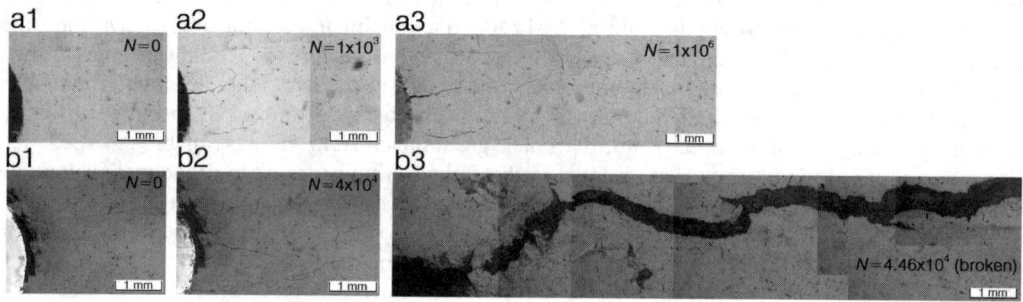

Fig. 5 Examples of observation of crack growth in the case of holed specimen; (a) σ_a=32 MPa (non-broken), (b) σ_a=32.5 MPa (broken). (Now, the shape of hole edge was clear, but it look likes bad conditions in these figures because of the film pasted)

Conclusions

The relationship between the crack growth length and fatigue life were examined by using net stress σ. The main results obtained are as follows,
1. The specimen order of high fatigue limit σ_w was the holed specimen, slit specimen and blunt notched specimen because of the fracture constraint.
2. Crack growth behavior was different between the holed specimen and blunt notched specimen. The critical length of crack where the crack did not grow at $N=1\times10^6$ in the case of holed specimen was longer than that in the case of blunt notched specimen. Such the crack growth behavior was related to the order of fatigue limit of specimen used.

References

[1] Information on http://www.boeing.com
[2] S.A. Setyabudi, C. Makabe, M. Fujikawa, T. Tohkubo: Journal of Solid Mechanics and Materials Engineering Vol. 5 (2011), p. 640.
[3] A. Murdani, C. Makabe, M. Fujikawa: Carbon Vol. 47 (2009), p. 3355.

Influence of Adhesive Layer on Actuation of Lamb Wave Signals

Z. Sharif Khodaei[1,a], Liu Qu[1,b] and M. H. Aliabadi[1,c]

[1]Department of Aeronautics, Imperial College London, South Kensington Campus,
Roderic Hill Building, Exhibition Road, SW7 2AZ, London, UK.

[a]z.sharif-khodaei@imperial.ac.uk, [b]liu.qu10@imperial.ac.uk, [c]m.h.aliabadi@imperial.ac.uk,

Keywords: Structural Health Monitoring, Ultrasonic Guided Waves, Piezoelectric transducer.

Abstract. In this work, Lamb wave generation and propagation have been modelled in composite plates. Actuation and acquisition of signals when the PZT transducers are tied to the structure or bonded with an adhesive layer are investigated. The effect of adhesive thickness and actuation frequency of Lamb wave have been examined.

Introduction

Although composite materials have been widely used in industry due to the excellent mechanical properties and low density, the invisible defects caused by the impact during the service life reduce the safety. Non-Destructive Techniques (NDT) such as thermography and ultrasonic techniques have been used for damage detection due to the timely testing, easy to conduct and relative low cost. However traditional techniques can be conducted before the assembly of the component but not during the service life of the structure without it being disassembled. Therefore, recent techniques based on sensor data are increasingly being evaluated for monitoring the health of composite structures.

The Structural Health Monitoring (SHM) techniques have the potential to reduce the maintenance activities and operational costs as well as to improve the aircraft safety and reliability. SHM techniques are aimed at moving away from the current schedule-based maintenance to a reliable condition-based maintenance based on monitoring the structure in real time. SHM methodologies are established by comparing the actual state of the structure to a pristine baseline case [1]. SHM techniques can be used in passive sensing for impact detection and classification [2] and in active sensing for damage detection and characterisation [3, 4]. The active sensing techniques use sensorised panels to measure the response of the structure in real time and conduct a diagnosis analysis. One methodology for detecting damage in composite structures is using Ultrasonic Guided (UG) waves [3]. Lamb waves are a specific type of UG waves. Lamb wave method was first regarded as an effective damage detection technique at 1961 when Worlton studied the characteristics of Lamb wave in terms of phase velocity to frequency and the thickness of the *Al* and *Zr* plate [5]. Even though analytical models are available for numerical modeling the wave propagation in composite plates [6], most of the analytical formulations are simplified and they assume tie connections between the plate and the PZT transducers. The influence of adhesive have been considered by few researchers [7]. To ensure correct damage detection technique based on numerical actuation and acquisition of Lamb waves, the accuracy of the numerical model needs to be evaluated. In this paper, the influence of adhesive layer on the actuation and sensing of Lamb waves are investigated.

Piezoelectric transducers

Based on piezoelectric (PZT) effects, PZT transducers have been the major application for generating and receiving Lamb waves. By applying an electric field, PZT material will produce strain change and similarly, when strain is applied, it will respond with a change in voltage. Unlike other smart materials such as shape memory alloys, PZTs have wide range of frequency responses, low power consumption, low cost and light weight [8]. There are two different transducer shapes that can be used: rectangular and circular. Circular PZT patches are used in this work. The numerical models were simulated by the commercial FE code ABAQUS. Due to the absence of piezoelectric elements in ABAQUS Explicit, the transducer models are based on the SMART FE model developed in [4] where the voltage will be applied as corresponding radial displacement in the FE model.

Influence of adhesive layer

The effect of adhesive layer is evaluated on actuation frequencies of PZT transducers surface mounted on a CFRP plate with a layer of CW2400 adhesive. Two PZT transducers are placed symmetrically in the middle of the plate, 100 mm apart from each other. The PZT sensor modeled in this work is a circular patch made from PZT PKI 402. The geometry for the CFRP plate is $500\ mm \times 500\ mm \times 2\ mm$ (see Figure 1). The stacking sequence for the plate is $[45_2/-45_2/0_2/90_2]_s$ and the material properties are shown in Table 1 to Table 3.

Table 1 Material properties of CFRP plate

E_{11} (GPa)	E_{22} (GPa)	E_{33} (GPa)	G_{12} (GPa)	G_{13} (GPa)	G_{23} (GPa)	V_{12}	V_{13}	V_{23}	ρ (kg/m^3)	Geometry (m^3)
153.7	9.49	9.49	4.26	4.26	3.44	0.295	0.295	0.381	1528	$0.5 \times 0.5 \times 0.002$

Table 2 Effective material propertied of PKI 402 PZT patch

Young's modulus (GPa)	Poisson's ration	Density (kg/m^3)	Relative dielectric constant K_3	Piezoelectric charge coefficient d_{31} (m/V)	Radius (mm)	Thickness (mm)
72.5	0.31	7800	1280	130×10^{-12}	3.4	0.5

Table 3 Material properties for adhesive Chemtronics CW2400

Young's modulus (GPa)	Shear modulus (GPa)	Poisson's ratio	Density (kg/m^3)
2.60	1.00	0.30	1100

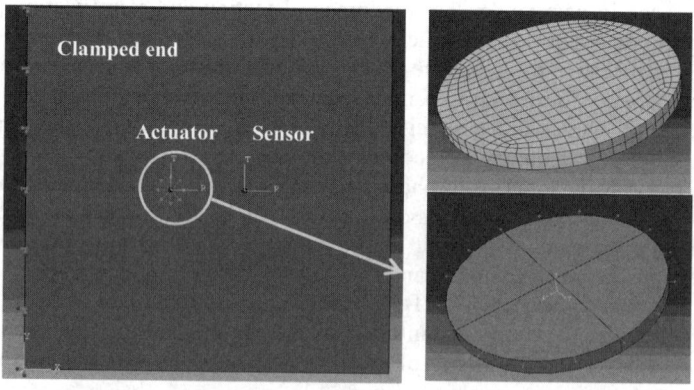

Figure 1 Geometry of CFRP plate and PZT trasnducers

The thickness of adhesives varies from 0.01 mm to 0.10 mm in the frequency range from 300 kHz to 600 kHz. The mesh size and the time increment are governed by the geometry of the plate, the material properties and the actuating frequency. A dynamic Finite element simulation was performed. The developed SMART elements were used for generating and sensing of the Lamb wave signals.

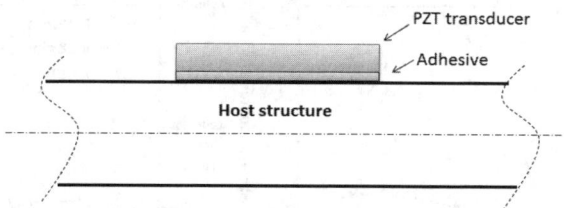

Figure 2 The adhesive bonded PZT transducer on the plate

The adhesive layer has been modeled with isotropic material properties, once with solid elements and once with cohesive elements with traction-separation behavior. To our interest, the amplitude of the sensor readings defer for the two different models of the adhesive layer (Figure 3 (a)).

A range of actuation frequency versus adhesive layer thickness were evaluated and plotted in Figure 3 (b). It can be seen that for different frequencies, the effect of adhesive layer is very different. Moreover the strains responses in the PZT sensor for different thickness of adhesive layer defer from the case were the PZT transducers are simply tied to the plate (0 mm adhesive layer). Only for some combination of frequency and thickness the response is the same as the case with no adhesive layer. This influence needs to be taken into account when simulating Lamb wave actuation and propagation numerically.

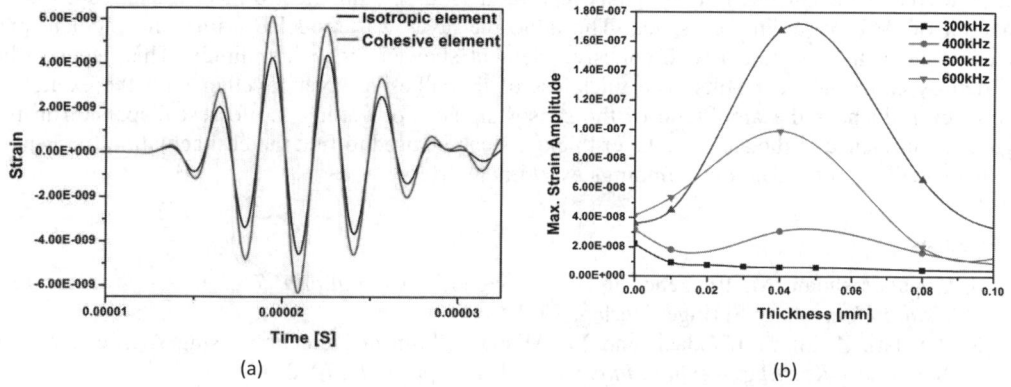

Figure 3 Influence of adhesive layer on acquisition signal (a) Solid and cohesive element comparison (b) Maximum strain vs. adhesive thickness

The reason for the strain decrease with increasing thickness is due to the shear-lag model which will reduce the strain transmitted between the PZT and the CFRP plate at the low frequency [9]. Strain transmission will be less effective with increasing thickness of adhesive, thus resulting in the reduction for the strain amplitude [7]. Due to the resonant phenomenon, the vibration of the PZT transducer is less constrained with higher thickness of adhesive and thus more energy can be triggered at the actuator. Therefore at higher frequencies, the amplitude of the sensors signal will increase at about the resonant frequency which is about 500 kHz for the CW 2400 adhesive. At around 500 kHz, the 0.04 mm-thick adhesive layer can provide the highest strain amplitude; see Figure 4 (a).

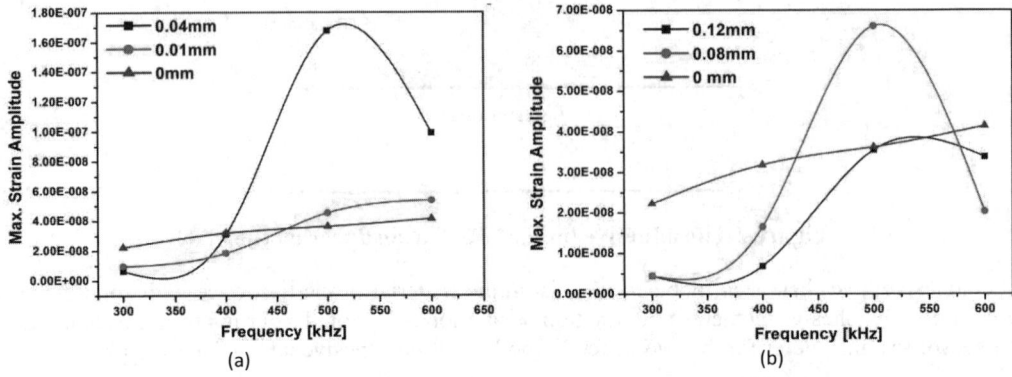

Figure 4 Maximum strain amplitude for different frequencies vs. adhesive thickness (a) 0 - 0.04 mm (b) 0.08 – 0.12 mm

Conclusion

The influence of adhesive layer on propagation of Lamb wave in composite plates has been evaluated. The propagation of guided waves was modeled numerically. Actuation and sensing of Lamb waves were carried out by analytical models based on electro-mechanical constitutive equation of PZT wafer in free space. The adhesive layer was modeled with solid element and cohesive element. The model with cohesive element showed higher amplitude. This needs to be verified by experimental results. The thickness of the adhesive layer together with the excitation frequency influences the amplitude of the sensor signal. For getting the lowest dispersion in the signal, the influence of the adhesive layer thickness can be used to find the best actuation frequency. Next step will be to validate these findings experimentally.

References

[1] S. Gopalakrishnan, M. Ruzzene, and S. Hanagud, *Computational Techniques for Structural Health Monitoring*: Springer Verlag, 2011.
[2] M. Ghajari, Z. Sharif Khodaei, and M. Aliabadi, "Impact Detection Using Artificial Neural Networks," *Key Engineering Materials,* vol. 488, pp. 767-770, 2012.
[3] Z. Sharif Khodaei and M. H. Aliabadi, "Damage Identification Using Lamb Waves," *Key Engineering Materials,* vol. 452, pp. 29-32, 2011.
[4] Z. Sharif Khodaei, R. Rojas-Diaz, and M. Aliabadi, "Lamb-Wave Based Technique for Impact Damage Detection in Composite Stiffened Panels," *Key Engineering Materials,* vol. 488, pp. 5-8, 2012.
[5] D. Worlton, "Experimental confirmation of Lamb waves at megacycle frequencies," *Journal of Applied Physics,* vol. 32, pp. 967-971, 1961.
[6] Z. Su and L. Ye, *Identification of Damage Using Lamb Waves: From Fundamentals to Applications*: Springer Verlag, 2009.
[7] S. Ha and F. K. Chang, "Adhesive interface layer effects in PZT-induced Lamb wave propagation," *Smart Materials and Structures,* vol. 19, p. 025006, 2010.
[8] Z. Su, L. Ye, and Y. Lu, "Guided Lamb waves for identification of damage in composite structures: a review," *Journal of Sound and Vibration,* vol. 295, pp. 753-780, 2006.
[9] S. Ha, A. Mittal, K. Lonkar, and F. K. Chang, "Adhesive layer effects on temperaturesensitive lamb waves induced by surface-mounted pzt actuators," 2009, pp. 2221-2233.

Evaluation of Setting Time in Ultra High Performance Concrete

Sung-Wook Kim[1,a], Jung-Jun Park[1,b], Doo-Youl Yoo[2,c], and Young-Soo Yoon[2,d]

[1] Structural Engineering Research Division, Korea Institute of Construction Technology, Goyang, 411-712, Korea

[2] School of Civil, Environmental and Architectural Engineering, Korea University, Seoul, 136-713, Korea

[a]swkim@kict.re.kr, [b]jjpark@kict.re.kr, [c]dooyoul@korea.ac.kr, [d]ysyoon@korea.ac.kr

Keywords: Ultra-High-Performance Concrete, Setting Time, Chemical Shrinkage

Abstract. Ultra high performance concrete (UHPC), characterized by a high strength and high ductility, is also subjected to large shrinkage due to its low water-to-binder ratio and its large content in high fineness materials. The large amount of autogenous shrinkage of UHPC can induce crack on structural member when it was restrained with reinforcement and form. However, shrinkage of UHPC in plastic state is not generating confining stress, which is the main cause of initial crack. Normally, the setting time in concrete is an index to distinguish shrinkage which occur confining stress or not. An estimation of setting time is conducted in compliance with ASTM C 403 till now however, that test standard reveals error of results due to discordance of test condition as following with concrete type. This study therefore evaluated setting time of UHPC through the modified test method which was proposed by KICT. Test results and analyses proved a discrepancy of setting time between ASTM and proposed method. The proposed method put faith in evaluation of setting time in accordance with UHPC.

Introduction

Ultra High performance Concrete shows a large amount of shrinkage due to low water-to-binder ratio (W/B) and an abundant quantity of binders. UHPC is a cementitious composite reducing the W/B ratio down to 0.2 and admixing high fineness admixtures and high-elastic steel fiber. However, its low W/B compared to conventional concretes, the admixing of large quantities of admixtures and the absence of coarse aggregates lead to large autogenous shrinkage and large risk of occurrence of shrinkage cracking at early age[1].

Since the shrinkage cracking behavior of concrete is influenced by the shrinkage development speed and size according to age as well as by the development of strength, stress relaxation and degree of restraint, this behavior is extremely complex. An accurate evaluation of setting time of UHPC is needed in order to estimate autogenous shrinkage strain as first step of prediction crack behavior of UHPC at early age. Especially, the surface of UHPC would be dried when it is exposed on the air after casting soon. The dried surface of specimen influenced setting time test result as earlier time. The importance of evaluation for setting time is provided because shrinkage before visco-elastic behavior does not make a restrained stress on the UHPC. Accordingly, this study performs setting test considering 3 types of treatment on the surface of UHPC

Materials and Mix Design

The test adopts type 1 Portland cement, fine aggregates with grain size smaller than 0.5 mm, filler with granulometry of 2 μm and 98% content of SiO_2, and silica fume (SF) with specific surface area of 200,000 cm^2/g. W/B ratio of 0.2 is used for the mix of UHPC. Polycarbonate superplasticizer is admixed to secure workability. Table 1 lists the mix proportions. This is one of normalized mix proportion on the level of 200MPa compressive strength. Steel fiber which is used on the mixture of UHPC was excluded in these tests.

Table 1. Mix proportion of UHPC (by weight ratio)

Nomenclature	W/B	Cement	SF	Filler	Sand	SP
UHPC	0.2	1	0.25	0.30	1.10	0.012

where SF = Silica Fume, SP = Superplasticizer

Test of Setting Time

Test of setting time of UHPC was carried out by ASTM C 403[2] which was similar to KS F 2436 [3] as following of Fig.1. The penetration time and depth in order to measure setting time were 10 seconds and 25mm relatively. When vicat spindle reached at 25mm during on 10seconds, the setting time after casting UHPC on test bowl was recorded. Generally, the initial setting time is defined when the internal stress of matrix by vicat spindle reach to $3.5N/mm^2$. Hence the final setting time is defined when the internal stress of matrix will be reached to $28.5N/mm^2$.

As following of ASTM C 403 test method, a curing membrane or an impermeable membrane is recommended to prevent evaporation on the surface of specimen. However, a membrane must be removed when vicat spindle puts into concrete. Whenever a membrane was removed on the specimen, the surface of specimen would be evaporated. The evaporation of surface on the specimen influences the stress of penetration resistance. Especially, the influence of evaporation will be more sensitive because W/B is very low in UHPC. In this study, 3 types of coverage which were an impermeable membrane, a membrane forming compound, and a paraffin oil applied in order to prevent evaporation. Fig.2 showed each test specimens installed in laboratory.

Fig. 1 Measuring of setting time on UHPC   Fig. 2 Specimens for setting time

Test Results and Discussion of Setting Time

Fig.3 revealed test results of setting time each conditions. The initial and final setting time in the specimens covered with paraffin oil on the top of specimen represented a middle value among 3 types of specimen. When test was carried by ASTM C 403 method out, setting time could be represented as Eq.(1).

$$Log(PR) = -b + aLog(t) \tag{1}$$

Where, PR is stress of penetration resistance, and a, b are constants of regression. The constants a and b were induced from setting time test results. Table 2 represented values of a, b.

Table 2. Comparison of setting time and a,b constants according to test methods

Type	a	b	R^2	Initial Set (hours)	Final Set (hours)
ASTM C 403	10.5742	24.0086	0.9930	5.60	6.82
Paraffin oil	12.2736	28.7182	0.9884	6.06	7.17
Membrane	10.1610	23.6676	0.9941	6.57	8.06

Test results indicated that when ASTM C 403 was applied, setting time revealed earlier than other tests. It proved that evaporation occured on the surface of UHPC. Hence, In case of using membrane forming compound as auxiliary measures to prohibit evaporation, setting time was measured later than other tests because the component matters included with membrane forming compound influenced to delay settlement of UHPC.
On the other hand, setting time in case of covering surface by paraffin oil revealed intermediate value. However there is no evidence that paraffin oil didn't accelerate or delay hydration of UHPC. One of test methods for verifying influence of agents on hydration of cementitious materials is a chemical shrinkage tset.

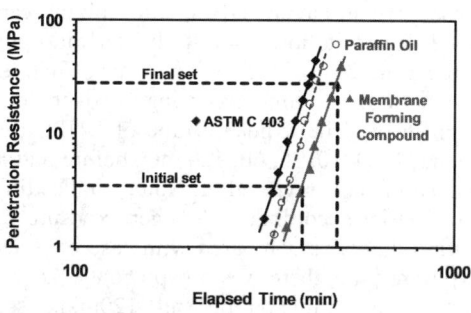

Fig.3 Comparison of setting time

Test Results of Chemical Shrinkage

Chemical shrinkage was measured in compliance with ASTM C 1608 (Procedure A)[4]. Cement paste was placed in 50mL Erlenmeyer flasks with a height of 10mm. Demineralized and distilled water was placed into the recipients without disturbing the paste. The quantity of water fixed with weight according to rate of W/C=40%,50%,60%, all 3cases, Fig. 4 shows test installation of chemical shrinkage. The volume change was measured by using a pipette with a minimum scale of 0.01mL. To prevent evaporation of water, a silicon rubber stopper was inserted into the flask and paraffin oil was dropped onto the top of the water. The flasks were left in a water bath at 23±1°C during the test. The position of meniscus in the pipettes was first recorded 1 hour after putting water. During the first 24 hours, the position of meniscus was recorded at intervals of 1 hour, and after that it was measured at every 1 to 24 hours.

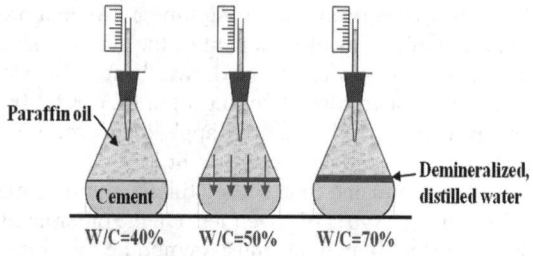

• 50mL Erlenmeyer flask
• Pipette with a minimum scale of 0.01mL

Fig.4 Test set-up for chemical shrinkage

Test Results and Discussion of Chemical Shrinkage

Fig.5 shows test results of chemical sjrinkage. Although W/C were different respectively, test results of chemical shrinkage revealed almost same values. It means that paraffin oil didn't influence on hydration of UHPC. The density of paraffin oil is around 0.85~0.94, it is lower than water of 1. Therefore, paraffin oil can not be distilled with water, and don't supply any matter for hydration.
In its final analysis, when paraffin oil uses by coverage material to prohibit evaporate mixing water, setting time can be evaluated accurately.

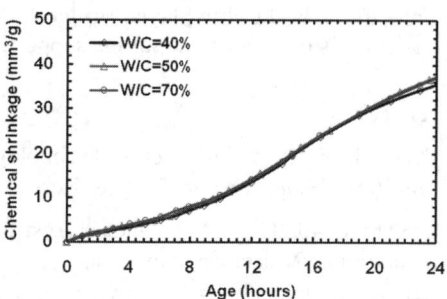

Fig.5 Test results of chemical shrinkage

Hence, the surface of UHPC after mixing can be easily dried due to its low water-to-binder ratio and its large content in high fineness materials. Fig.6 shows test results of setting time according to coverage moment of paraffin oil. The exposure time of UHPC after casting applied on 0, 10, 30, 60, 120 min. before putting paraffin oil. Test results revealed that initial and setting time was more earier according to longger exposure time. When setting time was compared with case of 0 and 120min. exposure time, there was a gap about 1 hour. In case of comparing with 60min. and 120min., setting time displayed almost same in Fig.6. The results represented that the exposed surface of UHPC on the air dried easily, and setting time of UHPC was influenced by conditions of surface.

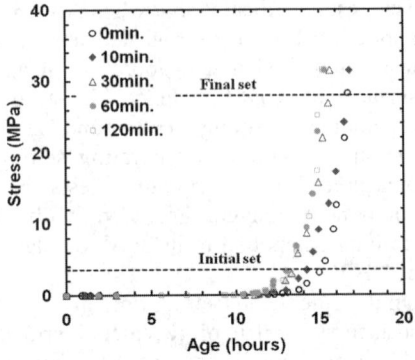

Fig.6 Setting time according to exposure

Conclusions

This study compared the setting time between a present test method and a proposed method. On the condition of 3types of treatment of the surface of UHPC, setting time tests were carried out, and the following conclusions can be drawn from the results.

    1) The test results of setting time of UHPC by ASTM C 403 can make an error. Although an impermeable membrane was applied to prohibit evaporation of water on the surface of UHPC, evaporation would be occurred a little.

    2) The initial and final setting time in the specimens covered with paraffin oil represented a middle value among 3 types of specimen. On the other hand, setting time by ASTM C 403 method was earlier than paraffin oil method, and a membrane forming compound method was measured later than that of method.

    3) The test results of chemical shrinkage represented that paraffin oil didn't influence on hydration of UHPC, when paraffin oil uses by coverage material to prohibit evaporation for mixing water, setting time of UHPC can be evaluated accurately.

    4) The large amount of autogenous shrinkage of UHPC can induce crack on structural member at early age. In order to distinguish shrinkage which occurs confining stress or not, setting time is an important index, and should be evaluated exactly.

Acknowledgements

This study was carried out as a partial research of the "Development of Ultra High Performance Concrete for Hybrid Cable Stayed Bridges" in the Korea Institute of Construction Technology. The authors express their gratitude for the support.

References

[1] Korea Institute of Construction Technology: *Development of Ultra High Performance Concrete for Hybrid Cable Stayed Bridges*, Research Report, KICT 2011-067 (2011).

[2] ASTM C 403/C 403M, Standard Test Method for Time of Setting of Concrete Mixtures by Penetration Resistance, American Society of Testing Materials, (2008).

[3] KS D 2436, Test Method for Time of Setting of Concrete Mixtures by Penetration Resistance, Korean Standards, (2007).

[4] ASTM C 1608, Standard Test Method for Chemical Shrinkage of Hydraulic Cement paste, American Society of Testing Materials, (2007).

Keyword Index

3D Simulation..193

A
Acoustic Emission...................................65, 141
Adhesion..65
AFOSM..101
Aggressive Environment..............................501
Air Plasma Sprayed Thermal
    Barrier Coating...13
Aircraft Landing..497
Airport Pavement..497
Alternating Temperature..............................421
Aluminum Alloy......................81, 213, 221, 605
Aluminum Alloy Thick Plate........................105
Analytical Solutions......................................601
Anchor Blocks...561
Anti-Plane Case..489
Artificial Crack..445
Artificial Neural Network (ANN)............565, 581
Attenuation..609
Austenitic Stainless Steel......................201, 217
Austin Moore...41
AZ91D Magnesium Alloy..................................9

B
Back Stress..441
Bamboo..609
Bending Property..49
Bent Beams..485
Bi-Material...445, 545
Bi-Material Wedge..93
Biodegradable Resin.......................................65
Boron Carbide...321
Boundary Element Method..............................1
Bridge..37, 593
Bridge Maintenance.....................................141
Brittle Elements..485
Brittle Fracture.......................................13, 489
Buckling..337
Buckling Critical Load Coefficient...............101
Buckling Reliability Analysis........................101

C
C/C Composites..613
Cable-Stayed Bridges...................................593
Calibration Spacing......................................365
Cantilever Specimens.....................................57
Car Body..453
Cast Aluminum Alloy....................................169
Cement..41
Central Crack..85, 89
CFRP...49
Chemical Shrinkage.....................................621

Circular Lining Structure.............................345
Clearance..161
Coalescence..613
Coatings..509
Cohesive..393
Cohesive Crack...513
Cohesive Zone..181
Cohesive Zone Model...................................117
Cold-Drawn Steel...553
Combination Method...................................157
Compact Tension Specimen........................549
Complex Potentials......................................545
Composite Laminate.............265, 389, 393, 521
Composite Materials.............................149, 225
Composite Panel...533
Composite Sandwich Panel.........................457
Composites........................45, 73, 197, 261,
    337, 565, 593
Composites Delamination............................409
Compression Test.................................457, 561
Compressive Test..385
Concrete..153, 209
Concrete Bridge..145
Concrete Composite
    Strengthening Method..........................145
Conformal Mapping.....................................293
Conservation Laws.......................................301
Constraint Effect..237
Construction Monitoring.............................577
Contact Angle...49
Corner Cracks..509
Correlation..421
Corrosion..505
Corrosion Behavior......................................129
Corrosion Resistance......................................9
Corrugated Core Sandwich.........................117
Crack..397, 529
Crack Arrest by Electromagnetic Heating...405
Crack Closure...517
Crack Deflection..221
Crack Growth..................121, 181, 209, 597
Crack Growth Behavior...............................613
Crack Initiation......................................69, 469
Crack Shape Prediction...............................177
Crack-Tip Field...297
Cracked Flattened Brazilian
    Disk (CFBD)......................................85, 89
Creep Conditions..297
Creep Failure..245
Creep Property...53
Cross Stiffened Plate....................................361
CTOD-R Curves..549
CTOD Test..425

Cu-Ni Alloy 325
Cumulative Damage 153
Current Density 405
Curved Multi-Pass Welding Line 349
Curved Steel Bridges 133
Cut-Out 457
Cyclic Creep 441
Cyclic Temperature 429
Cylindrical Inclusion 305
CZM 45

D
Damage 253, 265, 397, 541, 589
Damage Analysis 373
Damage Detection 433, 569, 581
Damage Diagnosis 477
Damage Mechanics Model 69, 469
Damage Model 393
Damage Monitoring 141
Damage Tolerance 389
Damping Characteristics 605
Damping Coefficient 609
Davies-Zhang 533
Debonding 65
Deck 37, 593
Defect 21
Deformation 201, 357
Deformation Modes 261
Delamination 121, 233, 393, 509, 533
Delamination Threshold Load 521
Density Function 413
Deposition Process 269
Detecting Method 597
Deterministic Spectra 537
Diffusion Coefficient 317
Dislocation Distribution Technique 517
Dislocation Technique 189
Dislocation Zones 33
Dispersion 433
Double Ellipsoid Heat Source 349
Double Network Hydrogel 193
DP Steel 69
Dual Boundary Element Method 17
Dual Form 313
Ductile Damage 469
Dugdale Model 489
Dynamic 261
Dynamic Response 573
Dynamic Strain Monitoring 145
Dynamic Stress Concentration
 Factor (DSCF) 293, 305

E
Effective Properties 73
Effective Stress 297

Eigen Value 445
Ejection 401
Elastic Deformation 57
Elastic Modulus 609
Elastic-Plastic Deformation 57
Elastic-Plastic-Model 349
Elastodynamics 601
Electro-Brush Plated Nano-Al_2O_3/Ni
 Composite Coating 277
Electro-Mechanical Impedance 569
Electro-Thermal Explosion Spraying 269
Electromagnetic Shielding Effectiveness 437
Electromechanical Impedance 581
Element-Free Galerkin Method 17
Elongation 421
Emergency Break-Away 257
Engine 257
Epoxy Adhesive Layer 61, 205
Existing Concrete Bridge 141
Experiment 285, 397
Experimental Modeling 33
Experimental Study 477
Exponential Cohesive Zone Model 409
Extended Finite Element Method 413
Extreme Temperature 185

F
Failure 45, 429, 541
Failure Behavior 469
Failure Initiation 61
Failure Model 373
Failure Probability 361
Fastener Load Distribution 285
Fatigue 37, 141, 153, 165, 217, 253, 333, 517, 597
Fatigue Behavior 169
Fatigue Crack 81
Fatigue Crack Growth 17, 221, 505
Fatigue Crack Growth Rate 137
Fatigue Life 145, 417
Fatigue Limit 613
Fatigue Property 241
FDM 573
FE Analysis 201, 541
FE Simulation 281
FEA 289, 337, 473
FEM 25, 73, 497
FEM Simulation 469
Fiber Fracture 65
Fiber-Reinforced Composite 541
Fiber Reinforced Concrete 185
Fibre Bridging 121
Finite Deformation 313
Finite Element 37
Finite Element Analysis (FEA) 241, 385
Finite Element Method (FEM) 225

Finite Element Model (FEM) 133
Finite Volume Method 185
Foam 289
Force Measurement System 13
Formality 69
Forming Limit Diagram 69
Fracture 261, 513, 585, 589
Fracture Constraint 613
Fracture Mechanics 173, 209
Fracture Toughness 29, 117, 233, 425, 609
Fragmentation 401
Freezing-Thawing Environment 153
Friction Stir Processing 169
Friction Stir Weld 129
Friction Stress 441
FRP 593

G
Galvanic Corrosion 325
Gas Dynamics 185
Generalized Fracture Mechanics 61
Generalized Stress Intensity Factor 545
GFRP Bridge Deck 249
GLARE 121
Global-Local 533
Grain Size 201
Green Composite 53, 65
Green's Function 293, 345
Ground Motion 481
Guided Waves 433

H
Harmonic Tests 477
HDPE 473
Heat Affected Zone (HAZ) 585
High Cycle Fatigue 113, 213
High Grade Line Pipe 425
High Strength Steel 325
High-Temperature Structural Analysis 465
High-Temperature
  Structure Analysis 5, 461
Higher-Order Terms 529
Hip 41
Honeycomb Sandwich 113
Hopkinson Bar Test 377
Hybrid Joint 161
Hyperelasticity 25
Hypervelocity Impact 97, 397, 401

I
Impact 289, 385, 497, 565
Impact Damage 365
Impact Detection 581
Impact Load 481
Impact Location Detection 365

Inconel 605
Inelastic Zone 529
Influence Coefficient 389
Initiated Crack 545
Initiation Life 369
Inserting Pin 597
Intensity Factors 229
Interface 65
Interface Fracture 509
Interface Properties 269
Interface Slip 357
Interfacial Crack 229
Interfacial Debond 117
Interfacial Performance 409
Interfacial Strength 49
Intergranular Fracture 273
Interlaminar Shear Strength 249
Interphase 173
Inverse Problems 189
Inversed Torsion Pendulum Method 605
Inverted Torsion Pendulum 609

J
J-Integral 181, 313
Johnson-Cook Model 377
Joint 45

K
K_{res} 125
K_t Effect 505
Knee-Point 389

L
Large Mass 565
Laser Cladding 77
Laser Peening 125
Laser Surface Treatment 241
Lattice Model 485
Load due to Explosion 185
Load Test 453
Local Axial Symmetry Initial Defects 101
Localization 69
Low Cycle Fatigue 441
Low Tensile Stress 273, 317
Low-Velocity Impact 265, 393, 521
Lubricant Additives 329

M
Magnesium Alloy 165, 605
Matching Asymptotic 489
Matching Pursuit 433
Material Homogenization 1
Material Parameters 289
Material Point Method 97, 109, 513
Material Recycling 473

Maximum Entropy Method 361
Mechanical 261
Mechanical Properties 137, 277, 321, 381
Mechanical Strength 437
Medium-Scale Process Heat
    Exchanger (PHE) 5
Mesh Bumper 401
Mesh Refinement.. 413
Mesh Size 485
Meshless Method.. 601
Metal Magnetic Memory 77
Micro-Cantilever Test 13
Micro Debonding Test....................................... 49
Micro Material... 165
Micro Mechanical Testing............................... 165
Micro-Scale Testing 57
Micro-Sized Mechanical Testing..................... 585
Microcracks 589
Microelectronics... 509
Micromechanics 73
Microstructural Modelling................................... 1
Microstructural Modification 169
Microstructure 81, 505
Microstructure of Nugget 129
Middle Tension Specimen................................ 549
Mixed-Mode 89
Mixed Mode I, II Crack 341
Mode I 297
Mode Mixity... 509
Modelling 149
Modified Hill Yield Criterion.......................... 341
Molecular Chain Network Model.................... 193
Monitoring Fatigue Cracks.............................. 189
Monte Carlo Method 157
MSD 369
Multi-Bolted Joint 285
Multi-Material Coupling 109
Multilevel 197
Multiobjective.. 197
Multiple Cracks 345
Multiscale Modelling....................................... 589
Multiscale Simulation...................................... 309

N
Nano-Indentation.. 309
Nanocrystalline Nickel 57
Natural Fiber... 53, 65
Natural Frequency 253, 453
Near-Crack-Tip Fields..................................... 529
Near-Field Explosion....................................... 361
Necking 193
Necuron 21
Neodymium 277
Nickel-Based Single Crystal............................ 341
Niobium Film 9

Nitriding.. 217
Non-Circular Cavity 293
Non-Contact 3D Deformation
    Measurement 493
Non-Crimp Fabric (NCF) 493
Non-Equilibrium Grain-Boundary
    Segregation (NGS)............................. 273, 317
Non-Linear Matrix FEM.................................. 173
Nondestructive Evaluation................................ 21
Notch 545
Numerical 41
Numerical Analysis 285
Numerical Method... 85
Numerical Optimization 377
Numerical Simulation............. 105, 109, 517, 577

O
Old Metal Bridge 137
Open-Hole Tension... 373
Optimisation 197
Orthotropic... 37
Orthotropy.. 545
Out-of-Plane Distortional Fatigue 133
Overload Retardation Effect............................ 553

P
Palygorskite Powders....................................... 329
Parameters 133
Paris Curve 553
Particulate Composite.. 25
Path Independence 181
Path-Independent Integrals.............................. 301
Persistent Slip Band.. 213
Phosphorus... 273, 317
Piezo-Magnetic Media..................................... 301
Piezoelectric Transducer.................................. 617
PLA... 53
Plastic Zone 341
Plastic Zone Size 553
Plate-Like Structure... 365
Point Load.. 177
Polycrystalline Materials 1
Polyisocianurate Foams..................................... 29
Polymer Particulate Composites...................... 173
Polymer Pressure Pipes 177
Post-Tensioning.. 561
Power Function Curve..................................... 417
PP 49
Pressure-Sensitive Dilatant Material 353
Pressureless Sintering...................................... 321
Primary Defects 573
Prior Corrosion 81
Probabilistic Fracture Mechanics 157
Probabilistic Spectra 537
Process Heat Exchanger (PHE) 465

Process Heat Exchanger
  (PHE) Prototype 461
Prosthesis 41
Pulse Current 405
Push-Out Test 61, 205
Pylon 257

Q
Quasi-Brittle Fracture 529
Quasi-Static Growing Crack 353
Quasistatic Incubation Tests 501

R
R-Ratio 553
Radial Basis Function 601
Rapeseed Oil 29
Real World Storage 421
Refractories 589
Remanufacturing 77
Residual Compressive Strength 265
Residual Fatigue Life 157
Residual Life 113
Residual Strength 369
Residual Stress 105, 125, 281, 349
Resonance 253
Restrained Shrinkage 449
Resultant Stress-Strain Curve 245
Risk Analysis 537
Rock Salts 245
Rupture 13
RVE Analysis 497

S
Safety Evaluation 141
Sandwich Plate System 337
Scattering of SH-Wave 345
Security Coefficient 113
Sedimentation 577
SEM *In Situ* Technology 81
SEM Analysis 249
Semi-Space 305
Sensibility 289
Setting Time 621
SH-Wave Scattering 305
Shaking Table 477
Shank-Hole 161
Shear Behavior 557
Shear Bond Strength 205
Shielding Effect 173
Ship Structure 337
Shock Wave 97
Silicon 57
Simplified Procedure 333
Simulated Concrete Environment 249
Simulation 401, 429

Size Effect 165, 309, 485
Small Crack 213
Small Mass 565
Small-Scale Gas Loop 5, 461, 465
Soft-Storey Failure 481
Solution Treatment 217
Spall Fracture 97
Spatial Oblique-Elliptical
  Embedding Crack 405
SPH 257
Stainless Steel 549
Stainless Steel Mesh 397
Stainless Steel Net 437
Steady-State 5
Steel-Concrete Joint 61, 205
Steel-Concrete-Steel Sandwich Beam 357
Steel Frame Building 481
Steel Plate 145
Steel Surface 329
Steel Tank 477
Stiffened Panel 565
Stiffening Plate 337
Stiffness 145
Stitch 233, 493
Stitched 45
Stitching Parameter 493
Stop-Hole 597
Strain Control 441
Strain Field 353
Strain-Induced Martensitic
  Transformation 217
Stress 41, 453
Stress Analysis 133, 453
Stress Concentration 597
Stress Extrapolation Method 445
Stress Field 353
Stress Intensity 597
Stress Intensity
  Factor (SIF) 17, 85, 89, 93, 177, 189
Stress Relaxation 449
Stress Singularity 445
Stress-Strain 429
Stress-Strain-Time 245
Strong Discontinuity 513
Structural Design 225
Structural Health Monitoring 569, 581, 617
Structural Steel 585
Submarine Pressure Structure 157
Subway Tunnel 573, 577
Super Singular Element 93
Superposition Technique 533
Surface Crack 185, 237
Surface Displacement 345
Surface Roughness 417
Surface Treatment 53

T

T_{11}-Stress 237
Taylor Impact Test 377
Temperature 341, 405
Tensile Behavior 149
Tensile Strength 449
Test Methods 381
Theoretical Analysis 105
Theoretical Expression 357
Thermal Bridge 473
Thermal Damage 77, 241
Thermal Shock Resistance 589
Thermo-Mechanical Stress 93
Thermography 21
Thick Composites 381
Thin Layers 509
Three-Dimensional Finite
    Element Analysis 237
Three Point Bend 201
Through-Thickness 381
Ti-6Al-4V Alloy 441
Ti6Al4V 501, 505
Time-Reversal Method 365
Titanium 325
Titanium Diboride 321
Toughness 137
Transversely Isotropic Piezoelectric 229
Trimaran 333
Trip Condition 5
Two Dissimilar Piezoelectric Media 293
Two-State Integral 545

U

UHPC 561
Ultimate Bearing Capacity 145
Ultimate Compressive Load 385
Ultra High Performance Concrete ... 449, 557, 621
Ultrasonic Guided Waves 581, 617
Underground Structures 33
Underwater Explosion 109
Unit Cell Model 25
Unreinforced Masonry 537

V

Very High Temperature
    Reactor (VHTR) 5, 461, 465
Vibration State 253
Viscoelastic 473

W

Water Concentration 609
WC/Co Coating 269
Wear Failure 329
Web Gaps 133
Wedge-Splitting Test 209
Weibull Statistics 149
Welding 281
Welding Condition 349
Williams Series 529
Wind 45
Wind Turbine Blade 225
Wing Skin 369
Wood Fiber 437
Woven Composite 373

Z

Z-Pin 233

Author Index

A

Abe, D. 73
Ai, C.S. 429
Aimmanee, S. 541
Aiyappa, D. 45
Aliabadi, M.H. 1, 17, 73, 197, 565, 569, 581, 601, 617
Amarandei, M. 21, 41
An, H. 101
An, H.P. 441
An, W.G. 101
Andersons, J. 29
Apicella, A. 197, 581
Asahina, T. 605, 609
Asifa, K. 281, 333
Avila-Haro, J.A. 537
Avila, J.A. 525

B

Bacarreza, O. 73, 197
Bai, Y. 217
Bao, R. 221
Baragetti, S. 501, 505
Barbat, A.H. 537
Benedetti, I. 1
Benhadj-Djilali, R. 45
Berdich, K. 21
Boel, V. 61, 205, 209
Bogdan, L. 41
Borrelli, R. 533
Boštík, J. 33
Boukellif, R. 189
Buonsanti, M. 497
Burkacki, D. 477

C

Cābulis, U. 29
Caputo, F. 161, 533
Cen, Y. 413
Chang, Z.L. 357
Chen, C.L. 365
Chen, H.B. 417
Chen, W.D. 97, 109
Chen, X. 261
Chen, X.Y. 361
Chen, Y. 521
Chen, Z.Q. 149
Cheng, J. 133
Cheng, L.N. 389
Cho, J.R. 593
Cho, K.H. 593
Choi, D.B. 453
Choi, E.S. 557, 561

Choi, Y.S. 561
Cui, H. 233
Cui, X.F. 9, 269, 277
Cui, Y.P. 573, 577

D

Darnbrough, J.E. 57
De Corte, W. 37, 61, 205, 209
De Schutter, G. 61, 205, 209
Ding, B. 421, 429
Dong, F. 573, 577
Dong, J. 573, 577
Dong, S.M. 85, 89
Duan, L. 137
Duan, M.G. 289

E

Eliáš, J. 485
Enkelhardt, A. 41
Enoki, S. 493

F

Faur, N. 41
Feng, Y.R. 425
Ferdous, M.S. 597
Flewitt, P.E.J. 13, 57, 201
Foglia, C. 505
Fu, H.B. 317
Fu, K. 249
Fu, Y.D. 273
Fu, Y.M. 405
Fujikawa, M. 613

G

Gan, W.X. 437
Gao, Y.Y. 101
Gerosa, R. 505
Ghajari, M. 565, 581
Ghasemnejad, H. 45
González-Drigo, J.R. 537
González, B. 553
Guan, G.S. 397, 401
Gubeljak, N. 549
Guiamatsia, I. 513
Guo, W.G. 373, 377
Guo, Y.L. 409
Guo, Y.Z. 261

H

Ha, Y. 397
Hagiwara, Y. 65
Halpern, L. 489
Haritos, G. 521

He, J. 361
He, Y. 157
Hei, B.P. 345
Helincks, P. 61, 205
Henneberg, D. 589
Hong, S.D. 5
Hu, H.T. 253
Hu, J.Y. 137
Hu, Y.Y. 105
Hua, L. 77, 241
Huang, F. 429
Huang, J.D. 437
Huang, Y.Y. 437
Hutař, P. 25, 177

J
Jagarinec, D. 549
Jang, Y.S. 457
Jankowski, R. 477, 481
Jansseune, A. 37
Jensen, H.M. 509
Jia, Y.X. 409
Jin, G. 9, 269, 277
Joh, C. 557, 561
Judt, P. 181

K
Kakiuchi, T. 165, 169
Kanazawa, K. 493
Katayama, T. 493
Katogi, H. 49, 53
Kayamori, Y. 585
Khurram, S. 281, 333
Kim, B.S. 557, 593
Kim, J.G. 453
Kim, J.S. 561
Kim, S.T. 593
Kim, S.W. 449, 621
Kimyong, C.Y. 541
Klusák, J. 61, 205, 545
Kong, C.W. 457
Kong, D.S. 373
Koo, D.H. 453
Korte, S. 209
Kotoul, M. 545
Kucharczyk, P. 469
Kun, L. 21
Kwak, I.J. 557
Kwark, J.W. 561

L
Lamanna, G. 161
Landes, J. 549
Lee, J.W. 457, 557, 561
Lee, Y.S. 457

Lei, J. 229
Leng, L. 93
Leonardi, G. 497
Li, B. 265
Li, C.K. 317
Li, D. 293
Li, D.Y. 577
Li, H. 281, 349
Li, H.L. 305
Li, H.Q. 445
Li, H.Y. 433
Li, J. 337
Li, L. 149, 281, 349
Li, Q.F. 9, 269, 273, 277, 325, 329
Li, X. 265
Li, X.D. 81
Li, Y.G. 117
Li, Y.L. 233, 253, 257, 261, 365
Li, Y.T. 381, 441, 445
Li, Y.Z. 265, 285
Lian, J.H. 69
Liang, W.Y. 297
Liu, A.D. 321
Liu, D. 13
Liu, E.B. 9, 277
Liu, L.P. 145
Liu, P.F. 69
Liu, Q. 617
Liu, X.D. 285
Liu, Y.Y. 321
Liu, Z.G. 81
Liu, Z.M. 301, 313
Lu, J. 113, 117
Luo, G. 381

M
Ma, C.H. 393
Ma, J.W. 245
Ma, Q.R. 425
Ma, Y.E. 121, 125, 129
Mahalingam, S. 57, 201
Majer, Z. 173
Makabe, C. 597, 613
Mao, J.Z. 313
Marigo, J.J. 489
Marşavina, L. 21
Máša, B. 25
Matejka, L. 473
Matějka, L. 473
Matos, J.C. 553
Matsuda, M. 585
Medolago, A. 501
Meng, Q.H. 297, 309
Migda, W. 481
Mimura, K. 193

Miyamoto, S. 53
Miyazaki, T. 597
Mizuno, S. 165
Mu, Z.T. 81
Münstermann, S. 69, 469

N
Náhlík, L. 25, 177
Nakagaito, A.N. 65
Nakajima, M. 217
Nakamura, Y. 217
Nes, C.S. 41
Nguyen, G. 513
Nie, J.Q. 409
Nishida, S. 585
Niu, R.T. 401
Noguchi, H. 213

P
Papadopoulos, F. 45
Park, G.S. 453
Park, H.Y. 5
Park, J.J. 449, 621
Park, S.Y. 593
Pěnčík, J. 473
Ping, X.C. 93
Predan, J. 549
Prochazka, P.P. 185
Profant, T. 545
Pu, D.D. 397
Pujades, L.G. 537

Q
Qiao, Y.J. 317, 321
Qiu, X. 421
Qu, J. 341
Qu, P. 409
Qu, T.J. 385

R
Ren, H.L. 333
Riccio, A. 533
Ricoeur, A. 181, 189, 589
Riku, I. 193
Rustamov, I. 225

S
Scaramuzzino, F. 533
Schwankl, M. 569
Scoppelliti, F. 497
Seitl, S. 61, 209, 529
Sekine, M. 605
Sellitto, A. 533
Šestáková, L. 529
Setyabudi, S.A. 613

Ševčík, M. 177
Ševeček, O. 545
Shang, L. 337
Shao, T.M. 9
Shapriya, R. 45
Sharif Khodaei, Z. 565, 569, 581, 617
Shen, Z. 521
Sheng, H.J. 137
Shi, P. 421
Shikama, T. 213
Shimada, Y. 585
Shimizu, T. 217
Shioya, T. 605, 609
Shterenlikht, A. 201
Shu, H. 285
Sobek, J. 529
Song, H.Y. 301, 313
Song, K.N. 5, 461, 465
Song, T.S. 157, 293
Song, X.H. 101
Soprano, A. 161
Sotirchos, E. 45
Spārniņš, E. 29
Stirna, U. 29
Su, C.W. 437
Sun, W.X. 273
Sun, W.Z. 573, 577
Suzuki, K. 217
Szigyarto, I. 21

T
Takagi, H. 65
Takahashi, Y. 213
Takashima, K. 585
Takemura, K. 49, 53
Tan, S.G. 261
Tanaka, K. 493
Tang, L.Q. 353
Tian, L. 141
Tian, W. 77, 241
Tiu, W. 517, 521
Toguri, T. 605
Toribio, J. 553
Tozaki, Y. 169
Tuyet, D.T.B. 489

U
Uematsu, Y. 165, 169, 217
Uthaisangsuk, V. 541

V
Vargas, Y.F. 537
Veluri, B. 509
Veselý, V. 529
Vořechovský, M. 485

W

Wang, C.L. 325
Wang, C.S. 133, 137, 141, 145
Wang, F. 149
Wang, H.T. 425
Wang, J.L. 253, 405
Wang, L.M. 329
Wang, Q.J. 377
Wang, Q.Y. 85, 89
Wang, R.F. 441
Wang, S.N. 369
Wang, W. 369
Wang, X.Y. 261
Wang, Z.F. 105
Wang, Z.H. 125
Wang, Z.Q. 297, 309
Wang, Z.Y. 245
Wechsatol, W. 541
Wei, M.S. 145
Weiglová, K. 33
Weimer, C. 569
Wen, P.H. 17, 601
Wu, J.H. 325
Wu, S.H. 93
Wu, Y. 337

X

Xia, P.X. 357
Xiao, Q.H. 153
Xiao, W.L. 417
Xie, W. 237
Xiong, Q.R. 425
Xu, B.S. 329
Xu, F. 289, 377, 393
Xu, H. 433
Xu, H.N. 345
Xu, Y. 329
Xu, Y.G. 517, 521
Xu, Y.Z. 517, 521
Xu, Z.Z. 425
Xue, Q.C. 337
Xue, W.C. 249

Y

Yan, S.L. 133, 137
Yan, X.Z. 369
Yang, L.H. 341
Yang, P.F. 113, 117
Yang, W.M. 97, 109
Yang, Y. 305, 345, 353
Yang, Z.J. 309
Yang, Z.L. 345, 353
Yao, Z.H. 265, 285
Yin, Y. 417
Yoo, D.Y. 449, 621
Yoon, S.C. 453
Yoon, Y.S. 449, 621
Yu, H.L. 329
Yuan, F.G. 365
Yuan, Q.P. 437
Yun, S. 121

Z

Zeng, C. 77, 241
Zeng, Z. 261
Zeng, Z.Y. 377
Zhai, M.S. 141
Zhang, B. 329
Zhang, C.L. 429
Zhang, D. 385, 389
Zhang, F. 97, 109
Zhang, H.B. 245
Zhang, J.X. 425
Zhang, J.Y. 221
Zhang, M.J. 293
Zhang, T. 221
Zhang, X. 257
Zhang, X.Q. 373
Zhang, Z.Y. 157
Zhao, X.C. 221
Zhao, Z.Q. 129
Zhen, C.B. 333
Zheng, L.J. 405
Zheng, X.T. 381, 385, 389
Zheng, Y. 349
Zhong, H.J. 229
Zhou, H.M. 405
Zhou, J.S. 301
Zhou, Y.F. 293
Zhu, G.W. 409
Zhu, S.F. 225
Zou, G.P. 113, 117, 337, 341, 357
Zouhar, M. 177